University of
Hertfordshire

College Lane, Hatfield, Herts. AL10 9AB

**Learning and Information Services**

For renewal of Standard and One Week Loans,
please visit the web site **http://www.voyager.herts.ac.uk**

This item must be returned or the loan renewed by the due date.
The University reserves the right to recall items from loan at any time.
A fine will be charged for the late return of items.

# Nutritional Aspects of Bone Health

This book is dedicated to the memory of Mrs Linda Edwards, Director of the National Osteoporosis Society (1986–2002) whose compassion for osteoporosis sufferers, commitment to their welfare and capacity for hard work shall remain an inspiration to us, always.

The editors and author royalties from the sale of this book have been donated to the National Osteoporosis Society.

National Osteoporosis Society
Camerton, Bath BA2 0PJ
Tel: 01761 471771
Helpline: 0845 450 0230
Website: www.nos.org.uk
Email: info@nos.org.uk

# Nutritional Aspects of Bone Health

Edited by

**Susan A. New**
*School of Biomedical and Molecular Sciences, University of Surrey, Guildford, UK*

**Jean-Philippe Bonjour**
*Department of Rehabilitation and Geriatrics, University Hospital, Geneva, Switzerland*

advancing the chemical sciences

ISBN 0-85404-585-6

A catalogue record for this book is available from the British Library

Published by The Royal Society of Chemistry,
Thomas Graham House, Science Park, Milton Road,
Cambridge CB4 0WF, UK

Registered Charity Number 207890

For further information see our web site at www.rsc.org

Typeset by Alden Bookset, Northampton, UK
Printed and Bound by TJ International Ltd, Padstow, Cornwall, UK

# Foreword

The skeleton is a remarkable organ. Fossil records remind us of its persistence long after life is over. However in life the skeleton is a vital structure undergoing remarkable changes in shape and structure during growth, and subject to continual processes of removal and renewal throughout life. All these processes are under the control of genetic, hormonal, nutritional and other factors.

Osteoporosis is the commonest disorder of bone. Commonly quoted figures state that at least one in three women will have a fracture over the age of 50, and more than one in twelve men. Osteoporosis will present one of the major challenges to health-care systems in the coming decades. In women the loss of oestrogen at the menopause is the major change leading to loss of bone, but in men the causes are less obvious. However in both sexes many factors contribute, and there is a strong interplay between genetic, nutritional and other environmental influences.

Diet plays an important role in the development and maintenance of bone health throughout life. There are many clinical disorders associated with faulty nutrition, ranging from the obvious effects of deficiency of key nutrients, especially Vitamin D and calcium, to the many other dietary components that affect the skeleton in obvious or more subtle ways.

In this book, *Nutritional Aspects of Bone Health*, two eminent leaders in this field, Susan New and Jean-Philippe Bonjour, have assembled a world-class group of contributors to produce the first textbook devoted specifically to this topic. There are 32 chapters covering the range of relevant topics. The book will appeal to many audiences, from undergraduates to the many specialists who work on these topics, as well as others interested in medical and health matters.

This is an outstanding collection of informative chapters which will have wide appeal. The editors and authors have generously agreed to donate proceeds from sales of this book to the National Osteoporosis Society (UK) a charity devoted to meeting the needs of patients suffering from this often distressing and disabling condition.

I strongly recommend this book to everyone with an interest in this important area of human health.

Graham Russell
*Norman Collisson Professor of Musculoskeletal Sciences,*
*University of Oxford,*
*The Botnar Research Centre,*
*Oxford University Institute of Musculoskeletal Sciences,*
*Nuffield Orthopaedic Centre, Headington, Oxford, UK*

# Preface

It is with great honour that we introduce our book, *Nutritional Aspects of Bone Health*. We have assembled together the world's leading researchers in this field to formulate the first-ever textbook on the subject of nutrition and bone. Each and every chapter includes current and cutting edge science underpinning the role of diet in the development and maintenance of bone health throughout the lifecycle and in the prevention of osteoporosis in later life.

The book is divided into six core Sections and includes a total of 32 chapters with contributions from 66 authors/co-authors. Section One defines bone health and discusses the fundamental areas of epidemiology of osteoporosis, genetic and non-nutritional influences, the role of nutrition in other metabolic bone diseases and the assessment of diet in individuals and populations. Section Two covers the established nutrients for bone health protection and osteoporosis prevention, namely calcium and vitamin D. Section Three details the effect of macronutrients (protein) and macrominerals (phosphorus, sodium and potassium) on bone. Section Four covers the influence of micronutrients on the skeleton and includes a detailed discussion of vitamin K, vitamin A, magnesium, trace elements as well as foods groups, isoflavones, alcohol and caffeine. Section Five discusses a number of important areas including use of twin studies for investigating the impact of nutrition on bone health, nutrient:gene interactions affecting the skeleton, nutrient/bone health relationships in different population groups, the effect of weight reduction and eating disorders on skeletal integrity as well as bone health issues during pregnancy and lactation. Section Six focuses on the critical areas of fracture risk reduction, cost-effectiveness of nutritional therapies for osteoporosis prevention and key nutritional strategies for optimising bone health in individuals and populations.

This Preface would not be complete without acknowledgement of a number of key individuals, whose efforts have made this book possible. Firstly, to Leatherhead Food Publishing, and in particular Mrs Victoria Emerton and Mr Matthew Pfleger, who initially identified the need for such a book and for their grace and sincerity in handing over the publishing reins to the Royal Society of Chemistry (RSC). To the staff at the RSC for their tremendous enthusiasm and total commitment to the book. In particular, Commissioning Editors, Mrs Janet Freshwater and Dr Robert Eagling; Marketing Manager, Mr Christopher Marshall; Books Editor, Mr Tim Fishlock; Production Co-ordinator, Ms Katrina Turner and Marketing Promotions Executive, Yolaine Bennett. Specific and sincere thanks for invaluable advice are due to Dr David Bender (University College London) and Dr Jane Morgan (University of Surrey). Finally, a very special thank you is owed to Professor

John Dickerson (Emeritus Professor, University of Surrey) whose assistance and advice throughout all stages of book production has been an absolute inspiration.

We do so hope that you will find our book to be a great source of information and that you will have much enjoyment in reading it.

**Susan Alexandra New**                                    **Jean-Philippe Bonjour**
*University of Surrey*                                        *University Hospital Geneva*

# Contents

**Section Five**

**Section Six**

CHAPTER 1

# An Overview of Osteoporosis

DAVID M. REID

Professor of Rheumatology, University of Aberdeen, Department of Medicine and Therapeutics, Medical School, Foresterhill, Aberdeen, AB25 2ZD, UK, Email: d.m.reid@abdn.ac.uk

**Key Points**
- Osteoporosis is a systemic skeletal disease with loss of bone mineral leading to excess fracture risk
- The diagnosis of osteoporosis is based on measurements of axial bone mineral density, but there are a number of other techniques which also predict future fracture
- Cost-effective intervention with pharmaceutical preparations is based on the diagnosis of osteoporosis, but in addition increasingly there is a need to consider the absolute fracture risk
- Nutritional and other non-pharmaceutical measures may be a useful and potentially cost-effective adjunct to pharmaceutical treatments or an alternative preventative measure
- More research is required to determine the appropriate targets for pharmaceutical and particularly nutritional interventions

Osteoporosis is defined as a skeletal disorder characterized by compromised bone strength predisposing a person to an increased risk of fracture[1]. Bone strength primarily reflects the integration of bone density and bone quality, a somewhat nebulous concept, which includes the biochemical integrity of bone[2]. While fractures of the vertebral body, distal forearm (Colles) and proximal femur are considered to be characteristic, it is important to realize that not all fractures of these bone sites are due to osteoporosis and *low trauma* fractures of most skeletal sites, except arguably the ankle,[3] can be associated with low bone mass.

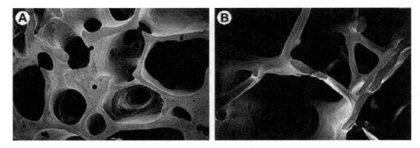

**Figure 1.1**  *Scanning electron micrograph of the trabecular structure of cancellous bone
from a normal (A) subject and a patient with osteoporosis (B)*
Reproduced with permission of Dr David Dempster and the Journal of
Bone and Mineral Research. (D.W. Dempster, The contribution of trabecular architecture to cancellous bone quality. *J. Bone Miner. Res.*, 2000,
**15**(1), 20–24.)

As not *all* fractures of even classical sites are associated with low bone mass,
it is important to have a practical definition, which can be used *in vivo*. Architectural features of bone can only be assessed *in vitro* by three-dimensional
computed tomography (CT) techniques and by scanning electron micrographs,
which demonstrate marked differences in the trabecular structure of cancellous
bone in normal and osteoporotic subjects (Figure 1.1).

However, such three-dimensional techniques cannot yet be used *in vivo*
and accordingly a working group of the World Health Organization (WHO)
has defined osteopenia and osteoporosis according to what can be measured, that is bone mineral content (BMC) or bone mineral density (BMD)
(Table 1.1).[4]

Use of this definition has allowed the prevalence of osteoporosis at different
skeletal sites to be determined for women in the Western world (Table 1.2) and
these prevalence figures are roughly equivalent to the prevalence rates of hip
and other fractures at various ages (see Chapter 3).

**Table 1.1.**  *The definition of osteoporosis as suggested by a Working Group of the
World Health Organisation[4]*

---

*Normal*
Women with BMC or BMD greater than 1 SD below the young normal mean (*T*-score)

*Osteopenia*
Women with BMC or BMD less than 1 SD and more than 2.5 SD below the young
normal mean (*T*-score)

*Osteoporosis*
Women with BMC or BMD more than 2.5 SD below the young normal mean

*Severe (established) osteoporosis*
Women with BMD more than 2.5 SD below the young normal mean and fragility
fractures

---

**Table 1.2.** *Prevalence of osteoporosis in Western women (World Health Organization, 1994[4])*

| Age range | Osteoporosis at any site (%) | Osteoporosis at hip (%) |
|-----------|------------------------------|-------------------------|
| 30–49 | 0 | 0 |
| 50–59 | 14.8 | 3.9 |
| 60–69 | 21.6 | 8.0 |
| 70–79 | 38.5 | 24.5 |
| 80 + | 70.0 | 47.6 |
| 50 + | 30.3 | 16.2 |

The term 'osteopenia' as used by the WHO can be confused with radiological osteopenia, that is bones which appear demineralized on X-ray. As the BMC/BMD term 'osteopenia' has not subsequently been found to have clinical relevance, it has been largely ignored of late. The WHO definition only applies to women, although recent reviews have suggested that applying the same definition to men based on a male normative range would have the same utility,[5] although this is not universally accepted.[6]

The need for a diagnosis to be based on the assessment of BMC or BMD has encouraged further development of bone mass measurement tools of which there are now a confusing profusion.

# Measurement of Bone Mass

In the last four decades, there has been an increasing interest in capturing numerically the obvious appearance of bone demineralization (loss of bone mineral), which may be an inevitable accompaniment of ageing of the skeleton[7]. This interest was initially spurned by the observation that there could be up to 30% reduction in bone mass before it became radiologically evident. This headlong rush to develop new techniques has been given impetus by the decision by the WHO working group to define osteoporosis by BMC or density (see above). The various techniques, which are still in use in some guise in the new millennium, are summarized in Table 1.3.

Initially, techniques used simple measures of cortical thickness (radio-grammetry) measured laboriously and imprecisely from radiographs using a ruler and Vernier calipers. However the field was revolutionized in the 1960s when Sorenson and Cameron[8] first described the use of photon absorptiometry to measure BMC of the forearm. The physics of the various techniques described since the initial observations are beyond the scope of this chapter, but photon absorptiometry has become the standard technique of assessment today with the initial radioisotope sources used to produce photons being replaced with much more stable and reproducible X-ray sources. Peripheral BMC, usually assessed at the radius or os calcis was measured initially using the absorption of transmitted photons from a single energy iodine source (SPA). However, use of a single photon source required the measurement to be

**Table 1.3.** *Advantages and disadvantages of bone measurement techniques*

| Technique | Site | Advantages | Disadvantages |
|---|---|---|---|
| **Peripheral skeleton** | Heel, forearm | Generally reasonably inexpensive; fairly portable | Measure relatively metabolically inactive bone sites; limited value in monitoring treatment effects or bone loss |
| Single energy X-ray absorptiometry (pSEXA) | Radius, os calcis | Relatively inexpensive | Requires a water bath; has radiation source |
| Dual energy X-ray absorptiometry (pDEXA) | Radius, os calcis | Does not require water-bath; precise | Has radiation source |
| Computed tomography (pQCT) | Radius, tibia | Measures true bone density | Relatively expensive; precision less than pDEXA |
| Digital radiogrammetry | Metacarpals, radius | Well-tried technique based on digitized radiographs; precise; inexpensive | Measures only cortical bone |
| Transmission QUS | Os calcis | Inexpensive portable; may give data on bone structure; easy to use | Does not measure bone density directly |
| Reflectance QUS | Radius, phalanx, tibia, metatarsal | Multi-site option with some machines | Primarily a measurement of cortical bone; needs training to give good precision |
| **Axial Skeleton** | Spine, hip | Measures areas adjacent to sites of osteoporotic fracture | Not easily portable; needs extensive operator training and interpretation skills |
| Dual energy X-ray absorptiometry (DEXA) | Lumbar spine, proximal femur, total body, peripheral sites | 'Gold' standard technique; multi-detector scanners give reasonable image quality; good precision | Measures *areal* BMD alone; difficult to interpret spine scans in the elderly |
| Computed tomography (QCT) | Lumbar spine, proximal femur | Measures true bone density | High radiation dose; very expensive and heavily used equipment |
| Magnetic resonance imaging | Spine, hip, forearm | May give bone quality data | Does not measure bone density; not yet a standard technique |

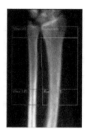 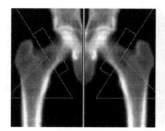 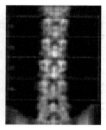

**Figure 1.2**  *Images from a Lunar Prodigy®, GE Medical Systems LUNAR, Madison, WI, USA (www.gemedicalsystems.com)*

carried out while the limb was immersed in water, which acted as a soft tissue equivalent thus allowing the transmission profile of water/soft tissue to be compared with significantly more dense bone.

The use of a water bath is somewhat inconvenient for peripheral bone sites and impossible for axial sites hence encouraging the development of 'dry systems', which currently use a 2-photon energy system replacing the original isotope photon source with X-rays, *viz.* dual energy X-ray absorptiometry (DEXA or DXA). This has become the standard clinical methodology for peripheral and axial measurements allowing assessment of areal BMD (BMC corrected for the assessed bone area) at sites of importance for fracture, including the forearm, lumbar spine and both proximal femurs (Figure 1.2).

The DEXA techniques thus allow a measurement of the mineral content of bone partially corrected for skeletal size, albeit only in two-dimensions. Accordingly, many authors have criticized the use of the term 'BMD' as assessed by DEXA as misleading, because it can only be considered to be areal BMD and incompletely corrects skeletal dimensions.[9] Use of additional correction equations to determine estimated volumetric BMD have been published and have value,[9,10] but it must be remembered that one of the main purposes of measurement of bone mass is in fracture prediction and here the data support the use of areal BMD as the preferred measurement. Assessment of true volumetric bone density is possible using CT technology; this technique has the disadvantage of requiring very expensive CT scanners and also involving high radiation exposure at least when used at axial sites. However, it has the advantage of being able to assess both cortical and trabecular bone either at peripheral sites, usually the radius, or at the important axial sites, most commonly the vertebral bodies.

For peripheral bone measurements, a new development has been the use of a non-ionizing radiation method, quantitative ultrasound (QUS). As can be seen from Table 1.3, this uses either transmission of propagated ultrasound through a peripheral bone or reflection of the ultrasound wave from primarily cortical bone. Transmission QUS allows measurements of attenuation of the broadband ultrasound beam and the speed of passage of the sound-wave across the bone along with manufacturer specific combined indices such as the inappropriately named 'stiffness', which is little to do with the elasticity of bone. Such indices correlate rather poorly *in vivo* with site-specific and distant-site BMD measurements, implying that the measurement is assessing other bone parameters as well as BMC and density. As QUS techniques do not measure bone density directly, they cannot be used to diagnose osteoporosis, as the definition of the condition is firmly dependent on the assessment of BMC or density. They do, however, predict those patients who subsequently fracture almost as well as BMD measurements and may give additional data on bone structure, a parameter poorly captured by the current methods of BMD measurement[11] An important potential use of bone mass equipment for the nutritionist is the assessment of body composition. Whole body DXA techniques allow measurement of bone and soft tissue using the two energies of X-rays assessed as part of the measurement protocol. Software splits the measurement of soft tissue into three compartments, *viz.*, (1) bone, (2) lean body mass and (3) total body fat, although to derive data for a three-compartment model accurately, it would theoretically be important to have beams of three different energies rather than two. For this reason, the technique has not received very wide-spread use and while giving some useful and fairly reproducible data,[12] different machines may give markedly different values.[13]

In summary, measurement of bone mass can be used theoretically to address the questions laid out in Table 1.4.

The requirement to use various techniques for diagnosing osteoporosis based on BMC or density clearly rules out the use of QUS or possibly radio-grammetry as potential diagnostic tools, but the more important question is whether the techniques predict future fracture. If, as seems likely, the diagnosis of osteoporosis and particularly intervention thresholds in the future becomes more allied to life-time or 10-year fracture risk,[14] then clearly as all techniques predict fracture, all will have future utility in diagnostic terms. It does not seem to matter at which site bone mass is assessed in terms of fracture prediction although site-specific assessment does show a slightly improved relative risk of future fracture compared with distant site assessment[15] (Table 1.5).

## Intervention Thresholds

In recent years, decisions on treatment with pharmaceutical preparations have become dependent on the presence of a diagnosis of osteoporosis, *i.e.* an axial BMD measurement below a *T*-score of (2.5. However, the publication of 10-year absolute fracture risks for men and women have shown that these are dependent not only on age as discussed in Chapter 2, but also on BMD and sex.[16] As can be seen from Figure 1.3, rates of fracture in women at any age are much higher than in men, but of more significance in terms of cost effective pharmaceutical intervention, the absolute risk of fracture at any one age in either sex is heavily dependent on the BMD measurement. This means that intervention with an expensive

**Table 1.4.** *Clinical uses of BMD measurements*

| Reason for use | QUS | Peripheral DEXA | Radiogrammetry | Peripheral QCT | Axial QCT | Axial (spine and hip) DEXA | Whole body DEXA |
| --- | --- | --- | --- | --- | --- | --- | --- |
| Diagnosis of osteoporosis | No | Yes | Possibly | Yes | Yes | Yes | No |
| Prediction of fracture | Yes | Yes | Yes | Yes | Yes | Yes | Yes |
| Targeting patients for axial DEXA | Yes | Possibly | Possibly | ? | ? | Not relevant | Not relevant |
| Target therapy | ? | ? | ? | ? | ? | Yes | No |
| Monitoring response to therapy | Probably not | Probably not | Probably not | Possibly | Yes, but high radiation | Yes but see text | No |
| Body composition | No | No | No | No | No | No | Yes |

**Table 1.5.** *Relative risks of fracture for each standard deviation reduction in age-standardized BMD*

| Site of measurement | Forearm fracture | Hip fracture | Vertebral fracture | All fractures |
|---|---|---|---|---|
| Distal radius | 1.7 (1.4–2.0) | 1.8 (1.4–2.2) | 1.7 (1.4–2.1) | 1.4 (1.3–1.6) |
| Hip | 1.4 (1.4–1.6) | 2.6 (2.0–3.5) | 1.8 (1.1–1.7) | 1.6 (1.4–1.8) |
| Lumbar spine | 1.5 (1.3–1.8) | 1.6 (1.2–2.2) | 2.3 (1.9–2.8) | 1.5 (1.4–1.7) |

preparation is unlikely to become cost effective in younger post-menopausal women or middle-aged men unless they have incredibly low BMD values. If nutrition and other lifestyle changes can be shown to be effective in fracture prevention (see Chapters 29 and 30) and such interventions turn out to be

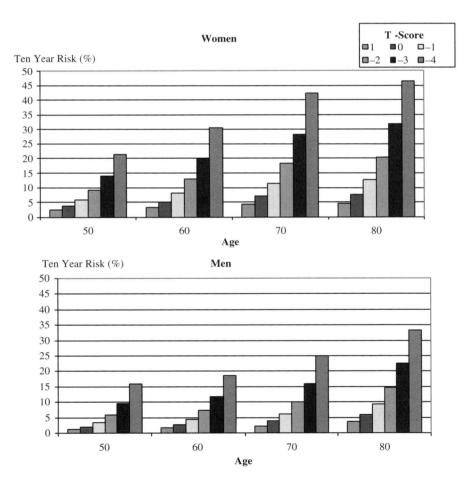

**Figure 1.3**   *Ten-year fracture risk according to age and BMD T-score in women and men*[16]

less expensive than the newer pharmaceutical preparations, then it may be that such preventative treatment could be targeted at younger middle-aged subjects of either sex.

Until cost effective intervention thresholds based on absolute fracture risk predictions are developed for each therapeutic intervention, the over-riding question is whether intervening at a particular bone threshold, for example a *T*-score of $< -2.5$, measured at a specific site, such as the total femur will be associated with fracture reduction. This has only been demonstrated with axial BMD and then only with the bisphosphonates.[17,18] Until such data become available with other intervention thresholds it is dangerous to assume that treating women with low QUS scores, for example, will necessarily translate into absolute fracture reduction. For that, if no other reason, recent guidelines have suggested that 2-site axial DEXA (usually spine and hip) is the current method of choice for diagnosis.[18–20] Targeting BMD on those with risk factors has been suggested by recent guidelines (Table 1.6), an approach recently endorsed by a meta-analysis of the literature.[21]

Many of these risk factors are historical and not necessarily evidence-based, but act as the best decision process available for case-finding, reflecting as they do many non-nutritional risk factors for osteoporosis (see Chapter 4). Although genetic factors are clearly the most determinant of bone mass (see Chapter 5), only a maternal hip fracture is accepted as an indication for BMD, reflecting the

**Table 1.6.** *Risk factors providing indications for the diagnostic use of bone densitometry[19]*

---

1. Presence of strong risk factors
   Oestrogen deficiency:
     Premature menopause (age <45 years)
     Prolonged secondary amenorrhoea
     Primary hypogonadism
   Corticosteroid therapy
     Oral Prednisolone to be used for three months or more[a]
     Maternal family history of hip fracture
     Low body mass index ($< 19$ kg m$^{-2}$)
   Other disorders associated with osteoporosis:
     Anorexia nervosa
     Malabsorption syndromes
     Primary hyperparathyroidism
     Post-transplantation
     Chronic renal failure
     Hyperthyroidism
     Prolonged immobilization
     Cushing's syndrome
2. Radiographic evidence of osteopenia and/or vertebral deformity
3. Previous fragility fracture, particularly of the hip, spine or wrist
4. Loss of height, thoracic kyphosis (after radiographic confirmation of vertebral deformities)

---

[a]Updated from guidelines specifically for the prevention and management of glucocorticoid induced osteoporosis.[22]

difficulty of clinically defining the condition of osteoporosis without bone mass assessment. As yet there are no cost-effective data available, which would argue in favour of population screening at any age, although guidelines in the US[23] do consider that assessment and treatment of all women over the age of 60 could be cost-effective, if a cost per quality associated life year of $30 000 is accepted. As the costs of the management of current fractures in the UK has been estimated at £1.7 billion per annum,[24] the potential saving that could be induced by a population-based approach to fracture reduction including targeted assessment of bone mass is likely to be considered in future years. What is universally accepted is that BMD measurements are not simple to interpret[25] and require an experienced operator.

## Targeting Prevention without Bone Mineral Measurements

As indicated above, it appears that for cost-effective use of bisphosphonates to prevent fracture, targeting treatment based on measurements of axial DEXA are essential. However, other therapies such as hormone replacement therapy may be effective in women regardless of their BMD.[26] As will be discussed in Chapter 8.2, calcium may have a role in fracture prevention regardless of the presence of osteoporosis, especially in the elderly and when used in association with vitamin D (Chapter 9). Such therapies may not to be targeted at those specifically with low bone mass. For example, calcium and vitamin D are recommended in fracture prevention in the frail elderly especially those at risk of hip fracture.[27] Recent evidence suggests that vitamin D supplementation may improve muscle strength and hence potentially reduce falls in those who are deficient. Therefore, it may be concluded that vitamin D supplementation will in the future need to be targeted at those who have been shown to be deficient. Similarly, wearing hip protectors may prevent hip fractures[28] and an obvious use may be to target the use of these garments in those at risk of hip fracture due to frequent falls.[29]

## Conclusions

This chapter has briefly reviewed the relationship between osteoporosis and fractures, the epidemiology of which is fully discussed in Chapter 2. While there are many techniques for measuring bone mass, only BMD measurements can be used to diagnose osteoporosis although other techniques are almost as predictive of fracture. Previously, intervention decisions have been based simply on a diagnosis of osteoporosis, but there is increasing recognition of the importance of the use of 10-year fracture risk to determine cost-effective intervention. The use of relatively expensive drug therapy may therefore require to be targeted at those with a high 10-year fracture risk including a diagnosis of osteoporosis. There is therefore a clear role for non-pharmaceutical interventions, such as nutrition in prevention and treatment in those at lower absolute risk. Further work is required to determine the most appropriate triggers for nutritional and other pharmaceutical interventions.

# References

1. Consensus Development Panel on Osteoporosis Prevention DaT, Osteoporosis prevention, diagnosis and therapy, *JAMA*, 2001, **285**, 785–795.
2. C.H. Turner, Biomechanics of bone: determinants of skeletal fragility and bone quality, *Osteoporosis Int.*, 2002, **13**, 97–104.
3. B.M. Ingle and R. Eastell, Site-specific bone measurements in patients with ankle fracture. *Osteoporos Int.*, 2002, **13**, 342–347.
4. Anonymous. Assessment of fracture risk and its application to screening for postmenopausal osteoporosis. Report of a WHO Study Group. [Review] [466 refs]. World Health Organization Technical Report Series 843: 1–129. 94.
5. J.A. Kanis, O. Johnell, A. Oden, C. De Laet and D. Mellstrom, Diagnosis of osteoporosis and fracture threshold in men, *Calcif Tissue Int.*, 2001, **69**, 218–221.
6. L.J. Melton III, E.S. Orwoll, R.D. Wasnich, Does bone density predict fractures comparably in men and women? *Osteoporos Int.*, 2001, **12**, 707–709.
7. H.K. Genant, K. Engelke, T. Fuerst, C.C. Gluer, S. Grampp, S.T. Harris, M. Jergas, T. Lang, Y. Lu, S. Majumdar, A. Mathur and M. Takada, Noninvasive assessment of bone mineral and structure: state of the art, *J. Bone Miner. Res.*, 1996, **11**, 707–730.
8. J.A. Sorenson and J.R. Cameron, A reliable in vivo measurement of bone-mineral content. *J. Bone Joint Surg. Am.*, 1967, **49**, 481–497.
9. A. Prentice, T.J. Parsons and T.J. Cole, Uncritical use of bone mineral density in absorptiometry may lead to size-related artifacts in the identification of bone mineral determinants, *Am. J. Clin. Nutr.*, 1994, **60**, 837–842.
10. N.F. Peel and R. Eastell, Diagnostic value of estimated volumetric bone mineral density of the lumbar spine in osteoporosis, *J. Bone Miner. Res.*, 1994, **9**, 317–320.
11. A Stewart and D.M. Reid, Quantitative ultrasound in osteoporosis, *Semin. Musculoskelet. Radiol.*, 2002, **6**, 229–232.
12. L.K. Bachrach, Dual energy X-ray absorptiometry (DEXA) measurements of bone density and body composition: promise and pitfalls, *J. Pediatr. Endocrinol. Metab.*, 2000, **13**(Suppl 2), 983–988.
13. L. Genton, D. Hans, U.G. Kyle, C. Pichard, Dual-energy X-ray absorptiometry and body composition: differences between devices and comparison with reference methods, *Nutrition*, 2002, **18**, 66–70.
14. J.A. Kanis, Diagnosis of osteoporosis and assessment of fracture risk. *Lancet*, 2002, **359**, 1929–1936.
15. D. Marshall, O. Johnell and H. Wedel, Meta-analysis of how well measures of bone mineral density predict occurrence of osteoporotic fractures, *BMJ*, 1996, **312**, 1254–1259.
16. J.A. Kanis, O. Johnell, A. Oden, A. Dawson, C. De Laet and B. Jonsson, Ten year probabilities of osteoporotic fractures according to BMD and diagnostic thresholds, *Osteoporos Int.*, 2001, **12**, 989–995.
17. S.R. Cummings, D.M. Black, D.E. Thompson, W.B. Applegate, E. Barrett-Connor, T.A. Musliner, L. Palermo, R. Prineas, S.M. Rubin, J.C. Scott, T. Vogt, R. Wallace, A.J. Yates and A.Z. LaCroix, Effect of alendronate on risk of fracture in women

with low bone density but without vertebral fractures: results from the Fracture Intervention Trial, *JAMA*, 1998, **280**, 2077–2082.

18. Royal College of Physicians BaTS, Update on pharmacological interventions and an algorithm for management, *Osteoporosis: Clinical Guidelines for Prevention and Treatment*, Royal College of Physicians, London, 2000.
19. Royal College of Physicians, *Osteoporosis: Clinical Guidelines for Prevention and Treatment*, Royal College of Physicians, London, 1999.
20. Management of Osteoporosis, Edinburgh: Scotttish Intercollegiate Guideline Network, 2003.
21. S.R. Cummings, D. Bates and D.M. Black, Clinical use of bone densitometry: scientific review, *JAMA*, 2002, **288**, 1889–1897.
22. Anonymous, *Glucocorticoid Induced Osteoporosis: Guidelines for Prevention and Treatment*, Royal College of Physicians, London, 2002.
23. D.M. Eddy, C.C. Johnston Jr., S.R. Cummings, B. Dawson-Hughes, R. Lindsay, I.I.I.L.J. Melton and C.W. Slemenda, Osteoporosis: review of the evidence for prevention, diagnosis, and treatment and cost-effectiveness analysis, status report, *Osteoporosis Int.*, **8** (suppl. 4), 1–82. see also p. 98.
24. D.J. Torgerson, C.P. Iglesias and D.M. Reid, The economics of fracture prevention. In *The Effective Management of Osteoporosis*, 2001, 111–121.
25. D.W. Bates, D.M. Black and S.R. Cummings, Clinical use of bone densitometry: clinical applications, *JAMA*, 2002, **288**, 1898–1900.
26. Writing Group for the Women's Health Initiative Investigators, Risks and benefits of estrogen plus progestin in healthy postmenopausal women, *JAMA*, 2002, **288**, 321–333.
27. Prevention and Management of Hip Fracture in Older People, Edinburgh: Scottish Intercollegiate Guidelines Network, 2002.
28. M.J. Parker, L.D. Gillespie and W.J. Gillespie, Hip protectors for preventing hip fractures in the elderly, [comment][update of Cochrane Database Syst Rev., 2000; (4): CD001255; PMID: 11034706], [Review] [31 refs], Cochrane (2), CD001255, 2001. Notes: COMMENTS: Comment in: Evid Based Nurs., 2002 Jan; 5(1): 23, PMID: 11915801.
29. G. Meyer, A. Warnke, R. Bender, I. Muhlhauser, Effect on hip fractures of increased use of hip protectors in nursing homes: cluster randomised controlled trial, *BMJ*, 2003, **326**, 76

CHAPTER 2

# Epidemiology and Public Health Impact of Osteoporosis

ANN SCOTT RUSSELL, ELAINE DENNISON and
CYRUS COOPER

MRC Environmental Epidemiology Unit, Southampton General Hospital,
Southampton SO16 6YD, UK, Email: graham.russell@ndos.ox.ac.uk

**Key Points**

- Osteoporotic fractures are associated with increased mortality and morbidity and are a considerable burden to the health care system.
- The annual cost to the health care system in the UK has been estimated at 1.7 billion pounds.
- The lifetime risk of a fragility fracture in the UK is 53.2% in women and 20.7% in men, at the age of 50.
- A 50 year old US White woman has a 13% chance of experiencing attributable functional decline after any osteoporotic fracture.
- It has been estimated that there were 1.66 million hip fractures worldwide in 1990.
- Predictions of an increase in hip fracture incidence to 6.26 million in 2050 have been made, largely as a result of demographic changes in the population.

## Introduction

Osteoporosis is a systemic skeletal disorder characterised by low bone mass and micro-architectural deterioration of bone tissue with a consequent increase in bone fragility and susceptibility to fracture.[1] These fragility fractures have devastating health consequences through their association with increased mortality and morbidity and are consequently a considerable burden to the health care system. Prospective studies indicate that the risk of osteoporotic fracture increases continuously as bone mineral density (BMD) declines with a 1.5- to 3-fold increase in risk of fracture for each standard deviation fall in BMD.[2] Other risk factors include increasing age, female sex, race, family history, hormonal status and steroid usage. Osteomalacia may co-exist, particularly in the elderly, when the disease may be due to chronic malabsorption from any cause, dietary deficiency of vitamin D, or reduced exposure to sunlight. Other groups at risk from osteomalacia include women from ethnic groups where exposure of the skin to sunlight is forbidden, particularly if the diet is poor in dairy products, and also adults with chronic renal failure or those taking anti-epileptic medication. Fractures due to osteoporosis are found in areas typified by large amounts of trabecular bone. Common fracture sites include the proximal femur, vertebrae and distal radius. This chapter reviews the epidemiology and the personal and economic burden associated with osteoporotic fracture.

## Absolute Risks of Fracture in Individuals

While most American women under the age of 50 years have normal BMD, 27% are osteopenic and 70% are osteoporotic at the hip, lumbar spine or forearm by the age of 80 years. Epidemiological studies from North America have estimated the remaining lifetime risk of common fragility fractures to be 17.5% for hip fracture, 15.6% for clinically diagnosed vertebral fracture and 16% for distal forearm fracture among White women aged 50 years. Corresponding risks among men are 6, 5 and 2.5% respectively. A recent British study using the General Practice Research Database[3] estimated the lifetime risk of any fracture to be 53.2% at the age of 50 among women, and 20.7% at the same age among men. Projected risks of fracture at various ages are displayed in Table 2.1. These fall to 26.8% at the age of 80 among women, and 9.6% at the same age among men. Site-specific lifetime risks at the age of 50 were as follows: women – radius/ulna 16.6%, femur/hip 11.4%, vertebral body 3.1%; men – radius/ulna 2.9%, femur/hip 3.1%, vertebral body 1.2%. Among women, the 10-year risk of any fracture increased from 9.8% at the age of 50 to 21.7% at the age of 80, while among men the 10-year risk remained fairly stable with advancing age (*i.e.* 7.1% at the age of 50 and 8.0% at the age of 80).

## Health Impact of Osteoporotic Fracture

All osteoporotic fractures are associated with significant morbidity, but both hip and vertebral fractures are also associated with excess mortality.

### Mortality

For fractures of the femur/hip and vertebral body, previous studies have shown clear evidence of excess mortality up to 5 years after fracture for both sexes, although fractures of the radius/ulna were only associated with slight excess

**Table 2.1.** *Estimated risks of fractures at various ages*

|  | Current age (years) | Any fractures (%) | Radius/ulna (%) | Femur/hip (%) | Vertebral (%) |
|---|---|---|---|---|---|
| *Lifetime risk* | | | | | |
| Women | 50 | 53.2 | 16.6 | 11.4 | 3.1 |
| | 60 | 45.5 | 14.0 | 11.6 | 2.9 |
| | 70 | 36.9 | 10.4 | 12.1 | 2.6 |
| | 80 | 28.6 | 6.9 | 12.3 | 1.9 |
| Men | 50 | 20.7 | 2.9 | 3.1 | 1.2 |
| | 60 | 14.7 | 2.0 | 3.1 | 1.1 |
| | 70 | 11.4 | 1.4 | 3.3 | 1.0 |
| | 80 | 9.6 | 1.1 | 3.7 | 0.8 |
| *10-Year risk* | | | | | |
| Women | 50 | 9.8 | 3.2 | 0.3 | 0.3 |
| | 60 | 13.3 | 4.9 | 1.1 | 0.6 |
| | 70 | 17.0 | 5.6 | 3.4 | 1.3 |
| | 80 | 21.7 | 5.5 | 8.7 | 1.6 |
| Men | 50 | 7.1 | 1.1 | 0.2 | 0.3 |
| | 60 | 5.7 | 0.9 | 0.4 | 0.3 |
| | 70 | 6.2 | 0.9 | 1.4 | 0.5 |
| | 80 | 8.0 | 0.9 | 2.9 | 0.7 |

Reproduced by permission from Van Staa *et al.*[3]

mortality amongst men (Table 2.2). While this may represent complications of the fracture and subsequent surgery for hip fractures, it is likely to reflect coexisting co-morbidity in vertebral fracture sufferers. It has been estimated that 8% of men and 3% of women over 50 years die whilst hospitalised for their hip fracture; mortality rates continue to rise over the subsequent months such that at 1-year mortality is 36% for men (higher for the very elderly) and 21% for women.[4] By 2 years after hip fracture, mortality rates decline back to baseline except in elderly patients and amongst men. The four main predictors for higher mortality appear to be: male sex, increasing age, coexisting illness and poor pre-fracture functional status.

Excess mortality after vertebral fracture appears to increase progressively after diagnosis of the fracture, as demonstrated in Figure 2.1.[5] This has been observed in studies based on clinically diagnosed vertebral deformities and in those using radiological morphometric approaches to classify vertebral deformity.[5] Impaired survival is more pronounced for vertebral fractures that follow moderate rather than severe trauma of which only 8% are actually attributable to osteoporosis (Figure 2.2). Five-year survival appears to be worse for men (72% 5-year-survival) than for women (84% 5-year-survival). In women, an excess risk of death from cardiovascular and pulmonary disease that rises with increasing number of vertebral fractures has been observed.[6]

**Table 2.2.** *Observed and expected survival (%) following a fracture among men and women aged >65 years*

| | | Radius/ulna (%) | | Femur/hip (%) | | Vertebral (%) | |
|---|---|---|---|---|---|---|---|
| | | *Observed* | *Expected* | *Observed* | *Expected* | *Observed* | *Expected* |
| Women | At 3 months | 98.2 | 98.6 | 85.6 | 97.7 | 94.3 | 98.4 |
| | At 12 months | 94.0 | 94.4 | 74.9 | 91.1 | 86.5 | 93.6 |
| | At 5 years | 75.5 | 73.8 | 41.7 | 60.9 | 56.5 | 69.6 |
| Men | At 3 months | 97.3 | 98.0 | 77.7 | 97.3 | 87.8 | 97.9 |
| | At 12 months | 89.6 | 92.4 | 63.3 | 90.0 | 74.3 | 91.8 |
| | At 5 years | 62.8 | 66.4 | 32.2 | 58.2 | 42.1 | 64.4 |

Reproduced with permission from Van Staa *et al.*[3]

## Morbidity

In the United States about 7% of survivors of all types of fragility fractures have some degree of permanent disability and 8% require long term nursing home care. Overall, a 50-year old US White woman has a 13% chance of experiencing attributable functional decline after any fracture.[7] Hip fracture invariably requires hospitalisation. The degree of functional recovery after this injury is age-dependent; in the US 14% of hip fracture patients in the 50–55 age group were discharged to nursing homes compared to 55% of those aged > 90 years.[4] The length of stay was also related to age. In addition, malnutrition, particularly of protein, may significantly delay recovery and is a common problem in the hospital environment. Premorbid status is also a strong predictor of outcome. In the United States, 25% of formerly independent people became at least partially dependent, 50% of those who were dependent pre-fracture were admitted to nursing homes, and those who were already in nursing homes remained there.[7] Results appear to be similar in France.[8] Hip fracture also has a significant effect on mobility; 1 year after hip fracture 40% were still unable to walk independently, 60% required assistance with at least one essential activity of daily living (*e.g.* dressing, bathing) and 80% were unable to perform at least one instrumental activity of daily living such as driving or shopping.[4]

The impact of a single vertebral fracture may be low, but multiple fractures cause progressive loss of height and kyphosis and severe back pain in the acute stages. The resultant loss of mobility can further exacerbate underlying osteoporosis leading to increased risk of further fractures.[9] The psychological impact of functional loss can cause depression and social isolation as well as loss of self-esteem.

Although good functional recovery post distal forearm fracture may be poor, reflecting complications such as reflex sympathetic dystrophy, neuropathies and post traumatic arthritis, mortality after Colles fracture does not deviate from the expected rate.

## Economic Cost

In the UK, the annual cost to the healthcare system from osteoporotic fracture cases has been estimated at 1.7 billion pounds, a significant contribution to the costs for Europe as a whole (13.9 billion euros), and substantially lower than those for the USA (17.9 billion dollars). Hip fractures account for over a third of the total figure, and reflect the cost of inpatient medical services and nursing home care. Expenditures are rising faster than the general rate of inflation and are a major source of concern to governmental leaders. Currently, research is focussing on the cost-effectiveness of treatment programmes for osteoporosis; at present this still remains difficult to assess, and in addition in many countries widespread use of expensive drugs may not be affordable.

# Fracture Epidemiology

Fracture incidence in the community is bimodal, with a peak in the young and elderly. In youth, fractures are usually associated with substantial trauma, occur in the long bones and are seen more frequently in males than females. Osteoporotic fractures characteristically occur in those areas of the skeleton with high amounts of trabecular bone after low or moderate trauma. There is increased frequency of fracture with age in both sexes, reflecting a combination of lower bone density with an increased tendency to fall in the elderly. Racial variation exists, with lower fracture rates observed in Black than in Caucasian or Asian populations, although geographic variations in fracture rates have been demonstrated even within countries, suggesting that environmental factors are also important in the pathogenesis of hip fracture.

The three sites most closely associated with osteoporotic fracture are the hip, spine and distal forearm. However the epidemiological characteristics vary with each site as will be discussed below.

## Hip Fracture

It has been estimated that worldwide there were 1.66 million hip fractures in 1990,[10] representing 1 197 000 fractures in females and 463 000 among males. In Western populations, among individuals above 50 years of age, there is a female preponderance of hip fracture with a female-to-male incidence ratio of approximately 2–1.

Hip fracture incidences vary considerably according to geographic area and race, and may vary widely within the same country. Within Europe, epidemiological studies report rates that vary up to 7-fold from country to country. Age- and sex-adjusted hip fracture rates are generally higher in Whites than in Black or Asian populations.[11] The highest recorded rates of hip fracture, after age-adjustment, come from Sweden and the northern US, intermediate rates are found in Asian populations; African people have the lowest rates. Men and women of African origin have similar hip fracture rates.[12] Some inter-racial differences may be explained by variations in reversible lifestyle factors such as low-milk consumption, cigarette smoking, lack of sunlight exposure, low BMI and physical activity,[13] but genetic factors may also be important. Various diseases associated with secondary osteoporosis and with falling seem to be a more important cause of hip fractures in men than in women.[14]

## Vertebral Fracture

Until recently, accurate epidemiological studies of vertebral fracture have been limited, reflecting the clinically silent nature of two-thirds of these fractures, in addition to a lack of consensus regarding the techniques that should be used to define vertebral deformity.[15] However, the advent of description of morphometric and semiquantitative visual techniques have now enabled a number of studies to report prevalence of vertebral fracture. Only about a third of all vertebral deformities noted on radiographs come to medical attention, and less than 10% necessitate admission to hospital.[16]

For example, in Rochester, MN the prevalence of any vertebral deformity was estimated at 25.3% per 100 women aged 50 years and above, whilst the incidence of a new deformity in this group was estimated at 17.8 per 1000 person-years.[17] In contrast, in the European Vertebral Osteoporosis Study, one in eight women and men aged 50 years and over had evidence of vertebral deformity. Prevalence of vertebral deformity increased steadily with age in both sexes although the gradient was steeper for women. There was a 3-fold variation in the occurrence of deformity across Europe and up to a 2-fold variation in centres within individual countries,[18] perhaps reflecting a combination of environmental and genetic factors. The risk of vertebral deformity among men was significantly elevated in those with high levels of physical activity suggesting the aetiological importance of trauma. In contrast, women with higher levels of customary physical activity have a reduced risk of deformity.[19]

In instances where comparable methods and definitions have been used in studies, the prevalence of vertebral fractures has been more similar across regions than seen for hip fractures.[20-22] For example, vertebral fracture prevalence among women in Hiroshima was 20–80% greater than for White women in Rochester, MN, USA despite lower hip fracture rates in the former.[20] Similarly, the risk of vertebral fractures among postmenopausal women in Beijing is only 25% lower than that among women in Rochester, MN, despite much lower hip fracture rates in the former.[21]

Only a quarter of vertebral fractures result from falls, and most are precipitated by routine daily activities such as bending or lifting light objects, reflecting the compressive load of such acts.

## Wrist Fracture

Distal forearm fracture almost always results as a consequence of a fall onto an outstretched hand. These fractures show a steep rise in incidence during the perimenopausal period among women, but a plateau thereafter. In men, there is no apparent increase in the incidence of wrist fracture with age. In White women, the incidence increases linearly between the ages of 40 and 65 and then stabilises, while in men the incidence remains constant between 20 and 80 years (Figure 2.3). A much stronger sex ratio exists for this fracture than for most others and this has been estimated to be 4:1 in favour of women. Although geographic variation exists, a partial explanation may be methodological considerations of case ascertainment, as less than 20% of forearm fracture patients are hospitalised. A winter peak is again demonstrated, but this probably is due to falls on icy surfaces. The plateau with age in women may be due to the mode of falls; later in life a woman is more likely to fall onto a hip than an outstretched hand as her neuromuscular co-ordination deteriorates.

## Other Fractures

The incidence of proximal humeral, pelvic, and proximal tibial fractures also rise steeply with age, and are greater in women than men. These are often termed frailty fractures, as they typically occur in women who are losing weight involuntarily.[23] Furthermore, there is some direct evidence that these fractures are associated with low BMD.[24] Three quarters of all proximal humerus fractures are due to moderate trauma, typically a fall from standing hight or less, and tend to be more common in women with poor neuromuscular function.[25]

# Clustering of Fractures

Previous vertebral deformities have been shown to increase the risk of subsequent vertebral deformities by 7- to 10-fold.[26] This is similar to the increase in risk of second hip fracture observed amongst those who have already sustained their first hip fracture. In Rochester, MN, residents aged 70 years or less with radiologically diagnosed vertebral deformities were followed up for the development of subsequent limb fractures. The standardised morbidity ratios of observed fractures to expected fractures for both sexes were 1.7 (95% CI 1.3–2.2) for the hip, 1.4 (95% CI 1.0–1.8) for the forearm and 1.5 (95% CI 1.3–1.8) for any limb fracture. This increased risk was more marked in subjects with vertebral deformities associated with moderate or minimal trauma than with severe trauma.[27]

The study of osteoporotic fractures, a prospective study of 9704 US women aged 65 years or older, has also investigated the relationship between prevalent vertebral deformity and incident osteoporotic fracture.[28] Prevalent vertebral deformity (assessed morphometrically) was associated with a 5-fold increased risk of sustaining a further vertebral deformity; the risk of hip fracture was increased 2.8-fold (95% CI 2.3–3.4) and the risk of any non-vertebral fracture 1.9-fold (95% CI 1.7–2.1), after adjustment for age and calcaneal BMD. Although there was a small increased risk of wrist fracture, this was not significant after adjustment for age and BMD.

The Rochester MN project has also been used to ascertain the ability of distal forearm fractures to predict future fractures.[29] Among residents who experienced their first distal forearm fracture aged 35 years or older, and excluding fractures that occurred on the same day as the index forearm fracture, hip fracture risk was increased 1.4-fold in women (95% CI 1.1–1.8) and 2.7-fold in men (95% CI 0.98–5.8). Excess risk in women was confined to those individuals who sustained their first forearm fracture at the age of 70 or greater. In contrast, vertebral fracture was increased at all ages, with a 5.2-fold (95% CI 4.5–5.9) increase in risk among women and a 10.7-fold (95% CI 6.7–16.3) increased risk among men.

# BMD and Fracture

The relationship between BMD and osteoporosis can be compared to that between blood pressure and stroke. Although low BMD is not a prerequisite for osteoporotic fracture, the risk of fracture is considerably elevated in the presence of low bone mass. Therefore, as with blood pressure, appropriate cut-off values can be defined so as to direct intervention towards 'at risk' individuals. BMD taken at different sites can be used to predict the future risk of fracture at the same, or other sites.

# Falls and Fracture

Distal forearm fracture and hip fracture usually follow a fall from standing high or low. The likelihood of fall increases with age, especially for elderly women. In one British study based in Oxford, one in three women had experienced a fall in the previous year in the 80–84 year age group. Above this age, half the women and one third of the men had experienced a fall in the preceding year.[30]

There are multiple factors that increase the likelihood of a fall, including use of medications such as antidepressants and benzodiazepines[31,32] and coexistent illness. Physical activity may be protective. In addition, only 1% of falls lead to a hip fracture, mainly due to the orientation of the fall. A fall sideways leads to direct impact on the greater trochanter and failure to break the fall with an outstretched hand.[33]

# Future Projections

Osteoporotic fractures represent a significant public health burden, that is set to rise in future generations. Life expectancy is increasing worldwide and it is estimated that the number of individuals aged 65 years and over will increase from the current figure of 323 million to 1555 million by the year 2050. These demographic changes alone can be expected to cause the number of hip fractures occurring worldwide to increase from 1.66 million in 1990 to 6.26 million in 2050. While half of all hip fractures among elderly people in 1990 took place in Europe and North America by 2050, the rapid ageing of the Asian and Latin American populations suggests that 75% of hip fractures will occur there. In addition, although age-adjusted rates appear to have levelled off in the northern part of the US and in the UK, rates continue to increase in countries such as Hong Kong and Finland. On the basis of current trends, hip fracture rates might increase in the UK from 46 000 in 1985 to 1 17 000 in 2016. A report from the Asian osteoporosis study has recently demonstrated moderate variation in hip fracture rates among Asian countries, with the highest rates in urbanised countries, suggesting that rapid economic development may prove important in rapidly increasing fracture rates.

Incidence for fractures at other skeletal sites have also risen during the last half century. Studies from Malmo, Sweden, have suggested age-specific secular increases for distal forearm, ankle, proximal humerus and vertebral fractures. These changes in vertebral fracture rate are particularly important as they point to an increasing prevalence of osteoporosis as an explanation for these trends. Similarly, a Finnish study has predicted three times as many osteoporotic fractures in the proximal humerus in Finland in 2030 than in 1998.

# Conclusions

Osteoporosis is a major public health problem through its association with fragility fracture, which typically occurs at the hip, spine and distal forearm. These fragility fractures increase in rate with advancing age, and are more common in women than men. In addition to this sex difference, racial variations in fracture rate are well recognised, with lower rates in Black than Caucasian or Asian populations. These fractures are associated with excess mortality (hip and vertebral fractures) and morbidity (all fractures) and hence represent a major public health burden. Future projections predict huge increases in fracture rate, particularly in developing countries as the population ages. Hence, the need for preventive strategies is becoming increasingly urgent.

# References

1. Consensus Development Conference, Prophylaxis and treatment of osteoporosis. *Osteoporosis Int.*, 1991, **1**, 114–117.
2. World health Organisation. Assessment of fracture risk and its application to screening for postmenopausal osteoporosis. *WHO Technical Report Series*: Geneva, WHO 1994.
3. T.P. Van Staa, E.M. Dennison, H.G.M. Leufkens and C. Cooper, Epidemiology of fractures in England and Wales, *Bone*, 2001, **29**, 517–522.
4. Hip fracture outcomes in people aged fifty and over: Mortality, Service Use, Expenditures, and Long-Term Functional Impairment. Washington, DC: Office of Technology Assessment, Congress of the United States; 1993. *US Department of Commerce Publication NTIS PB94107653* 1993.
5. C. Cooper, E.J. Atkinson, S.J. Jacobsen, W.M. O'Fallon and L.J. Melton III, Population-based study of survival after osteoporotic fractures. *Am. J. Epidemiol.*, 1993, **137**,1001–1005.
6. D.M. Kado, W.S. Browner, L. Palermo, M.C. Nevitt, H.K. Genant and S.R. Cummings, Vertebral fractures and mortality in older women: a prospective study, *Arch. Int. Med.*, 1999, **159**, 1215–1220.
7. E.A. Chrischilles, C.D. Butler, C.S. Davis and R.B. Wallace, A model of lifetime osteoporosis impact, *Arch. Int. Med.*, 1991, **151**, 2026–2032.
8. C. Baudoin, P. Fardellone, K. Bean, A. Ostertag-Ezembe and F. Hervy, Clinical outcomes and mortality after hip fracture: a 2-year follow-up study, *Bone*, 1996, **3**(Suppl), 149S–157S.
9. D.T. Gold, The clinical impact of vertebral fractures: quality of life in women with osteoporosis, *Bone*, 1996, **18**(Suppl. 3), S185–S189.
10. C. Cooper, G. Campion and L.J. Melton III, Hip fracture in the elderly: a worldwide projection, *Osteoporosis Int.*, 1992, **2**, 285–289.
11. S. Maggi, J.L. Kelsey, J. Litvak and S.P. Heyse, Incidence of hip fractures in the elderly: a cross-sectional study, *Osteoporosis Int.*, 1991, **1**, 232–241.

12. M.E. Farmer, L.R. White, J.A. Brody and K.R. Bailey, Race and sex differences in hip fracture incidence, *Am. J. Public Health*, 1984, **74**, 1374–1380.

13. O. Johnell, B. Gullberg, J.A. Kanis, E. Allander, L. Elffors, J. Dequeker, G. Dilsen, C. Gennari, A.L. Vaz, G. Lyritis, G. Mazzuoli, L. Miravet, M. Passeri, R.P. Cano, A. Rapado and C. Ribot, Risk factors for hip fracture in European women: the MEDOS study, *J. Bone Miner. Res.*, 1995, **10**(11), 1802–1814.

14. G. Poor, E.J. Atkinson, W.M. O'Fallon and L.J. Melton III, Predictors of hip fractures in elderly men, *J. Bone Miner. Res.*, 1995, **10**, 1900–1907.

15. C. Cooper, E.J. Atkinson, M.W. O'Fallon and J.L. Melton, Incidence of clinically diagnosed vertebral fractures: a population-based study in Rochester, Minnesota, 1985–1989. *J. Bone Miner. Res.*, 1992, **7**(2), 221–227.

16. C. Cooper and L.J. Melton III, Vertebral fracture: how large is the silent epidemic? *BMJ*, 1992, **304**, 793–794.

17. L.J. Melton, A.W. Lane, C. Cooper, R. Eastell, W.M. O'Fallon and B.L. Riggs, Prevalence and incidence of vertebral deformities. *Osteoporosis Int.*, 1993, **3**, 113–119.

18. T.W. O'Neill, D. Felsenberg, J. Varlow, C. Cooper, J.A. Kanis and A.J. Silman, The prevalence of vertebral deformity in European men and women: the European Vertebral Osteoporosis Study, *J. Bone Miner. Res.*, 1996, **11**(7), 1010–1017.

19. A.J. Silman, T.W. O'Neill, C. Cooper, J.A. Kanis and D. Felsenberg, Influence of physical activity on vertebral deformity in men and women: results from the European Vertebral Osteoporosis Study. *J. Bone Miner. Res.*, 1997, **12**, 813–819.

20. P.D. Ross, S. Fujiwara, C. Huang, J.W. Davis, R.S. Epstein, R.D. Wasnich, K. Kodama and L.J. Melton 3rd, Vertebral fracture prevalence in women in Hiroshima compared to Caucasians or Japanese in the US, *Int. J. Epidemiol.*, 1995, **24**, 1171–1177.

21. X. Ling, S.R. Cummings, Q. Mingwei, Z. Xihe, C. Xioashu, M. Nevitt and K. Stone, Vertebral fractures in Beijing, China: the Beijing osteoporosis project, *J. Bone Miner. Res.*, 2000, **15**, 2019–2025.

22. K.-S. Tsai, S.-J. Twu, R.-S. Chieng, R.S. Yang and T.K. Lee, Prevalence of vertebral fractures in Chinese men and women in urban Taiwanese communities. *Calcified Tissue Int.*, 1996, **59**, 249–253.

23. K.E. Ensrud, J. Cauley, R. Lipschutz and S.R. Cummings, Weight change and fractures in older women, *Arch. Intern. Med.*, 1997, **157**, 857–863.

24. D.G. Seeley, W.S. Browner, M.C. Nevitt, H.K. Genant, J.C. Gott and S.R. Cummings, Which fractures are associated with low appendicular bone mass in elderly women? The study of osteoporotic fractures research group, *Ann. Int. Med.*, 1991, **115**, 837–842.

25. J.L. Kelsey, W.S. Browner, D.G. Seeley, M.C. Nevitt and S.R. Cummings, Risk factors for fractures of the distal forearm and proximal humerus, *Am. J. Epidemiol.*, 1992, **135**, 477–489.

26. P.D. Ross, J.W. Davis, R. Epstein and R.D. Wasnich, Pre-existing fractures and bone mass predict vertebral fracture incidence, *Ann. Intern. Med.*, 1991, **114**, 919–923.

27. C. Cooper, M. Kotowicz, E.J. Atkinson, W.M. O'Fallon, B.L. Riggs and L.J. Melton, Osteoporosis, In *Risk of limb fractures among men and women with vertebral fractures*, Papapoulos Se, P. Lips, H.A.P. Pols, C.C. Johnson, P.D. Delmas, (eds.), Amsterdam, Elsevier, 1996, pp. 101–104.

28. D.M. Black, N.K. Arden, L. Palermo, J. Pearson, and S.R. Cummings, Prevalent vertebral deformities predict hip fractures and new vertebral deformities but not wrist fractures, *J. Bone Min. Res.*, 1999, **14**, 821–828.

29. M.T. Cuddihy, S.E. Gabriel, C.S. Crowson, W.M. O'Fallon and L.J. Melton, Forearm fractures as predictors of subsequent osteoporotic fractures. *Osteoporosis Int.*, 1999, **9**, 469–475.
30. S.J. Winner, C.A. Morgan and J.G. Evans, Perimenopausal risk of falling and incidence of distal forearm fracture, *Br. Med. J.*, 1989, **298**, 2486-2488.
31. B. Liu, G. Anderson, N. Mittmann, T. To, T. Axcell and N. Shear, Use of selective Serotonin-reuptake inhibitors or tricyclic antidepressants and risk of hip fractures in elderly people, *Lancet*, 1998, **351**, 1303–1307.
32. R.G. Cumming, Epidemiology of medication-related falls and fractures in the elderly, *Drugs Aging*, 1998, **12**, 43–53.
33. J. Pakkari, P. Kannus, M. Palvanen, A. Natri, J. Vainio, H. Aho, I. Vuori, M. Järvinen, Majority of hip fractures occur as a result of a fall and impact on the greater trochanter of the femur: a prospective controlled hip fracture study with 206 consecutive patients, *Calcified Tissue Int.*, 1999, **65**, 183–187.

CHAPTER 3

# Non-nutritional Risk Factors for Bone Fragility

OLOF JOHNELL

Department of Orthopaedics, Malmö University Hospital, SE-205 02 Malmö, Sweden, Email: carin.holmberg@orto.mas.lu.se

**Key Points**
- Risk factors can be used to predict fracture risk
- Some risk factors are also modifiable and interventions against these might reduce fracture rate
- Risk factors can be divided into major and minor factors
- Major factors include age, sex, bone mineral density (BMD), prior fracture
- To assess fracture risk, the absolute 10-year risk is the best measurement
- Combination of risk factors can be used to identify individuals with high-risk for fractures

Risk factors or protective factors for osteoporosis and fractures can be broadly divided into non-modifiable and modifiable risk factors. Non-modifiable risk factors, such as age, sex, *etc.*, are important when identifying high-risk patients based on their risk factor profile for various interventions. The modifiable risk factors, such as low bone mass, eye vision *etc.*, can, apart from identifying high-risk patients, be modified so that the patient will have less impact of the risk factor on the fracture risk. Other important questions are whether risk factors are the same for men and women, and the same in the years around the menopause as in the most elderly? Are risk factors the same all over the world? Most of these questions cannot be answered, but several of the risk factors seem to be the same in both men and women in the most part of the world where studies have been performed. It is also important to divide the risk factors into those important for osteoporosis and those important for fractures. It is important to identify the risk factors for osteoporosis – low

**Table 3.1.** *Risk factors for osteoporotic fractures*

| *Not modifiable* | *Possibly modifiable* |
|---|---|
| Advanced age | Low BMD |
| Female sex | Current cigarette use |
| Personal history of adult fracture | Low body weight |
| History of fracture in | Estrogen deficiency |
| first-degree relative | including menopause onset <45 years |
| Caucasian and Asian race | Alcoholism |
| Dementia | Oral glucocorticoid use |
| Poor health/frailty | Life long low calcium intake |
| | Recurrent falls |
| | Little or no physical activity |

Adapted from National Osteoporosis Foundation Physician's Guide to Prevention and Treatment of Osteoporosis, Belle Mead, NJ Excerpta Medica, Inc. 1998. Italics denote risk factors that are key factors for risk of hip fracture, independent of bone density.

bone mass – where we can offer BMD measurement, whereas risk factors for fractures are important to identify high-risk patients in whom intervention may modify the risk factors. Most risk factors are same for both osteoporosis and fracture. However, for some, such as height, there may be a discrepancy. In some studies, increasing height seems to be associated with higher bone mass, whereas in others it is associated with an increased risk of fracture. It is also important to identify independent risk factors – especially when treatment will be based on an absolute 10-year risk in the future – to identify risk factors that are independent of each other and can be used to accurately predict the patients' fracture risk. Another important aspect of this enquiry is the effect of the modification of risk factors in reducing the number of hip fractures. Examples of modifiable and non-modifiable risk factors are found in Table 3.1.

# Strong Risk Factors

## Age

Age is probably one of the most important risk factors, both for osteoporosis and for fractures. In the studies of the incidence of hip fractures there is a doubling of the incidence each 5-10 years, whereas in other fractures this age effect is less. In the study of osteoporotic fractures by multivariate analysis (Table 3.2),[1] age was significant. For 5 years the relative risk was 1.4.

It is concluded then, that advanced age is an important risk factor, both for men and women and BMD can only partly explain this age effect.

## Gender

The risk of fractures in most studies has been higher in women than in men. In some studies, it was found that 75% of all hip fractures occurred in women, but this partly depended on the lower life expectancy of men. In a study from

**Table 3.2.** *Multivariable models of risk factors for hip fracture with and without adjustment for fractures and calcaneal bone density among 9516 white women*

| Measurement (comparison or unit)[a] | Relative risk (95% confidence interval) | |
|---|---|---|
| | Base model[b] | Add fractures and bone density |
| Age (per 5 yr) | 1.5 (1.3–1.7) | 1.4 (1.2–1.6) |
| History of maternal hip fracture (*vs.* none) | 2.0 (1.4–2.9) | 1.8 (1.2–2.7) |
| Increase in weight since age 25 (per 20%) | 0.6 (0.5–0.7) | 0.8 (0.6–0.9) |
| Height at age 25 (per 6 cm) | 1.2 (1.1–1.4) | 1.3 (1.1–1.5) |
| Self-rated health (per 1-point decrease)[c] | 1.7 (1.3–2.2) | 1.6 (1.2–2.1) |
| Previous hyperthyroidism (*vs.* none) | 1.8 (1.2–2.6) | 1.7 (1.2–2.5) |
| Current use of long-acting benzodiazepines (*vs.* no current use) | 1.6 (1.1–2.4) | 1.6 (1.1–2.4) |
| Current use of anticonvulsant drugs (*vs.* no current use) | 2.8 (1.2–6.3) | 2.0 (0.8–4.9) |
| Current caffeine intake (per 190 mg/day) | 1.3 (1.0–1.5) | 1.2 (1.0–1.5) |
| Walking for exercise (*vs.* not walking for exercise) | 0.7 (0.5–0.9) | 0.7 (0.5–1.0) |
| On feet ≤4 hr/day (*vs.* >4 hr/day) | 1.7 (1.2–2.4) | 1.7 (1.2–2.4) |
| Inability to rise from chair (*vs.* no inability) | 2.1 (1.3–3.2) | 1.7 (1.1–2.7) |
| Lowest quartile for distant depth perception (*vs.* other three) | 1.5 (1.1–2.0) | 1.4 (1.0–1.9) |
| Low-frequency contrast sensitivity (per 1 SD decrease) | 1.2 (1.0–1.5) | 1.2 (1.0–1.5) |
| Resting pulse rate >80 beats/min (*vs.* ≤80 beats/min) | 1.8 (1.3–2.5) | 1.7 (1.2–2.4) |
| Any fracture since age of 50 (*vs.* none) | – | 1.5 (1.1–2.0) |
| Calcaneal bone density (per 1 SD decrease) | – | 1.6 (1.3–1.9) |

Source: NJEM.[1]

[a]For continuous variables, the relative risks are expressed as a change in risk for each specified change in the risk factor.

[b]Base-model values are based on proportional-hazards analysis with backward stepwise elimination. Best subset models yielded similar sets of risk factors, including the number of steps in a 360° turn and the functional-status score; some did not include low-frequency contrast sensitivity, long-acting benzodiazepine therapy, or walking for exercise.

[c]Health was rated as poor (1 point), fair (2 points), or good to excellent (3 points).

Rotterdam, men reached the same incidence level for hip fracture as women 5 years later. In a review paper, several studies including both men and women were compared.[2] In the MEDOS study, women had similar risk factors such as body mass index, recreational and physical activity, co-morbidity and secondary osteoporosis.[3,4] There was a significant difference in the interaction for mental score and functional activity, and for body weight there was significance for the level of the variable. However, these differences could only explain part of the difference in the incidence between men and women. BMI could explain 4% and in combination with calcium, recreational activity, work activity, sun exposure and previous fracture could explain 20% of the difference in incidence between men and women.[5] In the EVOS study,[6] risk factors for vertebral deformities were studied in men and women and for multiple deformities, in contrast to findings for single deformities, there was a similar pattern in men and women. In a metaanalysis, Klotzbuecher *et al.*[7] studied the predictive ability of prior fractures for new fractures and found only a few studies in men, but there seems to be a similar risk for new fractures in men and women with a prior fracture. For genetic factors: in the Rotterdam study polymorphism in COLIA 1 was found to be similar in men and women.[8]

One major problem in studying risk factors is that the value of the risk factors is different. An example is from the Copenhagen study,[9] where 1800 men and 1400 women were perspectively followed for hip fractures. The high alcohol intake was a risk factor for men, but women had a much lower intake. The same was found for physical activity where men in general have a higher physical activity than women. There is still a lack of more studies in men. This is one of the research topics for the next decade.

## Low BMD

BMD is described elsewhere in this book. However, low BMD is an important risk factor for fractures in men and women.[10] In a metaanalysis,[11] it was shown that for any fracture, at any site, 1SD decrease in BMD may predict new fractures with a relative risk of 1.6, whereas for specific sites it seems that measurements at the hip give the best prediction of hip fracture (Table 3.3).

Measurement of BMD can be done in different sites and with different techniques. At present, measurement at the hip with the DEXA technique is the gold standard. However, new techniques, such as ultrasound measurement at the heel, have shown prospective and predictive ability.

**Table 3.3.** *Age-adjusted relative increase in risk of fracture (with 95% confidence interval) in women for every 1 SD decrease in BMD (absorptiometry)*

| Site of measurement | Forearm fracture | Hip fracture | Vertebral fracture | All fractures |
|---|---|---|---|---|
| Distal radius | 1.7 (1.4–2.0) | 1.8 (1.4–2.2) | 1.7 (1.4–2.1) | 1.4 (1.3–1.6) |
| Hip | 1.4 (1.4–1.6) | 2.6 (2.0–3.5) | 1.8 (1.1–2.7) | 1.6 (1.4–1.8) |
| Lumbar spine | 1.5 (1.3–1.8) | 1.6 (1.2–2.2) | 2.3 (1.9–2.8) | 1.5 (1.4–1.7) |

**Table 3.4.** *Associations of prior and subsequent fractures*

| Prior fracture | Wrist | Vertebral | All | Hip | Pooled |
|---|---|---|---|---|---|
| Wrist | 3.3 (2.0–5.3) | 1.7 (1.4–2.1) | 2.4 (1.7–3.4) | 1.9 (1.6–2.2) | 2.0 (1.7–2.4) |
| Vertebral | 1.4 (1.2–1.7) | 4.4 (3.6–14.6) | 1.8 (1.7–1.9) | 2.3 (2.0–2.8) | 1.9 (1.7–2.3) |
| Other | 1.8 (1.3–2.4) | 1.9 (1.3–2.8) | 1.9 (1.3–2.7) | 2.0 (1.7–2.3) | 1.9 (1.7–2.2) |
| Hip | | 2.5 (1.8–3.5) | 1.9 | 2.3 (1.5–3.7) | 2.4 (1.9–3.2) |
| Pooled | 1.9 (1.3–2.8) | 2.0 (1.6–2.4) | 1.9 (1.6–2.2) | 2.0 (1.9–2.2) | 2.0 (1.8–2.1) |

The diagnosis of osteoporosis is based on a BMD measurement and defined as a *T* score of – 2.5 at the hip, spine or wrist with the DEXA technique. The WHO diagnosis is at present defined only for women.

BMD can predict new fractures in the same way as blood pressure can predict stroke and cholesterol can predict myocardial infarction.

## Prior Fractures

Many studies have revealed that one of the most important risk factors for a new fracture is to have a previous history of osteoporotic fracture. In a metaanalysis, Klotzbuecher *et al.*[7] found that RR was 2 to have a new fracture of any kind if one had had a prior fracture, with the exception for vertebral fracture where the RR for a new vertebral fracture was >4 (Table 3.4).

Several new studies have found the same results as in the metaanalysis. Two studies focused on the time sequence after a prior fracture. Lindsay *et al.*[12] found a very high number of new vertebral fracture cases during the first year after starting a drug trial. In a study from Sweden,[13] all hospitalized cases with vertebral fractures were followed up regarding hospitalization for new fractures and it was found that there was a sharp increase in new fractures just after discharge for the vertebral fracture. During the first 3 years there was an increase in new fractures as compared with the general population, but in parallel to the general population after 3 years, it still increased. This indicates that treatment should be started immediately after the prior fracture has occurred.

## Other Risk Factors

### Ethnicity

The incidence of hip fracture has been the highest in Caucasian women and lower for other ethnic groups. In the NHANES study,[14,15] the BMD was different in different ethnic groups with the lowest values in Caucasian women.

In the NORA study[16,17] in the US, the odds ratio for the diagnosis of osteoporosis was found to be increased for white women compared with Asian or Spanish American women.

This ethnic difference in fracture risk can only partly be explained by the difference in BMD and may be due to body weight.

# Heredity

There are several studies looking at different polymorphisms and other genetic aspects on osteoporosis and fractures. These studies are not discussed here. Heredity means that mother or father had had a fracture. In the SOF study, the RR was doubled if the mother had had a hip fracture.[1] This was even more pronounced if she had had a hip fracture before the age of 80 (RR = 2.7; 1.7–4.4) than if the hip fracture occurred after the age of 80 (RR = 1.6; 1.0–2.7). The EVOS study showed that if the mother had had a hip fracture, there was an increased risk of vertebral fracture in men, but no significant risk was found for women.[14] In a Finnish study, in a twin register there was a lower significance for heredity between the twins.[16]

Heredity for fracture is a risk factor for new fractures.

# Physical Inactivity

There are several studies on the relationship between physical activity and fracture rate.[3,4] Most studies seem to indicate that individuals with increased physical activity have a lower fracture rate when compared to those with low physical activity, both during leisure time and work. The type of physical activity is also important. First of all, it has to be fun otherwise the compliance will be extremely low and from epidemiological studies it also seems as if load bearing is of high benefit for reducing the fracture risk.

In this risk factor there seems to be an interaction with nutrition, *i.e.* those who are physically active also have a better food intake.

# Smoking

In epidemiological studies it has been shown that BMD is lower in men and women who are smokers.[18] In studies from the US, smoking has been shown to be a risk factor both for men and women.[19] However, the finding is not consistent, for, in a Danish prospective[9] study, smoking was related to a RR of 1.36 in women and 1.59 in men to have a hip fracture.

Tobacco smoking seems to give a low BMD and in some studies an increased fracture risk, both in men and women. There might be an interaction with lean women to exaggerate the risk.

# Body Height, Body Weight and BMI

A few studies have shown that increased body height is related to an increased risk of hip fractures as in the MEDOS,[3,4] where the tallest quintile in women had an increased risk. This was also found in a Norwegian study.

In several studies, low body weight was related to increased fracture risk and low BMD. In the SOF study,[1] comparing the lowest quartile with the highest quartile of body weight, the RR for hip fracture was 1.93 (1.34–2.90). However, after adjustment for BMD at the hip the relationship was not significant.

In several studies, the BMI has been shown to be related to hip fracture risk. The problem is that during aging height is reduced, either through kyphosis, disc degeneration or vertebral fractures, *etc.*, thus the BMI is changed due to this height difference.

It is also interesting to know whether there is a cut-off for body weight or BMI when the fracture risk is steeply increased. It is important to find this cut-off point. Another problem is that body weights are different in different countries. In the US, the body weight is higher than in Europe and in Asia the body weight is lower than in Europe. Hence, the cut-off must be region-specific for body weight.

## Increased Propensity to Fall

This is an important risk factor, probably more important in the most elderly since most fractures occur after a fall. Most studies are on hip fractures and the type of fall is important. Several studies have shown that neuromuscular dysfunction is important, as in the EPIDOS study.[20] The risk factors for falling seem to be somewhat different from fractures. To reduce the impact of falling, treatment like hip protectors should be discussed.

## Low Sun Exposure

This risk factor may be a proxy for vitamin D production in the skin. In the MEDOS study[3] it was found that sun exposure was also an important risk factor, when adjusting other risk factors, both for men and women. Thus, reduced sun exposure seems to increase the hip fracture risk, probably as a result of induced vitamin D deficiency, which is discussed elsewhere in this book.

## Low Visual Acuity

Several studies[1] have shown that reduced visual acuity is associated with an increased number of hip fractures and also lighting at home is an important factor to take into account when trying to prevent hip fractures. It is possible to change both sight and lighting to modify the importance of this risk factor.

Risk factors involving calcium, vitamin D and nutrition are discussed elsewhere.

## Other Risk Factors

There are a number of other risk factors that are discussed, such as coffee and alcohol intake, increased cadmium intake, increased vitamin A intake, *etc.* All these seem to be important but have to be documented more clearly in several detailed studies.

## Secondary Osteoporosis

One of the strong risk factors for osteoporosis and fractures is the use of glucocorticoids. Several studies have shown that oral intake of glucocorticoids is associated with an increased risk especially of vertebral fractures and also of hip fractures. In a detailed study conducted by a database in the UK,[21] it was found that in those with a high oral intake of glucocorticoids the risk of having a vertebral fracture was 2.3, hip fractures 2.1 and somewhat lower for other fractures. In two randomized trials there has not been any significant difference in BMD after the use of lower doses of inhaled steroids.[22,23]

The question is whether this important risk factor is independent of BMD. It seems that the use of glucocorticoids ever is less dependent on BMD than current use of oral steroids. Statistically, it has been discussed whether a dose of 5–7.5 mg of prednisolon daily poses a significantly increased risk. The problem is that different individuals have different sensitivity to the use of oral glucocorticoids. For inhaled steroids it seems that in doses $< 800$ μg/day there is no significantly increased risk of osteoporosis or fractures.

There are several diseases associated with osteoporosis and fractures that are discussed.

# Combination of Risk Factors – Risk Assessment

Until the present time the intervention against osteoporosis has been decided mainly on BMD measurement and the cut-off has been chosen as the WHO T score of $< -2.5$. In future, an absolute risk cut-off based on an absolute 10-year risk, will be used and therefore the combination of risk factors, including BMD is important. In a study from SOF in women $> 65$ years, 17 risk factors were studied in a multivariate analysis and it was found that in a combination of at least two risk factors, but normal BMD, there was an incidence of 1.1 (0.5–1.6) hip fracture per 1 000 women, whereas in those who had five risk factors, but normal BMD, the incidence was 19 (15–22).[1] If BMD was added to those who had at least five risk factors and also the lowest tertile of BMD, the incidence was 27 (20–34). Thus it is important to combine risk factors to find the women with the highest absolute risk.

In the MEDOS study the attributable risk was studied,[3,4] *i.e.*, the proportion of hip fractures that could be avoided if the risk factors were eliminated and if there was a causal relationship between the risk factor and hip fracture. The finding was that if those with the lowest milk consumption had started consuming enough milk, then 4% of the hip fractures are avoided. Likewise, 10% of the hip fractures could be avoided if those with less physical activity increased their physical activity. In another study from the SOF,[24] the seven strongest risk factors were used to construct an index and the variables were age, BMD (*T* score), fracture after the age 50, body weight $< 57$ kg, smoking and use of arms at the arm chair rise test. This index could be used with and without BMD and was associated with an increase in hip fractures. It was also tested for vertebral fractures. This index was also validated in the EPIDOS

**Table 3.5.** *Ten-year probability (%) of any osteoporotic fractures in men and women according to gradient of risk (RR/SD) at the ages shown*

| RR | Age (years) | | | | | | | |
|---|---|---|---|---|---|---|---|---|
| | 50 | 55 | 60 | 65 | 70 | 75 | 80 | 85 |
| *Men* | | | | | | | | |
| 1.0 | 3.3 | 3.6 | 4.7 | 5.5 | 7.0 | 9.9 | 12.6 | 11.4 |
| 2.0 | 6.5 | 7.2 | 9.1 | 10.7 | 13.5 | 18.7 | 23.1 | 21.1 |
| 3.0 | 9.6 | 10.5 | 13.3 | 15.5 | 19.4 | 26.4 | 31.9 | 29.4 |
| 4.0 | 12.6 | 13.8 | 17.3 | 20.1 | 24.9 | 33.2 | 39.3 | 36.5 |
| 5.0 | 15.5 | 16.9 | 21.1 | 24.4 | 30.0 | 39.2 | 45.5 | 42.6 |
| 6.0 | 18.3 | 19.9 | 24.8 | 28.5 | 34.6 | 44.5 | 50.8 | 47.8 |
| *Women* | | | | | | | | |
| 1.0 | 5.8 | 7.2 | 9.6 | 12.4 | 16.1 | 18.7 | 21.5 | 20.7 |
| 2.0 | 11.3 | 13.8 | 18.2 | 23.3 | 29.4 | 33.5 | 37.4 | 36.1 |
| 3.0 | 16.5 | 20.0 | 26.0 | 32.6 | 40.4 | 45.2 | 49.2 | 47.6 |
| 4.0 | 21.4 | 25.7 | 33.1 | 40.8 | 49.5 | 54.6 | 58.1 | 56.4 |
| 5.0 | 26.0 | 31.0 | 39.4 | 47.9 | 57.2 | 62.1 | 64.8 | 63.1 |
| 6.0 | 30.3 | 35.9 | 45.2 | 54.2 | 63.5 | 68.1 | 70.0 | 68.3 |

study and the algorithm was also possible to use in this cohort. Otherwise, the problem is that several cohorts have to be used to find the true combination of risk factors since there might be bias in one single cohort.

An estimate on the use of combinations of risk factors and its effect on the 10-year risk has been presented[25,26] and in Table 3.2 it is shown the absolute 10-year risk for different BMD cut-offs (Table 3.5) and for different RR.

In future we will probably choose intervention based on the absolute fracture risk, using several risk factors like age, sex, BMD, previous fractures, use of glucocorticoids, *etc.*[27]

# Conclusions

Risk factors are important as they serve several purposes; one is to identify high-risk patients for an intervention or a BMD measurement. Some risk factors are modifiable and intervention against these risk factors might reduce the fracture rate. The major risk factors are age, sex, BMD, prior fracture *etc.* Instead of using the relative risk, we should change and do assessment of fracture risk and calculate the 10-year risk of having a fracture for each of the risk factors. For the risk assessment in practical care it is important to do a new index based on a combination of risk factors.

# References

1. S.R. Cummings, M.C. Nevitt, W.S. Browner, K. Stone, K.M. Fox, K.E. Ensrud, J. Cauley, D. Black and T.M. Vogt, Risk factors for hip fracture in white women. Study of osteoporotic fractures research group. *New Engl. J. Med.*, 1995, **332**, 814–815.

2. J.M. Kaufman, O. Johnell, E. Abadie, S. Adami, M. Audran, B. Avouac, W.B. Sedrine, G. Calvo, J.P. Devogelaer, V. Fuchs, G. Kreutz, P. Nilsson, H. Pols, J. Ringe, L. Van Haelst and J.Y. Reginster, Background for studies on the treatment of male osteoporosis: state of the art. *Ann. Rheum. Dis.,* 2000, **59**, 765–772.

3. O. Johnell, B. Gullberg, J.A. Kanis, E. Allander, L. Elffors, J. Dequeker, G. Dilsen, C. Gennari, A. Lopes Vaz and G. Lyritis, Risk factors for hip fracture in European women: the MEDOS study. Mediterranean osteoporosis study. *J. Bone Miner. Res.,* 1995, **10**, 1802–1815.

4. J. Kanis, O. Johnell, B. Gullberg, E. Allander, L. Elffors, J. Ranstam, J. Dequeker, G. Delsen, C. Gennari, A.L. Vaz, G. Lyritis, G. Mazzuoli, L. Miravet, M. Passeri, R Pererz Cano, A. Rapado and C. Ribot, Risk factors for hip fracture in men from southern Europe: the MEDOS study. Mediterranean osteoporosis study. *Osteoporosis International*, 1999, **9**, 45–54.

5. O. Johnell, J. Kanis and G. Gullberg, Mortality, morbidity, and assessment of fracture risk in male osteoporosis. *Calcified Tissue Int.,* 2001, **69**, 182–184.

6. A.A. Ismail, T.W. O'Neill, C. Cooper and A.J. Silman, The European Vertebral Osteoporosis Study Group, Risk factors for vertebral deformities in men: relationship to number of vertebral deformities. *J. Bone Miner. Res.,* 2000, **15**, 278–283.

7. C.M. Klotzbuecher, P.D. Ross, P.B. Landsman, T.A. Abbot III and M. Berger, Patients with prior fractures have an increased risk of future fractures: a summary of the literature and statistical synthesis. *J. Bone Miner. Res.,* 2000, **15**, 721–739.

8. B.L. Landahl, S.H. Ralston, S.F.A. Grant and E.F. Eriksen, An Sp1 binding site polymorphism in the COLIA1 gene predicts osteoporotic fractures in both men and women. *J. Bone Miner. Res.,* 1998, **13**, 1384–1389.

9. S. Høidrup, M. Grønbæk, A. Gottschau, J. Bruun Lauritzen and M. Schroll, The Copenhagen Centre for Prospective Population Studies, Alcohol intake, beverage preference, and risk of hip fracture in men and women. *Am. J. Epidemiol.,* 1999, **149**, 993–1001.

10. J.A. Kanis, O. Johnell, A. Odén, C. De Laet and D. Mellström, Diagnosis of osteoporosis and fracture threshold in men. *Calcified Tissue Int.,* 2001, **69**, 218–221.

11. D. Marshall, O. Johnell and H. Wedel, Meta-analysis of how well measures of bone mineral density predict occurrence of osteoporotic fractures. *Br. Med. J.,* 1996, **312**, 1254–1259.

12. R. Lindsay, S.L. Silverman, C. Cooper, D.A. Hanley, I. Barton, S.B. Broy, A. Licata, L. Benhamou, P. Geusens, K. Flowers, H. Stracke and E. Seeman, Risk of new vertebral fracture in the year following a fracture. *J. Am. Med. Assoc.,* 2001, **285**, 320–323.

13. O. Johnell, A. Odén, F. Caulin and J.A. Kanis, Acute and long-term increase in fracture risk after hospitalisation for vertebral fracture. *Osteoporosis Int.,*2001, **12**, 207–214.

14. M.N. Diaz, T.W. O'Neill and A.J. Silman, The influence of family history of hip fracture on the risk of vertebral deformity I men and women: the European vertebral osteoporosis study. *Bone*, 1997, **20**, 145–149.

15. A.C. Looker, E.S. Orwoll, C.C. Johnston Jr., R.L. Lindsay, H.W. Wahner, W.L. Dunn, M.S. Calvo, T.B. Harris and S.P. Heyse, Prevalence of low femoral bone density in older US adults from NHANES III. *J. Bone Miner. Res.,* 1997, **12**, 1769–1771.

16. P. Kannus, M. Palvanen, J. Kaprio, J. Parkkari and M. Koskenvuo, Genetic factors and osteoporotic fractures in elderly people: prospective 25 year follow up of a nationwide cohort of elderly Finnish twins. *Br. Med. J.,* 1999, **319**, 1334–1337.

17. E.S. Siris, P.D. Miller, E. Barrett-Connor, K.G. Faulkner, L.E. Wehren, T.A. Abbott, M.L. Berger, A.C. Santora and L.M. Sherwood, Identification and fracture outcomes of undiagnosed low bone mineral density in postmenopausal women: results from the National osteoporosis risk assessment. *J. Am. Med. Assoc.*, 2001, **286**, 2815–2822.
18. G. Jones, C. White, T. Nguyen, P. Sambrook, P. Kelly and J. Eisman, Cigarette smoking and vertebral body deformity. *J. Am. Med. Assoc.*, 1995, **274**, 1834–1835.
19. A. Paganini-Hill, A. Chao, R.K. Ross and B.E. Henderson, Exercise and other factors in the prevention of hip fracture: the leisure world study. *Epidemiology*, 1991, **2**, 16–25.
20. P. Dargent-Molina, F. Favier, H. Grandjean, C. Baudoin, A.M. Schott, E. Hausherr, P. J. Meunier and G. Breart, Fall-related factors and risk of hip fracture: the EPIDOS prospective study. *Lancet*, 1996, **348**, 145–149.
21. T.P. Van Staa, H.G. Leufkens, L. Abenhaim, B. Zhang and C. Cooper, Use of oral corticosteroids and risk of fractures. *J. Bone Miner. Res.*, 2000, **15**, 993–1 000.
22. O. Johnell, R. Pauwels, C.G. Löfdahl, L.A. Laitinen, D.S. Postma, N.B. Pride and S.V. Ohlsson, Bone mineral density in patients with chronic obstructive pulmonary disease treated with budesonide Turbuhaler. *Eur. Resp. J.*, 2002, **19**, 1058–1063.
23. A.E. Tattersfield, G.I. Town, O. Johnell, C. Picado, M. Aubier, P. Braillon and R. Karlstrom, Bone mineral density in subjects with mild asthma randomised to treatment with inhaled corticosteroids or non-corticosteroid treatment for two years. *Thorax*, 2001, **56**, 272–278.
24. D.M. Black, M. Steinbuch, L. Palermo, P. Dargent-Molina, R. Lindsay, M.S. Hoseyni and O. Johnell, An assessment tool for predicting fracture risk in postmenopausal women. *Osteoporosis Int.*, 2001, **12**, 519–528.
25. J.A. Kanis, O. Johnell, A. Odén, C. De Laet, B. Jönsson and A. Dawson, Ten-year risk of osteoporotic fracture and the effect of risk factors on screening strategies. *Bone*, 2002, **30**, 251–258.
26. J.A. Kanis, O. Johnell, A. Odén, A. Dawson, C. De Laet and B. Jönsson, Ten year probabilities of osteoporotic fractures according to BMD and diagnostic thresholds. *Osteoporosis Int.*, 2001, **12**, 989–995.
27. J.A. Kanis, D. Black, C. Cooper, P. Dargent, B. Dawson-Huges, C. De Laet, P. Delmas, J. Eisman, O. Johnnell, B. Jonsson, L. Melton, A. Oden, S. Papapoulos, H. Pols, R. Rizzoli, A. Silman and A. Tenenhouse, International Osteoporosis Foundation, National osteoporosis foundation. *Osteoporosis Int.*, 2002, **13**, 527–536.

CHAPTER 4

# Genetic Susceptibility to Osteoporosis

FIONA E.A. MCGUIGAN and STUART H. RALSTON

Department of Medicine and Therapeutics, University of Aberdeen Medical School, University of Aberdeen, Aberdeen AB25 2ZD, UK, Email: f.mcguigan@abdn.ac.uk

**Key Points**

- Genetic factors play an important role in the regulation of bone mineral density (BMD) and other traits relevant to the pathogenesis of osteoporotic fracture.
- The genes responsible for several monogenic bone diseases, such as osteoporosis-pseudoglioma syndrome, high bone mass (HBM) syndrome, and osteopetrosis have now been identified, but it remains to be seen whether polymorphic variations in these genes contribute to the regulation of BMD in normal populations.
- The genes which regulate BMD in the normal population are poorly understood, but a number of candidate loci have been identified by linkage studies in man.
- Genetic mapping studies in mice have identified many chromosomal regions which are linked to BMD. Many of these are conserved in different mouse strains suggesting that they may also be important regulators of BMD in humans.
- Population based studies and case-control studies have identified polymorphisms in several candidate genes that predispose to fracture and are associated with BMD. These include the vitamin D receptor, collagen type I alpha 1 gene and estrogen receptor.
- Studies on the genetic basis of osteoporosis offers the prospect of developing genetic markers for the assessment of fracture risk and identifying new molecules that could act as targets for drug design.

# Genetics and Osteoporosis

Genetic factors play a key role in the pathogenesis of osteoporosis. Twin studies have shown that up to 85% of the population variance in bone mineral density (BMD) at the hip and spine is genetically determined, with a polygenic pattern of inheritance.[1–4] Twin studies have shown that genetic factors play a role in regulating hip axis length,[5] quantitative ultrasound properties of bone,[6] body mass index and muscle mass[7] and age of menarche and menopause.[8–10] Biochemical markers of bone turnover have also been found to be genetically controlled with heritability estimates of 29% for serum osteocalcin, 74% for bone specific alkaline phosphatase and 60–65% for serum PTH and 1,25 dihydroxyvitamin D.[11–13] The role of genetic factors in the pathogenesis of fracture has also been studied. Population based studies have consistently shown that family history of hip fracture is a risk factor for fracture incidence, independent of BMD[14] with an estimated heritability of 25–35%.[15] This figure is much lower than the heritability of the skeletal phenotypes which predispose to fracture, probably because of the importance of fall-related factors in determining fracture risk.

# Identifying Genes which Predispose to Osteoporosis

Identifying the genes responsible for regulation of bone mass and other determinants of osteoporotic fracture risk is made difficult by the fact that the traits in question are determined by a complex interaction of genetic, metabolic and environmental factors. Moreover, current evidence suggests that in normal subjects, the effect of individual genes is small and influenced by an interaction with environmental factors and by gene–gene interactions. The most common approaches employed to dissect out the genetic contribution to complex diseases are (i) linkage analysis in families, (ii) allele-sharing methods in sib-pairs, (iii) association studies in unrelated subjects and (iv) linkage studies in mice.

## Linkage Analysis in Families

Linkage studies may be used to search either the entire genome, or chromosomal regions harbouring genes known to regulate quantitative traits such as bone mass or skeletal geometry. These regions are called quantitative trait loci or QTLs. Linkage studies involve genotyping a large number of polymorphic markers, typically spread at 5–10 centimorgan (cM) intervals throughout the genome. Such searches are based on the assumption that relatives within a family, sharing a common phenotype, will share one or more chromosomal segments containing the gene or genes of interest that are responsible for that phenotype.

The results of linkage studies are expressed in LOD (log of the odds) scores, which estimate the ratio of the odds that the candidate locus is linked (as opposed to unlinked) to the trait under study. By convention, LOD scores of above + 3.0 are taken as significant evidence of linkage; those above + 1.9 are taken as evidence 'suggestive' of linkage and those below −2.0 are taken to exclude linkage. An advantage of linkage studies is that they are statistically robust and unlikely to give false-positive results. A disadvantage is that they are less suitable for the analysis of complex traits that are genetically heterogeneous and have low statistical power to detect genes that have modest effects on BMD.

## Allele Sharing in Sib-pairs

Allele sharing methods involve testing whether affected relatives inherit a certain chromosomal region more often than expected under random Mendelian segregation. Within a pedigree, it is determined how often the area of interest is inherited from a common ancestor (identical-by-descent (IBD)). Even in the presence of confounding factors (such as phenocopy, incomplete penetrance and genetic heterogeneity in the population) relatives who are affected should share an excess of alleles compared to unaffected individuals. Even if it is impossible to determine IBD it is usually possible to determine if they share alleles at a series of markers *i.e.*, they are identical-by-state (IBS). Similar to linkage analysis in pedigrees, allele sharing studies are generally performed as a genome wide search, but can also be conducted using markers within specific candidate genes or loci. One advantage of this method is, that unlike classical linkage analysis, a disease model need not be specified making it particularly suitable for the analysis of complex traits such as BMD. A disadvantage is that large numbers of sib-pairs are necessary to obtain adequate statistical power. It is also important to note that LOD scores obtained using this method must be at least + 0.5 units higher than the LOD score values suggesting linkage obtained by classical linkage analysis.[16]

## Candidate Gene Studies

This approach has been most widely used in studying the genetic basis of osteoporosis. Candidate gene studies relate allelic variants of the gene under consideration to BMD or osteoporotic fracture in a population or a case-control study. While association studies are relatively easy to perform and can be powered to detect small effects, one particular consideration is whether the association between the candidate gene and BMD is real or an artefact resulting from confounding factors. Associations can also occur as the result of linkage disequilibrium (LD) with a causal gene situated nearby on the same chromosome. LD refers to the phenomenon whereby genes, which lie close together, tend to be inherited together from one generation to the next. Current evidence suggests that the average extent of LD ranges from 60 to 350 Kb in the human genome, although occasionally significant LD can extend for up to 1 Mb or more.[17] The transmission disequilibrium test (TDT) is a special type of association study that examines the frequency with which individuals inherit alleles, suspected to cause disease, from a heterozygous parent.[18] In a TDT analysis, the transmitted allele acts as the 'case' and the non-transmitted allele acts as the 'control', which makes the TDT immune to the confounding effects of population stratification. These advantages are offset by the fact that the TDT can only be used where parents are available who are heterozygous for the marker of interest. This limits the applicability of the method to the study of late-onset diseases such as osteoporosis and consequently few studies have as yet been performed.[19]

## Linkage Studies in Mice

Genetic linkage studies in experimental animals have long been used in the identification of genes responsible for complex traits and over recent years, many such studies have been performed in an attempt to identify loci involved in the regulation of BMD in mice.[20] The approach used involves crossing laboratory strains of mice with low and high bone density. The resulting (F1) generation typically have intermediate BMD because they receive a set of high-BMD alleles from one parent and a set of low-BMD alleles from the other. By interbreeding offspring from the (F1) generation, a second (F2) generation of mice is established whose BMD levels vary, because of segregation of the alleles that regulate BMD in the F2 offspring. A genome wide search is then performed in the F2 generation and the inheritance of strain-specific alleles is related to BMD. This approach has several advantages. The influence of confounding factors can be minimised through a carefully controlled environment and the statistical power of the study is extremely high due to the large numbers of progeny that can be generated. Also, fine mapping of identified loci can be achieved by back crossing mice which inherit a locus for regulation of BMD into the background strain and selecting offspring which retain the phenotype for further fine mapping studies.[20] Although this procedure can be time consuming, it is a very good way of narrowing down the critical interval containing the gene of interest. Possibly the main disadvantage of genetic mapping studies in mice is the fact that genes which regulate BMD in mice may not necessarily be the same as those that regulate BMD in man.

# Mapping Osteoporosis and Bone Related Phenotypes

## Monogenic Bone Disease Genes

Spectacular progress has been made in identifying genes for rare monogenic bone diseases (Table 4.1). A genome wide search in a family with unusually high bone mass resulted in the identification of a candidate locus on chromosome 11q12-13. This locus is of special interest since it harbours the gene causing HBM syndrome and osteoporosis pseudoglioma syndrome (OPS)– an autosomal recessive disease, characterised by severe juvenile osteoporosis. Mutations of the lipoprotein-related receptor 5 (Lrp-5) gene are now known to be responsible for both diseases. An activating (gain in function) mutation

**Table 4.1.** *Candidate loci and genes for rare bone diseases identified by linkage analysis*

| Disease | Locus | Candidate gene |
|---|---|---|
| High bone mass | 11q12 | LRP-5 |
| Osteoporosis-pseudoglioma syndrome | 11q12 | LRP-5 |
| Autosomal recessive osteopetrosis | 11q12 | TCIRG1 |
| Autosomal recessive osteopetrosis | 16p13 | CLCN7 |
| autosomal dominant osteopetrosis | 16p13 | CLCN7 |
| Camaruti Engelmann disease | 9q13 | TGFβ1 |
| Sclerosteosis/Van Buchem's disease | 17q12 | SOST |

in a so-called propeller motif of the gene is responsible for HBM[21] whilst inactivating (recessive) mutations of the gene are responsible for the OPS.[22] Autosomal recessive osteopetrosis (or infantile osteopetrosis) has also been mapped to this locus.[23] Recessive osteopetrosis is a rare disorder characterised by abnormally dense bones due to defects in osteoclast function. Inactivating mutations of the osteoclast specific TCIRG1 gene, which encodes a subunit of the osteoclast proton pump, have been shown to be responsible for this disease.[24] In contrast, autosomal dominant osteopetrosis (Albers-Schönberg disease) the most common form of osteopetrosis, has been found to be caused by specific heterozygous mutations in the voltage-gated chloride channel CLCN7 gene (16p13.3) presumably by exerting a dominant negative effect on chloride channel function.[25] Interestingly, a patient with autosomal recessive osteopetrosis was also identified as having homozygous inactivating mutations within this gene.[25,26]

Sclerosteosis is an autosomal-recessive disorder of bone remodelling, characterised by bone overgrowth of the calvaria, base of the skull and tubular bones. Unlike osteopetrosis, which is attributed to abnormal osteoclast function, sclerosteosis is primarily a disorder of the osteoblast.[27] Both sclerosteosis and van Buchem's disease map to the same region of chromosome 17q12-q21.[28] Recent work has shown that both disorders are due to mutations affecting the SOST gene.[29,30] Recently, it has been shown that mutations in the latency-activating peptide (LAP) domain of the TGF beta-1 gene (19q13.1) are responsible for Camurati–Engelmann disease (CED), a progressive diaphyseal dysplasia characterised by hyperosteosis and sclerosis of the diaphysis of the long bones.[31] There is evidence that some genes which cause monogenic bone disease, such as TGFB-1,[32] TCIRG1 and LRP-5 may contribute to regulation of BMD in the normal population.[33]

## Quantitative Trait Loci for Regulation of BMD in Man

Linkage studies in normal sib-pairs and in extended families with osteoporosis have been used to identify loci that are linked to BMD. Koller and co-workers used non-parametric sib-pair linkage analysis to determine if the HBM/osteoporosis pseudoglioma locus on chromosome 11q12-13 might contribute to the genetic variation in BMD in normal subjects. In a study of Caucasian and African–American premenopausal siblings, significant evidence of linkage was found with positive LOD scores (LOD +3.50) observed in both ethnic groups. However, it was noted that the effect was greater among Caucasian women, raising the possibility that the genes responsible for regulation of bone mass may vary in different ethnic groups.[33] A subsequent genome wide search in the same population showed a LOD score of +3.86 at chromosome 1q21-23 for lumbar spine BMD and other lodscores suggestive of linkage were observed on chromosome 5q33-35 (LOD score +2.23 with femoral neck BMD) and chromosome 6p11-12 (LOD score +2.13 with lumbar spine BMD). Linkage to 11q12-13 was also supported, as previously, although the peak LOD score was decreased to +2.16 when the study sample was increased to 595 sibling pairs.[34] Possible explanations for this finding include genetic heterogeneity, with true linkage in a subset of the population, or a false-positive result in the 11q12-13 linkage study. Linkage studies in the same population have also identified multiple loci for regulation of femoral neck geometry.[35] These include a locus on chromosome 5q for hip axis length (LOD +4.3); a locus on chromosome 4q showing linkage to both hip axis length and mid-femur width (LOD +3.9 and +3.5, respectively) and a locus on chromosome 17q linked to femur head width (LOD +3.6).

Devoto and colleagues conducted a genome search in an extended sib-pair study, to screen for chromosomal regions involved in the determination of bone density.[36] Three regions with LOD scores between +2.5 and +3.5 on chromosomes 2p23-p24, 4q32-34 and 1p36 were identified and another locus on chromosome 11q had a LOD score of +2.08. Further studies by the same group subsequently confirmed evidence of linkage to the 1p36 locus in a second set of sib-pairs.[37] Another genome wide scan, performed by Niu, in Chinese sib-pairs searched for loci that regulated forearm BMD. In this study, evidence suggestive of linkage was found on chromosome 2p21.1-24 (lodscore +2.15), which overlaps with the region identified by Devoto as being linked to spine BMD. A further region showing possible linkage was identified on chromosome 13q21-34 (maximum lodscore +1.67).[38]

Duncan and colleagues conducted a sib-pair linkage study using polymorphic markers in the vicinity of 23 candidate genes implicated in the regulation of bone mass.[39] Using a total of 64 microsatellite markers they found strong suggestive evidence of linkage at the PTH receptor 1 locus in chromosome 3p (Lodscores between +2.7 and +3.5 depending on the method of analysis). Moderate evidence for linkage (lodscore above +1.7) was observed at loci for COLIA1 (17q21.31-22.15), epidermal growth factor (4q25), VDR (12q13) and IL-6 (7p21). A Japanese study confirmed moderate evidence for linkage of low BMD at the IL-6 locus when microsatellites in the IL-6 (7p21), IL-6 receptor (1q21), calcium receptor (3q13.3-q21) and matrix Gla protein (12p13.1-p12.3) genes were analysed.[40] In another study, no evidence of linkage could be demonstrated between BMD and a number of microsatellite markers flanking the VDR gene (12q12-14).

## Quantitative Trait Loci for Regulation of BMD in Mice

Linkage studies using inbred strains of mice have identified several loci that regulate bone mass. Klein and colleagues performed a genome search in 24 inbred strains of mice with varying BMD (generated by crossing high-BMD and low-BMD strains). Eight loci showed evidence of suggestive linkage to bone mass and two loci on chromosomes 7 and 14 showed definite linkage.[41] In a further study, two loci with linkage to peak bone mass were identified corresponding to human chromosomes 1q42, 7p15, 6p22 and 17q11-q22 and a locus on the X chromosome linked to BMD.[42] These loci are distinct from those identified by Klein, indicating that different genes may regulate bone mass in different strains of mouse. Beamer and colleagues identified four loci which were linked to BMD on chromosomes 1, 5, 13 and 15.[43] Similar studies corroborate the role of these loci in the regulation of bone mass.[44] Yershov and colleagues, using recombinant congenic (RC) mice, have also identified QTL for parameters such as bone strength and bone biomechanics localised on chromosomes already known to contain genes involved in the regulation of bone mass.[45]

Currently, QTL for BMD regulation have been identified on almost all mouse chromosomes and many of these QTL are shared across mouse strains implying that they contain conserved genes that could be important for regulation of BMD in other species.[44] Fine mapping studies of these loci are currently in progress to narrow down the regions of interest and identify the genes responsible. When this has been achieved, the next step will be to determine if the genes for regulation of BMD in mice are also important regulators of BMD in humans and to separate the functional from the non-functional polymorphisms within these genes.

## Candidate Gene Studies

This approach has been most widely used in studying the genetic basis of osteoporosis, with most research focussing on cytokines and growth factors, which regulate bone turnover, those that encode components of bone matrix and those that encode receptors for calciotropic hormones (Table 4.2). A particular problem associated with this type of study, is that often, due to small sample size or other confounding factors, conflicting results are obtained in different populations (as can be seen in some of the studies, discussed below). To address this issue, some researchers are exploring the use of meta-analysis to gain an accurate estimate of effect size for different polymorphisms.[46–48] Individual candidate genes that have been implicated in the regulation of bone mass or osteoporotic fractures are discussed in detail below.

### *Vitamin D Receptor*

Vitamin D, by interacting with its receptor, plays an important role in calcium homeostasis by regulating bone cell growth and differentiation, intestinal

**Table 4.2.** *Chromosome loci for candidate genes*

| Candidate gene | Gene locus |
|---|---|
| Tumor necrosis factor receptor | 1p36.3 |
| Methylenetetrahydrofolate reductase | 1p36.3 |
| Osteocalcin | 1q25-q31 |
| Interleukin-1 receptor antagonist | 2q13 |
| Interleukin-1 β | 2q14 |
| Peroxisome proliferator-activated receptor γ | 3p25 |
| Ca sensing receptor | 3q13.3-q21 |
| α2-HS-glycoprotein (AHSG) | 3q27-q29 |
| Glucocorticoid receptor | 5q31 |
| Estrogen receptor alpha | 6q25.1 |
| Interleukin-6 | 7p21 |
| Calcitonin receptor | 7q21.3 |
| WRN | 8p12-p11.2 |
| Osteoprotegerin | 8q24 |
| Parathyroid hormone | 11p15 |
| P57 kip2 | 11p15.5 |
| Vitamin D receptor | 12q13 |
| Insulin-like growth factor-1 | 12q22-q24.1 |
| Klotho | 13q12 |
| Aromatase | 15q21.1 |
| Type I collagen | 17q21.3-q22 |
| Apolipoprotein E | 19q13 |
| Transforming growth factor beta-1 | 19q13.1-13.3 |
| Androgen receptor | Xq11-q12 |

calcium absorption and parathyroid hormone secretion. Morrison and colleagues identified three common polymorphisms, in strong LD, situated in the 3′ region of the gene, between exons 8 and 9 and recognised by the restriction enzymes *BsmI, ApaI* and *TaqI*. These were found to be associated with circulating levels of the osteoblast-specific protein osteocalcin and bone mass in twin studies and a population-based study.[49] Other studies of VDR in relation to bone mass have been conflicting however; and moreover, a recent family based study showed no evidence of linkage between the VDR locus and BMD.[50] A meta-analysis of 16 studies, published in 1996, concluded that the VDR genotype was associated with relatively modest effects on BMD (amounting to a difference of about 0.15–0.20 Z score units between genotypes) and that the effect on bone mass diminished with age.[47] Studies of VDR in relation to osteoporotic fracture have also been conflicting. Possible explanations for the discrepancies between studies include differences in sample size, ethnicity and the behavioural characteristics of the study subjects. Another possible explanation is that the effects of VDR genotype on bone loss and bone mass may be modified by both calcium and vitamin D intake,[51–54] a fact which needs to be considered when targeting individuals for vitamin D and calcium based treatment regimes.

The mechanisms by which these polymorphisms affect VDR function are unclear. Although the coding sequence of VDR is unchanged, reporter gene constructs have shown haplotype-specific effects in reporter assays, suggesting that they may act as markers for RNA stability, because they are in LD with a polymorphic poly A tract in the 3′ UTR of the VDR mRNA or with functional polymorphisms elsewhere in the VDR gene.

A polymorphism in exon 2, creating an alternative translational start site, resulting in the production of two isoforms of the VDR protein, which differ in length by three amino acids, has also been described.[55] Evidence for functional differences between the isoforms has been conflicting, as is the association between bone density and fracture.[56-58] There is no evidence of LD between this and the intron 8/exon 9 polymorphisms.

Another polymorphism has been identified in the promoter of VDR at a binding site for the transcription factor Cdx-2 (a homeodomain protein-related caudal), which has been associated with BMD in Japanese subjects.[59] The Cdx-2 polymorphism appears to be functional since it has been shown to influence DNA protein binding and to modulate gene expression in reporter assays.[59] Further studies of this polymorphism in relation to BMD are awaited with interest.

## Type I Collagen

The genes encoding type I collagen (COLIA1 and COLIA2) are important candidates for the pathogenesis of osteoporosis. Grant and colleagues[60] described a common polymorphism affecting a binding site for the transcription factor Sp1 in the first intron of COLIA1 which was more prevalent in osteoporotic patients than in controls. Positive associations between the COLIA1 Sp1 polymorphism, bone mass and osteoporotic fractures were subsequently reported in several populations.[61-64] A recent meta-analysis showed that the COLIA1 genotype conferred differences in BMD of approximately 0.15 Z score units per copy of the 's' allele and an increase in fracture risk of approximately 62% per copy of the 's' allele.[48] COLIA1 Sp1 alleles have also been associated with other phenotypes relevant to osteoporosis including post-menopausal bone loss,[65] femoral neck geometry[66] and response to etidronate therapy.[67]

Significant ethnic differences have been noted in the population prevalence of the 's' allele (it is common in Caucasian populations, but rare in Africans and Asians). In view of this, it has been suggested that this may explain in part, the observed racial differences in osteoporotic fracture risk.[68] The mechanism by which the Sp1 polymorphism predisposes to osteoporosis has been investigated by Mann and colleagues[48] who found that the 's' allele had increased affinity for Sp1 protein binding and was associated with elevated allele-specific transcription in heterozygotes. These abnormalities were accompanied by increased production of the alpha 1 chain of collagen by osteoblasts cultured from 'Ss' heterozygotes, resulting in an increased ratio of the alpha 1 to alpha 2 chains, reflecting the presence of alpha 1 homotrimer formation. Biomechanical testing of bone samples from 'Ss' heterozygotes showed reduced bone strength compared with 'SS' homozygotes and a slight reduction in mineralisation of bone. The data suggest that the COLIA1 Sp1 polymorphism is a functional variant, which has adverse effects on bone composition and mechanical strength. Haplotype analysis suggests that susceptibility to fracture is probably driven by the Sp1 polymorphism rather than other polymorphisms within the COLIA1 gene.[69] However, two polymorphisms that are in LD with the Sp1 polymorphism, have recently been identified in the promoter of COLIA1 and are also associated with BMD.[70] Further work will be required to assess the functional significance of these polymorphisms and determine if they interact with the Sp1 polymorphism to regulate BMD and bone fragility. From a clinical viewpoint, the COLIA1 polymorphism may be of value as a marker of osteoporotic fracture risk, since it predicts fractures independent of BMD and interacts with BMD to enhance fracture prediction.[71]

## Estrogen Receptor Alpha

Estrogen is known to increase bone mass and inhibit bone turnover and as such, is a potentially important candidate gene for osteoporosis. Genetic variations of the ER gene have been found to be related to the onset of natural menopause and the risk of surgical menopause[72] and a protein coding mutation in the gene has been observed in a man suffering from severe osteoporosis.[73] An association has been reported between a TA repeat polymorphism in the ER promoter and bone mass in Japanese, American and European populations[74–77] and other investigators have reported positive associations between *PvuII* and *XbaI* polymorphisms in the first intron of the gene and bone mass.[78–80] These polymorphisms are in strong LD both with each other and with the TA repeat promoter polymorphism. Single nucleotide polymorphisms in exon 8 (six nucleotides from a stop codon) and exon 1 (29 nucleotides downstream from the putative start codon) of the gene have recently been identified in a Thai population of postmenopausal women and found to be associated with osteoporotic fracture. The exon 1 polymorphism is in LD with *PvuII* and correlates to the skeletal response after long-term estrogen treatment.[81,82] The mechanism by which these polymorphisms influence bone mass or ER function is as yet unclear.

## Transforming Growth Factor Beta-1

Transforming growth factor beta 1 (TGFβ-1) is a strong candidate gene for regulation of bone mass in view of its abundance in bone, its potent effects on both osteoblast and osteoclast activity and the fact that mutations in the LAP region of the gene cause the sclerotic bone dysplasia, Camurati-Engelmann disease (CED).[83] It has also been implicated as a mediator of the skeletal effects of estrogen.[84] Several polymorphisms have been described; these include a polymorphism within exon 1, causing a protein coding change in the signal peptide, which has been shown to be associated with BMD. Two promoter polymorphisms are also associated with circulating TGFβ levels.[85] While this is consistent with the suggested role of TGFβ in bone remodelling, the significance of these observations for regulation of bone mass remains to be fully determined, since these three polymorphisms are in LD.[32] Another rare polymorphism has been described close to the splice site in exon 5, although it is not yet known if this influences splicing.[86] In summary, there is good evidence to suggest that TGFβ-1 is a candidate gene for the regulation of bone mass, but the mechanisms by which polymorphisms of the gene influences TGFβ function are unclear.

## Interleukin-6

Interleukin-6 (IL-6) is known to act as a modulator of osteoclast differentiation and function[87] and may mediate some of the effects of estrogen on bone.[88] An association has been found with BMD in both pre-and postmenopausal women from a Caucasian population.[89] Another CT rich minisatellite has also been reported to be associated with wrist BMD in a Japanese Population.[90,91] These associations may be mediated by LD with polymorphisms in the 5′ promoter region of the gene, which have been associated with bone mass and bone turnover[92] since at least one of these polymorphisms is known to have a regulatory effect on gene expression.[93]

## Apolipoprotein E

Apolipoprotein E plays an important role in the transport of vitamin K from the intestine to target sites such as liver and bone. Its relevance to the pathogenesis of osteoporosis lies in the fact that vitamin K is an essential co-factor for carboxylation of the osteoblast specific protein, osteocalcin and other matrix carboxylated proteins. Of the three common alleles identified, the APOE*4 allele has been associated, in a study of US women, with an increased risk of fracture at both the hip and wrist, that is not explained by bone density, incidence of falls or impaired cognitive function.[94] While an association with reduced BMD at the lumbar spine has been reported in Japanese women,[95] and an increase in the rate of bone loss,[96] the evidence from studies is conflicting although there is a suggestion that the effects of the polymorphism may be modified by estrogen status.

## Osteocalcin

Osteocalcin is the most abundant non-collagenous protein in bone, but its precise function is unknown. Osteocalcin is synthesised by osteoblasts, and a proportion of this is released into the circulation, providing a biochemical marker of osteoblast activity. A polymorphism has been identified in the promoter region of the gene, in an area containing a number of important regulatory elements (including glucocorticoid repression and osteoblast specific regulatory elements) in addition to those necessary for osteocalcin gene expression. This polymorphism has been found to be associated with low bone mass in Japanese, Chinese and Swedish populations.[97-99] The mechanism by which osteocalcin polymorphisms could influence bone mass remain unclear, but gene knockout studies have shown that mice which are deficient in osteocalcin have increased bone mass, raising the possibility that it may act as a negative regulator of osteoblast function.[100]

## Calcitonin Receptor

Calcitonin exerts inhibitory effects on osteoclast function and stimulates urinary calcium excretion by interacting with its receptor, which is highly expressed on osteoclasts. Different isoforms of the calcitonin receptor (CTR) result from gene splicing and differential expression of these isoforms may explain the variable responsiveness to calcitonin in patients with high turnover metabolic bone diseases. The CTR has seven potential transmembrane domains and a coding polymorphism has been identified in intracellular domain 4.[101] The polymorphism, resulting in an amino acid substitution, has been found to be associated with bone mass in an Italian population.[102,103] It is not yet clear how the polymorphism affects bone mass, but it has been speculated that the amino acid change could cause a conformational change in the receptor and alter receptor signalling.

## Parathyroid Hormone

The parathyroid hormone (PTH) gene is located on the short arm of chromosome 11, an area already identified in linkage studies as associated with the regulation of bone mass.[33,104] A polymorphism in intron 2 has been described in a Japanese population, where it was found to be associated with reduced bone mass and increased bone turnover.[105] It is also associated with bone geometry, in particular the cross-sectional cortical area and a slower reduction in radial cortical area with increasing age.[106] The mechanism by which this polymorphism influences bone mass is not yet clear, since it lies in a non-regulatory region of the gene.

## Interleukin-1 Receptor Antagonist

The IL-1 receptor antagonist (IL-1ra) competitively inhibits IL-1β (by binding to the IL-1 receptor without inducing any effects) and a decrease in IL-1ra is one consequence of estrogen deficiency. In ovariectomised animals, osteoclast formation and bone loss is blocked following treatment with IL-1ra. Consequently, this gene is a possible candidate for the regulation of bone mass and postmenopausal bone loss.

Nine polymorphisms, all in LD, have been identified throughout the gene. An 86 bp VNTR in intron 3 has been found to be associated with differential rates of bone loss at the lumbar spine.[107,108] Another study found an association between those genotypes with a high number of repeats and an increased odds ratio of osteoporotic fracture.[109] This VNTR may be functionally significant since the sequence contains three potential transcription factor binding sites. Further studies are necessary to determine if there is a relationship with BMD and fracture risk in other populations.

## Osteoprotegerin

Osteoprotegerin (OPG) is a soluble receptor for RANKL and therefore a competitive inhibitor of osteoclast differentiation and activity, making it a strong candidate gene for bone metabolism regulation and osteoporosis susceptibility. A total of 12 polymorphisms have been identified in the gene, five of which are in complete LD. In a Danish case-control study, these polymorphisms were over represented in individuals with osteoporotic fractures of the spine, and a haplotype for fracture risk has been suggested.[110] In this study and an Irish study,[111] polymorphisms of the OPG gene were not found to be associated with BMD or biochemical markers of bone turnover. The mechanism by which these polymorphisms affect fracture risk are yet to be determined and further studies are necessary to confirm the findings in other populations.

## Tumour Necrosis Factor Receptor

One of the strongest candidate loci for regulation of BMD in man is on chromosome 1p36. This was initially identified by linkage analysis in sib-pairs[36] and was subsequently confirmed by variance components linkage analysis in a separate set of families.[37] The tumour necrosis factor receptor superfamily member 1B (TNFRSF1B) is a strong positional candidate gene within this region. The TNFRSF1B gene encodes a 75 kDa tumour necrosis factor receptor, which is highly expressed in osteoclast precursors and plays an important role in mediating the effects of TNF on osteoclastogenesis. A number of polymorphisms have been identified in this gene. These include a microsatellite repeat in intron 4, a SNP in exon 6 and three polymorphisms in the 3' UTR.[112] Spotila reported evidence of an allelic association between specific haplotypes of the 3' UTR region of TNFRSF1B and spine BMD, but the study population was small. Recent studies by Albagha have confirmed an association between the 3' UTR polymorphisms and BMD in a large population based study, although the risk haplotype differed from that reported by Spotila.[113] The mechanism of action is as yet unclear, but it is possible that sequence variations in this region of the gene could affect mRNA structure and stability.

## Peroxisome Proliferator-activated Receptor Gamma

The role of peroxisome proliferator-activated receptor gamma (PPAR-γ) in bone metabolism is as yet unclear, although it is speculated that it may be involved in osteoblast differentiation. Expressed predominantly in adipose tissue, it is a member of the nuclear receptor superfamily of ligand dependent transcription factors. Its main function is the regulation of glucose homeostasis, lipid homeostasis and adipocyte differentiation from precursors common also to osteoblasts. A polymorphism in exon 6 of the gene has previously been found to be associated with plasma leptin levels in obese individuals and recently has also been found to be associated with reduced bone density in a Japanese population of postmenopausal women.[114] Further studies are necessary to determine the functional effects of this polymorphism and its effect in other populations.

## Androgen Receptor

A repeat polymorphism $(AGC)_n$ coding for a polyglutamine tract in the activation domain of the X-linked androgen receptor has been described in exon 1 of the gene. Previous studies have found an association between length variations of the polymorphism and receptor function and it is of interest that large expansions of the tract have been found to cause X-linked spinocerebellar muscular atrophy.[115] In an American population of pre and perimenopausal women, these variations were found to be associated with BMD at both the hip and spine;[75] however, in an elderly male Belgian population, no association between repeat length and BMD or bone turnover was found.[116] Further studies are necessary to clarify the nature of the relationships observed.

## Methylenetetrahydrofolate Reductase

Homocysteine is a sulfur-containing amino acid generated as an intermediate product of methionine metabolism. Homocystinuria is a rare autosomal recessive disease, which in addition to mental retardation and skeletal abnormalities, manifests itself in early onset osteoporosis as a result of impaired collagen cross-linking. This disease is characterised by elevated concentrations of plasma homocysteine and is caused by a deficiency of cystathionine β synthase or a mutant form of methylenetetrahydrofolate reductase (MTHFR), the enzyme which catalyses the reduction of 5,10-methylenetetrahydrofolate to 5-methyltetrahydrofolate. A C/T polymorphism in the gene, resulting in an amino acid substitution $(Ala_{222}Val)$ has been identified in the coding region of the gene. This is a functional polymorphism, which has previously been associated with reduced enzyme activity and increased thermolability.[117] In a Japanese study of postmenopausal women, this polymorphism has been associated with reduced spine and total body BMD, concomitant with elevated levels of plasma homocysteine.[118] Further studies in other populations are necessary to determine its role in the pathogenesis of osteoporosis.

## α2-HS-glycoprotein (AHSG)

AHSG is a non-collagenous serum-derived protein of bone matrix with a high affinity for hydroxyapatite, which plays a role in the recruitment of osteoclast precursors. A polymorphism with two common alleles (resulting from an amino acid change at two positions within the gene) has been identified. It has been found to be associated with reduced height, BMD and reduced calcaneal broadband ultrasound attenuation (BUA) in different populations in addition to lower estradiol levels in postmenopausal women.[119–121]

## Calcium Sensing Receptor

The calcium-sensing receptor (CaSR), expressed in the parathyroid gland, thyroid C cells and kidney, regulates calcium homeostasis by responding to decreases in extracellular calcium levels by secreting PTH and stimulating renal calcium reabsorbtion. As such, it is an attractive candidate gene for osteoporosis susceptibility. A CA repeat polymorphism[122] in the gene has been associated with bone density in a Japanese cohort of women and a common polymorphism, causing an amino acid change at position 986 in the intracellular tail of the CaSR gene is known to be associated with serum calcium concentration in healthy individuals. Its association with BMD, however, is inconclusive, with contradictory results in different studies.[123–125]

## Insulin-like Growth Factor-1

In addition to stimulating skeletal growth, Insulin-like growth factor-1 (IGF-1) has a role in osteoblast and osteoclast function and bone matrix protein synthesis. Previously, circulating levels of IGF-1 have been demonstrated to have a heritable component[126] and low levels of IGF-1 have been implicated in the pathogenesis of male osteoporosis.[127] In a recent study, a dinucleotide repeat polymorphism, 1 kb upstream from the transcriptional start site, has been associated with serum IGF-1 concentrations and bone density in American males with osteoporosis[128] although similar studies in women have been inconclusive. The mechanism of action is far from clear and further investigations are necessary to determine if the observed results are due to LD with polymorphisms elsewhere in the gene.[129]

## Other Potential Candidate Genes

Polymorphisms of several other candidate genes have been associated with BMD. For the most part these have been studied in single populations and further work will be required to confirm or refute their role in the pathogenesis of osteoporosis. Briefly, these are outlined below.

The cyclin-dependent kinase (CDK) inhibitor gene P57 plays an important role in osteoblast proliferation and differentiation. A small association study in 154 postmenopausal women showed that a four-amino-acid insertion/deletion polymorphism was related to total body BMD in Japanese subjects.[130] The rare autosomal recessive disease Werner syndrome (WS), which is characterised by premature ageing, is caused by inactivating mutations in the WRN gene, a member of the RecQ family of helicase genes. Recent work by Ogata and colleagues, showed evidence of an association between spine BMD and a C/T polymorphism at position 4330 in the gene causing a cysteine to arginine amino acid substitution.[131] The same authors described an association between a CA repeat polymorphism downstream of exon 5 in the Klotho gene and Lumbar Spondylosis. The Klotho gene, like the WRN gene is believed to play an important role in age-related disorders, including osteoporosis. An association with BMD was also demonstrated but only in a subset of elderly women.[132]

A polymorphism has been described in exon 2 of the glucocorticoid receptor (GR), which causes an asparagine to serine amino acid change at codon 363. This polymorphism has been associated with altered sensitivity to glucocorticoids and in one study was associated with BMD, although the result just failed to reach statistical significance.[133] Finally, a tetranucleotide repeat polymorphism has been described within intron 4 of the aromatase (CYP19) gene which has been associated with BMD in Italian women[134] and polymorphisms of the interleukin 1beta (IL-1β) gene have been implicated as a predictor of bone loss in inflammatory bowel disease (IBD)[109,135] although Langdahl and colleagues found no association between IL-1b polymorphisms and BMD in Danish subjects.[109]

## Implications for Clinical Practice

Genetic markers will also be used as diagnostic tools in the assessment of individuals at risk of developing osteoporotic fractures. We already know that the BMD *T*-score value of −2.5, which is widely used as a treatment threshold for osteoporosis, identifies only a small proportion of individuals in the community who actually suffer fractures. Genetic markers of bone fragility or bone loss could be used alongside bone density measurements to help target preventative therapies to those individuals who are at risk of fracture. The most promising marker identified so far in this respect is the COLIA1 Sp1 polymorphism, which seems to predict fractures independent of BMD,[71] although it is likely that further markers of bone fragility will also be identified in the future. Another use of genetic profiling would be to distinguish treatment responders from non-responders and to identify patients who might be at risk of developing unwanted side effects.[67] This is an area which has been little studied in the bone field, but is one which is well developed in areas such as oncology and has tremendous future potential in all spheres of medicine.[136]

# Conclusions

Genetic factors play an important role in regulating BMD and other determinants of osteoporotic fracture risk. Osteoporosis is a relatively late-onset disease and increasingly, much of our knowledge of the genetic regulation of bone mass comes from the study of monogenic bone diseases such as OPS and the HBM syndrome, which occur as the result of mutations in a single gene – the lipoprotein receptor-related protein 5 (LRP-5) gene.

Population-based and case-control studies have identified a number of candidate genes that are associated with BMD or osteoporotic fracture; however, the functional effects of most of these polymorphisms are not fully understood.

Linkage studies in mice and man have identified loci on a number of chromosomes showing definite or probable linkage to BMD and a future challenge will be to define the genes responsible for BMD regulation and investigate relevance of mouse BMD QTL in humans.

From a clinical standpoint, advances in knowledge about the genetic basis of osteoporosis are important, since they offer the prospect of developing genetic markers for the assessment of fracture risk and the opportunity to identify molecules that will be used as targets for the design of new drugs for the prevention and treatment of bone disease.

# References

1. D.M. Smith, W.E. Nance, K.W. Kang, J.C. Christian and C.C. Johnston, Genetic factors in determining bone mass, *J. Clin. Invest.*, 1973, **52**, 2800–2808.
2. J.C. Christian, P.L. Yu, C.W. Slemenda and C.C. Johnston, Heritability of bone mass: a longitudinal study in aging male twins, *Am. J. Human Genet.*, 1989, **44**, 429–433.
3. N.A. Pocock, J.A. Eisman, J.L. Hopper, M.G. Yeates, P.N. Sambrook and S. Eberl, Genetic determinants of bone mass in adults: a twin study, *J. Clin. Invest.*, 1987, **80**, 706–710.
4. R. Gueguen, P. Jouanny, F. Guillemin, C. Kuntz, J. Pourel and G. Siest, Segregation analysis and variance components analysis of bone mineral density in healthy families, *J. Bone Miner. Res.*, 1995, **12**, 2017–2022.
5. K.G. Faulkner, S.R. Cummings, D. Black, L. Palermo, C.C. Gluer and H.K. Genant, Simple measurement of femoral geometry predicts hip fracture: the study of osteoporotic fractures, *J. Bone Miner. Res.*, 1993, **8**, 1211–1217.
6. N.K. Arden and T.D. Spector, Genetic influences on muscle strength, lean body mass, and bone mineral density: a twin study, *J. Bone Miner. Res.*, 1997, **12**, 2076–2081.
7. T.V. Nguyen, G.M. Howard, P.J. Kelly and J.A. Eisman, Bone mass, lean mass, and fat mass: same genes or same environments? *Am. J. Epidemiol.*, 1998, **147**, 3–16.
8. J. Kaprio, A. Rimpela, T. Winter, R.J. Viken, M. Rimpela and R.J. Rose, Common genetic influences on BMI and age at menarche, *Human Biol.*, 1995, **67**, 739–753.
9. D.Z. Loesch, R. Huggins, E. Rogucka, N.H. Hoang and J.L. Hopper, Genetic correlates of menarcheal age: a multivariate twin study, *Ann. Human Biol.*, 1995, **22**, 470–490.
10. H. Snieder, A.J. MacGregor and T.D. Spector, Genes control the cessation of a woman's reproductive life: a twin study of hysterectomy and age at menopause, *J. Clin. Endocrinol. Metab.*, 1998, **83**, 1875–1880.
11. P.J. Kelly, J.L. Hopper, G.T. Macaskill, N.A. Pocock, P.N. Sambrook and J.A. Eisman, Genetic factors in bone turnover, *J. Clin. Endocrinol. Metab.*, 1991, **72**, 808–813.

12. A. Tokita, P.J. Kelly, T.V. Nguyen, J.C. Qi, N.A. Morrison, L. Risteli, P.N. Sambrook and J.A. Eisman, Genetic influences on type I collagen synthesis and degradation: further evidence for genetic regulation of bone turnover, *J. Clin. Endocrinol. Metab.*, 1994, **78**, 1461–1466.

13. D. Hunter, M. de Lange, H. Snieder, A.J. MacGregor, R. Swaminathan, R.V. Thakker and T.D. Spector, Genetic contribution to bone metabolism, calcium excretion, and vitamin D and parathyroid hormone regulation, *J. Bone Miner. Res.*, 2001, **16**, 371–378.

14. D.J. Torgerson, M.K. Campbell, R.E. Thomas and D.M. Reid, Prediction of perimenopausal fractures by bone mineral density and other risk factors, *J. Bone Miner. Res.*, 1996, **11**, 293–297.

15. H.W. Deng, W.M. Chen, S. Recker, M.R. Stegman, J.L. Li, K.M. Davies, Y. Zhou, H. Deng, R. Haney and R.R. Recker, Genetic determination of Colles' fracture and differential bone mass in women with and without Colles' fracture, *J. Bone Miner. Res.*, 2000, **15**, 1243–1252.

16. D.R. Nyholt, All LODs are not created equal, *Am. J. Human Genet.*, 2000, **67**, 282–288.

17. D.E. Reich, M. Cargill, S. Bolk, J. Ireland, P.C. Sabeti, D.J. Richter, T. Lavery, R. Kouyoumjian, S.F. Farhadian, R. Ward and E.S. Lander, Linkage disequilibrium in the human genome, *Nature*, 2001, **411**, 199–204.

18. R.S. Spielman, R.E. McGinnis and W.J. Ewens, Transmission test for linkage disequilibrium: the insulin gene region and insulin-dependent diabetes mellitus (IDDM), *Am. J. Human Genet.*, 1993, **52**, 506–516.

19. H.W. Deng and W.M. Chen, The power of the transmission disequilibrium test (TDT) with both case-parent and control-parent trios, *Genet. Res.*, 2001, **78**, 289–302.

20. R.F. Klein, A.S. Carlos, K.A. Vartanian, V.K. Chambers, E.J. Turner, T.J. Phillips, J.K. Belknap and E.S. Orwoll, Confirmation and fine mapping of chromosomal regions influencing peak bone mass in mice, *J. Bone Miner. Res.*, 2001, **16**, 1953–1961.

21. R.D. Little, J.P. Carulli, R.G. Del Mastro, J. Dupuis, M. Osborne, C. Folz, S.P. Manning, P.M. Swain, S.C. Zhao, B. Eustace, M.M. Lappe, L. Spitzer, S. Zweier, K. Braunschweiger, Y. Benchekroun, X. Hu, R. Adair, L. Chee, M.G. FitzGerald, C. Tulig, A. Caruso, N. Tzellas, A. Bawa, B. Franklin, S. McGuire, X. Nogues, G. Gong, K.M. Allen, A. Anisowicz, A.J. Morales, P.T. Lomedico, S.M. Recker, P. Van Eerdewegh, R.R. Recker and M.L. Johnson, A mutation in the LDL receptor-related protein 5 gene results in the autosomal dominant high-bone-mass trait, *Am. J. Human Genet.*, 2002, **70**, 11–19.

22. Y. Gong, R.B. Slee, N. Fukai, G. Rawadi, S. Roman-Roman, A.M. Reginato, H. Wang, T. Cundy, F.H. Glorieux, D. Lev, M. Zacharin, K. Oexle, J. Marcelino, W. Suwairi, S. Heeger, G. Sabatakos, S. Apte, W.N. Adkins, J. Allgrove, M. Arslan-Kirchner, J.A. Batch, P. Beighton, G.C. Black, R.G. Boles, L.M. Boon, C. Borrone, H.G. Brunner, G.F. Carle, B. Dallapiccola, A. De Paepe, B. Floege, M.L. Halfhide, B. Hall, R.C. Hennekam, T. Hirose, A. Jans, H. Juppner, C.A. Kim, K. Keppler-Noreuil, A. Kohlschuetter, D. LaCombe, M. Lambert, E. Lemyre, T. Letteboer, L. Peltonen, R.S. Ramesar, M. Romanengo, H. Somer, E. Steichen-Gersdorf, B. Steinmann, B. Sullivan, A. Superti-Furga, W. Swoboda, M.J. van den Boogaard, W. Van Hul, M. Vikkula, M. Votruba, B. Zabel, T. Garcia, R. Baron, B.R. Olsen and M.L. Warman, LDL receptor-related protein 5 (LRP5) affects bone accrual and eye development, *Cell*, 2001, **107**, 513–523.

23. C. Heaney, H. Shalev, K. Elbedour, R. Carmi, J.B. Staack, V.C. Sheffield and D.R. Beier, Human autosomal recessive osteopetrosis maps to 11q13, a position predicted by comparative mapping of the murine osteosclerosis (oc) mutation, *Human Mol. Genet.*, 1998, **7**, 1407–1410.

24. A. Frattini, P.J. Orchard, C. Sobacchi, S. Giliani, M. Abinun, J.P. Mattsson, D.J. Keeling, A.K. Andersson, P. Wallbrandt, L. Zecca, L.D. Notarangelo, P. Vezzoni and A. Villa, Defects in TCIRG1 subunit of the vacuolar proton pump are responsible for a subset of human autosomal recessive osteopetrosis, *Nature Genet.*, 2000, **25**, 343–346.

25. E. Cleiren, O. Benichou, E. Van Hul, J. Gram, J. Bollerslev, F.R. Singer, K. Beaverson, A. Aledo, M.P. Whyte, T. Yoneyama, M.C. deVernejoul and W. Van Hul, Albers-Schonberg disease (autosomal dominant osteopetrosis, type II) results from mutations in the ClCN7 chloride channel gene, *Human Mol. Genet.*, 2001, **10**, 2861–2867.

26. U. Kornak, D. Kasper, M.R. Bosl, E. Kaiser, M. Schweizer, A. Schulz, W. Friedrich, G. Delling and T.J. Jentsch, Loss of the ClC-7 chloride channel leads to osteopetrosis in mice and man, *Cell*, 2001, **104**, 205–215.

27. S.A. Stein, C. Witkop, S. Hill, M.D. Fallon, L. Viernstein, G. Gucer, P. McKeever, D. Long, J. Altman, N.R. Miller, S.L. Teitelbaum and S. Schlesinger, Sclerosteosis: neurogenetic and pathophysiologic analysis of an American kinship, *Neurology*, 1983, **33**, 267–277.

28. W. Van Hul, W. Balemans, E. Van Hul, F.G. Dikkers, H. Obee, R.J. Stokroos, P. Hildering, F. Vanhoenacker, G. Van Camp and P.J. Willems, Van Buchem disease (hyperostosis corticalis generalisata) maps to chromosome 17q12-q21, *Am. J. Human Genet.*, 1998, **62**, 391–399.

29. W. Balemans, N. Patel, M. Ebeling, E. Van Hul, W. Wuyts, C. Lacza, M. Dioszegi, F.G. Dikkers, P. Hildering, P.J. Willems, J.B. Verheij, K. Lindpaintner, B. Vickery, D. Foernzler and W. Van Hul, Identification of a 52 kb deletion downstream of the SOST gene in patients with van Buchem disease, *J. Med. Genet.*, 2002, **39**, 91–97.

30. M. Brunkow, J. Gardner, J. Van Ness, B. Paeper, B. Kovacevich, S. Proll, J.E. Skonier, L. Zhao, P.J. Sabo, Y. Fu, R.S. Alisch, L. Gillett, T. Colbert, P. Tacconi, D. Galas, H. Hamersma, P. Beighton and J. Mulligan, Bone dysplasia sclerosteosis results from loss of the sost gene product, a novel cystine knot-containing protein, *Am. J. Human Genet.*, 2001, **68**, 577–589.

31. K. Janssens, R. Gershoni-Baruch, N. Guanabens, N. Migone, S. Ralston, M. Bonduelle, W. Lissens, L. Van Maldergem, F. Vanhoenacker, L. Verbruggen and W. Van Hul, Mutations in the gene encoding the latency-associated peptide of TGF-beta1 cause Camurati–Engelmann disease, *Nature Genet.*, 2000, **26**, 273–275.

32. Y. Yamada, A. Miyauchi, Y. Takagi, M. Tanaka, M. Mizuno and A. Harada, Association of the C-509→T polymorphism, alone of in combination with the T869→C polymorphism, of the transforming growth factor-beta1 gene with bone mineral density and genetic susceptibility to osteoporosis in Japanese women, *J. Mol. Med.*, 2001, **79**, 149–156.

33. D.L. Koller, L.A. Rodriguez, J.C. Christian, C.W. Slemenda, M.J. Econs, S.L. Hui, P. Morin, P.M. Conneally, G. Joslyn, M.E. Curran, M. Peacock, C.C. Johnston and T. Foroud, Linkage of a QTL contributing to normal variation in bone mineral density to chromosome 11q12-13, *J. Bone Miner. Res.*, 1998, **13**, 1903–1908.

34. D.L. Koller, M.J. Econs, P.A. Morin, J.C. Christian, S.L. Hui, P. Parry, M.E. Curran, L.A. Rodriguez, P.M. Conneally, G. Joslyn, M. Peacock, C.C. Johnston and T. Foroud, Genome screen for QTLs contributing to normal variation in bone mineral density and osteoporosis, *J. Clin. Endocrinol. Metab.*, 2000, **85**, 3116–3120.

35. D.L. Kollers, G. Liu, M.J. Econs, S.L. Hui, P.A. Morin, G. Joslyn, L.A. Rodriguez, P.M. Conneally, J.C. Christian, C.C. Johnston Jr., T. Foroud and M. Peacock, Genome screen for quantitative trait loci underlying normal variation in femoral structure, *J. Bone Miner. Res.*, 2001, **16**, 985–991.

36. M. Devoto, K. Shimoya, J. Caminis, J. Ott, A. Tenenhouse, M.P. Whyte, L. Sereda, S. Hall, E. Considine, C.J. Williams, G. Tromp, H. Kuivaniemi, L. Ala-Kokko, D.J. Prockop and L.D. Spotila, First-stage autosomal genome screen in extended pedigrees suggests genes predisposing to low bone mineral density on chromosomes 1p, 2p and 4q, *Eur. J. Human Genet.*, 1998, **6**, 151–157.

37. M. Devoto, C. Specchia, H.H. Li, J. Caminis, A. Tenenhouse, H. Rodriguez and L.D. Spotila, Variance component linkage analysis indicates a QTL for femoral neck bone mineral density on chromosome 1p36, *Human Mol. Genet.*, 2001, **10**, 2447–2452.

38. T. Nui, C. Chen, H. Cordell, J. Yang, B. Wang, Z. Wang, Z. Fang, N.J. Schork, C.J. Rosen and X. Xu, A genome-wide scan for loci linked to forearm bone mineral density, *Human Genet.*, 1999, **104**, 226–233.

39. E.L. Duncan, M.A. Brown, J. Sinsheimer, J. Bell, A.J. Carr, B.P. Wordsworth and J.A. Wass, Suggestive linkage of the parathyroid receptor type 1 to osteoporosis, *J. Bone Miner. Res.*, 2000, **14**, 1993–1999.

40. N. Ota, S.C. Hunt, T. Nakajima, T. Suzuki, T. Hosoi, H. Orimo, Y. Shirai and M. Emi, Linkage of interleukin 6 locus to human osteopenia by sibling pair analysis, *Human Genet.*, 1999, **105**, 253–257.

41. R.F. Klein, S.R. Mitchell, T.J. Phillips, J.K. Belknap and E.S. Orwoll, Quantitative trait loci affecting peak bone mineral density in mice, *J. Bone Miner. Res.*, 1998, **13**, 1648–1656.

42. M. Shimizu, K. Higuchi, B. Bennett, C. Xia, T. Tsuboyama, S. Kasai, T. Chiba, H. Fujisawa, K. Kogishi, H. Kitado, M. Kimoto, N. Takeda, M. Matsushita, H. Okumura, T. Serikawa, T. Nakamura, T.E. Johnson and M. Hosokawa, Identification of peak bone mass QTL in a spontaneously osteoporotic mouse strain, *Mammalian Genome*, 1999, **10**, 81–87.

43. W.G. Beamer, L.R. Donahue, C.J. Rosen and D.J. Baylink, Genetic variability in adult bone density among inbred strains of mice, *Bone*, 1996, **18**, 397–403.

44. H. Benes, R.S. Weinstein, W. Zheng, J.J. Thaden, R.L. Jilka, S.C. Manolagas and R.J. Shmookler Reis, Chromosomal mapping of osteopenia-associated quantitative trait loci using closely related mouse strains, *J. Bone Miner. Res.*, 2000, **15**, 626–633.

45. Y. Yershov, T.H. Baldini, S. Villagomez, T. Young, M.L. Martin, R.S. Bockman, M.G. Peterson and R.D. Blank, Bone strength and related traits in HcB/Dem recombinant congenic mice, *J. Bone Miner. Res.*, 2001, **16**, 992–1003.

46. J.P. Ioannidis, E.E. Ntzani, T.A. Trikalinos and D.G. Contopoulos-Ioannidis, Replication validity of genetic association studies, *Nature Genet.*, 2001, **29**, 306–309.

47. G.S. Cooper and D.M. Umbach, Are vitamin D receptor polymorphisms associated with bone mineral density? A meta-analysis, *J. Bone Miner. Res.*, 1996, **11**, 1841–1849.

48. V. Mann, E.E. Hobson, B. Li, T.L. Stewart, S.F. Grant, S.P. Robins, R.M. Aspden and S.H. Ralston, A COL1A1 Sp1 binding site polymorphism predisposes to osteoporotic fracture by affecting bone density and quality, *J. Clin. Invest.*, 2001, **107**, 899–907.

49. N.A. Morrison, J.C. Qi, A. Tokita, P. Kelly, L. Crofts, T.V. Nguyen, P.N. Sambrook and J.A. Eisman, Prediction of bone density from vitamin D receptor alleles, *Nature*, 1994, **367**, 284–287.

50. R.Y. Zee, R.H. Myers, M.T. Hannan, P.W. Wilson, J.M. Ordovas, E.J. Schaefer, K. Lindpaintner and D.P. Kiel, Absence of linkage for bone mineral density to chromosome 12q12-14 in the region of the vitamin D receptor gene, *Calcif. Tissue Int.*, 2000, **67**, 434–439.

51. E.A. Krall, P. Parry, J.B. Lichter and B. Dawson-Hughes, Vitamin D receptor alleles and rates of bone loss: influence of years since menopause and calcium intake, *J. Bone Miner. Res.*, 1995, **10**, 978–984.

52. D.P. Kiel, R.H. Myers, L.A. Cupples, X.F. Kong, X.H. Zhu, J. Ordovas, E.J. Schaefer, D.T. Felson, D. Rush, P.W. Wilson, J.A. Eisman and M.F. Holick, The BsmI vitamin D receptor restriction fragment length polymorphism (bb) influences the effect of calcium intake on bone mineral density, *J. Bone Miner. Res.*, 1997, **12**, 1049–1057.

53. S. Ferrari, R. Rizzoli, D. Slosman and J.-P. Bonjour, Do dietary calcium and age explain the controversy surrounding the relationship between bone mineral density and vitamin D receptor gene polymorphisms? *J. Bone Miner. Res.*, 1998, **13**, 363–370.

54. W.C. Graafmans, P. Lips, M.E. Ooms, J.P.T.M. van Leeuwen, H.A.P. Pols and A.G. Uitterlinden, The effect of vitamin D supplementation on the bone mineral density of the femoral neck is associated with vitamin D receptor genotype, *J. Bone Miner. Res.*, 1997, **12**, 1241–1245.

55. H. Arai, K.-I. Miyamoto, Y. Taketani, H. Yamamoto, Y. Iemori, K. Morita, T. Tonai, T. Nishisho, S. Mori and E. Takeda, A vitamin D receptor gene polymorphism in the translation initiation codon: effect on protein activity and relation to bone mineral density in Japanese women, *J. Bone Miner. Res.*, 1997, **12**, 915–921.

56. C. Gross, T.R. Eccleshall, P.J. Malloy, M.L. Villa, R. Marcus and D. Feldman, The presence of a polymorphism at the translation initiation site of the vitamin D receptor gene is associated with low bone mineral density in postmenopausal Mexican–American women, *J. Bone Miner. Res.*, 1997, **12**, 1850–1856.

57. S.S. Harris, T.R. Eccleshall, C. Gross, B. Dawson-Hughes and D. Feldman, The vitamin D receptor start codon polymorphism (FokI) and bone mineral density in premenopausal American black and white women, *J. Bone Miner. Res.*, 1997, **12**, 1043–1048.

58. L. Gennari, L. Becherini, R. Mansani, L. Masi, A. Falchetti, A. Morelli, E. Colli, S. Gonnelli, C. Cepollaro and M.L. Brandi, FokI polymorphism at translation initiation site of the vitamin D receptor gene predicts bone mineral density and vertebral fractures in postmenopausal Italian women, *J. Bone Miner. Res.*, 1999, **14**, 1379–1386.

59. H. Arai, K.I. Miyamoto, M. Yoshida, H. Yamamoto, Y. Taketani, K. Morita, M. Kubota, S. Yoshida, M. Ikeda, F. Watabe, Y. Kanemasa and E. Takeda, The polymorphism in the caudal-related homeodomain protein Cdx-2 binding element in the human vitamin D receptor gene, *J. Bone Miner. Res.*, 2001, **16**, 1256–1264.

60. S.F.A. Grant, D.M. Reid, G. Blake, R. Herd, I. Fogelman and S.H. Ralston, Reduced bone density and osteoporosis associated with a polymorphic Sp1 site in the collagen type I alpha 1 gene, *Nature Genet.*, 1996, **14**, 203–205.

61. B.L. Langdahl, S.H. Ralston, S.F.A. Grant and E.F. Eriksen, An Sp1 binding site polymorphism in the COLIA1 gene predicts osteoporotic fractures in men and women, *J. Bone Miner. Res.*, 1998, **13**, 1384–1389.

62. R.W. Keen, K.L. Woodford-Richens, S.F.A. Grant, S.H. Ralston, J.S. Lanchbury and T.D. Spector, Polymorphism at the type I collagen (COLIA1) locus is associated with reduced bone mineral density, increased fracture risk and increased collagen turnover, *Arthritis Rheum.*, 1999, **42**, 285–290.

63. C. Roux, M. Dougados, L. Abel, G. Mercier and G. Lucotte, Association of a polymorphism in the collagen I α 1 gene with osteoporosis in French women, *Arthritis Rheum.*, 1998, **41**, 187–188.

64. M. Bernad, M.E. Martinez, M. Escalona, M.L. Gonzalez, C. Gonzalez, M.V. Garces, M.T. Del Campo, E. Martin Mola, R. Madero and L. Carreno, Polymorphism in the type I collagen (COLIA1) gene and risk of fractures in postmenopausal women, *Bone*, 2002, **30**, 223–228.

65. S.S. Harris, M.S. Patel, D.E. Cole and B. Dawson-Hughes, Associations of the collagen type I alpha1 Sp1 polymorphism with five- year rates of bone loss in older adults, *Calcif. Tissue Int.*, 2000, **66**, 268–271.

66. A.M. Qureshi, F.E.A. McGuigan, D.G. Seymour, J.D. Hutchison, D.M. Reid and S.H. Ralston, Association between COLIA1 Sp1 alleles and femoral neck geometry, *Calcif. Tissue Int.*, 2001, **69**, 67–72.

67. A.M. Qureshi, R.J. Herd, G.M. Blake, I. Fogelman and S.H. Ralston, COLIA1 Sp1 polymorphism predicts response of femoral neck bone density to cyclical etidronate therapy, *Calcif. Tissue Int.*, 2002, **70**, 158–163.

68. S. Beavan, A. Prentice, B. Dibba, L. Yan, C. Cooper and S.H. Ralston, Polymorphism of the collagen type I alpha 1 gene and ethnic differences in hip-fracture rates, *New Engl. J. Med.*, 1998, **339**, 351–352.

69. F.E.A. McGuigan, D.M. Reid and S.H. Ralston, Susceptibility to osteoporotic fracture is determined by allelic variation at the Sp1 site, rather than other polymorphic sites, at the COLIA1 locus, *Osteoporosis Int.*, 2000, **11**, 338–343.

70. N. Garcia-Giralt, X. Nogues, A. Enjuanes, J. Puig, L. Mellibovsky, A. Bay-Jensen, R. Carreras, S. Balcells, A. Diez-Perez and D. Grinberg, Two new single nucleotide polymorphisms in the COLIA1 upstream regulatory region and their relationship with bone mineral density, *J. Bone Miner. Res.*, 2002, **17**, 384–393.

71. F.E.A. McGuigan, G. Armbrecht, R. Smith, D. Felsenberg, D.M. Reid and S.H. Ralston, Prediction of osteoporotic fractures by bone densitometry and COLIA1 genotyping: a prospective, population-based study in men and women, *Osteoporosis Int.*, 2001, **12**, 91–96.

72. A.E. Weel, A.G. Uitterlinden, I.C. Westendorp, H. Burger, S.C. Schuit, A. Hofman, T.J. Helmerhorst, J.P. van Leeuwen and H.A. Pols, Estrogen receptor polymorphism predicts the onset of natural and surgical menopause, *J. Clin. Endocrinol. Metab.*, 1999, **84**, 3146–3150.

73. E.P. Smith, J. Boyd, G.R. Frank, H. Takahashi, R.M. Cohen, B. Specker, T.C. Williams, D.B. Lubahn and K.S. Korach, Estrogen resistance caused by a mutation in the estrogen-receptor gene in a man, *New Engl. J. Med.*, 1994, **331**, 1056–1061.

74. M. Sano, S. Inoue, T. Hosoi, Y. Ouchi, M. Emi, M. Shiraki and H. Orimo, Association of estrogen receptor dinucleotide repeat polymorphism with osteoporosis, *Biochem. Biophys. Res. Commun.*, 1995, **217**, 378–383.
75. M. Sowers, M. Willing, T. Burns, S. Deschenes, B. Hollis, M. Crutchfield and M. Jannausch, Genetic markers, bone mineral density and serum osteocalcin levels, *J. Bone Miner. Res.*, 1999, **14**, 1411–1419.
76. L. Becherini, L. Gennari, L. Masi, R. Mansani, F. Massart, A. Morelli, A. Falchetti, S. Gonnelli, G. Fiorelli, A. Tanini and M.L. Brandi, Evidence of a linkage disequilibrium between polymorphisms in the human estrogen receptor alpha gene and their relationship to bone mass variation in postmenopausal Italian women, *Human Mol. Genet.*, 2000, **9**, 2043–2050.
77. H.Y. Chen, W.C. Chen, H.D. Tsai, C.D. Hsu, F.J. Tsai, C.H. Tsai, Relation of the estrogen receptor alpha gene microsatellite polymorphism to bone mineral density and the susceptibility to osteoporosis in postmenopausal Chinese women in Taiwan, *Maturitas*, 2001, **40**, 143–150.
78. S. Kobayashi, S. Inoue, T. Hosoi, Y. Ouchi, M. Shiraki and H. Orimo, Association of bone mineral density with polymorphism of the estrogen receptor gene, *J. Bone Miner. Res.*, 1996, **11**, 306–311.
79. H. Mizunuma, T. Hosoi, H. Okano, M. Soda, T. Tokizawa, I. Kagami, S. Miyamoto, Y. Ibuki, S. Inoue, M. Shiraki and Y. Ouchi, Estrogen receptor gene polymorphism and bone mineral density at the lumbar spine of pre- and postmenopausal women, *Bone*, 1997, **21**, 379–383.
80. O.M. Albagha, F.E.A. McGuigan, D.M. Reid and S.H. Ralston, Estrogen receptor alpha gene polymorphisms and bone mineral density: haplotype analysis in women from the United Kingdom, *J. Bone Miner. Res.*, 2001, **16**, 128–134.
81. B. Ongphiphadhanakul, S. Chanprasertyothin, P. Payattikul, S. Saetung, N. Piaseu, L. Chailurkit and R. Rajatanavin, Association of a G2014A transition in exon 8 of the estrogen receptor-alpha gene with postmenopausal osteoporosis, *Osteoporosis Int.*, 2001, **12**, 1015–1019.
82. B. Ongphiphadhanakul, R. Rajatanavin, S. Chanprasertyothin, N. Piaseu, L. Chailurkit, R. Sirisriro and S. Komindr, Estrogen receptor gene polymorphism is associated with bone mineral density in premenopausal women but not in postmenopausal women, *J. Endocrinol. Invest.*, 1998, **21**, 487–493.
83. K. Janssens, R. Gershoni-Baruch, N. Guanabens, N. Migone, S. Ralston, M. Bonduelle, W. Lissens, L. Van Maldergem, F. Vanhoenacker, L. Verbruggen and W. Van Hul, Mutations in the gene encoding the latency-associated peptide of TGF-beta1 cause Camurati-Engelmann disease, *Nature Genet.*, 2000, **26**, 273–275.
84. J.A. Robinson, B.L. Riggs, T.C. Spelsberg and M.J. Oursler, Osteoclasts and transforming growth factor-beta: estrogen-mediated isoform-specific regulation of production, *Endocrinology*, 1996, **137**, 615–621.
85. D.J. Grainger, J. Percival, M. Chiano and T.D. Spector, The role of serum TGF-beta isoforms as potential markers of osteoporosis, *Osteoporosis Int.*, 1999, **9**, 398–404.
86. B.L. Langdahl, J.Y. Knudsen, H.K. Jensen, N. Gregersen and E.F. Eriksen, A sequence variation: 713-8delC in the transforming growth factor-beta 1 gene has higher prevalence in osteoporotic women than in normal women and is associated with very low bone mass in osteoporotic women and increased bone turnover in both osteoporotic and normal women, *Bone*, 1997, **20**, 289–294.

87. G.D. Roodman, Interleukin-6: An osteotropic factor? *J. Bone Miner. Res.*, 1992, **7**, 475–476.

88. S.H. Ralston, Analysis of gene expression in human bone biopsies by polymerase chain reaction: evidence for enhanced cytokine expression in postmenopausal osteoporosis, *J. Bone Miner. Res.*, 1994, **9**, 883–890.

89. R.E. Murray, F.E.A. McGuigan, S.F.A. Grant, D.M. Reid and S.H. Ralston, Polymorphisms of the interleukin-6 gene are associated with bone mineral density, *Bone*, 1997, **21**, 89–92.

90. K. Tsukamoto, N. Ohta, Y. Shirai and M. Emi, A highly polymorphic CA repeat marker at the human interleukin 6 receptor (IL6R) locus, *J. Human Genet.*, 1998, **43**, 289–290.

91. K. Tsukamoto, H. Yoshida, S. Watanabe, T. Suzuki, M. Miyao, T. Hosoi, H. Orimo and M. Emi, Association of radial bone mineral density with CA repeat polymorphism at the interleukin 6 locus in postmenopausal Japanese women, *J. Human Genet.*, 1999, **44**, 148–151.

92. N. Ota, T. Nakajima, I. Nakazawa, T. Suzuki, T. Hosoi, H. Orimo, S. Inoue, Y. Shirai and M. Emi, A nucleotide variant in the promoter region of the interleukin-6 gene associated with decreased bone mineral density, *J. Human Genet.*, 2001, **46**, 267–272.

93. D. Fishman, G. Faulds, R. Jeffery, V. Mohamed-Ali, J.S. Yudkin, S. Humphries and P. Woo, The effect of novel polymorphisms in the interleukin-6 (IL-6) gene on IL-6 transcription and plasma IL-6 levels, and an association with systemic-onset juvenile chronic arthritis, *J. Clin. Invest.*, 1998, **102**, 1369–1376.

94. J.A. Cauley, J.M. Zmuda, K. Yaffe, L.H. Kuller, R.E. Ferrell, S.R. Wisniewski and S.R. Cummings, Apolipoprotein E polymorphism: a new genetic marker of hip fracture risk – the study of osteoporotic fractures, *J. Bone Miner. Res.*, 1999, **14**, 1175–1181.

95. M. Shiraki, Y. Shiraki, C. Aoki, T. Hosoi, S. Inoue, M. Kaneki and Y. Ouchi, Association of bone mineral density with apolipoprotein E phenotype, *J. Bone Miner. Res.*, 1997, **12**, 1438–1445.

96. L.M. Salamone, J.A. Cauley, J. Zmuda, A. Pasagian-Macaulay, R.S. Epstein, R.E. Ferrell, D.M. Black and L.H. Kuller, Apolipoprotein E gene polymorphism and bone loss: estrogen status modifies the influence of apolipoprotein E on bone loss, *J. Bone Miner. Res.*, 2000, **15**, 308–314.

97. Y. Dohi, M. Iki, H. Ohgushi, S. Gojo, S. Tabata, E. Kajita, H. Nishino and K. Yonemasu, A novel polymorphism in the promoter region for the human osteocalcin gene: the possibility of a correlation with bone mineral density in postmenopausal Japanese women, *J. Bone Miner. Res.*, 1998, **13**, 1633–1639.

98. H.Y. Chen, H.D. Tsai, W.C. Chen, J.Y. Wu, F.J. Tsai and C.H. Tsai, Relation of polymorphism in the promotor region for the human osteocalcin gene to bone mineral density and occurrence of osteoporosis in postmenopausal Chinese women in Taiwan, *J. Clin. Lab. Anal.*, 2001, **15**, 251–255.

99. A. Gustavsson, P. Nordstrom, R. Lorentzon, U.H. Lerner and M. Lorentzon, Osteocalcin gene polymorphism is related to bone density in healthy adolescent females, *Osteoporosis Int.*, 2000, **11**, 847–851.

100. P. Ducy, C. Desbois, B. Boyce, G. Pinero, B. Story, C. Dunstan, E. Smith, J. Bonadio, S. Goldstein, C. Gundberg, A. Bradley and G. Karsenty, Increased bone formation in osteocalcin deficient mice, *Nature*, 1996, **382**, 448–452.

101. M. Nakamura, S. Morimoto, Z. Zhang, H. Utsunomiya, T. Inagami, T. Ogihara and K. Kakudo, Calcitonin receptor gene polymorphism in Japanese women:

correlation with body mass and bone mineral density, *Calcif. Tissue Int.*, 2001, **68**, 211–215.

102. L. Masi, L. Becherini, E. Colli, L. Gennari, R. Mansani, A. Falchetti, A.M. Becorpi, C. Cepollaro, S. Gonnelli, A. Tanini and M.L. Brandi, Polymorphisms of the calcitonin receptor gene are associated with bone mineral density in postmenopausal Italian women, *Biochem. Biophys. Res. Commun.*, 1998, **248**, 190–195.

103. L. Masi, L. Becherini, L. Gennari, E. Colli, R. Mansani, A. Falchetti, C. Cepollaro, S. Gonnelli, A. Tanini and M.L. Brandi, Allelic variants of human calcitonin receptor: distribution and association with bone mass in postmenopausal Italian women, *Biochem. Biophys. Res. Commun.*, 1998, **245**, 622–626.

104. M.L. Johnson, G. Gong, W. Kimberling, S. Recker, D.B. Kimmel and R.R. Recker, Linkage of a gene causing high bone mass to human chromosome 11 (11q12-13), *Am. J. Human Genet.*, 1997, **60**, 1326–1332.

105. T. Hosoi, M. Miyao, S. Inoue, S. Hoshino, M. Shiraki, H. Orimo and Y. Ouchi, Association study of parathyroid hormone gene polymorphism and bone mineral density in Japanese postmenopausal women, *Calcif. Tissue Int.*, 1999, **64**, 205–208.

106. G. Gong, M.L. Johnson, M.J. Barger-Lux and R.P. Heaney, Association of bone dimensions with a parathyroid hormone gene polymorphism in women, *Osteoporosis Int.*, 2000, **9**, 307–311.

107. R.W. Keen, K.L. Woodford-Richens, J.S. Lanchbury and T.D. Spector, Allelic variation at the interleukin-1 receptor antagonist gene is associated with early postmenopausal bone loss at the spine, *Bone*, 1998, **23**, 367–371.

108. J.K. Tarlow, A.I. Blakemore, A. Lennard, R. Solari, H.N. Hughes, A. Steinkasserer and G.W. Duff, Polymorphism in human IL-1 receptor antagonist gene intron 2 is caused by variable numbers of an 86-bp tandem repeat, *Human Genet.*, 1993, **91**, 403–404.

109. B.L. Langdahl, E. Lokke, M. Carstens, L.L. Stenkjaer and E.F. Eriksen, Osteoporotic fractures are associated with an 86-base pair repeat polymorphism in the interleukin 1-receptor antagonist gene but not with polymorphisms in the interleukin-1beta gene, *J. Bone Miner. Res.*, 2000, **15**, 402–414.

110. B.L. Langdahl, M. Carstens, L. Stenkjaer and E.F. Eriksen, Polymorphisms in the osteoprotegerin gene are associated with osteoporotic fractures, *J. Bone Miner. Res.*, 2002, **17**, 1245–1255.

111. F. Wynne, F. Drummond, K. O'Sullivan, M. Daly, F. Shanahan, M.G. Molloy and K.A. Quane, Investigation of the Genetic Influence of the OPG, VDR (Fok1), and COLIA1 Sp1 Polymorphisms on BMD in the Irish Population, *Calcif. Tissue Int.*, 2002, **71**(1).

112. L.D. Spotila, H. Rodriguez, M. Koch, K. Adams, J. Caminis and H.S. Tenenhouse, Association of a polymorphism in the TNFR2 gene with low bone mineral density, *J. Bone Miner. Res.*, 2000, **15**, 1376–1383.

113. O.M.E. Albagha, P.N. Tasker, F.E.A. McGuigan, D.M. Reid and S.H. Ralston, Linkage disequilibrium between polymorphisms in the human TNFRSF1B gene and their association with bone mass in perimenopausal women, *Human Mol. Genet.*, 2002, **11**, 2289–2295.

114. S. Ogawa, T. Urano, T. Hosoi, M. Miyao, S. Hoshino, M. Fujita, M. Shiraki, H. Orimo, Y. Ouchi and S. Inoue, Association of bone mineral density with a polymorphism of the peroxisome proliferator-activated receptor gamma gene: PPAR gamma expression in osteoblasts, *Biochem. Biophys. Res. Commun.*, 1999, **260**, 122–126.

115. T.G. Tut, F.J. Ghadessy, M.A. Trifiro, L. Pinsky and E.L. Yong, Long polyglutamine tracts in the androgen receptor are associated with reduced transactivation, impaired sperm production and male infertility, *J. Clin. Endocrinol. Metab.*, 1997, **82**, 3777–3782.

116. P.I. Van, S. Lumbroso, S. Goemaere, C. Sultan and J.M. Kaufman, Lack of influence of the androgen receptor gene CAG-repeat polymorphism on sex steroid status and bone metabolism in elderly men, *Clin. Endocrinol.*, 2001, **55**, 659–666.

117. P. Frosst, H.J. Blom, R. Milos, P. Goyette, C.A. Sheppard, R.G. Matthews, G.J. Boers, M. den Heijer, L.A. Kluijtmans, L.P. van den Heuvel, R. Rozen, A candidate genetic risk factor for vascular disease: a common mutation in methylenetetrahydrofolate reductase, *Nature Genet.*, 1995, **10**, 111–113.

118. M. Miyao, H. Morita, T. Hosoi, H. Kurihara, S. Inoue, S. Hoshino, M. Shiraki, Y. Yazaki and Y. Ouchi, Association of methylenetetrahydrofolate reductase (MTHFR) polymorphism with bone mineral density in postmenopausal Japanese women, *Calcif. Tissue Int.*, 2000, **66**, 190–194.

119. J.E. Eichner, J.A. Cauley, R.E. Ferrell, S.R. Cummings and L.H. Kuller, Genetic variation in two bone-related proteins: is there an association with bone mineral density or skeletal size in postmenopausal women? *Genet. Epidemiol.*, 1992, **9**, 177–184.

120. I.R. Dickson, R. Gwilliam, M. Arora, S. Murphy, K.T. Khaw, C. Phillips and P. Lincoln, Lumbar vertebral and femoral neck bone mineral density are higher in postmenopausal women with the alpha 2HS-glycoprotein 2 phenotype, *Bone Miner.*, 1994, **24**, 181–188.

121. J.E. Eichner, C.A. Friedrich, J.A. Cauley, M.I. Kamboh, J.P. Gutai, L.H. Kuller and R.E. Ferrell, Alpha 2-HS glycoprotein phenotypes and quantitative hormone and bone measures in postmenopausal women, *Calcif. Tissue Int.*, 1990, **47**, 345–349.

122. K. Tsukamoto, H. Orimo, T. Hosoi, M. Miyao, N. Ota, T. Nakajima, H. Yoshida, S. Watanabe, T. Suzuki and M. Emi, Association of bone mineral density with polymorphism of the human calcium-sensing receptor locus, *Calcif. Tissue Int.*, 2000, **66**, 181–183.

123. I. Takacs, G. Speer, E. Bajnok, A. Tabak, Z. Nagy, C. Horvath, K. Kovacs and P. Lakatos, Lack of association between calcium-sensing receptor gene 'A986S' polymorphism and bone mineral density in Hungarian postmenopausal women, *Bone*, 2002, **30**, 849–852.

124. M. Eckstein, I. Vered, S. Ish-Shalom, A.B. Shlomo, A. Shtriker, N. Koren-Morag and E. Friedman, Vitamin D and calcium-sensing receptor genotypes in men and premenopausal women with low bone mineral density, *Israel Med. Assoc. J.*, 2002, **4**, 340–344.

125. M. Lorentzon, R. Lorentzon, U.H. Lerner and P. Nordstrom, Calcium sensing receptor gene polymorphism, circulating calcium concentrations and bone mineral density in healthy adolescent girls, *Eur. J. Endocrinol.*, 2001, **144**, 257–261.

126. A.G. Comuzzie, J. Blangero, M.C. Mahaney, S.M. Haffner, B.D. Mitchell, M.P. Stern and MacCluer JW, Genetic and environmental correlations among hormone levels and measures of body fat accumulation and topography, *J. Clin. Endocrinol. Metab.*, 1996, **81**, 597–600.

127. E.S. Kurland, C.J. Rosen, F. Cosman, D. McMahon, F. Chan, E. Shane, R. Lindsay, D. Dempster and J.P. Bilezikian, Insulin-like growth factor-I in men with idiopathic osteoporosis, *J. Clin. Endocrinol. Metab.*, 1997, **82**, 2799–2805.

128. C.J. Rosen, E.S. Kurland, D. Vereault, R.A. Adler, P.J. Rackoff, W.Y. Craig, S. Witte, J. Rogers and J.P. Bilezikian, Association between serum insulin growth factor-I (IGF-I) and a simple sequence repeat in IGF-I gene: implications for genetic studies of bone mineral density, *J. Clin. Endocrinol. Metab.*, 1998, **83**, 2286–2290.

129. I. Takacs, D.L. Koller, M. Peacock, J.C. Christian, S.L. Hui, P.M. Conneally, C.C. Johnston Jr., T. Foroud and M.J. Econs, Sibling pair linkage and association studies between bone mineral density and the insulin-like growth factor I gene locus, *J. Clin. Endocrinol. Metab.*, 1999, **84**, 4467–4471.

130. T. Urano, T. Hosoi, M. Shiraki, H. Toyoshima, Y. Ouchi and S. Inoue, Possible involvement of the p57(Kip2) gene in bone metabolism, *Biochem. Biophys. Res. Commun.*, 2000, **269**, 422–426.

131. N. Ogata, M. Shiraki, T. Hosoi, Y. Koshizuka, K. Nakamura and H. Kawaguchi, A polymorphic variant at the Werner helicase (WRN) gene is associated with bone density, but not spondylosis, in postmenopausal women, *J. Bone Miner. Metab.*, 2001, **19**, 296–301.

132. N. Ogata, Y. Matsumura, M. Shiraki, K. Kawano, Y. Koshizuka, T. Hosoi, K. Nakamura, M. Kuro-O and H. Kawaguchi, Association of klotho gene polymorphism with bone density and spondylosis of the lumbar spine in postmenopausal women, *Bone*, 2002, **31**, 37–42.

133. N.A. Huizenga, J.W. Koper, P. de Lange, H.A. Pols, R.P. Stolk, H. Burger, D.E. Grobbee, A.O. Brinkmann, F.H. De Jong and S.W. Lamberts, A polymorphism in the glucocorticoid receptor gene may be associated with and increased sensitivity to glucocorticoids in vivo, *J. Clin. Endocrinol. Metab.*, 1998, **83**, 144–151.

134. L. Masi, L. Becherini, L. Gennari, A. Amedei, E. Colli, A. Falchetti, M. Farci, S. Silvestri, S. Gonnelli and M.L. Brandi, Polymorphism of the aromatase gene in postmenopausal Italian women: distribution and correlation with bone mass and fracture risk, *J. Clin. Endocrinol. Metab.*, 2001, **86**, 2263–2269.

135. A. Nemetz, M. Toth, M.A. Garcia-Gonzalez, T. Zagoni, J. Feher, A.S. Pena and Z. Tulassay, Allelic variation at the interleukin 1beta gene is associated with decreased bone mass in patients with inflammatory bowel diseases, *Gut*, 2001, **49**, 644–649.

136. A.D. Roses, Pharmacogenetics and the practice of medicine, *Nature*, 2000, **405**, 857–865.

# CHAPTER 5

# Rickets, Osteomalacia and other Metabolic Bone Diseases: Influence of Nutrition

JOHN MORLEY PETTIFOR

MRC Mineral Metabolism Research Unit, Department of Paediatrics, Chris Hani Baragwanath Hospital and University of the Witwatersrand, P O Bertsham 2013, South Africa Email: pettiforjm@medicine.wits.ac.za

## Key Points

- This chapter covers the metabolic bone diseases, other than osteoporosis, caused by nutritional factors. The nutrients discussed include vitamin D, calcium, phosphorus, and fluoride.
- Nutritional rickets and osteomalacia are diseases, typically occurring at the extremes of life, due primarily to vitamin D deficiency as a result of inadequate sunlight exposure. Thus, infants and toddlers and the elderly, or the infirm are at risk unless precautions are taken to ensure an adequate vitamin D intake or sunlight exposure.
- Although rickets was thought to be nearly eradicated in a number of developed countries, it has seen a resurgence in minority communities and immigrant families in the USA and Europe. The disease remains a major public health problem in a number of developing countries, despite effective and cheap means of controlling the incidence.
- Low dietary calcium intakes in children living in developing countries may result in rickets and bone deformities without the associated vitamin D deficiency. The disease occurs after weaning, when calcium intakes fall as a result of the low calcium content of the diet that consists mainly of cereal staples without dairy products. Dietary calcium supplementation results in a rapid improvement in the radiological and biochemical changes, while the bone deformities may improve over several years.

- Nutritional phosphate deficiency, as a cause of metabolic bone disease, is an uncommon problem except in the breastfed, very-low-birth-weight (VLBW) infant, in whom the low phosphorus content of breast-milk is unable to meet the requirements of the rapidly growing tissues and bones.
- Endemic fluorosis, which results in dental changes, osteosclerosis, ligamentous calcification, bone deformities, joint stiffness, and rickets and osteomalacia, has been described in a number of developing countries, such as India, following the prolonged ingestion of bore-hole or well water with elevated fluoride concentrations. In children, the disease may manifest solely as an increase in the prevalence of knock-knees in the community.

# Introduction

Metabolic bone diseases are those conditions affecting bone, which have their origins distant from bone and therefore manifest as general diseases affecting bone turnover, formation or resorption. Thus, these diseases may result in generalised osteoporosis, osteomalacia/rickets or osteosclerosis. The nutritional factors, which are responsible for these generalised bone diseases, range from macronutrients such as protein deficiency, to micronutrients and minerals such as calcium and phosphorus deficiencies, fluoride excess and vitamin D deficiency. Much of the book will deal with nutrients and their role in bone health, particularly as they relate to fracture risk and the development of osteoporosis in adults. This chapter will concentrate on those nutrients responsible for metabolic bone diseases other than osteoporosis. There will naturally be some overlap with other chapters, but the emphasis in this chapter will be to steer away from osteoporosis and to provide a paediatric bias, as rickets and osteomalacia are more common during childhood than in adulthood. The nutrients, which will be discussed in detail, are vitamin D, calcium and phosphorus, and fluoride.

# Vitamin D Deficiency

It is over 70 years since vitamin D was isolated and its role in the pathogenesis of rickets highlighted. Since that time considerable advances have been made in our understanding of its physiology, and in particular of its role in bone and mineral homeostasis, but also of the broader actions of vitamin D and its metabolites in such functions as immune modulation[1] and cell differentiation (for a detailed discussion of vitamin D and its actions see Ref. [2]).

## The Physiology of Vitamin D

Vitamin D is a seco-steroid that really should not be considered a nutrient or a vitamin as most foods contain very little of this vitamin and the content in normal diets of most populations is insufficient to meet the normal requirements.[3] Rather, vitamin D behaves more like a prohormone, as it is physiologically inactive in its natural form, and must first be hydroxylated by the liver and then by the kidney before becoming active. The production of the active metabolite (or hormone) is tightly controlled in the kidney through feedback mechanisms *via* parathyroid hormone (PTH) and phosphorus concentrations particularly.

Vitamin D exists in two natural forms, vitamin $D_2$ or ergocalciferol (plant vitamin D), and vitamin $D_3$ or cholecalciferol. The former is synthesised in plants from the precursor sterol, ergosterol, while the latter is formed in the skin of animals from 7-dehydrocholesterol, both under the influence of ultraviolet irradiation. It is the dermal synthesis of vitamin $D_3$ that is of paramount importance in maintaining vitamin D sufficiency in man, unless the diet is vitamin D fortified.[4] The two forms of vitamin D are thought to have very similar actions in man, although there are suggestions that they may have different pharmacokinetics.[5]

A number of factors are important in influencing the production of vitamin $D_3$ in the skin. These include the amount of ultraviolet B radiation reaching the skin, the degree of melanin pigmentation in the skin, and the amount of precursor available in the skin.[4] The amount of UV light reaching the skin is influenced by the degree of atmospheric pollution, the thickness of the ozone layer and the zenith angle of the sun, which in turn is dependent on latitude, season and time of day.[6] Thus, vitamin D deficiency is more common in countries of high latitude such as in northern Europe, and in air-polluted large cities. Further, clothing and sunscreens effectively prevent UV radiation from reaching the skin, and extensive skin coverage by clothing is thought to be a major contributor to the high prevalence of vitamin D deficiency in the Middle East.[7] Recently, the use of sunscreens in Asian communities in the United Kingdom has been suggested as a possible factor contributing to their high prevalence of vitamin D deficiency.[8] Melanin pigmentation, which naturally absorbs solar UV radiation, thus reducing the amount available to convert 7-dehydrocholesterol to previtamin D, may be an important contributor to the pathogenesis of rickets in Black and Asian families living in countries where sunlight exposure and UV radiation are limited, as occurs in northern European countries.[9] Finally, there is evidence that as individuals age, the amount of 7-dehydrocholesterol (provitamin D) in the skin declines progressively, resulting in a reduced ability to form vitamin D under the influence of UV radiation.[10]

Once formed in the skin, vitamin D attached to the vitamin D binding protein is transported to the liver where it is hydroxylated to 25-hydroxy-vitamin D (25-OHD).[11] Vitamin D is also stored in adipose tissue and muscle, thus preventing rapid fluctuations in the availability of substrate for the further metabolism of vitamin D. 25-OHD is the major circulating form of the vitamin and serum levels are a good indication of the vitamin D status of an individual (see sections later). At physiological levels, it is thought that 25-OHD is biologically inactive, although some researchers have suggested it may have a direct role in intestinal calcium absorption.[12] In the kidney, 25-OHD is further hydroxylated to the physiologically active metabolite, 1,25-dihydroxyvitamin

D (1,25-$(OH)_2$D), under the influence of 25-hydroxyvitamin D 1α-hydroxylase in the proximal renal tubule. 1,25-$(OH)_2$D synthesis is stimulated by the elevated levels of PTH, hypocalcaemia and hypophosphataemia, while synthesis is suppressed by hyperphosphataemia.

The physiological effects of 1,25-$(OH)_2$D on bone and mineral homeostasis are principally directed towards maintaining normocalcaemia through increasing intestinal calcium absorption and to a lesser extent by increasing bone turnover and resorption. Considerable research into the mechanisms of action of vitamin D on intestinal calcium absorption has taken place over the past 20 years. Vitamin D is generally considered to be the major if not the only hormone or factor controlling dietary calcium absorption.[13] Details of the mechanisms by which 1,25-$(OH)_2$D increases intestinal calcium absorption are beyond the scope of this chapter, but the reader is referred to a number of excellent reviews.[14–16] 1,25-$(OH)_2$D *via* its interaction with the intracellular vitamin D receptor and vitamin D responsive genes stimulates intestinal calcium absorption by increasing the production of a number of intracellular proteins, including calbindin-$D_{9k}$.

Although there is considerable debate about the magnitude and role of vitamin D dependent active intestinal calcium absorption, as opposed to passive paracellular diffusion, in maintaining normal calcium balance,[16,17] the primary consequence of vitamin D deficiency is a marked impairment of intestinal calcium absorption, which results in an inability to maintain normocalcaemia. This sets in train a whole series of secondary consequences ending in the development of rickets/osteomalacia if it persists for long enough. Impaired calcium absorption leads to a fall in serum ionised calcium concentration, which in turn stimulates PTH secretion. Increased PTH levels have several effects: in the kidney PTH increases tubular calcium reabsorption and decreases phosphate reabsorption, resulting in conservation of calcium, but loss of phosphate. Further, it stimulates 1α-hydroxylase activity in the proximal renal tubular cells increasing the production of 1,25-$(OH)_2$D if sufficient substrate (25-OHD) is present. In bone, PTH increases bone turnover by stimulating not only osteoblastic activity, but also bone resorption. Persistent elevation of PTH levels results in a net loss of bone with the development of osteoporosis (Figure 5.1).

## The Consequences of Vitamin D Deficiency on Bone

Vitamin D deficiency thus leads to perturbations in calcium homeostasis resulting in a fall in ionised calcium concentrations and secondary hyperparathyroidism. The net effect of an increase in PTH secretion is an attempt to restore normocalcemia through increasing intestinal calcium absorption, renal calcium retention and bone resorption. Serum phosphorus concentrations tend to remain constant or fall due to the increased urinary loss secondary to hyperparathyroidism.

The consequence of these changes is the development of rickets/osteomalacia in children and osteomalacia is adults. Rickets is characterised by a delay in or failure of mineralisation at the growth plate, whilst osteomalacia is

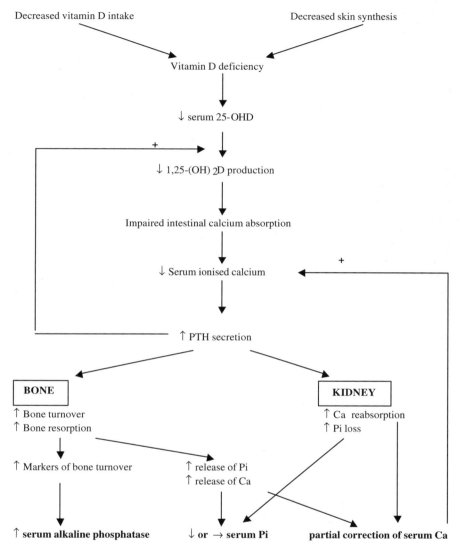

**Figure 5.1** *The biochemical changes associated with vitamin D deficiency*

characterised by an impairment of mineralisation of preformed osteoid at sites of bone remodelling on the trabecular bone surface and of osteoid deposition at the endosteal or periosteal bone surfaces.[18] The net effect of the impaired mineralisation at the growth plate is for the growth plate to widen and to distort under pressure from weight bearing and muscle forces. Further, the trabecular and cortical bone is poorly mineralised with an increase in the extent and thickness of osteoid seams lining the bone surfaces. Secondary hyperparathyroidism increases bone resorption resulting in thinning of the trabecular and cortical bone leading to a reduction in mineralised bone volume. The mechanism for the failure of mineralisation of the preformed osteoid or cartilage matrix is unclear, but it is thought to be primarily due to a lack of mineral substrate (calcium and/or phosphorus) at the mineralisation front as a consequence of hypocalcaemia and/or hypophosphataemia, rather than a direct effect of vitamin D deficiency *per se*.

At the growth plate, vitamin D deficiency is associated with a number of other changes as well, besides the failure of mineralisation. It has been shown that vitamin D metabolites regulate chondrocyte proliferation, metabolism, differentiation and maturation, as well as alter events in the extracellular matrix. In vitamin D deficiency, the prehypertrophic and hypertrophic zones of cartilage are markedly increased with abnormal invasion of metaphyseal blood vessels into the zone. Furthermore, lipid metabolism in the growth plate is altered and alkaline phosphatase activity in the hypertrophic zone is reduced. Recently, studies have shown that cartilage cells within the growth plate are able to metabolise 25-OHD to $1,25\text{-}(OH)_2D$ and $24,25\text{-}(OH)_2D$. The role of this local synthesis is unclear, although it has been suggested that $24,25\text{-}(OH)_2D$ may have a role in the early stages of endochondrial maturation, while $1,25\text{-}(OH)_2D$ may influence later maturation.[19]

Vitamin D metabolites, in particular $1,25\text{-}(OH)_2D$, also have direct effects on the development of osteomalacia, as $1,25\text{-}(OH)_2D$ has been shown to influence osteoblast function and thus the development of osteomalacia through its effects on matrix maturation and mineral transport.[18]

## The Clinical Features of Vitamin D Deficiency – Rickets/Osteomalacia

The clinical features of rickets/osteomalacia result from a combination not only of the effects of vitamin D deficiency on bone and growth plate structure, but also through its effects on muscle and mineral homeostasis. The latter may include features of hypocalcaemia such as tetany, convulsions, apnoeic attacks and stridor, while hypophosphataemia may compound the effects of vitamin D deficiency on muscle function leading to hypotonia and a proximal myopathy. These features present as delayed gross motor milestones in the infants and young children and as difficulty in getting out of chairs or climbing stairs in the older individuals.[20] They may also contribute to the risk of falling and fracturing in the elderly.[21,22]

The bony changes that occur in vitamin D deficiency are most apparent in the growing child whose growth plates are open, as deformities of the long bones develop rapidly with weight bearing. In the infant, the typical changes include a delay in closure of the cranial fontanelles, frontal and parietal bossing, the development of craniotabes (softening of the skull behind the ear) and a delay in eruption of the teeth. The chest may develop beading of the costochondral junctions (rachitic rosary), indrawing of the ribs along the line of the diaphragmatic attachments (Harrison's sulcus), and narrowing of the lateral diameter of the upper chest (violin case deformity). In the young infant, the long bone deformities tend to occur at the distal third of the forearms and the lower leg. Enlargement of the wrists and knees due to the splaying of the metaphyses and growth plate may also be noted. In the older infant and toddler, the lower limb deformities become more prominent as weight-bearing places extra strains on growth plates. The parents often notice progressively increasing bowlegs as the child starts to stand and walk. In the older child, knock-knees are more common and occasionally a windswept deformity of the knees may occur (a valgus deformity of one leg and varus deformity of the other). Rarely seen except in longstanding and severe vitamin D deficiency, are pelvic deformities, which may result in the narrowing of the pelvic outlet and obstructed labours in women.[23]

Osteomalacia in the adult is often difficult to diagnose, as the symptoms are typically insidious in onset and deformities do not occur unless associated with minimal trauma fractures. The disease typically presents with diffuse bone pain and proximal muscle weakness, which is often diagnosed initially as arthritis. Only when minimal trauma fractures occur and the patient is investigated with X-rays and blood tests may osteomalacia be suspected.

## Radiological Changes

The radiological changes that occur in rickets are characteristic of the disease, but are generally not helpful in differentiating the various causes of rickets from one another.[24] The pathognomonic changes occur at the growth plates and metaphyses of rapidly growing bones, especially at the wrist and the knee although changes may be noted at other sites as well, including the costochondral junctions, the proximal femur and distal tibia. The early sign is a widening of the growth plates, but as the disease progresses the distal ends of the metaphyses lose the well-defined zone of provisional calcification and become cupped, frayed and splayed (Figure 5.2). The cortices of the long bones become thin and may show periosteal new bone formation and features of secondary hyperparathyroidism may be noted (loss of the lamina dura around the teeth and the development of subperiosteal erosions particularly along the shafts of the phalanges). Deformities associated with bowing of the long bones are also seen. In severe rickets, pelvic deformities, such as protrusio acetabulae and narrowing of the pelvic outlet, may occur.

Osteomalacia is much more difficult to diagnose radiographically as the only feature may be a loss of bone density, giving the appearance of osteoporosis. Careful attention to the trabecular pattern may detect a coarsening and sparseness of the trabecular bone. Although not pathognomonic of osteomalacia, the finding of Looser's zones is very suggestive of the disease. These are radiolucent lines running perpendicular to and through the cortex, but not extending across the entire shaft. They occur typically in the femoral neck, the pubic rami, the scapula and ribs.

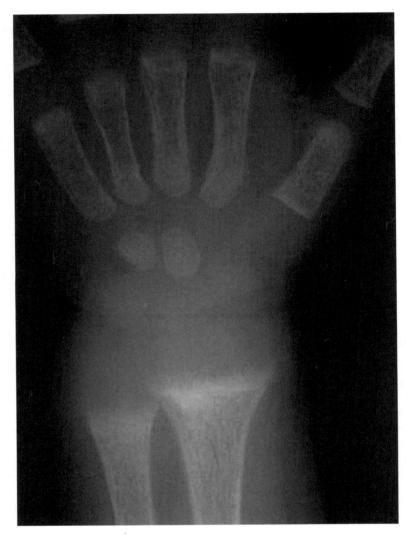

**Figure 5.2**  *Radiographic changes at the wrist of an infant with vitamin D deficiency rickets. The widening of the growth plate, and splaying and fraying of the distal ulna and radial metaphyses are clearly visible*

## Biochemical Changes

The biochemical changes associated with vitamin D deficiency rickets or osteomalacia are related mostly to the perturbation in calcium homeostasis and to the increased bone turnover consequent on vitamin D deficiency.[25] Thus hypocalcaemia, hypophosphataemia and elevated PTH values are typical. Markers of bone turnover are increased, thus alkaline phosphatase and

C-terminal telopeptides of collagen values are increased in serum and hydroxyproline and pyridinoline excretion in the urine is raised. The characteristic biochemical feature of vitamin D deficiency is a reduced circulating level of 25-OHD. However, serum concentrations of the active metabolite of vitamin D, 1,25-$(OH)_2$D, may be low, normal or even elevated.

## The Diagnosis of Vitamin D Deficiency

As mentioned above, the hallmark of vitamin D deficiency is a reduced serum level of 25-OHD, as the latter is the major circulating form of the vitamin. Nevertheless, there is considerable controversy surrounding the normal range of the metabolite and what levels constitute vitamin D deficiency or insufficiency.[26] Vitamin D is required to maintain normal calcium homeostasis, thus vitamin D sufficiency should be defined as that level, which is necessary to maintain serum calcium values without perturbing the homeostatic mechanisms. However, the maintenance of normal calcium homeostasis is also dependent on other factors besides vitamin D, such as the dietary calcium intake and its bioavailability, the ability of the intestine to absorb calcium, and the calcium requirements of the organism.

A number of different terms have been introduced to define the vitamin D status of an individual; vitamin D insufficiency is associated with a disturbance in normal calcium homeostasis as manifested by an elevation in PTH and/or 1,25-$(OH)_2$D levels without overt clinical disease, while vitamin D deficiency is a more severe form of the disease and is associated with features of rickets/osteomalacia.[27] It should be apparent that there is no clear-cut separation between these divisions as 25-OHD levels are a continuum from vitamin D deficiency through to vitamin D sufficiency and toxicity, and the development of biochemical or clinical features is dependent not only on factors such as intestinal calcium absorption and the calcium demands of the individual, but also on the duration of a particular 25-OHD level and the ability of the individual to produce 1,25-$(OH)_2$D. Furthermore, the actual serum levels of 25-OHD, which define vitamin D deficiency, insufficiency or sufficiency, are in dispute as a number of different assays, using different techniques to measure 25-OHD, have been employed to assess the vitamin D status of individuals. These assays show considerable variation in their so-called normal reference ranges.[28,29]

There has been renewed interest over the last decade in defining vitamin D sufficiency particularly in the elderly,[30] as a number of studies have shown that vitamin D supplementation may reduce the incidence of minimal trauma fractures and improve motor power in this age group.[20,21,31] The problems associated with defining the upper threshold value for vitamin D insufficiency have been highlighted by McKenna and Freaney[26] and Lips.[30] This threshold level varies from 25 to over 100 nmol $L^{-1}$ depending on which criteria are used to establish the diagnosis and possibly on the assay method employed.[28,32,33] Table 5.1 outlines the proposed staging of vitamin D deficiency/insufficiency in

**Table 5.1.** *Proposed staging of vitamin D insufficiency and deficiency in the elderly*[30]

| Stage | Serum 25-OHD (nmol L$^{-1}$) | Serum PTH increase | Bone histology |
|---|---|---|---|
| Vitamin D insufficiency | 25 – 50 | 15% | Normal or high turnover |
| Moderate vitamin D deficiency | 12.5–25 | 15–30% | High turnover |
| Severe vitamin D deficiency | <12.5 | >30% | Mineralisation defect, osteomalacia |

the elderly as proposed by Lips.[30] He uses a threshold value of 50 nmol L$^{-1}$ for the diagnosis of vitamin D insufficiency, but there are researchers who believe that this value is too low,[34,35] particularly in the elderly.

The public health relevance of defining vitamin D insufficiency in the elderly is that there is increasing evidence to indicate that prolonged mildly elevated PTH levels associated with insufficiency lead to increased bone turnover and a reduction in bone mass[36] and thus may be a major factor in hastening the onset of osteoporosis and minimal trauma fractures in this age group. Furthermore, there is mounting evidence that the elderly in many countries suffer from a high prevalence of vitamin D insufficiency,[37] even though in some countries vitamin D deficiency might be relatively uncommon.[38] Although the role of vitamin D in the pathogenesis of osteoporosis is discussed extensively in other chapters in this book, and thus will not be dwelt on here, it is necessary to highlight the point that the early phase of vitamin D deficiency osteomalacia is characterised only by an increase in bone turnover.[18] Most researchers would agree that vitamin D deficiency in the adult is associated with 25-OHD levels of < 10–2 ng mL$^{-1}$ (25–30 nmol L$^{-1}$).

In the paediatric literature, vitamin D deficiency is also generally considered to be present if 25-OHD levels fall below 10–12 ng mL$^{-1}$ (25–30 nmol L$^{-1}$)[39] as the majority of reported cases of untreated vitamin D deficiency rickets have values below this cut-off[40–43] and often below 4–5 ng mL$^{-1}$ (10–12.5 nmol L$^{-1}$). A number of studies have also investigated the association of 25-OHD levels with altered calcium homeostasis in both neonates and older children and adolescents.[44–46] In neonates, values below 12 ng mL$^{-1}$ (30 nmol L$^{-1}$) are associated with elevated PTH and low serum calcium values,[44] while in older children seasonal fluctuations in 25-OHD levels are associated with changes in PTH values.[45] In adolescents, it appears that PTH values only rise when 25-OHD levels fall below 12 ng mL$^{-1}$ (30 nmol L$^{-1}$).[47]

Using a rise of 1,25-(OH)$_2$D levels following an administration of 25-OHD as indicative of vitamin D insufficiency, Docio and coworkers[46] have suggested a threshold 25-OHD value of between 12 and 20 ng per mL (30–50 nmol L$^{-1}$). Thus, it appears that normal calcium homeostasis is maintained in children with serum 25-OHD values above 12–20 ng mL$^{-1}$ (30–50 nmol L$^{-1}$), which are substantially lower than the threshold suggested in the elderly. It should, however, be emphasised that these data apply only to children, who are on western type diets with relatively good calcium intakes.

## Prevention

Prevention of rickets/osteomalacia is dependent on ensuring an adequate intake of vitamin D. Table 5.2 lists the current recommended intakes of calcium and vitamin D for the United Kingdom[48] and North America.[49] It should be noted that there are a number of differences between the recommendations for the two regions. The United Kingdom recommendations assume that normal healthy adults do not require a dietary intake of vitamin D, as they should receive adequate sunlight exposure and thus UV irradiation to maintain normal circulating levels of 25-OHD. The North American recommendations suggest an increase in dietary intake in the elderly, taking cognisance of the research suggesting that vitamin D formation in the skin is impaired in the elderly and that they may require higher circulating 25-OHD levels to maintain normal calcium homeostasis. Further, many elderly are infirm and thus spend little time out of doors, reducing their dermal formation of vitamin D even more.

# Low Dietary Calcium Intakes

Adult physicians and endocrinologists in the developed world have long been interested in the role of dietary calcium intake in the pathogenesis of osteoporosis and minimal trauma fractures (see Chapter 9). However, little attention has been paid to its role in the exacerbation of other metabolic bone diseases, in particular rickets/osteomalacia, until relatively recently.

Kooh and coworkers[50] and Maltz and coworkers[51] in the early 1970s were the first to provide convincing evidence that low dietary calcium intakes in infants could lead to rickets in the face of an adequate vitamin D intake. Since that time, further isolated reports have appeared.[52] However, it is only in the last 20 years that the importance of low dietary calcium intakes in the pathogenesis of nutritional rickets in older children living in tropical and sub-tropical climates has been appreciated.[53] Studies of children with rickets living in rural communities in South Africa were the first to draw attention to the possibility of low dietary calcium intakes being responsible for the disease in otherwise healthy children exposed to adequate amounts of sunlight (Figure 5.3).[54–56] These studies not only found that the children with active rickets had 25-OHD levels within the normal range, but that they also had elevated serum concentrations of 1,25-$(OH)_2$D, a combination of biochemical changes indicative of an inadequate intestinal calcium absorption resulting from factors other than vitamin D deficiency. Furthermore, dietary calcium intakes in children with biochemical changes indicative of disturbed calcium homeostasis were approximately 200 mg per day (approximately 25% of

**Table 5.2.** Recommended dietary intakes of calcium and vitamin D

| Age | Vitamin D (µg/d) | | Calcium (mg/d) | |
| --- | --- | --- | --- | --- |
| | UK recommendations | North American recommendations | UK recommendations | North American recommendations |
| 0–6 months | 8.5 | 5 | 525 | 210 |
| 7 months–3 years | 7 | 5 | 525 (7–12 m) <br> 350 (1–3 years) | 270 (7–12 m) <br> 500 (1–3 years) |
| 4–50 years | 0 | 5 | 450–550 (4–10 years) <br> 800–1000 (11–18 years) <br> 700 (18–50 years) | 800 (4–8 years) <br> 1300 (9–18 years) <br> 1000 (19–50 years) |
| 50 + yrs | 10 (61 + years) | 10 (51–70 years) <br> 15 (71 + yrs) | 700 | 1200 |
| Pregnancy and lactation | 10 | 5 | + 550 for lactation | + 300 (if <19 years) |

Table adapted from References 48 and 49.

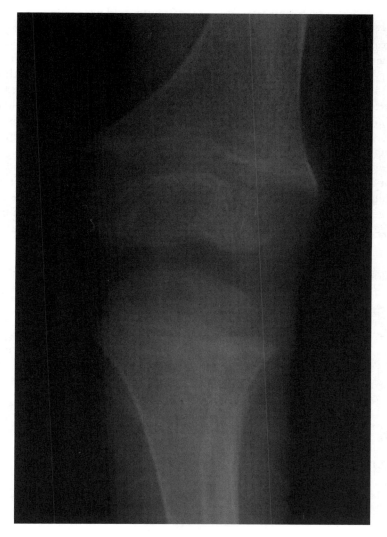

**Figure 5.3**   *Radiographic changes at the knee of a 6-year old child with dietary calcium deficiency rickets. The changes are similar to those seen in vitamin D deficiency with widening of the growth plates and splaying and fraying of the femoral and tibial metaphyses*

RDA), which were significantly lower than normal age-matched controls living in the same community.[57] Dietary calcium supplementation resulted in an improvement not only of the biochemical and radiological features of rickets, but also of the histological features of osteomalacia on bone biopsy.

Since that time, studies in Nigeria and Bangladesh have suggested that the problem is more widespread than initially thought.[58–63] Unlike the South African children, who came from rural areas of the country and were generally between the ages of 4 and 15 years, the Nigerian children are urban and tend to be younger with a mean age of presentation of approximately four years or younger.[59,60] Another difference between the South African and Nigerian children is the finding that control children in Nigeria have similar calcium intakes to those of affected children (approximately 200 mg per day).[60] Despite being unable to find a difference in dietary calcium content between the affected and control children, calcium supplementation of the diet was shown not only to heal those with rickets, but also to be significantly more effective than vitamin D therapy in treating the disease.[61]

The positive therapeutic response to calcium supplementation in the face of similar calcium intakes between affected and control children raises the possibility that factors other than calcium content might play a role in the pathogenesis of the disease in this group of children (Figure 5.4). It is possible that the concentration of inhibitors of calcium absorption, such as oxalates and

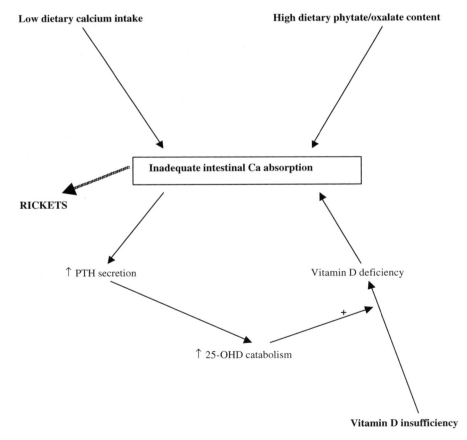

**Figure 5.4**  *The pathogenesis of rickets due to low dietary calcium intakes*

phytates in the diet, might be important.[53] Further, it has been hypothesised that children at risk of developing rickets in the face of low dietary calcium intakes may be genetically less well able to adapt to the stress induced by the calcium deficiency.[64] Studies to investigate these possibilities are currently under way, but at present no leads have been found.

The diets of children who develop dietary calcium deficiency rickets are typical of many children living in conditions of poverty in tropical and sub-tropical climates. The diet shows little variety or variability and consists mainly of maize or rice based staple (both of which are low in calcium content), with an almost total lack of dairy products. It is thus not surprising that studies from South Africa suggest that the children with active rickets and bone deformities represent the tip of an iceberg. Biochemical evidence of perturbations in calcium homeostasis and increased bone turnover is much more common in communities from which children with calcium deficiency rickets have been identified than in those communities with normal children.[65] Further, a study of appendicular bone mass in children suggested that children living in a community in which dietary calcium deficiency was common, had lower radial bone mass, but wider bones than urban children.[66] Thus, it is apparent that dietary calcium deficiency in children presents a continuum of changes from mild elevations of alkaline phosphatase through reductions in bone mass to frank rickets.

The level of reduced calcium intake necessary to induce these changes is unclear, as factors other than the dietary calcium content may play important roles in determining the fraction of calcium absorbed from the diet. Nevertheless, one study suggested that intakes of approximately 400 mg per day were associated with normal biochemistry, while intakes of approximately 200 mg per day were associated with biochemical changes indicative of dietary calcium deficiency.[57]

Dietary calcium deficiency as a cause of rickets has been almost exclusively described in children in developing countries, who are outside the infant and toddler age groups. However, this may not be the true situation, as very recently a study from the Eastern seaboard of the USA has provided evidence to indicate that a high percentage of cases of rickets in mainly African–American children have features suggestive of dietary calcium deficiency, in that they have normal 25-OHD and elevated 1,25-$(OH)_2$D levels and respond to calcium supplements. It is suggested the reason for dietary calcium deficiency rickets developing in an industrialised country is that many children are now being weaned onto fruit juices rather than cow's milk preparations and thus have low dietary calcium intakes once breast-feeding ceases. Further studies would need to be conducted to determine how widespread a problem this is.

This chapter has so far dealt with vitamin D deficiency and dietary calcium deficiency as separate entities. As can be seen in Figure 5.4, both causes of rickets are thought to act through a common path by reducing intestinal calcium absorption below levels necessary to maintain normal calcium homeostasis. Low dietary calcium intakes or reduced dietary calcium bioavailability may also precipitate rickets through another mechanism and that is the induction of vitamin D deficiency (Figure 5.4). Over 80 years ago Mellanby[67] showed that the onset and severity of rickets could be exacerbated by the addition of unrefined cereals to the diet of vitamin D deficient dogs. More recently, similar effects were produced by the addition of maize to the vitamin D deficient diet of captive baboons.[68] It has taken some 70 years since the observations of Mellanby for scientists to provide a rational explanation for these findings. Elegant experiments in rats[69–71] were able to show that low dietary calcium intakes or diets high in phytate resulted in hyperparathyroidism and elevated 1,25-$(OH)_2$D levels, which increased the catabolism of 25-OHD and reduced its half-life, thus increasing the requirements of vitamin D needed to maintain vitamin D sufficiency. Intestinal calcium malabsorption or high fibre diets in man have been shown to have similar effects on the half-life of 25-OHD[72,73]. It appears that the increased catabolism of 25-OHD is mediated through elevated levels of 1,25-$(OH)_2$D, which have been shown to correlate inversely with the half-life of 25-OHD, irrespective of whether the elevated levels were obtained through endogenous or exogenous means[74,75]. It is thus apparent that low calcium containing diets or diets high in oxalates or phytates, which inhibit the absorption of calcium, might precipitate rickets directly as described in the earlier section or indirectly through exacerbating vitamin D deficiency. It is this latter mechanism, which has been proposed as being responsible for the high prevalence of rickets in the Asian community in the United Kingdom.[76]

Thus vitamin D deficiency and dietary calcium deficiency should be seen as being at either end of a continuum in the pathogenesis of rickets.[77] It is highly possible that the vast majority of cases of rickets are as a result of a combination of both causes to a greater or lesser degree. The breastfed infant with vitamin D deficiency rickets and the Nigerian child with dietary calcium deficiency represent the two ends of the spectrum, but the majority of children with rickets fit between these two extremes.

# Endemic Fluorosis

In the early 1960s, sodium fluoride was introduced as an experimental drug for the treatment of osteoporosis.[78] Over the next 30 years, numerous investigators reported on the effects of fluoride therapy on bone mass and several studies investigated its effect on fracture rates. Although there was some initial enthusiasm, as there was a lack of other drugs for the treatment of osteoporosis, which had positive effects on osteoblastic activity, its use has fallen into disrepute because of its side effects (mainly the development of histological osteomalacia, stress fractures and bone pain) and the lack of convincing evidence that it reduced fracture rates.[79]

Fluoride is thought to produce its effects on bone by a direct effect on osteoblast function and through its incorporation into the hydroxyapatite crystal increasing its size and reducing its solubility. Fluoride also has a stimulatory effect on osteoblast proliferation and differentiation.[80]

The effects of chronic fluoride ingestion had been known for a number of decades prior to the use of fluoride as an experimental drug. Endemic fluorosis had been described from many areas of the world, such as South Africa, India, Saudi Arabia, Tanzania, China and Japan, in the 1940s and 1950s. The major cause of endemic fluorosis is the ingestion of borehole or well water with fluoride concentrations above 3–4 ppm, although other sources, such as industrial contamination, may also produce clinical effects. The earliest sign of fluorosis is the mottling and staining of dental enamel as this occurs at relatively low levels of chronic ingestion and at a time when the dental enamel is being laid down. Thus, it is seen in the primary and secondary teeth of children and adults who have lived in an endemic fluorosis area all their lives. Dental staining does not occur in children or adults who have moved into an area after their teeth have erupted. Dental fluorosis has been noted in children living in areas where the fluoride content of water is approximately 1 ppm. Although the incorporation of small amounts of fluoride into dental enamel is important for the prevention of dental caries, excessive fluoride results not only in unsightly staining, but also pitting and chipping of the teeth.

The skeletal effects of fluorosis characteristically occur only after a prolonged residence (10–15 years) in an endemic area[81] with water fluoride levels of 3 ppm or more, although the signs and symptoms may develop more rapidly in children. Typically, symptoms develop insidiously with generalised bone pains, stiffness and limitation of joint movement, and rigidity of the spine.[82] Neurological symptoms due to spinal cord compression may occur in very severe and long standing cases. In children, the presentation is frequently that of genu valgum, occurring particularly in young male adults and adolescents,[83,84] however, in areas with very high water fluoride concentrations features similar to those described in adults may be noted.[85,86] An association between the development of genu valgum and low dietary calcium intakes in areas of endemic fluorosis has been suggested.[84]

The radiological changes are very characteristic; they include axial osteosclerosis, periosteal new bone formation, calcification of interosseous ligaments and muscular attachments, and the development of exostoses and osteophytes.[81,87] Peripheral osteoporosis may also be present. Features typical of rickets and osteomalacia have also been described (Figure 5.5).[88] Appendicular bone density measurements in children living in an endemic fluorosis region in South Africa confirmed the radiological finding of osteoporosis at the distal radius. Further, the cortical diameter was increased suggesting that chronic hyperparathyroidism induced by fluoride ingestion caused increased endosteal bone resorption and periosteal new bone formation.[89]

Bone histology in patients with endemic fluorosis typically reveals increased cortical and trabecular bone thickness with disordered lamellar orientation. The Haversian systems are enlarged and poorly formed. The surfaces are lined by plumb active looking osteoblasts, while increased osteoid thickness and surface suggestive of osteomalacia may be noted. In a number of patients, features of hyperparathyroidism may be noted on bone histology with increased resorption surfaces and osteoclast numbers.[90]

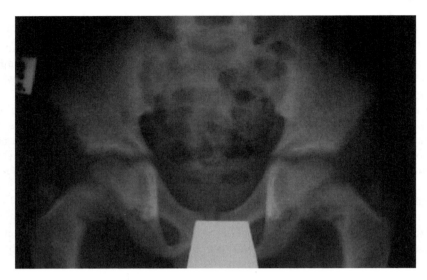

**Figure 5.5**   *Radiographic changes in a 15-year old boy suffering from endemic fluorosis. The pelvic bones are sclerotic with coarsening of the trabecular pattern. The femoral necks show features of rickets and osteomalacia (bilateral Looser's zones)*

Generally, there is little biochemical evidence of a disturbance in calcium homeostasis besides an elevation of alkaline phosphatase values in subjects living in areas of endemic fluorosis, however, hypocalcaemia and secondary hyperparathyroidism have been described in a minority of subjects.[81,88] Urine calcium excretion is generally reduced.

## Phosphorus Deficiency

The role of phosphorus in bone health is discussed in detail in chapter 12. This section will briefly discuss the role of dietary phosphorus deficiency in the development of rickets and osteomalacia.

Dietary phosphorus deficiency is unlikely to occur in man eating a normal diet, as phosphorus is a ubiquitous element found in almost all foods. In fact, there is a concern that the modern diet might result in excessive phosphate intake, particularly as a result of the increasing consumption of phosphate containing carbonated drinks and reduced calcium diets in many developed countries.[91]

The one situation, in which dietary phosphorus deficiency is thought to be primarily responsible for metabolic bone disease, is that of the breast-milk fed VLBW infant (<1500 g at birth). *In utero*, approximately 80% of skeletal mineral in the foetus is deposited between 25 weeks gestation and term during which time phosphorus accretion rates of between 2.1 and 2.5 mmol per day are achieved. Thus, the preterm infant is born during a rapid phase of bone growth and mineralisation, which is often compromised postnatally, as nutrient intakes are frequently poor during the first weeks of life because of low concentrations of various nutrients in breast-milk, illness and poorly developed gastrointestinal function. Although the features may be less severe than in those fed breast-milk, preterm infants fed with soy-based formulas and standard infant formulas may also show radiological and biochemical signs of dietary phosphorus deficiency.[92]

Von Sydow[93] in classic studies in the 1940s showed clearly that premature infants fed breast-milk developed rickets, which was unresponsive to vitamin D. He further suggested that the disease was a consequence of phosphorus deficiency and that an increase in calcium and phosphorus supply might prevent the condition. Since that time numerous studies have verified the conclusions and have expanded our understanding of the problem.

Breast-milk, when fed at intakes of 180–200 mL per kg per day to a VLBW infant, only provides approximately 25% of the daily requirement of phosphorus and approximately 50% of the calcium requirement.[94] It is thus not surprising that phosphorus deficiency manifests in a large proportion of VLBW infants in the first 6–8 weeks of life.

Although phosphorus deficiency is a major factor in the pathogenesis of metabolic bone disease in the very small preterm infant, a number of other factors might also play to a greater or lesser extent a role in individual patients. The condition is aggravated by low calcium intakes in the breast-milk fed infant, by the use of diuretics, which increase calcium loss, by reduced oral intake of nutrients in general including vitamin D, and by the use of drugs such as corticosteroids, which inhibit bone formation.[95,96]

Metabolic bone disease in VLBW infants manifests radiologically as a spectrum of appearances from under mineralised bones (osteopenia) to severe rickets and fractures. Osteopenia is seen in the majority (up to 75%) of VLBW infants[97] and particularly in those less than 1000 g at 3 weeks of age, while the features of rickets may be found in as high as 55%.[95] Nearly a quarter of infants less than 1500 g were noted to have fractures in one study.[95] Numerous studies using single photon absorptiometry and dual-energy X-ray absorptiometry have more accurately confirmed the radiological diagnosis of osteopenia in preterm infants fed with either breast-milk or standard infant formulas during the first months of life.[98–100]

Biochemically, the characteristic features of phosphorus deficiency in the VLBW infant are hypophosphataemia, elevated alkaline phosphatase values,[101] normocalcaemia, hypercalciuria and reduced urinary phosphate excretion. Typically, the biochemical changes reach their peak between 6 and 10 weeks postnatally. Markers of bone turnover (bone-specific alkaline phosphatase, C-terminal propeptide of type-I collagen, C-terminal telopeptides of type-I collagen) are increased in premature infants compared to values in full-term infants.[102]

The long-term consequences of metabolic bone disease in the neonatal period have not been well studied. However, Lucas and co-workers[103] found that preterm infants with evidence of metabolic bone disease during the neonatal period were shorter than their biochemically normal peers at

18 months of age. Further studies suggest that these children remain shorter than their peers at 9–12 years of age.[104] As far as the consequences of metabolic bone disease on bone mass are concerned, studies suggest that bone mineral content in children who had had metabolic bone disease during their early postnatal life, is similar to that of normal controls after adjusting for differences in current weight and height.[105–107]

The prevention of metabolic bone disease in VLBW infants has been an area of much research over the past 20 years. Early studies demonstrated that the addition of phosphate supplements to breast-milk fed preterm infants improved serum phosphorus values and reduced urinary calcium excretion, while serum calcium values remained within the normal range as did serum PTH values, although they increased.[108] More importantly, phosphorus supplementation improved calcium retention.[109] More recently, specially designed preterm infant formulas and breast-milk fortifiers have been introduced to increase not only the calcium and phosphorus content of the diet, but also that of other electrolytes, vitamins and protein.[110] The latest recommendations for calcium and phosphorus content of milk formulas designed for preterm infants are as follows: a calcium-to-phosphorus ratio of between 1.7:1 and 2.0:1; a calcium content of 123–185 mg per 100 kcal, and a phosphorus content of 82–109 mg per 100 kcal.[110]

# Conclusions

This chapter has provided an overview of a number of nutritional factors, of which a deficiency or excess may result in metabolic bone diseases other than osteoporosis. Despite our increased understanding of bone and mineral physiology and the availability of effective and often cheap methods to prevent these diseases, they remain as problems in a number of developing countries, in particular. In developed countries, special attention must be paid in eradicating vitamin D deficiency from the immigrant and minority communities. With a generally ageing population in many areas of the world, the number of elderly, who are likely to have subclinical vitamin D deficiency and thus be at risk of developing muscle weakness and an increased incidence of falls and fractures, is huge and places an enormous burden on health facilities and budgets. Thus, continued attention must be given to ensure that effective programmes to address these problems are continually sought.

# References

1. H.F. DeLuca and M.T. Cantorna, Vitamin D: its role and uses in immunology, *FASEB J.*, 2001, **15**(14), 2579–2585.
2. D. Feldman, F.H. Glorieux and J.W. Pike, *Vitamin D*, Academic Press, San Diego, 1997.
3. B. Dawson-Hughes, Vitamin D and calcium: recommended intake for bone health, *Osteoporos Int.*, 1998, **8** (Suppl. 2), S30–S34.

4. M.F. Holick, Environmental factors that influence the cutaneous production of vitamin D, *Am. J. Clin. Nutr.*, 1995, **61**(Suppl. 3), 638S–645S.

5. H.M. Trang, D.E.C. Cole, L.A. Rubin, A. Pierratos, S. Siu and R. Vieth, Evidence that vitamin D$_3$ increases serum 25-hydroxyvitamin D more efficiently than does vitamin D$_2$, *Am. J. Clin. Nutr.*, 1998, **68**, 854–858.

6. J.M. Pettifor, G.P. Moodley, F.S. Hough, H. Koch, T. Chen, Z. Lu and M.F. Holick, The effect of season and latitude on *in vitro* vitamin D formation by sunlight in South Africa, *S. Afr. Med. J.*, 1996, **86**, 1270–1272.

7. M.H. Gannage-Yared, R. Chemali, N. Yaacoub and G. Halaby, Hypovitaminosis D in a sunny country: relation to lifestyle and bone markers, *J. Bone Miner. Res.*, 2000, **15**(9), 1856–1862.

8. S. Zlotkin, Vitamin D concentrations in Asian children living in England. Limited vitamin D intake and use of sunscreens may lead to rickets, *Br. Med. J.*, 1999, **318**(7195), 1417.

9. M.F. Holick, J.A. MacLaughlin and S.H. Doppelt, Factors that influence the cutaneous photosynthesis of previtamin D$_3$, *Science*, 1981, **211**, 590–593.

10. J. MacLaughlin and M.F. Holick, Aging decreases the capacity of human skin to produce vitamin D3, *J. Clin. Invest.*, 1985, 76, 1536–1538.

11. G. Jones, S.A. Strugnell and H.F. DeLuca, Current understanding of the molecular actions of vitamin D, *Physiol. Rev.*, 1998, **78**(4), 1193–1231.

12. M.J. Barger-Lux, R.P. Heaney, S.J. Lanspa, J.C. Healy and H.F. DeLuca, An investigation of sources of variation in calcium absorption efficiency, *J. Clin. Endocrinol. Metab.*, 1995, **80**, 406–411.

13. I. Nemere and D. Larsson, Does PTH have a direct effect on intestine? *J. Cell Biochem.*, 2002, **86**(1), 29–34.

14. R.H. Wasserman, Vitamin D and the intestinal absorption of calcium and phosphorus, In *Vitamin D*, D. Feldman, F.H. Glorieux and J.W. Pike (eds.), Academic Press, San Diego, 1997, pp. 259–274.

15. L. Gueguen and A. Pointillart, The Bioavailability of dietary calcium, *J. Am. Coll. Nutr.*, 2000, **19**, 119S–136S.

16. C.C. McCormick, Passive diffusion does not play a major role in the absorption of dietary calcium in normal adults, *J. Nutr.*, 2002, **132**(11), 3428–3430.

17. R.P. Heaney, Vitamin D: role in the calcium economy, In *Vitamin D*, D. Feldman, F.H. Glorieux and J.W. Pike (eds.), Academic Press, San Diego, 1997, pp. 484–487.

18. A.M. Parfitt, Vitamin D and the pathogenesis of rickets and osteomalacia, In *Vitamin D*, D. Feldman, F.H. Glorieux and J.W. Pike (eds.), Academic Press, San Diego, 1997, pp. 645–662.

19. B.D. Boyan, D.D. Dean, V.L. Sylvia and Z. Schwartz, Cartilage and vitamin D: genomic and nongenomic regulation by 1,25(OH)$_2$D$_3$ and 24,25(OH)$_2$D$_3$, In *Vitamin D*, D. Feldman, F.H. Glorieux and J.W. Pike (eds.), Academic Press, San Diego, 1997, pp. 395–421.

20. H. Glerup, K. Mikkelsen, L. Poulsen, E. Hass, S. Overbeck, H. Andersen, P. Charles and E.F. Eriksen, Hypovitaminosis D myopathy without bio-chemical signs of osteomalacic bone involvement, *Calcif. Tissue Int.*, 2000, **66**(6), 419–424.

21. H.C. Janssen, M.M. Samson and H.J. Verhaar, Vitamin D deficiency, muscle function, and falls in elderly people, *Am. J. Clin. Nutr.*, 2002, **75**(4), 611–615.

22. H.A. Bischoff, H.B. Stahelin, W. Dick, R. Akos, M. Knecht, C. Salis, M. Nebiker, R. Theiler, M. Pfeifer, B. Begerow, R.A. Lew and M. Conzelmann, Effects of

vitamin D and calcium supplementation on falls: a randomized controlled trial, *J. Bone Miner. Res.*, 2003, **18**(2), 343–351.

23. J.P. Maxwell, Further studies in adult rickets (osteomalacia) and foetal rickets, *Proc. R. Soc. Med.*, 1934, **28**, 265–300.

24. J.E. Adams, Radiology of rickets and osteomalacia, In *Vitamin D*, D. Feldman, F.H. Glorieux and J.W. Pike (eds.), Academic Press, San Diego, 1997, 619–642.

25. K. Kruse, Pathophysiology of calcium metabolism in children with vitamin D-deficiency rickets, *J. Pediatr.*, 1995, **126**(5 pt 1), 736–741.

26. M.J. McKenna and R. Freaney, Secondary hyperparathyroidism in the elderly: means to defining hypovitaminosis D, *Osteoporos Int.*, 1998, **8** (Suppl. 2) S3–S6.

27. M. Peacock, P.L. Selby, R.M. Francis, W.B. Brown and L. Hordon, Vitamin D deficiency, insufficiency, sufficiency and intoxication. What do they mean? In *Vitamin D. A Chemical, Biochemical and Clinical Update*, A.W. Norman (ed.), Walter de Gruyter, Berlin, 1985, pp. 569–570.

28. R. Vieth, Problems with direct 25-hydroxyvitamin D assays, and the target amount of vitamin D nutrition desirable for patients with osteoporosis, *Osteoporosis Int.*, 2000, **11**(7), 635–636.

29. P. Lips, M.C. Chapuy, B. Dawson-Hughes, H.A.P. Pols and M.F. Holick, An international comparison of serum 25-hydroxyvitamin D measurements, *Osteoporos Int.*, 1999, **9**, 394–397.

30. P. Lips, Vitamin D deficiency and secondary hyperparathyroidism in the elderly. Consequences for bone loss and fractures and therapeutic implications, *Endocr. Rev.*, 2001, **22**(4), 477–501.

31. B. Dawson-Hughes, S.S. Harris, E.A. Krall and G.E. Dallal, Effect of calcium and vitamin D supplementation on bone density in men and women 65 years of age or older *N. Engl. J. Med.*, 1997, **337**(10), 670–676.

32. A. Malabanan, I.E. Veronikis and M.F. Holick, Redefining vitamin D insufficiency, *Lancet*, 1998, **351**(9105), 805–806.

33. A.G. Need, M. Horowitz, H.A. Morris and B.E.C. Nordin, Vitamin D status: effects on parathyroid hormone and 1,25-dihydroxyvitamin D in postmenopausal women, *Am. J. Clin. Nutr.*, 2000, **71**(6), 1577–1581.

34. M.C. Chapuy, P. Preziosi, M. Maamer, S. Arnaud, P. Galan, S. Hercberg and P.J. Meunier, Prevalence of vitamin D insufficiency in an adult normal population, *Osteoporosis Int.*, 1997, **7**(5), 439–443.

35. E.A. Krall, N. Sahyoun, S. Tannenbaum, G.E. Dallal and B. Dawson-Hughes, Effect of vitamin D intake on seasonal variations in parathyroid hormone secretion in postmenopausal women, *N. Engl. J. Med.*, 1989, **321** (26), 1777–1783.

36. S.S. Harris, E. Soteriades and B. Dawson-Hughes, Secondary hyperparathyroidism and bone turnover in elderly blacks and whites, *J. Clin. Endocrinol. Metab.*, 2001, **86**(8), 3801–3804.

37. S.H. Scharla, Prevalence of subclinical vitamin D deficiency in different European countries, *Osteoporosis Int.*, 1998, **8** (Suppl. 2), S7–S12.

38. J.C. Gallagher, H.K. Kinyamu, S.E. Fowler, B. Dawson-Hughes, G.P. Dalsky and S.S. Sherman, Calciotropic hormones and bone markers in the elderly, *J. Bone Miner, Res.*, 1998, **13**(3), 475–482.

39. J.M. Pettifor, What is the optimal 25(OH)D level for bone in children? In *Vitamin D Endocrine System: Structural, Biological, Genetic and Clinical Aspects*, A.W. Norman, R. Bouillon and M. Thomasset (eds.), University of California, Riverside (California), 2000, pp. 903–907.

40. T. Markestad, S. Halvorsen, K. Seeger Halvorsen, L. Aksnes and D. Aarskog, Plasma concentrations of vitamin D metabolites before and during treatment of vitamin D deficiency rickets in children, *Acta Paediatr. Scand.*, 1984, **73**, 225–231.

41. K.M. Goel, E.M. Sweet, R.W. Logan, J.M. Warren, G.C. Arneil and R.A. Shanks, Florid and subclinical rickets among immigrant children in Glasgow, *Lancet*, 1976, **1**, 1141–1145.

42. M. Garabedian, M. Vainsel, E. Mallet, H. Guillozo, M. Toppet, R. Grimberg, T.M. Nguyen and S. Balsan, Circulating vitamin D metabolite concentrations in children with nutritional rickets, *J. Pediatr.*, 1983, **103**, 381–386.

43. J.M. Pettifor, J.M. Isdale, J. Sahakian and J.D.L. Hansen, Diagnosis of subclinical rickets, *Arch. Dis. Child.*, 1980, **55**, 155–157.

44. F. Zeghoud, C. Vervel H. Guillozo, O. Walrant-Debray, H. Boutignon and M. Garabedian, Subclinical vitamin D deficiency in neonates: definition and response to vitamin D supplements, *Am. J. Clin. Nutr.*, 1997, **65**, 771–778.

45. J. Guillemant, S. Cabrol, A. Allemandou, G. Peres and S. Guillemant, Vitamin D-dependent seasonal variation of PTH in growing male adolescents, *Bone*, 1995, **17**(6), 513–516.

46. S. Docio, J.A. Riancho, A. Perez, J.M. Olmos, J.A. Amado and J. Gonzales-Macias, Seasonal deficiency of vitamin D in children: a potential target for osteoporosis-preventing strategies? *J. Bone Miner. Res.*, 1998, **13**(4), 544–548.

47. J. Guillemant, H.-T. Le, A. Maria, A. Allemandou, G. Peres and S. Guillemant, Wintertime vitamin D deficiency in male adolescents: effect on parathyroid function and response to vitamin $D_3$ supplements, *Osteoporosis Int.*, 2001, **12**, 875–879.

48. Department of Health, In *Nutrition and Bone Health*, 49th edn, The Stationary Office, London, 1998.

49. Standing Committee on the Scientific Evaluation of Dietary Reference Intakes, Food and Nutrition Board. Institute of Medicine, In *Dietary Reference Intakes for Calcium, Phosphorus, Magnesium, Vitamin D, and Fluoride*, National Academy Press, Washington, 1997.

50. S.W. Kooh, D. Fraser, B.J. Reilly, J.R. Hamilton, D. Gall and L. Bell, Rickets due to calcium deficiency, *N. Engl. J. Med.*, 1977, **297**, 1264–1266.

51. H.E. Maltz, M.B. Fish and M.A. Holliday, Calcium deficiency rickets and the renal response to calcium infusion, *Pediatrics*, 1970, **46**, 865–870.

52. W. Proesman, E. Legius and E. Eggermont, Rickets due to calcium deficiency, In *Proceedings of the Symposium on Clinical Disorders of Bone and Mineral Metabolism*, Mary Ann Liebert, New York, 1988, 15pp.

53. A.K. Bhattacharyya, Nutritional rickets in the tropics, In *Nutritional Triggers for Health and in Disease*, A.P. Simopoulos (ed.), Karger, Basel, 1992, pp. 140–197.

54. J.M. Pettifor, P. Ross, J. Wang, G. Moodley and J. Couper-Smith, Rickets in children of rural origin in South Africa: is low dietary calcium a factor? *J. Pediatr.*, 1978, **92**, 320–324.

55. J.M. Pettifor, F.P. Ross, R. Travers, F.H. Glorieux and H.F. DeLuca, Dietary calcium deficiency: a syndrome associated with bone deformities and elevated serum 1,25-dihydroxyvitamin D concentrations, *Metab. Bone Rel. Res.*, 1981, **2**, 301–305.

56. P.J. Marie, J.M. Pettifor, F.P. Ross and F.H. Glorieux, Histological osteomalacia due to dietary calcium deficiency in children, *N. Engl. J. Med.*, 1982, **307**, 584–588.

57. C. Eyberg, J.M. Pettifor and G. Moodley, Dietary calcium intake in rural black South African children. The relationship between calcium intake and calcium nutritional status, *Hum. Nutr. Clin. Nutr.*, 1986, **40C**, 69–74.

58. F. Okonofua, D.S. Gill, Z.O. Alabi, M. Thomas, J.L. Bell and P. Dandona, Rickets in Nigerian children: a consequence of calcium malnutrition, *Metabolism*, 1991, **40**, 209–213.

59. E.E. Ekanem, D.E. Bassey and M. Eyong, Nutritional rickets in Calabar, Nigeria, *Ann. Trop. Paediatr.*, 1995, **15**(4), 303–306.

60. T.D. Thacher, P.R. Fischer, J.M. Pettifor, J.O. Lawson, C. Isichei and G.M. Chan, Case-control study of factors associated with nutritional rickets in Nigerian children, *J. Pediatr.*, 2000, **137**, 367–373.

61. T.D. Thacher, P.R. Fischer, J.M. Pettifor, J.O. Lawson, C.O. Isichei, J.C. Reading and G.M. Chan, A comparison of calcium, vitamin D, or both for nutritional rickets in Nigerian children, *N. Engl. J. Med.*, 1999, **341**(8), 563–568.

62. P.R. Fischer, A. Rahman, J.P. Cimma, T.O. Kyaw-Myint, A.R. Kabir, K. Talukder, *et al.*, Nutritional rickets without vitamin D deficiency in Bangladesh, *J. Trop. Pediatr.*, 1999, **45**(5), 291–293.

63. L.M. Oginni, M. Worsfold, O.A. Oyelami, C.A. Sharp, D.E. Powell and M.W. Davie, Etiology of rickets in Nigerian children, *J. Pediatr.*, 1996, **128**, 692–694.

64. P.R. Fischer, T.D. Thacher, J.M. Pettifor, L.B. Jorde, T.R. Eccleshall and D. Feldman, Vitamin D receptor polymorphisms and nutritional rickets in Nigerian children, *J. Bone Miner. Res.*, 2000, **15**, 2206–2210.

65. J.M. Pettifor, F.P. Ross, G.P. Moodley and E. Shuenyane, Calcium deficiency in rural black children in South Africa – a comparison between rural and urban communities, *Am. J. Clin. Nutr.*, 1979, **32**, 2477–2483.

66. J.M. Pettifor and G.P. Moodley, Appendicular bone mass in children with a high prevalence of low dietary calcium intakes, *J. Bone Miner. Res.*, 1997, **12**(11), 1824–1832.

67. E. Mellanby, An experimental investigation on rickets, *Lancet*, 1919, **1**, 407–412.

68. M.R. Sly, W.H. van der Walt, D. Du Bruyn, J.M. Pettifor and P.J. Marie, Exacerbation of rickets and osteomalacia by maize: a study of bone histomorphometry and composition in young baboons, *Calcif. Tissue Int.*, 1984, **36**, 370–379.

69. M.R. Clements, L. Johnson and D.R. Fraser, A new mechanism for induced vitamin D deficiency in calcium deprivation, *Nature*, 1987, **325**, 62–65.

70. B.P. Halloran, D.D. Bikle, M.J. Levens, M.E. Castro, R.K. Globus and E. Holton, Chronic 1,25-dihydroxyvitamin $D_3$ administration in the rat reduces serum concentration of 25-hydroxyvitamin D by increasing metabolic clearance rate, *J. Clin. Invest.*, 1986, **78**, 622–628.

71. B.P. Halloran and M.E. Castro, Vitamin D kinetics in vivo: effect of 1,25-dihydroxyvitamin D administration, *Am. J. Physiol.*, 1989, **256**, E686–E691.

72. A.J. Batchelor, G. Watson and J.E. Compston, Changes in plasma half-life and clearance of $^3$H-25-hydroxyvitamin $D_3$ in patients with intestinal malabsorption, *Gut*, 1982, **23**, 1068–1071.

73. A.J. Batchelor and J.E. Compston, Reduced plasma half-life of radio-labelled 25-hydroxyvitamin $D_3$ in subjects receiving a high-fibre diet, *Br. J. Nutr.*, 1983, **49**, 213–216.

74. M.R. Clements, M. Davies, D.R. Fraser, G.A. Lumb, B. Mawer and P.H. Adams, Metabolic inactivation of vitamin D is enhanced in primary hyperparathyroidism, *Clin. Sci.*, 1987, **73**, 659–664.

75. M.R. Clements, M. Davies, M.E. Hayes, C.D. Hickey, G.A. Lumb, E.B. Mawer and P.H. Adams, The role of 1,25-dihydroxyvitamin D in the mechanism of acquired vitamin D deficiency, *Clin. Endocrinol.*, 1992, **37**, 17–27.

76. M.R. Clements, The problem of rickets in UK Asians, *J. Hum. Nutr. Diet.*, 1989, **2**, 105–116.
77. J.M. Pettifor, Privational rickets: a modern perspective, *J. R. Soc. Med.*, 1994, **87**, 723–725.
78. C. Rich and J. Ensinck, Effect of sodium fluoride on calcium metabolism of human beings, *Nature*, 1961, **191**, 184–185.
79. B.L. Riggs, W.M. O'Fallon, A. Lane, S.F. Hodgson, H.W. Wahner, J. Muhs, *et al.*, Clinical trial of fluoride therapy in postmenopausal osteoporotic women: extended observations and additional analysis, *J. Bone Miner. Res.*, 1994, **9**, 265–275.
80. M. Kassem, L. Mosekilde and E.F. Eriksen, Effects of fluoride on human bone cells in vitro: differences in responsiveness between stromal osteoblast precursors and mature osteoblasts, *Eur. J. Endocrinol.*, 1994, **130**, 381–386.
81. S.P.S. Teotia and M. Teotia, Endemic fluorosis in India: a challenging national health problem, *J. Assoc. Physicians India*, 1984, **32**, 347–352.
82. A. Singh, S.S. Jolly, B.C. Bansal and C.C. Mathur, Endemic fluorosis, *Medicine (Baltimore)*, 1963, **42**, 229–246.
83. K.A.V.R. Krishnamachari and B. Sivakumar, Endemic genu valgum: a new dimension to the fluorosis in India, *Fluoride*, 1976, **9**, 185–200.
84. K.A.V.R. Krishnamachari and K. Krishnaswamy, An epidemiological study of the syndrome of genu valgum among residents of endemic areas for fluorosis in Andhra Pradesh, *Ind. J. Med. Res.*, 1974, **62**, 1415–1423.
85. M. Teotia, S.P.S. Teotia and K.B. Kunwar, Endemic skeletal fluorosis, *Arch. Dis. Child*, 1971, **46**, 686–691.
86. A. Mithal and A.K. Singh, Endemic fluorosis and calcium intake, In *Nutrition and Bone Development*, J.-P. Bonjour and R.C. Tsang (eds.), Nestle Workshop Series, Nestec Ltd. Vevey/Lippincott-Raven, Philadelphia, 1999, vol. 41, pp. 213–229.
87. S.P.S. Teotia, M. Teotia and N.P.S. Teotia, Skeletal fluorosis: roentgenological and histopathological study, *Fluoride*, 1976, **9**, 91–98.
88. J.M. Pettifor, C.M. Schnitzler, F.P. Ross, G.P. Moodley, Endemic skeletal fluorosis in children: hypocalcemia and the presence of renal resistance to parathyroid hormone, *Bone Miner.* 1989, **7**, 275–288.
89. J.M. Pettifor, Calcium deficiency and impaired bone mass acquisition, In *Nutrition and Bone Development*, J.-P. Bonjour and R.C. Tsang (eds.), Nestle Workshop Series, Nestec Ltd. Vevey/Lippincott-Raven, Philadelphia, 1999, vol. 41, pp. 199–211.
90. S.P.S. Teotia and M. Teotia, Secondary hyperparathyroidism in patients with endemic fluorosis, *Br. Med. J.*, 1973, **1**, 637–640.
91. M.S. Calvo, Dietary phosphorus, calcium metabolism and bone, *J. Nutr.*, 1993, **123**, 1627–1633.
92. P.B. Kulkarni, R.T. hall, P.G. Rhodes, M.B. Sheehan, J.C. Callenbach, D.R. Germann, *et al.*, Rickets in very low-birth-weight infants, *J. Pediatr.*, 1980, **96**, 249–252.
93. C. Von Sydow, A study of the development of rickets in premature infants, *Acta Paediatr. Scand.*, 1946, **33**(Suppl. 2), 3–122.
94. J.M. Pettifor, Rickets in breast-fed and artificially fed infants, In Vitamins and Minerals in Pregnancy and Lactation, H. Berger (ed.), Nestle Nutrition Workshop Series, Raven Press, New York, 1988, **16**, 359–379.
95. M.C. Backstrom, A.L. Kuusela and R. Maki, Metabolic bone disease of prematurity, *Ann. Med.*, 1996, **28**(4), 275–282.

96. S. Kurl, K. Heinonen and E. Lansimies, Effects of prematurity, intrauterine growth status, and early dexamethasone treatment on postnatal bone mineralisation, *Arch. Dis. Child Fetal Neonatal Ed.*, 2000, **83**(2), F109–F111.

97. L.S. Hillman, N. Hoff, S. Salmons, L. Martin, W. McAlister and J. Haddad, Mineral homeostasis in very premature infants: serial evaluation of serum 25-hydroxyvitamin D, serum minerals, and bone mineralization, *J. Pediatr.*, 1985, **106**, 970–980.

98. S.D. Minton, J.J. Steichen and R.C. Tsang, Bone mineral content in term and preterm appropriate-for-gestational-age infants, *J. Pediatr.*, 1979, **95**, 1037–1042.

99. F.R. Greer and A. McCormick, Bone growth with low bone mineral content in very low birth weight premature infants, *Pediatr. Res.*, 1986, **20**, 925–928.

100. F. Pohlandt and N. Mathers, Bone mineral content of appropriate and light for gestational age preterm and term newborn infants, *Pediatrics*, 1989, **71**, 383–388.

101. S.J. Gross, Growth and biochemical response of preterm infants fed human milk or modified infant formula, *N. Engl. J. Med.*, 1983, **308**, 237–241.

102. Y. Shiff, A. Eliakim, R. Shainkin-Kestenbaum, S. Arnon, M. Lis and T. Dolfin, Measurements of bone turnover markers in premature infants, *J. Pediatr. Endocrinol. Metab.*, 2001, **14**, 389–395.

103. A. Lucas, O.G. Brooke, B.A. Baker, N. Bishop and R. Morley, High alkaline phosphatase activity and growth in preterm neonates, *Arch. Dis. Child.*, 1989, **64**, 902–909.

104. M.S. Fewtrell, T.J. Cole, N.J. Bishop and A. Lucas, Neonatal factors predicting childhood height in preterm infants: evidence for a persisting effect of early metabolic bone disease? *J. Pediatr.*, 2000, **137**, 668–673.

105. S. Kurl, K. Heinonen, E. Lansimies and K. Launiala, Determinants of bone mineral density in prematurely born children aged 6–7 years, *Acta Paediatr.*, 1998, **87**(6), 650–653.

106. I. Helin, L.A. Landin and B.E. Nilsson, Bone mineral content in preterm infants at age 4 to 16, *Acta Paediatr. Scand.*, 1985, **74**(2), 264–267.

107. M.C. Backstrom, R. Maki, A.L. Kuusela, H. Sievanen, A.M. Koivisto, M. Koskinen, R.S. Ikonen and M. Maki, The long-term effect of early mineral, vitamin D, and breast milk intake on bone mineral status in 9- to 11-year-old children born prematurely, *J. Pediatr. Gastroenterol. Nutr.*, 1999, **29**(5), 575–582.

108. L. Sann, B. Loras, L. David, F. Durr, C. Simonnet, P. Baltassat, *et al.*, Effect of phosphate supplementation to breast fed very low birthweight infants on urinary calcium excretion, serum immunoreactive parathyroid hormone and plasma 1,25-dihydroxy-vitamin D concentration, *Acta Paediatr. Scand.*, 1985, **74**(5):664–668.

109. J. Senterre, G. Putet, B.L. Salle and J. Rigo, Effects of vitamin D and phosphorus supplementation on calcium retention in preterm infants fed banked human milk, *J. Pediatr.*, 1983, **103**, 305–307.

110. C.J. Klein, Nutrient requirements for preterm infant formulas, *J. Nutr.*, 2002, **132**(6), 1395S–1577.

CHAPTER 6

# Assessment of Dietary Intake and Nutritional Status

## GAIL R. GOLDBERG

Nutrition and Bone Health Research Group, MRC Human Nutrition Research, Elsie Widdowson Laboratory, Fulbourn Road, Cambridge CB1 9NL, UK, Email: gail.goldberg@mrc-hnr.cam.ac.uk

**Key Points**

- Valid information about the intake of foods and nutrients and their relationships with health in individuals and populations is very important.
- There are many nutrients (and other components of foods and drinks) and foods that of are of particular interest and importance to those working in bone health.
- It is often wrongly assumed that dietary intake measurements are simple to undertake, and the data straightforward to analyse and interpret.
- This chapter summarises the main approaches to assessing dietary intake in studies of nutrition and health. Emphasis is placed on the advantages and disadvantages of different methods and their appropriate use.
- Some of the many other factors that should be considered when planning and designing studies are also outlined.
- The importance of critically evaluating dietary intake data before drawing any conclusions from the studies of nutrition and (bone) health is stressed.
- The acquisition of valid dietary intake data presents many challenges to nutritionists, dietitians and other health professionals, physiologists, psychologists, and statisticians.
- There is plenty of scope for fruitful collaborations across different disciplines.

# Introduction

"The measurement of the habitual food intake of an individual must be among the most difficult tasks a physiologist can undertake[1]"

"What foods do people habitually eat? A dilemma for nutrition, an enigma for psychology[2]"

Even if they do not purchase or prepare food, everyone understands about eating. Unlike for many clinical, physiological and biochemical assessments, complex and expensive equipment is not required to assess dietary intake. Samples do not have to be collected and analysed under rigorous conditions, so it is often wrongly assumed that assessing dietary intake is the 'easy' part of a study. However, the same rigour needs to be applied as with any other aspects of a study. Researchers need to ask themselves what they need to know and why, the methods used have to be appropriate; thought needs to be given to 'quality control', accuracy and precision, and data analysis and interpretation. The study duration, the sample size, and characteristics of the subjects under investigation will be important determinants of the method used, as will the other measurements and investigations to be undertaken, and resource and funding implications.

The above quotes encapsulate the challenge of acquiring and interpreting good dietary intake data. The valid measurement of food intake is essential to studies of nutrition and health, whether they be national or international observational surveys or epidemiological investigations, or smaller scale studies (*e.g.* evaluating the effects of an education programme on food choice, or collecting baseline data before an intervention). For example, if the intake of a nutrient is over or underestimated, this might lead to wrong estimates of the prevalence of inadequate intake. Between-subject differences in reporting will create variable bias that may distort the relationship between nutrient intake and health, and finally, biased intake data may lead to testing of false hypotheses (and therefore wasted effort and resources) or the unnecessary rejection of other data.

The main reasons for the need to obtain information about the food intake of individuals, groups or populations are to assess the adequacy of the diet; to aid the diagnosis and treatment of diet-related conditions; and to study the relationships between dietary intake and physiological function, disease, or risk factors or markers associated with a disease. With respect to the methods that are available, consideration has to be given to the sample size, logistical implications, subject characteristics and the quality of data required.

Small-scale research studies ($n < 100$ subjects) are the most demanding of quality data. The ability to measure the intakes of individuals with adequate precision is important. The interest may focus on intake during a specific period of study, rather than on the average intake over a long time. Epidemiological studies, where sample size can range from hundreds to thousands, focus on comparisons between groups of subjects (*e.g.* those with a high *vs.* low intakes of a nutrient or food) or on correlations between food or nutrient intake and outcome measures (*e.g* osteoporotic fracture). The primary interest in epidemiological studies is measuring intake averaged over long periods of time, commonly referred to as 'habitual' or 'usual' intake. Good precision of data at the individual level is less important. Assessing food intake in order to offer practical clinical or dietetic advice on a one-to-one basis can be the least or the most demanding of data. A purely qualitative record of the foods eaten may suffice, or it may be necessary to obtain a precise measure of an individual's nutrient intake.

Many of the nutrients and other constituents that are of particular interest in the field of bone health are discussed elsewhere in this volume. The nutrient data that can be obtained from dietary assessments can be broadly broken down into:

- Total energy intake.
- Macronutrient intakes – protein (Chapters 10.1 and 10.2) fats, carbohydrates, alcohol (Chapter 20).
- Micronutrient intakes – calcium (Chapter 8), vitamin D (Chapter 9), phosphorus (Chapter 11), sodium (Chapter 12), potassium (Chapter 13.1), magnesium (Chapter 15), and vitamins A and K (Chapters 14 and 17).
- Other constituents of foods and drinks – isoflavones (Chapter 19), caffeine (Chapter 20), water (Chapter 13.2).

The main (UK) food sources of these constituents are given in Table 6.1. It should be noted that the situation might well be different in other countries both because the major foods eaten are different and/or eaten with different frequencies.

**Table 6.1.** *Nutrients and other constituents of food and drinks of relevance to bone health* (adapted from[15])

| Constituent | Main sources in the UK diet |
| --- | --- |
| Alcohol | Alcoholic beverages – wines, beers, spirits, liqueurs, premixed drinks. |
| Caffeine | Coffee, tea, colas, 'stimulant drinks', chocolate. |
| Calcium | Milk and milk products are a major provider; other sources include white and brown flour and bread, pulses, dark green leafy vegetables (and, if eaten regularly, dried fruit, nuts and seeds, and the soft bones found in canned fish). |
| Copper | Although shellfish and liver are particularly rich in copper, the main sources in the British diet are meat, bread and other cereal products, and vegetables. |
| Fluoride | Tea is a major source in the British diet. Other sources include fish and water. |

**Table 6.1.** cont.

| Constituent | Main sources in the UK diet |
|---|---|
| Iron | Meat and meat products are a rich source of well-absorbed iron. Other important sources are cereal products, particularly bread and breakfast cereals, but also other products made from fortified flour; and vegetables. Also found in eggs, beans, and lentils, potatoes and dried fruit. |
| Magnesium | Widespread; main sources are cereals, cereal products, *e.g.* bread (particularly wholegrain/wholemeal), and green vegetables. Some also found in milk, and a small amount is contributed by meat and potatoes. Nuts and seeds are quite rich in magnesium. |
| Manganese | Tea is a major source; other sources include wholegrain cereals, bread, vegetables, nuts, and seeds. |
| Phosphorus | Milk, cheese, meat, fish, and eggs are good sources. |
| Isoflavones | Soya beans and soya products (*e.g.* textured vegetable protein, tofu, tempeh and soya drinks), other pulses, seeds (*e.g.* linseed), nuts, grains. |
| Potassium | All foods except sugars, fats and oils. Particularly abundant in vegetables, potatoes, fruit, (especially bananas) and juices. It is also found in bread, fish and nuts and seeds. Meat and milk also contribute to intake. |
| Protein | The amount in food varies, but the main sources include meat, fish, eggs and dairy foods, cereals, nuts, and pulses (peas, beans, and lentils). |
| Riboflavin | Milk and milk products, especially milk, and fortified breakfast cereals are the main sources in the British diet. Meat, and meat products provide smaller quantities. |
| Sodium | Used as a preservative and a flavouring agent. Sodium and chloride are comparatively low in all unprocessed foods. Sodium in the diet comes mainly from processed foods where 'salt' has been added, *e.g.* bacon, sausages and other meat products, canned vegetables, butter, margarine and spreads, cheese, bread, many savoury snack foods, and some breakfast cereals. 'Salt' can also be added during cooking or at the table. Sodium is also present in monosodium glutamate and sodium bicarbonate, and in some medicinal products *e.g.* antacids. |
| Vitamin A | As retinol in foods from animal sources – in milk, fortified margarine, butter, cheese, egg yolk, liver, and fatty fish. As carotenoids in foods from plant sources – milk, carrots, tomatoes, dark green vegetables. Orange-coloured fruits, *e.g.* mango and apricots. Beta-carotene can be converted to retinol in the body. There is statutory fortification of margarine in the UK, and voluntary fortification of reduced fat spreads. |
| Vitamin C | Richest sources are citrus fruit, citrus fruit juices, kiwi fruit, and soft fruits. Other sources include green vegetables, other fruit, peppers, and potatoes. |

**Table 6.1.** cont.

| Constituent | Main sources in the UK diet |
| --- | --- |
| Vitamin D | Found in foods in two main forms, mostly as cholecalciferol and in small amounts as ergocalciferol. Also made by the action of ultra violet rays on the skin and this is the most important source for the majority of people since few foods (fortified margarine, oily fish, egg yolk, fortified breakfast cereals) contain significant amounts of vitamin D. There is statutory fortification of margarine in the UK. |
| Vitamin K | Found in foods from both plant and animal sources and is also synthesised by gut bacteria. Dark green leafy vegetables are the richest source, but it is also found in other vegetables, fruit, dairy products, vegetable oils, cereals, and meat. |
| Zinc | Meat, meat products and milk and its products, bread, and other cereal products (especially wholegrain) are the major providers. Other sources include eggs, fish, beans and lentils, nuts, sweetcorn, and rice. |

# Methods of Dietary Assessment

There are basically four approaches to measuring food intake: diet records, diet recall, diet history, and food frequency questionnaire (FFQ). For more details about the different methodologies, their uses, advantages and disadvantages see reviews by Bingham[3] and by Rutishauser and Black.[4] There is also a checklist available for the methods sections of papers involving dietary investigations for authors, reviewers, and editors of journals.[5] The large and increasing literature on the relative and absolute validity of different dietary assessment methods is discussed in a recent, very comprehensive review and references therein.[6]

## Diet Records

The subject is asked to keep a detailed record, on specified days, of all items of food and drink at the time of consumption. The number of days' record is typically seven, but is often less (*e.g.* two week days and a weekend day). Weighed records are favoured where kitchen scales are a common item of household equipment, and recipes are given by weight (*e.g.* in the UK). Estimated records are favoured where standard cups and spoons are used and recipes are quoted in these measures (*e.g.* in the USA). Records are coded, and the data calculated using food composition tables.

### Weighed Record

This is the 'gold standard' method, and often the one of choice for small-scale research studies. The subject is required to weigh each item of food and drink at the time of consumption. Food is weighed 'as served' and leftovers and plate waste is also weighed. The advantages of weighed records are that they can provide good information on individual subjects and on individual foods, meal patterns can be categorised and the number of days studied can be varied. However, only a limited time period is covered, there is substantial burden on the subject, compliance may be poor and subjects may alter food patterns to simplify the recording, and thus may introduce bias into the results.

## Estimated Record

The procedure is as same as for the weighed record, but portions are described in household measures (*e.g.* cups, spoons), in dimensions, or in number of items of predetermined size. Diagrams or photographs may be used to help quantify portions.[7,8] Investigators then have to assign weights to the portions before calculating nutrient intake from food tables.[9] Because weights are estimated the errors are greater than for weighed records. These errors may be random, but could also be biased. However, the respondent's burden is less than for the weighed record and compliance may be better. Investigators have to decide whether to use average portion weights, to differentiate between small, medium and large portions, or to assess portion sizes for every individual. For some foods, subjects may be able to report portion sizes relatively easily (*e.g.* the number and size of bread slices). For some foods, such as eggs or apples, questions might be included about the number eaten per week and an average value could be used per item. For other foods, such as meats, rice and pasta, which cannot be quantified in 'units', an average portion weight could be used or portion sizes could be assessed by using descriptions, photographs or food models.[7-9]

## Menu Record

This is a qualitative record of foods eaten with no quantification of portion sizes. The record may be analysed as frequencies of consumption of different foods or by assigning 'average' weights to portions. Assigning weights should not bias the observed mean, but precision is reduced at the individual level. The method has the potential for collecting information about foods eaten and meal patterns over a prolonged period of time.

## Diet Recall

The subject is asked to recall the actual food and drink consumed in the past on specified days, usually the immediate past 24 h (24-h recall). Portions are quantified as in estimated records. The main advantages of 24-h recall is that there is minimal subject burden, they can be conducted by a single short interview or by telephone, and large numbers of subjects can be studied.

However, additional error is introduced because of the dependence on memory. Diet recalls provide no information on individuals; although 24-h recalls can be repeated at intervals over time to improve precision at the individual level.

## Diet History

This is a face-to-face interview in which the investigator questions the subject about past food intake. A 24-h recall is first taken and then each meal and inter-meal period is considered in turn to identify and quantify possible alternative menus. The interview may be open-ended or fully structured. A checklist of foods is usually used to probe for missing items. The method requires an experienced interviewer who is very knowledgeable about local foods and meal patterns and customary portion sizes. The interview takes 60–90 min to conduct properly and demands high levels of concentration and communication skills from both the interviewer and the respondent. If the subject is not usually involved in shopping and cooking, they may not be able to adequately answer questions about their food intake, so someone may need to accompany them. The supposed advantage of a diet history is that it can refer to an extended period of time and is presumed to measure 'habitual' intake, but there is evidence that the near past is better remembered than the distant past. The disadvantages are large random errors, the method depends on memory and on complex cognitive tasks such as the respondents' perception of their dietary pattern (which may be distorted) and their ability to conceptualise portions (which may be poor) and the method is unsuitable for erratic meal patterns. An additional problem is that subjects may tell the interviewer what they think he/she wishes to know rather than the true answer.

## Food Frequency (and Amount) Questionnaires (FFQ)

In this method, the subject is given a pre-printed list of foods with options to indicate how often each is eaten. Portion sizes might be included. The form is for self-completion and so can be issued and returned by post. The list of foods can be of variable length and depends on the aims of the study. The technique is designed for, and only suitable for, epidemiological scale studies and each questionnaire must be developed for the specific study aims and study population. To give an extreme theoretical example, a questionnaire designed to assess calcium intake in adolescents in the UK could not be used at a later date to assess protein intakes of elderly subjects in the USA! The advantages of FFQs is that they can be used with very large numbers, there is only moderate subject burden and forms can be designed for computer scanning for data entry. The disadvantages are the dependence on memory and ability to convert very variable dietary patterns into frequencies of consumption over usually over a long period of time. It should also be remembered that some subjects may not be fluent in the written or spoken language of the questionnaire. Precision at the individual level is poor. FFQs might be more suitable than diet records for items that might not be consumed very often, but are important sources of nutrients, *e.g.* oily fish (containing essential fatty acids) or if the nutrients of interest are present in a few easily identifiable foods, *e.g.* vitamins A, C, and D. Whether a particular questionnaire may be considered valid will be largely dependent on the purpose of the study. For example, in case-control or intervention studies the mean intakes of groups are compared. The questionnaire must therefore be able to satisfactorily estimate group mean intakes of the nutrients of interest. In cross-sectional studies to investigate whether there are associations between diet and disease risk factors such as bone mineral density, blood lipids or blood pressure, the method chosen needs to be able to rank individuals by their intakes for the foods or nutrients of interest. The food sources of each nutrient within the population being studied must be taken into account. Criteria for foods are that they should be eaten reasonably often by an appreciable number of individuals and that they should have a substantial content of the nutrient(s) of interest. If the study requires that subjects be classified into tertiles or quintiles according to their nutrient intakes then, in order to differentiate between individuals, foods should be included for which there is likely to be reasonable variation in consumption between individuals.

Many factors need to be considered when developing a questionnaire, whether it is to be used to estimate nutrient intakes or for other data required for nutrition studies (*e.g.* recording basic subject information such as date of birth and sex, to record measurements such as height and weight, and to collect information such as smoking habit, medical history, attitudes or food intake). These have been discussed in detail by Fehily and Johns.[10]

# Other Elements of Study Design

To some extent the overall requirements and design of a given study will determine what dietary assessment method is used. However, other issues also have to be considered. Dietary intake is a derived measurement; it is food intake that is the primary measure. Hence, in addition to the nutrient(s) of interest, attention also needs to be paid to foods, meals, portion sizes, and whether the methods used are culturally appropriate, with respect to eating habits, dietary patterns, and food composition data.

## How Many Days

The number of days needed to obtain a good measure of 'habitual' intake in an individual depends on the level of precision required and the component of interest. For example, a 21-day record is needed to measure the energy intake (EI) of an individual to 10%, a 7-day record gives a precision of between 15 and 20%. The within-subject coefficient of variation of daily intake for the macronutrients is similar to that for energy (20–26%). For micronutrients, it is between 30 and 40% for those such as iron, magnesium, and the B vitamins which are widely distributed in foods, about 60% for vitamin C and very much greater for nutrients such as vitamin A that are in found in large amounts in few foods [A.E. Black. The use of recommended daily allowances to assess dietary adequacy, *Proc. Nutr. Soc.*, 1986, **45**, 369–381.].

## Cultural and Geographical Considerations

Attempts to use some of the dietary assessment methods outlined above will not be appropriate in some settings. For example, if meals are traditionally consumed directly from a shared bowl of cooked food, as in many parts of Africa, then asking subjects to weigh and record individual items of food onto a plate or bowl before consumption, or doing so on their behalf is inappropriate. Hence, other methods need to be employed.[11] Major sources of nutrients of interest, possibly very different to those investigators may be accustomed to also have to be considered. For example, though in many countries, dairy products are the major sources of calcium, both because of the nutrient composition of these foods, and the quantities and frequency with which they are consumed, different foods may be the major sources in other parts of the world. For example, various species of fish of commercial importance have been found to be important sources of calcium and phosphorus in Venezuela.[12] Ho *et al.* has found that in Chinese women about 50% of the calcium source was from vegetables and 22% from dairy products.[13] Prentice *et al.* analysed raw ingredients, snack foods, and prepared dishes in rural Gambia. They also collected information about the contribution of mineral-rich seasonings and made efforts to discover unusual sources that might not be perceived as food by the subject or the observer. The main contributors to daily calcium intake were leaves, fishes, cereals, groundnuts, and local salt.[14] In many countries the availability of foodstuffs varies according to season, and so this should be factored in to research proposals. Because some foodstuffs are specific to particular countries and cultures, they are less likely to be included in widely available food composition tables and databases. Vitamin and mineral content of foods is likely to vary depending on the region they were grown or produced, cooking methods may be specific to particular countries, and the nearest equivalent widely used in food tables may not be appropriate. For all these reasons, it might be important to use or develop local or national food composition data.[25] Finally, the composition and source of water used for drinking and for preparation of foods will vary according to geographical location. For example, in hard water regions or areas where copper piping is used, drinking water may provide a substantial amount of calcium and copper respectively. Consumption of bottled mineral waters may also need to be taken into account.

## Categorising Nutrients and Foods

How foods are grouped or classified is also important. The following two examples are taken from Fehily and Johns.[10] "Should information about 'tomatoes' be asked for in a separate question and include cooked, raw and sauces, or should it be included with 'salads'?" "Should a question be asked about 'beef' – all forms including minced beef or should there be a number of questions: 'roast beef', 'beef stew', 'minced beef', 'beef burgers', *etc*.?" Groupings will depend on the hypotheses being tested; how similar the nutrient compositions of the foods are, and the likely ability of subjects to assess how often they eat the foods. Subjects may be better at estimating how often they have 'meals containing beef' than estimating the frequency of consumption for each form of beef. Another consideration is whether subjects may count the same food twice when completing a questionnaire. For example, questions about 'milk' may apply to that taken in tea and coffee and that eaten with breakfast cereals. Another example is 'fruits and vegetables'; how will these be categorised, if at all ? For example, divided into green leafy vegetables, citrus fruits, *etc*. Although in a culinary sense potatoes are vegetables, with respect to food groups, nutritionists classify them as 'starchy carbohydrates' grouped together with foods such as bread and pasta. If account is not taken of this, then the frequency of consumption or the weights of 'fruits and vegetables' may be very inaccurate and any comparisons between different studies may be potentially very misleading. Although water does not contribute to energy and macronutrient intake, the mineral content may be important. So, it might be useful to know about the source of tap water and consumption of bottled mineral waters. Finally, if caffeine intake is of interest then it is important to ensure distinctions are made between regular and decaffeinated versions of coffee, tea, and soft drink.

## Conversion of Food Consumption Data to Nutrient Data

How food consumption data are to be converted to nutrient intake data is another important consideration. For example, with respect to milk, provision should be made to include its different types (whole milk, semi-skimmed, skimmed) and diet records should be checked and details clarified with the subject. Whilst different types of milk may make only small differences to contributions to, for example calcium, and protein intakes, they might have a large impact on the reported intake of total energy, fat, and vitamins A and D. If the research topic of interest is 'protein', provision should probably be made, when analysing the data, to distinguish between animal and plant sources, and even different amino acids.

## Impact of Cooking Methods, Food Processing, Food Fortification and Supplements

Whether a given vegetable is boiled, steamed or microwaved, whether it is fresh, frozen or canned, or cooked with or without added salt, will have little impact on energy and macronutrient intake, but may well affect measurements of micronutrient intake. Manufactured products may also change over time (sometimes very rapidly), and so food composition tables or portion size information may not be sufficiently up to date. Examples of foods which might be affected in this way include commonly consumed items such as snacks, and breakfast cereals that may be fortified with vitamins and/or minerals. Traditionally, in western countries dairy foods provide the major source of calcium and are a source of the fat-soluble vitamins A and D. In many countries, for many years these vitamins have been added to margarines, and calcium added to white flour. However, more recently many other foods have been fortified with a range of nutrients, for example, breakfast cereals with iron, calcium and folic acid, and soya products (*e.g.* soya drink) fortified with calcium to bring them more in line with the nutrients that dairy products provide. The production of 'functional foods is increasing, so for example some brands of orange juice, bottled water, and milk have added calcium. Thus to obtain as complete a record as possible of, for example, calcium intake, an awareness of other less obvious foods sources is necessary. Vitamin and mineral supplementation is also increasing, and again questions should be asked about these, and any prescribed or over-the-counter medication that might affect the nutrients of interest, especially if status (*e.g.* from analysis of blood or urine samples) is also being assessed. It is often useful to ask subjects to preserve labels and packaging if possible.

## Food Groups and Meal Patterns

In recent years, nutrition epidemiologists have been increasingly interested in dietary patterns, not simply in nutrient intakes. For example, there is now a substantial literature, which suggests that diets high in plant-based foods (fruits, vegetables, and starchy carbohydrates) may protect against chronic diseases such as CVD and cancer.[15] Dairy foods have long been associated with bone health, at least in western countries, because of the calcium they provide. More recently, there has been interest in the acid–base balance effects of fruits and vegetables.

Meal patterns are also of increasing interest. They may give insights into other aspects of a subject's lifestyle and behaviour, and may also be important if issues relating to bioavailibility and absorption are of interest. For example, concerns have been expressed about the possible adverse effects of consuming iron fortified breakfast cereals with milk. Studies of this nature are often confined to detailed metabolic investigations, but if findings are to be extrapolated to a 'real life' situation, then information about meal patterns may be important. Excess dietary protein from either animal or plant proteins might be detrimental to bone health, but any effects will be modified by other nutrients in the food and the total diet. The effects of protein on urinary calcium and bone metabolism (see Chapter 10.2) are modified by other nutrients found in that protein food source. For example, the high amount of calcium in milk compensates for urinary calcium losses generated by milk protein. Similarly, the high potassium levels of plant protein foods, such as legumes and grains, will decrease urinary calcium. The hypocalciuric effect of the high levels of phosphate associated with the amino acids of meat at least partially offsets the hypercalciuric effect of the protein. Other food and dietary constituents such as vitamin D, isoflavones in soy, caffeine, and added salt also have effects on bone. Many of these other components are considered in the potential renal acid load of a food or diet, which predicts its effect on urinary acid and thus calcium (see Chapter 10.2).

## Collection of Subsidiary Data

Other information, which depending on a given study, would be useful or essential to collect is listed in Table 6.2. The data might be important for using to assess the validity of dietary intake assessments and/or might provide valuable insights into any relationships between intake and other parameters being measured in a study.

# Evaluation of Dietary Intake Data

Nutrient intake is a derived measurement. The primary measurement is of the foods eaten, and there are thousands of foods and food products. The choices made by individuals vary widely in terms of foods eaten, and the daily pattern of meals and snacks. Foods chosen vary in kind and quantity from day to day, week to week and season to season, so the energy and nutrient intakes also vary widely. Having decided the method(s) to be used in an investigation, and giving consideration to all the information that should be collected, the data must then be critically evaluated. It is all to easy to process the reported food intake through a nutrient analysis programme, obtain reported nutrient intakes calculated to several decimal places, and assume that the data are valid and provide a true record of intake.

The terminology applied to the critical evaluation of food intake is essentially the same as that for any other analytical method:

- Validity/accuracy – does the method measure the 'true' value?
- Precision, repeatability, reliability – does the method give the same answer every time?
- Random errors – operate in all directions and contribute to poor precision (confidence limits)
- Systematic errors – operate in one direction and introduce bias

Ideally, a method would give a valid and precise result, but data can also be invalid, but precise, or valid but imprecise! A particular survey technique may induce method-specific behavioural alterations in the subjects' actual and reported intakes. These are difficult to quantify, and so the true validity of different dietary assessment methods is unknown. The distinction must be made between validity at the group level (the mean value) and validity at the individual level (correct ranking). A method may provide a valid mean, but

**Table 6.2.** *Subsidiary information that should or could be collected in dietary intake studies*

| Information | Examples |
| --- | --- |
| Personal data | Date of birth<br>Medical history<br>Medication |
| Subject characteristics | Sex, height, weight, body mass index, parity, socio-economic status, geographical location, ethnic background |
| Lifestyle, social and cultural factors | Occupation and leisure activities<br>Weight change<br>Smoking habits<br>Alcohol consumption |
| Assessment of energy expenditure or physical activity | Doubly labelled water measurements of total energy expenditure, heart-rate monitoring, accelerometers or physical activity questionnaires |
| Attitudes to food, dieting/weight history<br>Subjects random or self-selected | |
| Motivation for participating in study<br>Time of year studied | |
| Psychological characteristics (*e.g.* using questionnaires derived from the area of eating disorders) | Dutch Eating Behaviour Questionnaire (DEBQ) with scales for restraint, emotional and external eating<br>The Three Factor Eating Questionnaire (TFEQ) with scales for restraint, disinhibition and hunger |
| External validation | Doubly labelled water (for EI)<br>Urinary nitrogen (for protein intake) |
| Nutritional status | *e.g.* assessed from analysis of urine or blood samples |

rank poorly due to poor precision or variable bias across subjects. Alternatively, a method may give an invalid mean, but nevertheless rank subjects correctly if precision is good and bias is similar across subjects.

As illustrated above, sources of error and bias can be due to:

- Food composition tables.
- Estimated portion size.
- Variation with time, poor memory.
- Deliberate mis-reporting.
- Lack of understanding/inadequate explanation.
- Study protocols interfering with normal eating behaviour.

- Subjects knowingly making omissions or changes because of the nuisance.
- Subjects changing their habits, for example, no second helpings, avoiding snacks, starting a weight reducing diet.
- Subjects do not eat and/or report foods perceived as unhealthy or bad.

Three concepts are fundamental to understanding the limitations of dietary assessment: habitual intake, validity, and precision.

## Habitual Intake

The habitual intake of an individual is their intake averaged over a prolonged period of time (weeks or months rather than days). For energy, it is the intake that maintains weight stability. For other nutrients, it can be the intakes required to produce a steady physiological state and hence influence nutritional status and health in both the short and the long term. Habitual intake is the value that studies of diet and health would ideally measure; however, it is a largely hypothetical concept, because intake varies widely from day to day, and weekly or monthly variation can also be significant.

## Validity (accuracy)

When investigators are questioning if their dietary intake data are valid they should ask themselves if the method measures what they believe they are measuring, and does it give an observed estimate of intake that is close to the true intake? A valid dietary assessment faithfully reports 'true' intake undistorted by systematic errors (these operate in one direction and thus introduce bias into the results). They distort the data and may, for example, lead to estimates of mean intake that are too high or too low. A valid diet record is a complete and accurate record of all food consumed on specified days, in which the subject eats and records exactly what they ate during the period of study and this is what they would have eaten if no investigator had intervened. A valid diet recall is a complete and accurate recall of all food and drink consumed on specified day(s). A valid diet history or FFQ accurately reflects typical food consumption over a designated period of time, undistorted by behavioural patterns or false memory. A valid diet report is one in which the subject reports past food intake without conscious or subconscious distortion. Records of poor validity may be a result of underreporting, undereating, overreporting, or overeating during the measurement period.

## Precision (repeatability, reproducibility, reliability)

When investigators are questioning if their dietary intake data are valid they should ask themselves if the method gives the same answer on repeated applications. A precise technique is one that yields the same answer on repeated administrations. Precision may be expressed in various ways. For example: mean absolute difference, mean difference as a percentage of overall mean intake, coefficient of variation of the differences within individuals, correlation coefficient, percentage of individuals classified in the same quantile on different occasions, and the 95% confidence limits of repeated measurements. Precision is low when there are large random errors. Poor precision does not affect the estimate of the mean, but prevents correct ranking of individuals. Because food intake varies widely with time, precision of dietary assessment at the individual level is poor even when repeated surveys show good agreement for mean intake. It should also be remembered that just because a method gives the same answer on more than one occasion, it does not necessarily mean that answer is correct.

## Critical Evaluation of EI Data

Between the 1930s and 1970s, the 'validity' of dietary assessment techniques was tested by comparing one method with another, but without any way of testing, which if any, method was valid. The so-called validation studies simply compared the results of one method with another. The weighed dietary record was often assumed to be the gold standard, and the validity of other methods was evaluated by comparison. These studies are actually studies of 'relative validity' rather than studies of validity that use external markers of intake. There were no independent markers of intake other than direct observation (*e.g.* in institutions and in detailed metabolic studies). So, until relatively recently dietary intakes were reported as if valid, and the interpretation of links between intake and health were based, often erroneously, on this assumption. In the 1980s and 1990s the advent of external markers of intake made it possible to test assumptions about validity. In contrast to the micronutrients there are no biochemical biomarkers of energy intake (EI). All three methods of validation of EI assessments assume that EI has to equal energy expenditure (EE) when weight is stable. The use of doubly labelled water (DLW) to measure total free-living EE in humans was central to disclosing the widespread problem of invalid dietary intake data. Initially, the observations were made in obese subjects[16,17] and later in most other subject groups.[18,19] After a further decade of work it is now generally recognised that self-reports of food intake underestimate food and nutrient intake. The problem was identified by several external markers of intake: comparisons between self reported EI and that required to maintain body weight in long term metabolic studies; comparisons between reported nitrogen/protein intake and 24-h nitrogen excretion; comparisons between reported EI and total free-living EE measured by doubly labelled water; comparison of reported EI and EE both expressed as multiples of BMR. The last approach was first developed by using 'fundamental principles of energy physiology' combined with statistical considerations, to derive 'cut-off' limits.[20,21] In this technique, mean EI is expressed as a multiple of the mean BMR estimated from equations and is compared with the presumed mean EE of the population, which is also expressed as a multiple of the BMR. The ratio EE:BMR is usually referred to as the physical activity level (PAL). Because at the group level weight may be regarded as stable in the timescale of a dietary assessment, the validity of reported EI could be evaluated by comparing it with either measured EE or an estimate of the energy requirement of the population. Values of EI:BMR falling below the 95% confidence limit of agreement between these two measures signify the presence of underreporting. The equation used to derive cut-offs calculates the lower 95% confidence limit of EI:BMR assuming a given PAL requirement, below which it is unlikely that the mean intake represents either habitual intake for weight maintenance or a random low intake.[21] It makes allowance for the errors associated with the number of subjects, the duration of the dietary assessment and variation in each of food intake, BMR and physical activity. In the original publication, the cut-off was calculated assuming an energy requirement of 1.55xBMR, and it was demonstrated conclusively that underreporting was widespread.[20] A PAL of 1.55xBMR was selected as the basis for comparison because it is the value defined by FAO/WHO/UNU as that which represents a sedentary level of EE. However, subsequent analysis of nearly 600 DLW measurements has shown that this is a conservative figure. In 16 age–sex groups, except those over 75-year old, the mean PAL in free-living people was > 1.55.[22,23] Thus, the extent of underreporting of EI based on this figure is almost certainly underestimated.

Many other workers in various ways have since followed this approach, and more recently Black has re-stated the principles underlying the cut-offs, examined each parameter used in the equations and the values to be inserted. She has also provided guidance for the application of the cut-offs and commented on their usefulness and limitations.[24] Cut-offs have been used by numerous authors to identify individual underreporters in different dietary databases to explore the variables associated with underreporting. These studies have been comprehensively reviewed by Livingstone and Black.[6] They explore the characteristics of underreporters and the biases in estimating nutrient intake and in describing meal patterns associated with underreporting, and also examined some of the problems for the interpretation of data introduced by underreporting and particularly by variable underreporting across subjects. Future directions for research were also identified.

## Implications of Under-reporting of EI on the Assessment of Other Nutrients and Food Intake

Validation against EE identifies only the bias in reporting EI. However, if the estimated EI is significantly lower than probable EE, as judged by knowledge of the lifestyle and physical activity of the subject, then food intake has been underreported. Energy is, of course, a surrogate measure for the total quantity of food eaten. Because the amount of any nutrient obtained is related to the quantity of food eaten (the greater the total quantity of food, the greater the intake of any nutrient), evaluating the validity of the reported EI provides a check on the overall quality of dietary intake data. Thus, if total energy intake is underestimated, it is likely that the intakes of other nutrients correlated with EI (the macronutrients, most minerals, and the B vitamins) are also underestimated. This may, for example, lead to overestimation of the proportion of the population with deficient intake or distortion of the associations between nutrient intake and disease outcome. A recent review discusses studies that have examined whether this reflects underreporting of the diet as a whole or whether there is bias in estimating nutrient intakes through altered food choices and/or selective reporting of foods or meals or snacks.[6] It was concluded that under-reporting of food intake is a selective rather than a general phenomenon, with the evidence pointing to differences in reporting for macro- and micronutrients, foods, and meal patterns.

## Conclusions

There are a number of different approaches that can be used to assess dietary intake. None are perfect. The method chosen must be suitable for the overall design of an investigation and the scientific questions to be addressed.

Dietary intake data must be critically evaluated before conclusions can be drawn from the results. Evaluating the validity of reported EI provides a valuable check on the general quality of the dietary data in any study.

In order to evaluate the reported EI of any one individual or a group of subjects, it is essential to obtain as much information as possible about lifestyle and physical activity.

Subjects' characteristics and behaviour, the method of dietary assessment chosen, other measurements included in a study which add to subject burden, study personnel, and indeed participation in a study or survey itself, all have to be taken into account.

Mis-reporting of intake is a serious problem in studies of nutrition and health. The phenomenon is the marked bias to the underestimation of food intake when the information is derived from self-reported dietary assessments.

The generally high correlation between micronutrient and total EI has implications for estimating the proportion of a population with suboptimal intakes.

Uncovering of the problems associated with systematic bias in dietary assessments does not mean that dietary studies should be abandoned. Study design and the handling of flawed data present a challenge to nutritionists, dietitians and other health professionals, physiologists, psychologists, and statisticians.

Researchers in bone health who have little or no knowledge or expertise in dietary intake methodology and validation should aim to collaborate with those who do. The outcomes of well-designed studies can only prove beneficial to all parties and interests.

# Acknowledgements

The author has drawn heavily on recent work by her friends and colleagues Dr Alison Black and Professor Barbara Livingstone.

# References

1. J.S. Garrow, In *Energy Balance and Obesity in Man*, 1st edn, North Holland Publishing Company, Amsterdam, 1974.
2. J. Blundell, What foods do people habitually eat? A dilemma for nutrition, an enigma for psychology, *Am. J. Clin. Nutr.*, 2000, **71**, 3–5.
3. S.A. Bingham, The dietary assessment of individuals: methods, accuracy, new techniques and recommendations, *Nutr. Abs. Rev. (Series A)*, 1987, **57**, 705–742.
4. I. Rutishauser and A. Black, Measuring food intake, In *Introduction to human nutrition*, M. Gibney, E. Vorster and F. Kok (eds.), Blackwell Publishing on behalf of the Nutrition Society, Oxford, 2002.
5. M. Nelson, B.M. Margetts and A.E. Black, Check-list for the methods section of dietary investigations, *J. Human Nutr. Dietet.* 1993, **6**, 79–83.
6. M.B.E. Livingstone and A.E. Black, Markers of the validity of reported energy intake, *J. Nutr.*, 2003, **133**, 895S–920S.
7. M. Nelson and J. Haraldsdottir, Food photographs: practical guidelines I. Design and analysis of studies to validate portion size estimates, *Public Health Nutr.*, 1998, **1**, 219–230.
8. M. Nelson and J. Haraldsdottir, Food photographs: practical guidelines II. Development and use of photographic atlases for assessing food portion size, *Publ. Health Nutr.*, 1998, **1**, 231–237.
9. H. Crawley, Food portion sizes, London: MAFF, HMSO, 1988.
10. A. Fehily and A. Johns, Designing questionnaires for nutrition research www.tsis@tinuviel.u-net.com.
11. G.J. Hudson, Food intake in a west African village. Estimation of food intake from a shared bowl, *Br. J. Nutr.*, 1995, **73**, 551–569.
12. P.I. Corser, G.T. Ferrari, Y.B. de Martinez, E.M. Salas and M.A. Cagnasso, Proximal analysis, fatty acids profile, essential amino acids and mineral content in 12 species of fishes of commercial importance in Venezuela, *Arch Latinoam Nutr.*, 2000, **50**, 187–194.
13. S. Ho, P. Leung, R. Swaminathan, C. Chan, S. Chan, Y. Fan and R. Lindsay, Determinants of bone mass in Chinese women aged 21–40 years II. Pattern of dietary calcium intake and association with bone mineral density, *Osteoporosis Int.* 1994, **4**, 167–175.
14. A. Prentice, M.A. Laskey, J. Shaw, G. Hudson, K. Day, L. Jarjou, B. Dibba and A.A. Paul, The calcium and phosphorus intakes of rural Gambian women during pregnancy and lactation, *Br. J. Nutr.*, 1993, **69**, 889–896.
15. British Nutrition Foundation, *Plants: diet and health. The Report of the British Nutrition Foundation Task Force,* G.R. Goldberg (ed.), Blackwell Publishing, Oxford, 2003.
16. L.G. Bandini, D.A. Schoeller, H.N. Cyr and W.H. Dietz, Validity of reported energy intake in obese and nonobese adolescents, *Am. J. Clin. Nutr.*, 1990, **52**, 421–425.

17. A.M. Prentice, A.E. Black, W.A. Coward, H.L. Davies, G.R. Goldberg, P.R. Murgatroyd, J. Ashford, M. Sawyer and R.G. Whitehead, High levels of energy expenditure in obese women, *Br. Med. J.*, 1986, **292**, 983–987.
18. M.B.E. Livingstone, A.M. Prentice, J.J. Strain, W.A. Coward, A.E. Black, M.E. Barker, P.G. McKenna and R.G. Whitehead, Accuracy of weighed dietary records in studies of diet and health, *Br. Med. J.*, 1990, **300**, 708–712.
19. A.E. Black, A.M. Prentice, G.R. Goldberg, S.A. Jebb, S.A. Bingham, M.B.E. Livingstone and W.A. Coward, Measurements of total energy expenditure provide insights into the validity of dietary measurements of energy intake, *J. Am. Dietetic Assoc.*, 1993, **93**, 572–579.
20. A.E. Black, G.R. Goldberg, S.A. Jebb, M.B.E. Livingstone, T.J. Cole and A.M. Prentice, Critical evaluation of energy intake data using fundamental principles of energy physiology. 2. Evaluating the results of dietary surveys, *Eur. J. Clin. Nutr.*, 1991, **45**, 583–599.
21. G.R. Goldberg, A.E. Black, S.A. Jebb, T.J. Cole, P.R. Murgatroyd, W.A. Coward and A.M. Prentice, Critical evaluation of energy intake data using fundamental principles of energy physiology. 1. Derivation of cut-off limits to identify under-recording, *Eur. J. Clin. Nutr.*, 1991, **45**, 569–581.
22. A.E. Black, Physical activity levels from a meta-analysis of doubly labelled water studies for validating energy intake as measured by dietary assessment, *Nutr. Rev.*, 1996, **54**, 170–174.
23. A.E. Black, W.A. Coward, T.J. Cole and A.M. Prentice, Human energy expenditure in affluent societies: an analysis of 574 doubly-labelled water measurements, *Eur. J. Clin. Nutr.*, 1996, **50**, 72–92.
24. A. Black, Critical evaluation of energy intake using the Goldberg cut-off for energy intake: basal metabolic rate. A practical guide to its calculation, use and limitations, *Int. J. Obesity*, 2000, **24**, 1119–1130.
25. C.J. Prynne, A.A. Paul, B. Dibba and L. Jarjou, Gambian Food Records: a new framework for computer coding. *J. Food Compos. Anal.* 2002, **15**, 349–357.

CHAPTER 7

# Nutritional Aspects of Bone Growth: An Overview

JEAN-PHILIPPE BONJOUR, PATRICK AMMANN,
THIERRY CHEVALLEY, SERGE FERRARI
and RENÉ RIZZOLI

Division of Bone Diseases [World Health Organization Collaborating Center for Osteoporosis Prevention], Department of Rehabilitation and Geriatrics, University Hospital, Geneva 1206, Switzerland,
E-mail: jean-philippe.bonjour@medecine.unige.ch

**Key Points**

- Peak bone mass is an important determinant of osteoporotic fracture risk.
- Bone mass accumulation from infancy to post-puberty involves interrelated actions of genetic, endocrine, mechanical and nutritional factors.
- Like standing height, bone mineral mass during growth follows a trajectory that can be influenced by environmental factors.
- Increasing calcium intakes and/or mechanical loading can shift the age-bone mass trajectory upward.
- Protein intake, probably by influencing IGF-1, can also positively influence bone mass accrual.

## Introduction

The extent to which variations in the intakes of some specific nutrients by healthy, apparently well-nourished, children and adolescents affect bone mass accumulation, has received increasing attention over the last 15 years. This particular interest essentially stems from the growing awareness of two related important evidences: (1) fragility fractures occurring during adult life are a major public health problem; (2) the amount of bone acquired at the end of the growth phase is an important determinant of the risk of fragility fractures of which the occurrence exponentially increases from the sixth decade of life on.[1-3] These fragility fractures are mainly due to osteoporosis. The definition of osteoporosis as endorsed by the World Health Organization states that it is *a systemic skeletal disease characterized by low bone mass and micro-architectural deterioration of bone tissue, with a consequent increase in bone fragility and susceptibility to fracture risk.* Osteoporosis is one of the most serious diseases facing the aging population. It affects one in three women and one in eight men during their lives. Most subjects who suffer from the disease are in the last third of their lives. Obviously, osteoporosis is widespread, and as the world's population ages, more and more people will suffer from this debilitating and sometimes fatal disease. Therefore, it is essential to develop a worldwide strategy for osteoporosis management and prevention. Over the last 25 years, measures of prevention have been aimed mainly at reducing the bone loss occurring at the menopause and/or that related to aging. Another preventive strategy to avoid osteoporotic fractures in later life is to build up the strongest bone possible during infancy, childhood and adolescence. The aim of this strategy is to achieve at the end of the growth period the maximum genetically attainable bone mass by modifying environmental determinants, particularly nutritional factors and physical activity. With respect to the impact of nutrients on bone mineral mass acquisition most studies have focused on the intake of calcium. However, other nutrients such as proteins also should be considered. Before describing specifically the impact of nutrition, it appears important to consider how bone growth trajectory in youth is to a large extent at the origin of fragility fractures in adulthood.

## Characteristics of Bone Mass Development

The mechanical strength of bone depends of several structural elements, which follow distinct development trajectories from the origin of intrauterine life to the end of the skeletal growth process, *i.e.*, when peak bone mass is attained by the end of the second decade. The size of the bone, the amount of bony tissue within the periosteal envelope and its space distribution, *i.e.*, the micro- and macro-architecture, and the degree of mineralization of the organic matrix are the most important structural elements which determine the resistance to mechanical loading. During development, the increase in areal bone mineral density (aBMD) is essentially due to a rise in bone size, which results in a proportional augmentation in the amount of mineralized tissue within the periosteal envelope. In contrast, the volumetric (v) BMD increases very little from infancy to the end of the growth period. In either gender the main bone strength determinants follow a trajectory similar to that recorded for standing height. In healthy girls longitudinal examination at 4.5 years interval, before and during pubertal maturation, of lumbar spine development shows that the Z-scores of aBMD, BMC, vBMD, as estimated by calculating bone mineral apparent density (BMAD), as well as vertebral body width and height are highly correlated, with correlation coefficients ranging from 0.70 to 0.82, as compared to 0.85 for standing height. In the lumbar spine, the gender difference observed when peak bone mass is attained consists mainly in a greater vertebral body diameter in the frontal plane of males as compared to females. This gender-related structural differentiation, which does not attenuate with ageing, is certainly an important macro-architectural determinant of the difference in fragility vertebral fracture incidence observed between female and male subjects in later life. It probably also plays an important role in the vertebral fracture risk within each gender, as demonstrated in postmenopausal women with equally low trabecular vBMD, those with vertebral fracture displaying a smaller vertebral body cross-sectional area.[4] The gender difference in either aBMD or BMC observed at the level of the radial or femoral diaphysis once peak bone mass is attained appears to be also essentially due to a greater gain in bone size in males than females during pubertal maturation. A very recent study comparing bone variables (BMC, aBMD and vBMD) in opposite-sex twins corroborate this notion.[5]

## Bone Mass Acquisition During Puberty

Before puberty, no substantial gender difference has been reported in bone mass of the axial (lumbar spine) or appendicular (*e.g.* radius and femur) skeleton. There is no evidence of a gender difference in bone mass at birth; the vBMD actually appears to be similar in female and male newborns.[6] This absence of sex differences in bone mass is maintained until the onset of pubertal maturation. During puberty, the gender difference in bone mass becomes expressed. This difference appears to be due mainly to a more prolonged bone maturation period in males than in females, with a larger resulting increase in bone size and cortical thickness. Puberty affects bone size much more than it does vBMD.[1,6] There is no significant sex difference in volumetric trabecular density at the end of pubertal maturation (for review, see Ref. 1). Nevertheless, the bone mass accumulation rate at the lumbar spine and femoral neck levels increases 4- to 6-fold over a 3- and 4-year period in females and males, respectively; the change in bone mass accumulation rate is less marked in long bone diaphysis.[7] There is an asynchrony between the gain in standing height and the growth of bone mineral mass during pubertal maturation.[7–9] This phenomenon may be responsible for the occurrence of a transient bone fragility in adolescence that may contribute to the higher incidence of fracture that occurs when the dissociation is maximal between the rates of statural growth and mineral mass accrual.[10,11]

## Time of Peak Bone Mass Attainment

In adolescent females, gain in bone mass declines rapidly after menarche; no further statistical gains are observed 2 years later at least in sites such as lumbar spine or femoral neck.[7] In adolescent males, the gain in BMD or BMC that is accelerated particularly from 13–17 years declines markedly thereafter, although it remains significant between 17–20 years in both lumbar spine BMD and BMC and in midfemoral shaft BMD; in contrast, no significant increase is observed for femoral neck BMD.[7] In subjects who reached pubertal stage P5 and grew less than 1 cm year$^{-1}$, a significant bone mass gain persisted in males but not in females.[7] This suggests the existence of an important sex difference in the magnitude and/or duration of the so-called 'consolidation' phenomenon that contributes to PBM.

The maximal vBMD of the lumbar vertebral body as measured by quantitative computed tomography is achieved soon after menarche; no differences are observed between mean values of 16 and 30 year-old subjects.[12] This is in keeping with numerous observations indicating that bone mass does not increase significantly from the third to the fifth decade. Nevertheless, a few studies suggest that bone mass acquisition may still be substantial during the third and fourth decades. In any case, the balance of published data does not sustain the concept that bone mass at any skeletal site, in either genders and in any ethnic geographic population group, continues to accumulate through the fourth decade.[6]

## Peak Bone Mass Variance

At the beginning of the third decade, there is a large variability in the normal values of the aBMD in axial and appendicular skeleton.[1] This is the outcome of the broad range of the trajectories followed by the various bone components during development. As already mentioned peak bone mass is considered as important as postmenopausal or age-dependent bone loss in the risk of fragility fractures. Information from epidemiological studies suggest that an increase by 10%, *i.e.*, an increase by about one standard deviation (SD) in peak bone mass would reduce the fracture risk by 50%. This quantitative relationship prompted the interest for exploring ways to influence positively the trajectory of bone development. The large variance in PBM, which is observed at sites particularly susceptible to osteoporotic fractures such as lumbar spine and femoral neck, is barely reduced after correction for standing height and does not appear to increase substantially during adult life.[13] The height-independent broad variance in bone mass that is present before puberty appears to increase further during pubertal maturation at sites such as the lumbar spine and femoral neck.[7,9,13] Note that, in young healthy adults, the biological variance in lumbar spine BMC is 4–5 times larger than that of standing height. It is also important to add that the variance in standing height does not increase during puberty.[9,13]

## Calcium Phosphate Metabolism During Growth

Several physiological functions influence bone mass accumulation during growth. Animal studies have identified physiological mechanisms that sustain increased bone mineral demand in relation to variations in growth velocity. In this context, two adaptive mechanisms affecting calcium phosphate metabolism appear to be particularly important, namely the increase in the plasma concentration of 1,25-dihydroxyvitamin $D_3$ (calcitriol), and the stimulation of the renal tubular reabsorption of inorganic phosphate (Pi). Elevations in the production and plasma levels of calcitriol enhance the capacity of intestinal epithelium to absorb calcium and Pi. Increases in the tubular reabsorption of Pi result higher extracellular concentrations. Without these two concerted adaptive responses, growth and mineralization would not be optimal. Note that the increase in tubular Pi reabsorption is not mediated by a rise in either the renal production of calcitriol or its plasma level.[14]

These adaptive mechanisms appear to be essential to cope successfully with the increased bone mineral demand of the pubertal growth spurt.[6] Indeed, marked increases in the plasma concentrations of calcitriol and Pi in relation with bone mineral mass gain occur during pubertal maturation. The insulin-like growth factor I (IGF-I) is responsible presumably for the renal stimulation of calcitriol production and maximum tubular Pi reabsorption rate (TmPi/GFR) in response to the increased calcium and Pi demands associated with the acceleration in bone growth.[8,14,15] In humans, the plasma level of IGF-I rises transiently during pubertal maturation, and reaches a peak during mid-puberty, the maximal level thus occurring at earlier chronological ages in females than in males.[16] Increased plasma levels of IGF-I, calcitriol and Pi are correlated with the elevations in indices of bone appositional rates such as those represented by alkaline phosphatase and osteocalcin (for review, see Ref. 6). The role of IGF-I in calcium phosphate metabolism during pubertal maturation in relation with nutrients that are essential for bone growth is illustrated in Figure 7.1. Note that the increases in plasma concentrations of gonadal sex hormones and adrenal androgens (dehydroepiandrosterone and androstenedione), which occur before and during pubertal maturation, do not seem to be in synchrony with the accelerated bone mass gain. As recently reviewed, the interaction between the growth hormone-IGF-I axis and sex steroids is quite complex.[17] This axis can be influenced by the intake of some nutrients, particularly proteins, as discussed below, and maybe calcium.

# Nutrition as a Determinant of Bone Mass Acquisition

Nutrition is one of the many factors that influence bone mass accumulation during growth. The list of other determinants classically includes heredity, sex, endocrine factors (sex steroids, calcitriol, IGF-I), mechanical forces (physical activity, body weight) and exposure to risk factors. Quantitatively, the most prominent determinants appear to be genetically related. These factors appear to be inter-related. However, robust prospective data on these putative inter-relations are still lacking. Some observations suggest that the response to specific bone acting nutrients, such as calcium, inorganic phosphate, vitamin D – an essential nutrient when its cutaneous production is insufficient–, and proteins may be modulated by certain gene polymorphisms, or by the degree of physical activity, or still by factors that determine the onset of pubertal maturation. Likewise, the possibility of some synergistic effect of specific bone acting nutrients remains nowadays hypothetical. Obviously, this uncertainty

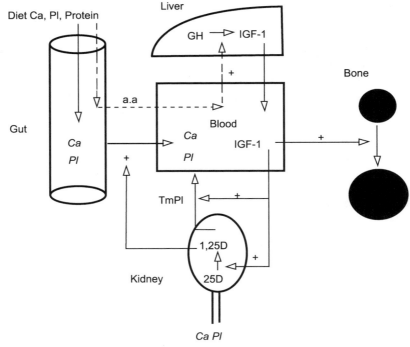

**Figure 7.1** *Relation between essential nutrients for bone mass accrual, insulin-like growth factor-I (IGF-I), and calcium phosphate metabolism during growth. The hepatic production of IGF-I is under the positive influence of growth hormone (GH) and essential amino acids (a.a.). IGF-I exerts a direct action on bone growth. In addition, at the kidney level, IGF-I increases both the 1,25-dihydroxyvitamin D (1,25D) conversion from 25-hydroxyvitamin D (25D) and the maximal tubular reabsorption of Pi (TmPi). By this dual renal action IGF-I favours a positive calcium and phosphate balance as required by the increased bone mineral accrual. See text for further details*

does not apply to the well-documented action of vitamin D on the intestinal absorption of calcium and inorganic phosphate.

The extent to which variations in the intakes of some nutrients by healthy, apparently well-nourished, children and adolescents affect bone mass accumulation, particularly at sites susceptible to osteoporotic fractures, has received increasing attention over the last 10 years. Most studies have focused on the intake of calcium. However, other nutrients such as proteins also should be considered.

## Calcium

It is usually accepted that increasing calcium intake during childhood and adolescence will be associated with a greater bone mass gain and thereby a higher PBM. However, a literature survey of the relationship between dietary calcium

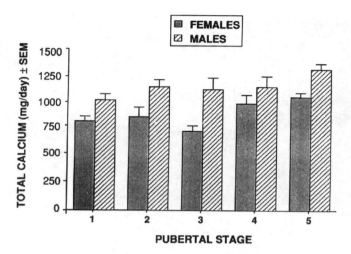

**Figure 7.2** *Calcium intakes in relation with pubertal maturation in both female and male subjects aged 9–19 years. The mean calcium intake from dairy, vegetal and mineral sources was recorded in two five-day diet diaries at one-year interval in 193 healthy adolescents (96 females and 97 males), aged 9–19 years. The method and corresponding macronutrient intakes are described in Ref. 57*

and bone mass indicates that some but not all observational studies have found a positive correlation between these two variables.[6,18–20] In our own cross-sectional study, a significant positive relationship between total calcium intake, as determined by two 5-day diaries (Figure 7.2), and the bone mass accrual was found in females in the pubertal subgroup P1–P4, but not in the P5 subgroup (Figure 7.3).[6] Furthermore, when results were analyzed by taking into account the influence of age and pubertal maturation, the relationship between the absolute values of the calcium intake and the gain in BMD Z-score suggested that calcium may be more important before than it is during pubertal maturation.[6]

Several intervention studies have been carried out in children and adolescents.[21–28] Overall, these studies indicate greater bone mineral mass gain in children and adolescents receiving calcium supplementation over periods varying from 12 to 36 months. The benefit of supplemental calcium has been greater in the appendicular that in the axial skeleton.[20–28] Thus, in pre-pubertal children, calcium supplementation is more effective on cortical appendicular bone (radial and femoral diaphysis) than on axial trabecular rich bone (lumbar spine) or in the hip (femoral neck, trochanter).[20,22,26] The skeleton appears to be more responsive to calcium supplementation before the onset of pubertal maturation.[20,22,26] As intuitively expected, this benefit may be particularly substantial at the end of the intervention period in children with a relatively low calcium intake.[26] In 8-year-old pre-pubertal girls with a spontaneously low calcium intake, increasing calcium intakes from about 700 to 1400 mg augmented the mean gain in aBMD of six skeletal sites by 58% as compared to the placebo group, after 1 year of supplementation (Figure 7.4).

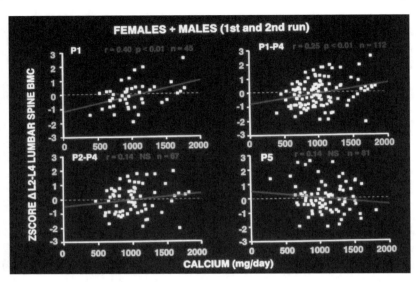

**Figure 7.3**  *Relation between calcium intakes and changes in lumbar BMC in pre-, peri-*
*and post-pubertal female and male adolescents. Mean calcium intakes from*
*dairy, vegetal and mineral sources were recorded in two 5-day diet diaries at*
*1-year interval in 193 healthy adolescents (96 females and 97 males) aged 9–*
*19 years. The relation with pubertal maturation is presented in Figure 7.2.*
*Each dot corresponds to the change in BMC adjusted for age and gender (Z*
*score). The BMC data are from Ref. 7. A positive correlation was found in*
*pre-pubertal (P1), but neither in peri-pubertal (P2–P4) nor in post-pubertal*
*(P5) subjects*

This difference corresponds to a gain of 0.24 SD; if sustained over a period of 4
years, such an increase in calcium intake could augment mean aBMD by 1 SD.
Thus, calcium supplementation could modify the bone growth trajectory and
thereby increase PBM. Interventions limited to the first period of life may
modify the trajectory of bone mass accrual. Thus, a vitamin D supplement
of 400 IU (10 µg) given daily to female infants for an average of one year
was associated with a significant increase in aBMD measured at the age of
7–9 years.[29] In this study the aBMD difference between the vitamin
D-supplemented and non-supplemented groups was most significant at the
femoral neck, trochanter and radial metaphysis.[29] Such an observation would
be compatible with the 'programming' concept, according to which environmental stimuli during critical periods of early development can provoke long-
lasting modifications in structure and function.[30,31]

The calcium salt used to supplement diets may modulate the nature of the
bone response. The response to a calcium phosphate salt from milk extract
appears to differ from those recorded with other calcium supplements. Indeed,
the positive effect on aBMD of a milk-extracted calcium phosphate salt
supplement was associated with an increase in the projected bone area at
several sites of the skeleton and a slight increase in statural height.[26] This type

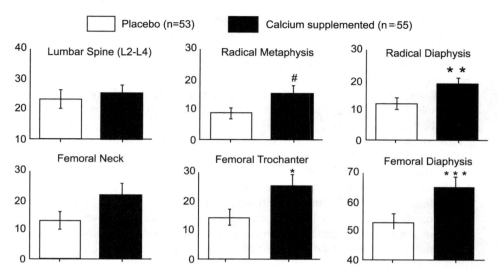

**Figure 7.4** *Influence of calcium enriched foods on bone mineral accrual in pre-pubertal girls. Bars represent means ± SEM of gains in aBMD measured at six skeletal sites in girls consuming food enriched (supplemented) or not (placebo) in calcium during 48 weeks. The calcium enriched foods provided a supplement of 850 mg day$^{-1}$. # p = 0.075; *p = 0.037; **p = 0.015; ***p = 0.007 for differences between the two groups. Adapted from Ref. [26]*

of response was not usually observed when calcium was given as citrate maltate salts,[22,23] carbonate alone[24,25] or carbonate combined with gluconate lactate.[27] Interestingly, in another study, the response to the calcium phosphate salt supplement was similar to the response to whole milk supplementation.[32] However, in this study, the positive effect on bone size could be ascribed to other nutrients contained in whole milk, whereas in the other one, the tested calcium-enriched foods had the same energy, lipid, and protein content as those given to the placebo-group.[26] Note that a very recent trial carried out in 16–18 years old boys indicates that calcium supplementation using a salt other than that contained in milk could also affect bone mass and size as reflected by an increase in standing height.[33] Interestingly this positive effect of calcium carbonate supplementation was associated with an increase in the circulating level of IGF-1.[34,35] These responses to calcium carbonate would suggest that the bone response to calcium supplementation is dependent upon various factors including the spontaneous calcium and protein (see below) intakes, the pubertal maturation as well as the age at the time of the intervention.

It is important to consider whether or not gains resulting from specific interventions are lost after their discontinuation. The answer remains uncertain. It could depend on the type of bone effect which is achieved. As just mentioned, in pre-pubertal girls supplemented with milk calcium phosphate salt, the increase in aBMD appeared to be associated with an increase of bone size;[26] one year after discontinuing the intervention, differences in the gain in aBMD and in the size of some bones were still detectable, at the limit of statistical significance.[26] More recently, we observed that this difference was still present three and half years after discontinuation of the supplementation.[36] These results need additional confirmation by longer term follow-up of the cohort, ideally until PBM has been attained. Nevertheless, they apparently differ from results obtained with other calcium salt supplements. As a matter of fact, calcium given in other forms to pre- or peri-pubertal girls does not appear to modify bone size,[22–25,27] nor to induce a persistent effect after stopping the intervention.[37–39,50,51] This comparative analysis suggests that the effects observed on the aBMD or BMC gain with citrate maltate salts or carbonate alone[22–25] could be related primarily to an increment in the volumetric density resulting from an inhibition of bone remodeling. However, this interpretation is probably too simplistic, since other factors may also play a role in the maintenance of the effect. Thus, as discussed above for the response combining an increase in both bone mass and size, the persistence of the difference in midshaft radius aBMD was also observed 12 and 24 months after the end of intervention in a recent study in boys supplemented with a calcium salt other than that contained in milk.[40]

Despite a positive effect on mean aBMD gain, there is still a wide inter-individual variability in the response to calcium supplementation. It is quite plausible that part of the variability in the bone gain response to calcium supplementation is related to genetic factors. It has been suggested that vitamin D receptor (VDR) gene polymorphisms may modulate the response to calcium supplementation.[41] Other candidate genes could be involved in the variability in the bone response to dietary calcium. Only prospective studies involving initial genotypic or haplotypic stratification followed by randomization in calcium supplemented and placebo groups could assess the functional role, if any, of certain candidate gene polymorphisms in the variability of the bone response.

The possibility that physical activity could modulate the bone response to dietary calcium supplementation during growth has been considered in infants, children and adolescents.[42–45] Overall, the results suggest an interaction, the effect of increased physical activity exerting a positive effect on bone growth to the extent calcium intake is high. At moderately low calcium intake the effect may not be positive. Thus, in a longitudinal study in infants 6–18 months of age, *i.e.*, during rapid bone growth, loading of the skeleton was associated with a reduced increase in total body BMC in the presence of a moderately low calcium intake.[42] In young children aged 3–5 years, either calcium supplement or gross motor activity increased bone mass accrual as compared to either placebo or fine motor activity.[44] Furthermore, the bone response to calcium supplement was greater in children with gross than fine physical activity.[44] In another recent study in 8–9 years old girls, greater gains in bone mass at skeletal loaded sites were observed when moderate exercise was combined with calcium supplementation.[45] Thus, the positive interaction of calcium supplementation and increased physical activity appears to be region specific. This regional specificity suggests that the effect of physical activity alone or combined to calcium supplementation is not merely due to an indirect influence on the energy intake, which in turn would positively affect bone mass acquisition.

## Proteins

Various experimental and clinical observations point to the existence of a relationship between the level of protein intake and either calcium phosphate metabolism or bone mass, or even osteoporotic fracture risk.[46,47] Multiple animal and human studies indicate strongly that low protein intake *per se* could be particularly detrimental to both the acquisition of bone mass and the conservation of bone integrity with aging. Undernutrition, including inadequate supplies of energy and protein during growth can impair bone development severely. Studies in experimental animals indicate that isolated protein deficiency leads to reduced bone mass and strength without histomorphometric evidence of osteomalacia.[48–50] Thus, an inadequate protein supply appears to play a central role in the pathogenesis of the delayed skeletal growth and reduced bone mass that is observed in undernourished children.

Low protein intake could be detrimental to skeletal integrity by lowering the production of IGF-I. Indeed, the hepatic production and plasma concentration of IGF-I are under the influence of dietary protein.[51] Protein restriction was shown to reduce circulating IGF-I by inducing resistance to the hepatic action of growth hormone. In addition, protein restriction appears to decrease the anabolic actions of IGF-I on some target cells. In this regard, it is important to note that growing rats maintained on a low protein diet failed to restore growth when IGF-I was administered at doses sufficient to normalize its plasma concentration.[52]

Variations in the production of IGF-I could explain some of the changes in bone and calcium phosphate metabolism that have been observed in relation to dietary protein intake. Indeed, the plasma level of IGF-I is related closely to the growth rate of the organism. In humans, circulating IGF-I rises progressively from 1 year of age to reach peak values during puberty. As depicted in Figure 7.1, this factor appears to play a key role in calcium phosphate metabolism during growth by stimulating two kidney processes: Pi transport and the production of calcitriol.[15,53] Furthermore, IGF-I is considered as an essential factor for bone longitudinal growth, as it stimulates proliferation and differentiation of chondrocytes in the epiphyseal plate.[54] It also plays a role on trabecular and cortical bone formation. In experimental animals, administration of IGF-I also affects bone mass positively, increasing the external diameter of long bone, probably by enhancing the process of periosteal apposition.[55] Therefore, during adolescence a relative deficiency in IGF-I or a resistance to its action may result in a reduction in the skeletal longitudinal growth, and impaired width- or cross-sectional bone development.

In 'well' nourished children and adolescents, the question arises whether or not variations in the protein intake within the 'normal' range as depicted in Figures 7.5 can influence skeletal growth and thereby modulate the influence of genetic determinants on peak bone mass attainment[56,59,60]. In the relationship between bone mass gain at the lumbar and femoral levels[7,8] and protein intake,[57] it is not surprising to find a positive correlation between these two variables. Like for the calcium intake,[36] the association appears to be particularly significant in pre-pubertal children. Indeed, in healthy pre-pubertal subjects, independently of the intake of calcium, a relatively low protein diet is associated with a reduced gain in aBMD or content (BMC) at both femoral and spinal levels (Figure 7.6). These results suggest that relatively high protein intakes could favor bone mass accrual during childhood. Nevertheless these prospective observational results should not be interpreted as evidence for a causal relationship between bone mass gain and protein intake. Indeed, it is quite possible that protein intake which is related to the overall amount of ingested calories, is to a large extent determined by growth requirements during childhood and adolescence. Only interventional studies testing different levels of protein intakes in otherwise isocaloric diets could eventually determine the quantitative relationship between protein intake and bone mass acquisition during childhood and adolescence. Finally, very recent data suggest that in healthy pre-pubertal boys the response to calcium supplementation can be influenced by the spontaneous protein intake.[58] It is possible that the individual calcium requirement for optimal bone mass accrual could be less at high

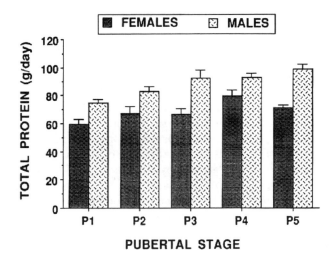

**Figure 7.5** *Protein intakes in relation with pubertal maturation in both female and male subjects aged 9–19 years. Mean protein intakes from dairy and vegetal sources were recorded in two 5-day diet diaries at 1-year interval in 193 healthy adolescents (96 females and 97 males), aged 9–19 years. The method and corresponding macronutrient intakes are described in Ref. 57*

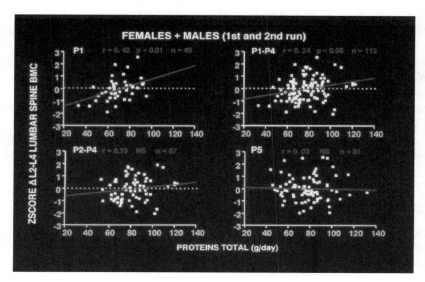

**Figure 7.6** *Relation between protein intakes and change in lumbar BMC in pre-, peri- and post-pubertal female and male adolescents. Mean protein intakes from dairy and vegetal sources were recorded in two 5-day diet diaries in 193 subjects aged from 9 to 19 years (96 females and 97 males) at 1-year interval. The relation with pubertal maturation is presented in Figure 7.5. The BMC data are from Ref. 7. Each dot corresponds to the change in BMC adjusted for age and gender (Z score). A positive correlation was found in pre-pubertal (P1), but neither in peri-pubertal (P2–P4) nor in post-pubertal (P5) subjects*

protein intake. The possible positive interaction between protein and calcium intake deserved to be investigated in the perspective of increasing peak bone mass by modifying bone trophic nutrients.

## Conclusions

Peak bone mass is an important determinant of osteoporotic fracture risk. This explains the current interest in exploring ways to increase PBM as a primary strategy to prevent osteoporosis. Bone mineral mass accumulation from infancy to post-puberty is a complex process involving interrelated actions of genetic, endocrine, mechanical and nutritional factors. From birth to the end of the second decade, the increase in bone mass and strength is due essentially to increments in bone size. Like standing height, bone mineral mass during growth follows a trajectory corresponding to a given percentile or SD from the mean. Nevertheless, this trajectory can be influenced by environmental factors. On the negative side, various chronic diseases and/or their treatment can shift this trajectory downward. On the positive side, and most important in the context of primary prevention of adult osteoporosis, prospective randomized controlled trials strongly suggest that increasing calcium intakes or mechanical loading can shift the age-bone mass trajectory upward. Pre-puberty appears to be an opportune time for obtaining a substantial benefit by increasing the calcium intake and/or physical activity. Further studies should demonstrate that these changes remain substantial by the end of the second decade and thus are translated in a greater PMB. In long-term evaluations of consequences of modifying environments, it will be of critical importance to assess whether any change in densitometric or morphometric bone variables observed at PBM confer a greater resistance to mechanical strain.

## Acknowledgements

Research by the authors of this chapter is supported by the Swiss National Science Foundation (Grants Nos 32-49757.96, 32-58880.99 and 32-58962.99) and by Nestec Ltd, Cerin, Novartis and Institut Candia.

## References

1. J.P. Bonjour, G. Theintz, F. Law, D. Slosman and R. Rizzoli, Peak bone mass, *Osteoporos. Int.*, 1994, **4**(Suppl 1), 7–13.
2. E. Seeman and J.L. Hopper, Genetic and environmental components of the population variance in bone density, *Osteoporos. Int.*, 1997, 7(Suppl 3), S10–S16.
3. B.L. Riggs and L.J. Melton III, *Osteoporosis. Etiology, Diagnosis and Management*, Raven Press, New York, 1988.
4. V. Gilsanz, M.L. Loro, T.F. Roe, J. Sayre, R. Gilsanz and E.E. Schulz, Vertebral size in elderly women with osteoporosis. Mechanical implications and relationship to fractures, *J. Clin. Invest.*, 1995, **95**(5), 2332–2337.
5. V. Naganathan and P. Sambrook, Gender differences in volumetric bone density: a study of opposite-sex twins, *Osteoporos. Int.*, 2003, **14**(7), 564–569.
6. J.P. Bonjour and R. Rizzoli, Bone acquisition in adolescence, In *Osteoporosis*, R.F.D. Marcus, J. Kelsey (eds.), Academic Press, San Diego, CA, 2001, pp. 621–638.
7. G. Theintz, B. Buchs, R. Rizzoli, D. Slosman, H. Clavien, P.C. Sizonenko and J.P. Bonjour, Longitudinal monitoring of bone mass accumulation in healthy adolescents: evidence for a marked reduction after 16 years of age at the levels of lumbar spine and femoral neck in female subjects, *J. Clin. Endocrinol. Metab.*, 1992, **75**(4), 1060–1065.
8. J.P. Bonjour, G. Theintz, B. Buchs, D. Slosman and R. Rizzoli, Critical years and stages of puberty for spinal and femoral bone mass accumulation during adolescence, *J. Clin. Endocrinol. Metab.*, 1991, **73**(3), 555–563.
9. P.E. Fournier, R. Rizzoli, D.O. Slosman and G. Theintz and J.P. Bonjour, Asynchrony between the rates of standing height gain and bone mass accumulation during puberty, *Osteoporos. Int.*, 1997, **7**(6), 525–532.
10. D.A. Bailey, J.H. Wedge, R.G. McCulloch, A.D. Martin and S.C. Bernhardson, Epidemiology of fractures of the distal end of the radius in children as associated with growth, *J. Bone Joint Surg. Am.*, 1989, **71**(8), 1225–1231.

11. L.A. Landin, Fracture patterns in children. Analysis of 8,682 fractures with special reference to incidence, etiology and secular changes in a Swedish urban population 1950–1979, *Acta Orthop. Scand. Suppl.*, 1983, **202**, 1–109.

12. V. Gilsanz, D.T. Gibbens, M. Carlson, M.I. Boechat, C.E. Cann and E.E. Schulz, Peak trabecular vertebral density: a comparison of adolescent and adult females, *Calcif. Tissue Int.*, 1988, **43**(4), 260–262.

13. P.E. Fournier, R. Rizzoli, D.O. Slosman, B. Buchs and J.P. Bonjour, Relative contribution of vertebral body and posterior arch in female and male lumbar spine peak bone mass, *Osteoporos. Int.*, 1994, **4**(5), 264–272.

14. J.P. Bonjour, J. Caverzasio and R. Rizzoli, Homeostasis of inorganic phosphate and the kidney, In *Rickets. vol. 21. Nestlé Nutrition Workshop Series*, Ed. Glorieux, F. H., New York, Raven Press. 1991, pp. 35–46.

15. J. Caverzasio and J.P. Bonjour, IGF-I, a key regulator of renal phosphate transport and 1,25-Dihydroxyvitamine D3 production during growth, *News Physiol. Sci.*, 1991, **6**, 206–210.

16. L.E. Underwood, A.J. D'Ercole and J.J. Van Wyk, Somatomedin-C and the assessment of growth, *Pediatr. Clin. North. Am.*, 1980, **27**(4), 771–782.

17. P. Szulc, E. Seeman and P.D. Delmas, Biochemical measurements of bone turnover in children and adolescents, *Osteoporos. Int.*, 2000, **11**(4), 281–294.

18. D. Teegarden, R.M. Lyle, W.R. Proulx, C.C. Johnston and C.M. Weaver, Previous milk consumption is associated with greater bone density in young women, *Am. J. Clin. Nutr.*, 1999, **69**(5), 1014–1017.

19. C.M. Weaver, Calcium requirements of physically active people, *Am. J. Clin. Nutr.*, 2000, **72**(2 Suppl), 579S–584S.

20. K.S. Wosje and B.L. Specker, Role of calcium in bone health during childhood, *Nutr. Rev.*, 2000, **58**(9), 253–268.

21. V. Matkovic, D. Fontana, C. Tominac, P. Goel and C.H. Chesnut, III, Factors that influence peak bone mass formation: a study of calcium balance and the inheritance of bone mass in adolescent females, *Am. J. Clin. Nutr.*, 1990, **52**(5), 878–888.

22. C.C. Johnston Jr., J.Z. Miller, C.W. Slemenda, T.K. Reister, S. Hui, J.C. Christian and M. Peacock, Calcium supplementation and increases in bone mineral density in children, *N. Engl. J. Med.*, 1992, **327**(2), 82–87.

23. T. Lloyd, M.B. Andon, N. Rollings, J.K. Martel, J.R. Landis, L.M. Demers, D.F. Eggli, K. Kieselhorst and H.E. Kulin, Calcium supplementation and bone mineral density in adolescent girls, *JAMA*, 1993, **270**(7), 841–844.

24. W.T. Lee, S.S. Leung, S.H. Wang, Y.C. Xu, W.P. Zeng, J. Lau, S.J. Oppenheimer and J.C. Cheng, Double-blind, controlled calcium supplementation and bone mineral accretion in children accustomed to a low-calcium diet, *Am. J. Clin. Nutr.*, 1994, **60**(5), 744–750.

25. W.T. Lee, S.S. Leung, D.M. Leung, H.S. Tsang, J. Lau and J.C. Cheng, A randomized double-blind controlled calcium supplementation trial, and bone and height acquisition in children, *Br. J. Nutr.*, 1995, **74**(1), 125–139.

26. J.P. Bonjour, A.L. Carrie, S. Ferrari, H. Clavien, D. Slosman, G. Theintz and R. Rizzoli, Calcium-enriched foods and bone mass growth in prepubertal girls: a randomized, double-blind, placebo-controlled trial, *J. Clin. Invest.*, 1997, **99**(6), 1287–1294.

27. C.A. Nowson, R.M. Green, J.L. Hopper, A.J. Sherwin, D. Young, B. Kaymakci, C.S. Guest, M. Smid, R.G. Larkins and J.D. Wark, A co-twin study of the effect of calcium supplementation on bone density during adolescence, *Osteoporos. Int.*, 1997, **7**(3), 219–225.

28. B. Dibba, A. Prentice, M. Ceesay, D.M. Stirling, T.J. Cole and E.M. Poskitt, Effect of calcium supplementation on bone mineral accretion in Gambian children accustomed to a low-calcium diet, *Am. J. Clin. Nutr.*, 2000, **71**(2), 544–549.
29. S.A. Zamora, R. Rizzoli, D.C. Belli, D.O. Slosman and J.P. Bonjour, Vitamin D supplementation during infancy is associated with higher bone mineral mass in prepubertal girls, *J. Clin. Endocrinol. Metab.*, 1999, **84**(12), 4541–4544.
30. D.J. Barker, Intrauterine programming of adult disease, *Mol. Med. Today*, 1995, **1**(9), 418–423.
31. C. Cooper, C. Fall, P. Egger, R. Hobbs, R. Eastell and D. Barker, Growth in infancy and bone mass in later life, *Ann. Rheum. Dis.*, 1997, **56**(1), 17–21.
32. J. Cadogan, R. Eastell, N. Jones and M.E. Barker, Milk intake and bone mineral acquisition in adolescent girls: randomised, controlled intervention trial, *BMJ*, 1997, **315**(7118), 1255–1260.
33. A. Prentice, S.J. Stear, F. Ginty, S.C. Jones, L. Mills and T.J. Cole, Calcium supplementation increases height and bone mass of 16–18 year old boys, *J. Bone Miner. Res.*, 2002, **17**(Suppl 1), S397.
34. F. Ginty, S.J. Stear, S.C. Jones, D. Stirling, J. Bennett, A. Laidlaw, T.J. Cole and A. Prentice, Impact of calcium supplementation on markers of bone and calcium metabolism in 16–18 year old boys, *J. Bone Miner. Res.*, 2002, **17**(Suppl 1), S178.
35. F. Ginty, A. Prentice, A. Laidlaw, L. McKenna, S. Jones, S. Stear and T.J. Cole, Calcium carbonate supplementatin increases IGF-1 in older adolescent boys and girls, In *Fifth International Symposiuum on Nutritional Aspects of Osteoporosis*, Lausanne, Switzerland, 2003 p. 4.
36. J.P. Bonjour, T. Chevalley, P. Ammann, D. Slosman and R. Rizzoli, Gain in bone mineral mass in prepubertal girls 3.5 years after discontinuation of calcium supplementation: a follow-up study, *Lancet*, 2001, **358**(9289), 1208–1212.
37. W.T. Lee, S.S. Leung, D.M. Leung and J.C. Cheng, A follow-up study on the effects of calcium-supplement withdrawal and puberty on bone acquisition of children, *Am. J. Clin. Nutr.*, 1996, **64**(1), 71–77.
38. W.T. Lee, S.S. Leung, D.M. Leung, S.H. Wang, Y.C. Xu, W.P. Zeng and J.C. Cheng, Bone mineral acquisition in low calcium intake children following the withdrawal of calcium supplement, *Acta Paediatr.*, 1997, **86**(6), 570–576.
39. C.W. Slemenda, M. Peacock, S. Hui, L. Zhou and C.C. Johnston, Reduced rates of skeletal remodeling are associated with increased bone mineral density during the development of peak skeletal mass, *J. Bone Miner. Res.*, 1997, **12**(4), 676–682.
40. B. Dibba, A. Prentice, M. Ceesay, M. Mendy, S. Darboe, D.M. Stirling, T.J. Cole and E.M. Poskitt, Bone mineral contents and plasma osteocalcin concentrations of Gambian children 12 and 24 mo after the withdrawal of a calcium supplement, *Am. J. Clin. Nutr.*, 2002, **76**(3), 681–686.
41. S.L. Ferrari, R. Rizzoli, D.O. Slosman and J.P. Bonjour, Do dietary calcium and age explain the controversy surrounding the relationship between bone mineral density and vitamin D receptor gene polymorphisms?, *J. Bone Miner. Res.*, 1998, **13**(3), 363–370.
42. B.L. Specker, L. Mulligan and M. Ho, Longitudinal study of calcium intake, physical activity, and bone mineral content in infants 6–18 months of age, *J. Bone Miner. Res.*, 1999, **14**(4), 569–576.
43. D. Courteix, Jaffré C., E. Lespessailles and C. Benhamou, Cumulative effects of calcium supplementation and physical activity on bone accretion in prepubertal period: a randomized double-blind placebo-controlled trial, *J. Bone Miner. Res.*, 2002, **17**(Suppl. 1), S136.

44. B. Specker and T. Binkley, Randomized trial of physical activity and calcium supplementation on bone mineral content in 3- to 5-year-old children, *J. Bone Miner. Res.*, 2003, **18**(5), 885–892.

45. S. Iuliano-Burns, L. Saxon, G. Naughton, K. Gibbons and S.L. Bass, Regional specificity of exercise and calcium during skeletal growth in girls: a randomized controlled trial, *J. Bone Miner. Res.*, 2003, **18**(1), 156–162.

46. E.S. Orwoll, The effects of dietary protein insufficiency and excess on skeletal health, *Bone*, 1992, **13**(4), 343–350.

47. R. Rizzoli and J.P. Bonjour, Determinants of peak bone mass and mechanisms of bone loss, *Osteoporos Int.*, 1999, **9**(Suppl 2), S17–S23.

48. P. Ammann, S. Bourrin, J.P. Bonjour, J.M. Meyer and R. Rizzoli, Protein undernutrition-induced bone loss is associated with decreased IGF-I levels and estrogen deficiency, *J. Bone Miner. Res.*, 2000, **15**(4), 683–690.

49. S. Bourrin, P. Ammann, J.P. Bonjour and R. Rizzoli, Dietary protein restriction lowers plasma insulin-like growth factor I (IGF-I), impairs cortical bone formation, and induces osteoblastic resistance to IGF-I in adult female rats, *Endocrinology*, 2000, **141**(9), 3149–3155.

50. S. Bourrin, A. Toromanoff, P. Ammann, J.P. Bonjour and R. Rizzoli, Dietary protein deficiency induces osteoporosis in aged male rats, *J. Bone Miner. Res.*, 2000, **15**(8), 1555–1563.

51. W.L. Isley, L.E. Underwood and D.R. Clemmons, Dietary components that regulate serum somatomedin-C concentrations in humans, *J. Clin. Invest.*, 1983, **71**(2), 175–182.

52. J.P. Thissen, S. Triest, M. Maes, L.E. Underwood and J.M. Ketelslegers, The decreased plasma concentration of insulin-like growth factor-I in protein-restricted rats is not due to decreased numbers of growth hormone receptors on isolated hepatocytes, *J. Endocrinol.*, 1990, **124**(1), 159–165.

53. J. Caverzasio, C. Montessuit and J.P. Bonjour, Stimulatory effect of insulin-like growth factor-1 on renal Pi transport and plasma 1,25-dihydroxyvitamin D3, *Endocrinology*, 1990, **127**(1), 453–459.

54. E.R. Froesch, C. Schmid, J. Schwander and J. Zapf, Actions of insulin-like growth factors, *Annu. Rev. Physiol.*, 1985, **47**, 443–467.

55. P. Ammann, R. Rizzoli, K. Muller, D. Slosman and J.P. Bonjour, IGF-I and pamidronate increase bone mineral density in ovariectomized adult rats, *Am. J. Physiol.*, 1993, **265**(5 Pt 1), E770–E776.

56. S. Ferrari, R. Rizzoli and J.P. Bonjour, Genetic aspects of osteoporosis, *Curr. Opin. Rheumatol.*, 1999, **11**(4), 294–300.

57. H. Clavien, G. Theintz, R. Rizzoli and J.P. Bonjour, Does puberty alter dietary habits in adolescents living in a western society?, *J. Adolesc. Health*, 1996, **19**(1), 68–75.

58. T. Chevalley, S. Ferrari, D. Hans, D. Slosman, M. Fueg, J.P. Bonjour and R. Rizzoli, Protein intake modulates the effect of calcium supplementation on bone mass gain in prepubertal boys, *J. Bone Miner. Res.*, 2002, **17**(Suppl. 1), S172.

59. R. Rizzoli, J.-P. Bonjour and S.L. Ferrari, Osteoporosis, genetics and hormones, *J. Mol. Endocrinol.*, 2001, **26**(2), 79–94.

60. S. Yakar and C.J. Rosen, From mouse to man: redefining the role of insulin-like growth factor-I in the acquisition of bone mass, *Exp. Biol. Med. (Maywood)*, 2003, **228**(3), 245–252.

# Role of Calcium in Maximizing Peak Bone Mass Development

VELIMIR MATKOVIC, ZELJKA CRNCEVIC-ORLIC and
JOHN D. LANDOLL

Osteoporosis Prevention and Treatment Center and Bone and Mineral
Metabolism Laboratory, The Ohio State University, Columbus, Ohio 43210,
The Department of Endocrinology, Medical Faculty, University of Rijeka,
Rijeka, Croatia, Email: matkovic.1@osu.edu

**Key Points**

- Calcium is an essential threshold nutrient.
- Calcium intake thresholds differ with various phases of growth.
- Calcium deficiency during early growth leads to rickets or could be associated with cortical osteopenia and bone fragility fracture during puberty.
- Calcium intake below the requirement level during bone modeling and skeletal consolidation could lower peak bone mass in the population.

## Introduction

During the last 20 years, the focus of nutrition research and recommendations for children and adolescents has shifted from the prevention of nutritional deficiencies to the early establishment of recommended diets to prevent chronic diseases. These priorities may eventually lead to dietary guidelines for the prevention and treatment of those conditions by targeting predisposed individuals early in life.[1] This, in particular, applies to the role of calcium in the primary prevention of osteoporosis. Calcium intake influences bone accretion during growth and ultimately this may affect peak bone mass.

**Table 8.1.1.** *Age of the inflection point for various bone parameters with estimates of bone variable at the inflection point and R-squared of each model (Mean ± SE)*

| ROI | N | $R^2$ | Inflection point age (years) | Estimate of bone variable |
|---|---|---|---|---|
| Height | 234 | 0.692 | 16.25±0.04 | 162.8 cm |
| TBBMC | 231 | 0.624 | 18.33±0.07 | 2432 g |
| TBBMD | 231 | 0.587 | 18.70±0.08 | 1.11 g cm$^{-2}$ |
| Skull | 231 | 0.642 | 21.77±0.12 | 2.29 g cm$^{-2}$ |
| Fem. Neck | 232 | 0.425 | 17.23±0.07 | 1.04 g cm$^{-2}$ |
| Wards | 232 | 0.328 | 18.49±0.09 | 1.02 g cm$^{-2}$ |
| Trochanter | 232 | 0.229 | 16.72±0.12 | 0.86 g cm$^{-2}$ |
| L$_{2-4}$ BMC | 231 | 0.657 | 18.79±0.07 | 48.05 g |
| L$_{2-4}$ BMD | 231 | 0.630 | 18.45±0.07 | 1.18 g cm$^{-2}$ |
| L$_3$ Body Vol. | 228 | 0.583 | 19.18±0.10 | 16.46 cm$^{-3}$ |
| L$_3$ Body BMC | 229 | 0.482 | 20.02±0.10 | 4.20 g |
| L$_3$ Body-LAT. | 229 | 0.274 | 23.97±0.19 | 0.77 g cm$^{-2}$ |
| L$_3$ Body-MID | 229 | 0.237 | 23.15±0.24 | 0.73 g cm$^{-2}$ |
| L$_3$ Body DENS.[a] | 227 | 0.131 | 27.18±0.06 | 0.257 g cm$^{-3}$ |
| Radius | 223 | 0.673 | 17.82±0.07 | 0.65 g cm$^{-2}$ |
| Wrist | 222 | 0.407 | 22.32±0.18 | 0.34 g cm$^{-2}$ |

ROI, regions of interest; TBBMC, total body bone mineral content; TBBMD, total body bone mineral density; BMC, bone mineral content; BMD bone mineral density.
[a]True bone mineral density (g cm$^{-3}$).[5]

Peak bone mass is generally defined as the highest level of bone mass achieved as a result of normal growth. Peak bone mass is important because, together with age-related loss later on, it is one of the two principal factors determining bone mass and fracture resistance late in life.[2] The timing of peak bone mass has been considered by various authors to occur from ages as early as 17–18[3] to as late as 35 years,[4] however, timing of peak bone mass depends on skeletal site[5] (Table 8.1.1). In some reports, it is considered to last for only a brief moment before the decline or age-related loss begins, while in others the peak seems to a plateau, lasting several years. As shown in one study, both patterns are correct, but for different skeletal regions.[5] The results of this study indicated that most of the bone mass is accumulated by late adolescence (~18 years) (Table 8.1.1).

# Calcium and Peak Bone Mass

Calcium is the most common mineral in the human body in which it has almost the same relative abundance as in the Earth's crust. There are actually six stable isotopes of calcium of which [40]Ca is the most common (97%) and [46]Ca the least abundant (0.003%). Calcium is present in variable amounts in all the foods and water we consume, although the main sources are dairy products and vegetables. As 99% of the total body calcium is in the skeleton, the rate of growth and skeletal development have a significant impact on calcium retention, which in turn has a profound effect on the dietary requirement for calcium. Calcium requirement is much higher during skeletal modeling than during the bone consolidation phase when longitudinal bone expansion no longer exists and periosteal bone expansion is at its minimum. During the pubertal growth spurt teenagers may develop a transient osteopenia due to increased intracortical porosity,[6] presumably resulting from a high demand for calcium. Endosteal bone apposition with concomitant increase in bone mineral density proceeds during the bone consolidation phase. However, this process requires less calcium as compared to the bone modeling phase with a concomitant decline in calcium requirement.

Because calcium is a ubiquitous element, it has been difficult to determine if calcium deficiency diseases exist; this particularly applies to growth. Except for an isolated report that calcium deficiency can cause rickets,[7] low calcium intake may not have any immediate deleterious effect on the skeleton of young persons. Some studies indicate that calcium intake may play a role in the prevention of bone fragility fractures during growth.[8,9] However, the definitive data to support the above observations are still lacking. In a study in Palma de Mallorca, Spain, a significant difference in the fracture rate was found when cities with a high calcium content in their water (282 mg $L^{-1}$) were compared with those with a lower calcium content (86 mg $L^{-1}$).[10] The peak incidence of the forearm fractures in children coincides with the pubertal growth spurt and, presumably, osteopenia of growth.[11] The association between calcium intake and bone fragility fractures in children ought to be confirmed due to the growing number of teenagers in the United States. This segment of the population is expected to increase steadily during the following two decades; the total number of teenagers is expected to reach 30.8 million by the year 2010 (U.S. Census, 1994). Secondary to this change the number of fractures specific for puberty is expected to increase as well.

It is possible, however, that calcium deficiency during skeletal formation could decrease the peak bone mass and therefore increase the risk of fracture later in life. This was first shown in the ecological study from Croatia where two rural communities with a high level of physical activity but with different dietary habits over a lifetime were compared.[12] The study found a difference in bone mass and hip fracture rates. It appeared that both populations were losing bone with age at about the same rate, but those who started with more bone, ended up having a higher bone mass and lower incidence of hip fractures. The results of this research directly supported the hypothesis of Newton John and Brian Morgan[13] for the pathogenesis of osteoporosis and fractures in the population. The differences in bone mass and fracture rates were attributed primarily to calcium intake. The differences in bone mass between the communities were established at an early age (30 years), implying that if calcium intake is important, it may be during growth and skeletal development that it has its greatest impact. Results of a similar ecological study conducted in China on populations accustomed to different calcium intakes[14] confirmed the above finding and reiterated the importance of adequate nutrition for peak bone mass. Positive relations between the bone mineral density of adult women and their milk consumption in childhood, adolescence, or throughout life, found in several retrospective studies[15–19] further strengthen the hypothesis. Overall, it is likely that variations in calcium nutrition early in life may account for as much as a 5–10% difference in peak adult bone mass. Such a difference, although small, probably contributes to more than 25–50% of the difference in the hip-fracture rate later in life.[12]

All of the above-mentioned studies conducted in adults were unable to determine which developmental phase is the most important for the accumulation of maximal bone mass with regards to nutritional needs. This could be either the period of childhood, adolescence, or young adulthood. The assumption is that dietary calcium intake at/or above the threshold level is necessary throughout the entire bone modeling and consolidation phase (from childhood to young adulthood) if genetically predetermined peak bone mass is expected to be reached. Several clinical trials indicated that children and teenagers particularly, may benefit from higher calcium intake with further gain in bone mass.[20–29] However, all of the intervention studies either with calcium supplements or dairy products in children and adolescents completed to date were relatively short in duration (1–3 years) to address the issue of bone tissue adaptation to nutritional challenge leading to peak bone mass. The increase in bone mass observed in those studies could be explained to a large extent by the remodeling transient phenomenon.[30]

According to Heaney,[30] an intervention study should be long enough and of adequate sample size to allow for the assessment of the slope of the rate of change (gain) in the measured bone variable after the initial effect on the bone remodeling transient is expected to be completed (up to 40–60 weeks). In the continuation of the twin study[21] during the 4th year (calcium withdrawn, codes released) the difference in bone mass between the groups has been diminishing,[31] indicating that some of the gain has been lost as the result of the second transient. Similar findings were observed in two other studies.[32,33] Another contributing factor for the lack of difference between the groups could be due to a natural tendency of the placebo group to increase calcium intake to 'catch-up with skeletal mineralization'. It is, therefore, unclear from these short-term studies if a positive effect of a surplus intake of calcium on bone tissue can be maintained. In other studies, the effects of intervention were maintained 1–3 years after discontinuation of treatment.[29,34] This may be specific to the calcium source and/or the level of habitual dietary calcium intake in the population.

## Skeletal Development and Calcium Requirement

Longitudinal bone growth and the periosteal bone expansion are the driving forces in bone mass acquisition during puberty and under the influence of strains imposed on the bone tissue from muscle action and loads. The endosteal apposition seems to be of a secondary importance early in puberty and it could be sacrificed to accomplish the expansion of the bone volume.[6] If calcium intake is far below the calcium intake threshold required for optimal skeletal development, a permanent deficit in peak bone mass is expected. This could primarily be due to inadequate development of bone structures (thin cortex and/or reduced formation of trabecular network) within the expanding bone as an organ. Borderline calcium deficiency during growth may not result in the permanent deficit in bone mass at skeletal maturity (age 18 years) due to its effect on bone remodeling and the potential for recovery with the decline in calcium requirement associated with bone consolidation.[35] A long-term intervention study with calcium supplementation extending from childhood to young adulthood at various calcium intake levels seems to be mandatory to answer these important questions.

Calcium needs are greater during adolescence than in either childhood or young adulthood when comparing the maximal calcium retentions in the body. This results from the high velocity of growth during the peak of puberty as well as skeletal consolidation by late adolescence. These combined skeletal parameters exert a major influence on calcium requirements during adolescence. Puberty is an intensely anabolic period, with increases in height and weight, alterations in body composition resulting from increased lean body mass, changes in the quantity and distribution of body fat, and enlargement of many organ systems besides the skeleton. To adequately judge the requirements for calcium during this period of life the following features of growth must be addressed: the intensity and extent of the pubertal growth spurt; sexual differences in the timing of peak growth (boys develop later); and individual variations in the timing of the pubertal growth spurt.[36] All these aspects of growth probably have a profound effect on calcium and bone metabolism.[37-40] The growth spurt during adolescence contributes about 15% to adult height, and about 37% to the total body bone mineral content, including the spine.[5] The average gain in height is dramatically out-paced by the gain in total body calcium, as a result of periosteal expansion and endosteal apposition of bone. Therefore, it is easy to conclude that nutrition plays a very important role in the doubling of body mass during puberty as well as in peak bone mass formation. Since protein and calcium requirements are closely related to the rapid increase in body mass as well as in skeletal mass, it is no surprise that peak nutritional requirements appear to occur during the years of maximum growth.[36] Although both adolescent males and females gain a significant amount of bone mineral, there are probably marked sex differences with respect to bone modeling and skeletal consolidation. By the age of 10 years, the mean height velocity is 5.5 cm year$^{-1}$ in girls, and it increases to an average peak of 8.5 cm year$^{-1}$ by the age of 12. Peak height velocity for boys starts at 12 years of age (5 cm year$^{-1}$) and reaches a maximum by the age of 14 (9.5 cm year$^{-1}$). Mean height velocity is close to zero by the age of 16 in girls, and by the age of 17 in boys.[41] By the age of 16 years, girls accumulate more than 90% of the bone mass of their premenopausal mothers.[20] In a cross-sectional study of the timing of peak bone mass in Caucasian females, peak adult height was present in 16-year old individuals, with a delay in peak bone mass from 1–4 years depending on the skeletal location.[5] Similar findings were obtained in a longitudinal study among adolescents from Geneva.[42] This delay between the cessation of longitudinal bone growth and the time period when

**Table 8.1.2.** *Total body calcium (TBCa) accumulation between ages 11 and 12 years in 93 teenage females during pubertal growth spurt*[58]

|           | TBCa accumulation (g year$^{-1}$) | TBCa accumulation (mg day$^{-1}$) |
|-----------|-----------------------------------|-----------------------------------|
| Mean      | $109 \pm 51$[a]                   | 299                               |
| Minimum   | 20                                | 55                                |
| Maximum   | 313                               | 858                               |

[a]Mean$\pm$SD.

most of the skeletal mass will be reached indicates ongoing skeletal consolidation by the process of endosteal apposition.[3,20,43–47] About 37% of the total bone mass could be attained between pubertal stages 2 and 4 and about 10–12% of adult peak bone mass could be accumulated in just one year during growth spurt (Table 8.1.2). The average annual change in stature between the ages 8–16 is 4 cm, which is ~2.4% of the peak adult height for women. The accumulation of total bone mineral between the age of 8 and 18 years is about 146 g year$^{-1}$, or 6% of the total body mineral of 2432 g. When this is translated into total body calcium (assuming 39% Ca in the hydroxyapatite crystal), 58 g of calcium is required for annual accretion, or 160 mg day$^{-1}$, to reach an average maximal total body calcium content of 949 g.[5] Skeletal calcium accretion almost doubles during the pubertal growth spurt.[48] Young females who mature from pubertal stage 2–4 over 12 months are capable of accumulating on average 128 g of calcium for which a positive calcium balance (excluding skin losses) of ~350 mg day$^{-1}$ is required.[49] The daily peak increment of calcium during the growth spurt is greater, occurs later, and lasts longer in boys than in girls. Such rapid skeletal modeling requires optimal intake and utilization of all bone minerals. High calcium absorption during rapid skeletal modeling is mediated by calcitriol $(1,25(OH)_2D_3)$.[50]

During young adulthood (ages 18–30 years) bones are no longer growing along the longitudinal axis. The cycle of bone resorption and formation continues at a slower rate than during adolescence. There is small but continuous periosteal envelope expansion with age, which can contribute to the net positive bone tissue balance at most of the skeletal locations, including the spine, forearm, shaft of the femur, metacarpal bones, and the skull. In addition, there is also a minimal endosteal apposition of bone mass by the beginning of young adulthood indicating continuing consolidation of skeletal mass.[5] The above events contribute to the bone mineral gain in young adult women during the third decade of life.[4] This gain was in the range of 4.8% for the forearm, 6% for the lumbar spine, and 12.5% for the total body. In this 5-year longitudinal observational study, dietary calcium (calcium/protein ratio) and physical activity strongly influenced mineral apposition. When this gain in the total body bone mineral content is translated into calcium content, approximately 150 g of calcium is being incorporated into the skeleton during this time interval. Net skeletal calcium accretion during this period could therefore be about $+41$ mg day$^{-1}$, a figure close to the one obtained from metabolic balance studies.[37] This indicates that a window of opportunity for the restoration of skeletal mass in young adults under adequate nutrition is possible, but seems to be closing by the age of 30 years.[4] The period of young adulthood is characterized by the decline in calcium absorption and calcium retention with a concomitant increase in calcium output.[37,39,51] An analysis of calcium balances showed that across all intake ranges from 500 to 1600 mg day$^{-1}$, calcium absorption and retention were substantially lower in young adults than in children and adolescents. On the other hand, calcium output (urine and faeces) was the highest. Young adults are in positive calcium balance of about 100 mg day$^{-1}$. When skin losses of about 60 mg day$^{-1}$ (see Ref. 52) are subtracted from this amount, this leaves 40 mg of calcium for bone accretion. To provide this small amount of extra calcium for bone building and to match higher urinary calcium output, with the body's decreasing ability to absorb it, calcium intake among young adults should be at or above the recently established threshold of 960 mg day$^{-1}$.[51] The current dietary reference intakes for calcium for young adults in age range 19–30 years are set up at 1000 mg day$^{-1}$.

Surveys in the US reveal that adolescent girls are less likely to meet the current recommended dietary levels for calcium than are teenage boys, whose intake is higher and comes close to achieving recommended intakes. By contrast, calcium intake in girls starts to decline at the time of puberty.[53] Actually, the ratio of calcium intake and the potential for maximal calcium retention is lowest for adolescent females, indicating a status of relative calcium deficiency. The problem is further compounded by individual variability in body size and pubertal development, making age and sex alone poor predictors of individual calcium needs. This could be partially corrected by using skeletal age and/or pubertal staging markers of sexual development. Of all the anthropometric variables skeletal age emerged as the most significant predictor for almost all of the bone mass parameters.[54] Since skeletal age is relatively easy to assess and requires a relatively low-cost technique, it could have implications in the assessment of bone mineral status in the young population; this is particularly important for clinical trials and fracture studies during growth.

## Calcium Intake Thresholds

Calcium is an essential threshold nutrient. An intake threshold applies to the level below which skeletal accumulation is a function of intake, and above which skeletal accumulation is constant, irrespective of further increases in intake. This indicates that retained calcium below the threshold level cannot saturate the skeletal mass, while maximal saturation can be achieved at the threshold level and above. At fully adequate intakes, bone deposited would depend not upon intake, but upon what was programmed into the growth process; hence further increases in intake would not result in further skeletal retention. Thus, below a certain threshold level of dietary calcium intake, young persons will not be able to reach genetically predetermined peak bone mass, while adults will be losing bone tissue at a faster rate than is necessary.

**Table 8.1.3.** *Calcium intake thresholds and balances for growing individuals* [51]

| Age group (years) | Threshold intake ($mg \ day^{-1}$) | Threshold balance ($mg \ day^{-1}$) |
|---|---|---|
| 0–1 | 1090 | +503 |
| 2–8 | 1390 | +246 |
| 9–17 | 1480 | +396 |
| 18–30 | 957 | +114 |

A recent analysis of calcium balances during various stages of human development indicated that threshold behavior occurs in humans.[51] This is illustrated in Table 8.1.3 where the data for threshold intakes are presented for each age group from childhood to young adulthood. Net positive bone formation contributes to the constant demand for calcium throughout the developmental process. There is a significant positive relationship between calcium intake and retention for all growing individuals. Higher calcium intakes are associated with greater calcium retention in the body and lower calcium intakes, in spite of the relatively high absorption efficiency during growth, are associated with lower calcium retention in the body.

Attainment of peak bone mass requires maximal positive balance between calcium intake and obligatory losses of calcium in feces, urine, and sweat. If we assume that normal children do not have calcium malabsorption, then dietary calcium intake and urinary excretion of calcium seem to be the most important determinants of calcium retention in the body. During the period of most rapid growth, calcium excretion in the urine is practically unrelated to calcium intake.[20,37] A simple linear regression of urinary calcium on calcium intake in 381 young teenage females shows a weak positive association ($p < 0006$), with $R^2$ of 2.9% (linear equation of the relationship is: $Ca_{urine} = 0.02252 * Ca_{intake} + 1.5261$[55]. In a smaller group of adolescents on controlled diets over a range from 800 to 2100 mg $day^{-1}$, calcium intake explained only 6% of the variability in urinary calcium.[56] Body weight and age seem to be the principal determinants of urinary calcium excretion during early growth. A stronger relationship between urinary calcium and calcium intake exists during childhood when the rate of growth declines[37] and is definitely more pronounced in adults.[57] The explanation is that rapidly growing individuals retain the absorbed calcium in the skeleton rather than excreting it in the urine. Urinary calcium is only expected to rise after the skeletal compartment is being saturated with calcium at intakes at/above the threshold level with the concomitant increase in the filtered load of calcium. A recent study of urinary calcium excretion in 484 young teenage females provided a large enough data set to reevaluate the calcium intake threshold concept based on the urinary calcium–calcium intake relationship. The slope for the regression line for the subset of data below the calcium intake threshold (1480 mg $day^{-1}$) was 17 mg of urinary calcium per 1 g of calcium intake, while the regression line above the calcium intake threshold had a slope of 33 mg urinary calcium per 1 g of extra calcium intake. The same is also indicated by the boxplots of urinary calcium excretion (mg day$^{-1}$) according to the quintiles of dietary calcium intake. It is evident that there is a more drastic increase in urinary calcium excretion in quintile 5; calcium intake above 1500 mg day$^{-1}$. The above analysis reaffirms the threshold concept for calcium nutrition for adolescents. At ~1500 mg day$^{-1}$ dietary calcium intake, the skeleton of an average teenager will be saturated with calcium; after that set point there will be no further increase in skeletal retention of calcium and urinary calcium will start to rise more rapidly.[58]

## Calcium and other Nutrients

Renal excretion of calcium is believed to be regulated by hormones (PTH and estrogens) and is also influenced by dietary protein and sodium. High intake of salt increases the obligatory calcium loss in the urine in adults, as well as in children. There is a substantially stronger relationship between urinary sodium and urinary calcium ($R^2 = 17.4\%$). The linear equation for the regression of urinary calcium on urinary sodium is: $Ca_{urine} = 0.01154 \times Na_{urine} + 0.823$. For 100 mmol of sodium excreted there will be 1.15 mmol of calcium excreted.[55] The same relationship between the two elements exists in adults.[59] Low calcium intake and high obligatory calcium loss in the urine, potentiated by sodium intake during growth may reduce calcium retention in the body with a concomitant reduction in peak bone mass. This suggests that the calcium requirement for adolescents may be influenced by salt intake. The diet of a typical young American is low in calcium and high in sodium.[55]

Vitamin D deficiency causes rickets in children and osteomalacia in adults, and also is related to calcium malabsorption and secondary hyperparathyroidism in the very elderly. Seasonal deficiency of vitamin D has been recently described in children from Spain with a corresponding increase in serum parathyroid hormone (PTH).[60] Out of 51 normal children studied in winter 31% had levels of 25-hydroxyvitaminn D [25(OH)D] below 12 ng mL$^{-1}$, and 80% had levels lower than 20 ng mL$^{-1}$. To what extent this may have implications with regard to skeletal mineralization and peak bone mass acquisition is unknown. Vitamin D deficiency is usually the result of inadequate exposure to sunlight and/or low dietary intake of vitamin D. Vitamin D facilitates calcium absorption from the diet. It stimulates active transport by inducing the synthesis of calcium-binding protein in intestinal mucosal cells. This function is particularly important for adaptation to low intakes. Most of the absorbed calcium, however, comes from the passive transport mechanism, which is not completely understood, and is not dependent on vitamin D.[61] The proportion of absorption by the two mechanisms varies with intake; at high calcium intakes it is likely that active transport contributes relatively little to the total absorbed load. Nevertheless, vitamin D status can influence absorptive performance and influence calcium requirement.[61] To meet these high calcium requirements adolescents have higher calcium absorption than children and young adults.[37,40,45] Such a high calcium absorption during these developmental phases is mediated by calcitriol.[50]

Phosphorus is also very important for bone health. Phosphate contributes about 50% of the weight of bone mineral and therefore must be present in adequate quantities in the diet, both to mineralize and to maintain the skeleton.[61] Phosphorus is present in relatively adequate quantities in the U.S. diet, and most of the recent concern of the nutrition community has centered around whether its presence in the diet might be excessive, with consequent development of secondary hyperparathyroidism which could potentiate bone loss. High phosphate intake and abnormally low calcium to phosphate ratio (*i.e.* 1:6) was implicated in the development of osteoporosis and secondary hyperparathyroidism in laboratory animals; however, this was due to a low calcium content of the experimental diets rather than the result of phosphorus surplus. In one study, an increase in the consumption of carbonated beverages has been shown to produce an increased incidence of fractures in adolescents, but this requires further confirmation.[62] Soft drinks have a tendency to displace more nutritious beverages (*e.g.* milk) from the diet.

Protein-calorie malnutrition during childhood can cause growth retardation and decreased formation of cortical bone,[63] and therefore can interfere with peak bone mass acquisition. This is probably mediated by IGF-I and leptin through its effect on reproductive function. Serum IGF-I is considered a biochemical marker of nutritional status, primarily protein intake. Excessive protein intake and increase in protein consumption above the recommended allowance level can be associated with hypercalciuria.[61] This, however, has not been confirmed in children who are in positive nitrogen balance reflective of protein synthesis during growth.[55]

## Primary Prevention Strategy

Since the optimal calcium intake during growth is important in maximizing an individual's genetic potential for peak bone mass, two leading national institutions in this country (National Science Foundation and National Institutes of Health) increased the daily calcium intake recommendations for individuals between the age of 11 and 24 years to 1300—1500 mg range.[64,65] However, adolescence is a critical and complex developmental period in which major biological, social, psychological, and cognitive changes occur. Those changes can influence teenagers' nutritional needs and status. Psychological changes involving the adolescent's search for independence and identity, desire for acceptance by peers, and preoccupation with physical appearance may affect eating habits, food choices, nutrient intake, and ultimately nutritional status. As young females reach their teen years most of them drink less milk due to fear of gaining weight, which leads to corresponding decline in calcium intake. For those who continue taking dairy products there is a concern that this drive for extra calcium intake may on the other hand perpetuate problems related to obesity and atherosclerosis.[1] The results of a 4-year longitudinal study of body composition in the group of young females who have been meeting their calcium intake standards from milk and dairy products as compared to those on calcium supplements, do not support the above view.

In fact, the dairy group did not gain more weight, increased BMI, or % body fat. In addition, there was no difference in serum leptin between the groups. The trend was opposite with lower accumulation of body fat in individuals who opted to drink more milk.[66] This suggests the importance of stimulating dietary habits of drinking milk early in childhood with constant encouragement during critical years of adolescence. For those who cannot meet dietary calcium intake standards through the consumption of dairy products, the alternative is to supplement the diets with calcium tablets. This will help to fulfill the goal of the primary prevention of osteoporosis among the young American as stressed by the Health People Act 2000.[67]

There is concern that such a high calcium intake may interfere with the utilization of other nutrients such as magnesium, zinc, and iron, which are also important for growth. Recent studies conducted in children showed that increasing the dietary calcium intake standards up to 1500 mg day$^{-1}$ does not adversely affect magnesium, zinc, selenium, and iron metabolism.[68–71]

# Conclusions

Sufficient exercise during childhood and adolescence, particularly during the prepubertal years, is more effective for increasing bone mass and strength than exercise in adulthood.[72] Its interaction with calcium could be even more important. Calcium seems to play a permissive role with regard to bone benefits of exercise.[73] The results of a recent intervention study with exercise and calcium indicate that skeletal sites prone to mechanical loading respond differently to calcium supplementation than non-loaded sites.[74] Exercise, therefore, in conjunction with good nutrition and calcium intake should play a powerful role with regard to primary prevention of osteoporosis.

# Acknowledgement

Supported in part by NIH RO1 AR40736-01A1, CRC-NIH M01-RR00034 and NRICGP/USDA.

# References

1. G. Miller, J.K. Jarvis and L.D. McBean, In *Handbook of Dairy Foods and Nutrition*, CRC Publ. Inc., Boca Raton, Florida, 1995.
2. R.P. Heaney and V. Matkovic, Inadequate peak bone mass, In *Osteoprosis: Etiology, Diagnosis and Management*, 2nd edn, B.L. Riggs and L.J. Melton (eds.), Raven Press, New York, 1995, 115–131.
3. J.P. Bonjour, G. Theintz, B. Buchs, D. Slosman and R. Rizzoli, Critical years and stages of puberty for spinal and femoral bone mass accumulation during adolescence, *J. Clin. Endocrinol. Metab.*, 1991, **73**, 555–563.
4. R.R. Recker, K.M. Davies, S.M. Hinders, R.P. Heaney, M.R. Stegman and D.B. Kimmel, Bone gain in young adult women, *JAMA*, 1992, **268**, 2403–2408.

5. V. Matkovic, T. Jelic, G.M. Wardlaw, J.Z. Ilich, P.K. Goel *et al.*, Timing of peak bone mass in Caucasian females and its implication for the prevention of osteoporosis. Inference from a cross-sectional model, *J. Clin. Invest.*, 1994, **93**, 799–808.

6. A.M. Parfitt, The two faces of growth: benefits and risks to bone integrity, *Osteoporosis Int.*, 1994, **4**, 382–398.

7. J.M. Pettifor, F.P. Ross, R. Travers, F.H. Glorieux and H.F. DeLuca, Dietary calcium deficiency; a syndrome associated with bone deformities and elevated serum 1,25-dihydorxyvitamin D concentrations, *Metab. Bone Rel. Res.*, 1981, **2**, 301–305.

8. G.M. Chan, M. Hess, J. Hollis and L.S. Book, Bone mineral status in childhood accidental fractures, *Am. J. Dis. Child.*, 1984, **139**, 569–570.

9. A. Goulding, R. Cannan, S.M. Williams, E.J. Gold, R.W. Taylor and N.J. Lewis-Barned, Bone mineral density in girls with forearm fractures, *J. Bone Miner. Res.*, 1998, **13**, 143–148.

10. S. Verd Vellespir, J. Dominguez Sanches, M. Gonzales Quintial, M. Vidal Mas, A.C. Soler Mariano *et al.*, Asociacion entre el contenido en calcio de las aguas de consumo y las fracturas en los ninos, *An. Esp. Pediatr*, 1992, **37**, 461–465.

11. D.A. Bailey, J.H. Wedge, R.G. McCulloch, A.D. Martin and S.C. Benhardson, Epidemiology of fractures of the distal end of the radius in children as associated with growth, *J. Bone Joint. Surg.*, 1989, **71A**, 8, 125–130.

12. V. Matkovic, K. Kostial, I. Simonovic, R. Buzina, A. Brodarec and B.E.C. Nordin, Bone status and fracture rates in two regions of Yugoslavia, *Am. J. Clin. Nutr.*, 1979, **32**, 540–549.

13. H.F. Newton-John and B.D. Morgan, The loss of bone with age: osteoporosis and fractures, *Clin. Orthop.*, 1970, **71**, 229–232.

14. J.F. Hu, X.H. Zhao, J.B. Jia, B. Parpia and T.C. Campbell, Dietary calcium and bone density among middle-aged and elderly women in China, *Am. J. Clin. Nutr.*, 1993, **58**, 219–227.

15. R.B. Sandler, C. Slemenda, R.E. LaPorte, J.A. Cauley, M.M. Schramm *et al.*, Postmenopausal bone density and milk consumption in childhood and adolescence, *Am. J. Clin. Nutr.*, 1985, **42**, 270–274.

16. J.A. Cauley, J.P. Gutai, L.H. Kuller, D. LeDonne, R.B. Sandler *et al.*, Endogenous estrogen levels and calcium intakes in postmenopausal women. Relationships with cortical bone measures, *JAMA*, 1988, **260**, 3150-3155.

17. L. Halioua and J.J.B. Anderson, Lifetime calcium intake and physical activity habits: independent and combined effects on the radial bone of healthy premenopausal Caucasian women, *Am. J. Clin. Nutr.*, 1989, **49**, 534–541.

18. S. Murphy, K.T. Khaw, H. May and J.E. Compston, Milk consumption and bone mineral density in middle aged and elderly women, *BMJ*, 1994, **308**, 939–941.

19. S. Soroko, T.L. Holbrook, S. Edelstein and E. Barrett-Connor, Lifetime milk consumption and bone mineral density in older women, *Am. J. Public Health*, 1994, **84**, 1319–1322.

20. V. Matkovic, D. Fontana, C. Tominac, P. Goel and C.H. Chesnut, Factors which influence peak bone mass formation: a study of calcium balance and the inheritance of bone mass in adolescent females, *Am. J. Clin. Nutr.*, 1990, **52**, 878–888.

21. C.C. Johnston, Jr., J.Z. Miller, C.W. Slemenda, T.K. Reister, S. Hui *et al.*, Calcium supplementation and increases in bone mineral density in children, *N. Engl. J. Med.*, 1992, **327**, 82–87.

22. T. Lloyd, M.B. Andon, N. Rollings, J.K. Martel, R.J. Landis *et al.*, Calcium supplementation and bone mineral density in adolescent girls, *JAMA*, 1993, **270**, 841–844.

23. W.T.K. Lee, S.S.F. Leung, S.F. Wang, Y.C. Xu, W.P. Zeng *et al.*, Double-blind, controlled calcium supplementation and bone mineral accretion in children accustomed to a low-calcium diet, *Am. J. Clin. Nutr.*, 1994, **60**, 744–750.

24. G.M. Chan, K. Hoffman and M. McMurray, Effect of dairy products on bone and body composition in pubertal girls, *J. Pediatr.*, 1995, **126**, 551–556.

25. J.P. Bonjour, A.L. Carrie, S. Ferrarri, H. Clavien, D. Slosman *et al.*, Calcium-enriched foods and bone mass growth in prepubertal girls: a randomized, double-blind, placebo-controlled, trial, *J. Clin. Investig.*, 1997, **99**, 1287–1294.

26. J. Cadogan, R. Eastell, N. Jones and M.E. Barker, Milk intake and bone mineral acquisition in adolescent girls: randomised, controlled intervention trial, *BMJ*, 1997, **315**, 1255–1260.

27. C.A. Nowson, R.M. Green, J.L. Hopper, A.J. Sherwin, D. Young *et al.*, A co-twin study of the effect of calcium supplementation on bone density during adolescence, *Osteoporosis Int.*, 1997, **7**, 219–225.

28. B. Dibba, A. Prentice, M. Ceesay, D.M. Stirling, T.J. Cole and E.M.E. Poskitt, Effect of calcium supplementation on bone mineral accretion in Gambian children accustomed to a low-calcium diet, *Am. J. Clin. Nutr.*, 2000, **71**, 544–549.

29. M.J. Merrilees, E.J. Smart, N.L. Gilchrist, C. Frampton, J.G. Turner *et al.*, Effects of dairy food supplements on bone mineral density in teenage girls, *Eur. J. Nutr.*, 2000, **39**, 256–262.

30. R.P. Heaney, Interpreting trials of bone-active agents, *Am. J. Med.*, 1995, **98**, 329–330.

31. C. Slemenda, M. Peacock, S. Hui, L. Zhou and C.C. Johnston Jr., Reduced rates of skeletal remodeling are associated with increased bone mineral density during the development of peak skeletal mass, *J. Bone Miner. Res.*, 1997, **12**, 676–682.

32. W.T.K. Lee, S.S.F. Leung, D.M.Y. Leung and J.C.Y. Cheng, A follow-up study on the effect of calcium-supplement withdrawal and puberty on bone acquisition of children, *Am. J. Clin. Nutr.*, 1996, **64**, 71–77.

33. T. Lloyd, N. Rollings, M.B. Andon, D.F. Eggli, E. Mauger and V. Chinchilli, Enhanced bone gain in early adolescence due to calcium supplementation does not persist in late adolescence, *J. Bone Miner. Res.*, 1996, **11**, S154.

34. B. Dibba, A. Prentice, M. Ceesay, M. Mendy, S. Darboe *et al.*, Bone mineral contents and plasma osteocalcin concentrations of Gambian children 12 and 24 mo after the withdrawal of a calcium supplement, *Am. J. Clin. Nutr.*, 2002, **76**, 681–686.

35. V. Matkovic, N.E. Badenhop-Stevens, J.D. Landoll, P. Goel and B. Li, Long Term effect of calcium supplementation and dairy products on bone mass of young females, *J. Bone Miner. Res.*, 2002, **17**, S172.

36. E.J. Gong and F.P. Heald, Diet, nutrition, and adolescence, In *Modern Nutrition in Health and Disease*, 8th edn, M.E. Shils, J.A. Olson and M. Shike, (eds.), Lea & Febiger, Philadelphia, 1994, 759–769.

37. V. Matkovic, Calcium metabolism and calcium requirements during skeletal modeling and consolidation of bone mass, *Am. J. Clin. Nutr.*, 1991, **54**, S245–S260.

38. A. Blumsohn, R.A. Hannon, R. Wrate, J. Barton, A.W. Al-Dehaimi *et al.*, Biochemical markers of bone turnover in girls during puberty, *Clin. Endocrinol.*, 1994, **40**, 663–670.

39. C.M. Weaver, B.R. Martin and M. Peacock, Calcium metabolism in adolescent girls, *Challenges Mod. Med.*, 1995, **7**, 123–128.

40. S.A. Abrams and J.E. Stuff, Calcium metabolism in girls: current dietary intakes lead to low rates of calcium absorption and retention during puberty, *Am. J. Clin. Nutr.*, 1994, **60**, 739–743.

41. J.M.H. Buckler, *A Reference Manual of Growth and Development*, Blackwell Scientific Publications, Oxford, 1981.

42. P.E. Fournier, R. Rizzoli, D.O. Slosman, G. Theintz and J.P. Bonjour, Asynchrony between the rates of standing height gain and bone mass accumulation during puberty, *Osteoporosis Int.*, 1997, **7**, 525–532.

43. C. Glastre, P. Braillon, L. David, P. Cochat, P.J. Meunier and P.D. Delmas, Measurement of bone mineral content of the lumbar spine by dual energy X-ray absorptiometry in normal children: correlations with growth parameters, *J. Clin. Endocrinol. Metab.*, 1990, **70**, 1330–1333.

44. G. Theintz, B. Buchs, R. Rizzoli, D. Slosman, H. Clavien *et al.*, Longitudinal monitoring of bone mass accumulation in healthy adolescents: evidence for a marked reduction after 16 years of age at the levels of lumbar spine and femoral neck in female subjects, *J. Clin. Endocrinol. Metab.*, 1992, **75**, 1060–1065.

45. C.M. Weaver, B.R. Martin and K.L. Plawecki, Differences in calcium metabolism between adolescent and adult females, *Am. J. Clin. Nutr.*, 1995, **61**, 577–581.

46. C.M. Weaver, M. Peacock, B.R. Martin, K.L. Plawecki and G.P. McCabe, Calcium retention estimated from indicators of skeletal status in adolescent girls and young women, *Am. J. Clin. Nutr.*, 1996, **64**, 67–70.

47. S.A. Abrams, K.O. O'Brien and J.E. Stuff, Changes in calcium kinetics associated with menarche, *J. Clin. Endocrinol. Met.*, 1996, **81**, 2017–2020.

48. A.D. Martin, D.A. Bailey, H.A. McKay and S. Whiting, Bone mineral and calcium accretion during puberty, *Am. J. Clin. Nutr.*, 1997, **66**, 611–615.

49. M. Skugor, J.Z. Ilich, N.E. Badenhop, J.D. Landoll, L.A. Nagode and V. Matkovic, Influence of puberty on body composition and bone mass over a 1-year period, *J. Bone Miner. Res.*, 1997, **12**, S252.

50. J.Z. Ilich, N.E. Badenhop, T. Jelic, A.C. Clairmont, L.A. Nagode and V. Matkovic, Calcitriol and bone mass accumulation in females during puberty, *Calcif. Tissue Int.*, 1997, **61**, 104–109.

51. V. Matkovic and R.P. Heaney, Calcium balance during human growth: evidence for threshold behavior, *Am. J. Clin. Nutr.*, 1992, **55**, 992–996.

52. P. Charles, F. Taagehoj Jensen, L. Mosekilde and H. Hvid Hansen, Calcium metabolism evaluated by 47Ca kinetics: estimation of dermal calcium loss, *Clin. Sci.*, 1983, **65**, 415–422.

53. K.H. Fleming and J.T. Heimbach, Consumption of calcium in the U.S.: Food sources and intake levels, *J. Nutr.*, 1994, **124**, 1426S–1430S.

54. J.Z. Ilich, T.A. Hangartner, M. Skugor, A.F. Roche, P. Goel and V. Matkovic, Skeletal age as a determinant of bone mass in young females, *Skel. Radiol.*, 1996, **25**, 431–439.

55. V. Matkovic, J.Z. Ilich, M.B. Andon, L.C. Hsieh, M.A. Tzagournis *et al.*, Urinary calcium, sodium, and bone mass of young females, *Am. J. Clin. Nutr.*, 1995, **62**, 417–425.

56. L.A. Jackman, S.S. Millane, B.R. Martin, O.B. Wood, G.P. McCabe *et al.*, Calcium retention in relation to calcium intake and postmenarcheal age in adolescent females, *Am. J. Clin. Nutr.*, 1997, **66**, 327–333.

57. B.E.C. Nordin and D.H. Marshall, Dietary requirements for calcium, In *Calcium in Human Biology*. ILSI Human Nutrition Reviews, B.E.C Nordin, (ed.), Springer-Verlag, Berlin, 1988, 447–471.

58. V. Matkovic, J.Z. Ilich and M. Skugor, Calcium intake and skeletal formation, *Challenges Mod. Med.*, 1995, **7**, 129–145.

59. B.E.C. Nordin, A.G. Need, H.A. Morris, M. Horowitz, The nature and significance of the relationship between urinary sodium and urinary calcium in women, *J. Nutr.*, 1993, **123**, 1615–1622.

60. S. Docio, J.A. Riancho, A. Perez, J.M. Olmos, J.A. Amado and J. Gonzales-Macias, Seasonal deficiency of vitamin D in children: a potential target for osteoporosis-preventing strategies, *J. Bone Miner. Res.*, 1998, **13**, 544–548.

61. R.P. Heaney, Nutrition and bone mass, *Phys. Med. Rehab. Clin. N Am.*, 1995, **6**(3), 551–556.

62. G. Wyshak and R.E. Frisch, Carbonated beverages, dietary calcium, the dietary calcium/phosphorus ratio, and bone fractures in girls and boys, *J. Adolesc. Health*, 1994, **15**, 210–215.

63. S.M. Garn, *The Earlier Gain and the Later Loss of Cortical Bone*, Charles C Thomas Publ., Springfield, IL, 1970.

64. Dietary Reference Intakes. Food and Nutrition Board, Institute of Medicine, National Academy Press, Washington, DC, 1997.

65. NIH Concensus Development Panel, Optimal calcium intake, *JAMA*, 1994, **272**, 1942–1948.

66. N.E. Badenhop, J.Z. Ilich, M. Skugor, J.D. Landoll and V. Matkovic, Changes in body composition and serum leptin in young females with high *vs.* low dairy intake, *J. Bone Miner Res.*, 1997, **12**, S487.

67. U.S. Department of Health and Human Services, Public Health Service, *Healthy People 2000, National Health Promotion and Disease Prevention Objectives*, Jones and Bartlett Publ., Boston, 1992, 1–153.

68. M.B. Andon, J.Z. Ilich, M.A. Tzagournis and V. Matkovic, Magnesium balance in adolescent females consuming a low or high calcium diet, *Am. J. Clin. Nutr.*, 1996, **63**, 950–953.

69. A.A. McKenna, J.Z. Ilich, M.B. Andon, C. Wang and V. Matkovic, Zinc balance in adolescent females consuming a low- or high-calcium diet, *Am. J. Clin. Nutr.*, 1997, **65**, 1460–1464.

70. D. Holben, A.M. Smith, E.J. Ha, J.Z. Ilich and V. Matkovic, Selenium (Se) absorption, balance, and status in adolescent females throughout puberty, *FASEB J.*, 1996, **10**, A532.

71. J.Z. Ilich, A.A. McKenna, N.E. Badenhop, A.C. Clairmont, M.B. Andon *et al.*, Iron status, menarche, and calcium supplementation in young females, *Am. J. Clin. Nutr.*, 1998, **68**, 880–887.

72. P. Kannus, H. Haapasalo, M. Sankelo, H. Sievanen, M. Pasanen, A. Heinonen *et al.*, Effect of starting age of physical activity on bone mass in the dominant arm of tennis and squash players, *Ann. Intern. Med.*, 1995, **123**, 27–31.

73. B.L. Specker, Evidence for an interaction between calcium intake and physical activity on changes in bone mineral density, *J. Bone Miner. Res.*, 1996, **11**, 1539–1544.

74. S. Iuliano-Burns, L. Saxon, G. Naughton, K. Gibbons and S.L. Bass, Regional specificity of exercise and calcium during skeletal growth in girls: a randomized controlled trial, *J. Bone Miner. Res.*, 2003, **18**, 156–162.

CHAPTER 8.2

# Role of Calcium in Reducing Postmenopausal Bone Loss and in Fracture Prevention

BESS DAWSON-HUGHES

Calcium and Bone Metabolism Laboratory, Jean Mayer USDA Human Nutrition Research Center on Aging, Tufts University, 711 Washington St., Boston, MA 02111, USA, Email: hughesb@hnrc.tufts.edu

**Key Points**

- Calcium is a substrate for bone formation and it is also an antiresorptive agent.
- Calcium can lower the bone-remodeling rate by 10 – 20% in older adults
- At intakes of 1200 mg per day, total body retention of calcium is maximal in adults.
- Calcium causes positive changes in bone mineral density (BMD) of about 2% in adults.
- The effect of calcium on fracture rates is not well defined, but calcium together with vitamin D lowers fracture rates in elderly subjects.
- Calcium appears to interact favorably with at least two antiresorptive therapies for osteoporosis, estrogen and calcitonin.
- To meet the National Academy of Sciences calcium intake recommendation of 1200 mg per day for men and women over age 50, many individuals would need supplements or fortified foods.

## Introduction

Calcium is the major mineral component of bone. It is present in hydroxyapatite crystals, which also contain phosphorus and water. In addition to providing structural support, the skeleton also serves as a reservoir of calcium that may be drawn upon to support the blood concentration of ionized calcium. There is little disagreement that dietary calcium is important in adults, but there is disagreement about the optimal intake of calcium. The aim of this review is to consider the calcium intake requirement in view of the evidence from calcium balance studies and from randomized controlled trials with end points of change in BMD and change in fracture incidence.

## Absorption and Impact on Endocrine System

Calcium is absorbed by two major mechanisms – active transport and passive diffusion. Active transport requires 1,25-dihydroxyvitamin D, the active form of vitamin D that acts on the enterocyte nuclear receptors to initiate the synthesis of calcium binding protein. The active transport process involves the movement of calcium from the intestinal lumen into the enterocyte and then out on the serosal side.

Passive diffusion involves the movement of calcium between enterocytes. It is driven by the gut luminal:serosal calcium concentration gradient. Figure 8.2.1 shows the relationship between dietary calcium intake and calcium absorption.[1] The steeper slope at low calcium intakes reflects 1,25-hydroxyvitamin D mediated active transport; this transport mechanism becomes saturated at calcium intake of around 500 mg per day. The slower rise in absorbed calcium at higher intake levels reflects additional passively absorbed calcium.

When calcium intake declines, there is an adaptation in the form of increased calcium absorption efficiency. The lower intake is associated with a decline in absorbed calcium and an accompanying subtle decline in the circulating

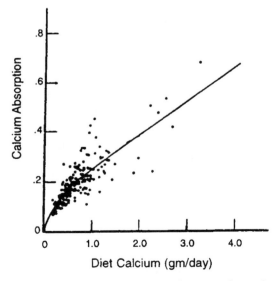

**Figure 8.2.1**   *The relationship between dietary calcium intake and calcium absorption* (Source: from Heaney *et al.*,[1] with permission)

ionized calcium concentration. This decline stimulates the release of parathyroid hormone (PTH). PTH acts in several ways to return the ionized calcium to its normal set point. First, PTH promotes the production of 1,25-hydroxyvitamin D and thus indirectly stimulates intestinal calcium absorption. Unfortunately, this adaptation is only partial. PTH also promotes renal reabsorption of filtered calcium, a positive adaptation. Finally, PTH draws on the skeletal reservoir of calcium by promoting bone resorption, a not-so-favorable adaptation from the perspective of bone health.

There are many determinants of calcium absorption. Intake is the most important. Aging, particularly after age 60, is associated with a significant decline in calcium absorption.[2] There is also an hereditary component in the form of vitamin D receptor alleles.[3,4] Other determinants include smoking (inhibitory), diet composition (*e.g.* chelates present in spinach and phytate in wheat inhibit absorption; sugar enhances absorption). Estrogen promotes absorption by enhancing production of 1,25-dihydroxyvitamin D and also by direct effects on the gut.[5] Calcium absorption is greater in summer than winter presumably because season influences skin synthesis of vitamin D. Low fractional calcium absorption has been associated with an increased risk of hip fracture in older women with low dietary calcium intakes.[6]

# Dietary Calcium – Mechanisms of Action

Calcium may be thought of as merely a passive substrate needed to support the bone formation phase of bone remodeling in adults or it may be regarded as both substrate, and, at higher intake levels, an anti-resorptive agent.

## Calcium as Substrate

Typically, about 5 mmol (200 mg) of calcium is removed from the skeleton and replaced every day. To restore this loss, one would need to consume about 600 mg of calcium (since calcium is not very efficiently absorbed). This intake estimate is an approximation since some of the absorbed calcium would be excreted by sweat, urine, and feces and thus not available for deposition in bone and some of the resorbed calcium will be recycled back into bone at new remodeling sites. The amount of calcium needed to provide 5 mmol of substrate might be thought of as a subsistence requirement. Several studies have demonstrated a significant impact of increasing calcium intake among those with very low usual calcium diets to the subsistence level. We have reported that supplementation with 500 mg per day of calcium had a greater positive effect on the change in BMD among postmenopausal women with self-selected calcium intakes under 400 mg per day than among those with intakes in the range of 400–650 mg per day.[7] In a large case-controlled study of hip fracture risk in women in Europe,[8] fracture risk declined as calcium intake rose to an estimated milk score of 4.6 (equivalent to a total calcium intake of 500 mg per day, assuming that

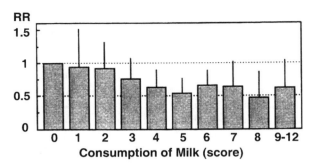

**Figure 8.2.2**  *In a large case-controlled study in European women, hip fracture risk declined as milk score rose to 4.6 (equivalent to a total calcium intake of about 500 mg per day)*
(Source: from Johnell et al.,[8] with permission)

milk accounts for about half of all dietary calcium as has been reported) (Figure 8.2.2). Since only 10% of the women studied had calcium intakes above 500 mg per day, this study does not allow one to examine carefully the impact of calcium at higher intake levels because of sample size limitations.[8] If replacing obligatory calcium losses were the only consideration, the average calcium requirement might be about 600 mg (and the associated recommended dietary intake about 800 mg per day).

## Calcium as Anti-resorptive Agent

It has recently been recognized that a high bone-remodeling rate is an independent risk factor for fracture.[9] Postulated mechanisms for this are that a high remodeling rate causes: (1) more perforations of trabeculae and thus greater architectural deformity, and (2) incomplete mineralization at new remodeling sites. Currently, approved treatments for osteoporosis, all anti-resorptive agents, may rely more on lowering the remodeling rate (typical reductions are 40–60%) for their anti-fracture efficacy than on increasing BMD (typical increases are 3–5%).

Dietary calcium at sufficiently high levels, usually 1,000 mg per day or more, lowers the bone-remodeling rate by about 10–20% in older men and women.[10–13] The degree of suppression (and also the degree of BMD gain) appears to be dose-related, as illustrated by Elders who treated postmenopausal Dutch women with either 1000 or 2000 mg of supplemental calcium.[13] The calcium induced suppression persists throughout the intervention periods of 3–4 years; however, the suppression of remodeling is readily reversible. When supplementation (in this case 500 mg of calcium and 700 IU of vitamin D) was discontinued after 3 years in one of these studies,[12] the remodeling rate returned to pre-treatment levels within 1–2 years.[14]

# Approaches to Assessing Impact of Dietary Calcium on the Skeleton

## Intake Associated with Maximal Calcium Retention

Calcium balance studies have often been performed with the aim of identifying the intake associated with the zero balance, or the intake at which calcium is neither lost nor gained from the body referred to above as the subsistence level. Balance studies may also be used to try to identify the intake associated not with zero balance, but with the most favorable balance that can be achieved as a result of increasing calcium intake. Increasing intake above the level associated with maximal calcium retention would result in more calcium being absorbed (by passive diffusion); however, that calcium would be excreted rather than retained.

To be useful in identifying the intake associated with maximal calcium retention, balance studies must have large sample sizes and include subjects with a wide range and a balanced distribution of calcium intakes. Most studies have not included enough subjects with high enough calcium intakes to be used for determining the point of maximal retention, with one notable exception.[15] Spencer and colleagues[15] performed balance studies in 181 men aged 34–71 years on six different calcium intake levels ranging from 234 to 2320 mg per day. Diets were supplemented with either calcium gluconate or with milk and balance periods were quite generous, averaging 20–38 days. The results are summarized in Table 8.2.1. Calcium absorption from gluconate and milk was

**Table 8.2.1.** *Calcium balances at different calcium intakes*

| Calcium intake (mg/day) | No. of studies | Average days per studyl | Ca balance (mg/day) | Significance[a] |
|---|---|---|---|---|
| Calcium gluconate studies | | | | |
| 234[b] | 22 | 37 | −95 | − |
| 804 | 67 | 36 | +22 | 0.005 |
| 1230 | 7 | 29 | +106 | 0.001 |
| 1431 | 13 | 35 | +104 | NS |
| 2021 | 12 | 34 | +147 | NS |
| 2320 | 14 | 38 | +139 | NS |
| Milk studies | | | | |
| 810 | 22 | 33 | +10 | − |
| 1248 | 15 | 35 | +106 | <0.001 |
| 1467 | 9 | 33 | +147 | NS |

Source: from Spencer *et al.*, with permission.[15]

[a]Significance in comparing calcium balance with the balance at previous calcium intake.
[b]Low calcium intake without the addition of calcium gluconate. All higher calcium intakes were achieved by adding calcium gluconate tablets to the constant low-calcium diet.

similar. Calcium retention increased significantly with increasing intake up to a maximum of about 1200 mg per day, in both the supplement and milk studies. Note that the zero balance point in this study occurred at the lower intake of 800 mg per day.

There are two reports of balance studies in women with calcium intakes over 1000 mg per day.[16,17] Both found positive correlations between intake and balance, and along with Spencer *et al.*[15] support the conclusion that increasing intake up to about 1200 mg per day is associated with increasing skeletal calcium retention. The studies in women do not exclude the possibility that even higher intakes could result in greater retention, presumably as a result of further suppression of the remodeling rate. If short-term balance studies define longer-term patterns, then greater retention would signify greater bone mass.

## Calcium and Changes in Bone Mineral Density

In the last 10–15 years, many randomized, controlled calcium-intervention trials have been reported. As indicated in a meta-analysis of 13 trials, calcium consistently induced significant gains (or slowed loss) at multiple skeletal sites in the first year.[18] In this analysis, the mean treatment-group difference between calcium and placebo ranged from 0.6 at the distal forearm to 3% at the spine, with a difference at the femoral neck of 2.6%. Mean differences after the first year were still positive, but greatly attenuated. The relatively strong initial response to calcium results from closure of remodeling space that occurs over the first 12–18 months. A more recent meta-analysis of 15 trials reported that calcium causes positive mean percentage changes (from baseline) of 2.05% for total body, 1.66% for lumbar spine, 1.64% for the hip, and 1.91% for the distal radius.[19]

Time since menopause may influence the impact of calcium on percentage changes in BMD. In the first 5–8 years after menopause, rapid bone loss occurs as a result of declining estrogen levels. The increased bone resorption that accompanies estrogen deficiency provides calcium to the blood and other extracellular space. In turn, PTH levels decline, 1,25-dihydroxyvitamin D levels decline, and the signal to absorb calcium declines. As would be expected, additional calcium intake does not entirely reverse this sequence and cannot be relied upon to prevent early menopausal bone loss, although in several studies it does attenuate it.[13,20] Calcium is generally more effective in older post-menopausal women,[7] and in large enough doses, it can reverse age-related increases in PTH and in bone remodeling.[21] In one trial, the effects of calcium from milk powder and supplements on changes in BMD in older post-menopausal women were compared and found to be similar.[22]

Studies have evaluated the impact of stopping calcium supplementation on changes in BMD. In older women and men who stopped taking supplements at the end of a 3-year trial, the BMD gains were lost over the following 1–2 years.[14] This loss was the inverse of the changes induced by supplementation, both in magnitude and timing.

## Calcium and Fracture Risk

Data are beginning to emerge on the impact of calcium[11,23-25] and the combination of calcium and vitamin D[12,26,27] on fracture rates. The recent Shea meta-analysis[19] found that calcium alone (*vs.* placebo) tended to lower risk of vertebral fractures (RR 0.77 [CI 0.54–1.09]) but not non-vertebral fractures (RR 0.86 [CI 0.43–1.72]). Studies reporting more than five fractures, including those using both calcium and vitamin D, are summarized in Table 8.2.2. These studies range from 18 months to four years in duration. The majority found significantly fewer fractures in the calcium ( $\pm$ vitamin D) than in the placebo arm. The largest of these studies, by Chapuy *et al.*[26] is likely to provide the most precise estimate of the impact of combined supplementation. Chapuy found a 25% decrease in hip and other non-vertebral fracture rates. But, this population may have been more likely than others to benefit from supplementation because of their low self-selected calcium intake of about 500 mg per day, their low initial 25-hydroxyvitamin D levels (mean of 25 nmol $L^{-1}$ in a subset[28]), and their advanced age which is usually associated with a higher turnover rate. Thus, a 25% reduction with combined calcium and vitamin D supplementation may be an optimistic expectation in less vulnerable populations.

## Calcium as Adjunctive Therapy

For ethical reasons, recent randomized, controlled trials testing the anti-fracture efficacy of anti-resorptive therapies (alendronate, risedronate, raloxifene, and calcitonin) have given calcium and often also vitamin D, to both the control and intervention groups. This allows one to define the impact of the drug in calcium- and vitamin D-replete patients. It does not, however, define the impact of drug treatment in calcium and vitamin D insufficient subjects.

There are several suggestions that calcium may interact favorably with antiresorptive therapies. In a meta-analysis, Nieves *et al.*[29] examined BMD responses of early postmenopausal women to hormone replacement therapy in relation to calcium intake. Among the 31 HRT intervention trials that met inclusion criteria for the analysis, 20 added calcium to both study arms (bringing mean total intake to 1183 mg per day) and 11 did not (mean calcium intake 563 mg per day). The BMD gains at the spine, hip, and forearm were greater in the women who increased their calcium intakes than in those who took the HRT without added calcium. This is consistent with the concept that calcium enables estrogen to be more effective in building BMD. In the large, observational Mediterranean Osteoporosis Study (MEDOS) in southern Europe, use of the antiresorptive drug nasal calcitonin was associated with a non-significant decrease in vertebral fracture risk (rr = 0.78 [$CI_{95}$ 0.48, 1.27], $P = 0.318$) as was use of calcium alone (rr = 0.82 [$CI_{95}$ 0.63, 1.07], $P = 0.149$).[30] Use of calcitonin and calcium together, however, was associated with a significant reduction in vertebral fracture risk (rr = 0.63 [$CI_{95}$ 0.44, 0.90], $P = 0.012$). Again, this suggests that the effects of calcium and another antiresorptive therapy may be additive.

**Table 8.2.2.** Calcium supplement trials and fracture incidence

| Study | Population | N enrolled | Mean age | Calcium intake (mg day$^{-1}$) | | Persons with new fracture | | Statistically significant difference |
|---|---|---|---|---|---|---|---|---|
| | | | | Diet | Supplement | N | Site | |
| Calcium | | | | | | | | |
| Recker[23] | Women; half with prior vertebral fracture | 251 | 74 ± 7 | 433 | 1200 | | | |
| With prior fracture | | | | | | 36 | Vertebra | Yes |
| No prior fracture | | | | | | 25 | Vertebra | No |
| Chevalley[11] | Men and women with no recent fracture | 93 | 72 ± 6 | 600 | 800 | 10 | Vertebra | Yes |
| Reid[24] | Women three or more years after menopause, no prior fracture | 86 | 59 ± 5 | 700 | 1000 | 9 | All sites | Yes |
| Riggs[25] | Healthy women | 236 | 66 ± 3 | 700 | 1600 | 40 | All sites | No |
| Calcium + Vitamin D | | | | | | | | |
| Chapuy[26] | Institutionalized elderly women | 3270 | 84 ± 6 | 500 | 1200[a] | 315 | Hip | Yes |
| | | | | | | 563 | All non-vertebral | Yes |
| Chapuy[27] | Ambulatory institutionalized women | 583 | 85 ± 7 | 557 | 1200[a] | 48 | Hip | ~[c] |
| Dawson-Hughes[12] | Healthy men and women aged 65 and older | 445 | 71 ± 5 | 700 | 500[b] | 37 | All non-vertebral | Yes |

[a]Plus 800 IU of vitamin D.
[b]Plus 700 IU of vitamin D.
[c]$P = 0.07$.

# Recommended Intake of Calcium

Intake recommendations vary enormously worldwide. Recommendations by the U.S. National Academy of Sciences are among the highest. The recommended intake of calcium for women and men over age 50 is 1200 mg per day.[31] This intake is higher than their previous recommendation of 800 mg per day. The increase was made on the basis of evidence that a higher intake was likely to improve calcium retention and bone mass and reduce the risk of fracture.

# Sources of Calcium

Consuming more natural calcium-rich foods, calcium-fortified foods, or supplements can increase calcium intake. Natural food sources of calcium are preferred because calcium-rich foods provide a variety of important nutrients that are not present in supplements. Calcium fortification of high-quality food items such as orange juice and soy beverages can be an effective strategy for raising calcium intakes. Although absorption of calcium from calcium-fortified soy drinks may not be quite equivalent to absorption of calcium from cow's milk (75 *vs.* 100%),[32] the amount of calcium added can of course be adjusted.

Calcium supplements are important for many individuals who are not willing or able to meet the calcium requirement from food sources. Many compounds including calcium acetate, carbonate, citrate, citrate malate, glubionate, lactate, lactogluconate, and tricalcium phosphate, and tricalcium phosphate are available. Absorbability of supplements is fairly comparable when the amounts of calcium absorbed per gram of supplement are compared under similar test conditions (similar test doses, test meals, *etc.*). In healthy individuals, calcium from carbonate, the most widely used supplement, is better and more consistently absorbed if it is taken with meals rather than during fasting.[33] This is especially important in patients with decreased gastric acid production.[34]

For all individuals who take more than 500–600 mg per day of calcium from supplements, the dose should be split to improve calcium absorption.[35] A dose of calcium at bedtime has been recommended to suppress the normal nocturnal rise in circulating PTH level that may increase nighttime bone resorption.[36]

# Conclusions

Balance data in men suggest that a calcium intake of 1200 mg per day may be optimal and available BMD data from supplement trials in women are compatible with this conclusion. There are few studies of the effect of calcium on fracture rates and there are no such studies in women with starting calcium intakes as high as 1200 mg per day to verify or refute conclusions from the balance and BMD studies. Additional large studies are needed to strengthen the evidence that increasing calcium intake will lower risk of fracture in older adults. Calcium appears to enhance the effect of estrogen on changes in BMD and the effect of calcitonin on hip fracture rates. Interactions of calcium with other therapies for osteoporosis have not been reported. In 1997, the National Academy of Sciences increased the recommended intake of calcium from 800 to 1200 mg day for men and women over age 50.[31] To meet this recommendation, many individuals would need to take calcium supplements or fortified foods.

# Acknowledgements

This material is based on the work supported by a grant (AG10353) from the National Institutes of Health and by the U.S. Department of Agriculture, under agreement No. 58-1950-9001. Any opinions, findings, conclusions, or recommendations expressed in this publication are those of the authors, and do not necessarily reflect the view of the U.S. Department of Agriculture.

# References

1. R.P. Heaney, P.D. Saville and R.R. Recker, Calcium absorption as a function of calcium intake, *J. Lab. Clin. Med.*, 1975, **85**(6), 881–890.
2. J.R. Bullamore, R. Wilkinson, J.C. Gallagher, B.E. Nordin and D.H. Marshall, Effect of age on calcium absorption, *Lancet*, 1970, **2**(7672), 535–537.
3. N.A. Morrison, J.C. Qi, A. Tokita, P.J. Kelly, L. Crofts, T.V. Nguyen, P.N. Sambrook and J.A. Eisman, Prediction of bone density from vitamin D receptor alleles, *Nature*, 1994, **367**(6460), 284–287.
4. E.A. Krall, P. Parry, J.B. Lichter and B. Dawson-Hughes, Vitamin D receptor alleles and rates of bone loss: influences of years since menopause and calcium intake, *J. Bone Min. Res.*, 1995, **10**(6), 978–984.
5. C. Gennari, D. Agnusdei, P. Nardi and R. Civitelli, Estrogen preserves a normal intestinal responsiveness to 1,25-dihydroxyvitamin D3 in oophorectomized women, *J. Clin. Endocrinol. Metabol.*, 1990, **71**(5), 1288–1293.
6. K.E. Ensrud, T. Duong, J.A. Cauley, R.P. Heaney, R.L. Wolf, E. Harris and S.R. Cummings, Low fractional calcium absorption increases the risk for hip fracture in women with low calcium intake. Study of osteoporotic fractures research group, *Ann. Int. Med.*, 2000, **132**(5), 345–353.
7. B. Dawson-Hughes, G.E. Dallal, E.A. Krall, L. Sadowski, N. Sahyoun and S. Tannenbaum, A controlled trial of the effect of calcium supplementation on bone density in postmenopausal women, *N. Engl. J. Med.*, 1990, **323**(13), 878–883.
8. O. Johnell, B. Gullberg, J.A. Kanis, E. Allander, L. Elffors, J. Dequeker, G. Dilsen, C. Gennari, V.A. Lopes and G. Lyritis, Risk factors for hip fracture in European women: the MEDOS Study. Mediterranean Osteoporosis Study, *J. Bone Miner. Res.*, 1995, **10**(11), 1805–1815.
9. P. Garnero, E. Hausherr, M.C. Chapuy, C. Marcelli, H. Grandjean, C. Muller, C. Cormier, G. Breart, P.J. Meunier and P.D. Delmas, Markers of bone resorption predict hip fracture in elderly women: the EPIDOS prospective study, *J. Bone Miner. Res.*, 1996, **11**(10), 1531–1538.

10. B. Riis, K. Thomsen and C. Christiansen, Does calcium supplementation prevent postmenopausal bone loss? A double-blind, controlled clinical study, *N. Engl. J. Med.*, 1987, **316**(4), 173–177.

11. T. Chevalley, R. Rizzoli, V. Nydegger, D. Slosman, C.H. Rapin, J.P. Michel *et al.*, Effects of calcium supplements on femoral bone mineral density and vertebral fracture rate in vitamin-D-replete elderly patients, *Osteoporosis Int.*, 1994, **4**(5), 245–252.

12. B. Dawson-Hughes, S.S. Harris, E.A. Krall and G.E. Dallal, Effect of calcium and vitamin D supplementation on bone density in men and women 65 years of age or older, *N. Engl. J. Med.*, 1997, **337**(10), 670–676.

13. P.J. Elders, J.C. Netelenbos, P. Lips, F.C. van Ginkel, E. Khoe, O.R. Leeuwenkamp, W.H. Hackeng and P.F. van der Stelt, Calcium supplementation reduces vertebral bone loss in perimenopausal women: a controlled trial in 248 women between 46 and 55 years of age, *J. Clin. Endocrinol. Metabol.*, 1991, **73**(3), 533–540.

14. B. Dawson-Hughes, S.S. Harris, E.A. Krall and G.E. Dallal, Effect of withdrawal of calcium and vitamin D supplements on bone mass in elderly men and women, *Am. J. Clin. Nutr.*, 2000, **72**(3), 745–750.

15. H. Spencer, L. Kramer, M. Lesniak, M. De Bartolo, C. Norris and D. Osis, Calcium requirements in humans. Report of original data and a review, *Clin. Orthopaed. Related Res.*, 1984, **184**, 270–280.

16. C. Hasling, P. Charles, F.T. Jensen and L. Mosekilde, Calcium metabolism in postmenopausal osteoporosis: the influence of dietary calcium and net absorbed calcium, *J. Bone Miner. Res.*, 1990, **5**(9), 939–946.

17. P.L. Selby, Calcium requirement – a reappraisal of the methods used in its determination and their application to patients with osteoporosis, *Am. J. Clin. Nutr.*, 1994, **60**(6), 944–948.

18. D. Mackerras and T. Lumley, First- and second-year effects in trials of calcium supplementation on the loss of bone density in postmenopausal women, *Bone*, 1997, **21**(6), 527–533.

19. B. Shea, G. Wells, A. Cranney, N. Zytaruk, V. Robinson, L. Griffith, Z. Ortiz, J. Peterson, J. Adachi, P. Tugwell and G. Guyatt, VII. Meta-analysis of calcium supplementation for the prevention of postmenopausal osteoporosis, *Endocrine Rev.*, 2002, **23**(4), 552–529.

20. J.F. Aloia, A. Vaswani, J.K. Yeh, P.L. Ross, E. Flaster and F.A. Dilmanian, Calcium supplementation with and without hormone replacement therapy to prevent postmenopausal bone loss, *Ann. Intern. Med.*, 1994. **120**(2), 97–103.

21. W.R. McKane, S. Khosla, K.S. Egan, S.P. Robins, M.F. Burritt and B.L. Riggs, Role of calcium intake in modulating age-related increases in parathyroid function and bone resorption, *J. Clin. Endocrinol. Metabol.*, 1996, **81**(5), 1699–1703.

22. R. Prince, A. Devine, I. Dick, A. Criddle, D. Kerr, N. Kent, R. Price and A. Randell, The effects of calcium supplementation (milk powder or tablets) and exercise on bone density in postmenopausal women, *J. Bone Miner. Res.*, 1995, **10**(7), 1068–1075.

23. R.R. Recker, S. Hinders, K.M. Davies, R.P. Heaney, M.R. Stegman, J.M. Lappe and D.B. Kimmel, Correcting calcium nutritional deficiency prevents spine fractures in elderly women, *J. Bone Miner. Res.*, 1996, **11**(12), 1961–1966.

24. I.R. Reid, R.W. Ames, M.C. Evans, G.D. Gamble and S.J. Sharpe, Long-term effects of calcium supplementation on bone loss and fractures in postmenopausal women: a randomized controlled trial, *Am. J. Med.*, 1995, **98**(4), 331–335.

25. B.L. Riggs, W.M. O'Fallon, J. Muhs, M.K. O'Connor, R. Kumar and L.J. Melton III, Long-term effects of calcium supplementation on serum parathyroid hormone

level, bone turnover, and bone loss in elderly women, *J. Bone Miner. Res.*, 1998, **13**(2), 168–174.

26. M.C. Chapuy, M.E. Arlot, P.D. Delmas and P.J. Meunier, Effect of calcium and cholecalciferol treatment for three years on hip fractures in elderly women, *Br. Med. J.*, 1994, **308**(6936), 1081–1082.

27. M.C. Chapuy, R. Pamphile, E. Paris, C. Kempf, M. Schlichting, S. Arnaud, P. Garnero and P.J. Meunier, Combined calcium and vitamin D3 supplementation in elderly women: confirmation of reversal of secondary hyperparathyroidism and hip fracture risk: the Decalyos II study, *Osteoporosis Int.*, 2002, **13**(3), 257–264.

28. P.J. Meunier, M.C. Chapuy, M.E. Arlot and F. Dubouef, Vitamin D and calcium: their roles in pathophysiology and prevention of hip fractures, In *Nutritional Assessment of Elderly Populations. Bristol-Myers Squibb/Mead Johnson Nutrial Symposia*, I.H. Rosenberg (ed), Raven Press, New York, 1995, 265–76.

29. J.W. Nieves, L. Komar, F. Cosman and R. Lindsay, Calcium potentiates the effect of estrogen and calcitonin on bone mass: review and analysis, *Am. J. Clin. Nutr.*, 1998, **67**(1),18–24.

30. J.A. Kanis, O. Johnell, B. Gullberg, E. Allander, G. Dilsen, C. Gennari, V.A. Lopes, G.P. Lyritis, G. Mazzuoli and L. Miravet, Evidence for efficacy of drugs affecting bone metabolism in preventing hip fracture, *Br. Med. J.*, 1992, **305**(6862), 1124–1128.

31. Standing Committee on the Scientific Evaluation of Dietary Reference Intakes, In *Dietary Reference Intakes: Calcium, Phosphorus, Magnesium, Vitamin D, and Fluoride*, National Academy Press, Washington, D.C, 1997.

32. R.P. Heaney, M.S. Dowell, K. Rafferty and J. Bierman, Bioavailability of the calcium in fortified soy imitation milk, with some observations on method, *Am. J. Clin. Nutr.*, 2000, **71**(5), 1166–1169.

33. R.P. Heaney, K.T. Smith, R.R. Recker and S.M. Hinders, Meal effects on calcium absorption, *Am. J. Clin. Nutr.*, 1989, **49**(2), 372–376.

34. R.R. Recker, Calcium absorption and achlorhydria, *N. Engl. J. Med.*, 1985, **313**(2), 70–73.

35. J.A. Harvey, M.M. Zobitz and C.Y. Pak, Dose dependency of calcium absorption: a comparison of calcium carbonate and calcium citrate, *J. Bone Miner. Res.*, 1988, **3**(3), 253–258.

36. M.S. Calvo, R. Eastell, K.P. Offord, E.J. Bergstralh and M.F. Burritt, Circadian variation in ionized calcium and intact parathyroid hormone: evidence for sex differences in calcium homeostasis, *J. Clin. Endocrinol. Metabol.*, 1991, **72**(1), 69–76.

CHAPTER 8.3

# Issues Concerning Calcium Absorption and Bone Health

## STEVEN A. ABRAMS

U.S. Department of Agriculture/Agricultural Research Service, Children's Nutrition Research Center, Department of Pediatrics, Baylor College of Medicine and Texas Children's Hospital, Houston, Texas 77030, USA, Email: sabrams@bcm.tmc.edu

*This chapter is dedicated to the four astronauts (Michael Anderson, Dave Brown, Laurel Clark and Ilan Ramon) who were study subjects for the calcium experiments as well as the entire STS-107 crew who perished in the conduct of scientific research.*

**Key Points**

- Calcium absorption can be measured using balance or isotopic methods
- Stable isotopic techniques are the most widely applicable and preferred method for measuring calcium absorption at present
- Increased absorption of calcium from infant formulas compared to human milk has not been shown to be beneficial
- Calcium absorption peaks during pubertal development and then rapidly decreases
- Calcium absorption may be related to identifiable genetic polymorphisms
- Calcium absorption increases during pregnancy but not substantially during lactation
- Dietary enhancers of calcium absorption such as prebiotics may be useful in improving mineral status
- Calcium absorption is impaired by bed-rest and microgravity

## Introduction

The role of calcium absorption in bone health has been the subject of innumerable clinical research studies over the past several decades. Increasing absorbed calcium is viewed as a practical and oftentimes, marketable, approach to increasing the calcium available to the body for bone formation or prevention of bone loss. Calcium absorption is determined by numerous factors, however, not all of them readily controllable. Factors difficult to control include the effects of genetics (include gender and ethnicity as well as putative specific genetic polymorphisms), puberty and medical therapies (*e.g.* steroid use). More readily controllable are the chemical form and solubility of the calcium and other dietary factors, which can either enhance or decrease calcium absorption.[1]

It is important to begin by actually considering the meaning of the term 'calcium absorption'. First, for nutritional purposes, calcium absorption is generally used interchangeably with the term 'calcium bioavailability'. This is because calcium absorbed from oral forms is essentially entirely available for 'utilization' by the body, especially the bones.[2] In this chapter the two terms will be used interchangeably, as they are in most of the primary research sources.

Another basic issue to consider is the difference between what is often referred to as 'net calcium absorption' and 'true calcium absorption'. The former term, derived from results from classical calcium balance studies, includes calcium that has been secreted into the intestine and then excreted in the stools (endogenous faecal excretion). Whereas this amount is positively correlated with calcium intake, it is negatively correlated with calcium absorption fraction.[3] This is because as absorption fraction increases, more of the secreted calcium is reabsorbed before being excreted. The term 'true calcium absorption' refers to the primarily unidirectional flow of calcium from an ingested source and is determined by tracer techniques.

# Relationship Between Calcium Absorption and Calcium Retention

Calcium absorption is an important regulator of balance and one that may, in some circumstances be positively or negatively affected by dietary or medical interventions. In general, increasing calcium absorption will lead to an increase in calcium retention. For example, in healthy children, the slope of the relationship between calcium absorption and urinary excretion is about 0.1–0.15, indicating that about 10–15% of absorbed calcium is excreted in the urine (Figure 8.3.1). In healthy adults, this fraction is somewhat higher, about 40–50% (Figure 8.3.2) and there is a much more highly significant relationship between absorption and urinary excretion. Therefore, if the total amount of calcium absorbed is increased (by increasing intake, or efficiency of absorption) a fraction (0.1–0.5) will be lost in the urine. Most of the rest will be retained, and presumably used for bone mineralization (Abrams and Heaney, unpublished observations).

# Measurement of Calcium Absorption

There are numerous methods for the assessment of calcium absorption. Most of these ultimately can be divided into those that utilize radioactive or

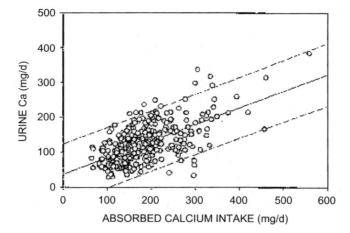

**Figure 8.3.1** *Relationship between absorbed calcium and urinary calcium in estrogen-replete adults. In this case Ca(urine) = 0.48 (absorbed calcium) + 36* (Figure courtesy of Dr. Robert Heaney).

stable tracers of calcium (or strontium) to measure calcium absorption and those that do not rely on tracer techniques. In addition, in growing infants and children, calcium absorption over a relatively prolonged period of time can be estimated using longitudinal whole body Dual-Energy X-ray Absorptiometry (DXA) measurements. We will first consider the non-tracer techniques.

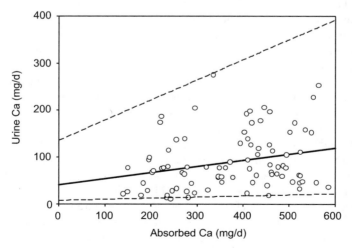

**Figure 8.3.2** *Relationship between absorbed calcium and urinary calcium in young adolescent females. In this case Ca(urine) = 0.13(absorbed calcium) + 41.*

# Non-Tracer Techniques

The most widely used method for determining calcium absorption is a mass balance. In a mass balance study, the net absorption of a nutrient is calculated by measuring the difference between mineral input from the diet and total faecal mineral output.[4] The faecal output of mineral includes both unabsorbed dietary mineral and endogenous faecal excretion. These two sources of mineral that appear in the faeces cannot be distinguished by a classical balance study.

Calcium absorption has also been estimated from the calciuric response to an oral calcium load. In this case, after a large oral calcium dose, one measures the rate at which calcium appears in the urine or changes in serum levels of ionized and total calcium or other bone mineral markers. This test may be useful for comparing some calcium sources, but cannot provide an accurate measure of dietary calcium absorption and is less preferable to tracer or mass balance methods.[5,6]

# Tracer Techniques

In the single-isotope technique, a single calcium isotope is given orally usually with a calcium load. A complete faecal collection is carried out until virtually all the unabsorbed oral tracer is recovered. The fraction of administered tracer that was absorbed is calculated from the difference between the amount of the oral dose and the faecal isotope recovery.

This technique has the benefit that the calculated absorption represents only the dietary component of the element that is absorbed and does not include endogenous secretory losses. It also does not require use of an intravenous (IV) infusion. A disadvantage to the use of this oral tracer approach is that extended faecal collections are required and the accuracy of the results depends on the completeness of the faecal collection.[1]

An alternate approach is to use a dual-tracer technique to measure calcium absorption (Table 8.3.1). In this technique, one calcium isotope is given orally and a different isotope is given intravenously.[1,7] With radioisotopes, a variation on this approach is to give $^{45}$Ca orally on one day and the same tracer intravenously a few weeks later. Although varying clinical protocols have been used for this method, we prefer to administer the tracer orally with a meal and a dietary source of calcium such as milk or calcium-fortified juice. In our protocols, early in the morning of the study, subjects are instructed to empty their bladders, and given breakfast. Toward the end of the meal, the subjects are given an isotope of calcium that has been premixed (and allowed to equilibrate in the refrigerator for 12–24 h) with the milk or juice. After breakfast, a different calcium isotope is administered intravenously over 2–3 min.[1,8–10] In some cases, especially if usual daily calcium absorption is being determined, it may be best to give isotope with several calcium-containing meals during the day in the approximate proportion of the total calcium in the meals.[11]

**Table 8.3.1.** *Mineral stable isotopes frequently used in pediatric nutritional research*

| Isotope studies | Natural abundance (%) | Typical dose for mineral absorption |
|---|---|---|
| $^{42}$Ca | 0.65 | 1–2 mg IV (3–6 mg if given orally) |
| $^{44}$Ca | 2.08 | 3–6 mg IV (10–15 mg if given orally) |
| $^{46}$Ca | 0.0032 | 15–20 μg given orally |

Isotope costs (typical costs as of 2003). $^{42}$Ca: $60 per mg (90–94% enriched); $^{44}$Ca: $18 per mg (95–99% enriched); $^{46}$Ca: $3 per μg of $^{46}$Ca. Typically $^{46}$Ca is provided as approximately 5% enriched material. So to provide 20 μg of $^{46}$Ca requires 20/0.05 = 400 μg of 5% enriched material. This material typically costs $150 per mg for the 5% enriched isotope. Therefore, one can calculate a typical cost of 150/0.05 = $3000 per mg = $3 per μg of 'pure' $^{46}$Ca.

After administration of the tracers, a complete 24-h urine collection is carried out. The relative fractions of the oral and the IV tracer doses in this 24-h urine pool are measured. This ratio represents the fraction of the oral tracer dose that was absorbed and is presumed to represent the fraction of the dietary calcium absorbed from that load. Spot determinations of urine or serum isotope levels of the tracers may also be used. However, for calcium, this method may not be as accurate as that determined from complete 24-h collections.[12] Furthermore, the spot serum method requires relative timing of the oral and the IV doses. In the 24-h urine method, because absorption is calculated from total urinary isotope recovery, it is not necessary to exactly sequence the time of administration of the oral and IV isotopes.

The single- and dual-tracer stable isotope techniques can be carried out using either stable or radioactive tracers. Studies using strontium as a marker for calcium have also been used.[13] Issues regarding the choice between stable or radioactive tracers, and between calcium and strontium relate to safety, cost, analytical availability and subject acceptance of the different isotopes. In our research group's view, only stable isotopes of calcium may ethically be used for children or during pregnancy and lactation. Currently, the cost of the isotopes for a typical absorption study in an average-sized adult is approximately $100 per study indicating that cost may not ultimately be the major issue in making this choice, even in healthy adults.

## Whole Body Bone Mineral Measurements

In the rapidly growing skeleton, substantial changes in whole body bone mineral content occur during a period of 6–12 months. Because of the consistent relationship between whole body bone mineral and whole body calcium content,[14] it is possible to calculate the mean calcium retention during a period of time based on such data. It is clearly best to use longitudinal data in individuals for this calculation, but cross-sectional data have also been used.[15] A limitation of this approach is that it determines retention, not absorption, but in children, this is not a major problem due to the primary importance of absorption in determining balance and the rapid growth of the skeleton. Furthermore, it only determines the mean retention between the age points, so it cannot distinguish dietary interactions or readily allow for direct comparisons of different calcium sources. Still, this is a useful adjunct to the other techniques especially in studies during puberty or infancy.

# Calcium Absorption from Human Milk Compared to Infant Formula

The challenge of determining the role of calcium absorption in bone health begins with infants. In this case, two potentially conflicting guidelines come into consideration. The first is the global understanding that human milk is an optimal nutritional source as a single food source in the first six months of life and as a primary milk source throughout the first year of life.[16] For calcium, this means that dietary requirements for calcium are based primarily on the intake of calcium from their mother's milk[17,18] as relatively little calcium is derived from most infant foods.

In potential conflict with this, however, is the demonstration that infant formulas have been shown to lead to slightly greater bone mineralization in both the whole body and some skeletal regions than those obtained in human milk-fed babies. However, this increased bone mineral content may not persist even during later infancy.[18,19]

There is a general consensus that calcium is well absorbed from human milk, with values for net calcium retention of about 50% of intake.[20,21] Calcium absorption values from infant formulas are highly variable due to the various carbohydrate, protein and mineral sources of these formulas. Although it is generally stated that calcium bioavailability is lower from formulas than from human milk, this may not always be the case. Early findings may have been related to the greater concentration of calcium in infant formulas than in human milk.[18,20] In general, however, values of 30–40% absorption are typical for cow's milk-based infant formulas or whole cow's milk.[22]

Since the Infant Formula Act in the United States, and numerous expert committee recommendations, advise or require higher concentrations of calcium in formulas than human milk,[17,18,22] calcium bioavailability comparisons at identical calcium concentration in human milk and formula have not been performed. However, several studies[23,24] have shown fractional calcium absorption from infant formulas very similar to the value for human milk.

These data indicate that it is possible for formula-fed infants to exceed the calcium absorbed and bone mass accumulated during the first 6 months of life by human milk-fed infants. It remains unknown, however, whether that is a worthwhile goal. There are no data to support any long-term benefit, in terms of either increased peak bone mass or the prevention of osteoporosis, to be gained by exceeding calcium absorption or bone mass in early infancy.[25] Animal data further support the idea that increases in calcium intake in early childhood do not have beneficial effects on long-term bone mass.[26] Ultimately, long-term research must be done before it is appropriate to advocate or target a higher calcium retention or bone mass accretion in artificially fed infants relative to the human milk-fed standard.

# Effects of Puberty, Ethnicity, and Other Factors on Calcium Absorption

## Pubertal Changes

Population and age-related variability in calcium absorption is largely related to factors that are not readily controlled such as pubertal status, ethnicity, and genetic factors including polymorphisms of the vitamin D receptor (VDR) related genes as well as other genes controlling estrogen and growth hormone.[27]

Calcium absorption increases substantially during puberty and rapidly decreases post-puberty.[28] In a longitudinal multiethnic study, we found a significant increase in the utilization of calcium associated with the onset of the physical development of puberty. This change was evidenced by increased calcium absorption, and kinetically determined rates of bone calcium deposition.[29] These findings support the importance of advocating adequate mineral nutrition in girls during early puberty.

We found that increases in calcium absorption and deposition were associated with maturation of the hypothalamic–pituitary axis as measured by a rise in the gonadotropin-releasing hormone-simulated luteinizing hormone (LH) level. This increase occurred after initial increases in estradiol in most girls. Small increases in estradiol are probably the earliest marker of puberty and occur before physical changes are observed. This is to be expected, as estradiol stimulates these physical changes. In this study, we found that changes in calcium metabolism were concomitant with physical changes of puberty and were likely the consequence of these hormonal changes. Recently, McKay and co-workers[30] have reported the changes in total body bone mineral content in girls using a DXA technique. From their data, the calculated increments in calcium gain are 110 mg day$^{-1}$ at age 10 and 140 mg day$^{-1}$ at age 11. Of note is that they reported a maximum increment of 260 mg day$^{-1}$ at age 12–13. Although this value is slightly higher than the peak value of 212 mg day$^{-1}$ reported by Martin *et al.*[15] for girls, it supports our data that calcium absorption peaks relatively early in puberty in girls. These more recent values for peak rates of calcium gain are very similar to the estimates derived from rates of weight change by Leitch and Aitken in the 1950s.[31]

## Ethnicity

There are marked ethnic differences in peak bone mass and the incidence of osteoporosis between African–Americans and Caucasians. We have demonstrated that African–American girls absorb more calcium than Caucasians, at similar calcium intakes.[9] This difference is not found in adults however, where calcium absorption is reported similar in African–American and Caucasians. A lower daily urinary calcium excretion has been reported in both adolescent and adult African–Americans compared with Caucasians.[32,33] There are fewer data comparing other ethnic groups in their ability to absorb calcium. We found no difference in prepubertal Mexican–American and Caucasian children in calcium absorption.[34] Differences in calcium absorption between population groups of Asian origin and other population groups have not been systematically studied.

## Genetic Polymorphisms and Calcium Absorption

Numerous recent studies have focused on identifying the mechanisms of the relationship between genetics and osteoporosis by evaluating specific genetic markers and their relationship to bone mass. One of the unknown aspects of this important relationship is the phenotypic mechanism by which the various candidate genes act. It is probable that calcium absorption, excretion and/or bone formation/resorption are altered in subjects with genotypes related to low bone mass. Because of the importance of puberty in bone mass, it is likely that an effect can be seen during pubertal development or sooner. Ultimately, early identification of abnormalities in calcium metabolism may lead to earlier dietary or other interventional approaches.

However, there are few data on the relationships in either adults or children between calcium absorption and putative genetic markers of bone mineral status. A single study by Dawson-Hughes *et al.*[35] found a significant difference in calcium absorption associated with VDR (Bsm-1 site) genotype in adults, but the effect was not large and has not been verified in other studies.

We reported[27] a significant relationship between polymorphisms of the VDR *Fok 1* genotype and calcium absorption. Children with the *FF* genotype absorbed on average 115 mg day$^{-1}$ more calcium than those with the *ff* genotype. We found that pubertal status, but not ethnicity, was associated with the amount of calcium absorbed in this study population. The association of genotype with calcium absorption was significant both when prepubertal children were analyzed separately and when all subjects were analyzed together with pubertal status included in the statistical model as a co-variate. Furthermore, we found the *FF* genotype to be associated with greater bone mineral density in the study subjects.

## Pregnancy

Calcium absorption is significantly increased during pregnancy as demonstrated in an early cross-sectional comparison by Heaney and Skillman[36] and confirmed in a longitudinal study by Ritchie *et al.*[37] The magnitude of increased absorption is quite high, representing an increase of 50% or more of usual absorptive levels. The additional absorbed calcium is comparable to the amount of calcium needed by the fetus so that there is little if any maternal skeletal loss during pregnancy.

During lactation however, calcium absorption declines and is either similar to, or only slightly greater than prior to pregnancy. Therefore, calcium must primarily be derived from the maternal bones to provide for the needed calcium from breast milk. It is of interest that studies have failed to document a significant maternal bone benefit to supplementing mothers with high amounts of calcium during pregnancy and lactation. Post-lactation, the catch-up in bone mineral may be related in part to a small increase in calcium absorption, but this is unlikely to be the primary mechanism of bone mineral catch-up in adults.[18]

## Aging

Many studies have documented an age-related decrease in calcium absorption in adults. In women, there is an abrupt decrease in calcium absorptive efficiency at menopause as shown in a study by Wishart *et al.*[38] In that 18-month longitudinal study, absorption was decreased in those with hormonal changes or irregular menses suggesting early menopause. Calcium absorption, however, was not decreased in women of similar ages who did not have evidence of early menopause.

In addition, an age-related decrease in calcium absorption appears to occur independent of the menopause as well.[39] It has been estimated that the overall decline in women of absorptive efficiency at age 60 is about 20–25% of their absorption fraction at age 40. The etiology of the change is less certain, but may be related, in part, to lower vitamin D status. However, variations in serum 1,25 dihydroxyvitamin D concentration do not seem to account for most of the variance in the age-related change in calcium absorption,[40] but women may develop resistance to 1,25 dihydroxyvitamin D as the age[41] which may contribute to their decreased calcium absorption.

Of importance is that this decreased calcium absorption may be directly related to an increased fracture risk.[42] Ensrud *et al.*[43] recently reported that women over age 69 with a low fractional absorption of calcium had an increased risk of hip fracture. They found that among women whose dietary calcium intake was less than 400 mg day$^{-1}$, those who had fractional calcium absorption at or below the median value of 32.3% had over twice the risk of hip fracture compared with those who had greater levels of calcium absorption.

# Use of Modifiers of Calcium Absorption (*e.g.* Highly Bioavailable Salts, Prebiotics)

The low population intakes relative to dietary requirements has led to the use of calcium supplements, usually in the form of pills, by a substantial proportion of the population. This has led, on competitive ground, to a substantial interest in identifying the most bioavailable supplement sources of calcium. Not unexpectedly, however, this has also led to competitive claims regarding these supplements the merits of which may not always be clear.

For example, it has been suggested based on a meta-analysis of a series of studies that calcium citrate and/or calcium citrate malate are more bioavailable than the less-expensive calcium carbonate.[44] The medical literature on this topic is difficult to evaluate due to large variations in the methods used for assessing calcium absorption as well as the subject population and size of supplement used. If there is a difference between these sources, it is likely, especially for calcium citrate, to be of small magnitude.[2] It is important to realize, however, in recommending a single supplement that the magnitude of difference in absorption may more readily be overcome by increasing the intake at a lower cost than purchasing a more expensive supplement. This perspective, well known to those providing iron supplements as part of global relief efforts, may be lost on the consumer. For example, using numbers that are typical for these types of supplements, if a 500 mg calcium supplement has 24% absorption, then providing 600 mg of a much lower priced supplement with 20% absorption would have the same effect.[2] Furthermore, other issues, including patient compliance, taste, calcium load, presence of absorption inhibitors such as phytates, and timing of dosing may all affect the long-term net calcium absorbed more than a small difference in bioavailability.[5]

## Prebiotics

An alternative dietary strategy to enhancing net absorbed calcium is to identify dietary strategies that enhance the calcium bioavailability of the whole diet and that have other health benefits. For example, functional foods including prebiotics such as non-digestible oligosaccharides (NDO) may be of benefit.

The addition of NDO to the diet may have numerous health benefits including a decreased incidence of some malignancies, improved glucose tolerance, improvement in blood lipid profiles, and a decreased risk of cardiovascular disease. Several NDO products have been shown to increase calcium absorption in experimental animals,[45,46] but data from human subjects have been less clear.

The current literature (Table 8.3.2) has involved studies of relatively large intakes of oligosaccharides, with short adaptation periods. Few data were available in adolescents, in whom the beneficial effects, if any, of NDO might be particularly important.[47–50]

The varying results between some earlier studies may also reflect methodological differences, or they may suggest that the benefit, if any, of NDO differs amongst different populations. Whether these differences might relate to genotypic variations (genetic polymorphisms), ethnicity, the subjects normal calcium intake, vitamin D status, the intake of other NDOs in the diet, or some other factor remains to be seen.

**Table 8.3.2.**  *Results of previous studies of calcium absorption from prebiotics*

|  | Coudray et al. 1997[7] | Van den Heuvel et al. 1998[48] |  | Van den Heuvel et al. 1999[49] |
|---|---|---|---|---|
| Oligosaccharide (g day$^{-1}$) | 40 (Inulin) | 15 (Inulin) | 15 (Oligo-fructose) | 15 g day$^{-1}$ Oligofructose |
| $N =$ | 9 | 12 |  | 12 |
| Calcium intake (mg day$^{-1}$) | ~850 | 955 |  | ≈1250 mg day$^{-1}$ |
| Adaptation period (days) | 20 | 21 |  | 9 |
| Method | Metabolic balance | Stable isotope 24 h urine |  | Stable isotope 36 h urine |
| *Ca absorption (mean±SD)* |  |  |  |  |
| Placebo (%) | 21.3±12.5 | 28.1±14.9 | 28.1±14.9 | 47.8±16.4 |
| Oligosaccharide (%) | 33.7±12.1 | 25.8±8.0 | 26.3±6.6 | 60.1±17.2 |
| *p*-value | <0.01 | NS | NS | <0.05 |

We recently completed a study of the effect of 8 g day$^{-1}$ of synergy1 (an NDO composed of a mixture of long- and short-chain length molecules) on calcium absorption in young girls (aged 11–13.9 years).[51] Subjects received in random order 8 g day$^{-1}$ of NDO (either synergy1 or oligofructose) and placebo (sucrose), added to a diet providing approximately 1200–1300 mg day$^{-1}$ calcium. Calcium absorption was measured after 21 days adaptation to the NDO or placebo using a stable isotope method. We found a significant increase in calcium absorption whilst consuming NDOs. Calcium absorption was significantly higher when subjects consumed 8 g day$^{-1}$ Synergy 1 than when consuming placebo, but no significant benefit was seen from 8 g day$^{-1}$ of oligofructose (Table 8.3.3). The increase in calcium absorption (32.3–38.2%) represents a relative increase of more than 18%. A change of this magnitude is highly clinically significant. Of importance, this effect was seen at a relatively high calcium intake, where simply increasing the amount of calcium in the diet would be unlikely to significantly increase calcium absorption.

**Table 8.3.3.**  *Results from a recent study[51] of calcium absorption from prebiotics*

|  | Calcium absorption (%) | p-value, vs. sucrose |
|---|---|---|
| Study I |  |  |
| Sucrose | 30.9±10.0 |  |
| Oligofructose | 31.8±9.3 | 0.75 |
| Study II |  |  |
| Sucrose | 32.3±9.8 |  |
| Fortified inulin | 38.2±9.8 | 0.007 |

The mechanism by which oligosaccharides might increase calcium absorption is not known. NDOs resist digestion in the human gut, but are fermented to volatile fatty acids in the colon.[52] These fatty acids may have a local effect in the colon by reducing the pH and increasing solubility of mineral in the aqueous phase of the colonic contents[53] permitting higher absorption of minerals in the colon, a site where little calcium absorption normally occurs. Alternatively the NDO, volatile fatty acids, or some other mediator may have a trophic effect on the gut,[46] improving overall 'gut health'. Such an effect is possible *via* a number of hormonal mediators that may be increased following regular consumption of NDO. Such an effect could increase calcium absorption throughout the length of the gastro-intestinal tract. Finally, NDOs may alter the composition of the colonic bacterial flora[53] and this might lead, directly or indirectly, to a change in mineral absorption of overall gut health.

## Other Factors Affecting Calcium Absorption

Disease processes can affect fractional absorption. Primary hyperparathyroidism has been associated with an increase in calcium absorption although this is probably an indirect effect related to increased 1,25 dihydroxyvitamin D concentrations in this condition.[54]

Of considerable importance are the well-described adverse effects of corticosteroid use on bone health. In particular, diseases related to increased cortisol levels or in which corticosteroids are administered orally or intravenously have been shown to decrease calcium absorption. This effect has been demonstrated in healthy adults given dexamethasone.[55]

We have found much lower rates of calcium absorption in adolescents with anorexia nervosa, even when placed on a high calcium diet than in healthy adolescents. In children with juvenile rheumatic conditions, we have also found low rates of calcium absorption. As these are likely to affect the whole diet, they can have a substantial effect on bone health and bone mass and are not easily dealt with by increasing calcium intake.[56,57]

The effect of cystic fibrosis on calcium absorption has recently been evaluated. Aris *et al.*[58] found that young adults with cystic fibrosis had low rates of calcium absorption that were not entirely resolved by therapy with pancreatic enzymes. They hypothesize that low vitamin D levels may be the etiology of the low absorption. Low vitamin D levels may also decrease calcium in neurological disorders including cerebral palsy.

Calcium absorption is decreased during long-term space flight[59] that is likely related to the lack of gravitational effects on the skeleton. This effect is in addition to bone resorption, which occurs during space flight. Although this condition is not encountered frequently in the clinical setting, it has profound implications for the on-going exploration of space, either for prolonged stays on the International Space Station, or for further manned exploration of the inner planets. Furthermore, insights from this extreme state may provide insights in to the bone loss that accompanies bed rest and prolonged inactivity such as may occur post-operatively. Whether the effect of weightlessness occurs rapidly upon entering low gravity was under investigation in an experiment conducted on the U.S. space shuttle mission STS-107. Four of the astronauts (Michael Anderson, David Brown, Laurel Clark, and Ilan Ramon) who perished on February 1, 2003 had participated in a calcium absorption and kinetic study which would have provided the first information regarding the nature and extent of decreased calcium absorption and increased bone resorption in the first days after launch into space. A recent study conducted in space demonstrated a marked decrease in calcium absorption towards the end of a 20-day mission.[13] Supporting this is the finding in long-term bed-rest studies of a decrease in gastrointestinal tract calcium absorption during prolonged immobility.[60]

A recent study in Australia[61] has suggested that in post-menopausal women, suppression of the parathyroid hormone–vitamin D axis from cigarette smoking is associated with decreased calcium absorption. This is similar to a finding of lower calcium absorption in elderly men and women[62] in the United States. Although the magnitude of the decrease in absorption was relatively small in both cases, as it likely affects the whole diet and is a prolonged effect, could have substantial impact on bone health.

## Conclusions

The measurement of calcium absorption, although traditionally performed using mass balances may not more readily be performed using stable isotopes. Calcium absorption is increased in some population groups, such as African-Americans and increases during pubertal development. It also goes up markedly during pregnancy to meet the fetal calcium needs. In developing plans to enhance bone mineralization, the use of dietary enhancers of calcium absorption such as prebiotics may be useful and is currently being evaluated. Decreases in calcium absorption are related to the aging process, smoking and long-term inactivity such as is associated with bed-rest or space travel.

## Acknowledgement

This work is a publication of the U.S. Department of Agriculture (USDA)/ Agricultural Research Service (ARS) Children's Nutrition Research Center, Department of Pediatrics, Baylor College of Medicine and Texas Children's Hospital, Houston, TX. This project has been funded in part with federal funds from the USDA/ARS under Cooperative Agreement number 58-6250-6-001. Contents of this publication do not necessarily reflect the views or policies of the USDA, nor does mention of trade names, commercial products, or organizations imply endorsement by the U.S. Government.

# References

1. S.A. Abrams, Using stable isotopes to assess mineral absorption and utilization by children, *Am. J. Clin. Nutr.*, 1999, **70**(6), 955–964.
2. R.P. Heaney, Factors influencing the measurement of bioavailability, taking calcium as a model, *J. Nutr.*, 2001, **131**(4 Suppl), 1344S–1348S.
3. R.P. Heaney and R.R. Recker, Determinants of endogenous fecal calcium in healthy women, *J. Bone Miner. Res.*, 1994, **9**(10), 1621–1627.
4. V. Matkovic and R.P. Heaney, Calcium balance during human growth: evidence for threshold behavior, *Am. J. Clin. Nutr.*, 1992, **55**(5), 992–996.
5. P. Charles, Calcium absorption and calcium bioavailability, *J. Intern. Med.*, 1992, **231**(2), 161–168.
6. R.P. Heaney, S.D. Dowell, J. Bierman, C.A. Hale and A. Bendich, Absorbability and cost effectiveness in calcium supplementation, *J. Am. Coll. Nutr.*, 2001, **20**(3), 239–246.
7. F. Bronner, Experimental studies of calcium absorption in man, *Nutritio Dieta*, 1962, **3**, 22–31.
8. S.A. Abrams, M.A. Grusak, J. Stuff and K.O. O'Brien, Calcium and magnesium balance in 9–14-y-old children, *Am. J. Clin. Nutr.*, 1997, **66**(5), 1172–1177.
9. S.A. Abrams, K.O. O'Brien, L.K. Liang and J.E. Stuff, Differences in calcium absorption and kinetics between black and white girls age 5–16 years, *J. Bone Miner. Res.*, 1995, **10**(5), 829–833.
10. S.A. Abrams and J.E. Stuff, Calcium metabolism in girls: current dietary intakes lead to low rates of calcium absorption and retention during puberty, *Am. J. Clin. Nutr.*, 1994, **60**(5), 739–743.
11. R. Eastell, N.E. Vieira, A.L. Yergey and B.L. Riggs, One-day test using stable isotopes to measure true fractional calcium absorption, *J. Bone Miner. Res.*, 1989, **4**(4), 463–468.
12. A.L. Yergey, S.A. Abrams, N.E. Vieira, A. Aldroubi, J. Marini and J.B. Sidbury, Determination of fractional absorption of dietary calcium in humans, *J. Nutr.*, 1994, **124**(5), 674–682.
13. A. Zittermann, M. Heer, A. Caillot-Augusso, P. Rettberg, K. Scheld, C. Drummer, C. Alexandre, G. Horneck, D. Vorobiev and P. Stehle, Microgravity inhibits intestinal calcium absorption as shown by a stable strontium test, *Eur. J. Clin. Investig.*, 2000, **20**(12), 1036–1043.
14. K.J. Ellis, R.J. Shypailo, A. Hergenroeder, M. Perez and S. Abrams, Total body calcium and bone mineral content: comparison of dual-energy X-ray absorptiometry (DXA) with neutron activation analysis (NAA), *J. Bone Miner. Res.*, 1996, **11**(6), 843–848.
15. A.D. Martin, D.A. Bailey, H.A. McKay and S. Whiting, Bone mineral and calcium accretion during puberty, *Am. J. Clin. Nutr.*, 1997, **66**(3), 611–615.
16. Committee on Nutrition, American Academy of Pediatrics, Breastfeeding and the use of human milk, *Pediatrics*, 1997, **100**, 1035–1039.
17. Committee on Nutrition, American Academy of Pediatrics, Calcium requirements of infants, children, and adolescents, *Pediatrics*, 1999, **104**(5 pt 1), 1152–1157.
18. Institute of medicine food and nutrition board's standing committee on the scientific evaluation of dietary intervals. Calcium. In *Dietary Reference Intervals for Calcium, Phosphorus, Magnesium, Vitamin D and Fluoride*, National Academy Press, Washington DC, 1997, pp. 71–146.

19. B.L. Specker, A. Beck, H. Kalfwarf and M. Ho, Randomized trial of varying mineral intake on total body bone mineral accretion during the first year of life, *Pediatrics*, 1997, **99**(6), E12.

20. S.J. Fomon and S.E. Nelson, Calcium, phosphorus, magnesium, and sulfur, In *Nutrition of Normal Infants*, S.J. Fomon (ed.), Mosby-Year Book, Inc, St. Louis, 1993, pp. 192–218.

21. S.A. Abrams, J. Wen and J.E. Stuff, Absorption of calcium, zinc and iron from breast milk by 5- to 7-month-old infants, *Pediatr. Res.*, 1997, **41**(3), 384–390.

22. Life Sciences Research Office (LSRO) report. Assessment of nutrient requirements for infant formulas, *J. Nutr.*, 1998, **128**(11 Suppl):i–iv, 2059S–2293S.

23. S.A. Abrams, I.J. Griffin and P.M. Davila, Calcium and zinc absorption from lactose-containing and lactose-free infant formulas, *Am. J. Clin. Nutr.*, 2002 (In Press).

24. C.L. Lifschitz and S.A. Abrams, Addition of rice cereal to formula does not impair mineral bioavailability, *J. Pediatr. Gastroenterol. Nutr.*, 1998, **26**(2), 175–178.

25. G. Jones, M. Riley and T. Dwyer, Breastfeeding in early life and bone mass in prepubertal children: a longitudinal study, *Osteoporosis Int.*, 2000, **11**, 146–152.

26. R.I. Gafni, E.F. McCarthy, T. Hatcher, J.L. Meyers, N. Inoue, C. Reddy, M. Weise, K.M. Barnes, V. Abad and J. Baron, Recovery from osteoporosis through skeletal growth: early bone mass acquisition has little effect on adult bone density, *FASEB J.*, 2002, **16**(7), 736–738.

27. S.K. Ames, K.J. Ellis, S.K. Gunn, K.C. Copeland and S.A. Abrams, Vitamin D receptor gene *Fok*1 polymorphism predicts calcium absorption and bone mineral density in children, *J. Bone Miner. Res.*, 1999, **14**(5), 740–746.

28. F. Bronner and S.A. Abrams, Development and regulation of calcium metabolism in healthy girls, *J. Nutr.*, 1998, **128**(9), 1474–1480.

29. S.A. Abrams, K.C. Copeland, S.K. Gunn, C.M. Gundberg, K.O. Klein and K.J. Ellis, Calcium absorption, bone accretion and kinetics increase during early pubertal development in girls, *J. Clin. Endocrinol. Metab.*, 2000, **85**(5), 1805–1808.

30. H.A. McKay, D.A. Bailey, R.L. Mirwald, S. Davison and R.A. Faulkner, Peak bone mineral accrual and age at menarche in adolescent girls: a 6-year longitudinal study, *J. Pediatr.*, 1998, **133**(5), 682–687.

31. I. Leitch and F.C. Aitken, The estimation of calcium requirement: a re-examination, *Nutr. Abstr. Rev.*, 1959, **29**, 393–409.

32. N.H. Bell, A.L. Yergey, N.E. Vieira, M.J. Oexmann and J.R. Shary, Demonstration of a difference in urinary calcium, not calcium absorption, in black and white adolescents, *J. Bone Miner. Res.*, 1993, **8**(9), 1111–1115.

33. B. Dawson-Hughes, S.S. Harris, S. Finneran and H.M. Rasmussen, Calcium absorption responses to calcitriol in black and white premenopausal women, *J. Clin. Endocrinol. Metab.*, 1995, **80**(10), 3068–3072.

34. S.A. Abrams, K.C. Copeland, S.K. Gunn, J.E. Stuff, L.L. Clark and K.J. Ellis, Calcium absorption and kinetics are similar in 7- and 8-year-old Mexican-American and Caucasian girls despite hormonal differences, *J. Nutr.*, 1999, **129**(3), 666–671.

35. S.S. Harris, T.R. Eccleshall, C. Gross, B. Dawson-Hughes and D. Feldman, The vitamin D receptor start codon polymorphism (FokI) and bone mineral density in premenopausal American black and white women, *J. Bone Miner. Res.*, 1997, **12**(7), 1043–1048.

36. R.P. Heaney and T.G. Skillman, Calcium metabolism in normal human pregnancy, *J. Clin. Endocrinol. Metab.*, 1971, **33**(4), 661–670.

37. L.D. Ritchie, E.B. Fung, B.P. Halloran, J.R. Turnland, M.D. Van Loan and D.E. Cann, A longitudinal study of calcium homeostasis during human pregnancy and lactation and after resumption of menses, *Am. J. Clin. Nutr.*, 1998, **67**(4), 693–701.

38. J.M. Wishart, F. Scopacase, M. Horowitz, H.A. Morris, A.G. Need, P.M. Clifton and B.E. Nordin, Effect of perimenopause on calcium absorption: a longitudinal study. *Climacteric*, 2000, **3**(2), 102–108.

39. R.P. Heaney, R.R. Recker, M.R. Stegman and A.J. Moy, Calcium absorption in women: relationships to calcium intake, estrogen status, and age. *J. Bone Miner. Res.*, 1989, **4**(4), 469–475.

40. H.K. Kinyamu, J.C. Gallagher, J.M. Prahl, H.F. Deluca, K.M. Petranick and S.J. Lanspa, Association between intestinal vitamin D receptor, calcium absorption, and serum 1,25 dihydroxyvitamin D in normal young and elderly women. *J. Bone Miner. Res.*, 1997, **12**(6), 922–928.

41. S. Pattanaungkul, B.L. Riggs, A.L. Yergey, N.E. Vieira, W.M. O'Fallon and S. Khosla, Relationship of intestinal calcium absorption to 1,25-dihydroxyvitamin D [1,25(OH)2D] levels in young *vs.* elderly women: evidence for age-related intestinal resistance to 1,25(OH)2D action, *J. Clin. Endocrinol. Metab.*, 2000, **85**(11), 4023–4027.

42. R.P. Heaney, S.A. Abrams, B. Dawson-Hughes, A. Looker, R. Marcus, V. Matkovic and C. Weaver, Peak bone mass, *Osteoporosis Int.*, 2000, **11**(12), 985–1009.

43. K.E. Ensrud, T. Duong, J.A. Cauley, R.P. Heaney, R.L. Wolf, E. Harris and S.R. Cummings, Low fractional calcium absorption increases the risk of hip fracture in women with low calcium intake, *Ann. Intern. Med.*, 2000, **132**(5), 345–353.

44. K. Sakhaee, T. Bhuket, B. Adams-Huet and D.S. Rao, Meta-analysis of calcium bioavailability: a comparison of calcium citrate with calcium carbonate, *Am. J. Therapeut.*, 1999, **6**(6), 313–321.

45. R. Brommage, C. Binacua, S. Antille and A.L. Carrie, Intestinal calcium absorption in rats is stimulated by dietary lactulose and other resistant sugars, *J. Nutr.*, 1993, **123**(12), 2186–2194.

46. C. Rémésy, M.A. Levrat, L. Gamet and C. Demigne, Cecal fermentations in rats fed oligosaccharides (inulin) are modulated by dietary calcium level, *Am. J. Physiol.*, 1993, **264**(5 pt 1), G855–G862.

47. C. Coudray, J. Bellanger, C. Castiglia-Delavaud, C. Rémésy, M. Vermorel and Y. Rayssignuier, Effect of soluble or partly soluble dietary fibres supplementation on absorption and balance of calcium, magnesium, iron and zinc in healthy young men, *Eur. J. Clin. Nutr.*, 1997, **51**(6), 375–380.

48. E.G. van den Heuvel, G. Schaafsma, T. Muys and W. van Dokkum, Non-digestible oligosaccharides do not interfere with calcium and nonheme-iron absorption in young, healthy men, *Am. J. Clin. Nutr.*, 1998, **67**(3), 445–451.

49. E.G. van den Heuvel, T. Muys, W. van Dokkum and G. Schaafsma, Oligofructose stimulates calcium absorption in adolescents, *Am. J. Clin. Nutr.*, 1999, **69**(3), 544–548.

50. C. Coudray and S.J. Fairweather-Tait, Do oligosaccharides affect the intestinal absorption of calcium in humans? *Am. J. Clin. Nutr.*, 1998, **68**(4), 921–923.

51. I.J. Griffin, P.M. Davila and S.A. Abrams, Non-digestible oligosaccharides and calcium absorption in girls with adequate calcium intakes, *Br. J. Nutr.*, 2002, **87**(Suppl. 2), 187–191.

52. J. Van Loo, J. Cummings, N. Delzenne, H. Englyst, A. Franck, M. Hopkins, N. Kok, G. Macfarlane, D. Newton, M. Quigley, M. Roberfroid, T. van Vliet and E. van den Heuvel, Functional food properties of non-digestible oligosaccharides: a consensus report from the ENDO project (DGXII AIRII-CT94-1095), *Br. J. Nutr.*, 1999, **81**(2), 121–132.

53. M.B. Roberfroid, Concepts in functional foods: the case of inulin and oligofructose, *J. Nutr.*, 1999, **129**(7 Suppl.), 1398S–1401S.

54. P. Charles, L. Mosekilde and F.T. Jensen, Primary hyperparathyroidism: evaluated by $^{47}$Calcium kinetics, calcium balance and serum bone-Gla-protein, *Eur. J. Clin. Investig.*, 1986, **16**(4), 277–283.

55. B.L. Wajchenberg, V.G. Pereira, J. Kieffer and S. Ursic, Effect of dexamethasone on calcium metabolism and 47Ca kinetics in normal subjects, *Acta Endocrinol. (Copenhagen)*, 1969, **61**(1), 173–192.

56. M.D. Perez, S.A. Abrams, L. Loddeke, R. Shypailo and K.J. Ellis, Effects of rheumatic disease and corticosteroid treatment on calcium metabolism and bone density in children assessed 1 year after diagnosis, using stable isotopes and dual-energy X-ray absorptiometry, *J. Rheumatol.*, 2000, **27**(Suppl. 58), 38–43.

57. M.D. Perez, S.A. Abrams, G. Koenning, J. Stuff, K.O. O'Brien and K.J. Ellis, Mineral metabolism in children with dermatomyositis, *J. Rheumatol.*, 1994, **21**(12), 2364–2369.

58. R.M. Aris, G.E. Lester, S. Dingman and D. Ontjes, Altered calcium homeostasis in adults with cystic fibrosis, *Osteoporosis Int.*, 1999, **10**(2), 102–108.

59. S.M. Smith, M.E. Wastney, B.V. Morukov, I.M. Larina, L.E. Nyquist, S.A. Abrams, E.N. Taran, C.Y. Shih, J.L. Nillen, J.E. Davis-Street, B.L. Rice and H.W. Lane, Calcium metabolism before, during, and after a 3-month space flight: kinetic and biochemical changes, *Am. J. Physiology*, 1999, **277**(1 pt 2), R1–R10.

60. A. LeBlanc, V. Schneider, E. Spector, H. Evans, R. Rowe, H. Lane, L. Demers and A. Lipton, Calcium absorption, endogenous excretion, and endocrine changes during and after long-term bed rest, *Bone*, 1995, **16**(4 Suppl.), 301S–304S.

61. A.G. Need, A. Kemp, N. Giles, H.A. Morris, M. Horowitz and B.E. Nordin, Relationships between intestinal calcium absorption, serum vitamin D metabolites and smoking in postmenopausal women, *Osteoporos Int.*, 2002, **13**(1), 83–88.

62. E.A. Krall and B. Dawson-Hughes, Smoking increases bone loss and decreases intestinal calcium absorption, *J. Bone Miner. Res.*, 1999, **14**(2), 215–220.

CHAPTER 9

# Vitamin D and Fracture Prevention

IAN R. REID

Department of Medicine, University of Auckland, Private Bag 92019, Auckland, New Zealand, Email: i.reid@auckland.ac.nz

**Key Points**

- In humans, most vitamin D (calciferol) is synthesised in the skin as a result of exposure to sunlight. Therefore, vitamin D deficiency is mainly seen in those having little sunlight exposure (*e.g.* because of frailty or style of dress) and in those with dark skin.
- Vitamin D has little bioactivity, and is hydroxylated in the liver to form 25-hydroxyvitamin D, which is bioactive. It is further hydroxylated in the kidney to form 1,25-dihydroxyvitamin D, which is much more active, though it circulates in much lower concentrations than 25-hydroxyvitamin D. Thus, both metabolites contribute to the biological activity of the vitamin D system.
- Vitamin D status is assessed by measurement of serum 25-hydroxyvitamin D concentrations.
- The vitamin D system maintains circulating calcium concentrations by stimulating both intestinal calcium absorption and bone resorption.
- Vitamin D deficiency results in a small decline in the circulating calcium concentration, which leads to an increase in parathyroid hormone (PTH) levels and accelerated bone loss.
- Supplementation with calcium and vitamin D consistently reduces the incidence of fractures in older populations.
- Vitamin D supplementation has little effect on bone density in normal, younger, postmenopausal women who are not vitamin D deficient.

# Introduction

Vitamin D is misnamed in that it is not a vitamin. This problem leads to fundamental confusion regarding its role in health and disease. Vitamins are usually defined as essential dietary constituents. While vitamin D can occur in the diet, endogenous production (in the skin) is quantitatively much more important in most individuals. Therefore, vitamin D should be regarded as a substrate necessary for the synthesis of a family of hormones primarily recognized for their regulation of calcium metabolism. As a result, vitamin D status is not really a nutritional issue, but is determined by sunlight exposure and supplement use. The parent compounds are metabolised to a number of other steroid-like molecules (secosteroids) of varying, and sometimes uncertain, biological activity. Some of these compounds are important regulators of calcium metabolism, though they are now being recognized as having other important activities as well. Some are available as pharmaceuticals, and play important roles in the management of conditions such as vitamin D deficiency, chronic renal failure, hypophosphataemia and hypocalcaemia. Novel pharmaceuticals are being developed within this class, which may find applications in other conditions. While the focus of this chapter is fracture prevention with vitamin D, this area is difficult to fully appreciate without an understanding of the relationships and mechanisms of action of the various members of this hormone family.

## Structure and Regulation of Synthesis of Vitamin D and Its Metabolites

Figure 9.1 shows the chemical structure of vitamin $D_3$, which is the naturally occurring vitamin D in animals. This is a 27-carbon secosteroid, which is derived from 7-dehydrocholesterol as a result of the action of ultraviolet (UV) light. The ultraviolet B spectrum (wavelength 290–320 nm) results in the rupture of the bond between carbon 9 and carbon 10 in 7-dehydrocholesterol, leading ultimately to the production of vitamin $D_3$, or cholecalciferol. A similar process takes place in some plants, leading to the conversion of ergosterol to vitamin $D_2$ (also known as ergocalciferol), which has a methyl group attached to carbon 24, and is widely used as a vitamin D supplement. The term 'calciferol' refers to both chole- and ergo-calciferols. With sustained exposure to sunlight, there is increased production of inactive vitamin D metabolites, providing a mechanism for preventing vitamin D intoxication. Cutaneous vitamin D production is related to the intensity of UVB irradiation, so diminishes with increasing latitude. It is also diminished by skin pigmentation (particularly in individuals of African and Indian origin) and by advancing age. Covering of the skin, whether with clothing or sunscreens, also provides an effective barrier to vitamin D production.

Vitamin $D_2$ and vitamin $D_3$ have usually been regarded as having comparable biological effects, though some recent studies have suggested that

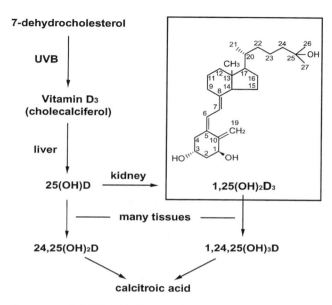

**Figure 9.1** *Structure of 1,25(OH)₂D, and the principal pathways for its production and metabolism*

metabolites in the vitamin D₃ series may have greater biological potency. They are sometimes measured in international units (IU), there being 40 IU per μg. The calciferols are fat-soluble and are absorbed primarily in the jejunum and ileum in a process involving bile acids. Thus, any disease state associated with malabsorption of fat is likely to be associated with malabsorption of vitamin D. However, vitamin D is relatively uncommon in an unsupplemented diet, occurring in fish oils, egg yolk, butter and liver. In the United States, supplementation of food (particularly dairy products) with vitamin D is common, though it has been demonstrated that actual levels of vitamin D in some foods are widely different from those claimed on the packaging.

The first step in the activation of vitamin D is its 25-hydroxylation, to form 25-hydroxyvitamin D [25(OH)D]. This takes place in the parenchymal cells of the liver. There appears to be very little regulation of the 25-hydroxylase, and only in very advanced liver failure does its activity become limiting to normal vitamin D metabolism.

25(OH)D formed in the liver returns to the circulation bound to vitamin D-binding protein. If it is to be further hydroxylated, this takes place in the proximal tubular cells of the kidney. Further activation of 25(OH)D is accomplished by 25-hydroxyvitamin D-1α-hydroxylase, which is up-regulated by PTH and down-regulated by its product, 1,25-dihydroxyvitamin D (1,25(OH)₂D, or calcitriol). Phosphate concentration is also an important regulator of 1α-hydroxylase activity, which is suppressed by high phosphate concentrations and stimulated by low levels. Calcium concentration impacts on

1α-hydroxylase activity mainly through its effect on circulating levels of PTH, though there is some evidence that it can directly influence enzyme activity also. There is also extra-renal synthesis of the hormone, particularly during pregnancy, when the placenta produces 1,25(OH)$_2$D.

An alternative fate for 25(OH)D, is to be hydroxylated on carbon 24, to produce 24,25(OH)$_2$D. This conversion is mediated by vitamin D-24-hydroxylase. This enzyme can also 24-hydroxylate 1,25(OH)$_2$D, and is regarded as the principal pathway for deactivating vitamin D metabolites. As a consequence, it is widely expressed throughout the body, particularly in vitamin D target tissues, including the proximal and distal tubules of the kidney. 1,25(OH)$_2$D is the principal regulator of 24-hydroxylase activity. In the renal proximal tubule, PTH inhibits 24-hydroxylase activity, though it may have an opposite effect in the distal tubule. The induction of 24-hydroxylase by 1,25(OH)$_2$D also leads to accelerated catabolism of 25(OH)D. This explains the low levels of this metabolite seen in patients treated with calcitriol, those suffering from primary hyperparathyroidism, and those with very low calcium intakes. After 24-hydroxylation, vitamin D metabolites undergo further hydroxylations and side-chain cleavage, resulting in the production of water-soluble calcitroic acid, which has no biological activity.

# Actions of Vitamin D and Its Metabolites

## Mechanisms of Action

The classic mechanism of action of vitamin D metabolites on their target tissues is similar to that of other steroid hormones and is mediated by the vitamin D receptor (VDR), a member of the steroid hormone receptor super family. Both 1,25(OH)$_2$D and 25(OH)D bind to this receptor, though the affinity of 1,25(OH)$_2$D for the receptor is a 1000-fold higher. However, it should be remembered that 25(OH)D circulates in concentrations a 1000-fold higher than those of 1,25(OH)$_2$D, so both may contribute to receptor activation. The vitamin D metabolites bind to the VDR which then forms a heterodimeric complex with the retinoic acid X receptor (RXR). This complex then interacts with vitamin D response elements associated with a variety of genes.

In addition to this classic mechanism of action of vitamin D metabolites, rapid responses have been observed that are thought to occur too rapidly to be accounted for by a genomic mechanism of action. These rapid responses include changes in intracellular calcium and a rapid transport of calcium from the intestinal lumen to the circulation, known as transcaltachia. The structure–activity relationships of vitamin D metabolites in stimulating these rapid effects is different from that for the classic genomic effects of vitamin D, and some of these effects have been demonstrated in cells lacking the VDR. These observations have raised the possibility of a different VDR present on the cell membrane.

# Effects on Calcium Metabolism

$1,25(OH)_2D$ is one of the classic calcitropic hormones, its primary function being to elevate serum calcium concentrations. Probably the main way in which this is achieved is by increasing the absorption of calcium in the small intestine. VDRs are found throughout the small intestine but are most abundant in the duodenum. Administration of calcitriol to animals or humans leads to increases in intestinal calcium absorption, though the precise mechanism of this is not fully understood. $1,25(OH)_2D$ regulates production of a calcium binding protein, calbindin D 8k, within the intestinal epithelial cell. This is mediated by a vitamin D response element in the gene for calbindin D9k. It is thought that this calcium binding protein facilitates transport of calcium through the intestinal epithelial cells. $1,25(OH)_2D$ may also facilitate entry of calcium from the intestinal lumen into the epithelial cell, possibly through a channel, and it increases activity of an ATP-dependent calcium pump at the basolateral membrane. $1,25(OH)_2D$ also stimulates intestinal phosphate absorption, though this mostly takes place in the distal small bowel. Thus, it also has an important role in the maintenance of serum phosphate concentrations.

The direct effects of $1,25(OH)_2D$ on bone work in concert with those already described in the intestine, to maintain or increase serum calcium concentrations. $1,25(OH)_2D$ acts on osteoblasts and their precursors, causing the production of RANK-L which binds to pre-osteoclasts to stimulate their development into osteoclasts. This leads to an increase in osteoclastic bone resorption. $1,25(OH)_2D$ also directly stimulates alkaline phosphatase activity and the production of osteopontin and osteocalcin by osteoblasts. While it has been suggested that $1,25(OH)_2D$ may directly influence skeletal mineralization, the balance of evidence suggests that this occurs indirectly, as a result of vitamin D effects on serum calcium and phosphate concentrations.

$1,25(OH)_2D$ also acts on the parathyroid glands, where its binding to the VDR directly decreases expression of the PTH gene. This genomic effect is reinforced by the action of $1,25(OH)_2D$ to increase serum calcium and thus, diminish PTH secretion. The direct effects of vitamin D metabolites on PTH secretion are particularly important in the management of the secondary hyperparathyroidism of chronic renal failure. They may also be important in the prevention and treatment of primary hyperparathyroidism, since this condition may be exacerbated by the vitamin D deficiency (low levels of 25(OH)D) with which it is frequently associated.

When vitamin D deficiency develops, circulating concentrations of PTH increase, leading to increased bone resorption. This accelerates bone loss in the elderly, and probably increases the risk of fractures. In more severe deficiency (serum 25(OH)D 20 nmol $L^{-1}$), the increase in bone resorption is not able to maintain normal levels of serum calcium and phosphate, so normal bone mineralization is interfered with, resulting in osteomalacia. In children, this manifests itself as rickets, with reduced linear growth and deformities around the growth plates.

## Effects on Other Tissues

1,25(OH)$_2$D receptors are expressed in a wide variety of tissues other than those classically involved in mineral and bone homeostasis. In keratinocytes and in some white blood cells or their precursors, 1,25(OH)$_2$D has anti-proliferative and pro-differentiation effects. Thus, in psoriasis the hyper-proliferative state of skin cells appears to be able to be controlled with the use of vitamin D analogues. There is also experimental evidence that some leukaemic cell lines show similar responses to vitamin D metabolites. As a result, there has been an interest in using 1,25(OH)$_2$D for treating leukaemias and other tumours, such as those of the breast, colon and prostate. Despite the existence of promising preliminary results in these areas for a number of years, the vitamin D metabolites are not yet established as having a therapeutic role in any of these malignant conditions. Vitamin D metabolites may also have direct actions on muscle cells, adipocytes, immune regulation, and on endocrine tissues, such as the pancreatic beta cell.

# Assessment of Vitamin D Status

25(OH)D is the principal circulating vitamin D metabolite, and it is the entity that should be assessed when determining an individual's vitamin D status. Because ingested or endogenously produced calciferols are converted to 25(OH)D with very little regulation, serum levels of this metabolite accurately reflect both excess and deficiency states. 25(OH)D circulates bound to vitamin D-binding protein so any condition associated with hypoproteinaemia may produce falsely depressed levels of 25(OH)D.

When assessing vitamin D status, it is important to consider whether the 'normal' range – which varies with latitude – can be regarded as optimal. This has recently been addressed by Malabanan *et al.*,[1] who demonstrated that vitamin D supplementation only suppressed PTH levels in subjects whose baseline serum 25-hydroxyvitamin D was less than 50 nmol L$^{-1}$. This suggests that levels below this threshold are sub-optimal, and are likely to result in increased PTH concentrations and, thus, increased bone resorption. Some cross-sectional studies relating PTH levels to those of 25(OH)D suggest that 25(OH)D concentrations as high as 100 nmol L$^{-1}$ may be necessary to minimise serum PTH. However, achievement of such high vitamin D levels would require almost universal supplementation.

Measurements of serum 1,25(OH)$_2$D are also widely available, but should be little used. Their value is mainly in elucidating the cause of difficult cases of hypercalcaemia, since levels of this metabolite are high when hypercalcaemia is a direct consequence of overproduction of 1,25(OH)$_2$D. This occurs in some granulomatous conditions and lymphomas. The widespread impression that 1,25(OH)$_2$D is the only biologically active vitamin D metabolite frequently leads to inappropriate use of vitamin D assays. This belief results in the expectation that levels of 1,25(OH)$_2$D are all that influences calcium metabolism, whereas this is clearly not the case. As vitamin D deficiency develops, serum 25(OH)D declines. In response to this, secondary hyperparathyroidism develops, and with it there are increases in serum 1,25(OH)$_2$D. Thus, the paradoxical situation can develop of a patient having clinical and histological evidence of osteomalacia, in whom 1,25(OH)$_2$D is either normal or supranormal. Conversely, overdosage with vitamin D will elevate 25(OH)D immediately, but will only push 1,25(OH)$_2$D above the normal range with much more severe intoxication.

# Replacement Doses of Vitamin D in Postmenopausal Osteoporosis

In recent years, there has been an increasing recognition that vitamin D deficiency is common in the elderly, particularly those who are no longer fully independent and therefore less exposed to sunlight. The problem is often greater at higher latitudes, though can also occur in very hot climates where the sun is often avoided because of the heat. Vitamin D deficiency leads to secondary hyperparathyroidism and a resulting increase in bone loss.

## Biochemical and Bone Density Effects

The most physiological way of replacing vitamin D in the housebound elderly is to expose them to sunlight. This approach has been described by Reid et al.[2] Fifteen elderly rest home residents were randomly assigned to three groups: (i) no intervention; (ii) spending 15 min per day outdoors; (iii) spending 30 min per day outdoors. The study was carried out during spring at a latitude of 37° south. There were dose-related increases in serum 25(OH)D levels which had not plateaued by the end of the 1 month study (Figure 9.2). In subjects spending 30 min daily outdoors, serum 25(OH)D levels increased by more than 30%. The changes in the intervention groups were associated with significant increases in intestinal strontium absorption (used as a surrogate for intestinal calcium absorption) and declines in serum alkaline phosphatase activity. Circulating concentrations of 1,25(OH)$_2$D did not change during the study period. These data imply that modest changes in vitamin D status in the frail elderly can produce significant beneficial effects on calcium metabolism.

Larger changes in circulating levels of 25(OH)D can be produced with oral supplementation, and this has recently been reviewed by Vieth.[3] The dose–response is relatively flat up to intakes of several thousand IU day$^{-1}$, as demonstrated in Figure 9.3. In young men receiving oral treatment for 8 weeks, daily calciferol doses of 1000, 10 000, and 50 000 IU increased serum 25(OH)D concentrations by 13, 137, and 883 nmol L$^{-1}$, respectively.[4] Honkanen et al.[5] demonstrated a doubling of 25(OH)D levels in elderly subjects receiving 1800 IU of cholecalciferol daily over an 11 week winter period, during which time the control group showed declines of 30–50% in circulating levels of this metabolite. Comparable effects have been demonstrated by others, and have

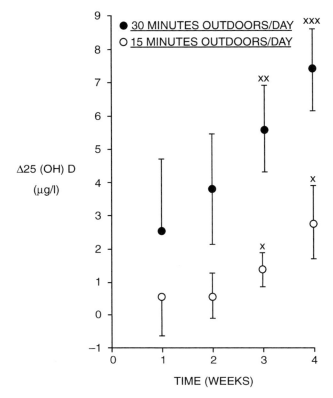

**Figure 9.2**   *The effect of spending time outdoors daily, for the periods indicated, on serum 25-hydroxyvitamin D concentrations in elderly retirement home residents* (From Ref. [2], used with permission)

been associated with reductions in serum PTH concentrations.[6,7] Similar effects can be produced with a number of different dosing regimens. Zeghoud *et al.*[8] raised very low concentrations of 25(OH)D to 25 nmol $L^{-1}$ with 3-monthly oral administration of 100 000 IU of cholecalciferol. Heikinheimo *et al.*[9] found that annual intramuscular injection of 150 000 IU of ergocalciferol achieved a similar end. Vitamin D deficiency can be effectively treated with 500 000 IU of calciferol, either given in a single dose or spread over days or weeks.[10] The addition of calciferol to foods such as milk also results in maintenance of normal vitamin D status.[11,12] Some studies have suggested that cholecalciferol results in larger increments in circulating 25(OH)D levels than does ergocalciferol.[13,14]

The effects of physiological vitamin D supplementation on bone density have been studied in a number of contexts. In normal, early postmenopausal women, it appears to have small if any effects on bone density, probably because such populations are already optimally supplied with the compound. For example, Christiansen *et al.* followed radial bone mineral content in early

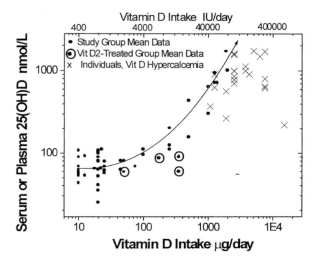

**Figure 9.3.** *Dose–response relationship between daily vitamin D intake and mean 25(OH)D concentration. The solid points show mean results for groups of adults consuming the indicated doses of vitamin D, usually vitamin $D_3$. Results for groups of adults that were definitely consuming vitamin $D_2$, are shown by the circled points. The results represented by 'X' are for individuals who had hypercalcaemia as a result of vitamin D consumption. The toxic intakes were vitamin $D_2$, not $D_3$*
(From R. Vieth, in *Nutritional Aspects of Osteoporosis*, P. Burckhardt, R. Heaney, B. Dawson-Hughes (eds), Academic Press, San Diego, 2001, pp. 173–195. Used with permission)

postmenopausal women over a 2-year period during which the subjects received either vitamin $D_3$ or placebo.[15] There was no difference in rates of bone loss between the groups. Studies in Finnish women have produced similar results.[16,17]

In otherwise normal women whose vitamin D status is marginal, bone density increases with calciferol treatment. Nordin *et al.*[18] conducted a randomised placebo-controlled trial of vitamin $D_2$ 15 000 IU weekly in normal women aged 65–74 years. Metacarpal cortical area was monitored over a 2-year treatment period. Serum concentrations of 25(OH)D were elevated to the young normal range with this intervention and there was an upward trend in bone area in the treated subjects, whereas a significant decline occurred in the control group. The between-groups difference was significant ($P < 0.01$). Ooms *et al.*[19] carried out a randomised controlled trial of calciferol 400 IU daily *vs.* placebo, and found an increase in serum 25(OH)D concentrations from 27 to 62 nmol $L^{-1}$, suppression of serum PTH concentrations, and beneficial effects on femoral neck bone mineral density of about 2% at 2 years. Dawson-Hughes *et al.*[20] showed a benefit to bone density at the femoral neck of 1.5% over 2 years from the use of a daily calciferol supplement of 700 IU. This increased serum 25(OH)D concentrations to 100 nmol $L^{-1}$, in comparison with concentrations of 66 nmol $L^{-1}$ in the control group. In a separate study, this group found a similar benefit in the lumbar spine.[21] Baeksgaard *et al.*[22] found increases in spinal and femoral neck densities of about 1% at 2 years in healthy women (average age 63 years) randomised to receive calciferol 560 IU day$^{-1}$ or placebo. An uncontrolled study of patients with vitamin D deficiency suggested increases of up to 4% in spine and hip bone density could be achieved with vitamin D replacement.[10]

## Fracture Prevention with Vitamin D

There have now been two large studies of the effects of calciferol administration alone on fractures in the elderly. Heikinheimo *et al.*[23] studied almost 800 elderly subjects in Finland. Approximately two thirds were living in their own homes and the remainder were in a home for the elderly. About a quarter of the subjects were males and the mean age of the subjects was between 86 and 87 years. Subjects were randomised to receive 150 000 IU vitamin D2 annually (in 1 of the 5 years 300 000 IU was given) or to act as controls. Follow-up was from 2 to 5 years, the mean follow-up period being just over 3 years. Circulating levels of 25(OH)D were 31 and 14 nmol $L^{-1}$ in the control subjects living independently or in a municipal home, respectively. These were normalized by the intervention (respective means, 49 and 45 nmol $L^{-1}$). There was a fall in serum alkaline phosphatase activity only in the year during which the double dose of calciferol was given. Symptomatic fractures (confirmed by radiographs) were the principal end-point of the study. Fracture numbers were reduced by 25% in the vitamin D-treated subjects ($P = 0.03$).

In contrast, Lips *et al.*[24] showed no difference in fracture incidence in 2578 independently living men and women over the age of 70 years randomised to receive calciferol 400 IU day$^{-1}$ or placebo over a period of up to 3.5 years. Mean serum 25(OH)D concentration in the third year of the study was 23 nmol $L^{-1}$ in the placebo group and 60 nmol $L^{-1}$ in the vitamin D group.

There have also been three large studies of combined treatment with calcium and calciferol in the elderly. In the 3-year study of Chapuy *et al.*[25,26] more than 3000 women aged 69–106 years living in institutions for the elderly were randomly allocated to take placebo or 1.2 g of elemental calcium plus 800 IU of vitamin $D_3$ daily. There were 32% fewer non-vertebral fractures in those receiving active treatment ($P = 0.02$) and 43% fewer hip fractures ($P = 0.04$). There were more than 350 fractures in total, 99% of which resulted from a fall. The data suggest that there was an increase in fracture rate over time in the placebo-treated subjects and this was prevented by the therapeutic regimen used. Bone density measurements taken in a sub-set of the patients would be consistent with this. Proximal femoral bone mineral density increased 2.7% in those receiving active treatment, but declined 4.6% in the placebo group ($P < 0.001$). At the end of 3 years of treatment, the probabilities of non-vertebral fractures and hip fractures were reduced by 24 and 29%, respectively ($P < 0.001$), in those receiving active therapy.

Calcium-related biochemical indices were assessed in a sub-group of subjects in the Chapuy study. Baseline serum 25(OH)D concentrations were 33 and 40 nmol L$^{-1}$ in the placebo and active treatment groups, respectively. This implies that a substantial proportion of patients had significant vitamin D deficiency. Levels of 25(OH)D were stable in the placebo group throughout the study but rose to 100–105 nmol L$^{-1}$ in those receiving active therapy. This was accompanied by a fall in serum PTH concentrations from 54 to 30 pg mL$^{-1}$, whereas PTH levels rose by a small but statistically significant amount in the placebo-treated subjects. Baseline PTH concentrations were inversely related to femoral bone mineral density. Serum alkaline phosphatase activity remained stable in the active group but rose significantly in placebo-treated subjects. Serum concentrations of 1,25(OH)2D were stable in both groups over the trial period.

Dawson-Hughes *et al.*[27] have reported similar findings. They randomised 389 men and women aged over 65 years to treatment with either 500 mg of calcium plus 700 IU of calciferol per day, or to placebo. At the end of 3 years, there were non-vertebral fractures in 26 subjects in the placebo group and in 11 in the calcium-vitamin D group ($P = 0.02$). Thus, this study reproduces the results of the Chapuy study in an American cohort living at home.

Recently, the Chapuy group have repeated their original study, this time with only 583 women.[28] Again, they have demonstrated normalisation of circulating levels of 25(OH)D and PTH with the calcium/vitamin D intervention, a 2.7% increase in femoral neck bone density (relative to placebo), and a substantial reduction in hip fracture numbers, not quite reaching conventional levels of statistical significance in this smaller cohort (Figure 9.4).

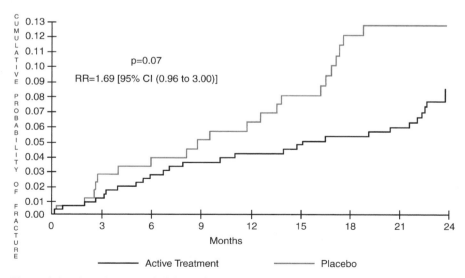

**Figure 9.4** *Cumulative probability of hip fracture in 583 elderly women randomised to calcium plus vitamin D or to placebo for 2 years*
(From Chapuy *et al.*, Osteoporos. Int., 2002, **13**, 257–264, used with permission)

The data from these studies indicate that vitamin D deficiency and secondary hyperparathyroidism are common in frail elderly subjects, and suggest that these changes contribute to the progressive reduction in bone density, which occurs in this age group. These biochemical abnormalities are reversible with physiological doses of vitamin D and calcium, which result in beneficial effects on bone density and, more importantly, on fracture rates. Not only does this lead to a substantial reduction in morbidity in the subjects, but is likely to be associated with a significant prolongation of life, since there was a 24% mortality amongst patients developing hip fractures in the first Chapuy study.

It remains uncertain which component of the calcium–vitamin D intervention contributed to the therapeutic benefit seen in the Chapuy and Dawson-Hughes studies or whether the combination of agents is necessary. The apparently contradictory results from the Heikinheimo and Lips studies do not help address this question. However, three recent studies have suggested an anti-fracture efficacy for calcium supplementation alone[29–31] indicating that a significant contribution from the calcium is likely.

## Falls Prevention with Vitamin D

Because of the actions of vitamin D on muscle, it has been suggested that some of its anti-fracture effects might be mediated by improved muscle function resulting in fewer falls. Two recent studies comparing the effects of calcium alone with calcium and vitamin D support this contention. Pfeifer *et al.*[32] randomised 148 healthy older women to 800 IU of cholecalciferol plus calcium 1200 mg day$^{-1}$, or to calcium alone, daily for 8 weeks. Mean serum 25(OH)D) at baseline was 10 µg L$^{-1}$ and increased to 26 µg L$^{-1}$ after 8 weeks of treatment. There was a reduction in the number of falls and in the number of people who fell in the treatment group compared to the control group at 1 year, and there was also a significant improvement in two out of three measures of body sway. Another study of 122 women has reported in an abstract that a significant reduction in falls occurred following administration of daily calcium and vitamin D supplements for 12 weeks.[33] Physical performance measures of mobility and strength also improved in the treatment group. However, when vitamin D alone has been compared with placebo[34] or when calcium plus vitamin D are compared with placebo,[25] no effect is found. Differences in the baseline frailty or vitamin D status of the study populations may account for some of these apparent discrepancies.

# High-Dose Vitamin D and Use of Potent Metabolites

The current practice of giving vitamin D in doses sufficient to restore serum 25(OH)D to the optimal range, contrasts with an earlier approach of using pharmacological doses of this agent to increase the efficiency of intestinal calcium absorption. Several studies indicated that this resulted in, at best, no benefit,[35,36] and there was some evidence of increased bone loss and fractures.

Such adverse outcomes could be explained in terms of the effect of vitamin D to increase osteoclast recruitment and thus accelerate bone resorption directly.

Somewhat analogously, the more potent vitamin D metabolites have been assessed as therapies for osteoporosis. However, results of controlled trials have been inconsistent. Studies in Danish women in their fifties[37] and their seventies[38,39] showed no beneficial effects of calcitriol on bone loss, and suggested that it accelerated the rate of vertebral height loss. Similar negative findings have been reported in osteoporotic women by others.[40,41] In contrast, Gallagher[42] and Aloia[43] observed increases in bone density in osteoporotic patients treated with calcitriol. The large study of Tilyard *et al.*[44] found fewer fractures in patients receiving this therapy in comparison with those treated with calcium alone. However, the number of patients with new fractures was stable over the 3 years of the study in patients receiving calcitriol, whereas it increased 3-fold in the calcium group. This pattern of response to calcium supplementation has not been observed in other studies and suggests there is something atypical about this group of calcium-treated patients, which renders them unsuitable as a reliable comparator. There is a similar inconsistency in the data with alfacalcidol (reviewed in Ref. [45]) though the observation that a number of positive studies have come from Italy and Japan has led to a suggestion that there may be racial differences in responsiveness to these agents. Such differences could be related to differences in customary dietary calcium intakes or to differences in VDR gene alleles.

# Safety

The use of replacement doses of calciferol is a very safe intervention, as would be expected since the intention is to restore circulating levels of 25(OH)D to those which are present in the ambulant population. Thus, in the study of Chapuy in which more than 1600 women were treated with calciferol, the only individual who developed hypercalcaemia was subsequently found to have primary hyperparathyroidism. A similar zero-incidence of significant hypercalcaemia has been reported by other investigators using either continuous or intermittent low-dose regimens of calciferol administration.[5,8,46] In contrast, pharmacological doses of calciferol can produce severe hypercalcaemia, often of long duration and sometimes associated with renal failure.[47,48] Vieth[3] has recently reviewed this issue and concluded that the relationship of vitamin D dose to serum 25(OH)D concentration is relatively flat up to a daily calciferol dose of 10 000 IU (Figure 9.3). Possibly doses of up to 10 000 IU day$^{-1}$ are safe in individuals without conditions which predispose them to hypercalcaemia (*e.g.* primary hyperparathyroidism, acidosis), and fully documented cases of toxicity have only occurred with intakes of 40 000 IU day$^{-1}$ or more. The propensity of the potent vitamin D metabolites to cause hypercalcaemia and hypercalciuria is dependent on their dose and on dietary calcium intake.

Usually they can be used safely if their dosing is appropriate to calcium intake and if serum calcium is monitored.

## Use of Vitamin D Preparations in Other Conditions

Failure of the 1α-hydroxylation of 25(OH)D is probably the single most important contributor towards the hypocalcaemia of renal failure, and the resulting development of renal bone disease. Thus, the availability of 1,25(OH)$_2$D (calcitriol) and 1α-hydroxyvitamin D (alfacalcidol) as pharmaceuticals has revolutionised the management of calcium metabolism in renal failure. The development of hypercalcaemia can be dose-limiting when using vitamin D metabolites to reverse secondary hyperparathyroidism in renal failure. This has led to the development of synthetic vitamin D analogues, which suppress PTH secretion but have less effect on serum calcium.

The 1α-hydroxyvitamin D metabolites have also revolutionised the management of other conditions associated with either hypocalcaemia (such as hypoparathyroidism) or hypophosphataemia (*e.g.* X-linked-hypophosphataemia, oncogenic osteomalacia). Previously these conditions were managed with very large doses of calciferol. Because of its long half-life, any over-dosage resulted in sustained hypercalcaemia, often leading to renal failure.

As noted above, the anti-proliferative actions of the vitamin D metabolites are now being used in the management of psoriasis, and experimental work is continuing regarding their use in some cancers. There is also epidemiological evidence suggesting that high vitamin D levels are associated with a lower incidence of cardiovascular disease. The latter work is potentially subject to a number of biases (*e.g.* people who are more physically active and therefore at lower risk of cardiovascular disease, spend more time outdoors and therefore have better vitamin D status). Intervention studies will be necessary before the significance of these findings can be determined, and vitamin D supplementation is not a standard part of cardiovascular disease prevention at the present time.

## Conclusions

Though vitamin D deficiency was originally described in children as a privational disease resulting in rickets, it is now most commonly a problem of the frail elderly. It accelerates postmenopausal bone loss and probably contributes to the acceleration of fracture rates in advanced old age. Its treatment is inexpensive, safe and straight forward, and results in 25–50% reductions in fracture rates. The most consistent evidence of anti-fracture efficacy is with the combined use of calcium and calciferol supplementation. The recognition of the problem of vitamin D deficiency in the elderly and the institution of calcium and vitamin D replacement represents the most cost-effective strategy currently known for fracture prevention.

# References

1. A. Malabanan, I.E. Veronikis and M.F. Holick, Redefining vitamin D insufficiency, *Lancet,* 1998, **351**, 805–806.
2. I.R. Reid, D.J.A. Gallagher and J. Bosworth, Prophylaxis against vitamin D deficiency in the elderly by regular sunlight exposure, *Age Ageing*, 1986, **15**, 35–40.
3. R. Vieth, Vitamin D supplementation, 25-hydroxyvitamin D concentrations, and safety, *Am. J. Clin. Nutr.*, 1999, **69**, 842–856.
4. M.J. Barger-Lux, R.P. Heaney, S. Dowell, T.C. Chen and M.F. Holick, Vitamin D and its major metabolites: serum levels after graded oral dosing in healthy men, *Osteoporosis Int.*, 1998, **8**, 222–230.
5. R. Honkanen, E. Alhava, M. Parviainen, S. Talasniemi and R. Monkkonen, The necessity and safety of calcium and vitamin D in the elderly, *J. Am. Geriatr. Soc.*, 1990, **38**, 862–866.
6. M.C. Chapuy, P. Chapuy and P.J. Meunier, Calcium and vitamin D supplements: effects on calcium metabolism in elderly people, *Am. J. Clin. Nutr.*, 1987, **46**, 324–328.
7. P. Lips, A. Wiersinga, F.C. Van Ginkel, M.J.M. Jongen, J.C. Netelenbos, W.H.L. Hackeng, P.D. Delmas and W.J.F. Van der Vijgh, The effect of vitamin D supplementation on vitamin D status and parathyroid function in elderly subjects, *J. Clin. Endocrinol. Metab.*, 1988, **67**, 644–650.
8. F. Zeghoud, A. Jardel, M. Garabedian, R. Salvatore and R. Moulias, Vitamin D supplementation in institutionalized elderly. Effects of vitamin D3 (100,000 IU) orally administered every 3 months on serum levels of 25-hydroxyvitamin D, *Rev. Rhum.*, 1990, **57**, 809–813.
9. R.J. Heikinheimo, M.V. Haavisto, E.J. Harju, J.A. Inkovaara, R.H. Kaarela, L.A. Kolho and S.A. Rajala, Serum vitamin D level after an annual intramuscular injection of ergocalciferol, *Calcif. Tissue Int.*, 1991, **49** Suppl, S87.
10. J.S. Adams, V. Kantorovich, C. Wu, M. Javanbakht and B.W. Hollis, Resolution of vitamin D insufficiency in osteopenic patients results in rapid recovery of bone mineral density, *J. Clin. Endocrinol. Metab.*, 1999, **84**, 2729–2730.
11. E.M. Keane, A. Rochfort, J. Cox, D. McGovern, D. Coakley and J.B. Walsh, Vitamin-D-fortified liquid milk–a highly effective method of vitamin D administration for house-bound and institutionalised elderly, *Gerontology*, 1992, **38**, 280–284.
12. M.J. McKenna, R. Freaney, R. Byrne, Y. McBrinn, B. Murray, M. Kelly, B. Donne and M. Obrien, Safety and efficacy of increasing wintertime vitamin d and calcium intake by milk fortification, *QJM*, 1995, **88**, 895–898.
13. L. Tjellesen, L. Hummer, C. Christiansen and P. Rodbro, Serum concentration of vitamin D metabolites during treatment with vitamin D2 and D3 in normal premenopausal women, *Bone Miner.*, 1986, **1**, 407–413.
14. J.L. Sebert, M. Garabedian, C. De Chasteigner and D. Defrance, Correction of vitamin D deficiency in the elderly: a comparative study of vitamin D2 and vitamin D3, *J. Bone Miner. Res.*, 1991, **6** (Suppl 1), S123.
15. C. Christiansen, M.S. Christensen, P. McNair *et al.*, Prevention of early postmenopausal bone loss: controlled 2-year study in normal females, *Eur. J. Clin. Invest.*, 1980, **10**, 273–279.
16. M. Komulainen, M.T. Tuppurainen, H. Kroger, A.M. Heikkinen, E. Puntila, E. Alhava, R. Honkanen and S. Saarikoski, Vitamin D and HRT – no benefit additional to that of HRT alone in prevention of bone loss in early postmenopausal

women – a 2.5-year randomized placebo-controlled study, *Osteoporosis Int.*, 1997, **7**, 126–132.

17. A.M. Heikkinen, M. Parviainen, L. Niskanen, M. Komulainen, M.T. Tuppurainen, H. Kroger and S. Saarikoski, Biochemical bone markers and bone mineral density during postmenopausal hormone replacement therapy with and without vitamin D-3 – a prospective, controlled, randomized study, *J. Clin. Endocrinol. Metab.*, 1997, **82**, 2476–2482.

18. B.E.C. Nordin, M.R. Baker, A. Horsman and M. Peacock, A prospective trial of the effect of vitamin D supplementation on metacarpal bone loss in elderly women, *Am. J. Clin. Nutr.*, 1985, **42**, 470–474.

19. M.E. Ooms, J.C. Roos, P.D. Bezemer, W.J.F. Vandervijgh, L.M. Bouter and P. Lips, Prevention of bone loss by vitamin D supplementation in elderly women: a randomized double-blind trial, *J. Clin. Endocrinol. Metab.*, 1995, **80**, 1052–1058.

20. B. Dawson-Hughes, S.S. Harris, E.A. Krall, G.E. Dallal, G. Falconer and C.L. Green, Rates of bone loss in postmenopausal women randomly assigned to one of two dosages of vitamin D, *Am. J. Clin. Nutr.*, 1995, **61**, 1140–1145.

21. B. Dawson-Hughes, G.E. Dallal, E.A. Krall, S. Harris, L.J. Sokoll and G. Falconer, Effect of vitamin D supplementation on wintertime and overall bone loss in healthy postmenopausal women, *Ann. Intern. Med.*, 1991, **115**, 505–512.

22. L. Baeksgaard, K.P. Andersen and L. Hyldstrup, Calcium and vitamin D supplementation increases spinal BMD in healthy, postmenopausal women, *Osteoporosis Int.*, 1998, **8**, 255–260.

23. R.J. Heikinheimo, J.A. Inkovaara, E.J. Harju, M.V. Haavisto, R.H. Kaarela, J.M. Kataja, A.M. Kokko, L.A. Kolho and S.A. Rajala, Annual injection of vitamin D and fractures of aged bones, *Calcif. Tissue Int.*, 1992, **51**, 105–110.

24. P. Lips, W.C. Graafmans, M.E. Ooms, P.D. Bezemer and L.M. Bouter, Vitamin D supplementation and fracture incidence in elderly persons – a randomized, placebo-controlled clinical trial, *Ann. Intern. Med.*, 1996, **124**, 400–406.

25. M.C. Chapuy, M.E. Arlot, F. Duboeuf, J. Brun, B. Crouzet, S. Arnaud, P.D. Delmas and P.J. Meunier, Vitamin D3 and calcium to prevent hip fractures in the elderly women, *N. Engl. J. Med.*, 1992, **327**, 1637–1642.

26. M.C. Chapuy, M.E. Arlot, P.D. Delmas and P.J. Meunier, Effect of calcium and cholecalciferol treatment for three years on hip fractures in elderly women, *Br. Med. J.*, 1994, **308**, 1081–1082.

27. B. Dawson-Hughes, S.S. Harris, E.A. Krall and G.E. Dallal, Effect of calcium and vitamin D supplementation on bone, density in men and women 65 years of age or older, *N. Engl. J. Med.*, 1997, **337**, 670–676.

28. M.C. Chapuy, R. Pamphile, E. Paris, C. Kempf, M. Schlichting, S. Arnaud, P. Garnero and P.J. Meunier, Combined calcium and vitamin D-3 supplementation in elderly women: confirmation of reversal of secondary hyperparathyroidism and hip fracture risk: the Decalyos II study, *Osteoporosis Int.*, 2002, **13**, 257–264.

29. R.R. Recker, D.B. Kimmel, S. Hinders and K.M. Davies, Anti-fracture efficacy of calcium in elderly women, *J. Bone Miner. Res.*, 1994, **9**(Suppl. 1), s154.

30. I.R. Reid, R.W. Ames, M.C. Evans, G.D. Gamble and S.J. Sharpe, Long-term effects of calcium supplementation on bone loss and fractures in postmenopausal women – a randomized controlled trial, *Am. J. Med.*, 1995, **98**, 331–335.

31. T. Chevalley, R. Rizzoli, V. Nydegger, D. Slosman, C.H. Rapin, J.P. Michel, H. Vasey and J.P. Bonjour, Effects of calcium supplements on femoral bone mineral density and vertebral fracture rate in vitamin-D-replete elderly patients, *Osteoporosis Int.*, 1994, **4**, 245–252.

32. M. Pfeifer, B. Begerow, H.W. Minne, C. Abrams, D. Nachtigall and C. Hansen, Effects of a short-term vitamin D and calcium supplementation on body sway and secondary hyperparathyroidism in elderly women, *J. Bone Miner. Res.*, 2000, **15**, 1113–1118.

33. H.A. Bischoff, H.B. Staehelin, W. Dick *et al.*, Fall prevention by vitamin D: a randomized controlled trial, *J. Bone Miner. Res.*, 2001, **16** (Suppl), s163.

34. W.C. Graafmans, M.E. Ooms, H.M.A. Hofstee, P.D. Bezemer, L.M. Bouter and P. Lips, Falls in the elderly: a prospective study of risk factors and risk profiles, *Am. J. Epidemiol.*, 1996, **143**, 1129–1136.

35. B.E.C. Nordin, A. Horsman, R.G. Crilly, D.H. Marshall and M. Simpson, Treatment of spinal osteoporosis in postmenopausal women, *Br. Med. J.*, 1980, **1**, 451–454.

36. B.L. Riggs, E. Seeman, S.F. Hodgson, D.R. Taves and W.M. O'Fallon, Effect of the fluoride/calcium regimen on vertebral fracture occurrence in postmenopausal osteoporosis, *N. Engl. J. Med.*, 1982, **306**, 444–450.

37. C. Christiansen, M.S. Christensen, P. Rodbro, C. Hagen and I. Transbol, Effect of 1,25-dihydroxy-vitamin D3 in itself or combined with hormone treatment in preventing postmenopausal osteoporosis, *Eur. J. Clin. Invest.*, 1981, **11**, 305–309.

38. G.F. Jensen, C. Christiansen and I. Transbol, Treatment of postmenopausal osteoporosis. A controlled therapeutic trial comparing oestrogen/gestagen,1,25-dihydroxy-vitamin D3 and calcium, *Clin. Endocrinol.*, 1982, **16**, 515–524.

39. G.F. Jensen, B. Meinecke, J. Boesen and I. Transbol, Does 1,25(OH)2 D3 accelerate spinal bone loss? *Clin. Orthop. Related Res.*, 1985, **192**, 215–221.

40. J.A. Falch, O.R. Odegaard, M. Finnanger and I. Matheson, Postmenopausal osteoporosis: no effect of three years treatment with 1,25-dihydroxycholecalciferol, *Acta Med. Scand.*, 1987, **221**, 199–204.

41. S.M. Ott and C.H. Chesnut, Calcitriol treatment is not effective in postmenopausal osteoporosis, *Ann. Intern. Med.*, 1989, **110**, 267–274.

42. J.C. Gallagher and D. Goldgar, Treatment of postmenopausal osteoporosis with high doses of synthetic calcitriol. A randomized controlled study, *Ann. Intern. Med.*, 1990, **113**, 649-655.

43. J.F. Aloia, A. Vaswani, J.K. Yeh, K. Ellis, S. Yasumura and S.H. Cohn, Calcitriol in the treatment of postmenopausal osteoporosis, *Am. J. Med.*, 1988, **84**, 401–408.

44. M.W. Tilyard, G.F. Spears, J. Thomson and S. Dovey, Treatment of postmenopausal osteoporosis with calcitriol or calcium, *N. Engl. J. Med.*, 1992, **326**, 357–362.

45. I.R. Reid, Vitamin D and its metabolites in the management of osteoporosis, In *Osteoporosis,* 2nd edn, R. Marcus, D. Feldman and J. Kelsey, (eds) Academic Press, San Diego, 2000.

46. D.J. Hosking, G.A. Campbell, J.R. Kemm, R.E. Cotton and R.V. Boyd, Safety of treatment for subclinical osteomalacia in the elderly, *Br. Med. J.*, 1984, **289**, 785–787.

47. M.S. Schwartzman and W.A. Franck, Vitamin D toxicity complicating the treatment of senile, postmenopausal, and glucocorticoid-induced osteoporosis, *Am. J. Med.*, 1987, **82**, 224–230.

48. R. Rizzoli, C. Stoermann, P. Ammann and J.P. Bonjour, Hypercalcemia and hyperosteolysis in vitamin D intoxication – effects of clodronate therapy, *Bone*, 1994, **15**, 193–198.

CHAPTER 10.1

# Effects of Dietary Protein Insufficiency on the Skeleton

RENÉ RIZZOLI, PATRICK AMMANN,
THIERRY CHEVALLEY and JEAN-PHILIPPE BONJOUR

Division of Bone Diseases [WHO Collaborating Center for Osteoporosis Prevention], Department of Rehabilitation and Geriatrics, University Hospital, Geneva, Switzerland, E-mail: Rene.Rizzoli@medecine.unige.ch

**Key Points**

- Dietary protein intake influences both bone mass acquisition and bone loss.
- Modifications in the GRF-growth hormone-IGF-I axis, associated with decreased bone formation, appear to be directly implicated in the low bone mass observed under a low protein diet.
- An increased bone resorption may also contribute to the deleterious effects of a low protein diet on bone mineral mass, bone architecture and bone strength. Thus, a low protein intake seems to be associated with a marked uncoupling between bone formation and bone resorption.
- Analyzing data obtained in various experimental models of rats and mice, the process leading to decreased bone strength induced by a low protein isocaloric diet appears to be characterized by skeletal site specificity, changes in bone remodeling rate, microarchitectural alterations, and bone envelop selectivity.
- These features are of major importance for the pathogenesis of protein deficiency-induced osteoporosis, and of its complications in elderly.

## Introduction

Undernutrition is often observed in elderly. It appears to be more severe in patients with hip fracture than in the general aging population.[1–3] Indeed, a state of undernutrition on admission is consistently documented in elderly patients with hip fracture.[3,4] In addition to an inadequate food intake during hospital stay, the state of undernutrition before admission can adversely influence the clinical outcome.[1,2,4,5] In hip fracture patients, in whom a lower femoral neck bone mineral density (BMD) at the level of the proximal femur has been demonstrated,[6] nutritional requirements are not met while the patients are in hospital although adequate quantities of food are offered.[2] Numerous studies have highlighted the handicaps, increased mortality rate and high costs associated with hip fracture.[7,8] A negative influence of undernutrition on hospital outcome can also be detected in non-fractured elderly, since poor nutritional status as assessed by the Mini Nutritional Assessment is associated with higher in-hospital mortality rate, higher rate of discharge to nursing homes and longer length of stay.[9] Among various nutrients, a low protein intake could be particularly detrimental for both the acquisition of bone mass and the conservation of bone integrity with aging.[10–13] Protein undernutrition can favor the occurrence of hip fracture by increasing the propensity to fall as a result of muscle weakness and of impaired movement coordination, by affecting protective mechanisms, such as reaction time or muscle strength,[1] and/or by decreasing bone mass.[12,14,15] Furthermore, a reduction in the protective layer of soft tissue padding decreases the force required to fracture an osteoporotic hip.[16–18]

## Bone Mineral Mass and Low Dietary Protein Intake

In numerous studies, a positive association between bone mass at various skeletal sites and spontaneous protein intake has been detected in pre- or postmenopausal women as well as in men (Table 10.1.1). In a survey carried out in hospitalized elderly patients, low protein intake was associated with reduced femoral neck areal BMD and poor physical performances.[14] The group with lower intakes and a lower BMD, particularly at the femoral neck level, had also a poor evolution of bicipital and quadricipital muscle strength and performance during the hospital stay, as indicated by the decreased capacity to walk and climb stairs, as compared with those with higher intakes, after 4 weeks of hospitalization.[14] Undernutrition can concern all kinds of nutrients and the specific role of a low protein intake besides low calorie consumption can be difficult to appraise in the elderly.[14] One of the key nutrients responsible for a beneficial effect on fracture outcome seems to be proteins.[19] Unadjusted BMD was greater in the group with the higher protein intake in a large series of data collected in the frame of the Study of Osteoporotic Fracture.[20] Besides numerous cross-sectional studies showing a positive association between bone mass and protein intake, a longitudinal follow-up in the frame of the Framingham study has demonstrated that the rate of bone mineral loss was inversely correlated to dietary protein intake.[21] Recently, spontaneous higher protein intake was associated with an increase in BMD in a group of patients receiving calcium supplements followed longitudinally[22] (Table 10.1.2). Several studies have failed to find any

**Table 10.1.1.** *Bone mineral density and protein intake (cross-sectional studies)*

| Positive association | | |
|---|---|---|
| Calvo (1998)[108] | 5900 Men and women | Proximal femur (< 50 years) |
| Chiu (1997)[109] | 258 Postmenopausal women | Lumbar spine |
| Cooper (1996)[110] | 72 Premenopausal women | Distal radius, proximal femur |
| Geinoz (1993)[14] | 74 Men and women | Femoral neck |
| Hirota (1992)[111] | 161 Premenopausal women | Forearm |
| Kerstetter (2000)[112] | 1822 Postmenopausal women | Proximal femur |
| Lacey (1991)[113] | 178 Pre- and postmenopausal women | Midradius |
| Lau (1998)[114] | 76 Postmenopausal women | Hip |
| Michaelsson (1995)[115] | 175 Pre- and postmenopausal women | Total, lumbar spine, proximal femur |
| Orwoll (1987)[116] | 92 Men | Midradius, lumbar spine |
| Teegarden (1998)[117] | 215 Premenopausal women | Midradius, lumbar spine |
| Tylavsky (1988)[118] | 375 Postmenopausal women | Mid- and distal radius |
| *No association* | | |
| Henderson (1995)[119] | 115 Premenopausal women | Lumbar spine, proximal femur, forearm |
| Mazess (1991)[120] | 200–300 Premenopausal women | Lumbar spine, forearm |
| New (1997)[121] | 994 Premenopausal women | Lumbar spine, proximal femur |
| Nieves (1995)[122] | 139 Premenopausal women | Lumbar spine, forearm |
| Wang (1997)[123] | 125 Postmenopausal women | Lumbar spine, proximal femur |
| *Negative association* | | |
| Anderson (1995)[23] | 220 Premenopausal women | Distal radius |
| Metz (1993)[34] | 38 Premenopausal women | Distal radius |

relationship between bone mineral mass and protein intake. In contrast, very few are the surveys in which low protein intake was accompanied with higher bone mass. In a cross-sectional study, a protein intake close to 2 g kg$^{-1}$ body weight was associated with reduced BMD only at one out of the two forearm sites measured in young college women.[23] In another study, BMD became lower with various adjustments, in the group with the higher animal-to-vegetable protein ratio.[20] Taken together, these results indicate that whereas a gradual decline in calorie intakes with age can be considered as an adequate adjustment to the progressive reduction in energy expenditure, the parallel reduction in protein intakes may be detrimental for maintaining the integrity and function of several organs or systems, including skeletal muscles and bone.

There is a longstanding suggestion that high protein intake could be harmful for bone, based on the positive association between urinary calcium excretion and protein intake. Thus, high protein intake would induce a negative calcium balance and, consequently, would favor bone loss.[24] However, further studies indicated that a reduction in dietary protein led to a decline in calcium

**Table 10.1.2.** *Bone mineral density changes and protein intake (longitudinal studies)*

| Rate of bone loss and protein intake | | |
| --- | --- | --- |
| *Inverse relationship (High intake → Low rate of bone loss)* | | |
| Dawson-Hughes (2002)[22] | 345 Men and women, 3 years | Femoral neck[a] |
| Freudenheim (1986)[124] | 67 Postmenopausal women, 4 years | Forearm |
| Hannan (2000)[21] | 615 Men and women (75 years), 4 years | Spine, femoral neck |
| *No association* | | |
| Recker (1992)[125] | 156 Premenopausal women, 3.4 years | Spine |
| Reid (1994)[126] | 122 Postmenopausal women, 2 years | Spine, hip |
| *Positive relationship (High intake → High rate of bone loss)* | | |
| Sellmeyer (2001)[20] | 1035 Postmenopausal women (>65 years) | Femoral neck[b] |

[a]In the group supplemented with calcium.
[b]Ratio animal/vegetable, adjusted for energy, calcium, protein, weight.

absorption and to secondary hyperparathyroidism.[25] There is some evidence that the favorable effect of increasing the protein intake on bone mineral mass as repeatedly observed in both genders requires an adequate supply of both calcium and vitamin D.[22,26–29]

The source of proteins, animal *vs.* vegetal, has been claimed to influence calcium metabolism. This appears to involve emotional belief rather than reliable scientific demonstration.[27] The hypothesis implies that animal proteins would generate more sulfuric acid from sulfur-containing amino acids than a vegetarian diet. Consequently, the nutrition-generated acid load would lead in healthy individuals to an increased bone dissolution. That animal protein in contrast to vegetal protein would be more detrimental for bone health is not supported by convincing experimental evidence. A vegetarian diet with protein derived from grains and legumes appears to deliver as many millimoles of sulfur per gram proteins as would a purely meat-based diet.[27] Several recent human studies do not suggest that the protective effect of protein on either bone loss or osteoporotic fracture is due to vegetable rather than animal protein.[21,22,30–32] In contrast with these consistent results, a recent study suggested that a high ratio of animal to vegetal protein was associated with a greater bone loss and higher risk of hip fracture.[20] An animal to vegetal protein ratio remains mechanistically difficult to interpret, since a same ratio can be achieved with a large variety of protein content, and thus of sulfur supply if the acid load theory was valid. More importantly, the statistically negative relationship between the animal to vegetal protein ratio and bone loss was obtained only after multiple adjustments, not only for age but also for energy intake, total calcium intake (dietary plus supplements), total protein intake, weight, current estrogen use, physical activity, smoking status and alcohol.[20] An inconsistency according to the way the data were analyzed makes difficult the generalization of these findings in terms of nutritional recommendations for bone health and osteoporosis prevention.[27]

## Fracture Risk and Low Dietary Protein Intake

Various studies, assessing the relationship between protein intake and bone metabolism[15,23,33–38] come to the conclusion that either a deficient or an excessive protein supply could negatively affect the balance of calcium. An indirect argument in favor of a deleterious effect of high protein intake on bone is that hip fracture appeared to be more frequent in countries with high protein intake of animal origin (Table 10.1.3).[39,40] But, as expected, the countries with the highest incidence are those with longest life expectancy, accounting for an elevated fracture incidence. In the large Nurse Health Study a trend for a hip fracture incidence inversely related to protein intake has been reported.[41] Similarly, hip fracture was higher with low energy intake, low serum albumin levels and low muscle strength in the NHANES I study.[42] In a prospective study carried out on more than 40 000 women in Iowa, higher protein intake was associated with a reduced risk of hip fracture.[31] The association was positive with dietary protein of animal origin. Similarly, a reduced relative risk of hip fracture was found with higher intake of milk.[43,44] A low plasma albumin level, which can reflect low nutritional intakes, has been repeatedly found in patients with hip fracture as compared to age-matched healthy subjects or patients with osteoarthritis.[2,5,45,46] In another survey, no association between hip fracture and non-dairy animal protein intake could be detected.[47] However, in this study, fracture risk was increased when a high protein diet was accompanied by a low calcium intake, in agreement with the requirement of sufficient calcium intake to detect a favorable influence of dietary protein on bone. In a longitudinal study, hip fracture incidence was positively related to a higher ratio animal-to-vegetal protein intake,[20] whereas protein of vegetable origin was rather protective.

**Table 10.1.3.** *Osteoporotic fracture and protein intake*

| | | |
|---|---|---|
| *Positive association* | | |
| Abelow (1992)[39] | Cross-cultural study | Hip (animal) |
| Feskanich (1996)[41] | 85 900 Women, 12 years | Forearm 1.2× (animal) |
| Frassetto (2000)[40] | Cross-cultural study | Hip |
| Meyer (1997)[47] | 39 787 Men and women | Hip (if low calcium in women) |
| Sellmeyer (2001)[20] | 1035 women, 7 years | Hip (animal/vegetable) |
| | | |
| *Inverse relationship (High intake → Low incidence)* | | |
| Huang (1996)[42] | 2565 women, 16 years | Hip |
| Munger (1999)[31] | 32 050 women, 3 years | Hip (animal > vegetable) |
| Sellmeyer (2001)[20] | 1035 women, 7 years | Hip (vegetable) |
| | | |
| *No association* | | |
| Feskanich (1996)[41] | 85 900 women, 12 years | Hip (animal) |

# Nutrition and Bone Mass Gain

Peak bone mass is defined as the amount of bony tissue present at the end of skeletal maturation (for review, see Refs. 15,48,49). During puberty, the accumulation rate in areal BMD at both the lumbar spine and femoral neck levels increases 4- to 6-fold over a 3- and 4-year period in females and males, respectively.[50,51] Puberty affects bone size much more than volumetric mineral density. In a prospective survey carried out in a cohort of female and male subjects aged 9–19 years, food intake was assessed twice, at a one-year interval, using a 5-day dietary diary method consisting in weighing all consumed foods.[52] In this adolescent cohort, we found a positive correlation between yearly lumbar and femoral bone mass gain and calcium or protein intake. The association remained statistically significant after adjustment for the alternate nutrient intake, and appeared to be significant mainly in prepubertal children, but not in those having reached a peri- or postpubertal stage.

Anorexia nervosa is a condition frequently observed in young women. BMD is reduced at several skeletal sites in most women with anorexia nervosa. Young women with anorexia nervosa are at increased risk of fracture later in life.[53] Abnormally low serum albumin levels ($\leq 36$ g $L^{-1}$ and a low body weight ($\leq 60\%$ of average body weight)) at the initial examination were variables best able to predict a lethal course.[54] Besides estrogen and calcium deficiency, low protein intake very likely contributes to the impaired bone acquisition or the accelerated bone loss observed in anorexia nervosa.

In athletes or ballet dancers intensive exercise can lead to hypothalamic dysfunction with delayed menarche and disturbances in menstrual cycles, and to bone loss.[55–59] Nutritional restriction can play an important role in the disturbance of the female reproductive system resulting from intense physical activity. Nutritional restriction could be more common when leanness confers an advantage for the athletic performance.[57] Insufficient energy intake with respect to energy expenditure is supposed to impair the secretion of GnRH and thereby to lead to a state of hypoestrogenism. However, the relative contribution of insufficient protein intake frequently associated with the low energy intake remains to be assessed. Indeed, experimental studies indicate that bone loss induced by isocaloric protein restriction in adult female rats is mediated by both dependent and independent sex-hormone deficiency mechanisms.[60–62]

# Pathogenesis of Bone Loss Associated with Low Dietary Protein Intake

IGF-I is an essential factor for longitudinal bone growth,[63] as it stimulates proliferation and differentiation of chondrocytes in epiphyseal plate. IGF-I also plays a role in trabecular and cortical bone formation. IGF-I can exert anabolic effects on bone mass not only during growth, but also during adulthood.[64–67] Furthermore, by its renal action on tubular reabsorption of phosphate and on the synthesis of calcitriol, through a direct action on renal cells,[68,69] IGF-I can be considered as an important controller of the intestinal absorption and of the extracellular concentration of both calcium and phosphate, the main elements of bone mineral. On the other hand, IGF-I can selectively stimulate the transport of inorganic phosphate across the plasma membrane in osteoblastic cell lines.[70,71]

Experimental and clinical studies suggest that dietary proteins, by influencing both the production and action of IGF-I, particularly the growth hormone (GH)–insulin-like growth factor (IGF) system, could control bone anabolism.[64,72] The hepatic production and plasma levels of IGF-I are under the influence of dietary proteins.[73–75] Protein restriction has been shown to reduce IGF-I plasma levels by inducing a resistance to the action of GH at the hepatic level,[76,77] and by an increase of IGF-I metabolic clearance rate.[78] Decreased levels of IGF-I have been found in states of undernutrition such as marasmus, anorexia nervosa, celiac disease or HIV infected patients.[73,74,79,80] Refeeding these patients led to an increase of IGF-I.[81,82] Elevated protein intake is able to prevent the decrease in IGF-I usually observed in hypocaloric states.[3,83] Protein restriction appears to render target organs less sensitive to IGF-I. When IGF-I was given to growing rats maintained under a low protein diet at doses normalizing its plasma levels, it failed to restore skeletal longitudinal growth.[84] The administration of pharmacological doses of IGF-I, producing a 5-fold increase in IGF-I circulating levels in adult rats, in an attempt to correct the negative influence of protein deficiency was without effects on bone if the protein intake was insufficient (Figure 10.1.1).[85]

Osteogenic cells not only express specific IGF-I receptors, but they can also be endowed with IGF-I producing machinery.[86] Regarding a possible influence of the local environment in proteins or amino acids on IGF-I production by bone cells, it has been found that the amino acids arginine or lysine increased IGF-I production and collagen synthesis in a mice osteoblastic cell line, on a time- and concentration-dependent manner.[87] This study underlines a possible influence of the local environment in proteins or amino acids on IGF-I production by bone cells, and suggests a potential role of locally produced IGF-I under the influence of extracellular amino acid concentration in the regulation of osteoblast function. Since undernutrition can concern all kinds of nutrients in the elderly,[2] and not only proteins, we developed an experimental model in adult female rats of selective protein deprivation with isocaloric low protein diets supplemented with identical amounts of minerals in order to study the specific influence of protein deficiency in the pathogenesis of osteoporosis (Figure 10.1.2).[62] This model enables the study of bone mineral mass, bone strength and bone remodeling. A decrease in BMD is observed at the level of skeletal sites formed by trabecular or cortical bone in animals fed 2.5% casein, but receiving the same amount of energy. This is associated with a marked and early decrease in plasma IGF-I by 40%. In this model, the decrease in bone mass and bone strength is related to an early inhibition of

A. Cortical bone: P- BFR
(µm/d)

B. Cancellous bone: BFR/BS
(µm3/µm2/d)

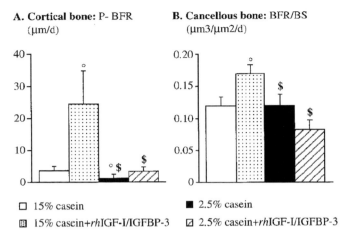

□ 15% casein                          ■ 2.5% casein
▦ 15% casein+*rh*IGF-I/IGFBP-3      ▨ 2.5% casein+*rh*IGF-I/IGFBP-3

**Figure 10.1.1**  *Bone formation rate (mean ± SEM) in the cortex (A) and trabeculae
(B). P-BFR (rate of periosteal bone formation) and BFR/BS (rate of
trabecular bone formation) were studied at mid-diaphysis and at the
secondary cancellous bone in the proximal tibia, respectively. The
isocaloric diets containing 15 or 2.5% casein, respectively, were given
for 24 days. The rats were given rhIGF-I/GFBP-3 for 10 days. °P < 0.05 as
compared to the group fed the control diet (15% casein); $P < 0.03 as
compared to the group fed the control diet (15% casein) and given the
IGF-I/GFBP-3 complex.*
(Data from Bourrin *et al.* Endocrinology, 2000, 141, 3149–3155, with
the permission of the editor)

bone formation, and a later acceleration of bone resorption.[62] Whereas a rapid
decrease in circulating IGF-I could account for the former, the latter might be
related to estrogen deficiency caused by protein undernutrition. Indeed, under
a low protein diet, the normal cycling in female rats disappeared, and was
recovered upon protein replenishment (Ammann *et al.* unpublished results).
However, an increase in bone resorption under a low protein diet was also
detected in ovariectomized animals, indicating the presence of a low protein
diet-dependent and sex hormone-independent component.[62] Histomorphometric analysis and biochemical markers of bone remodeling results indicate
that the low protein intake-induced decrease in bone mineral mass and bone
strength is related to an uncoupling between bone formation and resorption.[62]
Besides sex hormone deficiency, other mechanisms, including the effects of
circulating or locally released cytokines are likely.[88,89] Production and action of
TNFα play a central role in the accelerated bone loss caused by sex hormone
deficiency, as indicated by experiments carried out in transgenic mice
overexpressing TNFα receptor 1 protein, which blocks the effects of this
factor. Indeed, in these animals, the influence of ovariectomy[88] or orchidect-
omy[89] is prevented in transgenic animals overexpressing the receptor. To
address the issue of the accelerated bone loss occurring under a low protein

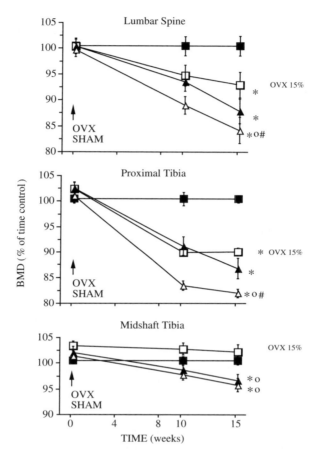

**Figure 10.1.2** *Effects of an isocaloric low protein diet and/or of oophorectomy on bone mineral density in adult female rats (mean ± SEM). Results are expressed as the mean ± SEM. \*P < 0.05 as compared to control animals; °P < 0.05 as compared to the oophorectomized group fed a normal-protein diet; #P < 0.05 as compared to the group fed a low protein diet.*
*(Data from Ammann et al. J. Bone Miner. Res. 2000, 15, 683–690, with the permission of the editor)*

diet, we used the model of transgenic mice, which overexpresses the soluble TNFα receptor. Blocking TNFα activity prevented the component of increased bone resorption induced by the isocaloric low protein diet, without modifying the alterations in bone formation.[90,91] Similarly, we also assessed whether interleukin-1 (IL-1) could be involved in this process. The effects of an isocaloric low protein diet was studied in transgenic mice overexpressing a IL-1 receptor antagonist, a situation in which IL-1 is prevented to exert its biological action. In this model, bone loss was identical in the IL-1 receptor-antagonist overexpressing mice and their negative.[90,91]

An isocaloric low protein (2.5% casein) diet decreases BMD and alters mechanical properties in male rats as well, in which protein deficiency induced cortical and trabecular thinning, in relation with a remodeling imbalance with impaired bone formation. This leads to a decrease in bone mineral mass and bone strength.[85] At an early time point (1 month), histomorphometry analysis shows that bone loss process is mainly related to a depressed bone formation.[85] In this gender too, some state of hypogonadism was associated with long-term isocaloric low protein diet. Adult male and female rats differed by the kinetics of the response to the isocaloric low protein diet, with changes occurring more slowly in males than in females.

GH secretion was evaluated by measuring GH pulsatility over a 24-h period. The amplitude of the pulses of GH was maintained, whereas the number of pulses, and thereby the area under the curve, appeared to be lowered under an isocaloric low protein diet.[92] Since there was an alteration of the GH–IGF-I bone axis in protein undernutrition with altered production of both hormones, decreased bone formation and increased bone resorption, and a marked increase in bone fragility, we investigated whether the administration of GH could reverse this process. Under an isocaloric low protein diet, the IGF-I response to GH appeared to be blunted, but the most striking finding was that GH was rather catabolic on bone, instead of anabolic, since there was a dose-dependent decrease of bone strength after 4 weeks of GH treatment in animals fed the isocaloric low protein diet.[93] We then tested the effects of protein replenishment by administering essential aminoacid supplements in the same relative proportion as in casein. These supplements caused an increase in IGF-I up to a level higher than in rats fed the control diet, increased biochemical bone formation and decreased markers of bone resorption, improved bone strength more than bone mineral mass, probably in relation with an increase in cortical thickness, as demonstrated by micro-quantitative computerized tomography.[93]

In addition to alterations in the control and action of the GH–IGF-I system, protein undernutrition can be associated with alterations of cytokines secretion, such as interferon gamma, tumor necrosis factor alpha (TNFα), or transforming growth factor beta.[94,95] TNFα and Interleukin-6 are generally increasing with aging.[96] In a situation of cachexia, such as in chronic heart failure, an inverse correlation between BMD and TNFα levels has been found,[97,98] further implicating a possible role of uncontrolled cytokines production in bone loss. Increased TNFα can be a crucial factor in the sex hormone deficiency-induced bone loss,[88] but it also plays a role in the target organ resistance to insulin, and possibly to IGF-I.[99] Along the same line, certain amino acids given to rats fed a low protein diet can increase the liver protein synthesis response to TNFα.[100] The amino acid oxidation rate was lower in children with kwashiorkor repleted with milk as compared with egg white, and protein breakdown and synthesis correlated inversely with TNFα levels.[101] Thus, the modulation by nutritional intakes of cytokines production and action[102] and the strong implication of various cytokines in the regulation of bone remodeling[103] suggest a possible role of certain cytokines in the nutrition–bone link.

# Effects of Correcting Protein Insufficiency

Taking into account these experimental and clinical observations, IGF-I could play a prominent role in the pathophysiology of osteoporosis, of osteoporotic fracture and of its complications. Under these conditions, a restoration of this altered system in elderly by protein replenishment is likely to favorably influence not only BMD, but also muscle mass and strength, since these two variables are important determinants of the risk of falling.[104,105]

Intervention studies using a simple oral dietary preparation that normalizes protein intake[2] can improve the clinical outcome after hip fracture. It should be emphasized that a 20 g protein supplement brought the intake from low to a level still below RDA (0.8 g kg$^{-1}$ body weight), avoiding thus the risk of an excess of dietary protein.[2] Follow-up showed a significant difference in the clinical course in the rehabilitation hospitals, with the supplemented patients doing better. The significantly lower rate of complication (bedsore, severe anemia, intercurrent lung or renal infections, 44 *vs.* 87%), and deaths was still observed at six months (40 *vs.* 74%).[2] The duration of hospital stay of elderly patients with hip fracture is not only determined by the present medical condition, but also by domestic and social factors.[4,7,106] In this study,[2] the total length of stay in the orthopedic ward and convalescence hospital was significantly shorter in supplemented patients than in controls (median: 24 *vs.* 40 days). It was then shown that normalization of protein intake, independently of that of energy, calcium and vitamin D, was in fact responsible for this more favorable outcome,[19] as shown in a randomized controlled trial, in which protein intake was the primary variable accounting for the better outcome which was recorded. In undernourished elderly with a recent hip fracture, an increase in the protein intake, from low to normal, can also be beneficial for bone integrity.[30] Indeed, in a double-blind, placebo-controlled study, protein repletion with 20 g protein supplement daily for 6 months as compared to an isocaloric placebo, produced greater gains in serum prealbumin, IGF-I, and IgM, and an attenuated proximal femur BMD decrease.[30] In this trial, all 82 patients (80.7 $\pm$ 1.2 years) were given 200 000 IU vitamin D once at baseline, and 550 mg day$^{-1}$ of calcium, starting within one week after an osteoporotic hip fracture. In a multiple regression analysis, baseline IGF-I concentrations, biceps muscle strength, together with protein supplements accounted for more than 30% of the variance of the length of stay in rehabilitation hospitals ($r^2 = 0.312$, $p < 0.0005$), which was reduced by 25% in the protein supplemented group. We recently completed short-term studies on the kinetics and determinants of the IGF-I response to protein supplements in two situations associated with low baseline IGF-I levels. In elderly with a recent hip fracture, we found that a 20 g day$^{-1}$ protein supplement increased serum IGF-I and IGF-binding protein-3 already by 1 week (Chevalley *et al.*, unpublished results). Finally, this normalization of protein intake was found to increase IGF-I, and even IgM concentrations.[30] Thus, the lower incidence of medical complications observed after a protein supplement[2,19] is also compatible with the hypothesis of IGF-I improving the immune status, as

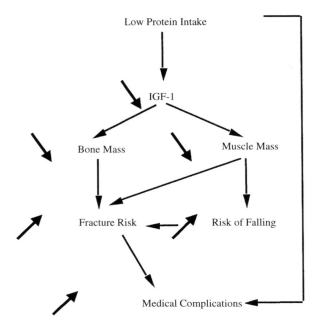

**Figure 10.1.3**   *Effects of protein deficiency in elderly subjects. Possible effects of IGF-I on muscle, bone fracture risk and incidence of postfracture medical complications*

this growth factor can stimulate the proliferation of immunocompetent cells and modulate immunoglobulin secretion.[107]

## Conclusions

In conclusion, there is a large body of evidences linking nutritional intakes, particularly protein undernutrition, to bone homeostasis and osteoporotic fractures (Figure 10.1.3). Sufficient dietary proteins are necessary for bone homeostasis, during growth as well as in elderly. Several mechanisms, among them the GH–IGF-I-target organ axis and various cytokines, are likely to be implicated. However, better understanding and elucidation of the mechanisms involved in low protein intake-induced decrease in bone mass are mandatory to appreciate the role of protein intakes in the prevention and treatment of osteoporosis.

## Acknowledgements

We thank Mrs. M. Perez for her secretarial assistance. The works from our group quoted were supported by the Swiss National Science Research Foundation (grants nos. 32-32415.91, 32-49757.96, 32-58880.99 and 3200B0-100714) and by Novartis-Nutrition, Berne, Switzerland.

# References

1. M.D. Bastow, J. Rawlings and S.P. Allison, Undernutrition, hypothermia, and injury in elderly women with fractured femur: an injury response to altered metabolism? *Lancet*, 1983, **1**(8317), 143–146.

2. M. Delmi, C.H. Rapin, J.M. Bengoa, P.D. Delmas, H. Vasey and J.P. Bonjour, Dietary supplementation in elderly patients with fractured neck of the femur, *Lancet*, 1990, **335**(8696), 1013–1016.

3. J.E. Jensen, T.G. Jensen, T.K. Smith, D.A. Johnston and S.J. Dudrick, Nutrition in orthopaedic surgery, *J. Bone Joint Surg. [Am.]*, 1982, **64**(9), 1263–1272.

4. M.D. Bastow, J. Rawlings and S.P. Allison, Benefits of supplementary tube feeding after fractured neck of femur: a randomised controlled trial, *Br. Med. J.*, 1983, **287**(6405), 1589–1592.

5. B.M. Patterson, C.N. Cornell, B. Carbone, B. Levine and D. Chapman, Protein depletion and metabolic stress in elderly patients who have a fracture of the hip, *J. Bone Joint Surg. [Am.]*, 1992, **74**(2), 251–260.

6. T. Chevalley, R. Rizzoli, V. Nydegger, D. Slosman, L. Tkatch, C.H. Rapin, H. Vasey and J.P. Bonjour, Preferential low bone mineral density of the femoral neck in patients with a recent fracture of the proximal femur, *Osteoporos. Int.*, 1991, **1**(3), 147–154.

7. M.A. Schurch, R. Rizzoli, B. Mermillod, H. Vasey, J.P. Michel and J.P. Bonjour, A prospective study on socioeconomic aspects of fracture of the proximal femur, *J. Bone Miner. Res.*, 1996, **11**(12), 1935–1942.

8. A. Trombetti, F. Herrmann, P. Hoffmeyer, M.A. Schurch, J.P. Bonjour and R. Rizzoli, Survival and potential years of life lost after hip fracture in men and age-matched women, *Osteoporos. Int.*, 2002, **13**(9), 731–737.

9. M.C. Van Nes, F.R. Herrmann, G. Gold, J.P. Michel and R. Rizzoli, Does the mini nutritional assessment predict hospitalization outcomes in older people? *Age Ageing*, 2001, **30**(3), 221–226.

10. S.M. Garn, M.A. Guzman and B. Wagner, Subperiosteal gain and endosteal loss in protein-calorie malnutrition, *Am. J. Phys. Anthropol.*, 1969, **30**(1), 153–155.

11. A.M. Parfitt, Dietary risk factors for age-related bone loss and fractures, *Lancet*, 1983, **2**(8360), 1181–1185.

12. R. Rizzoli, P. Ammann, T. Chevalley and J.P. Bonjour, Protein intake and bone disorders in the elderly, *Joint Bone Spine*, 2001, **68**, 383–392.

13. G. Schaafsma, E.C. van Beresteyn, J.A. Raymakers and S.A. Duursma, Nutritional aspects of osteoporosis, *World Rev. Nutr. Diet.*, 1987, **49**, 121–159.

14. G. Geinoz, C.H. Rapin, R. Rizzoli, R. Kraemer, D. Buchs, D. Slosman, J.P. Michel and J.P. Bonjour, Relationship between bone mineral density and dietary intakes in the elderly, *Osteoporos. Int.*, 1993, **3**(5), 242–248.

15. R. Rizzoli and J.P. Bonjour, Nutritional approaches to healing fractures in the elderly, In *The Aging Skeleton*, C.J. Rosen, J. Glowacki, and J.P. Bilezikian, (eds.), Academic Press, San Diego, 1999, pp. 399–409.

16. J.A. Grisso, J.L. Kelsey, B.L. Strom, G.Y. Chiu, G. Maislin, L.A. O'Brien, S. Hoffman and F. Kaplan, Risk factors for falls as a cause of hip fracture in women. The Northeast Hip Fracture Study Group, *N. Engl. J. Med.*, 1991, **324**(19), 1326–1331.

17. B. Vellas, R.N. Baumgartner, S.J. Wayne, J. Conceicao, C. Lafont, J.L. Albarede and P.J. Garry, Relationship between malnutrition and falls in the elderly, *Nutrition*, 1992, **8**(2), 105–108.

18. B.J. Vellas, J.L. Albarede and P.J. Garry, Diseases and aging: patterns of morbidity with age; relationship between aging and age-associated diseases, *Am. J. Clin. Nutr.*, 1992, **55**(Suppl. 6), 1225S–1230S.

19. L. Tkatch, C.H. Rapin, R. Rizzoli, D. Slosman, V. Nydegger, H. Vasey and J.P. Bonjour, Benefits of oral protein supplementation in elderly patients with fracture of the proximal femur, *J. Am. Coll. Nutr.*, 1992, **11**(5), 519–525.

20. D.E. Sellmeyer, K.L. Stone, A. Sebastian and S.R. Cummings, A high ratio of dietary animal to vegetable protein increases the rate of bone loss and the risk of fracture in postmenopausal women. Study of Osteoporotic Fractures Research Group, *Am. J. Clin. Nutr.*, 2001, **73**(1), 118–122.

21. M.T. Hannan, K.L. Tucker, B. Dawson-Hughes, L.A. Cupples, D.T. Felson and D.P. Kiel, Effect of dietary protein on bone loss in elderly men and women: the Framingham Osteoporosis Study, *J. Bone Miner. Res.*, 2000, **15**(12), 2504–2512.

22. B. Dawson-Hughes and S.S. Harris, Calcium intake influences the association of protein intake with rates of bone loss in elderly men and women, *Am. J. Clin. Nutr.*, 2002, **75**(4), 773–779.

23. J.J. Anderson and J.A. Metz, Adverse association of high protein intake to bone density, *Challenges Modern Med.*, 1995, **7**, 407–412.

24. R.P. Heaney and R.R. Recker, Effects of nitrogen, phosphorus, and caffeine on calcium balance in women, *J. Lab. Clin. Med.*, 1982, **99**(1), 46–55.

25. J.E. Kerstetter, K. O'Brien and K.L. Insogna, Dietary protein and intestinal calcium absorption, *Am. J. Clin. Nutr.*, 2002, **73**, 990–992.

26. R.P. Heaney, Calcium, dairy products and osteoporosis, *J. Am. Coll. Nutr.*, 2000, **19**(2 Suppl), 83S–99S.

27. R.P. Heaney, Protein intake and bone health: the influence of belief systems on the conduct of nutritional science, *Am. J. Clin. Nutr.*, 2001, **73**(1), 5–6.

28. R.P. Heaney, Protein and calcium: antagonists or synergists? *Am. J. Clin. Nutr.*, 2002, **75**(4), 609–610.

29. J. Bell and S.J. Whiting, Elderly women need dietary protein to maintain bone mass, *Nutr. Rev.*, 2002, **60**(10 pt 1), 337–341.

30. M.A. Schurch, R. Rizzoli, D. Slosman, L. Vadas, P. Vergnaud and J.P. Bonjour, Protein supplements increase serum insulin-like growth factor-I levels and attenuate proximal femur bone loss in patients with recent hip fracture. A randomized, double-blind, placebo-controlled trial, *Ann. Intern. Med.*, 1998, **128**(10), 801–809.

31. R.G. Munger, J.R. Cerhan and B.C. Chiu, Prospective study of dietary protein intake and risk of hip fracture in postmenopausal women, *Am. J. Clin. Nutr.*, 1999, **69**(1), 147–152.

32. J.H. Promislow, D. Goodman-Gruen, D.J. Slymen and E. Barrett-Connor, Protein consumption and bone mineral density in the elderly: the Rancho Bernardo Study, *Am. J. Epidemiol.*, 2002, **155**(7), 636–644.

33. J.P. Bonjour, M.A. Schurch and R. Rizzoli, Nutritional aspects of hip fractures, *Bone*, 1996, **18**(3 Suppl), 139S–144S.

34. J.A. Metz, J.J. Anderson and P.N. Gallagher Jr., Intakes of calcium, phosphorus, and protein, and physical-activity level are related to radial bone mass in young adult women, *Am. J. Clin. Nutr.*, 1993, **58**(4), 537–542.

35. E. Orwoll, M. Ware, L. Stribrska, D. Bikle, T. Sanchez, M. Andon and H. Li, Effects of dietary protein deficiency on mineral metabolism and bone mineral density, *Am. J. Clin. Nutr.*, 1992, **56**(2), 314–319.

36. E.S. Orwoll, The effects of dietary protein insufficiency and excess on skeletal health, *Bone*, 1992, **13**(4), 343–350.

37. R. Rizzoli and J.P. Bonjour, Determinants of peak bone mass and mechanisms of bone loss, *Osteoporos. Int.*, 1999, **9**(Suppl. 2), S17–S23.
38. R. Rizzoli, M.A. Schürch, T. Chevalley and J.P. Bonjour, Protein intake and osteoporosis, In *Nutritional Aspects of Osteoporosis*, P. Burckhardt, B. Dawson-Hughes and R.P. Heaney, (eds.), Springer, New York, 1998, pp. 141–154.
39. B.J. Abelow, T.R. Holford and K.L. Insogna, Cross-cultural association between dietary animal protein and hip fracture: a hypothesis, *Calcif. Tissue Int.*, 1992, **50**(1), 14–18.
40. L.A. Frassetto, K.M. Todd, R.C. Morris Jr. and A. Sebastian, Worldwide incidence of hip fracture in elderly women: relation to consumption of animal and vegetable foods, *J. Gerontol. A Biol. Sci. Med. Sci.*, 2000, **55**(10), M585–M592.
41. D. Feskanich, W.C. Willett, M.J. Stampfer and G.A. Colditz, Protein consumption and bone fractures in women, *Am. J. Epidemiol.*, 1996, **143**(5), 472–479.
42. Z. Huang, J.H. Himes and P.G. McGovern, Nutrition and subsequent hip fracture risk among a national cohort of white women, *Am. J. Epidemiol.*, 1996, **144**(2), 124–134.
43. J. Kanis, O. Johnell, B. Gullberg, E. Allander, L. Elffors, J. Ranstam, J. Dequeker, G. Dilsen, C. Gennari, A.L. Vaz, G. Lyritis, G. Mazzuoli, L. Miravet, M. Passeri, R. Perez Cano, A. Rapado and C. Ribot, Risk factors for hip fracture in men from southern Europe: the MEDOS study. Mediterranean Osteoporosis Study, *Osteoporos. Int.*, 1999, **9**(1), 45–54.
44. O. Johnell, B. Gullberg, J.A. Kanis, E. Allander, L. Elffors, J. Dequeker, G. Dilsen, C. Gennari, A. Lopes Vaz, G. Lyritis, G. Mazzuoli, L. Miravet, M. Passeri, R. Perez Cano, A. Rapado and C. Ribot, Risk factors for hip fracture in European women: the MEDOS Study. Mediterranean Osteoporosis Study, *J. Bone Miner. Res.*, 1995, **10**(11), 1802–1815.
45. C.H. Rapin, R. Lagier, G. Boivin, A. Jung and W. Mac Gee, Biochemical findings in blood of aged patients with femoral neck fractures: a contribution to the detection of occult osteomalacia, *Calcif. Tissue Int.*, 1982, **34**(5), 465–469.
46. D. Thiebaud, P. Burckhardt, M. Costanza, D. Sloutskis, D. Gilliard, F. Quinodoz, A.F. Jacquet and B. Burnand, Importance of albumin, 25(OH)-vitamin D and IGFBP-3 as risk factors in elderly women and men with hip fracture, *Osteoporos. Int.*, 1997, **7**(5), 457–462.
47. H.E. Meyer, J.I. Pedersen, E.B. Loken and A. Tverdal, Dietary factors and the incidence of hip fracture in middle-aged Norwegians. A prospective study, *Am. J. Epidemiol.*, 1997, **145**(2), 117–123.
48. J.P. Bonjour, T. Chevalley, P. Ammann, D. Slosman and R. Rizzoli, Gain in bone mineral mass in prepubertal girls 3.5 years after discontinuation of calcium supplementation: a follow-up study, *Lancet*, 2001, **358**(9289), 1208–1212.
49. J.P. Bonjour, G. Theintz, F. Law, D. Slosman and R. Rizzoli, Peak bone mass, *Osteoporos. Int.*, 1994, **4**(Suppl 1), 7–13.
50. G. Theintz, B. Buchs, R. Rizzoli, D. Slosman, H. Clavien, P.C. Sizonenko and J.P. Bonjour, Longitudinal monitoring of bone mass accumulation in healthy adolescents: evidence for a marked reduction after 16 years of age at the levels of lumbar spine and femoral neck in female subjects, *J. Clin. Endocrinol. Metab.*, 1992, **75**(4), 1060–1065.
51. J.P. Bonjour, G. Theintz, B. Buchs, D. Slosman and R. Rizzoli, Critical years and stages of puberty for spinal and femoral bone mass accumulation during adolescence, *J. Clin. Endocrinol. Metab.*, 1991, **73**(3), 555–563.

52. H. Clavien, G. Theintz, R. Rizzoli and J.P. Bonjour, Does puberty alter dietary habits in adolescents living in a western society? *J. Adolesc. Health*, 1996, **19**(1), 68–75.

53. A.R. Lucas, L.J. Melton III, C.S. Crowson and W.M. O'Fallon, Long-term fracture risk among women with anorexia nervosa: a population-based cohort study, *Mayo Clin. Proc.*, 1999, **74**(10), 972–977.

54. W. Herzog, H.C. Deter, W. Fiehn and E. Petzold, Medical findings and predictors of long-term physical outcome in anorexia nervosa: a prospective, 12-year follow-up study, *Psychol. Med.*, 1997, **27**(2), 269–279.

55. B.L. Drinkwater, K. Nilson, C.H. Chesnut III, W.J. Bremner, S. Shainholtz and M.B. Southworth, Bone mineral content of amenorrheic and eumenorrheic athletes, *N. Engl. J. Med.*, 1984, **311**(5), 277–281.

56. R. Marcus, C. Cann, P. Madvig, J. Minkoff, M. Goddard, M. Bayer, M. Martin, L. Gaudiani, W. Haskell and H. Genant, Menstrual function and bone mass in elite women distance runners. Endocrine and metabolic features, *Ann. Intern. Med.*, 1985, **102**(2), 158–163.

57. M.P. Warren and N.E. Perlroth, The effects of intense exercise on the female reproductive system, *J. Endocrinol.*, 2001, **170**(1), 3–11.

58. G. Gremion, R. Rizzoli, D. Slosman, G. Theintz and J.P. Bonjour, Oligo-amenorrheic long-distance runners may lose more bone in spine than in femur, *Med. Sci. Sports Exerc.*, 2001, **33**(1), 15–21.

59. B.R. Beck, J. Shaw and C.M. Snow, Physical activity and osteoporosis, In *Osteoporosis*, 2nd edn, R. Marcus, D. Feldman, and J. Kelsey, (eds.), Academic Press, San Diego, 2001, vol. 1, pp. 701–720.

60. P. Ammann, R. Rizzoli and J.P. Bonjour, Protein malnutrition-induced bone loss is associated with alteration of growth hormone-IGF-I axis and with estrogen deficiency in adult rats, *Osteoporos. Int.*, 1998, **8**(Suppl 3), 10.

61. P. Ammann, I. Garcia, J.P. Bonjour and R. Rizzoli, High expression of soluble tumor necrosis factor receptor-1 fusion protein prevents bone loss caused by testosterone deficiency, *Calcif. Tissue Int.*, 1999, **64**(Suppl 1), S26.

62. P. Ammann, S. Bourrin, J.P. Bonjour, J.M. Meyer and R. Rizzoli, Protein undernutrition-induced bone loss is associated with decreased IGF-I levels and estrogen deficiency, *J. Bone Miner. Res.*, 2000, **15**(4), 683–690.

63. E.R. Froesch, C. Schmid, J. Schwander and J. Zapf, Actions of insulin-like growth factors, *Annu. Rev. Physiol.*, 1985, **47**(8), 443–467.

64. C.J. Rosen and L.R. Donahue, Insulin-like growth-factor: potential therapeutic options for osteoporosis, *Trends Endocrinol. Metab.*, 1995, **6**, 235–241.

65. P. Ammann, R. Rizzoli, J.M. Meyer and J.P. Bonjour, Bone density and shape as determinants of bone strength in IGF-I and/or pamidronate-treated ovariectomized rats, *Osteoporos. Int.*, 1996, **6**(3), 219–227.

66. C. Bagi, M. van der Meulen, R. Brommage, D. Rosen and A. Sommer, The effect of systemically administered rhIGF-I/IGFBP-3 complex on cortical bone strength and structure in ovariectomized rats, *Bone*, 1995, **16**(5), 559–565.

67. C.M. Bagi, E. DeLeon, R. Brommage, D. Rosen and A. Sommer, Treatment of ovariectomized rats with the complex of rhIGF-I/IGFBP-3 increases cortical and cancellous bone mass and improves structure in the femoral neck, *Calcif. Tissue Int.*, 1995, **57**(1), 40–46.

68. J. Caverzasio and J.P. Bonjour, Insulin-like growth factor I stimulates Na-dependent Pi transport in cultured kidney cells, *Am. J. Physiol.*, 1989, **257**(5 pt 2), F712–F717.

69. J. Caverzasio, C. Montessuit and J.P. Bonjour, Stimulatory effect of insulin-like growth factor-1 on renal Pi transport and plasma 1,25-dihydroxyvitamin D3, *Endocrinology*, 1990, **127**(1), 453–459.

70. G. Palmer, J.P. Bonjour and J. Caverzasio, Stimulation of inorganic phosphate transport by insulin-like growth factor I and vanadate in opossum kidney cells is mediated by distinct protein tyrosine phosphorylation processes, *Endocrinology*, 1996, **137**(11), 4699–4705.

71. G. Palmer, J.P. Bonjour and J. Caverzasio, Expression of a newly identified phosphate transporter/retrovirus receptor in human SaOS-2 osteoblast-like cells and its regulation by insulin-like growth factor I, *Endocrinology*, 1997, **138**(12), 5202–5209.

72. E. Canalis and D. Agnusdei, Insulin-like growth factors and their role in osteoporosis [editorial], *Calcif. Tissue Int.*, 1996, **58**(3), 133–134.

73. J.P. Thissen, J.M. Ketelslegers and L.E. Underwood, Nutritional regulation of the insulin-like growth factors, *Endocr. Rev.*, 1994, **15**(1), 80–101.

74. J.P. Thissen, J.B. Pucilowska and L.E. Underwood, Differential regulation of insulin-like growth factor I (IGF-I) and IGF binding protein-1 messenger ribonucleic acids by amino acid availability and growth hormone in rat hepatocyte primary culture, *Endocrinology*, 1994, **134**(3), 1570–1576.

75. W.L. Isley, L.E. Underwood and D.R. Clemmons, Dietary components that regulate serum somatomedin-C concentrations in humans, *J. Clin. Invest.*, 1983, **71**(2), 175–182.

76. J.P. Thissen, S. Triest, M. Maes, L.E. Underwood and J.M. Ketelslegers, The decreased plasma concentration of insulin-like growth factor-I in protein-restricted rats is not due to decreased numbers of growth hormone receptors on isolated hepatocytes, *J. Endocrinol.*, 1990, **124**(1), 159–165.

77. M.J. VandeHaar, B.M. Moats-Staats, M.L. Davenport, J.L. Walker, J.M. Ketelslegers, B.K. Sharma and L.E. Underwood, Reduced serum concentrations of insulin-like growth factor-I (IGF-I) in protein-restricted growing rats are accompanied by reduced IGF-I mRNA levels in liver and skeletal muscle, *J. Endocrinol.*, 1991, **130**(2), 305–312.

78. J.P. Thissen and L.E. Underwood, Translational status of the insulin-like growth factor-I mRNAs in liver of protein-restricted rat, *J. Endocrinol.*, 1992, **132**(1), 141–147.

79. D.H. Sullivan and W.J. Carter, Insulin-like growth factor I as an indicator of protein-energy undernutrition among metabolically stable hospitalized elderly, *J. Am. Coll. Nutr.*, 1994, **13**(2), 184–191.

80. J.B. Pucilowska, M.L. Davenport, I. Kabir, D.R. Clemmons, J.P. Thissen, T. Butler and L.E. Underwood, The effect of dietary protein supplementation on insulin-like growth factors (IGFs) and IGF-binding proteins in children with shigellosis, *J. Clin. Endocrinol. Metab.*, 1993, **77**(6), 1516–1521.

81. D.R. Clemmons, M.M. Seek, and L.E. Underwood, Supplemental essential amino acids augment the somatomedin-C/insulin-like growth factor I response to refeeding after fasting, *Metabolism*, 1985, **34**(4), 391–395.

82. D.R. Clemmons, L.E. Underwood, R.N. Dickerson, R.O. Brown, L.J. Hak, R.D. MacPhee and W.D. Heizer, Use of plasma somatomedin-C/insulin-like growth factor I measurements to monitor the response to nutritional repletion in malnourished patients, *Am. J. Clin. Nutr.*, 1985, **41**(2), 191–198.

83. V.C. Musey, S. Goldstein, P.K. Farmer, P.B. Moore and L.S. Phillips, Differential regulation of IGF-1 and IGF-binding protein-1 by dietary composition in humans, *Am. J. Med. Sci.*, 1993, **305**(3), 131–138.

84. J.P. Thissen, S. Triest, B.M. Moats-Staats, L.E. Underwood, T. Mauerhoff, D. Maiter and J.M. Ketelslegers, Evidence that pretranslational and translational defects decrease serum insulin-like growth factor-I concentrations during dietary protein restriction, *Endocrinology*, 1991, **129**(1), 429–435.

85. S. Bourrin, P. Ammann, J.P. Bonjour and R. Rizzoli, Dietary protein restriction lowers plasma insulin-like growth factor I (IGF-I), impairs cortical bone formation, and induces osteoblastic resistance to IGF-I in adult female rats, *Endocrinology*, 2000, **141**(9), 3149–3155.

86. E. Canalis, T.L. McCarthy and M. Centrella, Growth factors and cytokines in bone cell metabolism, *Annu. Rev. Med.*, 1991, **42**, 17–24.

87. T. Chevalley, R. Rizzoli, D. Manen, J. Caverzasio and J.P. Bonjour, Arginine increases insulin-like growth factor-I production and collagen synthesis in osteoblast-like cells, *Bone*, 1998, **23**(2), 103–109.

88. P. Ammann, R. Rizzoli, J.P. Bonjour, S. Bourrin, J.M. Meyer, P. Vassalli and I. Garcia, Transgenic mice expressing soluble tumor necrosis factor-receptor are protected against bone loss caused by estrogen deficiency, *J. Clin. Invest.*, 1997, **99**(7), 1699–1703.

89. P. Ammann, S. Bourrin, J.P. Bonjour, F. Brunner, J.M. Meyer and R. Rizzoli, The new selective estrogen receptor modulator MDL 103,323 increases bone mineral density and bone strength in adult ovariectomized rats, *Osteoporos. Int.*, 1999, **10**(5), 369–376.

90. P. Ammann, C. Gabay, G. Palmer, I. Garcia and R. Rizzoli, Tumor necrosis factor alpha but not interleukine-1 is involved in protein undernutrition-induced bone resorption, *J. Bone Miner. Res.*, 2002, **17**(Suppl. 1), S205.

91. P. Ammann, M.L. Aubert, J.M. Meyer and R. Rizzoli, Protein undernutrition-induced bone resorption is dependent on tumor necrosis factor alfa (TNF), *Osteoporos. Int.*, 2002, **13**(Suppl 1), S5.

92. P. Ammann, I. Garcia, J.P. Bonjour and R. Rizzoli, Tumor necrosis factor-a (TNF) plays a prominent role in protein undernutrition-induced bone resorption, *J. Bone Miner. Res.*, 2001, **16**(Suppl 1), S147.

93. P. Ammann, A. Laib, J.P. Bonjour, J.M. Meyer, P. Ruegsegger and R. Rizzoli, Dietary essential amino acid supplements increase bone strength by influencing bone mass and bone microarchitecture in ovariectomized adult rats fed an isocaloric low-protein diet, *J. Bone Miner. Res.*, 2002, **17**(7), 1264–1272.

94. J. Chan, Y. Tian, K.E. Tanaka, M.S. Tsang, K. Yu, P. Salgame, D. Carroll, Y. Kress, R. Teitelbaum and B.R. Bloom, Effects of protein calorie malnutrition on tuberculosis in mice, *Proc. Natl Acad. Sci. USA*, 1996, **93**(25), 14857–14861.

95. G. Dai and D.N. McMurray, Altered cytokine production and impaired antimycobacterial immunity in protein-malnourished guinea pigs, *Infect. Immun.*, 1998, **66**(8), 3562–3568.

96. C.C. Spaulding, R.L. Walford and R.B. Effros, Calorie restriction inhibits the age-related dysregulation of the cytokines TNF-alpha and IL-6 in C3B10RF1 mice, *Mech. Ageing Dev.*, 1997, **93**(1–3), 87–94.

97. S.D. Anker, A.L. Clark, M.M. Teixeira, P.G. Hellewell and A.J. Coats, Loss of bone mineral in patients with cachexia due to chronic heart failure, *Am. J. Cardiol.*, 1999, **83**(4), 612–615.

98. S.D. Anker and A.J. Coats, Cardiac cachexia: a syndrome with impaired survival and immune and neuroendocrine activation, *Chest*, 1999, **115**(3), 836–847.

99. G.S. Hotamisligil, Mechanisms of TNF-alpha-induced insulin resistance, *Exp. Clin. Endocrinol. Diabetes*, 1999, **107**(2), 119–125.
100. R.F. Grimble, A.A. Jackson, C. Persaud, M.J. Wride, F. Delers and R. Engler, Cysteine and glycine supplementation modulate the metabolic response to tumor necrosis factor alpha in rats fed a low protein diet, *J. Nutr.*, 1992, **122**(11), 2066–2073.
101. M.J. Manary, D.R. Brewster, R.L. Broadhead, S.M. Graham, C.A. Hart, J.R. Crowley, C.R. Fjeld and K.E. Yarasheski, Whole-body protein kinetics in children with kwashiorkor and infection: a comparison of egg white and milk as dietary sources of protein, *Am. J. Clin. Nutr.*, 1997, **66**(3), 643–648.
102. R.F. Grimble, Nutritional modulation of cytokine biology, *Nutrition*, 1998, **14** (7–8), 634–640.
103. R.L. Jilka, Cytokines, bone remodeling, and estrogen deficiency: a 1998 update, *Bone*, 1998, **23**(2), 75–81.
104. A. Aniansson, C. Zetterberg, M. Hedberg and K.G. Henriksson, Impaired muscle function with aging. A background factor in the incidence of fractures of the proximal end of the femur, *Clin. Orthop.*, 1984, **191**, 193–201.
105. C. Castaneda, P.L. Gordon, R.A. Fielding, W.J. Evans and M.C. Crim, Marginal protein intake results in reduced plasma IGF-I levels and skeletal muscle fiber atrophy in elderly women, *J. Nutr. Health Aging*, 2000, **4**(2), 85–90.
106. D.H. Sullivan, G.A. Patch, R.C. Walls and D.A. Lipschitz, Impact of nutrition status on morbidity and mortality in a select population of geriatric rehabilitation patients, *Am. J. Clin. Nutr.*, 1990, **51**(5), 749–758.
107. C.J. Auernhammer and C.J. Strasburger, Effects of growth hormone and insulin-like growth factor I on the immune system, *Eur. J. Endocrinol.*, 1995, **133**(6), 635–645.
108. M.S. Calvo, C.N. Barton and Y.K. Park, Bone mass and high dietary intake of meat and protein: analyses of data from the Third National Health and Nutrition Examination Survey (NHANES III, 1988–94), *Bone*, 1998, **23**(Suppl.), S290.
109. J.F. Chiu, S.J. Lan, C.Y. Yang, P.W. Wang, W.J. Yao, L.H. Su and C.C. Hsieh, Long-term vegetarian diet and bone mineral density in postmenopausal Taiwanese women, *Calcif. Tissue Int.*, 1997, **60**(3), 245–249.
110. C. Cooper, E.J. Atkinson, D.D. Hensrud, H.W. Wahner, W.M. O'Fallon, B.L. Riggs and L.J. Melton 3rd, Dietary protein intake and bone mass in women, *Calcif. Tissue Int.*, 1996, **58**(5), 320–325.
111. T. Hirota, M. Nara, M. Ohguri, E. Manago and K. Hirota, Effect of diet and lifestyle on bone mass in Asian young women, *Am. J. Clin. Nutr.*, 1992, **55**(6), 1168–1173.
112. J.E. Kerstetter, A.C. Looker and K.L. Insogna, Low dietary protein and low bone density, *Calcif. Tissue Int.*, 2000, **66**(4), 313.
113. J.M. Lacey, J.J. Anderson, T. Fujita, Y. Yoshimoto, M. Fukase, S. Tsuchie and G.G. Koch, Correlates of cortical bone mass among premenopausal and postmenopausal Japanese women, *J. Bone Miner. Res.*, 1991, **6**(7), 651–659.
114. E.M. Lau, T. Kwok, J. Woo and S.C. Ho, Bone mineral density in Chinese elderly female vegetarians, vegans, lacto-vegetarians and omnivores, *Eur. J. Clin. Nutr.*, 1998, **52**(1), 60–64.
115. K. Michaelsson, L. Holmberg, H. Mallmin, A. Wolk, R. Bergstrom and S. Ljunghall, Diet, bone mass, and osteocalcin: a cross-sectional study, *Calcif. Tissue Int.*, 1995, **57**(2), 86–93.
116. E.S. Orwoll, R.M. Weigel, S.K. Oviatt, D.E. Meier and M.R. McClung, Serum protein concentrations and bone mineral content in aging normal men, *Am. J. Clin. Nutr.*, 1987, **46**, 614–621.

117. D. Teegarden, R.M. Lyle, G.P. McCabe, L.D. McCabe, W.R. Proulx, K. Michon, A.P. Knight, C.C. Johnston and C.M. Weaver, Dietary calcium, protein, and phosphorus are related to bone mineral density and content in young women, *Am. J. Clin. Nutr.*, 1998, **68**(3), 749–754.

118. F.A. Tylavsky and J.J. Anderson, Dietary factors in bone health of elderly lactoovovegetarian and omnivorous women, *Am. J. Clin. Nutr.*, 1988, **48**(3 Suppl), 842–849.

119. N.K. Henderson, R.I. Price, J.H. Cole, D.H. Gutteridge and C.I. Bhagat, Bone density in young women is associated with body weight and muscle strength but not dietary intakes, *J. Bone Miner. Res.*, 1995, **10**(3), 384–393.

120. R.B. Mazess and H.S. Barden, Bone density in premenopausal women: effects of age, dietary intake, physical activity, smoking, and birth-control pills, *Am. J. Clin. Nutr.*, 1991, **53**(1), 132–142.

121. S.A. New, C. Bolton-Smith, D.A. Grubb and D.M. Reid, Nutritional influences on bone mineral density: a cross-sectional study in premenopausal women, *Am. J. Clin. Nutr.*, 1997, **65**(6), 1831–1839.

122. J.W. Nieves, A.L. Golden, E. Siris, J.L. Kelsey and R. Lindsay, Teenage and current calcium intake are related to bone mineral density of the hip and forearm in women aged 30–39 years, *Am. J. Epidemiol.*, 1995, **141**(4), 342–351.

123. M.C. Wang, M. Luz Villa, R. Marcus and J.L. Kelsey, Associations of vitamin C, calcium and protein with bone mass in postmenopausal Mexican American women, *Osteoporos. Int.*, 1997, **7**(6), 533–538.

124. J.L. Freudenheim, N.E. Johnson and E.L. Smith, Relationships between usual nutrient intake and bone mineral content of women 35–65 years of age: longitudinal and cross-sectional analysis, *Am. J. Clin. Nutr.*, 1986, **44**, 863–876.

125. R.R. Recker, K.M. Davies, S.M. Hinders, R.P. Heaney, M.R. Stegman and D.B. Kimmel, Bone gain in young adult women, *JAMA*, 1992, **268**(17), 2403–2408.

126. I.R. Reid, R.W. Ames, M.C. Evans, S.J. Sharpe and G.D. Gamble, Determinants of the rate of bone loss in normal postmenopausal women, *J. Clin. Endocrinol. Metab.*, 1994, **79**(4), 950–954.

CHAPTER 10.2

# Excess Dietary Protein and Bone Health

LINDA K. MASSEY and SUSAN J WHITING

Food Science and Human Nutrition, Washington State University Spokane, Spokane WA 99210-1495, USA, Email: massey@wsu.edu
College of Pharmacy and Nutrition, University of Saskatchewan, Saskatoon SK S7N 5C9 Canada

**Key Points**

- Catabolism of sulfur-containing amino acids increases net renal acid excretion (NRAE), which in turn promotes urinary calcium excretion.
- Both purified animal and plant proteins increase urinary NRAE, and therefore urinary calcium.
- Increasing protein intake is not harmful but beneficial to bone health, in the range of usual dietary intakes of US postmenopausal women.
- Most prospective epidemiological studies find an adverse effect of high protein intakes on bone health only when dietary calcium is also low.
- Animal and plant foods' effects on bone are due to their non-protein constituents as well as protein.
- Animal foods are a more concentrated source of protein compared to plant foods.
- Dietary protein as foods is not a risk factor for bone loss when consumed in typical amounts.

Excess dietary protein has been considered a risk factor for osteoporosis, as it increases renal acid load. The catabolism of sulfur-containing amino acids is a major determinant of renal net acid excretion (RNAE), which in turn is strongly related to urinary calcium excretion. Bone is mobilized to neutralize acid and to maintain blood calcium levels. Experimental studies using purified protein sources show negative calcium balance from feeding either plant or animal protein, when phosphate intake is equal and calcium intake is low. Prospective studies provide information on bone mineral density and fracture incidence in relation to excess protein intake. However, data, while conflicting, suggest that a low protein intake may be of greater concern than excess protein, particularly in elderly women. The effects of a protein food on urinary calcium and bone metabolism are modified by other nutrients found in that food, such as calcium, potassium and phosphate. In a Western diet, animal protein foods are frequently consumed in excess in part because they are a concentrated energy source compared to plant-based foods, as well as being culturally preferred. However, there is little evidence to suggest that higher intakes of either animal-based protein or plant-based protein in an otherwise nutritionally balanced diet is detrimental to bone health.

# Introduction

Excess dietary protein may be harmful to bone health as the products of protein metabolism increase the amount of calcium that must be excreted through the kidney. In 2002 the Institute of Medicine (IOM) concluded that there was insufficient evidence to suggest a tolerable upper level for dietary protein.[1] It has long been recognized that an increase in protein consumption increases urinary calcium excretion over the entire range of protein intakes, from marginal to excess.[2] The IOM report, however, stated "the potential implications of high dietary protein for bone metabolism are not sufficiently unambiguous at present to make recommendations." While it has been difficult to define the amount of protein characterizing 'excess' dietary protein, the common perception is that the typical North American diet, high in protein and primarily animal-based, is sufficient to evoke detrimental changes in calcium metabolism, which result in bone loss and subsequent osteoporotic fractures.[3]

# Dietary Intakes of Protein Related to Osteoporosis

The incidence of hip fractures is higher in 'developed' countries. Many lifestyle changes occur during economic development; typically a decrease in physical activity and change in diet are observed. Dietary changes usually include an increase in animal foods at the expense of plant foods. Because of the associated increase in dietary animal protein and urinary calcium excretion, the increase in hip fracture incidence has frequently been attributed to the increase in animal protein. Frassetto *et al.*[3] found the cross-cultural relationship between hip fracture rates and dietary protein was positively related to animal protein intake and inversely related to vegetable protein intake. Even when non-Caucasian populations were removed from the data set to control for genetic differences in bone, this relationship held. These authors attempted to determine whether the source of protein was a factor, so they plotted the relationship between the ratio of vegetable to animal protein *vs.* hip fracture rate. This resulted in a non-linear relationship where countries having a vegetable:animal protein ratio over 2.0 had very low rates of hip fracture.

Figure 10.2.1 shows the relationship between protein type (animal or plant) as well as total protein intake (g day$^{-1}$ per capita), on the incidence of osteoporosis, in these 33 countries.[3] The higher protein intake in the second tertile is due to an increase in animal protein. The top two tertiles are not different in total protein, but there is a 7 g day$^{-1}$ increase in animal protein at the expense of plant protein. While protein appears to be a factor in hip fracture incidence, the large range within each tertile cannot be explained by protein intake alone. It should be noted these protein intake data were derived from food availability data, a dietary assessment method that over-estimates energy consumed in developed countries and underestimates it in less developed countries.[4]

Why people in Western countries consume more animal-based than plant-based protein, and why it may be consumed in excess, are complex issue.[5] Protein is a more costly source of energy than fat or carbohydrate, and low intakes of protein are common in areas of the world where food is more expensive relative to income. However, with modern agricultural practices, protein foods are relatively less costly in developed countries. In North America and other developed countries, plant proteins account for 65% of available food protein.[1] However, protein intakes in the US are two-thirds animal foods and only one-third from plant sources.[5] As shown in Table 10.2.1, the weight of food necessary to obtain 7 g protein (equivalent to that in one large egg) is much less if animal-based food is consumed compared to plant-based foods. The low energy to protein density of animal-based foods may be one reason why these foods are preferred.

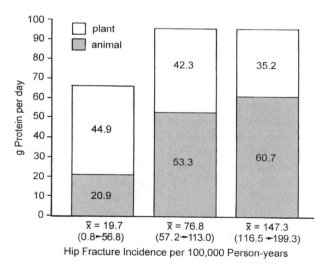

**Figure 10.2.1** *Dietary protein and hip fracture incidence*

**Table 10.2.1.** *Weight and energy of food portions containing 7 g of protein*

| Food | Weight (g) | Energy (kcal) | Energy/protein density (kcal g$^{-1}$ protein) |
|------|-----------|---------------|-----------------------------------------------|
| Chicken breast | 23 | 45 | 6.4 |
| Tuna | 24 | 28 | 4.0 |
| Hard cheese | 28 | 113 | 16.1 |
| Ground beef | 30 | 75 | 10.7 |
| Egg (1 large) | 56 | 83 | 11.9 |
| Peanuts, dry roasted | 30 | 176 | 24.1 |
| Soy beans, cooked | 42 | 73 | 10.4 |
| Peas | 100 | 84 | 12.0 |
| Macaroni, cooked | 147 | 207 | 29.6 |
| Brown rice, cooked | 300 | 337 | 48.1 |

# Mechanism of Protein-Induced Increase in Calcium Excretion

Urinary calcium excretion increases in response to ingestion of protein, from marginal to excess intakes.[2] Each 10 g increase in dietary protein increases urinary calcium by 6 mg, and doubling protein increases urinary calcium by 50%. An increase in urinary calcium excretion upon consumption of high protein diets has been observed in young and old adults[6–12] as well as adolescents.[13] The human observations are supported by findings in animal models such as the growing and adult rat.[14–16]

Dietary protein increases urinary calcium excretion, at least in part, by its effect on acid–base metabolism.[17–19] Dietary protein raises net acid excretion, and the resulting decrease in renal tubular reabsorption of calcium may lead to a negative calcium balance and subsequent bone loss if there is insufficient dietary calcium to maintain calcium balance.[6,8,9] Metabolism of amino acids leads to ammonium ion production. In addition, a molecule of sulfuric acid is potentially generated from each sulfur of cysteine and methionine, which contributes substantially to the rise in urinary acid excretion.[12,14] A change in blood pH has not been reported after high protein feeding[9]; however even small decreases in blood pH have been shown to activate bone resorption.[20] Urinary calcium excretion is strongly related to RNAE. From his meta-analysis, Lemann[21] concluded that urinary calcium increases by 0.035 mmol mEq$^{-1}$ RNAE increase; similarly urinary calcium increases 0.04 mmol (1.6 mg) per gram dietary protein increase. Dietary protein has also been shown to reduce the glomerular filtration rate (GFR) of calcium,[8,11] thus affecting the renal reabsorption of calcium.

Higher dietary protein intake appears to have a beneficial effect on bone healing. Two studies have shown that supplementation with 20 g protein improved fracture healing and reduced bone loss in elderly subjects.[22,23] In these studies, dietary protein intake was low before supplementation, 33 and 45 g, respectively. The mechanism seems to be at least partially through an increase in insulin-like growth factor-1 (IGF-1). IGF-1 stimulates proliferation and differentiation of osteoblasts, selectively stimulates the transport of phosphate across membranes of osteoblastic cells, and inhibits the degradation of collagen produced by osteoblasts.[24] Studies involving administration of recombinant human IGF-1 to osteoporotic patients suggest this hormone may play a unique role in fracture healing[25]; thus the effect of dietary protein on IGF-1 remains a promising area for further research.

# Experimental Studies

## Chronic Calcium Balance Studies

In many of the early studies isolated proteins (egg white, lactalbumin, casein and wheat gluten) were added to the basal diet (approximately 50 g protein) to provide a high protein (*i.e.* 112–150 g), low calcium (500 mg) diet.[7,10–12] One early study added protein as isolated beef protein.[8] Increasing dietary protein from 47 to 142 g doubled urinary calcium excretion,[7] and there was a concomitant change in GFR[8,11] that accounted for only a small part of the increase in urinary calcium observed. These experimental human studies demonstrated a rise in net acid excretion with protein consumption.[9,18,26] The sulfur amino acid content of the diet accounted for 90% of the RNAE[11] but only half of the hypercalciuria due to high protein feeding.[12] Replacing chloride with bicarbonate in the high protein diet, or adding potassium bicarbonate directly to the diet, reduced the RNAE and the calciuria; however, changes in blood pH were not evident.[6,9] When the phosphorus level of the lower protein control diet was equalized to the high protein diet, using either purified protein[10] or meat,[27,28] there was no hypercalciuria or change in calcium balance, in agreement with studies of phosphate administration reporting a hypocalciuric effect of phosphate.[29]

Several other human studies on calcium excretion and balance have been reported. Kitano *et al.*[30] reported that increasing protein from 1.0 to 2.0 g $kg^{-1}$ by adding beef, egg and milk increased urinary calcium and decreased calcium balance in six healthy young males. However, each study period was very short, and balance was measured on days 3–6 after only two days adaptation. In contrast, Moriguti *et al.*[31] found that increasing protein intake from 0.6 to 2.0 g $kg^{-1}$ $day^{-1}$ for 1 week by adding beef, milk, rice and beans so that plant:animal protein was unchanged at 1:1 did not increase urinary calcium in elderly men. The response of elderly men did not differ from young men with diets of 12 *vs.* 21% protein in a 3-week study reported by Pannemans *et al.*[32] In both groups urinary calcium did not change, and calcium balance was slightly improved in the elderly group. Finally, Jenkins *et al.*[33] found that increasing protein intake from 1.5 to 2.5 g $kg^{-1}$ $day^{-1}$ by replacing starch with wheat gluten in a control bread increased urinary calcium and N-teleopeptide during the last week of a 4 week study. Dietary calcium was high at 1580 mg $day^{-1}$ and calcium balance was unchanged. The difference in urinary calcium loss correlated with the serum anion gap as a marker of metabolic acid production, suggesting variable individual responses in this group of 20 men and women aged 35–71 years (mean 56). The high protein intake of the Jenkins *et al.* subjects exceeded the protein intakes of 98% of subjects in recent USDA survey data.[5] This study confirmed that not only animal proteins but also a plant protein, gluten, could increase urinary calcium if eaten in excess. Gluten contains 4.56% sulfur-containing amino acids, so has the potential to produce 86 mEq acid per 100 g of protein.

## Acute Studies

An acute load test developed by Allen and colleagues[34] has been useful for identifying postprandial effects of dietary protein on calcium excretion. Using the acute load protocol this group initially postulated a role of insulin secretion in protein-induced hypercalciuria[31] but further study revealed that insulin inhibition of calcium reabsorption could not explain the unique calciuria with high protein intake.[32] Whiting *et al.*[33] examined the effects of high protein with and without accounting for the rise in phosphate using the acute protocol. Results indicated urinary calcium excretion at 3 h postload was significantly higher when subjects were administered high protein compared to the moderate protein treatment having equivalent phosphorus, but not higher compared to moderate protein alone. Rises in urinary sulfate and RNAE were consistent with chronic studies. They also showed that the protein-induced hypercalciuria could be blunted by simultaneous administration of potassium bicarbonate.

A longer-term protocol developed by Kerstetter and co-workers to study protein effects on calcium metabolism involves 4-day experimental periods.[34,35] Subjects consume several levels of protein diets (0.7, 1.0 or 2.1 g kg$^{-1}$ day$^{-1}$) prepared from various foods; while protein intake varies 3-fold, phosphorus varies by only 50%. In comparing a high protein intake to low, this group has shown that urinary calcium increases, as does fractional and true calcium absorption.[34] Markers of bone resorption increase on the low protein intake but are not affected by the high protein regimen; however, there was a shift in the excretion of one marker indicating that high protein might influence bone turnover.[35] However, the effects of a sustained low *vs.* high protein intake have not yet been studied.

The experimental studies, with a few exceptions, examined the effect of high protein when only protein was increased. Most equalized the phosphorus levels so that the only difference between a high and low protein intake was the amino acid content of the increment in protein intake. While equalizing phosphorus decreased the hypercalciuria, a high protein intake from food changed many other variables that could affect calcium balance. Further, the short study durations have precluded any direct measurement of bone lone or fracture. Thus epidemiological studies looking at the effect of habitual high protein intakes on bone have been also examined using population studies.

# Epidemiological Observations

Relationships between bone mineral density or fracture incidence and dietary protein have been examined in both cross-sectional or prospective epidemiological studies. Because of the strength of the prospective approach, we will only discuss epidemiological studies with that design. A review of the recent studies on the effects of protein intake on bone mineral density or hip fracture incidence of middle-aged and elderly men and women[40] shows near consensus that increasing protein intake is not harmful, but actually beneficial to bone health in the range of usual intake of the women studied. Of the seven prospective studies relating dietary protein to bone health in older American women (Table 10.2.2), only Feskanich *et al.*[41] found a significant increase in fracture risk with higher protein intake. This risk was seen only for protein intake greater than 95 g day$^{-1}$, which corresponds to protein intakes of less than 10% of US women 50 years and older.[42] The other six studies report that subjects with higher protein intakes had reduced fracture risk,[43,44] higher BMD,[45] reduced BMD loss[46,47] or higher fracture only in the women in both high non-dairy protein/low calcium quartiles.[48] In these latter studies, mean protein intakes ranged between 68 and 79 g day$^{-1}$. Fewer studies of men[45–48] and only one on younger women[49] have been reported so patterns are not evident yet.

The debate concerning protein and its effect on bone has also raised the issue of the source of protein, with animal protein considered to provide more potential renal acid load (PRAL) than plant-based protein. In agreement with that, Sellmeyer *et al.*[44] found that a greater ratio of plant-based to animal protein was associated with reduced BMD loss and hip fracture risk in elderly women. However, as shown in Table 10.2.2, most of the recent studies of elderly women find no benefit of plant-based protein over animal protein.

# Variations in Protein Composition Which May Affect Bone Health

## Sulfur Amino Acid Content

Although vegetable proteins are considered to have poorer nutritional quality than animal proteins for humans, it is because they are imbalanced in the ratio of cysteine to methionine needed to meet requirements, not because they are lower in mEq sulfur per gram of protein. In a study comparing several protein sources and their effect on urinary calcium excretion, there was a positive relationship between sulfur content and calcium excretion.[14] When sulfur amino acids were added to a low protein diet, urinary calcium excretion rose.[12] Although animal proteins are commonly assumed to have a higher content of sulfur-containing amino acids per gram of protein, this is not always the case (Table 10.2.3). Some plant proteins have the potential of producing more mEq of sulfuric acid per gram of protein than some animal proteins. For example, wheat bread has a value of 58.1 while beef has a value of 53.0 mEq. Two common legumes,

**Table 10.2.2.** Summary of recent prospective studies examining the effect of total protein, animal protein and plant protein intakes on bone health in middle-aged and elderly men and women

| Study authors | Feskanich et al. (1996)[41] | Meyer et al. (1997)[48] | Munger et al. (1999)[43] | Hannan et al. (2000)[45] | Sellmeyer et al. (2001)[44] | Promislow et al. (2002)[47] | Dawson-Hughes et al. (2002)[46] |
|---|---|---|---|---|---|---|---|
| Duration | 12 years | 11.4 years | 3 years | 4 years | ~7 years | 4 years | 3 years |
| Number of subjects | 85, 900 women | 4916 women 4953 men | 32,006 women | 224 men 391 women | 1035 women | 572 women 388 men | 161 men 181 women |
| Subjects | Nurses' Health Study cohort | Norway (random) | Iowa (random) | Framingham cohort | Study of Osteoporosis Fractures cohort | Rancho Bernardo cohort | Boston volunteers |
| Age (years) at baseline | 35–59 | 35–49 (mean 47) | 55–69 (mean 61) | 75 (mean) 68–91 | >65 73 (mean) | 71 (mean) 55–92 | ≥65 70 (mean) |
| Average protein intake (g) | 80 | 49.5 women, 67.4 men | 78 | 68 | 50 | 71 women 74 men | 79 |
| Calcium intake (mg) | 718 | 589 women, 834 men | 1150 | 810 (all) | 853 | 797 women, 985 men | 1346 + Ca suppl 871 − Ca suppl |
| Calcium: protein (mg g$^{-1}$) | 9:1 | 12:1 | 15:1 | 11:1 | 17:1 | 14:1 | 17:1 |
| Conclusions | ↑p ↑Fx ↑ap ↑Fx ↑pp ↔ Fx | ↑ap[a] ↔Fx ↑ap and low Ca ↑Fx | ↑p ↓Fx ↑ap ↓Fx ↑pp ↑Fx | Women ↑p ↑BMD ↑ap ↑BMD ↑pp ↔ Bone loss | ↑ap:pp ↑hip Fx ↑ap:pp ↔BMD | Women ↑p ↔ BMD ↑ap ↑BMD ↑pp ↓BMD men ↑p ↔ BMD | ↑p ↑BMD ↑ap ↔BMD ↑pp ↔BMD |

[a]Non-dairy animal protein.

**Table 10.2.3.** *Potential acid as sulfate from sulfur amino acid content*[a]

| Food | Sulfur (mEq per 100 g protein) |
|---|---|
| Oatmeal | 82.2 |
| Egg | 79.6 |
| Walnuts | 73.8 |
| Pork | 73.0 |
| Wheat (whole) | 69.4 |
| White Rice | 68.0 |
| Barley | 67.6 |
| Tuna | 65.0 |
| Chicken | 65.0 |
| Corn | 61.4 |
| Beef | 59.4 |
| Milk | 54.8 |
| Cheddar | 46.2 |
| Soy | 39.8 |
| Peanuts | 39.6 |
| Millet | 31.2 |
| Almonds | 23.2 |
| Potato | 23.2 |

[a]Calculated from the amino acid composition published by Hands.[61]

soybeans and chickpeas, have values of 44.9 and 39.9 mEq, respectively. However, in a mixed diet containing a variety of both animal and plant protein foods, both are significant sources of sulfur containing amino acids.

## Phosphorus Content

Total phosphorus is similar in plant and animal foods, typically 8–20 mg per gram of protein.[1] However, in muscle protein foods from meat, poultry and fish, the phosphorus is found as phosphate bound to an amino acid side chain, which is released during digestion. In contrast, much of the phosphorus in plant foods is found as phytate, which is poorly digested, and therefore less phosphorus is absorbed. Most experimental studies of high protein-induced hypercalciuria corrected for the high phosphorus content of protein by adding phosphorus to the control diet,[6–12] thereby bypassing the hypocalciuric effect of phosphorus.

Many of the studies on the effect of increasing meat consumption on calcium balance have found no effect.[27,28] Hunt *et al.*[28] studied 14 postmenopausal women for seven weeks on three diets varying in meat content. The low meat diet had only 38.5 g meat compared to 289 g on the high meat diet. The low meat diet substituted fruit, sugars and oils for the meat energy. The third diet was the low meat diet supplemented with potassium, phosphate, iron and zinc salts. Calcium balance was not different during the last week of each diet. Urinary calcium was not different between the low and high meat diets, but was lower on the mineral supplemented diet. This last finding suggests that the higher phosphate intake of the high meat diet was an important factor, and its hypocalciuric effect compensated for the hypercalciuric effect of the higher protein intake. Since some fruit was substituted for meat energy, the higher potassium-base intake would also play a role, but unfortunately, dietary and urinary potassium data were not reported. However, consumption of meat *per se* does not appear to affect urinary calcium or calcium balance.

## Calcium

Some protein foods contain substantial amounts of calcium. Calcium in milk has been credited with compensating for any urinary calcium loss generated by milk protein.[49,50] The effects of dietary protein on calcium retention are not only related to urinary calcium excretion, but also to the amount of calcium absorbed, which in turn is related to dietary calcium intake. When the relationship of dietary calcium to calcium retention is plotted at various protein intakes, nearly all balances are negative when calcium intakes are below 800 mg per day. However, at 800 or 1400 mg of calcium, a range of both negative and positive balances was seen, with no obvious relationship to dietary protein intake.[2] These data suggest that higher protein intakes affect calcium retention most adversely when calcium intakes are simultaneously inadequate.

Promislow and co-workers examined whether calcium intake modified the protein–BMD relationships that were found for women.[47] At a high calcium intake, there was a negative effect of plant protein intake on BMD; at low calcium intakes there was a positive effect on total body BMD. This pattern of interaction between protein and calcium on BMD was not observed by Dawson-Hughes and Harris[46]; in their study, higher intakes of protein spared bone loss only in a calcium-supplemented group of elderly men and women, while those in the placebo group (with an average calcium intake of 871 mg) did not show any protein effect (positive or otherwise) on BMD. The calcium to protein ratio of subjects in prospective studies (Table 10.2.2) showing a protective effect of protein was higher than in the study showing an adverse effect, suggesting that the calcium:protein ratio may be important when considering protein effects on bone.

## Potassium and Base

The high potassium content of plant protein foods, such as legumes and grains, promotes retention of bone mineral.[53–55] Fruits and vegetables are generally recognized as sources of alkalinity that can neutralize acid from metabolism of sulfur amino acids.[54,55] Furthermore, plant foods may have additional benefit for bone health by mechanisms other than base potential. A recent study of onions, vegetable mixes, salads and herbs indicates inhibition of bone resorption by an unknown mechanism independent of base excess[56] when these foods are fed to rats.

# Potential Renal Acid Load (PRAL)

The urinary ions of a healthy adult eating a protein-rich diet are shown in Figure 10.2.2. NRAE can be calculated directly by subtracting the bicarbonate excretion from the combined ammonia and titratable acid; however, these are difficult to measure. An indirect calculation can be done by adding the remaining anions (Cl, S, P) and subtracting the remaining cations (Ca, Mg, Na, K). Remer and Manz[17] predicted the urinary composition from the diet by assuming an average absorption of each nutrient and assuming that all proteins have the same sulfur content per gram. Finally, when dietary sodium and chloride are equal (if their main source is dietary salt) these may be omitted. When body weight is known or assumed, the PRAL of foods and diet may be calculated. In a validation of this calculation, Remer and Manz showed that as dietary protein and PRAL increased, urine pH fell and urinary calcium increased.[18] Plant-based foods such as fruit and vegetables are generally recognized as being sources of alkalinity; while this is true for most fruits and vegetables, some plant-based proteins such as grains, legumes and nuts have PRAL values as high as meat[17] (Table 10.2.4).

There are several sources of misinterpretation of PRAL with respect to calcium metabolism. First, dietary salt is assumed to have no effect on urinary

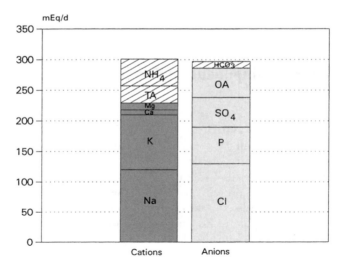

**Figure 10.2.2** *Typical urine ionogram (showing all quantitatively important urinary anions and cations) of a healthy adult consuming a protein-rich mixed diet. $NH_4$ = ammonium; $TA$ = titratable acid; $Mg$ = magnesium; $Ca$ = calcium; $K$ = potassium; $Na$ = sodium; $HCO_3$ = bicarbonate; $OA$ = organic acids; $SO_4$ = sulfate; $P$ = phosphorus; $Cl$ = chloride. The directly measured net acid excretion ($NAE_{direct}$) equals $TA + NH_4 - HCO_3$. Indirectly determined net acid excretion corresponds to the sum of the nonbicarbonate anions minus the sum of the mineral cations ($NAE_{direct} = (Cl + P + SO_4 + OA) - (Na + K + Ca + Mg)$)*

**Table 10.2.4.** *Calcium, potassium, phosphorus and PRAL of portions of food providing 7 g protein*[a]

| Weight (g) | Food | Ca (mg) | K (mg) | P (mg) | PRAL (mEq) |
|---|---|---|---|---|---|
| 28 | Hard cheese | 202 | 27 | 143 | 19.4 |
| 300 | Brown rice | 10 | 250 | 310 | 12.5 |
| 30 | Peanuts | 16 | 183 | 100 | 8.3 |
| 30 | Beef | 3 | 85 | 134 | 7.8 |
| 58 | Spaghetti | 15 | 146 | 110 | 6.5 |
| 78 | Lentils | 15 | 287 | 140 | 3.5 |
| 100 | Peas | 21 | 330 | 118 | 1.2 |
| 66 | Egg whites | 4 | 94 | 8 | 1.1 |
| 218 | Milk | 252 | 306 | 209 | 0.7 |
| 159 | Broccoli | 89 | 588 | 138 | −1.2 |
| 333 | Potato | 23 | 119 | 190 | −4.0 |

[a]From Remer and Manz.[17]

calcium, when in fact it does.[57] Second, all foods are assumed to have the same content of total sulfur amino acids per gram protein, while it actually varies over three-fold (Table 10.2.3). The lower bioavailability of calcium from oxalate and phytate salts is not considered, which is a significant effect in legume-based diets. Finally the positive benefit of dietary calcium on bone health, in the foods being compared using PRAL, is not taken into account. In spite of these shortcomings, PRAL is a useful tool for predicting NRAE of populations when dietary composition is known.[58]

## Other Dietary Components Affecting Bone

As indicated above, sodium chloride (table salt) has a hypercalciuric effect.[57] For healthy populations, each 100 mmol of salt increases urinary calcium 1 mmol (40 mg). Packard and Heaney[59] quantified the effects of protein and sodium on urinary calcium loss, and concluded that a high sodium intake would have a greater effect on calcium loss than a diet moderately high in protein. None of the studies in Table 10.2.2 reported whether those consuming higher protein (or animal protein) had better diets, *i.e.*, diets containing less high-salt, processed foods or more fruits and vegetables. Other food constituents such as vitamin D, isoflavones in soy, phosphate in meat, magnesium, and caffeine may also have modest effects on bone health. As dietary factors tend to exhibit colinearity with other components in the diet, dietary patterns should be considered. Research into dietary patterns indicate that a pattern predominantly fruit and vegetable-based is protective to bone[53] but that varying amounts of dairy or meat do not appear to have a large impact on bone in older adults. Interestingly, a diet that is predominant in candy (suggesting low nutrient density) impacts adversely on radius bone mineral density of women.[60] It is likely that the epidemiological studies of diet and bone, which report the specific effects of various single nutrients, are in reality measuring different dietary patterns.

# Conclusions

In a 'Western' diet, animal protein foods are more frequently consumed in excess of protein needs, partly because they are a concentrated calorie source compared to plant foods as well as culturally preferred. Increasing intake of purified proteins from either animal or plant sources increase RNAE, which in turn increases urinary calcium. The effects of excess dietary protein on urinary calcium and bone metabolism are reduced by other nutrients found in that protein source, such as calcium, phosphorus and potassium. Bone health is affected by other dietary constituents such as isoflavones, antioxidants, salt, oxalate, phytates and caffeine. Most calcium balance studies and prospective epidemiological surveys find that protein intakes in the forms and amounts commonly consumed have little effect on bone. Animal and plant foods may affect bone health differently, but these effects are due to other constituents, not protein. Except when consumed in extremely high amounts, excess dietary protein intake is not a risk factor for bone loss.

# References

1. Institute of Medicine. *Dietary Reference Intakes. Energy, Carbohydrate, Fiber, Fat. Fatty Acids, Cholesterol, Protein, and Amino Acids.* National Academy Press: Washington, DC, 2002. Available online at http://www.nap.edu
2. J.E. Kerstetter and L.H. Allen, Protein intake and calcium homeostasis, *Adv. Nutr. Res.*, 1994, **9**, 167–181.
3. L.A. Frassetto, K.M. Todd, R.C. Morris Jr., and A. Sebastian, Worldwide incidence of hip fracture in elderly women: Relation to consumption of animal and vegetable foods, *J. Gerontol.*, 2002, **55A**, M585–M592.
4. R.S. Gibson, *Principles of Nutritional Assessment*, Oxford University Press, New York, 1990, 21–25.
5. E. Smit, F.J. Nieto, C.J. Crespo and P. Mitchell, Estimates of animal and plant protein intake in US adults: results from the Third National Health and Nutrition Examination Survey, 1988–91, *J. Am. Dietetic Assoc.*, 1999, **99**, 813–820.
6. T.J. Green and S.J. Whiting, Potassium bicarbonate reduces high protein-induced hypercalciuria in adult men, *Nutr. Res.*, 1994, **14**, 991–1002.
7. Y. Kim and H.M. Linkswiler, Effect of level of protein intake on calcium metabolism and on parathyroid and renal function in the adult human male, *J. Nutr.*, 1979, **109**, 1399–1404.
8. A.A. Licata, Acute effects of increased meat protein on urinary electrolytes and cyclic adenosine monophosphate and serum parathyroid hormone, *Am. J. Clin. Nutr.*, 1981, **34**, 1779–1784.
9. J. Lutz, Calcium balance and acid–base status of women as affected by increased protein intake and by sodium bicarbonate ingestion, *Am. J. Clin. Nutr.*, 1984, **39**, 281–288.
10. S.A. Schuette and H.M. Linkswiler, Effects on Ca and P metabolism in humans by adding meat, meat plus milk, or purified proteins plus Ca and P to a low protein diet, *J. Nutr.*, 1982, **112**, 305–315.
11. S.A. Schuette, M.B. Zemel and H.M. Linkswiler, Studies of the mechanism of protein-induced hypercalciuria in older men and women, *J. Nutr.*, 1980, **110**, 305–315.

12. M.B. Zemel, S.A. Schuette, M. Hegsted and H.M. Linkswiler, Role of sulfur amino acids in protein-induced hypercalciuria in men, *J. Nutr.*, 1981, **111**, 545–552.
13. R. Schwartz, N.A. Woodcock, J.D. Blakely and I. MacKeller, Metabolic response of adolescent boys to two levels of dietary magnesium and protein. II. Effect of magnesium and protein level on calcium balance, *Am. J. Clin. Nutr.*, 1973, **26**, 519–523.
14. S.J. Whiting and H.H. Draper, The role of sulfate in the calciuria of high protein diets in adults rats, *J. Nutr.*, 1980, **110**, 212–222.
15. M.S. Calvo, R.R. Bell and R.M. Forbes, Effect of protein-induced calciuria on calcium metabolism and bone status in adult rats, *J. Nutr.*, 1982, **112**, 1401–1413.
16. S.L. Petito and J.L. Evans, Calcium status of the growing rat as affected by diet acidity from ammonium chloride, phosphate and protein, *J. Nutr.*, 1984, **114**, 1049–1059.
17. T. Remer and F. Manz, Potential renal acid load of foods and its influence on urine pH, *J. Am. Dietetic Assoc.*, 1995, **95**, 791–797.
18. T. Remer and F. Manz, Estimation of the renal net acid excretion by adults consuming diets containing variable amounts of protein, *Am. J. Clin. Nutr.*, 1994, **59**, 1356–1361.
19. L.A. Frassetto, K. Todd, R.C. Morris Jr., and A. Sebastian, Estimation of net endogenous noncarbonic acid production in humans from dietary protein and potassium contents, *Am. J. Clin. Nutr.*, 1998, **68**, 576–583.
20. U.S. Barzel, The skeleton as an ion exchange system: implications for the role of acid–base imbalance in the genesis of osteoporosis, *J. Bone Miner. Res.*, 1995, **10**, 1431–1436.
21. J. Lemann Jr., Relationship between urinary calcium and net acid excretion as determined by dietary protein and potassium: a review, *Nephron*, 1999, **81**(Suppl. 1), 18–25.
22. L. Tkatch, C-H. Rapin, R. Rizzoli, D. Slosman, V. Nydegger, H. Vasey, and J.-P. Bonjour, Benefits of oral protein supplementation in elderly patients with fracture of the proximal femur, *J. Am. Coll. Nutr.*, 1992, **11**(5), 519–525.
23. M.-A. Schurch, R. Rizzoli, D. Slosman, L. Vadas, P. Vergnaud and J.-P. Bonjour, Protein supplements increase serum insulin-like growth factor-I levels and attenuate proximal femur bone loss in patients with recent hip fracture, *Ann. Int. Med.*, 1998, **128**, 801–809.
24. E. Barrett-Connor and D. Goodman-Gruen, Gender differences in insulin-like growth factor and bone mineral density association in old age: The Rancho Bernardo Study, *J. Bone Miner. Res.*, 1998, **13**, 1343–1349.
25. S. Boonen, C. Rosen, R. Bouillon, A. Sommer, M. McKay and D. Rosen, *et al.*, Musculoskeletal effects of the recombinant human IGF-1/IGF binding protein-3 complex in osteoporotic patients with proximal femoral fracture: a double-blind, placebo-controlled pilot study, *J. Clin. Endocrinol. Metab.*, 2002, **87**, 1593–1599.
26. S.A. Schuette, M. Hegsted, M.B. Zemel and H.M. Linkswiler, Renal acid, urinary cyclic AMP, and hydroxyproline excretion as effected by level of protein, sulfur amino acid, and phosphorus intake, *J. Nutr.*, 1981, **111**, 2106–2116.
27. H. Spencer, L. Kramer and D. Osis, Effect of a high protein (meat) intake on calcium metabolism in man, *Am. J. Clin. Nutr.*, 1978, **31**, 2167–2180.
28. J.R. Hunt, S.K. Gallagher, L.K. Johnson and G.I. Lykken, High-versus low-mean diets: effects on zinc absorption, iron status, and calcium, copper, iron,

magnesium, manganese, nitrogen, phosphorus, and zinc balance in postmeno-pausal women, *Am. J. Clin. Nutr.*, 1995, **62**, 621–632.

29. H. Spencer, L. Kramer, D. Osis and C. Norris, Effect of phosphorus on the absorption of calcium and on the calcium balance in man, *J. Nutr.*, 1978, **108**, 447–457.

30. T. Kitano, T. Esashi, and S. Azami, Effect of protein intake on mineral (calcium, magnesium and phosphorus) balance in Japanese males, *J. Nutr. Sci. Vitaminol.*, 1988, **34**, 387–398.

31. D.J.A. Jenkins, C.W.C. Kendall, E. Vidgen, L.S.A. Augustin, T. Parker, D. Faulkner, R. Vieth, A.C. Vandenbroucke and R.G. Josse, Effect of high vegetable protein diets on urinary calcium loss in middle-aged men and women, *Eur. J. Clin. Nutr.*, 2003, **57**, 376–382.

32. D.L.E. Pannemans, G. Schaafsma and K.R. Westerterp, Calcium excretion, apparent calcium absorption and calcium balance in young and elderly subjects: influence of protein intake, *Br. J. Nutr.*, 1997, **77**, 721–729.

33. J.C. Moriguti, E. Ferriolli and J.S. Marchini, Urinary calcium loss in elderly men on a vegetable:animal (1:1) high-protein diet, *Gerontology*, 1999, **45**, 274–278.

34. L.H. Allen, R.S. Bartlett and G.D. Block, Reduction of renal calcium reabsorp-tion in man by the consumption of dietary protein, *J. Nutr.*, 1979, **109**, 1345–1350.

35. L.H. Allen, G.D. Block, R.J. Wood and G.F. Bryce, The role of insulin and para-thyroid hormone in the protein-induced calciuria of man, *Nutr. Res.*, 1981, **1**, 3–11.

36. M.G. Holl and L.H. Allen, Comparative effects of meals high in protein, sucrose, or starch on human mineral metabolism and insulin secretion, *Am. J. Clin. Nutr.*, 1988, **48**, 1219–1225.

37. S.J. Whiting, D.A. Anderson and S.J. Weeks, Calciuric effects of protein and potassium bicarbonate but not phosphate or sodium chloride can be detected acutely in adult females and males, *Am. J. Clin. Nutr.*, 1997, **65**, 1465–1472.

38. J.E. Kerstetter, K.O. O'Brien and K.L. Insogna, Dietary protein affects intestinal calcium absorption, *Am. J. Clin. Nutr.*, 1998, **68**, 859–865.

39. J.E. Kerstetter, M.E. Mitnik, C.M. Gundberg, D.M. Caseria, A.F. Ellison, T.O. Carpenter and K.L. Insogna, Changes in bone turnover in young women consuming different levels of dietary protein, *J. Clin. Endocrinol. Metab.*, 1999, **84**, 1052–1055.

40. J. Bell and S.J. Whiting, Elderly women need dietary protein to maintain bone mass, *Nutr. Rev.*, 2002, **60**, 337–341.

41. D. Feskanich, W.C. Willett, M.J. Stampfer and G.A. Colditz, Protein consump-tion and bone fractures in women, *Am. J. Epidemiol.*, 1996, **143**, 472–479.

42. J.W. Wilson, C.W. Enns, J.D. Goldman, K.S. Tippett, S.J. Mickle, L.E. Cleveland and P.S. Chahil, *Data tables: Combined Results from USDA's 1994 and 1995 Continuing Survey, of Food Intakes by Individuals and 1994 and 1995 Diet and Health Knowledge Survey*, ARS Food Surveys Research Group. U.S. Department of Commerce, National Technical Information Service, 1997, 25.

43. R.G. Munger, J.R. Cerhan and B. Chiu, Prospective study of dietary protein intake and risk of hip fracture in postmenopausal women, *Am. J. Clin. Nutr.*, 1999, **69**, 147–152.

44. D.E. Sellmeyer, K.L. Stone, A. Sebastian and S.R. Cummings, A high ratio of dietary animal to vegetable protein increases the rate of bone loss and the risk of fracture in postmenopausal women, *Am. J. Clin. Nutr.*, 2001, **72**, 118–122.

45. M.T. Hannan, K.L. Tucker, B. Dawson-Hughes, L.A. Cupples, D.T. Felson and D.P. Kiel, Effect of dietary protein on bone loss in elderly men and women: the Framingham osteoporosis study, *J. Bone Miner. Res.*, 2000, **15**, 2504–2512.

46. B. Dawson-Hughes and S.S. Harris, Calcium intake influences the association of protein intake with rates of bone loss in elderly men and women, *Am. J. Clin. Nutr.*, 2002, **75**, 773–779.

47. J. Promislow, D. Goodman-Gruen, D.J. Slymen and E. Barrett-Connor, Protein consumption and bone mineral density in the elderly: the Rancho Bernardo study, *Am. J. Epidemiol.*, 2002, **155**, 636–644.

48. H.E. Meyer, J.I. Pedersen, E.B. Loken and A. Tverdal, Dietary factors and the incidence of hip fracture in middle-aged Norwegians. A prospective study, *Am. J. Epidemiol.*, 1997, **145**, 117–123.

49. E.C.H. van Beresteijn, J.H. Brussaard and M. van Schaik, Relationship between the calcium-to-protein ratio in milk and the urinary calcium excretion in healthy adults – a controlled crossover study, *Am. J. Clin. Nutr.*, 1990, **52**, 142–146.

50. R.P. Heaney, Protein intake and the calcium economy, *J. Am. Dietetic Assoc.*, 1993, **93**, 1259–1260.

51. K.L. Tucker, M.T. Hannan, H. Chen, L.A. Cupples, P.W.F. Wilson and D.P. Kiel, Potassium, magnesium, and fruit and vegetable intakes are associated with greater bone mineral density in elderly men and women, *Am. J. Clin. Nutr.*, 1999, **69**, 727–736.

52. S.A. New, C. Boulton-Smith, D.A. Grubb and D.M. Reid, Nutritional influences on bone mineral density: a cross-sectional study in premenopausal women, *Am. J. Clin. Nutr.*, 1997, **65**, 1831–1839

53. S.A. New, S.P. Robins, M.K. Campbell, J.C. Martin, M.J. Garton, C. Boulton-Smith, D.A. Grubb, S.J. Lee and D.M. Reid, Dietary influences on bone mass and bone metabolism: further evidence of a positive link between fruit and vegetable consumption and bone health? *Am. J. Clin. Nutr.*, 2000, **71**, 142–151.

54. L.A. Frassetto, K. Todd, R.C. Morris Jr., and A. Sebastian, Estimation of net endogenous noncarbonic acid production in humans from dietary protein and potassium contents, *Am. J. Clin. Nutr.*, 1998, **68**, 576–583.

55. L. Frassetto, R.C. Morris Jr., D.E. Sellmeyer, K. Todd and A. Sebastian, Diet, evolution and aging – the pathophysiologic effects of the postagricultural inversion of the potassium-to-sodium and base-to-chloride ratios in the human diet, *Eur. J. Nutr.*, 2001, **40**, 200–213.

56. R.C. Mühlabauer, A. Lozano and A. Reinli, Onion and a mixture of vegetables, salads, and herbs affect bone resorption in the rat by a mechanism independent of their base excess, *J. Bone Miner. Res.*, 2002, **17**, 1230–1236.

57. L.K. Massey and S.J. Whiting, Dietary salt, urinary calcium and bone loss, *J. Bone Miner. Res.*, 1996, **11**, 731–736.

58. D.S. Michaud, R.P. Troiano, A.F. Subar, S. Runswick, S. Bingham, V. Kipnis and A. Schatzlcih, Comparison of estimated renal net acid excretion from dietary intake and body size with urine pH, *J. Amer. Diet. Assoc.*, 2003, **103**, 1001–1007.

59. P.T. Packard and R.P. Heaney, Medical nutrition therapy for patients with osteoporosis, *J. Am. Dietetic Assoc.*, 1997, **97**, 414–417.

60. K.L. Tucker, H. Chen, M.T. Hannan, L.A. Cupples, P.W.F. Wilson, D. Felson and D.P. Kiel, Bone mineral density and dietary patterns in older adults: the Framingham Osteoporosis Study, *Am. J. Clin. Nutr.*, 2002, **76**, 245–252.

61. E.S. Hands, *Nutrients in Food*, Williams & Wilkins,, Lippencott, 2000.

CHAPTER 11

# The Influence of Phosphorus on the Skeleton

MONA S. CALVO[1] and THOMAS O. CARPENTER[2]

[1]Center for Food Safety and Applied Nutrition, US Food and Drug Administration, Washington, DC, USA, Email: mona.calvo@cfsan.fda.gov
[2]Department of Pediatrics, Yale University School of Medicine, New Haven, CT, USA

**Key Points**

- Phosphorus homeostasis and physiological factors influencing hyper and hypophosphatemia
- Dietary intake of phosphorus and dietary requirement over life
- Endocrine regulation of plasma phosphorus homeostasis
- Mechanisms of control of plasma phosphorus homeostasis by sodium-phosphorus co-transporter proteins in renal tubule and intestinal epithelium
- Skeletal disorders of hypophosphatemia and phosphorus wasting
- Skeletal disorders of hyperphosphatemia associated with renal failure
- Endocrine and skeletal response to changing phosphorus intake and the low calcium to phosphorus intake of Western diets.

## Introduction: Phosphorus Homeostasis

The essential nutrient phosphorus is ubiquitous in its distribution in the food supply and in all life forms. Life as we know it has depended in many ways upon the element phosphorus, and has evolved complex mechanisms by which to utilize various forms for essential functions to the living organisms. Some of these functions relate to the physiology of the cell and take advantage of the chemical properties of the four oxygen per phosphorus molecule, phosphate, which renders this unique element stable in biological systems. Indeed, it has been suggested that the origins of life on earth were largely shaped by unleashing and concentrating phosphorus in the earth's crust.[1] Weathering of rock surfaces in the earth's crust occurring over time were thought to have exposed sufficient amounts of mineral to the atmosphere, such that water runoff would eventually carry phosphorus, probably as phosphate, or other compound of limited reactivity, into the seas. When sufficient phosphorus became deposited in concentrated areas, exposure of sufficient amounts of phosphate to early organic molecules was likely to occur. Subsequently molecular mechanisms must have arisen which could lead to an increasing complexity of these organic materials. What is clear is the indispensable quality of this element for life today.

The chemical properties of phosphorus in cytosolic solution provide a mechanism to mediate a variety of critical biological processes. Examples of this include the multiplicity of signaling events that occur upon activation of cell kinases, and the subsequent phosphorylation of proteins. The covalent attachment of a phosphate moiety to tyrosine residues in the insulin and IGF receptors results in activation of an intracellular cascade of events that mediate classic hormone signaling. Such an acute molecular 'switch' is able to coordinate cell signaling in numerous important pathways. Similarly, phosphorylation plays a major role in activating enzymes. The high-energy chemical bond of adenosine triphosphate utilizes phosphorus to provide the primary reservoir of rapidly accessible energy in the cell. Other properties of phosphorus are structural in nature: at the molecular level phosphorus plays an important role in DNA structure. With respect to the whole human organism, the organ with the greatest phosphorus content (in the chemical form of the crystalline bone, hydroxyapatite) is the human skeleton, our major structural organ.

Table 11.1 shows the basic physiologic functions of phosphorus (Pi) and its distribution in the body tissues which usually exists as phosphate ions ($HPO_4^{2-}$, $HPO_4^{-}$). The small percentage of phosphorus in the plasma and

**Table 11.1.** *Distribution and function of calcium and phosphate*

|  | *Calcium ions* | *Phosphate ions* |
|---|---|---|
| *Extracellular* |  |  |
| Concentration |  |  |
| Total in serum | $2.5 \times 10^{-3}$ M | $1.00 \times 10^{-3}$ M |
| Free | $1.2 \times 10^{-3}$ M | $0.85 \times 10^{-3}$ M |
| Functions | Bone mineral | Bone mineral |
|  | Blood coagulation | Acid-base balance |
|  | Membrane excitability |  |
| *Intracellular* |  |  |
| Concentration | $10^{-7}$ M | $1-2 \times 10^{-3}$ M |
| Functions | Signal for: | Structural role |
|  | Neuronal activation | High energy bond |
|  | Hormonal secretion | Regulation of proteins |
|  | Muscle contraction | by phosphorylation |

Adapted from Ref. [143].

extracellular fluid compartment is the key barometer of phosphorus homeostasis. This small body pool is the compartment from which phosphorus from intestinal absorption, soft tissue release, or bone resorption is ultimately cleared and excreted in urine or deposited in bone. The normal range of phosphorus concentration in this body pool is maintained through the hormonal regulation of intestinal absorption and renal excretion. Disruption of phosphorus homeostasis may result in hyperphosphatemia or hypophosphatemia, both of which may adversely affect bone. The primary goal of this chapter is to examine the effects of acute and chronic hypophosphatemia and hyperphosphatemia on bone health. To set the stage for this discussion, we first review some of the new findings concerning the physiologic factors that influence the normal fluctuations in serum phosphorus, the homeostatic mechanisms at the hormonal and tissue level that regulate phosphorus, and the potential impact of the changing phosphorus content of the Western diet.

## Fluctuations in Plasma Phosphorus

Blood phosphorus levels are subject to a wide day to day and within day variation due to intrinsic rhythms and the influence of diet content and hormones that fluctuate with food intake including insulin and epinephrine, and exercise.[2] Hypophosphatemia can result from tissue redistribution of phosphorus from extracellular to intracellular stores that occurs with carbohydrate loading or heavy exercise. Consequently, serum phosphorus is not considered a good indicator of body stores or nutritional status of phosphorus, since such levels can appear normal, even when some body stores are deplete.[3] Serum phosphorus levels also vary with age and gender.[4] The adult serum phosphorus range, expressed as elemental Pi, is 0.81–1.45 mmol $L^{-1}$ (2.5–4.5 mg $dL^{-1}$) and that for children is almost twice as high at 1.29–2.26 mmol $L^{-1}$ (4.0–7.0 mg $dL^{-1}$).[4] After the early adult years, phosphorus levels do not change in women, but decrease in elderly men,[5] however, significantly higher levels have been reported in women over 10 years past menopause, despite lower phosphorus intakes relative to their male counterparts.[6,7] Estrogen deficiency and estrogen's ability to suppress renal phosphorus reabsorption have been suggested as the underlying mechanisms for this marked gender difference in serum phosphorus in elderly men and women. Dick et al.[8] demonstrated that ovariectomy in rats was associated with a rise in plasma phosphorus and renal phosphorus reabsorption independent of any change in parathyroid hormone (PTH) which supports the suggested mechanism of estrogen inhibition of renal phosphorus reabsorption.

Phosphorus concentration in blood follows a pronounced circadian rhythm in old and young men and women.[9] The wide diurnal fluctuations in serum phosphorus follow a biphasic pattern with peak concentrations occurring in the late afternoon and evening after meals.[9,10] The general biphasic pattern remains even with changes in dietary phosphorus intake; however, changes in dietary phosphorus intake can influence the height of the first biphasic peaks.

Shifts of phosphorus between intracellular and extracellular compartments may create acute changes in the serum phosphorus level.[11] Indeed, in that phosphate is an intracellular ion as well as an extracellular ion, the serum phosphate level may not reflect current body stores of this ion. The movement of phosphorus as phosphate from the intracellular to extracellular space may occur in a variety of well-known clinical settings. In particular, the movement from the extracellular space to intracellular pools often occurs during nutritional recovery, as in provision of total parenteral nutrition to an emaciated patient. Acutely, when the correction of diabetic ketoacidosis occurs, phosphorus can rapidly shift into the intracellular space, decreasing extracellular phosphorus. Critical drops in serum phosphorus below 1.5 mg dL$^{-1}$ can be life-threatening in certain situations, and primarily manifest as severe muscle fatigue, and respiratory failure. This problem has also been noted in therapy for burn patients, the use of androgen, and alkalosis.

An interesting physiologic phenomenon repetitively observed in animal studies and in clinical settings is the reciprocal relationship between calcium and phosphorus. Because of the mass action equation, the concentration of available calcium and the concentration of available phosphorus drive the chemical reaction to hydroxyapatite in the skeleton. A manipulation in one of the divalent ions creates the reciprocal change in the other. For instance, upon provision of a calcium load in a normal individual, the serum phosphorus decreases, as a presumed consequence of driving the mass action equation mentioned above more rapidly in the direction of formation of hydroxyapatite without any increase in phosphorus supply. Likewise, a diminished calcium supply slows this reaction, thereby allowing phosphorus to accumulate. Hypophosphatemia often results in hypercalcemia, and exposure to an oral phosphate load often results in hypocalcemia. It should be noted that as a normal calcium × phosphorous product is probably necessary to drive the mass action equation to hydroxyapatite and effect bone mineralization, a high product creates risks for calcification of soft tissues.[12]

# Dietary Intake of Phosphorus and Dietary Requirements

## Multinational Guidelines and Cultural Differences in Phosphorus Intake

The total dietary intake of phosphorus is an important determinant of plasma phosphorus concentration. The total amount of phosphorus that is absorbed is influenced by the amount in the diet and its bioavailability.[13] The usual daily intake of phosphorus for the typical American diet varies with age and sex with more than half of the young and middle-aged men consuming 1600 mg day$^{-1}$ or more, while comparably aged women consume about 1000 mg day$^{-1}$.[14] This level of intake exceeds the US dietary reference intake (DRI) guidelines (Table 11.2) which are similar to those of other countries. Comparable phosphorus intakes are observed in other Western cultures such as the UK and Germany where men consume an estimated 1450 mg day$^{-1}$.[15] In contrast to

**Table 11.2.** *Multinational dietary guidelines for calcium and phosphorus intake in adults*

| | USA[a] | UK[b] | Thailand[c] | Germany[d] | Korea[e] | China[e] (Taiwan) | Japan[e] |
|---|---|---|---|---|---|---|---|
| Calcium (mg day$^{-1}$) | 1000 | 700 | 800 | 1000 | 700 | 600 | 600 |
| Phosphorus, (mg day$^{-1}$) | 700 | 550 | 800 | 700 | 550 | 600 | <1300 |
| Ca:Pi (wt:wt) | (1.42) | (1.27) | (1) | (1.42) | (1.27) | (1) | – |
| (moles:moles) | (1.15) | (1) | (0.8) | (1.15) | (1) | (0.8) | – |

[a]Recommended Dietary Allowance, RDA for Pi and Adequate intake (AI) for Ca from the Food and Nutrition Board. Ref. [3].
[b]Department of Health: Dietary Reference Values for Food, Energy and Nutrition for the United Kingdom, London: HMSO, 1991, 150–151.
[c]Committee on Health, Department of Health, Ministry of Public Health, The Government of Thailand: Recommendation of the Dietary Allowance for Thai People. Bankok; War Veterans Publishers, 1989.
[d]Deutsche Gesellschaft für Ernährung: Referenzwerte für die Nährstoffzufur. 1. Auffl. Frankfurt am Main: Umschau Braus GmbH. 2000, 154–168.
[e]Ref. [20].

the US, some European studies report baseline mean phosphorus intakes of 1400–1700 mg day$^{-1}$ in young adult women [16,17] or intakes of slightly >1000 mg day$^{-1}$ in perimenopausal women.[18] Further evidence that younger adults in Western cultures have relatively higher phosphorus intakes comes from a study in Greek children 2–14 years old. All children were reported to have higher phosphorus intakes than the dietary guidelines, half with intakes greater than 1500 mg day$^{-1}$.[19] Phosphorus intakes in Asian cultures are generally much lower, averaging 440–840 mg day$^{-1}$.[20,21]

Estimates of dietary phosphorus intake largely reflect the 'natural' phosphorus content of food which is richest in dairy, meats, poultry and fish, and grain products [14] and do not generally capture the phosphorus that is added to food during processing. In Western cultures, specifically in the US, additives can represent an average of 500 mg day$^{-1}$ of phosphorus per capita daily intake [22,23] and depending on food preferences these compounds can contribute from 300 mg to as much as a gram of phosphorus to individual daily intake. Asian and some other cultures that use less phosphate addition and less processing of their foods can over estimate their phosphorus intake when nutrient databases are used to estimate intake. This pattern of low phosphorus intake changes with the acceptance and import of processed foods from the West.[24] From 1960 to 1995, daily phosphorus intake increased in Japan approximately by 200 mg, but these estimates did not capture all the phosphate contributed by food additives; so the increase in phosphorus intake is actually much greater. Nutrient composition data bases do not account for all the food additives used in food processing; therefore, actual intakes of phosphorus are usually higher than reported.[25] Depending on individual food choices, phosphorus intakes can vary among individuals by

500–1000 mg day$^{-1}$.[26] The Joint FAO/WHO Expert Committee on Food Additives concluded that adverse effects were unlikely if daily total phosphorus consumption from all sources remains below 70 mg kg$^{-1}$ day$^{-1}$; however, this assumption of safe levels of phosphorus intake was based largely on animal data, not on observational studies in humans.[27]

Dietary guidelines for most countries also recognize the need for achieving a balance between dietary calcium and phosphorus intake and generally recommend a ratio of 1:1 on a molar basis and greater than 1:1 on a mg:mg basis as shown in Table 11.2. The relative importance of the Ca:Pi intake ratio to adult bone health was dismissed by the 1997 U.S. Food and Nutrition Board,[3] however, the committee lowered the adult 1989 RDA for phosphorus to 700 mg day$^{-1}$ for the 1997 RDA. Others [19,28–30] find utility in the Ca:Pi intake ratio as a means of gauging the imbalance between the intakes of these nutrients.

Western countries in general experience a great imbalance in the Ca:Pi ratio where calcium intake is low and phosphorus intake greatly exceeds the dietary guidelines. High protein intake (meat, poultry, fish and dairy) and the preference for highly processed convenience foods are well-known characteristics of the Western diet.[14] Both men and women in the US show an overall trend towards lower Ca:Pi intake ratios (0.6–0.7, mg:mg) or greater imbalance between calcium and phosphorus intake with increasing age. The Ca:Pi ratio is also indicative of nutrient poor diets. Ca:Pi ratios of low calcium consumers are significantly lower than those of high calcium consumers.[14]

A key factor contributing to the imbalance between calcium and phosphorus intake, notably in the US, is the decreased consumption of milk, a traditional source of both calcium and phosphorus in an ideal ratio for most dairy products that are not highly processed.[31] Recent survey data from the US (shown in Figure 11.1) demonstrates an alarming decrease in milk consumption by teens and young adults with a steady increase in soft drink consumption after age 10.[32] At the time of greatest physiologic demand for both of these nutrients to support rapid bone growth, American adolescents are abandoning milk, their richest, most balanced dietary source of calcium and phosphorus. Approximately 66% of the soft drinks consumed in the US are colas which supply phosphate and inorganic acid as phosphoric acid, but no calcium.[33,34] Soft drinks and soda ranked tenth in the order of food sources of phosphorus among US adults, contributing 2.1% of total daily adult phosphorus intake, while milk and cheese ranked first and second contributing 16.2 and 9.7%, respectively.[35] This is a surprisingly high contribution from one food source that is so void of nutrients.

The Ca:Pi intake ratio recommendations differ with age and are higher in infants more closely approximating the ratio in human milk. Feeding standard cow milk-based infant formulas (Ca:Pi molar ratio of ~1:1) have been reported to cause hypocalcaemia, convulsions and secondary hyperparathyroidism in the first month of life,[36] however such events were attributed to the high concentration of phosphorus in the formulas relative to the content of human milk and not specifically to the Ca:Pi ratio.

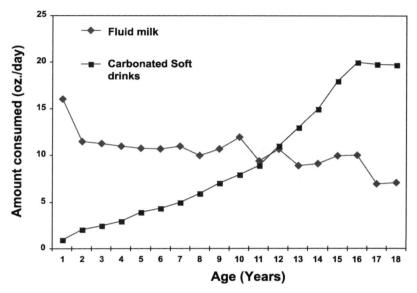

**Figure 11.1** *Daily Intake of Selected Beverage Categories by Age Group* From: Continuing Survey of Food Intake by Individuals, 1994–1995, 1998. (Redrawn from Ref. [32])

## Bioavailability of Phosphorus from Food

Bioavailability of the phosphorus source is another determinant of how much phosphorus is absorbed, but it is a greater consideration for animal feeds than it is for humans with the exception of some groups with unusual dietary habits. Phosphorus present as a component of phytic acid found in the cell walls of many grains is not bioavailable unless freed by treatment with phytase.[37] In the human food supply, this usually occurs when bread is leavened with yeast containing phytase. Another consideration is the rate of dissociation and absorption of the phosphorus source (salt or protein-bound) in the stomach. Phosphoric acid and orthophosphates, the most widely used phosphate food ingredients, are rapidly and efficiently absorbed, while protein-bound and polyphosphates must first be cleaved prior to absorption. Differences in the rate of absorption of these phosphate sources relative to the slower rate of calcium absorption influence how tissues and organs respond to fluxes in the plasma Ca × Pi product. Casein phosphopeptides and other protein-bound dietary sources would be expected to be associated with a phosphorus absorption rate closer to that of calcium, consequently rapid increases in serum phosphorus alone that may drive the mass action equation towards hypo-calcemia and soft tissue calcification would not be anticipated when cheese and dairy are the main sources of phosphorus. However, this could be the case with processed foods rich in orthophosphates or phosphoric acid since these dietary sources of phosphorus are so rapidly and efficiently absorbed. Interestingly, orthophosphates reportedly improve calcium balance while polyphosphates have a significantly negative effect when calcium intake is adequate.[38]

## Populations with Special Dietary Phosphorus Requirements: Premature Infants

Poor dietary phosphorus intake was once a common form of hypophosphatemic rickets in breast-fed, premature infants. Considerable mineral demands are imposed by the rapid skeletal growth normally occurring in utero late in gestation. Although these demands can be supplied in the usual transplacental route, they are in excess of what human breast milk can provide in the ex utero setting of a premature infant. Phosphorus demands are particularly at issue, because human breast milk contains relatively low amounts of phosphorus, as compared to most cow milk formulas. Supplementation with phosphorus-containing breast milk fortifiers have markedly reduced the incidence of this problem, however hypercalcemia and nephrocalcinosis have accompanied aggressive supplementation with such products, particularly in extremely small infants.[39] Nevertheless, secondary prevention of nutritional phosphopenic rickets by routine use of breast milk fortifier is recommended in premature infants. If rachitic disease related to nutritional phosphorus deprivation develops in the premature infant, it can be treated with 20–25 mg of elemental phosphorus per kg body weight per day, given as an oral supplement in 3–4 divided doses (Table 11.3).

# Hormonal Regulation of Phosphorus

## Hormonal Control of Plasma Phosphorus

The homeostatic mechanisms that maintain blood phosphorus levels are under the endocrine regulation of the PTH-vitamin D axis involving three key organs: kidney; small intestine; and bone. Phosphate ions are efficiently (60–75%) absorbed from the small intestine at all levels of dietary intake,[40] however, the kidney is the main regulatory organ in the maintenance of extracellular phosphorus homeostasis, normally reabsorbing up to 80% of the filtered load. While the kidney and small intestine regulate input and output of phosphorus from the plasma pool, the bone serves as the reservoir for both calcium and phosphorus.[31]

The calcitropic hormones which comprise the PTH-vitamin D axis include PTH, the active form of vitamin D, $1,25(OH)_2D$ or calcitriol, and calcitonin. Plasma phosphorus acts both directly and indirectly through PTH and calcitriol to regulate the production and catabolism of $1,25(OH)_2D$ and thus self regulates its own homeostasis. A low serum phosphorus concentration increases serum $1,25(OH)_2D$, while high serum phosphorus or a high phosphorus intake decreases circulating levels of $1,25(OH)_2D$. In situations where serum calcium is low, the calcium receptor on the parathyroid glands activates the secretion of PTH.[41] Increased circulating PTH stimulates the

**Table 11.3.** *Phosphorus preparations available for clinical use*

| Phosphorus preparations | Pi content | Na content |
|---|---|---|
| *Neutaphos*-powder (for mixing with liquid) | 250 mg per packet | 164 mg per packet |
| *Neutraphos*-K-powder (for mixing with liquid) | 250 mg per packet | 0 |
| *K-Phos Original*-uncoated tablet (to mix in liquid, acidifying) | 114 mg per tablet | 0 |
| *K-Phos MF*-coated tablet (mixing not required, acidifying) | 126 mg per tablet | 67 mg per tablet |
| *K-Phos #2* (double strength of K-Phos MF) | 250 mg per tablet | 133 mg per tablet |
| *K-Phos Neutral*-tablet (non-acidifying, mixing not required) | 250 mg per tablet | 298 mg per tablet |
| *Phospha-Soda*-solution (small doses may be given undiluted) | 127 mg ml$^{-1}$ | 152 mg ml$^{-1}$ |

renal proximal tubular cell 25(OH)D 1-hydroxylase enzyme activity and increases the circulating levels of calcitriol.[42] Acutely, PTH and 1,25(OH)$_2$D activate bone calcium and phosphorus mobilization and act on the kidney to increase calcium reabsorption from urine, while PTH alone stimulates increased renal excretion of phosphorus. High phosphorus intake can induce secondary hyperparathyroidism in animal models and humans.[43,44] The longer acting arm of this homeostatic mechanism involves the action of 1,25(OH)$_2$D on the intestine to stimulate the active transport of calcium and phosphorus restoring serum calcium levels to the normal range.[42]

Independent of serum calcium levels, low serum phosphorus can increase the activity of 25(OH)D-1-hydroxylase enzyme (1-OHlase) through mechanisms that are not entirely clear.[45,46] Recent findings in mice indicate that dietary phosphorus restriction regulates 1-OHlase gene expression in the proximal renal tubule.[47] Growth hormone and its active anabolic mediator, insulin-like growth factor-1 (IGF-1) are thought to play a role in maintaining the needed higher extracellular concentrations of phosphorus required during development for cellular growth and bone mineral deposition.[48]

In contrast to the effects of dietary phosphorus restriction on 1,25(OH)$_2$D, high dietary phosphorus (3000 mg Pi day$^{-1}$) reduces serum 1,25(OH)$_2$D levels by 30% in normal men relative to levels observed during normal phosphorus intake.[49] Earlier studies using oral phosphate therapy to treat hypercalciuria and hypercalcaemia were the first to demonstrate a negative association between high phosphorus intake and the active metabolites of vitamin D, even in the presence of increased circulating levels of PTH.[50,51] It is well established that phosphorus loading either acutely[16,52,53] or chronically[50,51,54,55] results in secondary hyperparathyroidism in humans as observed in animals fed high phosphorus diets.[56] An indirect mechanism where efficiently absorbed phosphorus precipitates an acute reduction in extracellular ionized calcium concentration is thought to be the main trigger for these events.

Several more recent studies have found that high phosphorus intake can attenuate the PTH-induced stimulation of 1,25(OH)$_2$D synthesis when the phosphate load is sufficiently high, and when the habitual Ca:Pi ratio in the diet is low and fed for a longer duration.[56] Portale *et al.* first suggested that dietary phosphorus finely regulates the renal production of plasma 1,25(OH)$_2$D at times pre-empting PTH stimulation.[49,57] This assessment is supported by data from a cross-sectional study of 275 healthy postmenopausal women where the relationship between PTH and 1,25(OH)$_2$D circulating levels were attenuated at high normal concentrations of serum phosphorus.[58] PTH was most strongly associated with 1,25(OH)$_2$D levels at mid-normal rather than low or high normal plasma phosphorus.

Direct stimulation of PTH secretion by phosphorus has also been proposed.[59,60,61] This direct effect of high levels of serum phosphorus on parathyroid gland function is probably independent of changes in ionized calcium and 1,25(OH)$_2$D, but is clearly dependent on a very high magnitude of hyperphosphatemia, well above the normal physiologic range. Under normal dietary conditions, serum phosphorus is too tightly regulated and would not be sufficiently high to stimulate PTH secretion directly. This effect of phosphorus is most evident in renal failure.

## Mechanisms of Endocrine Regulation of Phosphorus at the Cellular Level

Regulation of serum phosphorus is an important physiologic process involving a variety of factors. Dietary phosphorus is largely absorbed across the intestinal mucosa by passive diffusion, although this process can be upwardly modulated by 1,25(OH)$_2$D. Approximately 75% of dietary phosphorus is absorbed, independent of the amount ingested. As phosphorus is ample in most diets, dietary phosphorus deprivation is not common, and found in selected settings, such as in the unsupplemented, breast-fed premature infant, or where highly unusual dietary habits are practiced. Maintenance of total body phosphorus is largely dependent upon regulation of phosphorus transport at the level of the renal tubule. As the bulk of inorganic phosphate is filtered by the kidney, the regulatory mechanisms are most evident in the proximal renal tubule where phosphorus reabsorption is mediated by sodium–phosphate co-transporters, primarily the type II class of sodium–phosphate co-transporters.

## *NaPi2a*

The most abundant renal tubular phosphate transporter in mammalian kidney is the type II sodium–phosphate co-transporter, NaPi2a shown in Figure 11.2.[62] This transporter localizes to the brush border surface of proximal renal tubular cells, and its expression is regulated by PTH and dietary phosphorus. It is the rate limiting transporter for total body phosphorus homeostasis, and is a

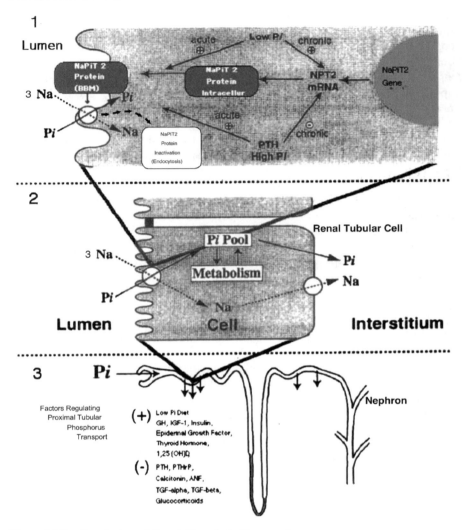

**Figure 11.2** *Regulation of Renal Phosphate Transport*

member of the type II class of transporters which serve specific epithelial phosphate uptake functions across these barrier cells. These molecules serve as co-transporters of phosphate and sodium across the apical membrane of the renal tubular cell in an electrogenic fashion. The inward downhill sodium gradient drives the process, and this is maintained by a basolateral Na-K ATPase which pumps sodium to the extracellular space. The NaPi2a transporter operates with 3 sodium: 1 phosphate stoichiometry and is stimulated under alkaline conditions. Thus by regulation of the numbers of functional NaPi2a molecules on the apical surface of the proximal renal tubule total body phosphorus is tightly controlled. One of the best known acute regulators of this process is PTH, which when applied to renal cells, results in rapid movement of the transporters from the apical surface *via* a clathrin coated vesicle to lysosomal pools where the transporters are rapidly degraded.[63] Dietary phosphorus intake also regulates the expression of protein at the transcriptional level. The degradation of these transporters after application of PTH, rather than storage in an inactive pool suggests that new protein synthesis would be required to restore active phosphate transport after withdrawal of PTH. Such a system seems highly inefficient, and the biologic rationale for this is unclear. On the other hand, there is some evidence that in intact tubules there is ability to restore transport independent of new protein synthesis, but the mechanism has not yet been elucidated.[64]

The NaPi2b isoform of the type II class of transporter is an intestinal sodium phosphate co-transporter that is regulated by phosphate intake and 1,25 $(OH)_2D$. This transporter differs slightly in sodium and pH dependence (stimulated by weak acid) from the 2a isoform. Two other classes of Na/Pi transporters are known: the type I co-transporters which are found in the renal proximal tubular apical membranes, are thought to have anion channel function and secretion roles. This transporter is also evident in brain, where it may mediate glutamate transport. The type III transporter is ubiquitously expressed, and is thought to serve cellular 'housekeeping' functions, not specialized epithelial ones. It is evident that the type II transporters are physiologically regulated and serve to maintain total body phosphate homeostasis.

Other numerous factors that regulate phosphate transport including IGF-1, growth hormone (through IGF-1), insulin, epidermal growth factor, thyroid hormone and $1,25(OH)_2D$, all of which stimulate tubular reabsorption of phosphate. PTHrP (like PTH), calcitonin, ANF, TGF-$\alpha$, TGF-$\beta$, and glucocorticoids inhibit renal phosphate transport directly or indirectly, through the availability of the high-affinity, low capacity NaPiT2a transporter.[64]

## NHERF

Other factors recently implicated in the regulation of phosphate transport are NHERF-1 and NHERF-2 (the sodium hydrogen exchanger regulatory factors-1 and 2). NHERF-1 and NHERF-2 are structurally related proteins that contain PDZ domains (specialized protein interaction domains) that may serve as a site for transporter binding. Recent information has been put forth demonstrating that NHERF-1 is an intermediary between the PTH receptor and the NaPiT2 protein, and important in trafficking NaPiT2 to the apical membrane of proximal tubular cells.[65] Furthermore, it appears that NHERF-2 directly binds to the PTH receptor and signals activation of phospholipase C, activating the PKC signaling pathway, as well as inhibition of adenylate cyclase.[66] NHERF-2 may account for certain tissue-specificity regarding the actions of PTH and PTHrP. Others have identified that PTH regulation of NaPiT2 availability utilizes an anchoring protein dependent on protein kinase A.[67]

## FGF23, FRP4, MEPE

Other factors that have been implicated in phosphorus homeostasis include those factors that have been isolated from oncogenic osteomalacia tumors[68] (Figure 11.3). Much recent work has focused on FGF23 (fibroblast growth factor 23), a novel member of the larger FGF family. This factor is over-expressed in oncogenic osteomalacia tumors, and is thought to mediate the hypophosphatemia seen in that syndrome. Accumulation of FGF23 in individuals with XLH (see below) due to mutations in PHEX, suggest an alternate mechanism by which accumulated FGF23 in the circulation can result in renal phosphate wasting.[69] Yet a third mechanism by which this factor is associated with hypophosphatemic disease relates to specific mutations of the molecule at arginine residues 176 or 179 adjacent to a furin enzyme recognition site.[70] This mutation appears to disrupt the catabolic pathway for the molecule thereby resulting in prolonged activation. When FGF23-transfected cells have been injected into nude mice, the subsequent development of hypophosphatemia and characteristic features of oncogenic osteomalacia occur.[71] It is not clear whether FGF23 is the direct mediator of phosphate wasting in the kidney or whether some metabolite or intermediary signal is responsible for the direct action on the proximal tubule. It is also unclear whether this molecule has direct effects on bone. Other molecules that have been associated with oncogenic osteomalacia associated tumors include: frizzled-related protein 4

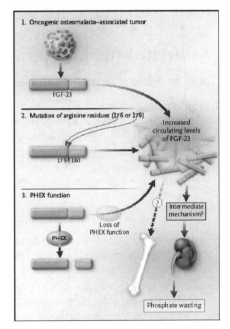

**Figure 11.3** *Proposed roles of fibroblast growth factor 23 (FGF-23) in the generation of phosphate wasting*

(FRP4) and MEPE (matrix extracellular phosphoglycoproteins). It is unclear what the roles of these two molecules are in the pathogenesis of oncogenic osteomalacia; however, preliminary data suggests that frizzled-related protein 4 may also mediate phosphate wasting.[68]

## PHEX

PHEX, (originally PEX), or a phosphate regulating gene with homologies to endopeptidase, is located on the X chromosome. PHEX is a member of a larger family of type II membrane proteins including neutral endopeptidases and the endothelin converting enzymes (ECE-1 and ECE-2), which degrade or activate peptide hormones.[72] These metalloenzymes are characterized by a short N-terminal intracellular portion, a single membrane-spanning domain, and a large C-terminal extra-cellular region that contains a zinc-binding motif consistent with a catalytic site.[73] PHEX is expressed in very low abundance in adult humans; its expression is relatively greater in bone than other tissues, yet is not expressed in kidney tissue.[73,74] Embryonic mice express PHEX during initial mineralization of certain bones with relative abundance by day 16 in calvaria, vertebrae, and teeth.[75] The protein in mature bone appears to be more abundant in osteocytes than in osteoblasts.[76] Its physiologic substrate is unknown at present. It appears likely that the PHEX gene product processes a factor important in phosphorus homeostasis, as loss-of-function mutations result in renal phosphate wasting. Thus abnormal processing of such a factor by PHEX is an oft-speculated mechanism for this abnormality in phosphorus homeostasis.

# Bone Disorders Associated with Hypophosphatemia

## Renal Tubular Defects in Phosphorus Homeostasis

As regulation of phosphate transport at the renal level contributes so dominantly to the maintenance of total body phosphorus homeostasis, it follows that many of the prototype disorders of inherited hypophosphatemia involve derangement of the renal mechanisms required to reclaim filtered phosphate from the renal tubular lumen.

## XLH

The prototype hypophosphatemic bone disease, X-linked hypophosphatemic rickets (XLH), was initially characterized by the inability of vitamin D treatment to significantly effect healing, and was referred to as 'vitamin D-resistant rickets'.[77] This designation continues in the common clinical lexicon, although it has been nearly 50 years since the recognition that abnormal phosphate homeostasis is the prominent pathophysiologic feature of the disease. Importantly, XLH should not be confused with the more recently characterized disorder of hereditary vitamin D resistance associated with mutations in the $1,25(OH)_2D_3$ receptor gene.[78] The latter disorder is often referred to as 'vitamin D-dependent rickets, type 2' in many medical textbooks, but newer reports utilize the biologically accurate descriptor 'hereditary resistance to vitamin D.' In order to develop a consistent and least confusing nomenclature, we recommend that 'X-linked hypophosphatemic rickets' or 'familial hypophosphatemic rickets' be employed for the hypophosphatemic disorder, instead of the older 'vitamin D resistant rickets.'

Several similar hypophosphatemic disorders of the skeleton have been described, and manifest the finding of inappropriate urinary excretion of phosphate; however, the most commonly identified etiology in this group is XLH. XLH is transmitted in X-linked dominant fashion, and is due to mutations in PHEX (see Figure 11.3.). The murine model for XLH, the *Hyp* mouse, has a 3' deletion in PHEX, and a similar model, the *Gy* mouse, has a 5' deletion in Phex.[79]

XLH typically presents in the second or third year of life. Leg bowing is the first noticeable sign, and is often attributed to physiologic bowing of childhood. Rickets (a bow or knock-knee deformity) is the prominent skeletal finding of a child with XLH. Lower extremities are more severely affected than the arms, ribs, or pelvis, which tend to be involved in severe forms of calciopenic rickets. Radiographic features of rickets are most apparent at the ends of long bones, and include flared metaphyses with frayed borders. The expansion of cartilage at the growth plate may give the appearance of a cup or bowl. The femora may develop bow defects in the anterior–posterior direction as well as the lateral, and patients may assume a lordotic posture as to compensate for the anterior–posterior bowing. In later childhood, asymmetric closure of the growth plate may occur, which can further exaggerate the bow defect, as one side of the extremity will continue to grow, and the other side cannot.[80] With fusion of the epiphyses and closure of the growth plate, the abnormalities are outgrown, although abnormal configuration of the epiphyses may persist into adulthood in severe cases. An important clinical feature of the epiphyseal abnormalities evident in XLH is the impaired lower extremity growth. Thus short stature in childhood, with an eventual outcome of a less than expected adult height, are of concern in the management of XLH. This feature of the disorder is variable; some have suggested growth hormone therapy for those children most severely affected with respect to poor lower extremity growth.

The characteristic microscopic finding of both trabecular and cortical bone is osteomalacia[81,82] Histomorphometric analysis reveals an increase in unmineralized bone matrix or osteoid. The total bone volume may be normal or even elevated.[83,84] Cellularity is usually reduced, however, in patients that manifest hyperparathyroidism; increased cellular parameters of bone turnover (*e.g.*, cell number, osteoblastic or osteoclastic surface) may be present.[84] Dynamic histomorphometry reveals a lag in mineralization time.[62] Increased bone density in adults with XLH has been reported, but may be related to the development of extra ossicles, osteophytes, or spinal stenosis. CT determination of cortical bone density is low in patients with XLH, despite increased cortical area.[85] Craniosynostosis may be observed as an infrequent occurrence in childhood; a more common finding is a scaphocephalic configuration of the skull. Optic canal stenosis has been reported.[86] Many patients develop other significant skeletal problems in the third and fourth decades of life. Vertebral abnormalities include thickening of spinous processes, fusion, thickening of facet joints, and spinal canal stenosis.[87] Calcification of tendons and ligaments occurs frequently in XLH, appearing at various ages, but unrelated to therapy.[88] Arthritis is common, and many patients complain of generalized bone pain in adulthood. Fractures also appear to occur with some frequency in adults with XLH, and we have anecdotally noted difficulty in fracture healing in adults with XLH. Finally, the presumption that the described histomorphometric finding of increased total bone volume does not necessarily indicate that affected patients are protected from the development of osteoporosis in later years.

Treatment of XLH is usually undertaken with the combination of oral phosphate salts (Table 11.3) and $1,25(OH)_2D_3$. This therapy must be monitored closely because of the frequent complications of secondary hyperparathyroidism and the potential for vitamin D intoxication. The therapy is helpful in terms of partial healing of the epiphyseal lesions, and a positive effect on stature in most cases, however this treatment does not entirely cure the disorder, nor prevent many of the later musculoskeletal complications.[77]

## *Autosomal Dominant Hypophosphatemic Rickets (ADHR)*

Another hypophosphatemic disorder, autosomal dominant hypophosphatemic rickets (ADHR), is evidently much less frequently occurring than XLH.[89] The ADHR phenotype is nearly identical to XLH, but appears to have a more variable penetrance, and postpubertal onset has been reported. Furthermore, two individuals have been reported to clearly demonstrate phosphate wasting in childhood, with resolution of this finding in early adulthood. ADHR has been found to be due to activating mutations of FGF23, on chromosome 12. Specific mutations in FGF23 stabilize a putative furin enzyme recognition site, thereby conferring potential stability to the encoded protein in ADHR. Such a protein may be inherently more potent, have more avid receptor binding, or be protected from normal degradation pathways. Therapy for this disorder is similar to that for XLH.

Given the apparent infrequency of ADHR, most children with sporadic occurrence of the classic XLH phenotype are probably due to spontaneous mutations in PHEX. Subsequent generations of such individuals usually transmit the disorder in X-linked dominant fashion. Incidence of XLH has been estimated between 1/10 000 to 1/20 000. In certain regions the disorder may be more frequent, and many affected individuals are unaware of their diagnosis. One study revealed that significant bone, joint, and back pain, dental abscesses, weakness, or hearing loss was present among 57 affected individuals in three large North Carolina families, yet less than a fourth of them were aware of their diagnoses, and only one was receiving any specific therapy.[90]

## Hereditary Hypophosphatemic Rickets with Hypercalciuria (HHRH)

HHRH is another variant of hypophosphatemic rickets described in a Bedouin tribe notable for its history of consanguinity.[91] Hypophosphatemia with renal phosphate wasting was evident with low renal tubular phosphate reabsorption. The serum alkaline phosphatase activity was elevated to levels typically observed in XLH. However, other parameters of mineral metabolism were distinctly different from those seen in patients with XLH: urinary calcium excretion was considerably elevated, and circulating levels of PTH were low. Another distinguishing feature of HHRH is the invariable finding of elevated circulating levels of $1,25(OH)_2D$, indicating that the normal $1\alpha$-hydroxylase stimulation of hypophosphatemia is intact, in contrast to XLH, in which the renal defect involves both the disruptions of the normal hypophosphatemia-$1\alpha$ hydroxylase axis and of renal tubular phosphate handling.

Osteomalacia and osteopenia have been noted in HHRH, also distinguishing the disorder from XLH in which normal, or even excess total bone may be present. Osteopenia has been observed in the three individuals that the authors have cared for with the disease, and it appears to be evident from an early age. Histological bone sections have demonstrated evidence of active bone resorption in one study,[91] but others have noted decreased numbers of osteoclasts.[92]

The consequences of the persistently elevated $1,25(OH)_2D$ levels include a state of intestinal calcium hyperabsorption, and an exaggerated rise in serum phosphorus following an oral phosphate load.[91] No apparent renal hypersensitivity to PTH is evident, as injection of PTH has resulted in appropriate decreases in tubular phosphate reabsorption and increases in urinary cyclic AMP excretion. In one of the authors' patients PTH and circulating $1,25(OH)_2D$ levels have been shown to decrease appropriately with intravenous calcium infusion, and $1,25(OH)_2D$ levels increase appropriately with PTH infusion, suggesting that the PTH/Ca regulation of the 1-OHlase axis is fully intact in HHRH.

As might be expected in the setting of increased calcium absorption, hypercalciuria and hyperphosphaturia, renal stones may occur in individuals with HHRH. In the original report of Tieder *et al.* renal stones were present only in the adult individual studied,[91] whereas renal stones have been present in both school-age and adolescent children affected with HHRH in our clinic.[92]

HHRH is usually managed by administration of oral phosphate salts alone (examples shown in Table 11.3). Attention to both the bone disease and hypercalciuria is important. The therapeutic strategy is to decrease the elevated circulating $1,25(OH)_2D$ levels, effecting a decrease in intestinal calcium absorption, and to provide an increase in ambient phosphorus levels as to enhance skeletal mineralization. Because of the unique pathophysiology, there is little risk of developing hyperparathyroidism, unlike with XLH in which hyperparathyroidism is notoriously problematic in view of phosphorus supplementation. Reports of long-term monitoring are limited, but in the authors' experience, improvements in bone pain and height velocity have occurred with therapy. Bone mineral density has improved, but one patient continues to have low bone density. The importance of generous hydration and avoidance of a high sodium intake is recommended.

In that the normal physiologic response to chronic hypophosphatemia can explain the spectrum of clinical and biochemical findings observed in patients with HHRH, it has been proposed that the defect is one specific to the regulation of epithelial renal tubular transport of phosphate. This differs from the pathophysiology of XLH where a defect proximal to isolated renal tubular transport is hypothesized, as other features of the disease are not easily explained by a solitary defect in apical phosphate transport. Therefore, candidate genes for the site of mutation in HHRH include NaPiT2, the major renal phosphate transporter. Two studies have examined whether HHRH maps to chromosome 5q35, where NaPiT2 is located. Jones *et al.*[93] examined four families (including the Bedouin tribe described above). Neither mutation in NaPiT2, nor evidence for linkage to the relevant area could be demonstrated. Van den Heuvel *et al.*[94] sequenced the entire open reading frame of the NaPiT2 gene in affected and unaffected members of another family, also finding no mutation in NaPiT2. Linkage disequilibrium studies in the Bedouin family also exclude association of HHRH with chromosome 6p22, the site of NaPiT1, a lesser abundant renal phosphate transporter. The phenotypic variability in the reported families has suggested that HHRH may be the manifestation of different genetic defects. It may be possible that some cases represent a two gene disorder with variable penetrance.

## *Oncogenic Osteomalacia*

Oncogenic osteomalacia (OO) is manifested by clinical features similar to those seen in XLH. The age of onset of the clinical features in OO are most consistent with an acquired phenotype, presenting at older ages, although children as young as 4–5 years of age have been reported with the disease.[95] XLH, on the other hand is often manifest by the second or third year of life. Characteristic hypophosphatemia and renal phosphate wasting are usually reported as presenting features in OO, as also seen in patients with XLH.[96–99] Serum phosphate may be evident in a lower range than usually observed in XLH, as can circulating $1,25(OH)_2D$ levels, but considerable overlap of these values between the two syndromes is evident upon examination of the bulk of reported cases. The patient with OO, however, frequently exhibits more severe symptoms, such as bone pain and muscle weakness, whereas these complaints are more tempered in the usual XLH patient. Adults with OO may present with fractures and proximal muscle weakness. Elevated serum alkaline phosphatase activity is usually present, and may be greater than that encountered in XLH. Serum calcium levels are usually normal, but low values have been reported; circulating PTH levels are usually normal, but have been variably reported as elevated or even low. Thus the typical clinical and biochemical phenotype is similar to that seen in XLH, but the degree of severity of the manifestations in OO is often greater.

Numerous pathologic diagnoses have been attributed to the causal tumors. Most reports have classified OO associated tumors as benign and of mesenchymal cell origin. Characteristic features noted in tabulated reviews of the syndrome include predominant pericytes, invoking the diagnosis of *hemangiopericytoma.*[97] Spindle cells and fibroblastic cells are also prominent features in other many reports; ossifying and non-ossifying fibromas often have described intraskeletal tumors causing OO. Despite this predominance of benign lesions, significant malignancies may be present with this syndrome, including prostatic carcinoma, oat cell carcinoma of the lung, and osteosarcoma. Although many of the OO-associated tumors are considered benign at the time of detection, their natural history is not well-documented, and malignant transformation is a concern in tumors not identified in timely fashion, or incompletely resected.

The presentation of the above described clinical features in an acquired setting, or in the absence of family history of XLH, raises the possibility of OO as the underlying disorder. A careful search for a possible tumor is important in the evaluation of such patients. The clinical course of this disease is dramatically affected by removal of the tumor, nearly always resulting in a rapid and complete response. The pathophysiology of OO has intrigued clinicians and physiologists for decades. The dramatic response to tumor removal provides evidence that some tumor-associated substance has direct renal tubule effects resulting in severe phosphate wasting. The bone disease has been explained by this severe hypophosphatemia. The inappropriate levels of circulating $1,25(OH)_2D$ in both XLH and OO, have suggested that at least one other renal tubular function was affected, such that both tubular phosphate transport and vitamin D-1$\alpha$ hydroxylation activity were aberrant in both syndromes. The similarity of findings in the two disorders and the recent finding that circulating levels of FGF23 are elevated in this condition provide substantial credibility to the hypothesis that a circulating factor mediates the pathophysiology of XLH and OO. With the advent of subtraction RNA technology, several candidates specifically expressed by OO tumors have been identified, including FGF23 itself;[100–102] see above, page 241.

## Fanconi Syndrome

Fanconi syndrome, a disorder of generalized proximal renal tubular dysfunction can be acquired or inherited. Typically glycosuria with normal blood glucose levels, phosphate and potassium wasting, renal tubular acidosis, and aminoaciduria occur. Uncommonly calcium excretion is excessive. One can develop Fanconi syndrome after ingestion of various toxins including heavy metals, drugs, including the aminoglycoside class of antibiotics, valproate, and with chemotherapeutic agents such as ifosfamide and cisplatin.

Fanconi syndrome may occur as an isolated inherited disorder, or in concert with other metabolic diseases such as cystinosis, Lowe's oculo-cerebro-renal syndrome, Wilson's disease, glycogen storage diseases, and hereditary fructose intolerance. The phosphate wasting component of Fanconi Syndrome results in a metabolic bone disease similar to XLH, and is usually treated with phosphate and vitamin D metabolites employing a similar treatment strategy. It is of interest that circulating 1,25(OH)$_2$D levels are often inappropriately normal in Fanconi syndrome, as occurs in XLH,[103] but this finding is variable. Often bicarbonate or citrate supplements are necessary, as renal tubular acidosis results in resistance to phosphate therapy.

# Bone Disorders Associated with Chronic Hyperphosphatemia and with High Phosphorus Intake, Low Dietary Calcium to Phosphorus Intake Ratio

## Osteodystrophy and Disorders of Calcification Associated with Hyperphosphatemia of End Stage Renal Disease

The most critical adaptive mechanism in the maintenance of phosphorus homeostasis lies in the ability of the kidneys to adjust the tubular reabsorption of phosphorus to changes in dietary phosphorus intake.[104,105] As discussed earlier, under normal conditions, renal tubular phosphorus reabsorption decreases when a high phosphorus diet is consumed. Under normal conditions, adaptive changes in phosphorus reabsorption are independent of PTH stimuli and reflect similar changes in the sodium-dependent phosphate transporters in the brush border membrane (BBM) of the convoluted proximal tubule of the nephron. Rapid adaptation is achieved by shuttling the co-transporter proteins between the BBM and inactive intracellular pool.

In early renal failure when nephron mass is reduced, phosphorus homeostasis is dependent upon increased PTH production, which maintains normal plasma phosphorus levels at the expense of elevated PTH.[105,106] With increased severity of renal disease, increasingly higher levels of PTH are needed to maintain phosphorus homeostasis. Such developments ultimately lead to hyperphosphatemia, secondary hyperparathyroidism, uremic bone disease, and soft tissue calcification.[108]

Hyperphosphatemia stimulates PTH production by several known mechanisms all of which are interrelated: (1) direct stimulation of PTH synthesis[59–61,107]; (2) precipitation of calcium phosphate in soft tissue resulting in hypocalcemia, the major stimulus for PTH secretion[106]; (3) inhibition of 1-OHlase activity resulting in a decrease 1,25-(OH)$_2$ D which disrupts the feed back inhibition of PTH gene transcription by the active metabolite of vitamin D and may result in a further rise in PTH.[108]

Renal osteodystropy is the term used to describe a wide spectrum of bone abnormalities that are associated with chronic renal failure. At one end of the spectrum is the high-turnover bone disorder of secondary hyperparathyroidism or *osteitis fibrosa cystica* which is a common skeletal consequence of end-stage renal disease.[107,109–111] Osteitis fibrosa cystica is characterized by increased osteoclastic bone resorption with extensive bone marrow fibrosis.[109] At the opposite end of the bone abnormality spectrum is adynamic bone disease, characterized by low bone turnover. Osteomalacia in adults and rickets in children may also occur due to impaired synthesis of 1,25(OH)$_2$D resulting in defective mineralization of osteoid.[109] Osteoporosis characterized by decreased bone density, thinning cortices and trabeculae may also occur. The spectrum of renal osteodystrophy has changed over time with technical advances in the treatment of end stage renal disease. Secondary hyperparathyroidism was the most prevalent type of bone disease observed in renal failure until just recently, when adynamic bone disease became the most commonly observed lesion in dialysis patients.[110] Using dynamic histomorphometric analyses of transiliac bone biopsies in end-stage renal failure patients not yet on dialysis, Spasovski *et al.*[110] demonstrated abnormal bone in 62% of the patients. Adynamic bone disease was the most prevalent type in this population appearing in 23% with notably high (12%) prevalence of osteomalacia and 18% with mixed osteodystrophy.

Renal osteodystrophy, although a serious disabling morbidity, is of less importance to the dialysis patient's over all health than the complications arising from the hyperphosphatemia induced soft tissue calcification that occur in end-stage renal failure. Hyperphosphatemia and elevated calcium × phosphorus product (Ca × Pi) promotes visceral and vascular calcification and is linked to increased cardiovascular mortality in renal failure.[112–116] The prevalence of cardiovascular disease is markedly higher than in the general population and cardiovascular disease accounts for almost half of all the deaths in dialysis patients. Serum phosphorus and calcium × phosphorus product have been significantly linked with mortality risk in chronic dialysis patients.[115] Jono *et al.*[114] showed dose-dependent increase in mineral deposition in human aortic smooth muscle cells cultured in media containing phosphorus levels comparable to those observed in hyperphosphatemia of chronic hemodialysis patients (1.14–2.0 mmol L$^{-1}$). They showed that the calcification of the aortic smooth muscle cells was mediated by a sodium-dependent phosphate co-transporter (identified as PIT-1, also called GLvr-1), a member of the novel type III sodium-dependent phosphate co-transporter family (see above, page 240). Their findings strongly support the hypothesis that elevated phosphorus can induce aortic smooth muscle cells to secrete a protein matrix capable of mineralization and that a sodium–phosphate co-transporter functions in cellular sensing of elevated phosphorus levels.

Since elevated phosphorus is a preventable risk factor for both bone disease and increased cardiovascular morbidity and mortality in end stage renal disease, effective strategies to reduce or maintain normal serum phosphorus levels are critical.[13] The dietary intake of phosphorus is an important determinant of dietary phosphorus restriction in the prevention and treatment of hyperphosphatemia. Dialytic removal of phosphate is not adequate to prevent hyperphosphatemia, consequently most dialysis patients need to take phosphate binders, preferably those that are calcium and aluminum free to avoid serious adverse effects.[117–119] However, these measures are not sufficient unless efforts are made to reduce the amount of phosphorus available for absorption. More attention needs to be given to the dietary management of phosphorus in dialysis patients.[13] Dietary phosphorus restriction has been shown to effectively slow the progression of renal insufficiency in humans[118,119] and animal models.[117]

There are several barriers to effectively reducing dietary phosphorus intake to levels recommended for dialysis patients. Because phosphorus and protein are inexorably linked in foods,[120] it is difficult to restrict dietary phosphate intake without inducing protein malnutrition, because daily phosphorus intakes of less than 750 mg correspond to a protein intake of less than 50 g.[121,122] The Dialysis Outcome Quality Initiative (DOQI) guidelines recommend minimal daily protein intakes of $1-1.2$ g $kg^{-1}$ $day^{-1}$ for dialysis patients,[13] however, this level of protein intake will produce hyperphosphatemia in chronic dialysis patients. The solution to effective dietary restriction of phosphorus lies in the careful selection of protein foods, selecting those with the best phosphorus to protein ratio.[121] The chronic dialysis patient also needs to become aware of the hidden sources of phosphate in their food choices that are added in processing.[13] A major barrier in accomplishing this is the general lack of information concerning the phosphorus content of foods on food labels. There is a universal need for accurate information on the phosphorus content of foods available on food labels to effectively manage chronic renal failure.

## Influence of Low Calcium to Phosphorus Intake Ratio of Western Diets on Bone Mass

As world populations increasingly adopt Western culture and diet, such as highly processed convenience foods, we see growing evidence of a disproportionate increase in phosphorus intake relative to calcium. Few cross-sectional studies in humans relate high phosphorus, low calcium intake to effects on bone mineral content or bone density, however this relationship is well established in animal models when calcium intake is both adequate and inadequate.[29,123] Table 11.4 reviews the small number of studies to date that have examined the relationship between phosphate intake or the Ca:Pi intake ratio of the diet and some measure of bone status such as bone mineral density or content (BMD/BMC). Of the more recent studies[18,124–127] only one[18] used food record, rather than food frequency questionnaires (FFQ) or food history to estimate phosphorus and calcium intake. Brot et al.[18] found a significant inverse association between Ca:Pi intake ratio and circulating levels of 1,25(OH)$_2$D. Further more, BMC and BMD at five bone sites were inversely related to serum 1,25(OH)$_2$D levels. Moreover, there was a significant positive association of 1,25(OH)$_2$D levels and markers of bone turnover, indicating that 1,25(OH)$_2$ D levels were associated with increased bone turnover and a reduced dietary Ca:Pi ratio. Brot et al. also found a positive association between the dietary Ca:Pi ratio and bone mineral density ($P < 0.0005$). A longitudinal study by Neville et al.[124] found mixed associations depending on age, however they also found a decline in the Ca:Pi intake ratio with

**Table 11.4.** *Cross-sectional studies of the relationship between phosphorus intake and bone health*

| Study method used to estimate nutrient intake | Population | Mean Pi intake (Ca:Pi ratio)[1] | Association between BMC/BMD[2] |
|---|---|---|---|
| Yano, 1985[128] | 1208 men | (0.5) | $P \leq 0.05$ distal radius |
| Food Record | 908 women | (0.5) | NS |
| Lukert, 1987[129] FFQ[3] | 29 elderly women | (0.85) | NS |
| and 24 h food record | 50 perimenopausal | (0.71) | $P \leq 0.04$ radius |
|  | 35 elderly men | (0.71) | NS |
| Tylavsky, 1988[130] FFQ | 88 vegetarian women | (0.72) | $P \leq 0.05$ Pi *vs.* mid-radius BMC |
|  | 278 omnivorous women | (0.72) |  |
| Metz, 1993[131] FFQ | 38 young women, | (0.72) | $P \leq 0.03$ Pi *vs.* d. radius BMC |
|  | 24–28 years of age |  | $P \leq 0.03$ Pi *vs.* d. radius BMD |
|  |  |  | $P \leq 0.03$ Pi *vs.* mid-radius BMC |
| Teergarden, 1998[125] FFQ | 215 women, | 1356 mg Pi day$^{-1}$ | $P \leq 0.005$ Pi *vs.* spine BMD |
|  | 24–28 years of age | (0.68) | NS |
| Brot, 1999[18] | 510 women, | 1079 mg Pi day$^{-1}$ | NS |
| 4–7 day food record | 45–58 years of age | (0.73) | $P \leq 0.005$ Ca: Pi ratio predictor of BMC/BMD in total body, spine, total femur, femur neck, trochanter |
| New, 2000[126] FFQ | 62 women, | 1536 mg Pi day$^{-1}$ | NS Pi *vs.* radius |
|  | 45–55 years of age |  | $P \leq 0.005$ negative relationship pyridionline[4] |
|  |  | (0.72) | Not measured |

cont.

**Table 11.4.** cont.

| Study method used to estimate nutrient intake | Population | Mean Pi intake (Ca:Pi ratio)[1] | Association between BMC/BMD[2] |
|---|---|---|---|
| Diet history, Neille, 2000[124] | Longitudinal study 238 men 205 women 15 and 20–25 years of age | (0.79→0.63) (0.79→0.64) | $P \leq 0.03$ positive correlation to adult Male spine $P \leq 0.02$ negative correlation to 15 year old female femur neck |
| Whiting, 2002[127] FFQ | 57 men, 39—42 years of age | 1741 mg Pi day$^{-1}$ (0.66) | $P \leq 0.01$ Pi vs. total body BMD $P \leq 0.05$ Pi vs. hip and spine BMD Not measured |

[1] Ca:Pi ratio, wt: wt.
[2] P value indicates level of significance of association between a measure of phosphorus intake (mg Pi day$^{-1}$) or calcium to phosphorus intake ratio (mg : mg) and a measure of bone mass (BMC/BMD) or measure of bone turnover (pyridinoline).
[3] FFQ = Food frequency questionnaire.
[4] Pyridinoline, a marker of bone resorption activity.

increasing age in teens similar to findings from the US population. Other more recent studies using FFQ[125–127] either found no relationship between phosphorus intake and bone[126] or observed a significant positive association.[125,127] With respect to the positive association between phosphorus intake and bone mass, findings from these recent as well as those from earlier studies[128–131] should be interpreted cautiously due to the use of FFQ to determine nutrient intake. In natural foods, phosphorus and protein are very strongly associated, a relationship that allows reasonably accurate estimation of phosphorus intake from knowledge of protein intake.[120] In addition, protein intake has a well established positive association with bone mass when calcium intake is adequate[132,133] Use of FFQ to estimate phosphorus intake is seriously flawed in that FFQ derived estimates of phosphorus intake most likely reflect protein intake and underestimate phosphorus intakes. This method of nutrient intake estimation cannot differentiate between the consumption of processed foods (high in phosphate) or unprocessed foods with lower phosphorus content.[127]

High phosphorus, low calcium diets do not generally increase serum phosphorus levels in fasted morning blood samples. To accurately observe both plasma phosphorus and PTH changes in response to phosphorus loading, multiple blood samples should be taken over the day.[54,55,134] Hyperphosphatemia was the criterion used to establish the U.S. Tolerable Upper Intake Limit for phosphorus (UL $= 4000$ mg day$^{-1}$);[3] however, no evidence was found to indicate hyperphosphatemia with intakes at or above this level probably due to the use of morning fasting blood measurements in most studies.

Experimental studies examining low calcium, high phosphorus consumption have produced hormonal changes that are not conducive to the development or maintenance of peak bone mass (Table 11.5). Elevated levels of serum PTH have been observed with high phosphorus feeding during both adequate and inadequate calcium intake. Acute,[16,43,53] and longer[17,19,49,50,52,54,55,135–140] exposure to oral phosphorus loading in the presence or absence of adequate calcium intake shows a rise in PTH. In contrast to short term or acute feeding studies, longer term studies[19,50,52,55,137] show a significant reduction or no change in $1,25(OH)_2D$, despite elevations in PTH. In these studies, chronic high phosphorus apparently pre-empted PTH induced stimulation of $1,25(OH)_2D$ synthesis, the adaptive mechanism needed for adequate calcium absorption and optimal bone accretion.

No prospective dietary studies have been of sufficient duration to accurately determine the direct effect of low calcium, high phosphate intake on bone mass accretion in young adults. Biochemical markers of bone turnover have been used to determine whether bone formation or resorption activities are influenced by high phosphorus feeding (Table 11.5) in chronic feeding studies of 4–7 weeks, but no longer. It is unclear how reliable such markers are under these study conditions since many of the bone turnover markers vary in their sensitivity to dietary changes and susceptibility to renal and hepatic degradation.[141] Several studies have shown declines[16,17] or no change in formation markers,[53,55] however most studies fail to show changes in resorption

**Table 11.5.** Calcitropic hormone and bone turnover response to high phosphate intake under conditions of adequate and low calcium intake in adults

| Study, Pi Source | Duration, Diet (Ca:Pi ratio)[3] | PTH[1] | | | Calcitriol[2] | | | PTH and Bone Markers % Change with high Pi intake |
|---|---|---|---|---|---|---|---|---|
| | | Control Diet | Test Diet | % Change | Control Diet | Test Diet | % Change | |
| **Adequate calcium intake** | | | | | | | | |
| Bell, 1977,[26] Food source | 14 weeks (Ca:Pi) | NA (0.7) | NA (0.33) | | NA (0.7) | NA (0.3) | | ↑22%cAMP* ↑20% Hydroxyproline*** |
| VandenBerg, 1980,[50] Phosphate salts | 14 days (Ca:Pi) | 16±2 (0.65) | 18±2 (0.25) | +12.5 | 57±13 (0.65) | 35±11 (0.25) | −38* | NA |
| Broadus, 1983,[136] Phosphate salts | 363 days (Ca:Pi) | 237 (0.5) | 332 (0.33) | +40 | 84±18 (0.5) | 56±14 (0.33) | −33*** | NA |
| Portale, 1986,[137] Phosphate salts | 10 days (Ca:Pi) | 12±2 (0.57) | 15±2 (0.28) | +25 | 38±3 (0.57) | 27±4 (0.28) | −29** | NA |
| Silverberg, 1986,[135] Phosphate salts | 5 days (Ca:Pi) | 14±2 (0.6) | 22±2 (0.2) | +57* | 35.1±45 (0.6) | 37.2±46 (0.2) | +6 NS | ↑54% osteocalcin* No change in urinary hydroxyproline |
| Portale, 1989,[49] Phosphate salts | 9 days (Ca:Pi) | 19±3 (1.36) | 23±4 (0.36) | +21 | 43±4 (1.36) | 29±4 (0.36) | −58** | NA |
| Silverberg, 1989,[52] Phosphate salts Normal women | 5 days (Pi load) | 15.4±2 (0) | 37.9±6 (2 g) | +250** | 34±4 (0) | 31±3 (2 g) | −9 NS | NA |

| Reference | Duration (Ca:Pi) | PTH[1] | PTH[1] | | Calcitriol[2] | Calcitriol[2] | | Comments |
|---|---|---|---|---|---|---|---|---|
| Bizik, 1996,[138] Food phosphorus | 7 days (Ca:Pi) | 27.2±6 (1.5) | 30.6±6 (0.75) | +12.5 | NA | NA | NA | No change in urinary hydroxyproline |
| Whybro, 1998,[139] Phosphate salts | 14 days (Ca:Pi) | 26.7±11 (1.0) | 33.7±12 (0.44) | +26** | NA | NA | NA | No change in osteocalcin or urinary N-telopeptide |
| Grimm, 1999,[17] Phosphate salts | 42 days (Ca:Pi) | 32.5±9 (0.88) | 43.1±17 (0.66) | +33 | 34.3±9 (0.88) | 39.1±13 (0.66) | +14 | NS ↓ osteocalcin |
| **Low calcium intake** | | | | | | | | |
| Calvo, 1988,[54] | 8 days ♂ | 13.8±0.7 (0.88) | 15.4±1 | +12*** | 62±6 (0.88) | 80.4±6 (0.25) | +30** | ↑ cAMP* |
| | ♀ | 12.6±0.8 (0.88) | 15.4±0.8 | +22**** | 78±7 (0.88) | 86.9±8 (0.25) | +11** | ↑ hydroxyproline* |
| Food phosphorus | (Ca:Pi) | | | | | | | |
| Calvo, 1990,[55] Food phosphorus | 28 days (Ca:Pi) | 26.9±2 (0.88) | 35.3±3.9 (0.25) | +31* | 62.2±306 (0.88) | 66.5±5.3 (0.25) | +7 NS | NS ↑ cAMP; NS ↑ hydroxyproline; NS osteocalcin |
| Barger-Lux, 1993,[140] Food phosphorus | 56 days (Ca:Pi) | 28 (0.63) | 34 (0.33) | +48* | NA | NA | NA | ↑ hydroxyproline** |

NA = Not analyzed; NS = Not statistically significant.
[1] Plasma PTH values presented as pg mL$^{-1}$ or pmol L$^{-1}$.
[2] Plasma calcitriol presented as pg mL$^{-1}$ or pmol L$^{-1}$.
[3] Ca:Pi ratio as mg:mg.
* Statistical comparisons made between normal or control diet vs. test diet with low Ca:Pi ratio.
*$P \leq 0.05$; **$P \leq 0.01$; ***$P \leq 0.001$.

markers despite evidence of an increase in PTH.[16,17,53,55,138,139] Of those studies that measured the pyridinoline cross links of collagen, a more sensitive and specific bone marker relative to hydroxyproline,[142] none showed an increase in these resorptive markers with the observed rise in PTH.[16,17,138,139] Of note, all these studies employed diets that were high in calcium as well as phosphorus, while more pronounced adverse effects were observed in those studies where the Ca:Pi intake ratio was very low and the calcium intake inadequate.[54,55,140] Americans typically consume diets that fail to meet the calcium intake guidelines, but exceed the intake guidelines for phosphorus. The health impact of high phosphorus consumption independent of calcium intake remains unclear. Significant phosphorus induced bone resorption may only occur when calcium intake is low, however this does not preclude the possibility that bone accretion may be impaired in adolescents and young adults with high phosphorus intakes independent of calcium status. This dietary pattern clearly merits further study since it is common in adolescents and young adults in Western cultures who have not reached peak bone mass.

## Conclusions

The homeostatic mechanisms that maintain serum phosphorus levels are under the endocrine regulation of the PTH– vitamin D axis with the primary point of regulation at the level of renal reabsorption of phosphorus. Recent findings show that serum phosphorus concentrations can directly influence PTH secretion, inhibition or stimulation of the synthesis of calcitriol, the active form of vitamin D, and can self-regulate its own renal excretion rate by controlling the internalization of the sodium–phosphate co-transporter proteins of the renal proximal tubular brush border. Disruption of phosphorus homeostasis may result in hyperphosphatemia or hypophosphatemia, both of which may adversely affect bone mineralization. Disorders in renal function that impair renal phosphate excretion or metabolic disorders that impair or prevent renal reabsorption of phosphorus can hinder bone mineralization and formation. Hypophosphatemic disorders generally result in severe mineralization defects of the skeleton and frequently result from the disruption of renal tubular reclamation of phosphorus. In end stage renal disease which results in severe hyperphosphatemia, normal or low levels of phosphorus intake can result in secondary hyperparathyroidism, uremic bone disease, and arterial calcification. As world populations increasingly adopt Western culture and diet, such as highly processed convenience and prepared foods, we see growing evidences of an increased phosphorus intake, relative to calcium. High phosphorus consumption, when calcium intake is low has been shown to result in secondary hyperparathyroidism and is associated with little to no change in bone resorption markers, whereas significant decreases in markers of bone formation occur, suggesting that a high phosphorus, low calcium intake could result in suboptimal maintenance of bone mass, and ultimately poor skeletal health.

# References

1. E.J. Griffith, C. Ponnamperuma and N.W. Gabel, Phosphorus, a key to life on the primitive earth. *Origins Life*, 1977, **8**, 71–85.
2. G.J. Kemp, A. Blumsohn and B.W. Morris, Circadian changes in plasma phosphate concentration, urinary phosphate excretion and cellular phosphate shifts, *Clin. Chem.*, 1992, **38**, 400–402.
3. Food Nutrition Board. Dietary Reference Intakes: Calcium, Phosphorus, Magnesium, Vitamin D, and Fluoride. Washington, DC, National Academy Press, 1997.
4. D.B. Endres and R.K. Rude, Mineral and bone metabolism. In *Teitz Clinical Chemistry*. C.A. Burtis, E.R. Ashwood (eds.), Philadelphia, W.B. Saunders, 1996, 1887–1972.
5. S.S. Sherman, B.W. Hollis and J.B. Tobin, Vitamin D status and related parameters in a healthy population: the effects of age, sex and season, *J. Clin. Endocrinol. Metab.*, 1990, **71**, 405–413.
6. R.L. Prince, A. Devine, R.I. Price, D.A. Gatteridge, D. Kerr, A. Criddle, A. Garcia-Webb and A. St. John, The effects of menopause and age on calcitropic hormones: a cross-sectional study of 655 healthy women aged 35 to 90, *J. Bone Miner. Res.*, 1995, **10**, 835–842.
7. C.J. Bates, A. Prentice and S. Finch, Gender differences in food and nutrient intakes and status indices from the National Diet and Nutrition Survey of People aged 65 years and over, *Eur. J. Clin. Nutr.*, 1999, **53**, 694–699.
8. I.M. Dick, A. St. John, S. Heal and R.L. Prince, The effect of estrogen deficiency on bone mineral density, renal calcium and phosphorus handling and calcitropic hormones in the rat, *Calcif. Tissue Int.*, 1996, **59**, 174–178.
9. M.S. Calvo, R. Eastell, K.P. Offord, E.J. Bergstralh and M.F. Burritt, Circadian variation in ionized calcium and intact parathyroid hormone: evidence for sex differences in calcium homeostasis, *J. Clin. Endocrinol. Metab.*, 1991, **72**, 69–76.
10. A.A. Portale, B.P. Halloran and R.C. Morris Jr., Dietary intake of phosphorus modulates the circadian rhythm in serum concentration of phosphorus, *J. Clin. Invest.*, 1987, **80**, 1147–1154.
11. T.O. Carpenter and L.L. Key, Disorders of the metabolism of calcium, phosphorus, and other minerals, In *Pediatric Textbook of Fluids and Electrolytes*, I. Ichikawa (ed.), Williams and Wilkins, Baltimore, 1990, pp. 237–268.
12. C.M. Giachelli, Ectopic calcification: new concepts in cellular regulation, *Z. Kardiol.*, 2001, **90**(3), 31–37.
13. J. Uribarri and M.S. Calvo, Hidden sources of phosphorus in the typical American diet: Does it matter in nephrology? *Seminars Dialysis*, 2003, **16**, 186–188.
14. M.S. Calvo and Y.K. Park, Changing phosphorus content of the U.S. diet: potential for adverse effects on bone, *J. Nutr.*, 1996, **126**, 1168S–1180S.
15. J. Gregory, K. Foster, H. Tyler and M. Wiseman, The Dietary and Nutritional Study of British Adults. London, HMSO, 1990, pp. 153–176.
16. M. Karkkainen and C. Lamberg-Allardt, An acute intake of phosphate increases parathyroid hormone secretion and inhibits bone formation in young women, *J. Bone Miner. Res.*, 1996, **11**, 1905–1912.
17. M. Grimm, A. Muller, G. Hein, R. Funfstuck and G. Jahreis, High phosphorus intake only slightly affects serum minerals, urinary pyridinium crosslinks and renal function in young women, *Eur. J. Clin. Nutr.*, 2001, **55**, pp. 153–161.
18. C. Brot, N. Jorgensen, O.R. Madsen, L.B. Jensen and O.H. Sorensen, Relationships between bone mineral density, serum vitamin D metabolites and calcium:

phosphorus intake in healthy perimenopausal women, *J. Intern. Med.*, 1999, **245**, 509–516.

19. E. Roma-Giannikou, D. Adamidis, M. Gianniou, R. Nikolara and N. Matsaniotis, Nutritional survey in Greek children: nutrient intake, *Eur. J. Clin. Nutr.*, 1997, **51**, 273–285.

20. Z-W. Zhang, S. Shimbo, K. Miyake, T. Wantanabe, H. Nakatsuka, N. Matsuda-Inoguchi, C-S. Moon, K. Higaashikawa and M. Ikeda, Estimates of mineral intakes using food composition tables vs measures by inductively-coupled plasma mass spectrometry: Part 1. calcium, phosphorus and iron, *Eur. J. Clin. Nutr.*, 1999, **53**, 226–232.

21. N. Matsuda-Inoguchi, S. Shimbo, S. Srianujata, O. Banjong, C. Chitchumroom-chokchai, T. Wantanabe, H. Nakatsuka, K. Higashikawa and M. Ikeda, Nutrient intake of working women in Bangkok, Thailand, as studied by total food duplicate method, *Eur. J. Clin. Nutr.*, 2000, **54**, 187–194.

22. J.L. Greger and M. Krystofiak, Phosphorus intake of Americans, *Food Technol.*, 1982, **36**, 78–84.

23. M.S. Calvo, Dietary considerations to prevent loss of bone and renal function, *Nutrition*, 2000, **16**, 564–566.

24. E. Takeda, K. Sakamoto, K. Yokota, M. Shinohara, Y. Taketani, K. Morita, H. Yamamoto, K. Miyamoto and M. Shibayama, Phosphorus supply per capita from food in Japan between 1960 and 1995, *J. Nutr. Sci. Vitaminol. (Tokyo)*, 2002, **48**, 102–108.

25. L.L. Oenning, J. Vogel and M.S. Calvo, Accuracy of methods estimating calcium and phosphorus intake in daily diets, *J. Am. Diet. Assoc.*, 1988, **88**, 1076–1080.

26. R.R. Bell, H.H. Draper, D.Y.M. Tzeng, H.K. Shin and G.R. Schmidt, Physiological responses of human adults to foods containing phosphate additives, *J. Nutr.*, 1977, **107**, 42–50.

27. M.L. Weiner, W.F. Salminen, P.R. Larson, R.A. Barter, J.L. Kranetz and G.S. Simon, Toxicological review of inorganic phosphates, *Food Chem. Technol.*, 2001, **39**, 759–786.

28. Scientific Committee for Food of the Commission of the European Communities (1993): Food-science and techniques: Reports of the Scientific Committee for Food: 31 series. Luxembourg.

29. M.S. Calvo, Dietary phosphorus, calcium metabolism, and bone, *J. Nutr.*, 1993, **123**, 1627–1633.

30. L. Sax, The Institute of Medicine's 'Dietary Reference Intake' for phosphorus: a critical perspective, *J. Am. College Nutr.*, 2001, **20**, 271–278.

31. J.J.B. Anderson, M.L. Sell, S.C. Garner and M.S. Calvo, Phosphorus, In *Present Knowledge of Nutrition*, B.A. Bowman, R.M. Russell (eds.), Washington, DC, IlSI Press, 2001, pp. 281–291.

32. G.C. Rampersaud, L.B. Bailey and G.P. Kauwell, National survey beverage consumption data for children and adolescents indicate the need to encourage a shift toward more nutritive beverages, *J. Am. Diet. Assoc.*, 2003, **103**, 97–100.

33. E.M. Pao, K.H. Fleming, P.M. Guenther and S.J. Mickle, Foods commonly eaten by individuals: amount per eating occasion. USDA Human Services Home Economics Research Report No. 44. Washington, DC, US Government Printing Office, 1982.

34. L.K. Massey and M.M. Strang, Soft drink consumption, phosphorus intake and osteoporosis, *J. Am. Diet. Assoc.*, 1982, **80**, 581–583.

35. A.F. Subar, S.M. Krebs-Smith, A. Cook and L.L. Kahle, Dietary sources of nutrients among adults, 1989 to 1991, *J. Am. Diet. Assoc.*, 1998, **98**, 537–547.
36. P.S. Venkataraman, R.C. Tsang, F.R. Greer, A. Noguchi, P. Laskarzewski and J.J. Steichen, Late infantile tetany and secondary hyperparathyroidism in infants fed humanized cow's milk formula: Longitudinal follow-up, *Am. J. Dis. Children*, 1985, **1939**, 664–668.
37. A. Vohra and T. Satyanarayana, Phytases: microbial sources, production, purification, and potential biotechnological applications, *Crit. Rev. Biotechnol.*, 2003, **23**, 29–60.
38. M.B. Zemel and H.M. Linkswiler, Calcium metabolism in the young adult male as affected by level and form of phosphorus intake and level of calcium intake, *J. Nutr.*, 1981, **11**, 315–324.
39. K.C. Moltz, P.G. Gallagher and R.E. Ehrenkranz *et al.*, Markedly elevated fractional calcium absorption in a very low birth weight infant: failure to adapt to calcium intake, *J. Bone Miner. Res.*, 1995, **10**(Suppl. 1) S511.
40. J. Lemann Jr., Intestinal absorption of calcium, magnesium and phosphorus, In *Primer on the Metabolic Bone Diseases and Disorders of Mineral Metabolism*, M.J. Favus (ed.), New York, Raven Press, 1993, pp. 46–50.
41. E.M. Brown, G. Gamba, D. Riccardi, M. Lombardi, R. Butters, O. Kifor, A. Sun, M.A. Hediger, J. Lutton and S.C. Herbert, Cloning and characterization of an extracellular $Ca^{+2}$ sensing receptor from bovine parathyroid, *Nature*, 1993, **366**, 575–580.
42. L.C. McCary and H.F. DeLuca, Functional Metabolism and Molecular Biology of Vitamin D action, In *Vitamin D: Physiology, Molecular Biology and Clinical Applications*, M.F. Holick (ed.), Totowa, NJ. Human Press, 1999, 39–56.
43. E. Reiss, J.M. Canterbury, M.A. Bercovitz and E.L. Kaplan, the role of phosphate secretion of parathyroid hormone in man, *J. Clin. Investig.*, 1970, **49**, 2146–2149.
44. H.H. Draper, Bone loss in animals, In *Advances in Nutritional Research*, H. Draper (ed.), New York. Plenum Press, 1994, pp. 79–106.
45. Y. Tanaka and H.F. DeLuca, The control of 25-hydroxyvitamin D metabolism by inorganic phosphorus, *Arch. Biochem. Biophys.*, 1973, **154**, 566–574.
46. W.J. Maierhofer, R.W. Gray and J. Lemann Jr., Phosphate deprivation increases serum 1,25(OH)$_2$ -vitamin D concentration in healthy men, *Kidney Int.*, 1984, **25**, 571–575.
47. M.Y.H. Zhang, X. Wang, J.T. Wang, N.A. Compagnone, S.H. Mellon, J.L. Olson, H.S. Tenenhouse, W.L. Miller and A.A. Portale, Dietary phosphorus transcriptionally regulates 25-hydroxyvitamin D-1-alpha hydroxylase gene expression in the proximal renal tubule, *Endocrinology*, 2002, **143**, 587–592.
48. J. Caverzasio and J.P. Bonjour, Growth factors and renal regulation of phosphate transport, *Pediatr. Nephrol.*, 1993, **7**, 802–806.
49. A.A Portale, B.P. Halloran and R.C. Morris Jr., Physiologic regulation of serum concentration of 1,25-(OH)$_2$ D$_3$ by phosphate in normal men, *J. Clin. Investig.*, 1989, **83**, 1494–1499.
50. C. Van den Berg, R. Kumar, D.M. Wilson, H. Health III and L.H. Smith, Orthophosphate therapy decreases urinary calcium excretion and serum 1,25-dihydroxyvitmain D concentrations in idiopathic hypercalciuria, *J. Clin. Endocrinol. Metab.*, 1980, **51**, 998–1001.
51. R.S. Goldsmith, J. Jowsey, W.J. Dube, B.L. Riggs, C.D. Arnaud and P.J. Kelley, Effects of phosphorus supplementation on serum PTH and bone morphology in osteoporosis, *J. Clin. Endocrinol. Metab.*, 1976, **43**, 523–532.

52. S.J. Silverberg, E. Shane, R.N. Luz del la Cruz, G.V. Segre, T.L. Clemens and J.P. Bilezikian, Abnormalities in parathyroid hormone secretion and 1,25-dihydroxyvitamin $D_3$ formation in women with osteoporosis, *New Engl. J. Med.*, 1989, **320**, 277–281.

53. K. Brixen, H.K. Nielsen, P. Charles and L. Mosekilde, Effects of a short course of oral phosphate treatment on serum parathyroid hormone (1–84) and biochemical markers of bone turnover: a dose response study, *Calcif. Tissue Int.*, 1992, **51**, 276–281.

54. M.S. Calvo, R. Kumar and H. Heath III, Elevated secretion and action of serum parathyroid hormone in young adults consuming high phosphorus, low calcium diets assembled from common foods, *J. Clin. Endocrinol. Metab.*, 1988, **66**, 823–829.

55. M.S. Calvo, R. Kumar and H. Heath III, Persistently elevated parathyroid hormone secretion and action in young women after four weeks of ingesting high phosphorus low calcium diets, *J. Clin. Endocrinol. Metab.*, 1990, **70**, 1334–1340.

56. M.S. Calvo, The effect of high phosphorus intake on calcium homeostasis, In *Advances in Nutritional Research*, H. Draper, New York, Plenum Press. 1994, pp. 183–207.

57. A.A. Portale, B.E. Booth, B.P. Halloran and R.C. Morris Jr., Effect of dietary phosphorus on circulating concentrations of 1,25-dihydroxyvitamin D and immunoreactive parathyroid hormone in children with moderate renal insufficiency, *J. Clin. Invest.*, 1984, **73**, 1580–1589.

58. B. Dawson-Hughes, S. Harris and G.E. Dallal, Serum ionized calcium, as well as phosphorus and parathyroid hormone, is associated with plasma 1,25-dihydroxyvitamin $D_3$ concentration in normal postmenopausal women, *J. Bone Miner. Res.*, 1991, **6**, 461–468.

59. J.C. Estepa, E. Aguilera-Tejero, I. Lopez, Y. Almaden, M. Rodriguez and A.J. Felsenfeld, Effect of phosphate on parathyroid hormone secretion *in vivo*, *J. Bone Miner. Res.*, 1999, **14**, 1848–1854.

60. Y. Almaden, A. Hernandez, V. Torregrosa, A. Canalejo, L. Sabate, L. Fernandez-Cruz, J.M. Campistol, A. Torres and M. Rodriguez, High phosphate level directly stimulates parathyroid hormone secretion and synthesis by human parathyroid tissue in vitro, *J. Am. Soc. Nephrol.*, 1998, **9**, 1845–1852.

61. A. Canalejo, A. Hernandez, Y. Almaden, A.T. Concepcion, A. Felsenfeld, A. Torres and M. Rodriguez, The effects of high phosphorus diet on parathyroid cell cycle, *Nephrol. Dialysis Transplant.*, 1998, **3**, 19–22.

62. H.M. Kronenberg, NPT2a - the key to phosphate homeostasis, *N. Engl. J. Med.*, 2002, **47**(13), 1022–1024.

63. H. Murer, N. Hernando, I. Forster and J. Biber, Molecular mechanisms in proximal tubular and small intestinal phosphate reabsorption (Plenary Lecture), *Mol. Membr. Biol.*, 2001, **18**, 3–11.

64. I.C. Forster, K. Köhler, J. Biber and H. Murer, Forging the link between structure and function of electrogenic cotransporters: the renal type IIa $Na^+/P_i$ cotransporter as a case study, *Progress Biophys. Mol. Biol.*, 2002, **80**, 69–108.

65. S. Shenolikar, J.W. Voltz, C.M. Minkoff, J.B. Wade and E.J. Weinman, Targeted disruption of the mouse NHERF-1 gene promotes internalization of proximal tubule sodium–phosphate cotransporter type IIa and renal phosphate wasting, *Proc. Natl. Acad. Sci. USA*, 2002, **99**(17), 11470–11475.

66. M.J. Mahon, M. Donowitz, C.C. Yun and G.V. Segre, $Na^+/H^+$ exchanger regulatory factor 2 directs parathyroid hormone 1 receptor signalling, *Nature*, 2002, **417**, 858–861.

67. S.J. Khundmiri, M.J. Rane and E.D. Lederer, Parathyroid hormone regulation of Type II sodium–phosphate cotransporters is dependent on an A kinase anchoring protein, *J. Biol. Chem.*, 2002, **278**(12), 10134–10141.

68. T.O. Carpenter, Oncologenic Osteomalacia - a complex dance of factors, *N. Engl. J. Med.*, 2003, **348**, 1705–1708.

69. K.B. Jonsson, R. Zahradnik and T. Larsson *et al.*, Fibroblast growth factor 23 in oncogenic osteomalacia and X-linked hypophosphatemia, *N. Engl. J. Med.*, 2003, **348**(17), 1656–1663.

70. K.E. White, G. Carn, B. Lorenz-Depiereux, A. Benet-Pages, T.M. Strom and M.J. Econs, Autosomal-dominant hypophosphatemic rickets (ADHR) mutations stabilize FGF-23, *Kidney Int.*, 2001, **60**, 2079–2086.

71. T. Shimada, S. Mizutani, T. Muto, T. Yoneya, R. Hino, S. Takeda, Y. Takeuchi, T. Fujita, S. Fukumoto and T. Yamashita, Cloning and characterization of FGF23 as a causative factor of tumor-induced osteomalacia, *Proc. Natl Acad. Sci., USA*, 2001, **98**(11), 6500–6505.

72. HYP-Consortium-1995. A gene (PEX) with homologies to endopeptidases is mutated in patients with X-linked hypophosphatemic rickets, *Nature Genet.*, 1995, **11**, 130–136.

73. H.S. Tenenhouse, X-linked hypophosphatemia: a homologous disorder in humans and mice, *Nephrol. Dialysis Transplant.*, 1999, **14**, 333–341.

74. L. Du, M. Desbarats, J. Viel *et al.*, cDNA cloning of the murine *Pex* gene implicated in X-linked hypophosphatemia and evidence for expression in bone, *Genomics*, 1996, **36**, 22–28.

75. A.F. Ruchon, M. Marcinkiewicz, G. Siegfried, H.S. Tenenhouse, L. DesGroseillers, P. Crine and G. Boileau, Pex mRNA is localized in developing mouse osteoblasts and odontoblasts, *J. Histochem. Cytochem.*, 1998, **46**, 459–468.

76. D. Miao, D. Panda, A. Karaplis and D. Goltzman, Localization of PHEX protein to osteocytes in adult mouse bone: implications for X-linked hypophosphatemia and osteoblast differentiation, *J. Bone Miner. Res.*, 1999, **14**, S475–S481.

77. T.O. Carpenter, New perspectives on the biology and treatment of X-linked hypophosphatemic rickets, *Pediatr. Clin. N. Am.*, 1997, **44**, 443–466

78. P.J. Malloy, J.W. Pike and D. Feldman, Hereditary 1,25-dihydroxyvitamin D resistant rickets, In *Vitamin D*, D. Feldman, F.H. Glorieux, J.W. Pike (eds.), San Diego, Academic Press, 1997, pp. 765–787.

79. T.M. Strom, F. Francis, B. Lorenz, A. Boddrich, M.J. Econs, H. Lehrach and T. Meitinger, Pex gene deletions in Gy and Hyp mice provide mouse models for X-linked hypophosphatemia, *Human Mol. Geneticist*, 1997, **6**, 165–171.

80. W.H. McAlister, G.S. Kin and M.P. Whyte, Tibial bowing exacerbated by partial premature epiphyseal closure in sex-linked hypophosphatemic rickets, *Radiology*, 1987, **162**, 461–463.

81. P.J. Marie and F.H. Glorieux, Histomorphometric study of bone remodeling in hypophosphatemic vitamin D-resistant rickets, *Metab. Bone Dis. Related Res.*, 1981, **3**, 31–38.

82. P.J. Marie and F.H. Glorieux, Bone histomorphometry in asymptomatic adult with hereditary hypophosphatemic vitamin D-resistant osteomalacia, *Metab. Bone Dis. Related Res.*, 1982, **4**, 249–253.

83. I.R. Reid, W.A. Murphy, D.C. Hardy *et al.*, X-linked hypophosphatemia: skeletal mass in adults assessed by histomorphometry, computed tomography, and absorptiometry, *Am. J. Med.*, 1991, **90**, 63–69.

84. W. Sullivan, T. Carpenter, F. Glorieux *et al.*, A prospective trial of phosphate and 1,25-dihydroxyvitamin D$_3$ therapy in symptomatic adults with X-linked hypophosphatemic rickets, *J. Clin. Endocrinol. Metab.*, 1992, **75**, 879–885.
85. A. Kovanlikaya, M.L. Loro, T.N. Hangarter, *et al.*, Osteopenia in Children: CT assessment, *Radiology*, 1996, **198**, 781–784.
86. K.S. Caldmeyer, R.R. Smith and M.K. Edwards-Brown, Familial hypophosphatemic rickets causing ocular calcification and optic canal narrowing, *Am. J. Nephrol. Res.*, 1995, **16**, 1252–1254.
87. I.R. Reid, D.C. Hardy and W.A. Murphy, X-linked hypophosphatemia: a clinical, biochemical and histopathologic assessment of morbidity in adults, *Medicine*, 1989, **68**, 336–352.
88. R.P. Polisson, S. Martinez, M. Khoury *et al.*, Calcification of entheses associated with X-linked hypophosphatemic osteomalacia, *N. Engl. J. Med.*, 1985, **313**, 1–6.
89. ADHR-consortium (2000). Autosomal dominant hypophosphataemic rickets is associated with mutations in FGF23. The ADHR Consortium, *Nature Genet.*, **26**, 345–348.
90. M.J. Econs, G.P. Samsa, M. Monger, *et al.*, X-linked hypophosphatemic rickets: a disease often unknown to affected patients, *Bone Miner.*, 1994, **24**, 17–24.
91. M. Tieder, D. Modai, R. Samuel, R. Arie, A. Halabe, I. Bab, D. Gabizon and U.A. Liberman, Hereditary hypophosphatemic rickets with hypercalciuria, *N. Engl. J. Med.*, 1985, **312**, 611–617.
92. C. Chen, T.O. Carpenter, N. Steg, R. Baron and C. Anast, Hypercalciuric hypophosphatemic rickets, mineral balance, bone histomorphometry, and therapeutic implications of hypercalciuria, *Pediatrics*, 1989, **84**, 276–280.
93. A. Jones, J. Tzenova, D. Frappier, M. Crumley, N. Roslin, C. Kos, M. Tieder, C. Langman W. Proesmans, T. Carpenter, A. Rice, D. Anderson, K. Morgan, T. Fujiwara and H. Tenenhouse, Hereditary hypophosphatemic rickets with hypercalciuria is not caused by mutations in the Na/Pi cotransporter NPT2 gene, *J. Am. Soc. Nephrol.*, 2001, **12**, 507–514.
94. L. van den Heuvel, K. Op de Koul, E. Knots, N. Knoers and L. Monnens, Autosomal recessive hypophosphataemic rickets with hypercalciuria is not caused by mutations in the type II renal sodium/phosphate cotransporter gene, *Nephrol. Dialysis Transplant.*, 2001, **16**, 48–51.
95. M. Reyes-Mugica, S.L. Arnsmeier, P.F. Backeljauw, J. Persing, B. Ellis and T.O. Carpenter, Phosphaturic mesenchymal tumor-induced rickets, *Pediatr. Devel. Pathol.*, 2000, **3**, 61–69.
96. M.K. Drezner, Tumor-induced rickets and osteomalacia, In *Primer on the Metabolic Bone Diseases and Disorders of Mineral Metabolism*, M. J. Favus (ed.), Lippincott-Raven, Philadelphia, 1999, pp. 331–337.
97. Z.S. Agus, Oncogenic hypophosphatemic osteomalacia, *Kidney Int.*, 1983, **24**, 113–123.
98. B. Eyskens, W. Proesmans, B. Van Damme, L. Lateur, R. Bouillon and M. Hoogmartens, Tumour-induced rickets: a case report and review of the literature, *Eur. J. Pediatr.*, 1995, **154**, 462–468.
99. D. Schapira, O. Ben Izhak, A. Nachtigal, A. Burstein, R.B. Shalom, I. Shagrawi and L.A. Best, Tumor-induced osteomalacia, *Seminar Arthritis Rheumatol.*, 1995, **25**, 35–46.
100. K.E. White, K.B. Jonsson, G. Carn, G. Hampson, T.D. Spector, M. Mannstadt, B. Lorenz-Depiereux, A. Miyauchi, I.M. Yang, O. Ljunggren, T. Meitinger, T.M. Strom, H. Juppner and M.J. Econs, The autosomal dominant hypopho-

sphatemic rickets (ADHR) gene is a secreted polypeptide overexpressed by tumors that cause phosphate wasting, *J. Clin. Endocrinol. Metab.*, 2001, **86**, 497–500.

101. A.E. Bowe, R. Finnegan, S.M. Jan de Beur, J. Cho, M.A. Levine, R. Kumar and S.C. Schiavi, FGF-23 inhibits renal tubular phosphate transport and is a PHEX substrate, *Biochem. Biophys. Res. Commun.*, 2001, **284**, 977–981.

102. S.M. Jan De Beur, J.R. O'Connell, R. Peila, J. Cho, Z. Deng, S. Kam and M.A. Levine, The pseudohypoparathyroidism type 1b locus is linked to a region including GNAS1 at 20q13.3, *J. Bone Miner. Res.*, 2003, **18**, 424–433.

103. D.T. Baran and T.W. Marcy, Evidence for a defect in vitamin D metabolism in a patient with incomplete Fanconi syndrome, *J. Clin. Endocrinol. Metab.*, 1984, **59**(5), 998–1001.

104. M. Logman-Adham, Adaptation to changes in dietary phosphorus intake in health and in renal failure, *J. Lab. Clin. Med.*, 1997, **129**, 176–188.

105. F. Llach and S.G. Massry, On the mechanism of secondary hyperparathyroidism in moderate renal insufficiency, *J. Clin. Endocrinol. Metab.*, 1985, **61**, 601–606.

106. K.J. Martin and E.A. Gonzalez, Strategies to minimize bone disease in renal failure, *Am. J. Kidney Dis.*, 2001, **38**, 1430–1436.

107. F. Llach and J. Bover, Renal Osteodystrophy, In *The Kidney, volume II*, B.M. Brenner (ed.), Philadelphia, W.B. Saunders Co., 1996, 2187–2273.

108. R.M. Edwards, Disorders of phosphate metabolism in chronic renal disease, *Curr. Opin. Pharmacol.*, 2002, **2**, 171–176.

109. V. Jevic, Imaging of renal osteodystrophy, *Eur. J. Radiol*, 2003, **46**, 85–95.

110. G.B. Spasovski, A.R.J. Bervoets, G.J.S. Behets, N. Ivafnovski, A. Sikole, G. Dams, M-M. Couttenye, M.E. Broe and P.C. D'Haese, Spectrum of renal bone disease in end-stage renal failure patients not yet on dialysis, *Nephrol. Dialysis Transplant.*, 2003, **18**, 1159–1166.

111. B.D. Kuizon and I.B. Salusky, Cell biology of renal osteodystrophy, *Pediatr. Nephrol.*, 2002, **17**, 777–789.

112. M. Cozzolino, A.S. Dusso and E. Slatopolsky, Role of calcium-phosphate product and bone-associated proteins on vascular calcification in renal failure, *J. Am. Soc. Nephrol.*, 2001 **12**, 2511–2516.

113. R. Narang, D. Ridout, C. Nonis and J.S. Kooner, Serum calcium, phosphorus and albumin levels in relation to angiographic severity of coronary artery disease, *Int. J. Cardiol.*, 1997, **60**, 73–79.

114. S. Jono, M.D. McKee, C.E. Murry, A. Shioi, Y. Nishizawa, K. Mori, H. Morii and C.M. Giachelli, Phosphate regulation of vascular smooth muscle cell calcification, *Circulation Res.*, 2000, **87**, e10–e17.

115. G.A. Block, S.T. Hulbert, N.W. Levin and F.K. Port, Association of serum phosphorus and calcium x phosphate product with mortality risk in chronic hemodialysis patients: a national study, *Am. Kidney Dis.*, 1998, **31**, 607–617.

116. T.B. Drüeke and S.G. Rostand, Progression of vascular calcification in uremic patients can it be stopped? *Nephrol. Dialysis Transplant.*, 2002, **17**,1365–1368.

117. S. Lopez-Hilker, A. Dusso, N. Rapp, K.J. Martin and E. Slatopolsky, Phosphorus restriction reverses hyperparathyroidism in uremia independent of changes in calcium and calcitriol, *Am. J. Physiol.*, 1990, **259**, F432–F437.

118. E. Slatopolsky, J. Finch, M. Denda, C. Ritter, M. Zhong, A. Dusso, P.N. Mac Donald and A.J. Brown, Phosphorus restriction prevents parathyroid gland growth. High phosphorus directly stimulates PTH secretion in vitro, *J. Clin. Investig.*, 1996, **97**, 2534–2540.

119. C.S. Ritter, D.R. Martin, Y. Lu, E. Slatopolsky and A.J. Brown, Reversal of secondary hyperparathyroidism by phosphate restriction restores parathyroid calcium-sensing receptor expression and function, *J. Bone Miner. Res.*, 2002, **17**, 2206–2213.

120. M. Boaz and S. Smetana, Regression equation predicts dietary phosphorus intake from estimate of dietary protein intake, *J. Am. Diet. Assoc.*, 1996, **96**, 1268–1270.

121. A. Cupisti, E. Morelli, C. D'Alessandro, S. Lupetti and G. Barsotti, Phosphate control in chronic uremia: don't forget diet, *J. Nephrol.*, 2003, **16**, 29–33.

122. M. Rufino, E. deBonis, M. Martin, S. Rebello, B. Martin, R. Miguel, M. Cobo, D. Hermandez, A. Torres and V. Lorenzo, Is it possible to control hyperphosphatemia with diet, without inducing protein malnutrition? *Nephrol. Dialysis Transplant.*, 1998, **13**, 65–67.

123. J.M. Pettifor, P.J. Marie, M.R. Sly, D.B. du Bruyn, F. Ross, J.M. Isdale, W. De Klerk and W.H. van der Walt, The effects of differing dietary calcium and phosphorus contents on mineral metabolism and bone histomorphometry in young vitamin D-replete baboons, *Calcif. Tissue Int.*, 1984, **36**, 668–676.

124. C.E. Neville, P.J. Robson, L.J. Murray, J.J. Strain, J. Twisk, A.M. Gallagher, M. McGuinnens, G.W. Cran, S.H. Ralston and C.A.G. Borehass, The effects of nutrient intake on bone mineral status in young adults, *Calcif. Tissue Int.*, 2002, **70**, 89–98.

125. D. Teegarden, R.M. Lyle, G.P. McCabe, L.D. McCabe, W.R. Proulx, K. Michon, A.P. Knight, C.C. Johnston and C.M. Weaver, Dietary calcium, protein and phosphorus are related to bone mineral density and content in young women, *Am. J. Clin. Nutr.*, 1998, **68**, 749–754.

126. S.A. New, S.P. Robins, M.K. Campbell, J.C. Martin, M.J. Garton, C. Bolton-Smith, D.A. Grubb, S.J. Lee and D.M. Reid, Dietary influences on bone mass and bone metabolism: further evidence of a positive link between fruit and vegetable consumption and bone health, *Am. J. Clin. Nutr.*, 2000, **71**, 142–151.

127. S.J. Whiting, J.L. Boyle, A. Thompson, R.L. Mirwald and R.A Faulkner, Dietary protein, phosphorus and potassium are beneficial to bone mineral density in adult men consuming adequate dietary calcium, *J. Am. College Nutr.*, 2002, **21**, 402–409.

128. K. Yano, L.K. Heilbrun, R.D. Wasnich, J.D. Hankin and J.M. Voegel, The relationhip between diet and bone mineral content of multiple skeletal sites in elderly Japanese-American men and women living in Hawaii, *Am. J. Clin. Nutr.*, 1985, **42**, 877–888.

129. B.P. Lukert, M. Carey, B. McCarty, S. Tiemann, L. Goodnight, M. Helm, R. Hassanein, C. Stevenson, M. Stofkopf and L. Doolan, Influence of nutritional factors on calcium-regulating hormones and bone loss, *Calcif. Tissue Int.*, 1987, **40**, 119–125.

130. F.A. Tylasky and J.J.B. Anderson, Dietary factors in bone health of elderly lacto-ovovegetarian and omnivorous women, *Am. J. Clin. Nutr.*, 1988, **48**, 842–9.

131. J.A. Metz, J.J.B. Anderson and P.N. Gallagher, Intake of calcium, phosphorus, and protein, and physical activity level are related to radical bone mass in young adult women, *Am. J. Clin. Nutr.*, 1993, **58**,537–542.

132. J.P. Bonjour, P. Ammann, T. Chevally and R. Rizzoli, Protein intake and bone growth, *Can. J. Appl. Physiol.*, 2001, **26**, S153–S166.

133. B. Dawson-Hughes and S. Harris, Calcium intake influences the association of protein intake with rates of bone loss in elderly men and women, *Am. J. Clin. Nutr.*, 2002, **75**, 773–779.

134. D.A. Smith and B.E.C. Nordin, The effect of high phosphorus intake on total and ultrafiltrable plasma calcium and phosphorus clearance, *Clin. Sci.*, 1964, **26**, 479–486.

135. S.J. Silverberg, E. Shane, T.L. Clemens, D.W. Dempster, G.V. Segre, R. Lindsay and J.P. Bilezikian, The effects of oral phosphate administration on major indices of skeletal metabolism in normal subjects, *J. Bone Miner. Res.*, 1986, **1**, 383–388.

136. A.E. Broadus, J.S. Magee, L.E. Mallette, R.L. Horst, L.R. Lang, P.S. Jensen, J.M. Gertner and R. Baron, A detailed evaluation of oral phosphate therapy in selected patients with primary hyperparathyroidism, *J. Clin. Endocrinol. Metab.*, 1983, **56**, 953–961.

137. A.A.Portale, B.P. Halloran, M.M. Murphy and R.C. Morris Jr., Oral intake of phosphorus can determine the serum concentration of 1,25dihydroxyvitamin D by determining its production rate in humans, *J. Clin. Investig.*, 1986, **77**, 7–12.

138. B.K. Bizik, W. Ding and F.L. Cerklewski, Evidence that bone resorption of young men is not increased by high dietary phosphorus obtained from milk and cheese, *Nutr. Res.*, 1996, **16**, 1143–1146.

139. A. Whybro, H. Jager, M. Barker and R. Eastell, Phosphate supplementation in young men: lack of effect on calcium homeostasis and bone turnover, *Eur. J. Clin. Nutr.*, 1998, **52**, 29–33.

140. M.J. Barger-Lux and R.P. Heaney, Effects of calcium restriction on metabolic characteristics of premenopausal women, *J. Clin. Endocrinol. Metab.*, 1993, **76**, 103–107.

141. J.A. Clowers, R.A. Hannon, T.S. Yap, N.R. Hoyle, A. Blumsohn and R. Eastell, Effect of feeding on bone turnover markers and its impact on biological variability of measurements, *Bone*, 2002, **30**, 886–890.

142. M.S. Calvo, D.R. Eyre and C.M. Gundberg, Molecular basis and clinical application of biological markers of bone turnover, *Endocrine Rev.*, 1996, **17**, 333–368.

143. F.R. Bringhurst, M.B. Demay and H.M. Kronenberg, Mineral metabolism: hormones and disorders of mineral metabolism, In *Williams Textbook of Endocrinology*, J.D. Wilson, D.W. Foster, H.M. Kronenberg and P.R. Larsen (eds.), Philadelphia, W.B. Saunders Co., 1998, pp. 1155–1209.

CHAPTER 12

# Sodium Effects on Bone and Calcium Metabolism

KEVIN D. CASHMAN [1,2] and ALBERT FLYNN[1]

[1]Department of Food and Nutritional Sciences, University College, Cork, Ireland, Email: k.cashman@ucc.ie
[2]Department of Medicine, University College, Cork, Ireland

**Key Points**

- Sodium-induced calciuria has been clearly demonstrated in many studies and the magnitude of this effect within the usual range of sodium intakes is significant in relation to overall calcium metabolism.
- The available evidence suggests that healthy individuals adapt to sodium-induced calciuria by a parathyroid hormone-mediated increase in intestinal calcium absorption.
- The intestinal adaptation does not appear to function in all individuals (*e.g.*, those with impaired parathyroid function, postmenopausal women with osteoporosis, as well as some healthy postmenopausal women) and even in those individuals who appear to adapt, the increase in net calcium absorption may not be sufficient to offset the increase in urinary calcium losses.
- The limited numbers of epidemiological studies, which have investigated the association of dietary or urinary sodium with bone mass and/or bone turnover have produced conflicting results.
- There is evidence in short-term studies of an association between high-salt intakes and increased bone resorption in postmenopausal women. However, whether this increased rate of bone resorption persists in the long-term and, more importantly, results in net bone loss has not been clearly established.
- There are still many unresolved questions regarding the impact of dietary sodium on bone health and further research is required.

# Introduction

Dietary sodium intake has long been known to influence urinary calcium excretion.[1,2] However, up to the early 1980s sodium was not included among factors believed to influence calcium requirements or the pathogenesis of osteoporosis.[3,4] In the past decade or so there has been a renewal of interest in the effect of sodium on calcium metabolism and the possible contribution of high dietary sodium intake to the development of osteoporosis.[5-9]

This article critically reviews the evidence for a possible aetiological role for dietary sodium in the development of osteoporosis through sodium–calcium interrelationships. This review will focus primarily on data from human studies, since studies on experimental animals (*e.g.*, rats) must be interpreted carefully, especially when extrapolating the results to humans.[5,6] Evidence from animal studies has been reviewed elsewhere.[5,6]

# Sodium Intake

Mean daily sodium intakes in adults in Western countries usually exceed 100 mmol (2.3 g), based on data from studies, which have assessed urinary sodium excretion (see Table 12.1), which is the preferred method for estimation of sodium intake. However, a large range in sodium excretion (0.9–246 mmol sodium excreted per day), and thus intake, has been reported.[10,11]

Since greater than 95% of dietary sodium is excreted in urine under normal conditions,[12] 24 h urinary sodium excretion is generally regarded as the best measure of total dietary sodium intake.[6,13] However, it should be noted that while single 24 h urinary values are adequate to estimate mean sodium intake of groups, multiple 24 h urine collections are needed to classify accurately individuals with respect to sodium intake because of the large day-to-day variation in salt intakes.[14,15]

It is important to note that habitual salt intake cannot be accurately determined using usual methods of dietary assessment such as dietary records, recalls, or food frequency questionnaires.[6] These assessment methods rarely account for salt added during cooking or at the table, while processed foods contain varying amounts of sodium for foods of, otherwise, similar nutrient content.[6] Despite these well-known drawbacks, some studies have still used such methods in examining the relationship between sodium intake and bone status (see section *Effect of sodium on bone metabolism and mass* below).

# Physiological Basis for Sodium–Calcium Interactions

Calciuria, sometimes referred to as hypercalciuria, has been defined as a 24 h urinary calcium excretion greater than 0.1 mmol $kg^{-1}$ body weight.[2] This condition is associated with a number of factors, including dietary factors such as sodium, protein and caffeine; drugs such as diuretics, antacids and antibiotics, and certain disease states such as thyrotoxicosis, primary

**Table 12.1.** *Estimates of average urinary sodium excretion (and thus sodium intakes) by adults in selected countries*

| Country[a] | Urinary sodium (mmol per 24 h)[b] |
|---|---|
| Argentina | 155.8 |
| Belgium | 144.4 |
| Brazil | 6.6 |
| Canada | 175.6 |
| Colombia | 201.4 |
| Denmark | 140.2 |
| Germany | 158.6 |
| Finland | 162.6 |
| Hungry | 198.3 |
| Iceland | 138.3 |
| India | 182.1 |
| Ireland[c] | 150.0 |
| Italy | 170.9 |
| Japan | 196.4 |
| Kenya | 56.8 |
| Malta | 169.8 |
| Mexico | 144.1 |
| The Netherlands | 150.6 |
| Papua New Guinea | 36.8 |
| People's Republic of China | 207.4 |
| Poland | 189.5 |
| Portugal | 181.9 |
| South Korea | 208.2 |
| Soviet Union | 161.7 |
| Spain | 178.9 |
| Taiwan | 141.4 |
| Trinidad and Tobago | 117.4 |
| UK | 152.7 |
| US | 141.8 |
| Zimbabwe | 140.5 |

[a]Adapted from Ref. [10].
[b]Intersalt study does not provide nationally representative data. Studies were carried out, where possible, in randomly selected subgroups. In some cases, *e.g.*, Brazil, data includes exceptional subgroups (Yanomamo Indians). Urinary sodium excretion values represent mean values calculated for a number of centers/regions in which measurements were made in men aged 20–39 years, men aged 40–59 years, women aged 20–39 years, and women aged 40–59 years and then averaged over age and sex groups.
[c]Data from Ref. [11].

hyperparathyroidism, calcium stone formers, and secondary cancer in bone.[5] Sodium-induced calciuria is believed to result from sodium–calcium interactions in the kidney.[16] Since there is no evidence for tubular secretion of either calcium or sodium in the nephron,[17] the primary determinant of renal excretion is glomerular filtration of the ions. The major proportion of both sodium and calcium in plasma appears to be filtered and urinary excretion of sodium and calcium is controlled by the extent of reabsorption of these ions from the nephron.[5] Normally, greater than 99% of filtered sodium and greater than 95% of filtered calcium is reabsorbed.[5] In quantitative terms, the location of the reabsorption is similar for both ions, namely 40–45, 43–45, 6–9 and 5%, for reabsorption at the proximal convoluted tubule, the loop of Henle, distal convoluted tubule, and collecting ducts, respectively, in the rat nephron.[18]

The dependence of urinary calcium excretion on urinary sodium excretion has been attributed to the existence of linked or common re-absorption pathways for both ions in the convoluted portion of the proximal tubule and thick ascending loop of Henle.[19,20] In contrast, there is no evidence for a similar relationship in the distal tubule or collecting ducts.[19]

There is a large variability in sodium-induced calciuria between individuals. For example, while some subjects show a significant response in terms of sodium-induced calciuria (sometimes referred to as 'sodium responsive' subjects), others show little or no response ('sodium-non-responsive').[5,15,21–23] The reasons for the lack of, or at least much reduced, dependency of urinary calcium excretion on urinary sodium excretion in the sodium-non-sensitive subjects are unclear.[23] It is tempting to suggest that such variation among individuals may be due, at least in part, to underlying genetic differences, predisposing certain groups of individuals to increased salt sensitivity. However, to date, this has received very little attention. Interestingly, Chen et al.[24] recently reported that a polymorphism in the vitamin D receptor (VDR) gene was associated with risk of calcium oxalate stone disease, a condition for which a high salt diet is a modifiable risk factor, suggesting a role for this genetic marker in renal calcium handling. Recently, the VDR gene polymorphism has also been reported to interact with high caffeine intake, which is another dietary calciuric factor, in determining the rate of bone loss in postmenopausal women.[25]

# Urinary Sodium and Calcium Excretion in Studies with Controlled Sodium Intake and Population Studies

Nordin et al.[16] proposed that for every 100 mmol of sodium excreted in urine there is approximately 1 mmol loss of urinary calcium in free-living, normocalciuric healthy populations. There is considerable evidence supporting a strong positive association between urinary sodium and calcium excretion in both young and adult free-living individuals of both sexes in different countries consuming their usual diet (see Table 12.2). Similarly, salt-loading studies show a strong positive correlation between urinary calcium and sodium excretion for adult men and women (see Table 12.3).

## Studies on Free-Living Iindividuals

Results from studies on free-living individuals suggest that the average increase in urinary calcium per 100 mmol increment in urinary sodium in healthy

**Table 12.2.** *Estimates of the incremental increase in urinary calcium excretion observed with 100 mmol increment in urinary sodium excretion in free-living individuals consuming their usual diets* [a]

| Subject characteristics (No. and gender) | Δ Urinary calcium/urinary sodium ratio (mmol per 100 mmol) | Reference |
|---|---|---|
| 46 M/F (Caucasian) | 2.4 | 26 |
| 50 M/F (Bantu) | 0.3 | 26 |
| 34 M/F | 1.3 | 83 |
| 25 M | 1.6 | 30 |
| 26 F | 2.6 | 30 |
| 38 F/M | 1.5 | 35 |
| 130 F/M | 0.7 | 84 |
| 484 M | 1.5 | 40 |
| 491 F | 1.3 | 40 |
| 467 F (Postmenopausal) | 1.3 | 29 |
| 58 F | 1.1 | 85 |
| 108 M | 1.0 | 85 |
| 46 M | 2.3 | 15 |
| 48 F | 1.5 | 15 |
| 2112 M | 2.8 | 86 |
| 1943 F | 3.0 | 86 |
| 39 M (Elderly) | 1.0 | 87 |
| 44 F (Postmenopausal) | 1.0 | 87 |
| 334 F | 0.9 | 55 |
| 170 F (Premenopausal) | 1.0 | 16 |
| 283 F (Postmenopausal) | 1.2 | 16 |
| 124 F (Postmenopausal) | 3.0 | 66 |

M, adult males and F, adult females.
[a]Adapted from Refs. [5,6].

individuals consuming their usual diet is in the range 0.2–3.0 mmol (78–120 mg) (Table 12.2), indicating considerable variation between groups examined. An example of such variation between ethnic groups is seen in the study of Modlin[26] in which Caucasian South Africans were found to excrete in urine, on average, 96 mg (2.4 mmol) calcium per 100 mmol increase in urinary sodium while the Bantu (a native South African tribes people) excreted only 12 mg (0.3 mmol). Such variation among ethnic groups may be due to underlying genetic differences, predisposing certain groups of individuals to increased salt sensitivity, but these variations could also be due to dietary differences.[5] Interestingly, Ojwang et al.[27] recently reported that South African Blacks had a significantly higher frequency of one of the VDR genotypes (namely, the TT *Taq* I VDR genotype group) compared to South African Caucasians, which may be associated with the lower incidence of osteoporosis in the former group. Whether this genetic factor interacts with sodium intake in determining bone mass in these population groups remains to be clarified.

One limiting factor in many of the studies carried out to date in relation to the calciuric response of free-living individuals to dietary sodium is the lack of

**Table 12.3.** *Estimates of the incremental increase in urinary calcium observed in healthy individuals with a direct increase in urinary sodium excretion of 100 mmol per day during controlled dietary studies*[a]

| Subject characteristics (No. and gender) | Δ Urinary calcium/urinary sodium ratio (mmol per 100 mmol) | Reference |
|---|---|---|
| 20 M | 2.0 | 88 |
| 2 M/F | 0.6 | 22 |
| 6 M | 0.7 | 20 |
| 11 M/F | 0.6 | 34 |
| 38 F/M | 1.5 | 35 |
| 12 M | 0.5 | 39 |
| 7 F (Osteoporotic) | 0.5 | 34 |
| 6 F | 2.1 | 40 |
| 5 M (Hypertensive) | 0.5 | 44 |
| 6 M | 0.9 | 15 |
| 6 F | 0.7 | 15 |
| 4 M | 1.5 | 89 |
| 10 F (Postmenopausal) | 0.6 | 36 |
| 17 F (Postmenopausal) | 1.0 | 90 |
| 8 M | 0.7 | 91 |
| 11 F | 0.9 | 48 |
| 7 F | 0.7 | 55 |
| 30 F (Postmenopausal) | 2.3 | 16 |
| 14 F (Postmenopausal) | 0.5 | 38 |
| 11 F (Premenopausal) | 0.5 | 37 |
| 11 F (Postmenopausal) | 0.7 | 37 |
| 29 F | 0.7 | 23 |
| 52 F (Postmenopausal) | 0.8 | 42 |

M, adult males and F, adult females.
[a]Adapted from Refs. [5,6].

control for factors known to influence calciuria which could potentially confound this relationship.[5,6] These include factors such as dietary calcium, protein, sulfur-containing amino acids, caffeine, potassium, phosphorus, and acid–base status,[6] as well as possible calciuric effects of non-dietary factors, such as diuretics, antacids or antibiotics.[28] The importance of controlling for such confounding variables is highlighted by the findings of Nordin and Polley[29] who found that a significant positive association between urinary sodium and calcium existed for a group of postmenopausal women ($n = 445$), but no significant association was observed for those individuals ($n = 174$) who consumed diets containing more than 1000 mg of calcium per day. Madden et al.[30] who observed a significant positive relationship between 24 h urinary excretion of sodium and calcium in 51 young adults consuming their usual diet, found that 24 h urinary calcium and sodium excretion were not significantly correlated with dietary calcium, protein, phosphorus or fibre intakes.

The median reported increase in urinary calcium excretion per 100 mmol increment in urinary sodium in these free-living populations was 1.3 mmol (Table 12.2).

## Studies with Controlled Sodium Intake

Table 12.3 lists and summarizes studies where salt intakes of healthy (or osteoporotic) adults were manipulated and urinary sodium and calcium excretion measured. The median reported increase in urinary calcium per 100 mmol increment in urinary sodium in these salt-loading studies was 0.7 mmol, lower than that in free-living populations (1.3 mmol). This discrepancy between the ratios may reflect a non-linear relationship over the ranges of sodium chloride intakes fed in loading studies, especially when the dietary intakes were extremely low or high.[6] In some metabolic studies, extremes of sodium intake were studied. For example, McCarron *et al.*[20] varied the dietary sodium intake of six adult males, who were consuming a fixed diet of 400 mg calcium per day, from 10 to 1500 mmol sodium per day. They found that mean urinary calcium excretion increased with progressive sodium loading and attained an average maximum of 262 mg per day (6.5 mmol per day) at 1500 mmol sodium per day. Within the normal physiological range of sodium intake (10–300 mmol per day) a linear relationship was observed between urinary sodium and calcium excretion, with urinary calcium excretion increased by 28 mg per day (0.7 mmol per day), on average, for each 100 mmol increment in dietary sodium intake. However, a non-linear relationship was observed over the entire range of sodium intakes examined (10–1500 mmol per day). Therefore, this study defined the limits of calcium excretion at extremes of sodium intake and indicated that the linear relationship between urinary sodium and calcium excretion, which is evident within the normal physiological range of sodium intakes, is not maintained when sodium intake is increased beyond the normal physiological limits. In addition, the urinary sodium excretions of some individuals partaking in these studies were likely to have been different from the habitual intake, so their urinary sodium–calcium relationships may not reflect the same relationship as individuals who had been adapted to that intake.[6] On that point, few researchers have considered it important to allow adequate time for adaptation to the change in dietary salt intake. While changes in sodium intake may not be fully reflected in urinary sodium for up to 5 days, with an average lag of 3–5 days,[31] many studies have been of very short duration (3–5 days).[5] Furthermore, dietary calcium has been changed from the usual intake in many salt-loading studies, again without consideration of whether adequate time for adaptation to this dietary change has been allowed. For example, Dawson-Hughes *et al.*[32] showed that when calcium intake was reduced from 50 to 7.5 mmol per day fractional calcium retention and serum parathyroid hormone (PTH) and 1,25 dihydroxyvitamin $D_3$ increased, urinary calcium excretion decreased, and all these parameters had reached a plateau by the second week, but not by the first week, after the diet was introduced.

There is some evidence that the magnitude of the sodium-induced calciuria is influenced by gender. For example, Dawson-Hughes *et al.*[33] reported that urinary sodium and calcium excretion were significantly correlated in elderly men ($r = 0.42$, $P = 0.0001$) and women ($r = 0.26$, $P = 0.0001$), but the regression coefficient was greater in men than in women (beta = 0.017 vs. 0.010, $P = 0.007$). Therefore, for a given sodium excretion, elderly men excrete more calcium than women.

A number of studies have investigated the influence of dietary calcium intake on the magnitude of the sodium-induced calciuria. Sodium-induced increases in urinary calcium excretion have been demonstrated in individuals with daily calcium intakes of 400 mg,[34,35] 500 mg,[23] 700–900 mg[15,22,36–38] and in excess of 1000 mg.[21,29,39,40] Castenmiller et al.[39] varied the level of dietary calcium intake (2.9 vs. 3.5 mmol/MJ per day), and the increases in urinary calcium loss per 100 mmol increment in urinary sodium were not markedly different, 0.30 vs. 0.33 mmol, respectively. However, the two dietary calcium levels were consumed by two different groups of six subjects. Goulding et al.[40] found that salt-induced increases in 24 h urinary calcium to creatinine ratios were similar in nine young women after consuming either 200 or 1500 mg of calcium per day for 4 days. Dawson-Hughes et al.[33] reported that in healthy elderly men and women, urinary sodium and calcium excretion are linked at moderate and high dietary calcium intakes, but not at low calcium intakes. Massey and Whiting[6] after reviewing the available data on the impact of dietary calcium, concluded that at this time, there is no evidence that the level of dietary calcium influences the magnitude of the salt-induced hypercalciuric response.

The magnitude of the calciuria elicited by increases in dietary sodium intake is at least comparable, but most likely, greater than the calciuric effects of other dietary constituents, such as protein and caffeine. For example, Schaafsma et al.[41] calculated that increasing the protein content of the diet from normal sources by 20 g per day results in an increase in urinary calcium of 20 mg per day and that an increase of 150 mg caffeine per day (equivalent to two cups of coffee) increases urinary calcium by 5 mg per day. In comparison, studies on either free-living subjects consuming their usual diet (Table 12.2) or healthy subjects participating in salt-loading studies (Table 12.3) indicate that an increase in dietary sodium intake of 100 mmol results in an average increase in urinary calcium excretion of 12–107 mg (0.3–2.7 mmol)/100 mmol increment in urinary sodium. Considering the reported range of sodium intake, as estimated from urinary sodium excretion, in human populations (0.9–246 mmol sodium per day[10,11]), the possible influence of dietary sodium intake on urinary calcium loss may be considerable.[5] To place this calcium loss in the context of bone health, a net deficit of only 1 mmol per day of calcium would result in losing one third of the calcium contained in the typical adult skeleton in just over two decades unless a compensatory increase in intestinal calcium absorption occurred[42]. This issue will be dealt with in later sections of this article (see sections *The effect of salt-induced calciuria on calcium metabolism* and *Effect of sodium on bone metabolism and mass* below).

# Calciuria – a Consequence of Sodium or Salt?

In much of the literature dealing with the relationship between sodium and calcium excretion in humans and the possible adverse effect of sodium on calcium balance and bone health, the words 'sodium' and 'salt' have been used interchangeably. While, in general, this should not lead to confusion, as most of the sodium consumed in our diet is as sodium chloride,[43] it should be clarified at this point that not all 'salts' or indeed 'sodiums' are equal in terms of their effect on urinary calcium loss. For example, unlike when sodium is consumed as a salt with chloride, sodium consumed as a salt with a metabolizable anion, such as bicarbonate or citrate appears to exert no adverse effect on calcium excretion.[6] Kurtz *et al.*[44] showed that ingestion of 240 mmol of sodium chloride per day increased urinary calcium excretion in salt-sensitive individuals but an equimolar amount of sodium citrate had no effect on urinary calcium excretion. Likewise, sodium bicarbonate does not increase urinary calcium when compared to sodium chloride.[45] The resulting fall in net acid excretion with bicarbonate or citrate as the anion may promote calcium retention and thus negate the sodium-induced calcium loss.[6]

Just as sodium not accompanied by chloride has no effect on urinary calcium,[44-46] chloride not accompanied by sodium, *e.g.*, potassium chloride, has no effect on urinary calcium excretion.[47,48] It is unclear whether chloride exerts an effect independent of sodium or whether chloride is as important as sodium in the development of hypercalciuria. In the remainder of this article, the use of the words 'sodium' and 'salt' will refer specifically to sodium chloride, unless specified otherwise.

# The Effect of Salt-Induced Calciuria on Calcium Metabolism

It has been suggested that the calciuria arising from increased sodium intake temporarily depresses serum ionized calcium concentration which, in turn, stimulates the release of PTH which may help restore plasma calcium, *via* increased 1,25 dihydroxyvitamin $D_3$, increasing intestinal absorption of calcium.[5] There is some evidence to support this suggested adaptive increase in calcium absorption arising from sodium-induced calciuria. For example, Breslau *et al.*[34] showed that supplementation for 10 days with 240 mmol per day of sodium chloride (from 10 to 250 mmol per day) increased the fractional intestinal absorption of $^{47}Ca$ (by 26%, on average) in eleven healthy young adults (six males and five females). Meyer *et al.*[22] also found that supplementation for 12 days with sodium chloride (from 10 to 250 mmol per day) increased (by 8% on average) the fractional intestinal absorption of $^{47}Ca$ in ten subjects. On the other hand, Breslau *et al.*[46] found no increase in intestinal calcium absorption in seven osteoporotic postmenopausal women when sodium intake was increased from 10 to 250 mmol per day while dietary calcium was maintained at 10 mmol per day. Similarly, both McParland *et al.*[36] and Evans *et al.*[37] reported a lack of effect of increasing dietary sodium on strontium absorption (an index of calcium absorption) in postmenopausal women, while Evans *et al.*[37] also found no effect of increasing dietary sodium on strontium absorption in premenopausal women. However, the precision of this method may have been too low to detect a change. Breslau *et al.*[34] showed that two patients with post-surgical hypoparathyroidism did not show an increase in calcium absorption when they were supplemented for 10 days with 240 mmol per day. The finding that calcium absorption increased in normal individuals given sodium chloride supplements, but not in subjects with hypoparathyrodism, would support a PTH-mediated mechanism.[6] However, the role of PTH in this adaptive process is debated. Serum PTH has been found to be significantly increased in association with sodium-induced calciuria when the level of sodium supplementation was within the normal physiological range (10–300 mmol per day) of sodium intakes.[20,34,37,49,50] However, despite similarity in study design (*i.e.*, subjects and sodium loads), the observed increment in serum PTH with the imposition of the high sodium diets differed considerably among studies.[20,34,49] Furthermore, yet other studies reported no effect of sodium chloride supplementation on serum PTH.[23,37,38,42] Interpretation of studies which examine PTH changes are made difficult by differences in sensitivity of the radioimmunoassays used, the pulsatile nature of serum PTH concentrations, the half-life of intact PTH (*i.e.*, 4 min) and the possible haemodilution effect of sodium loading.[6,51–54] Due to these difficulties with the measurement of PTH, some studies have measured urinary cyclic adenosine monophosphate (cAMP), a marker of PTH action on the kidney, instead. Urinary cAMP was elevated after salt-loading,[20,34,36,55] although not in all studies.[23,42]

While there is little doubt that increasing salt intake increases urinary calcium excretion, this phenomenon is of little significance if there is complete adaptation, *i.e.*, if net calcium absorption is increased by an amount sufficient to offset the increased urinary loss.[5] If, on the other hand, adaptation is not complete, the sodium-induced calciuria could lead to a negative calcium balance with implications for bone health. However, the effect of salt on calcium balance has received relatively little attention to date. While Goulding *et al.*[40] have suggested that the daily calcium requirement to maintain calcium balance is higher in individuals consuming a high sodium diet than in individuals consuming a low-sodium diet this has not been established in balance studies. Fujita *et al.*[56] found that, despite a marked increase in urinary calcium excretion, calcium balance was not significantly affected in eight healthy subjects in response to oral sodium chloride supplementation (100 mmol per day) together with frusemide (a diuretic agent) administration (80 mg per day) for a 8 day period. They observed significant increases in mean serum PTH and mean urinary cAMP excretion and a significant decrease in faecal calcium excretion (365 *vs.* 294 mg calcium per day). These results suggest that there was compensation for calciuria by increased calcium absorption mediated through serum PTH but that reduced endogenous calcium loss could also have contributed.[5] However, this latter effect seems not, as of yet, to have been investigated. This overall adaptive mechanism does not appear to function in all individuals (*e.g.*, those with impaired parathyroid function, postmenopausal women with osteoporosis, as well as some healthy postmenopausal women) and even in those individuals who appear to adapt, the increase in net calcium absorption may not be sufficient to offset the increase in urinary calcium losses.[5,42] Furthermore, the capacity for such adaptation may be limited by low dietary calcium intakes, poor vitamin D status, impaired renal function or poor intestinal calcium absorption.[5]

Besides acting on the kidney (*i.e.*, increasing renal calcium reabsorption) and intestine (*via* increased renal 1-α hydroxylase activity resulting in increased 1,25-dihydroxyvitamin $D_3$ synthesis), PTH may also act on bone to restore serum ionized calcium to normal levels,[40] especially when adaptation at the level of the intestine is not complete. The remainder of the article will examine the evidence for an effect of sodium on bone metabolism and status.

# Effect of Sodium on Bone Metabolism and Mass

## Studies in Animals

There is evidence that very high sodium intakes reduce bone mass and bone calcium content in rats (see reviews by Shortt and Flynn[5] and Massey and Whiting[6]). Numerous reports indicate that salt supplementation (typically 80 g $kg^{-1}$ diet) significantly increases urinary calcium excretion. There is also evidence that urinary hydroxyproline and pyridinium crosslinks, markers of bone resorption, are significantly increased and bone mass and bone calcium and phosphorus content and concentration are reduced in weanling, young adult or ovariectomized albino rats (usually of Sprague-Dawley or Wistar strain) given a salt-supplemented diet.[5,6,57] These effects are believed to be secondary to sodium-induced calciuria which is not compensated by increased calcium absorption or reduced endogenous loss.[5] However, there are limitations to the rat model for the study of dietary effects on bone metabolism as there are a number of differences in calcium and bone metabolism between rats and humans. These include a lower percentage of ingested calcium excreted in urine in young adult albino rats (~1%) than man (~16%); the renal concentrating capacity of the rat is about twice that of humans when expressed on a body-weight basis; rats lack lamellar bone and consequently have a limited capacity for bone remodelling; and, importantly, rats appear to be more susceptible than humans to the influence of dietary factors which increase the rate of bone resorption, *e.g.*, protein and phosphorus (see review by Shortt and Flynn[5]).

## Studies in Humans

In theory the impact of dietary sodium on human skeletal integrity could be evaluated by a variety of techniques, including bone histomorphometry, bone densitometry, fracture incidence or the measurement of the rate of bone turnover. Histomorphometry is highly invasive, expensive, has a long turnaround time, and is limited to a single skeletal site, the iliac crest. Intervention studies which use fracture incidence as an outcome measure are long-term studies requiring large numbers of subjects. Bone densitometry or measurement of bone mineral density (BMD), which can be measured by dual energy X-ray absorptiometry or related methods, is related to bone strength and is commonly used as an outcome in studies of bone health. BMD at any age is a predictor of bone strength and fracture risk at that age,[58,59] and in older adults is a predictor of future osteoporotic fracture risk.[60,61] While this is a precise and non-invasive method, it is slow to reveal changes.[62] The rate of bone remodeling or turnover can be determined by measurement of biochemical markers, which are relatively non-invasive and respond to intervention more rapidly than does densitometry.[62]

Biochemical markers that reflect the remodeling process and can be measured in blood or urine (see Table 12.4) fall into three categories: (a) enzymes or proteins that are excreted by cells involved in the remodeling process, (b) breakdown products generated in the resorption of old bone, and (c) byproducts produced during sythnesis of new bone.[62] Because of the phenomenon of coupling (*i.e.*, bone formation linked to bone resorption), these markers reflect the general process of bone turnover when bone is in steady state; however, markers are usually classified by the part of the remodeling process that they mainly reflect in acute situations (*i.e.*, resorption or formation). Because the process of resorption is shorter than the process of formation, resorption markers respond faster to changes in remodeling than do formation markers.[62] Increased bone turnover, as assessed by such biochemical markers, has been associated with increased risk of fracture. For example, in the OFELY study, baseline levels of markers of bone formation and bone resorption in the highest quartile were associated with increased risk of fracture in a five-year follow up of postmenopausal women.[63] The EPIDOS study showed correlations between high concentrations of bone resorption markers (CTx and Dpyr) and increased hip fracture risk similar in magnitude to that between low hip BMD and increased hip fracture risk.[64]

**Table 12.4.** *Biochemical markers of bone turnover*

|  | *Bone resorption* | *Bone formation* |
|---|---|---|
| Serum | Tartrate-resistant acid phosphatase (TRAP) Free γ-carboxy glutamic acid C-terminal pyridinoline cross-linked telopeptide of type 1 collagen (ICTP) | Alkaline phosphatase (total, bone specific) Osteocalcin Amino-terminal procollagen extension peptide (PINP) Carboxy-terminal procollagen extension peptide (PICP) |
| Urine | Calcium Hydroxyproline (total, free) Deoxypyridinoline (Dpyr) (total, free) Pyridinoline (Pyr) (total, free) N-telopeptides of collagen (NTx) C-telopeptides of collagen (CTx) Hydroxylysine glycosides | |

## Sodium and Bone Mass

The limited number of epidemiological studies which have investigated the association of dietary or urinary sodium with BMD in humans have produced conflicting results.[29,33,65–69] For example, in a 2-year longitudinal study of healthy postmenopausal women, urinary excretion of sodium was negatively correlated with changes in BMD at the introchanteric and hip sites.[66] The authors suggested that, based on their findings on regression analysis of the data, no bone loss occurred at the total hip when urinary sodium excretion was 92 mmol per day or less (approximately 2.1 g sodium intake). Nordin and Polley[29] found that forearm BMD (corrected for age and years since menopause) of 440 postmenopausal women was significantly and negatively correlated with urinary sodium excretion. On the other hand, Jones *et al.*[68] reported that while urinary sodium was correlated with urinary calcium in one hundred and fifty-four adults (34 males and 120 females, aged, 20–70 (mean 47) years), it was not associated with bone mineral content (BMC) or BMD (adjusted for sex, age, body weight and smoking) at the spine, femur or total body. Matkovic *et al.*[67] reported that urinary sodium was not directly associated with any measures of bone mass in early pubertal girls, although urinary sodium was the main determinant of urinary calcium. Similarly, Jones *et al.*[69] recently reported that, in a cross-sectional study of 330 pre-pubertal (eight-year-old) boys and girls, urinary sodium was not associated with BMD at any site. Greendale *et al.*[65] reported that dietary sodium intake was not related to bone loss in pre- or postmenopausal women measured 16 years later but there was a weak positive relationship of dietary sodium intake and BMD of the ultradistal radius in men; however, the sodium intake was assessed only by a single 24 h dietary recall at the initial examination, and sodium intake was not assessed at any time in the 16 year follow-up. In addition, 70% of the population studied also claimed to have reduced their salt intake over the time of the study, which may have contributed to the lack of association reported in the study.[6] Dawson-Hughes *et al.*[33] examined associations of urinary sodium excretion with urinary calcium and with BMD in healthy elderly (over age 65 years) men ($n = 249$) and women ($n = 665$). They found that, while urinary sodium and calcium excretion were positively correlated at moderate and high dietary calcium intakes, but not at low calcium intakes, urinary sodium was not correlated with BMD at the spine, hip, or whole-body.

Sodium intake has been reported to be associated with reduced BMD in patients with absorptive hypercalciuria[70] and in hypercalciuric calcium stone forming patients.[71] These patients may represent 'at risk' groups which are more susceptible to the negative effects of sodium on bone, and thus, may benefit from reducing salt intake.

## Sodium and Bone Metabolism

Besides the studies which investigated the impact of sodium intake on BMD, there are also studies which investigated the effect of dietary sodium on bone metabolism, or bone turnover. There is evidence that urinary hydroxyproline, a marker of bone resorption, is increased with increasing sodium intake in adults,[29,35,36,40,55,72–75] although Castenmiller et al.[39] found no association of sodium-induced calciuria with urinary hydroxyproline excretion. However, the suitability of urinary hydroxyproline as a marker of bone resorption has been questioned due to its lack of specificity and sensitivity and the changes in hydroxyproline excretion may not necessarily reflect changes in bone metabolism, but rather may be due to contributions from other sources, e.g., connective tissues or from an alteration in the liver catabolism of hydroxyproline.[38] Additionally, in a number of these studies, dietary hydroxyproline was not controlled and it has been well established that dietary hydroxyproline may affect urinary hydroxyproline excretion.[6,76].

Urinary pyridinium crosslinks of collagen are considered to be more reliable biochemical markers of bone resorption.[38,77] Itoh et al.[78] found a significant positive association between urinary deoxypyridinoline (Dpyr) and urinary sodium in Japanese women aged 50–79 years ($n = 216$), but not in women aged 20–49 years ($n = 205$) or in men aged 20–79 years ($n = 342$). Urinary sodium has also been shown to be positively associated with urinary Dpyr in 154 Tasmanian adults (34 males and 120 females, aged 20–70 (mean, 47) years), although there was no demonstratable association between bone mass and urinary sodium.[69]

Two controlled intervention studies have investigated the effects of dietary sodium on urinary pyridinium crosslinks of collagen in adult (premenopausal) women. Evans et al.[37] reported that urinary excretion of Dpyr was unaffected by increasing dietary sodium (i.e., from 50 to 300 mmol per day as salt supplements) for 7 days in premenopausal women (mean age 32 years) with average daily calcium intake of 860 mg, although urinary calcium excretion increased. Similarly, Ginty et al.[23] found that increasing sodium intake from 80 to 180 mmol per day for 14 days by dietary manipulation had no effect on urinary excretion of either pyridinoline (Pyr) or Dpyr in adult women (mean age 25 years) whose daily calcium intake was restricted to 500 mg, regardless of whether or not they were sodium-sensitive (i.e., showing a significant calciuric response to increased sodium intake). The lack of effect of sodium on urinary pyridinium crosslinks in the premenopausal women in these two studies suggests that the sodium-induced urinary calcium loss is compensated for by increased calcium absorption rather than increased bone resorption. Of particular note is the finding that in the study by Ginty et al.[23] the adaptive processes appeared to be adequate to protect bone, even though the calcium intake of the young women was restricted to 500 mg per day. Evidence for such adaptation is derived from the study of Breslau et al.[34] described previously, which showed that sodium supplementation of young men and women (mean age, 27 years) increased fractional calcium absorption by an average of 26%.

However, the capacity for such adaptation may be related to age and menopausal status.[42] Impaired adaptation may explain the increased urinary excretion of crosslinks of collagen in postmenopausal women in response to increased dietary sodium intake in some studies. For example,

Evans *et al.*[37] reported that, in postmenopausal women (mean age 57 years) with an average daily calcium intake of 750 mg, urinary excretion of Dpyr was higher following 7 days on a high sodium diet (300 mmol per day) than a low sodium diet (50 mmol per day). However, Lietz *et al.*[38] did not observe this in a similar study of postmenopausal women (mean age 62 years) with average daily calcium intake of 816 mg using a lower sodium load (60–170 mmol per day). Sellmeyer *et al.*[42] reported that when postmenopausal women, who received a daily supplement of 500 mg calcium in addition to their usual diet, and who were adapted to a low-salt diet (87 mmol sodium per day) for 3 weeks, were switched to a high-salt (225 mmol sodium per day) diet (achieved by sodium supplements and dietary manipulation) for a further 4 weeks, urinary N-telopeptides of collagen (NTx) levels, a sensitive and specific marker of bone resorption, were significantly increased.

Some of the above studies also assessed the effect of increasing sodium intake on the rate of bone formation, as assessed by biochemical markers, such as serum osteocalcin and bone-specific alkaline phosphatase. An increased rate of bone resorption and bone formation is indicative of an increased rate of bone turnover. McParland *et al.*[36] reported that in addition to an increased rate of bone resorption (as evidenced by increased urinary hydroxyproline excretion), serum osteocalcin levels were also increased in postmenopausal women supplemented with 100 mmol sodium per day.[36] Other sodium intervention studies reported a lack of effect of increasing sodium intake on serum osteocalcin and bone-specific alkaline phosphatase levels in premenopausal (osteocalcin and bone-specific alkaline phosphatase[23]) and postmenopausal (osteocalcin[37]) women. The lack of effect in these studies, however, could possibly be due to their relative short duration since bone formation, although coupled to resorption, is separated in time by approximately 6 weeks.[79]

Interestingly, two studies have reported that increasing sodium intake reduced serum osteocalcin levels in premenopausal[37] and postmenopausal women.[42] In the study by Evans *et al.*[37] reduced serum osteocalcin was observed in the absence of an effect on markers of bone resorption, while in the recent study by Sellmeyer *et al.*[42] reduced serum osteocalcin in response to increased dietary sodium was accompanied by an increase in the biochemical marker of bone resorption (urinary NTx). This raises the possibility that high sodium intake might be associated with uncoupling of the processes of bone resorption and formation, which would have potentially more serious adverse consequences for bone metabolism. Therefore, the effect of sodium on the rate of bone formation requires further investigation.

The above mentioned studies suggest that increasing sodium intake within the usual dietary range can increase bone resorption in postmenopausal women, even when calcium intake is adequate, due to maladaption of calcium absorption to sodium-induced calciuria. As mentioned earlier, an increased rate of bone turnover in older adults contributes to faster bone loss and is recognised as a risk factor for fracture.[63,64] However, it is worth noting that these studies were relatively short-term (typically less than the time needed for one bone remodeling cycle, *i.e.*, up to 6 months) and thus may not have allowed sufficient time for bone to reach a steady state after the sodium intervention. Larger studies with more long-term outcomes such as BMD and fracture will be needed to define the role of dietary sodium in postmenopausal bone loss and osteoporosis more completely. To date, there has not been any reported controlled intervention study of the effect of dietary sodium on BMD.

## Interactions of Sodium with Other Dietary Factors

There is evidence that sodium-induced calciuria, and its possible adverse effects on bone metabolism, may be modulated through interactions with other dietary factors, including calcium, alkaline salts of potassium and foods such as fruit and vegetables.

While there is no evidence that the level of dietary calcium influences the magnitude of the salt-induced calciuric response, it seems reasonable to assume that a sufficiently high dietary calcium intake could prevent the loss of bone that might occur with a high salt intake.[6] However, to date, this has not been tested experimentally. Goulding et al.[40] found that salt-induced increases in 24 h calcium/creatinine and hydroxyproline/creatinine ratios were the same in nine young women after consuming either 200 or 1500 mg of calcium per day for 4 days. However, this was a very short-term study and did not allow for sufficient time (at least 1–2 weeks) for adaptation to the altered calcium intake.[32] In addition, the study had the various limitations of using urinary hydroxyproline as a marker of bone resorption, as outlined previously. There has been some attempt to predict the levels of dietary calcium required to offset the potential negative impact of high sodium intake on bone. In the study of Devine et al.[66] regression analysis indicated that approximately 25 mmol per day (1000 mg) of dietary calcium would prevent bone loss at the hip in postmenopausal women ingesting 84 mmol per day (2 g sodium) salt. However, if salt intake rose from 84 to 127 mmol (3 g sodium per day), then calcium intake would have to rise to 37.5 mmol (1500 mg) to prevent loss of bone. A double-blind, randomised $2 \times 2$ factorial crossover design study of the impact of a high (5 g) vs. a low (1 g) sodium intake on calcium absorption (using stable isotopes), calcium metabolism (including endogenous calcium excretion) and bone metabolism (using urinary pyridinium crosslinks and NTx and serum osteocalcin and bone-specific alkaline phosphatase) in postmenopausal women consuming a low to moderate (500 mg) or high (1250 mg) daily level of calcium is currently underway as part of the EU Framework funded OSTEODIET project (see http://osteodiet.ucc.ie). This study will specifically address the issue of whether a high dietary calcium intake can overcome the loss of bone that may occur with a high salt intake.

Sellmeyer et al.[42] have suggested that dietary changes that reduce urinary calcium excretion could potentially prevent the calciuric effect of increased dietary sodium chloride. Alkaline salts of potassium (e.g., potassium bicarbonate, potassium citrate) have been shown to significantly reduce urinary calcium excretion in young men and both premenopausal and postmenopausal women,[45,48,80] even in the setting of a high sodium intake.[81]

Thus, Sellmeyer *et al.*[42] investigated the effect on urinary calcium and bone resorption of switching postmenopausal women, who were receiving a 500 mg calcium supplement in addition to their usual diet, and who were adapted to a low-salt diet (87 mmol sodium per day) for 3 weeks, to a high-salt diet (225 mmol sodium per day) plus either potassium citrate (90 mmol per day) or a placebo for a further 4 weeks. While the high sodium diet induced calciuria and led to increased urinary NTx levels relative to the low sodium diet, suggesting an increase in the rate of bone resorption, the potassium citrate attenuated the sodium-induced calciuria and associated elevation in bone resorption. Sellmeyer *et al.*[42] proposed that the dietary levels of alkaline salts of potassium which prevented the adverse effects of sodium in their study, are achievable by increasing intake of fruits and vegetables.

In support of this contention, a recent preliminary report of an ancillary study to the DASH (Dietary Approached to Stop Hypertension)-Sodium trial has shown that in comparison to the DASH control diet (a typical Western type diet), consumption of the DASH combination diet (high in fruit and vegetable and low-fat dairy produce) for 30 days reduced the levels of biochemical markers of bone formation (serum osteocalcin) and bone resorption (serum C-terminal telopeptide of type I collagen) at each of three dietary sodium intake levels (low, intermediate and high) in adult men and women.[82]

## Discussion

Sodium-induced calciuria has been clearly demonstrated in many studies and the magnitude of this effect within the usual range of sodium intakes is significant in relation to overall calcium metabolism.[5] There is also evidence of considerable inter-individual variation in the calciuric effect of sodium,[5,6,23] but more research is needed to understand the basis of this variation, including possible genetic influences. The available evidence suggests that healthy individuals adapt to sodium-induced calciuria by a PTH-mediated increase in intestinal calcium absorption.[5] However, this adaptive mechanism does not appear to function in all individuals (*e.g.*, those with impaired parathyroid, postmenopausal women with osteoporosis, as well as some healthy postmenopausal women) and, even in those individuals who appear to adapt, the increase in net calcium absorption may not be sufficient to offset the increase in urinary calcium losses.[5,42] Furthermore, the capacity for such adaptation may be limited by low dietary calcium intakes, poor vitamin D status, impaired renal function or poor intestinal calcium absorption.[5]

There is evidence in short-term studies of an association between high salt intakes and increased bone resorption (as indicated by hydroxyproline and the more sensitive and specific urinary pyridinum crosslinks of collagen and NTx) in postmenopausal women. However, whether this increased rate of bone resorption persists in the longer-term and, more importantly, results in net bone loss has not been clearly established and further research is required.

The possibility that a high dietary calcium intake could prevent the potential loss of bone that might occur with a high salt intake is the subject of ongoing research. Furthermore, there is some encouraging preliminary evidence that a diet high in fruit and vegetables may ameliorate the potential negative effects for bone of a high sodium intake[42,82]; however, more research is needed on this topic.

## Conclusions

Increasing salt intake within the usual range is associated with increased urinary calcium loss. However, the implications of this effect for bone health are still unclear. For example, although the balance of the evidence suggests that high salt consumption may adversely affect bone metabolism in postmenopausal women, the evidence of an association between sodium intake and bone loss is not clear and strong enough at present to justify interventions at the population level. Additional research is needed before specific recommendations about habitual sodium chloride intake can be made for the prevention or treatment of osteoporosis.

## References

1. J.C. Aub, D.M. Tibbetts and R. McLean, The influence of parathyroid hormone, urea, sodium chloride, fat and of intestinal activity upon calcium balance, *J. Nutr.*, 1937, **13**, 635–655.
2. A. Hills, D. Parsons, G. Webster, O. Rosenthal and H. Conover, Influence of renal excretion of sodium chloride upon the renal excretion of magnesium and other ions by human subjects, *J. Clin. Endocrinol. Metab.*, 1959, **19**, 1192–1211.
3. R.P. Heaney, J.C. Gallagher, C.C. Johnston, R. Neer, A.M. Parfitt and G.D. Whedon, Calcium nutrition and bone health in the elderly, *Am. J. Clin. Nutr.*, 1982, **36**, 986–1013.
4. H. Spencer, L. Kramer and D. Osis, Factors contributing to calcium loss in aging, *Am. J. Clin. Nutr.*, 1982, **36**, 776–787.
5. C. Shortt and A. Flynn, Sodium–calcium inter-relationships with specific reference to osteoporosis, *Nutr. Res. Rev.*, 1990, **3**, 101–115.
6. L.K. Massey and S.J. Whiting, Dietary salt, urinary calcium, and bone loss, *J. Bone Miner. Res.*, 1996, **11**, 731–736.
7. H. Burger, D.E. Grobbee and T. Drueke, Osteoporosis and salt intake, *Nutr. Metab. Cardiovasc. Dis.*, 2000, **10**, 46–53.
8. A.J. Cohen and F.J.C. Roe, Review of risk factors for osteoporosis with particular reference to a possible aetiological role of dietary salt, *Food Chem. Toxicol.*, 2000, **38**, 237–253.
9. H.E. De Wardener and G.A. MacGregor, Harmful effects of dietary salt in addition to hypertension, *J. Human Hypertens.*, 2002, **16**, 213–223.
10. Intersalt Cooperative Research Group Intersalt: an international study of electrolyte excretion and blood pressure. Results for 24 hour urinary sodium and potassium excretion, *Br. Med. J.*, 1988, **297**, 319–328.
11. A. Flynn, C. Shortt and P.A. Morrissey, Sodium and potassium intakes in Ireland, *Proc. Nutr. Soc.*, 1990, **49**, 323–332.

12. J. Schacter, P.H. Harper, M.E. Radin, A.W. Caggiula, R.H. McDonald and W.F. Diven, Comparison of sodium and potassium intake with excretion, *Hypertension*, 1980, **2**, 695–699.

13. S.A. Bingham, The dietary assessment of individuals; methods, accuracy, new techniques and recommendations, *Nutr. Abstr. Rev.*, 1987, **57**, 705–742.

14. F.C. Luft, N.S. Fineberg and R.N. Sloan, Estimating dietary sodium intake in individuals receiving a randomly fluctuating intake, *Hypertension*, 1982, **4**, 805–808.

15. C. Shortt, A. Madden, A. Flynn and P.A. Morrissey, Influence of dietary sodium intake on urinary calcium excretion in selected Irish individuals, *Eur. J. Clin. Nutr.*, 1988, **42**, 595–603.

16. B.E.C. Nordin, A. Need, H.A. Morris and M. Horowitz, The nature and significance of the relationship between urinary sodium and urinary calcium in women, *J. Nutr.* 1993, **123**, 1615–1622.

17. R.A.L. Sutton and J.H. Dirks, Renal handling of calcium: overview, In *Phosphate Metabolism*, S.G. Massry and E. Ritz (eds.), Plenum Press, New York, 1977, pp. 15–27.

18. W.N. Suki, Calcium transport in the nephron, *Am. J. Physiol.*, 1979, **237**, F1–F6.

19. L.D. Antoniou, G.M. Eisner, L.M. Slotkoff and L.S. Lilienfield, Relationship between sodium and calcium transport in the kidney, *J. Lab. Clin. Med.*, 1969, **74**, 410–420.

20. D.A. McCarron, L.I. Rankin, W.M. Bennett, S. Krutzik, M.R. McClung and F.C. Luft, Urinary calcium excretion at extremes of sodium intake in normal man, *Am. J. Nephrol.*, 1981, **1**, 84–90.

21. J.S. King, R. Jackson and B. Ashe, Relation of sodium intake to urinary calcium excretion, *Invest. Urol.*, 1964, **1**, 555–560.

22. W.J. Meyer, I. Transbol, F.C. Bartter and C. Delea, Control of calcium absorption: effect of sodium chloride loading and depletion, *Metabolism*, 1976, **25**, 989–993.

23. F. Ginty, A. Flynn and K.D. Cashman, The effect of short-term calcium supplementation on biochemical markers of bone metabolism in healthy young adults, *Br. J. Nutr.*, 1998, **80**, 437–443.

24. W.C. Chen, H.Y. Chen, H.F. Lu, C.D. Hsu and F.J. Tsai, Association of the vitamin D receptor gene start codon Fok I polymorphism with calcium oxalate stone disease, *Br. J. Urol. Int.*, 2001, **87**, 168–171.

25. P.B. Rapuri, J.C. Gallagher, H.K. Kinyamu and K.L. Ryschon, Caffeine intake increases the rate of bone loss in elderly women and interacts with vitamin D receptor genotypes, *Am. J. Clin. Nutr.*, 2001, **74**, 694–700.

26. M. Modlin, The aetiology of renal stone; a new concept arising from studies on a stone-free population, *Ann. R. Coll. Surg.*, 1967, **40**, 155–177.

27. P.J. Ojwang, R.J. Pegoraro, L. Rom and P. Lanning, Collagen Ialpha1 and vitamin D receptor gene polymorphisms in South African whites, blacks and Indians, *East Afr. Med. J.*, 2001, **78**, 604–607.

28. H. Spencer and L. Kramer, The calcium requirement and factors causing calcium loss, *Fed. Proc.*, 1986, **45**, 2758–2762.

29. B.E.C. Nordin and K.J. Polley, Metabolic consequences of the menopause: a cross-sectional, longitudinal and intervention study on 557 normal postmenopausal women, *Calcif. Tissue Int.*, 1987, **41**, S1–S59.

30. A. Madden, A. Flynn and F.M. Cremin, Relationship between dietary sodium intake and urinary calcium excretion, In *Research in Food Science and Nutrition*,

vol. 3, *Human Nutrition*, J.V. McLoughlin and B.M. McKenna (eds.), Boole Press, Dublin, 1983, pp. 30–31.

31. N.K. Hollenberg, Set point for sodium homeostasis: surfeit, deficit, and their implications, *Kidney Int.*, 1980, **17**, 423–429.

32. B. Dawson-Hughes, S. Harris, C. Kramich, G. Dallal and H.M. Rasmussen, Calcium retention and hormone levels in black and white women on high- and low-calcium diets, *J. Bone Miner. Res.*, 1993, **8**, 779–787.

33. B. Dawson-Hughes, S.E. Fowler, G. Dalsky and C. Gallagher, Sodium excretion influences calcium homeostasis in elderly men and women, *J. Nutr.*, 1996, **126**, 2107–2112.

34. N.A. Breslau, J.L. McGuire, J.E. Zerwekh and C.Y.C. Pak, The role of dietary sodium on renal excretion and intestinal absorption of calcium and on vitamin D metabolism, *J. Clin. Endocrinol. Metab.*, 1982, **55**, 369–373.

35. J. Sabto, M.J. Powell, M.J. Breidahl and F.W. Gurr, Influence of urinary sodium on calcium excretion in normal individuals: a definition of hypercalciuria, *Med. J. Aust.*, 1984, **140**, 354–356.

36. B.E. McParland, A. Goulding and A.J. Campbell, Dietary salt affects biochemical markers of resorption and formation of bone in elderly women, *Br. Med. J.*, 1989, **299**, 834–835.

37. C.E.L. Evans, A.Y. Chughtai, A. Blumsohn, M. Giles and R. Eastell, The effect of dietary sodium on calcium metabolism in premenopausal and postmenopausal women, *Eur. J. Clin. Nutr.*, 1997, **51**, 394–399.

38. G. Leitz, A. Avenell and S.P. Robins, Short-term effects of dietary sodium intake on bone metabolism in postmenopausal women measured using urinary deoxypyridinoline excretion, *Br. J. Nutr.*, 1997, **78**, 73–82.

39. J.J.M. Castenmiller, R.P. Mensink, L. Van der Heijden, T. Kouwenhoven, J.G.A.J. Hautvast, P.W. de Leeuw and G. Schaafsma, The effect of dietary sodium on urinary calcium and potassium excretion in normotensive men with different calcium intakes, *Am. J. Clin. Nutr.*, 1985, **41**, 52–60.

40. A. Goulding, H.E. Everitt, J.M. Cooney and G.F.S. Spears, Sodium and osteoporosis, In *Recent Advances in Clinical Nutrition*, M.L. Wahlqvist and A.S. Truswell (eds.), John Libbey, London, 1986, vol. 2, pp. 99–108.

41. G. Schaafsma, E.C.H. Van Beresteyn, J.A. Raymakers and S.A. Duursma, Nutritional aspects of osteoporosis, *World Rev. Nutr. Diet.*, 1987, **49**, 121–159.

42. D.E. Sellmeyer, M. Schloetter and A. Sebastian, Potassium citrate prevents increased urine calcium excretion and bone resorption induced by a high sodium chloride diet, *J. Clin. Endocrinol. Metab.*, 2002, **87**, 2008–2012.

43. A.J. Cohen and F.J.C. Roe, Evaluation of the aetiological role of dietary salt exposure in gastric and other cancers in humans, *Food Chem. Toxicol.*, 1997, **35**, 271–293.

44. T.W. Kurtz, H.A. Al-Bander and R.C. Morris, Jr, Salt sensitive essential hypertension in men. Is the sodium ion alone important? *N. Engl. J. Med.*, 1987, **317**, 1043–1048.

45. J. Lemann, Jr, R.W. Gray and J.A. Pleuss, Potassium bicarbonate, but not sodium bicarbonate, reduces urinary calcium excretion and improves calcium balance in healthy men, *Kidney Int.*, 1989, **35**, 688–695.

46. N.A. Breslau, K. Sakhaee and C.Y.C. Pak, Impaired adaptation to salt-induced urinary calcium losses in postmenopausal osteoporosis, *Trans. Assoc. Am. Physicians*, 1985, **98**, 107–115.

47. T.J. Green and S.J. Whiting, Potassium bicarbonate reduces high protein-induced hypercalciuria in adult men, *Nutr. Res.*, 1994, **14**, 991–1002.

48. R.R. Bell, M.M. Eldrid and F.R. Watson, The influence of NaCl and KCl on urinary calcium excretion in healthy young women, *Nutr. Res.*, 1992, **12**, 17–26.

49. F.L. Coe, J.J. Firpo, D.L. Hollandsworth, L. Segil, J.M. Canterbury and E. Reiss, Effect of acute and chronic metabolic acidosis on serum immunoreactive parathyroid hormone in man, *Kidney Int.*, 1975, **8**, 262–273.

50. M.B. Zemel, S.M. Gualdoni, M.F. Walsh, P. Komanicky, P. Standley, D. Johnson, W. Fitter and J.R. Sowers, Sodium excretion and plasma renin activity in normotensive black adults as affected by dietary calcium and sodium, *J. Hypertens.*, 1986, **4**, S364–S366.

51. W. Jubiz, J.M. Canterbury, E. Reiss, F.H. Tyler, J. Frailey, K. Bartholomew and M.A. Creditor, Circadian rhythm in serum parathyroid hormone concentration in human subjects: correlation with serum calcium, phosphate, albumin, and growth hormone levels, *J. Clin. Invest.*, 1972, **51**, 2040–2046.

52. E. Slatopolsky, K. Martin, J. Morrissey and K. Hruska, Current concepts of the metabolism and radioimmunoassay of parathyroid hormone, *J. Lab. Clin. Med.*, 1982, **99**, 309–316.

53. D.B. Magliaro, Pulsatile secretion of parathyroid hormone in humans, *Calcif. Tissue Int.*, 1983, **35**, 685.

54. J.M. Garel, Hormonal control of calcium metabolism during the reproductive cycle in mammals, *Physiol. Rev.*, 1987, **67**, 1–66.

55. E.L.P. Chan, C.S. Ho, D. MacDonald, S.C. Ho, T.Y.K. Chan and R. Swaminathan, Interrelationships between urinary sodium, calcium, hydroxy-proline and serum PTH in healthy subjects, *Acta Endocrinol.*, 1992, **127**, 242–245.

56. T. Fujita, J.C.M. Chan and F.C. Bartter, Effects of oral furosemide and salt loading on parathyroid function in normal subjects. Physiological basis for renal hypercalciuria, *Nephron*, 1984, **38**, 109–114.

57. A. Creedon and K.D. Cashman, The effect of high salt and high protein intake on calcium metabolism, bone composition and bone resorption in the rat, *Br. J. Nutr.*, 2000, **84**, 49–56.

58. A. Goulding, I.E. Jones, R.W. Taylor, S.M. Williams and P.J. Manning, Bone mineral density and body composition in boys with distal forearm fractures: a dual-energy X-ray absorptiometry study, *J. Pediatr.*, 2001, **139**, 509–515.

59. C.V. Oleson, B.D. Busconi and D.T. Baran, Bone density in competitive figure skaters. *Arch. Phys. Med. Rehabil.*, 2002, **83**, 122–128.

60. World Health Organization, *Assessment of Fracture Risk and Its Application to Screening for Postmenopausal Osteoporosis*, Geneva, World Health Organisation, 1994.

61. K.G. Faulkner, Bone matters: Are density increases necessary to reduce fracture risk? *J. Bone. Miner. Res.*, 2000, **15**, 183–187.

62. N.B. Watts, Clinical utility of biochemical markers of bone remodeling, *Clin. Chem.*, 1999, **45**, 1359–1368.

63. P. Garnero, E. Sornay-Rendu, B. Claustrat and P.D. Delmas, Biochemical markers of bone turnover, endogenous hormones and the risk of fractures in postmenopausal women: the OFELY study, *J. Bone Miner. Res.*, 2000, **15**, 1526–1536.

64. P. Garnero, E. Hausherr, M.C. Chapuy, C. Marcelli, H. Grandjean, C. Muller, *et al.*, Markers of bone resorption predict hip fracture in elderly women: the EPIDOS Prospective Study, *J. Bone Miner. Res.*, 1996, **11**, 1531–1538.

65. G.A. Greendale, E. Barrett-Connor, S. Edelstein, S. Ingles and R. Haile, Dietary sodium and bone mineral density: Results of a 16-year follow-up study, *J. Am. Geriatr. Soc.*, 1994, **42**, 1050–1055.

66. A. Devine, R.A. Criddle, I.M. Dick, D.A. Kerr and R.L. Prince, A longitudinal study of the effect of sodium and calcium intakes on regional bone density in postmenopausal women, *Am. J. Clin. Nutr.*, 1995, **62**, 740–745.

67. V. Matkovic, J.Z. Ilich, M.B. Andon, L.C. Hsieh, M.A. Tzagournis, B.J. Lagger and P.K. Goel, Urinary calcium, sodium, and bone mass of young females, *Am. J. Clin. Nutr.*, 1995, **62**, 417–425.

68. G. Jones, M.D. Riley and S. Whiting, Association between urinary potassium, urinary sodium, current diet, and bone density in prepubertal children, *Am. J. Clin. Nutr.*, 2001, **73**, 839–844.

69. G. Jones, T. Beard, V. Parameswaran, T. Greenaway and R. von Witt, A population-based study of the relationship between salt intake, bone resorption and bone mass, *Eur. J. Clin. Nutr.*, 1997, **51**, 561–565.

70. F. Pietschmann, N.A. Breslau and C.Y. Pak, Reduced vertebral bone density in hypercalciuric nephrolithiasis, *J. Bone. Miner. Res.*, 1992, **7**, 1383–1388.

71. L.A. Martini, L. Cuppari, F.A. Colugnati, D.M. Sigulem, V.L. Szejnfeld, N. Schor and I.P. Heilberg, High sodium chloride intake is associated with low bone density in calcium stone-forming patients, *Clin. Nephrol.*, 2000, **54**, 85–93.

72. A. Goulding, Fasting urinary sodium/creatinine in relation to calcium/creatinine and hydroxyproline/creatinine in a general population of women, *NZ Med. J.*, 1981, **93**, 294–297.

73. A. Goulding and B. McDonald, Intra-individual variability in fasting urinary calcium/creatinine and hydroxyproline/creatinine measurements, In *Recent Advances in Clinical Nutrition*, M.L. Wahlqvist and A.S. Truswell (eds.), John Libbey, London, 1986, vol. 2, pp. 312–313.

74. A.G. Need, H.A. Morris, D.B. Cleghorn, D.D. DeNichilo, M. Horowitz and B.E.C. Nordin, Effect of salt restriction on urine hydroxyproline excretion in postmenopausal women, *Arch. Intern. Med.*, 1991, **151**, 757–759.

75. R. Itoh and Y. Suyama, Sodium excretion in relation to calcium and hydroxy-proline excretion in a healthy Japanese population, *Am. J. Clin. Nutr.*, 1996, **63**, 735–740.

76. K. Kivirikko, Urinary excretion of hydroxyproline in health and disease, *Int. Rev. Connect. Tissue Res.*, 1970, **5**, 93–163.

77. P.D. Delmas, Clinical use of biochemical markers of bone remodelling in osteoporosis, *Bone*, 1992, **13**, S17–S21.

78. R. Itoh, Y. Suyama, Y. Oguma and F. Yokota, Dietary sodium, an independent determinant for urinary deoxypyridinoline in elderly women. A cross-sectional study on the effect of dietary factors on deoxypyridinoline excretion in 24-h urine specimens from 763 free-living healthy Japanese, *Eur. J. Clin. Nutr.*, 1999, **53**, 886–890.

79. F.F. Eriksen, H.J.G. Gundersen, F. Melsen and L. Mosekilde, Reconstruction of the formative site in iliac trabecular bone in 20 normal individuals employing a kinetic model for matrix and mineral apposition, *Metab. Bone Dis. Relat. Res.*, 1984, **5**, 243–252.

80. J. Lemann, Jr, J.A. Pleuss, R.W. Gray and R.G. Hoffmann, Potassium administration reduces and potassium deprivation increases urinary calcium excretion in healthy adults, *Kidney Int.*, 1991, **39**, 973–983.

81. R.C. Morris, A. Sebastian, A. Forman, M. Tanaka and O. Schmidlin, Normotensive salt-sensitivity: effects of race and dietary potassium, *Hypertension*, 1999, **33**, 18–23.
82. P. Lin, F. Ginty, L. Appel, L. Svetkey, A. Bohannon, D. Barclay, R. Gannon and U. Aickin, Impact of sodium intake and dietary patterns on biochemical markers of bone and calcium metabolism, *J. Bone Miner. Res.*, 2001, **16**, S511.
83. D.A. McCarron, P.A. Pingree, R.J. Rubin, S.M. Gaucher, M. Molitch and S. Krutzik, Enhanced parathyroid function in essential hypertension: a homeostatic response to a urinary calcium leak, *Hypertension*, 1980, **2**, 162–168.
84. A.G. Need, M.D. Guerin, R.W. Pain, T.F. Hartley and B.E.C. Nordin, The tubular maximum for calcium reabsorption: normal range and correction for sodium excretion, *Clin. Chim. Acta*, 1985, **150**, 87–93.
85. A.G. Wasserstein, P.D. Stolley and K.A. Soper, Case-control study of risk factors for idiopathic calcium urolithiasis, *Miner. Electrolyte Metab.*, 1987, **13**, 85–95.
86. H. Kesteloot and J.V. Joossens, The relationship between dietary intake and urinary excretion of sodium, potassium, calcium and magnesium: Belgian Interuniversity Research on Nutrition and Health, *J. Human Hypertens.*, 1990, **4**, 527–533.
87. R. Itoh, J. Oka, K. Yamada, Y. Suyama and K. Murakami, The interrelation of urinary calcium and sodium intake in healthy elderly Japanese, *Int. J. Vitam. Nutr. Res.*, 1991, **61**, 159–165.
88. M.J. Philips and J.N.C. Cooke, Relation between urinary calcium and sodium in patients with idiopathic hypercalciuria, *Lancet*, 1967, 1354–1357.
89. L.K. Law, R. Swaminathan and S.P.B. Donnan, Relationship between sodium excretion and calcium excretion in healthy subjects, *Med. Sci. Res.*, 1988, **16**, 643.
90. M. Zarkadas, R. Gourgeon-Reyburn, E.B. Marliss, E. Block and M. Alton-Mackey, Sodium chloride supplementation and urinary calcium excretion in postmenopausal women, *Am. J. Clin. Nutr.*, 1989, **50**, 1088–1094.
91. D.J. Kok, J.A. Iestra, J. Doorenbos and S.E. Papapoulos, The effects of dietary excesses in animal protein and in sodium on the composition and the crystallization kinetics of calcium oxalate monohydrate in urines of healthy men, *J. Clin. Endocrinol. Metab.*, 1990, **71**, 861–867.

CHAPTER 13.1

# Acid–Base Homeostasis and the Skeleton: Is there a Fruit and Vegetable Link to Bone Health?

SUSAN A. NEW

Centre for Nutrition and Food Safety, School of Biomedical and Molecular Sciences, University of Surrey, Guildford, Surrey GU2 7XH, UK, Email: s.new@surrey.ac.uk

**Key Points**

- The health-related benefit of high consumption of fruit and vegetables on a variety of diseases has been gaining increasing prominence in the literature over a number of years.
- The role that the skeleton plays in acid–base homeostasis is of considerable interest; with theoretical considerations of the role alkaline bone mineral may play in the defence against acidosis dating as far back as the late 19th century.
- Natural, pathological and experimental states of acid loading/acidosis have been associated with hypercalciuria and negative calcium balance and more recently, the detrimental effects of 'acid' from the diet on bone mineral have been demonstrated. At the cellular level, a reduction in extracellular pH has been shown to have a direct enhancement on osteoclastic activity, with the result of increased resorption pit formation in bone.
- A number of population-based studies, published in the last decade, have demonstrated a beneficial effect of fruit and vegetable/potassium intake on axial and peripheral bone mass and bone metabolism in men and women across the age ranges.
- Further support for a positive link between fruit and vegetable intake and bone health can be found in the results of the Dietary Approaches to Stopping Hypertension (DASH) and DASH-Sodium intervention trials.
- Future research should focus attention on four key areas:
  - (1) Intervention trials centred specifically on fruit and vegetables as the supplementation vehicle and assessing a wide range of bone health indices including fracture risk;

(2)  Experimental studies (at both the cellular, animal and human level) to determine whether there are other aspects of fruit and vegetables which are beneficial to bone metabolism, including key micronutrients and phytoestrogens;

(3)  Further definition of the relationship between net endogenous acid production and skeletal integrity and in particular whether very high protein intakes are detrimental to the skeleton in the absence of alkali-forming foods;

(4)  Re-analysis of existing dietary:bone mass/metabolism datasets to look in particular at the impact of 'dietary acidity' on the skeleton.

*One farmer says to me, "You cannot live on vegetable food solely, and it furnishes nothing to make bones with," and so he religiously devotes a part of his day to supplying his system with the raw material of bones; walking all the while he talks behind his oxen, which, with vegetable-made bones, jerk him and his lumbering plow along in spite of every obstacle.*
Henry David Thoreau: Walden

*Life is a struggle, not against sin, not against the money power, not against malicious animal magnetism, but against hydrogen ions*
Henry Louis Mencken (1919)

*these famous words by Mencken in the early 20th Century about the meaning of life and death, may also apply to the struggle of the healthy skeleton against the deleterious effects of retained acid!*
Kraut and Coburn (1994)

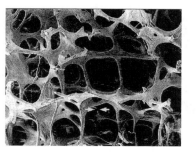

**Figure 13.1.1**   *Positive link between fruit and vegetables and bone health?*
Source: Proceedings of the First International Symposium on Osteoporosis, Editor: Uriel S. Barzel, New York, USA. 25–26th June 1969

# Importance of Nutrition to Bone Health Optimisation Throughout the Life-Cycle

Nutritional strategies for optimising bone health are important for a number of reasons: (i) nutrition is an exogenous factor and is thus amenable to change; (ii) identification of the key 'bone health nutrients' has relevant public health implications; (iii) a nutritional approach is far more popular with osteoporotic sufferers than drug intervention, a point of particular relevance given the poor, long-term compliance rates associated with a number of currently available treatments.[1] The health-related benefit of a high consumption of fruit and vegetables on a variety of diseases has been gaining increasing prominence in the literature over a number of years (Figure 13.1.2). A number of observational, experimental, clinical and intervention studies over the last decade have suggested a positive link between fruit and vegetable consumption and the skeleton.

## Acid–Base Homeostasis and Its Criticality to Health

Acid–base homeostasis is critical to health and it is well documented that extracellular fluid pH remains between 7.35–7.45. A major requirement therefore of our metabolic system is to ensure that hydrogen ion concentrations are maintained between 0.035 and 0.045 mEq $L^{-1}$.[2] The body's adaptive response involves three specific mechanisms: (i) buffer systems; (ii) exhalation of $CO_2$; and (iii) kidney excretion.

On a daily basis, humans eat substances that both generate and consume protons and as a net result, adult humans on a normal Western diet generate ~1 mEq per kg body weight of acid per day. Of course, the more acid precursors a diet contains, the greater the degree of systemic acidity.[3]

As humans become older, their overall renal function declines which includes their ability to excrete acid.[4] Hence, with increasing age, humans become

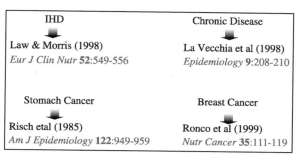

**Figure 13.1.2** *Examples of the importance of fruit and vegetables to the prevention of ischemic heart disease (IHD), cancer and other health outcomes*
With thanks to Dr Veronique Coxam (Aprifel, France) (2003) for assistance with information

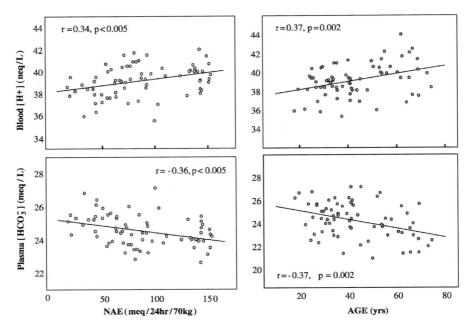

**Figure 13.1.3**  *Changes in acidity with ageing*
Reproduced with kind permission – Dr Lynda Frassetto (USA) (2003)

slightly but significantly more acidic.[5] As shown in Figure 13.1.3, blood $H^+$ concentrations increase with age and plasma bicarbonate levels decrease (Dr Lynda Frassetto, personal communication).

## Skeletal Link to Acid–Base Maintenance – Key Points

A number of points should be considered when examining the possible link between acid–base homeostasis and the skeleton:

(1) The theoretical considerations of the role of alkaline bone mineral may play in the defence against acidosis date back as far as the late 1880s/early 19th century,[6–8] with the fundamental concepts being established in the late 1960s/early 1970s. A number of studies published during this period provided evidence that in natural (*e.g.*, starvation), pathological (*e.g.*, diabetic acidosis) and experimental (*e.g.*, ammonium chloride ingestion) states of acid loading and acidosis, an association exists with both hypercalciuria and negative calcium balance.[9–10] The pioneering work of Lemann and Barzel showed extensively the effects of 'acid' from the diet on bone mineral in both man and animal.[11–12] As shown in Figures 13.1.4 and 13.1.5, pioneering work by Barzel (1969)[13] and Barzel and Jowsey (1969)[14] demonstrated that long-term ingestion of ammonium chloride caused a decrease in bone substances and the development of osteoporosis in the rat. Potassium bicarbonate prevented this bone loss.

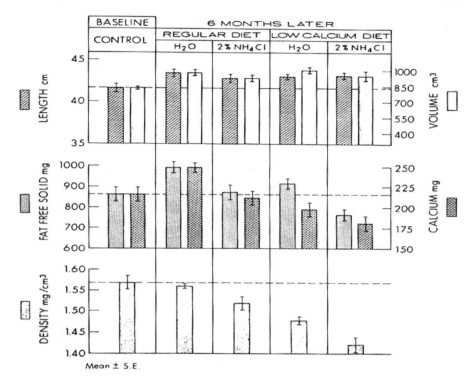

**Figure 13.1.4** *Long term ingestion of NH$_4$Cl caused a ↓ in bone substances and the development of osteoporosis in the rat*
From Ref. [13]

(2) There are clear mechanisms for a deleterious effect of acid on bone. Novel work in the 1980s by Arnett and Dempster[15] demonstrated a direct enhancement of osteoclastic activity following a reduction in extracellular pH. This effect was shown to be independent of the influence of parathyroid hormone[16–17] (Figure 13.1.6). Furthermore, osteoclasts and osteoblasts appear to respond independently to small changes in pH in the culture media in which they are growing.[18] There is evidence that a small drop in pH, close to the physiological range, causes a tremendous burst in bone resorption.[19–20] Metabolic acidosis has also been shown to stimulate resorption by activating mature osteoclasts already present in calvarial bone rather than by inducing formation of new osteoclasts[21] (Figure 13.1.7). It is considered that almost all the bone mineral release that occurs in response to acidosis is due to osteoclast activation, which results in increased resorption pit formation in bone (with the organic matrix being destroyed at the same time) (Dr TR Arnett, University College London, personal communication), although there is evidence that excess hydrogen ions directly induce physicochemical calcium release from bone.[22]

(3) The effect of dietary acidity on the skeleton needs only to be relatively small for there to be a large impact over time. In the 1960s, Wachman and

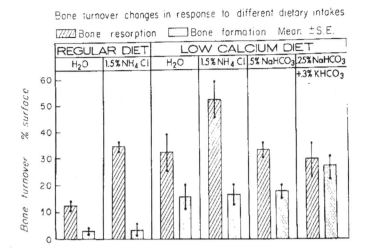

**Figure 13.1.5**  *Ingestion of $NH_4Cl$ increased bone resorption $NaHCO_3$ & $KHCO_3$ prevented the loss on bone*
From Ref. [14]

Bernstein put forward a hypothesis linking the daily diet to the development of osteoporosis based on the role of bone in acid–base balance and noted specifically that "the increased incidence of osteoporosis with age may represent, in part, the results of a life-long utilisation of the buffering capacity

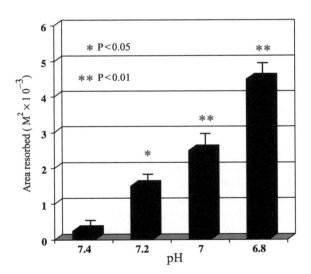

**Figure 13.1.6**  *Increase in osteoclastic activity with a reduction in extracellular pH. Mean values were significantly different from that at $pH = 7.4$: \*$P < 0.05$, \*\*$P < 0.01$*
From Ref. [15]

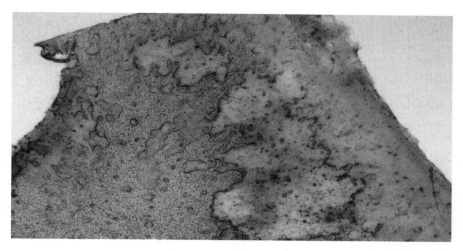

**Figure 13.1.7** *Resorption stimulated by HCO₃ – acidosis (pH = 7.01)*
Reproduced with permission – Dr TR Arnett, University College
London (2001)
From Ref. [21]

of the basic salts of bone for the constant assault against pH homeostasis".[23]
The net production of acid is related to nutrition and there is a gross
quantitative relation between the amount of acid produced (as reflected by
urine pH) and the amount of acid-ash consumed in the diet. When one
considers the extent of loss, if 2 mEq $kg^{-1}$ per body weight of Ca per day is
required to buffer about 1 mEq $kg^{-1}$ per body weight of fixed acid per day,
over 10 years (and assuming a total body Ca of approximately 1 kg), it would
account for 15% loss of inorganic bone mass in an average individual.[24]

# Vegetarianism and Skeletal Health: Potential Renal
Acid Load Concept

Following the recognition of the role that bone plays in acid–base balance and
the hypothesis linking diet to osteoporosis, it was proposed that long-term
ingestion of vegetable-based diets may have a beneficial effect on bone mineral
mass. The earlier studies (published before 1990), in general, appeared to
support the hypothesis, *i.e.*, bone mineral mass was found to be higher in the
vegetarian group compared to their omnivorous counterparts.[25–30] However,
there was a fundamental error in the interpretation of the photographic density
measurements in the first paper published by Ellis *et al.* (1972)[25] (*i.e.*, their
conclusions should have been the opposite of what they claimed)[31–34] and
subjects in several of the published studies were Seventh Day Adventists (SDA)
with a different lifestyle to that of the omnivorous group and is likely to have

**Table 13.1.1.** *Vegetarianism and bone health: summary of studies published in the 1990s*

| Author | Year | Source | Findings | Summary |
|--------|------|--------|----------|---------|
| Lloyd *et al.* | 1991 | *Am. J. Clin. Nutr.,* **54**:1005–1010 | No difference in BMD between groups | – |
| Tesar *et al.* | 1992 | *Am. J. Clin. Nutr.,* **56**:699–704 | No difference in BMD between groups | – |
| Reed *et al.* | 1994 | *Am. J. Clin. Nutr.,* **59**:1997–1202 | Bone loss rates similar | – |
| Chui *et al.* | 1997 | *Calcified Tissue Int.,* **60**:245–249 | BMD ↓ in vegan group | ✗ |
| Lau *et al.* | 1998 | *Eur. J. Clin. Nutr.,* **52**:60–64 | Hip BMD lower in vegetarian group | ✗ |

Reproduced with permission: Ref. [94].

been an important confounding influence (*e.g.*, the SDA group refrained from smoking and caffeine intake and physical levels were higher).

Studies published in the last decade suggest no differences in BMD between vegetarians and omnivores[35–36] (Table 13.1.1). In a 5-year prospective study of changes in radial bone density of elderly white American women, no differences were seen in bone loss rates between the lactoovovegetarians and the omnivorous group.[37] Furthermore, in the most recently published studies, bone mass was found to be significantly lower in the vegetable-based dietary groups,[38–39] although it is likely that protein 'undernutrition' may account for some of these differences.[40]

Very few studies have focused attention with respect to bone health on populations consuming a diet highly dependent on animal foods, particularly that of meat. Mazess and Mather (1974)[41] examined the bone mineral content of forearm bones in a sample of 217 children, 89 adults and 107 elderly Eskimo natives of the north coast of Alaska. After the age of 40 years, the Eskimos of both sexes were found to have a deficit of bone mineral in the order of magnitude between 10 and 15% relative to white standards. An even greater ageing bone loss was found in Canadian Eskimos.[42] Whilst the issue of 'dietary change' amongst the Eskimo population (particularly the utilisation of refined carbohydrates) was raised[43] and subsequently discussed. Clearly these findings of are considerable interest to the interaction between diet and bone in the regulations of systemic acid–base balance and further work in this area is clearly warranted.[44]

The potentially deleterious effect of specific foods on the skeleton has been a topic of some considerable recent debate.[45–48] Work by Remer and Manz (1995)[49] examining the potential renal acid loads (known as PRAL) of a variety of foods, have found that many grain products and some cheeses have a high PRAL level (Table 13.1.2). These foods, which are likely to be consumed in large quantities in lactoovovegetarians, may provide an explanation for the lack of a positive effect on bone health indices in studies comparing vegetarians *vs.* omnivores.[50]

**Table 13.1.2.** *Potential renal acid load (PRAL) values of a variety of foods and food groups. (From Ref. 49)*

| Food/Food Group | PRAL mEq/100g edible portion | Food/Food Group | PRAL mEq/100g edible portion |
|---|---|---|---|
| **Fruits and fruit juices** | | **Milk, dairy products and eggs** | |
| Apples | −2.2 | Milk (whole, pasteurised) | 0.7 |
| Bananas | −5.5 | Yoghurt (whole milk, plain) | 1.5 |
| Raisins | −21.0 | Cheddar cheese (reduced fat) | 26.4 |
| Grape Juice | −1.0 | Cottage cheese | 8.7 |
| Lemon Juice | −2.5 | Eggs (yolk) | 23.4 |
| **Vegetables** | | **Meat, meat products and fish** | |
| Spinach | −14.0 | Beef (lean only) | 7.8 |
| Broccoli | −1.2 | Chicken (meat only) | 8.7 |
| Carrots | −4.9 | Pork (lean only) | 7.9 |
| Potatoes | −4.0 | Liver sausage | 10.6 |
| Onions | −1.5 | Cod (fillets) | 7.1 |
| **Grain products** | | **Beverages** | |
| Bread (white wheat) | 3.7 | Coca Cola | 0.4 |
| Oat flakes | 10.7 | Coffee (infusion) | −1.4 |
| Rice (brown) | 12.5 | Tea (Indian infusion) | −0.3 |
| Spaghetti (white) | 6.5 | White wine | −1.2 |
| Cornflakes | 6.0 | Red wine | −2.4 |

# Evidence for a Beneficial Effect of Fruit and Vegetables/Alkali on Indices of Bone Health

## Observational Studies

A variety of population-based studies published in the latter part of the 20th century and more recently between 2001 and 2003 have demonstrated a beneficial effect of fruit and vegetable/potassium intake on indices of bone health in young boys and girls,[51–52] premenopausal women,[53–57] perimenopausal women,[55,58] postmenopausal women and elderly[53,59–64] men and women (Table 13.1.3).

## Aberdeen Prospective Osteoporosis Screening Study (APOSS) – Baseline and Longitudinal Findings

Baseline findings of the Aberdeen Prospective Osteoporosis Screening Study (APOSS) have shown specific associations between nutrients found in abundance in fruit and vegetables and axial and peripheral bone mass and markers of bone resorption. Women (*n* = 994) in the lowest quartile of intake of potassium, magnesium, fibre, vitamin C and β-carotene had significantly lower lumbar spine and femoral neck bone mineral density[54] (Figure 13.1.8). In a

**Table 13.1.3.** *Impact of fruit & vegetables on bone: a review of population-based studies showing a positive link*

| Author | Year | Source | Details | Findings |
|---|---|---|---|---|
| Eaton-Evans et al. | 1993 | Proc. Nutr. Soc., **52**:44A | 77 Females, 46–56 Yrs | ✓ Vegetables |
| Michaelsson et al. | 1995 | Calcified Tissue Int., **57**:86–93 | 175 Females, 28–74 Yrs | ✓ K Intake |
| New et al. | 1997 | Am. J. Clin. Nutr., **65**:1831–1839 | 994 Females, 45–49 Yrs | ✓ K, Mg, Fibre, Vitamin C |
| | | | | ✓ Past Intake: Fruit & Veg |
| Tucker et al. | 1999 | Am. J. Clin. Nutr., **69**:727–736 | 229 Males: 349 Females, 75 Yrs | ✓ K, Mg, Fruit & Vegetables |
| New et al. | 2000 | Am. J. Clin. Nutr., **72**: 142–151 | 62 Females, 45-54 Yrs | ✓ K, Mg, Fibre, Vitamin C |
| | | | | ✓ Past Intake: Fruit & Veg |
| Jones et al. | 2001 | Am. J. Clin. Nutr., **73**: 839–844 | 215 Boys: 115 Girls, ?Yrs | ✓ K, Urinary K |
| Chen et al. | 2001 | J. Bone Miner Res., **16**:S386 | 668 Females, >? years | ✓ Fruit |
| Miller et al. | 2001 | J. Bone Miner Res., **16**:S395 | ? Males, 50-91 Yrs | ✓ K, Mg |
| Stone et al. | 2001 | J. Bone Miner Res., **16**:S388 | 1075 Men, 65 Yrs and over | ✓ K, Lutein |
| New et al. | 2002 | Osteoporosis Int., (In the Press) | 164 Females, 55-87 Yrs | ✓ K, Fruit & Vegetables |

From Ref. [94].

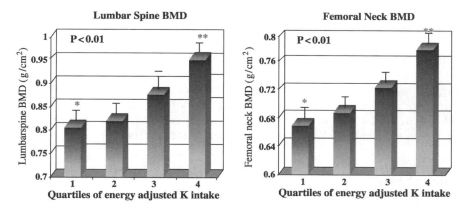

**Figure 13.1.8** *Fruit and vegetable intake and BMD in 994 women; baseline values are shown for the lumbar spine and femoral neck BMD from Study 1 of the Aberdeen Prospective* Osteoporosis Screening Study (APOSS) From Ref. [54]

second study, women ($n = 62$) with low intakes of these same nutrients were found to have lower forearm bone mass and higher bone resorption[55] (Figure 13.1.9), findings which were independent of important confounding factors. With financial assistance, initially from the Department of Health/MRC and more recently the Food Standards Agency (formerly MAFF), APOSS longitudinal is now the largest nutrition, genetic and bone health dataset currently available worldwide, involving ~5000 women. Preliminary analysis indicates a positive influence of alkaline forming foods on post-menopausal bone loss and bone turnover markers.[56,58] Further exploration of

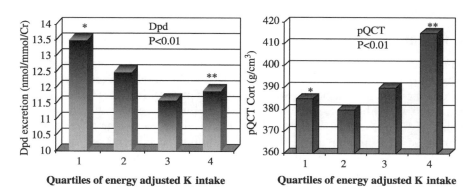

**Figure 13.1.9** *Fruit and vegetable intake and BMD in 62 women; baseline values are shown for the forearm bone mass and de-oxypridinoline excretion from Study 2 of the Aberdeen Prospective Osteoporosis Screening Study (APOSS)* From Ref. [55]

the data will enable determination of potential relationships between nutrient:gene interactions and bone health, with a specific focus on the role of the skeleton in acid–base maintenance.

In trying to clarify the size of the effect of fruit and vegetables/potassium intake on bone health, a recent systematic review on over 4500 subjects suggests a small (~0.9%) but significant effect on bone health (New and Torgerson, 2003, unpublished data). Much more detailed analysis is now urgently required with respect to which types of fruit and vegetables (including or excluding potatoes) have the most direct impact on the skeleton.


## Clinical Studies

The clinical application of the effect of normal endogenous acid production on bone is of considerable interest, with extensive work in this area by Lemann (at the subject level)[65–69] and Bushinsky (at the cellular level)[70–74]. Sebastian and colleagues[75] demonstrated that potassium bicarbonate administration resulted in a decrease in urinary calcium and phosphorus, with overall calcium balance becoming less negative (or more positive). Changes were also seen in markers of bone metabolism, with a reduction in urinary excretion of hydroxyproline (bone resorption) and an increased excretion of serum osteocalcin (bone formation). There has been some discussion concerning the level of protein consumed by the women in the study being higher than is typical of American women in this age group and a call for further studies to be undertaken with dietary protein being consumed at a lower level.[76] More recently, Buclin and colleagues have examined the effect of dietary modification on calcium and bone metabolism. The 'acid-forming' diet increased urinary calcium excretion by 74% and bone resorption, as measured by C-terminal peptide excretion by 19% in comparison to the alkali forming diet, both at baseline and after an oral calcium load.[77]


## Intervention Studies

Further support for a positive link between fruit and vegetable intake and bone health can be found in the results of the DASH and DASH-Sodium intervention trials (Dietary Approaches to Stopping Hypertension). In DASH, diets rich in fruit and vegetables were associated with a significant fall in blood pressure compared with baseline measurements. However, of particular interest to the bone field were findings that increasing fruit and vegetable intake from 3.6 to 9.5 daily servings decreased the urinary calcium excretion from 157 to 110 mm day$^{-1}$.[78] The authors suggested this was due to the "high fibre content of the diet possibly impeding calcium absorption". However, a more likely explanation put forward by Barzel (1997), was a reduction in the 'acid load' with the fruit and vegetable diet compared to the control diet.[79] This study is the first population based fruit and vegetable intervention trial showing a positive effect on calcium economy (albeit a secondary finding).

More recently, Lin *et al.* (2001)[80] have reported the findings of the DASH-Sodium trial. The impact of two dietary patterns on indices of bone metabolism were examined. The DASH diet emphasises fruits, vegetables, low-fat dairy products and is reduced in red meats and in this second DASH II trial, three levels of sodium intake were investigated (50, 100 and 150 nmol $L^{-1}$). Subjects consumed the control diet at the 150 mmol sodium intake per day levels for 2 weeks and were then randomly assigned to eat either the DASH diet or the control diet at all three sodium levels for a further 4 weeks in random order. The DASH diet, compared with the control diet was found to significantly reduce both bone formation (by measurement of the marker osteocalcin) by 8–10% and bone resorption (by measurement of the marker CTx) by 16–18% (Dr F Ginty, personal communication). Interestingly, sodium intake did not significantly affect the markers of bone metabolism. This is an important intervention study that shows a clear benefit of the high intake of fruit and vegetables on markers of bone metabolism. Research is now required to determine the long-term clinical impact of the DASH diet on bone health and fracture risk as well as clarification of the exact mechanisms involved with respect to this diet on skeletal protection.

# Dietary Balance and the Skeleton: Concept of Net Endogenous Acid Production (NEAP)

Determination of the acid–base content of diets consumed by individuals and populations is a useful way to quantify the link between acid–base balance and skeletal health. On a daily basis, humans eat substances that both generate and consume protons and as a net result, consumption of a normal Western diet is associated with chronic, low-grade metabolic acidosis.[3] The severity of the associated metabolic acidosis is determined, in part, by the net rate of endogenous noncarbonic acid production (NEAP) that varies with diet. Since 24-h urine collections are impractical for population-based studies, an alternative is to examine the net acid content of the diet. Frassetto *et al.* have found that the protein to potassium ratio predicts net acid excretion and in turn, net renal acid excretion predicts calcium excretion.[81] They propose a simple algorithm to determine the net rate of endogenous NEAP from considerations of the acidifying effect of protein (*via* sulfate excretion) and the alkalising effect of potassium (*via* provision of salts of weak organic acids) (Frassetto *et al.* 1998).

To examine this theory further, estimations of NEAP were calculated from the APOSS baseline and longitudinal datasets. In APOSS baseline, women with the lowest estimate of NEAP were found to have higher lumbar spine and femoral neck bone mineral density (Figure 13.1.9) and significantly lower urinary pyridinium cross link excretion (Figure 13.1.10).[82] Findings for bone resorption were mirrored in APOSS longitudinal. In 2323 women, adjusting for the key confounding factors, women with the lowest estimate of NEAP were found to have significantly lower pyridinoline and de-oxypyridinoline excretion.[83]

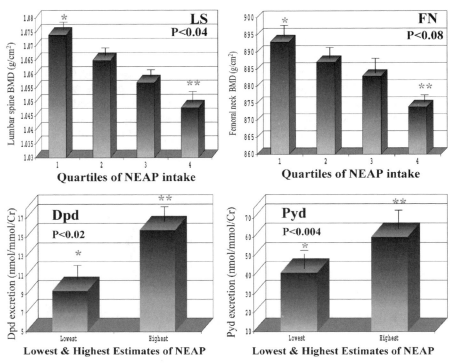

**Figure 13.1.10** *Association between net endogenous non-carbonic acid production (NEAP) and BMD in 994 women; baseline values are shown for the lumbar spine and femoral neck BMD from Study 1 of the Aberdeen Prospective Osteoporosis Screening Study (APOSS)*
From Ref. [82]

## Theoretical Considerations

The NEAP data from APOSS do not suggest that dietary protein is detrimental to bone health since even women in the lowest quartile of protein were consuming protein intakes well above the EAR (mean intake 82.5 g day$^{-1}$). Rather, these data indicate that dietary potassium (as the denominator in the NEAP algorithm) is the critical component, *i.e.*, diets that are characterised by less dietary acid (*i.e.*, closer to neutral) are associated with better indices of bone health. Although the variation in factors, including weight and height, account for most of the variation in BMD among subjects, estimates of NEAP still accounted for a statistically significant fraction of the BMD variation among subjects. Shifting from the top quartile to the lowest quartile of 'dietary acidity' as a group or from the top intake to the lowest intake of dietary acidity as an individual, resulted in a better bone mass. These findings concerning NEAP are critical since in adults, after the age of 30 years (approximately), both weight and height remain relatively stable; the majority of their influence in 'setting' bone mass is complete by this age. Thus in postmenopausal and/or elderly women, with one-time measure of BMD, weight and height influences on BMD manifest strongly among-subjects. However, NEAP, it can be postulated, continues to 'wear' away bone gradually and indefinitely after the age of 30 years, accelerating as the glomerular filtration rate falls with age (Professor Anthony Sebastian, personal communication). In other words, weight and height differences manifesting among subjects is not likely transfer their influence on BMD in each individual over time, but NEAP, as an on-going 'dynamic' influence on bone, presumably does continue its influence on bone mass in each individual subject over time.

## Ca/Alkali Supplements and Skeletal Integrity

Of growing interest in the literature is the recognition that calcium plays a critical role in the relationship between the balance of a beneficial *vs.* detrimental effect of protein on the skeleton.[84–86] It is possible that calcium supplements may be favourable to bone, not just through the additional mineral that they supply but also through their provision of additional alkali salts.[87] Since the relationship between dietary protein and skeletal health remains a controversial one, it might be prudent to suggest re-analysis of existing nutrition and bone health datasets to focus specifically on the effect of protein:potassium ratios and protein:calcium ratios on indices of bone health. It may be that in the absence of sufficient dietary alkali to neutralise the protein-derived acid, net calcium loss ensues and the anabolic drive of dietary protein on the bone matrix is ineffective in maintaining bone mineral density.

## Historical Diets

It should not be forgotten that it is generally believed that our modern diet is vastly different to that which early humans once consumed.[88] Estimates of the dietary content of pre-agricultural man estimate intakes of sodium to be 600 mg day$^{-1}$ and potassium to be at levels reaching 7000 mg day$^{-1}$, compared with intakes of sodium and potassium at levels of approximately 4000 and 2500 mg, respectively, in the UK, USA and Australia.[89–90]

## Other Components in Fruit and Vegetables which may be Important to Skeletal Integrity

It is also important to note that the positive associations found between fruit and vegetable consumption and bone may be due to some other, yet unidentified 'dietary' component rather than alkali-excess effect.[91] Convincing work by Muhlbauer and colleagues suggests that vegetables, herbs and salads commonly consumed in the human diet affect bone resorption in the rat by a mechanism that is not mediated by their base excess,[92] but possibly through pharmacologically active compounds which are currently being explored.[93]

# Conclusions

The role that the skeleton plays in acid–base homeostasis has been gaining increasing prominence in the literature and a number of observational, experimental, clinical and intervention studies over the last decade have indicated a positive link between fruit and vegetable consumption and bone.[94] Future research should focus on intervention trials centred specifically on fruit and vegetables as the supplementation vehicle and assessing a wide range of bone health indices including fracture risk as well as experimental studies (at both the cellular, animal and human level) to determine whether there are other aspects of fruit and vegetables which are beneficial to bone metabolism, including key micronutrients and phytoestrogens. If these studies provide concluding evidence, a 'fruit and vegetable'/alkali approach to optimising bone health may be a natural alternative therapy for osteoporosis prevention and treatment, with clear additional health-related benefits.

# Acknowledgements

The author would like to acknowledge the following individuals for the key roles they have played in helping and guiding during her different career stages thus far and to her understanding of acid–base homeostasis and the skeleton: Mrs. Linda Edwards, Director of the National Osteoporosis Society (1986–1992) for her dedication to osteoporosis sufferers; Professor David M. Reid FRCP, Director of APOSS, whose dedication to the field of osteoporosis and capacity for hard work shall always be an inspiration; Dr Simon P. Robins DSc, for the opportunity of working in his laboratory and experience the fascinating research into bone metabolism markers which he pioneered; Mr David A. Grubb, for his tremendous help with the dietary assessment methodology; Dr Helen M. Macdonald for her tireless work on APOSS longitudinal; senior colleagues in the field of Nutritional Science for their mentorship and motivation: Professor Peter Aggett OBE; Dr Jacki Bishop; Dr Judy Buttriss; Dr Clare Casey; Professor W. Philip T. James CBE; Professor Alan Jackson; Professor Ian Macdonald DSc; Professor Colin Mills DSc; Professor Joe Millward DSc; Dr Roger Whitehead CBE; Professor Christine Williams; Professor Martin Wiseman FRCP; to Professor Anthony Sebastian who remains a constant source of knowledge and inspiration as one of the main pioneers in the acid:base/skeletal link field; to Dr Lynda Frassetto for her outstanding work and willingness to discuss ideas and mechanisms; to International Colleagues in the field of metabolic bone disease: Dr Tim R Arnett (UK); Professor Uriel S Barzel (USA); Professor Jean-Philippe Bonjour (CH); Professor Peter Burckhardt (CH); Professor Bess Dawson-Hughes (USA); Dr Alan St.J. Dixon OBE; Professor Robert Heaney (USA); Professor R. Graham G. Russell (UK). And finally to my greatest supporters: my mother and father, Mrs. Lily Margaret Lanham (nee Topp) and Mr Victor Guy Lanham; my husband William John and our two children, Christian Stephen and Kristabel Pamela.

# References

1. S.A. New, Bone health: the role of micronutrients, *Br. Med. Bull.*, 1999, **55**, 619–633.
2. J. Green and R. Kleeman, Role of bone in regulation of systematic acid–base balance (Editorial Review), *Kidney Int.*, 1991, **39**, 9–26.
3. I. Kurtz, T. Maher, H.N. Hulter, M. Schambelan and A. Sebastian, Effect of diet on plasma acid–base composition in normal humans, *Kidney Int.*, 1983, **24**, 670–680.
4. L.A. Frassetto, R.C. Morris Jr and A. Sebastian, Effect of age on blood acid–base composition in adult humans: role of age-related renal functional decline, *Am. J. Physiol. (Renal Fluid Electrolyte Physiol. 40)*, 1996, **271**, F1114–F1122.
5. L.A. Frassetto and A. Sebastian, Age and systemic acid–base equilibrium: analysis of published data, *J. Gerontol.*, 1996, **51A**, B91–B99.
6. K. Goto, Mineral metabolism in experimental acidosis, *J. Biol. Chem.*, 1918, **36**, 355–376.
7. L. Irving and A.L. Chute, The participation of the carbonates of bone in the neutralisation of ingested acid, *J. Cell Comp. Physiol.*, 1933, **2**, 157.
8. F. Albright and E.C. Reifenstein Jr., *The Parathyroid Glands and Metabolic Bone Disease*, Williams & Wilkins, Baltimore, 1948, pp. 241–247.
9. M.M. Reidenberg, B.L. Haag, B.J. Channick, C.R. Schuman and T.G.G. Wilson, The response of bone to metabolic acidosis in man, *Metabolism*, 1966, **15**, 236–241.
10. C.F. Gastineau, M.H. Power and J.W. Rosevear, Metabolic studies of a patient with osteoporosis and diabetes mellitus: effects of testosterone enanthate and strontium lactate, *Proc. Mayo Clin.*, 1960, **35**, 105–111.
11. J. Lemann Jr., J.R. Litzow and E.J. Lennon, The effects of chronic acid loads in normal man: Further evidence for the participation of bone mineral in the defense against chronic metabolic acidosis, *J. Clin. Invest.*, 1966, **45**, 1608–1614.
12. U.S. Barzel, The role of bone in acid–base metabolism, In *Osteoporosis*, U.S. Barzel (ed.), Grune and Stratton, New York, 1970, pp. 199–206.
13. U.S. Barzel, The effect of excessive acid feeding on bone, *Calcif. Tissue Res.*, 1969, **4**, 94–100.
14. U.S. Barzel and J. Jowsey, The effects of chronic acid and alkali administration on bone turnover in adult rats, *Clin. Sci.*, 1969, **36**, 517–524.
15. T.R. Arnett and D.W. Dempster, Effect of pH on bone resorption by rat osteoclasts in vitro, *Endocrinology*, 1986, **119**, 119–124.
16. T.R. Arnett and D.W. Dempster, Perspectives: protons and osteoclasts, *J. Bone Miner. Res.*, 1990, **5**, 1099–1103.
17. T.R. Arnett, A. Boyde, S.L. Jones and M.L. Taylor, Effects of medium acidification by alteration of carbon dioxide or bicarbonate concentrations on the resorptive activity of rat osteoclasts, *J. Bone Miner. Res.*, 1994, **9**, 375–379.
18. N.A. Kreiger, N.E. Sessler and D.A. Bushinsky, Acidosis inhibits osteoblastic and stimulates osteoclastic activity in vitro, *Am. J. Physiol.*, 1992, **262**, F442–F448.
19. T.R. Arnett and M. Spowage, Modulation of the resorptive activity of rat osteoclasts by small changes in extracellular pH near the physiological range, *Bone*, 1996, **18**, 277–279.
20. D.A. Bushinsky, Metabolic alkalosis decreases bone calcium efflux by suppressing osteoclasts and stimulating osteoblasts, *Am. J. Physiol. (Renal Fluid Electrolyte Physiol.)*, 1996, **271**, F216–F222.
21. S. Meghji, M.S. Morrison, B. Henderson and T.R. Arnett, pH dependence of bone resorption: mouse calvarial osteoclasts are activated by acidosis, *Am. J. Physiol., Endocrinol. Metab.*, 2001, **280**, E112–E119.

22. D.A. Bushinsky, N.E. Sessler, R.E. Glena and J.D.B. Featherstone, Proton-induced physicochemical calcium release from ceramic apatite disks, *J. Bone Miner. Res.*, 1994, **9**, 213–220.

23. A. Wachman and D.S. Bernstein, Diet and Osteoporosis, *Lancet*, 1968, **I**, 958–959.

24. E.M. Widdowson, R.A. McCance and C.M. Spray, The chemical composition of the human body, *Clin. Sci.*, 1951, **10**, 113–125.

25. F.R. Ellis, S. Holesh and J.W. Ellis, Incidence of osteoporosis in vegetarians and omnivores, *Am. J. Clin. Nutr.*, 1972, **25**, 555–558.

26. A.G. Marsh, T.V. Sanchez, O. Micklesen, J. Keiser and G. Major, Cortical bone density of adult lacto-ovo-vegetarians and omnivorous women, *J. Am. Diet. Assoc.*, 1980, **76**, 148–151.

27. A.G. Marsh, T.V. Sanchez, F.L. Chaffee, G.H. Mayor and O. Mickelsen, Bone mineral mass in adult lactoovovegetarian and omnivorous males, *Am. J. Clin. Nutr.*, 1983, **83**, 155–162.

28. A.G. Marsh, T.V. Sanchez, O. Michelsen, F.L. Chaffee and S.M. Fagal, Vegetarian lifestyle and bone mineral density, *Am. J. Clin. Nutr.*, 1988, **48**, 837–841.

29. F. Tylavsky and J.J.B. Anderson, Bone health of elderly lactoovovegetarian and omnivorous women, *Am. J. Clin. Nutr.*, 1988, **48**, 842–849.

30. I.F. Hunt, N.J. Murphy, C. Henderson, V.A. Clark, R.M. Jacobs, P.K. Johnston and A.H. Coulson, Bone mineral content in postmenopausal women: comparison of omsnivores and vegetarians, *Am. J. Clin. Nutr.*, 1989, **50**, 517–523.

31. H.E. Meema, Photographic density versus bone density, *Am. J. Clin. Nutr.*, 1973, **26**, 687 (Letter).

32. F.R. Ellis, S. Holesh and T.A. Sanders, Osteoporosis in British vegetarians and omnivores, *Am. J. Clin. Nutr.*, 1974, **27**, 769–770.

33. H.E. Meema, What's good for the heart is not good for the bones? *J. Bone Miner. Res.*, 1996, **11**, 704 (Letter).

34. U.S. Barzel, Ne'ertheless, an acidogenic diet may impair bone, *J. Bone Miner. Res.*, 1996, **11**, 704 (Letter).

35. T. Lloyd, J.M. Schaeffer, M.A. Walker and L.M. Demers, Urinary hormonal concentrations and spinal bone densities of premenopausal vegetarian and nonvegetarian women, *Am. J. Clin. Nutr.*, 1991, **54**, 1005–1010.

36. R. Tesar, M. Notelovitz, E. Shim, G. Kauwell and J. Brown, Axial and peripheral bone density and nutrient intakes of postmenopausal vegetarian and omnivorous women, *Am. J. Clin. Nutr.*, 1992, **56**, 699–704.

37. J.A. Reed, J.B.B. Anderson, F.A. Tylavsky and P.N. Gallagher Jr., Comparative changes in radial bone density of elderly female lactoovovegetarians and omnivores, *Am. J. Clin. Nutr.*, 1994, **59**, 1197S–1202S.

38. J.F. Chiu, S.J. Lan, C.Y. Yang, P.W. Wang, W.J. Yao, L.H. Su and C.C. Hsieh, Long term vegetarian diet and bone mineral density in postmenopausal Taiwanese women, *Calcif. Tissue Int.*, 1997, **60**, 245–249.

39. E.M. Lau, T. Kwok, J. Woo and S.C. Ho, Bone mineral density in Chinese elderly female vegetarians, vegans, lactoovovegetarians and omnivores, *Eur. J. Clin. Nutr.*, 1998, **52**, 60–64.

40. R. Rizzoli, M.A. Schurch, T. Chevalley, P. Ammann and J.P. Bonjour, Protein intake and osteoporosis, In *Nutritional Aspects of Osteoporosis '97*, P. Burckhardt, B. Dawson-Hughes and R. P. Heaney (eds.), Proceedings of the Third International Symposium on Nutritional Aspects of Osteoporosis, Switzerland, Ares-Serono Symposia Publications, Italy, 1997, pp. 141–158.

41. R.B. Mazess and W.E. Mather, Bone mineral content of North Alaskan Eskimos, *Am. J. Clin. Nutr.*, 1974, **27**, 916–925.
42. R.B. Mazess and W.E. Mather, Bone mineral content in Canadian Eskimos, *Human Biol.*, 1975, **47**, 45.
43. G. Mann, Bone mineral content of North Alaskan Eskimos, *Am. J. Clin. Nutr.*, 1975, **28**, 566–567 (Letter).
44. S.A. New, Impact of Food Clusters on Bone, In *Nutritional Aspects of Osteoporosis '2000 (Fourth International Symposium on Nutritional Aspects of Osteoporosis, Switzerland, 1997)*, B. Dawson-Hughes, P. Burckhardt and R.P. Heaney (eds.), Challenges of Modern Medicine, Ares-Serono Symposia Publications, Academic Press, 2001, pp. 379–397.
45. D. Fox, Hard cheese, *New Scientist*, 2001.
46. S.A. New, H.M. Macdonald, D.M. Reid and A. St. J. Dixon, Hold the soda, *New Scientist*, 2002, **2330**, 54–55.
47. J. Plant and G. Tidey, *Understanding, preventing and overcoming osteoporosis*, Virgin Books Ltd, London, 2003.
48. S.A. New and R.F. Francis, Book review: Understanding, preventing and overcoming osteoporosis by Plant & Tidey, *Science in Parliament*, 2003 (in press).
49. T. Remer and F. Manz, Potential renal acid load of foods and its influence on urine pH, *J. Am. Diet. Assoc.*, 1995, **95**, 791–797.
50. S.A. New, Intake of fruit and vegetables – implications for bone health, *Proceedings of the Nutrition Society*, 2003 (in press).
51. F.A. Tylvasky, K. Holliday, R.K. Danish, C. Womack, J. Norwood, K.M. Ryder and L.S. Carbone, Fruit & vegetable intake is an independent predictor of bone mass in early-pubertal children, *J. Bone Miner. Res.*, 2002, **17**, S459.
52. G. Jones, M.D. Riley and S. Whiting, Association between urinary potassium, urinary sodium, current diet, and bone density in prepubertal children, *Am. J. Clin. Nutr.*, 2001, **73**, 839–844.
53. K. Michaelsson, L. Holmberg, H. Maumin, A. Wolk, R. Bergstrom and S. Ljunghall, Diet, bone mass and osteocalcin; a cross-sectional study, *Calcif. Tissue Int.*, 1995, **57**, 86–93.
54. S.A. New, C. Bolton-Smith, D.A. Grubb and D.M. Reid, Nutritional influences on bone mineral density: a cross-sectional study in premenopausal women, *Am. J. Clin. Nutr.*, 1997, **65**, 1831–1839.
55. S.A. New, S.P. Robins, M.K. Campbell, J.C. Martin, M.J. Garton, C. Bolton-Smith, D.A. Grubb, S.J. Lee and D.M. Reid, Dietary influences on bone mass and bone metabolism: further evidence of a positive link between fruit and vegetable consumption and bone health? *Am. J. Clin. Nutr.*, 2000, **71**, 142–151.
56. H.M. Macdonald, S.A. New, D.A. Grubb and D.M. Reid, Higher intakes of fruit and vegetables are associated with higher bone mass in perimenopausal Scottish women, *Proc. Nutr. Soc.*, 2002, **60**, 202A.
57. S. Whiting, *J. Nutr.*, 2002.
58. H.M. Macdonald, S.A. New, W.D. Fraser and D.M. Reid, Increased fruit and vegetable intake reduces bone turnover in early postmenopausal Scottish women, *Osteoporosis Int.*, 2002, **13**, S97.
59. J. Eaton-Evans, E.M. McIlrath, W.E. Jackson, P. Bradley and J.J. Strain, Dietary factors and vertebral bone density in perimenopausal women from a general medical practice in Northern Ireland, *Proc. Nutr. Soc.*, 1993, **52**, 44A (Abstract).

60. K.L. Tucker, M.T. Hannan, H. Chen, A. Cupples, P.W.F. Wilson and D.P. Kiel, Potassium and fruit and vegetables are associated with greater bone mineral density in elderly men and women, *Am. J. Clin. Nutr.*, 1999, **69**, 727–736.

61. Y. Chen, S.C. Ho, R. Lee, S. Lam and J. Woo, Fruit intake is associated with better bone mass among Hong Kong Chinese early postmenopausal women, *J. Bone Miner. Res.*, 2001, **16**(S1), S386.

62. D.R. Miller, E.A. Krall, J.J. Anderson, S.E. Rich, A. Rourke and J. Chan, Dietary mineral intake and low bone mass in men: The VALOR Study, *J. Bone Miner. Res.*, 2001, **16**(S1), S395.

63. K.L. Stone, T. Blackwell, E.S. Orwoll, J.C. Cauley, E. Barrett-Connor, R. Marcus, M.C. Nevitt and S.R. Cummings, The relationship between diet and bone mineral density in older men, *J. Bone Miner. Res.*, 2001, **16**(S1), S388.

64. S.A. New, R. Smith, J.C. Brown and D.M. Reid, Positive associations between fruit and vegetable consumption and bone mineral density in late postmenopausal and elderly women, *Osteoporosis Int.*, 2002, **13**, S77.

65. J. Lemann Jr., J.R. Litzow and E.J. Lennon, Studies of the mechanisms by which chronic metabolic acidosis augments urinary calcium excretion in man, *J. Clin. Invest.*, 1967, **46**, 1318–1328.

66. J. Lemann Jr., N.D. Adams and R.W. Gray, Urinary calcium excretion in humans, *N. Engl. J. Med.*, 1979, **301**, 535–541.

67. J. Lemann Jr., R.W. Gray, W.J. Maierhofer and H.S. Cheung, The importance of renal net acid excretion as a determinant of fasting urinary calcium excretion, *Kidney Int.*, 1986, **29**, 743–746.

68. J. Lemann Jr., R.W. Gray and J.A. Pleuss, Potassium bicarbonate, but not sodium bicarbonate, reduces urinary calcium excretion and improves calcium balance in healthy men, *Kidney Int.*, 1989, **35**, 688–695.

69. J. Lemann Jr., J.A. Pleuss, R.W. Gray and R.G. Hoffmann, Potassium administration increases and potassium deprivation reduces urinary calcium excretion in healthy adults, *Kidney Int.*, 1991, **39**, 973–983.

70. D.A. Bushinsky, N.S. Kreiger, D.I. Geisser, E.B. Grossman and F.L. Coe, Effects of bone calcium and proton fluxes in vitro, *Am. J. Physiol.*, 1983, **245**, F204–F209.

71. D.A. Bushinsky and N.E. Sessler, Critical role of bicarbonate in calcium release from bone, *Am. J. Physiol.* (*Renal Fluid Electrolyte Physiol.*), 1992, **263**, F510–F515.

72. D.A. Bushinsky, B.C. Lam, R. Nespeca, N.E. Sessler and M.D. Grynpas, Decreased bone carbonate content in response to metabolic, but not respiratory, acidosis, *Am. J. Physiol.* (*Renal Fluid Electrolyte Physiol.* 34), 1993, **265**, F530–F536.

73. D.A. Bushinsky, K. Gavrilov, J.M. Chabala and R. Levi-Setti, Metabolic acidosis decreases potassium content of bone, *J. Am. Soc. Nephrol.*, 1997, **7**, 1787.

74. D.A. Bushinsky, Decreased potassium stimulates bone resorption, *Am. J. Physiol.* (*Renal Fluid Electrolyte Physiol.*), 1997, **272**, F774–F780.

75. A. Sebastian, S.T. Harris, J.H. Ottaway, K.M. Todd and R.C. Morris Jr., Improved mineral balance and skeletal metabolism in postmenopausal women treated with potassium bicarbonate, *N. Engl. J. Med.*, 1994, **330**, 1776–1781.

76. R.J. Wood, Potassium bicarbonate supplementation and calcium metabolism in postmenopausal women: are we barking up the wrong tree? *Nutr. Rev.*, 1994, **52**, 278–280.

77. T. Buclin, M. Cosma, M. Appenzeller, A.F. Jacquet, L.A. Decosterd, J. Biollaz and P. Burckhardt, Diet acids and alkalis influence calcium retention on bone, *Osteoporosis Int.*, 2001, **12**, 493–499.

78. L.J. Appel, T.J. Moore, E. Obarzanek, W.M. Vallmer, L.P. Svetkey, F.M. Sacks, G.A. Bray, T.M. Vogt and J.A. Cutler, A clinical trial of the effects of dietary patterns on blood pressure, *N. Engl. J. Med.*, 1997, **336**, 1117–1124.
79. U.S. Barzel, Dietary patterns and blood pressure, *N. Engl. J. Med.*, 1997, **337**, 637 (Letter).
80. P. Lin, F. Ginty, L. Appel, L. Svetky, A. Bohannon, D. Barclay, R. Gannon and M. Aickin, Impact of sodium intake and dietary patterns on biochemical markers of bone and calcium metabolism, *J. Bone Miner. Res.*, 2001, **16**(S1), S511.
81. L. Frassetto, K. Todd, R.C. Morris Jnr and A. Sebastian, Estimation of net endogenous noncarbonic acid production in humans from dietary protein and potassium contents, *Am. J. Clin. Nutr.*, 1998, **68**, 576–583.
82. S.A. New, H.M. Macdonald, M.K. Campbell, J.C. Martin, M.J. Garton, S.P. Robins and D.M. Reid, Lower estimates of net endogenous noncarbonic acid production (NEAP) are positively associated with indices of bone health in pre/peri-menopausal women, *Am. J. Clin. Nutr.*, 2003 (in press).
83. H.M. Macdonald, S.A. New, W.D. Fraser and D.M. Reid, Estimates of NEAP are associated with increased bone turnover in early postmenopausal women: findings from APOSS longitudinal, *J. Bone Miner. Res.*, 2002, **17**, 1131.
84. R.P. Heaney, Excess dietary protein may not adversely affect bone, *J. Nutr.*, 1998, **128**, 1054–1057.
85. B. Dawson-Hughes and S.S. Harris, Calcium intake influences the association of protein intake with rates of bone loss in elderly men and women, *Am. J. Clin. Nutr.*, 2002, **75**, 773–779.
86. R.P. Heaney, Protein and calcium: antagonists or synergists? *Am. J. Clin. Nutr.*, 2002, **75**, 609.
87. S.A. New and D.J. Millward, Calcium, protein and fruit and vegetables as dietary determinants of bone health, *Am. J. Clin. Nutr.*, 2003, **77**, 1340–1341 (letter).
88. B.S. Eaton and M. Konner, Paleolithic nutrition. A consideration of it's nature and current implications, *N. Engl. J. Med.*, 1985, **312**, 283–290.
89. B.H. Patterson, G. Block and W.F. Rosenberger, Fruit and vegetables in the American diet: data from the NHANES II Survey, *Am J Public Health*, 1990, **80**, 1443–1449.
90. S. Finch, W. Doyle, C. Lowe, C.J. Bates, A. Prentice, G. Smithers and P.C. Clarke, *National Diet and Nutrition Survey: People Aged 65 Years and Over*, The Stationary Office, London, 1998.
91. M.S. Oh and J. Uribarri, Bone buffering of acid: fact of fancy? *J. Nephrol.*, 1996, **9**, 261–262.
92. R.C. Muhlbauer, A.M. Lozano and A. Reinli, Onion and a mixture of vegetables, salads and herbs affect bone resorption in the rat by a mechanism independent of their base excess, *J. Bone Miner. Res.*, 2002, **17**, 1230–1236.
93. R.C. Muhlbauer, R. Felix, A. Lozano, S. Palacio and A. Reinli, Common herbs, essential oils and monoterpenes potently modulate bone metabolism, *Bone*, 2003, (in press).
94. S.A. New, The role of the skeleton in acid–base homeostasis. The 2001 Nutrition Society Medal Lecture, *Proc. Nutr. Soc.*, 2002, **61**, 151–164.

# The Effect of Mineral Waters on Bone Metabolism and Bone Health

PETER BURCKHARDT

Department of Internal Medicine, University Hospital – CHUV, 1011 Lausanne, Switzerland, Email: peter.burckhardt@chuv.hospvd.ch

## Key Points

- Mineral waters vary considerably in their content of ions and are mostly of limited nutritional interest; but many are rich in ions that influence bone metabolism, especially calcium, which lowers PTH and resorption markers of bone immediately after absorption.
- When consumed over years, these waters contribute to bone health. But, of equal importance is their content in potassium and bicarbonate.
- Bicarbonate lowers bone resorption and calcium excretion and positivates calcium balance.
- Taken in form of mineral waters rich in bicarbonate, mineral waters also inhibit bone resorption. Potassium too is known to decrease calcium excretion. Therefore, potassium too might be a constituent of mineral waters that is crucial for bone health.
- There is some evidence to support the usefulness of mineral waters rich in bicarbonate and potassium.

## Introduction

The consumption of bottled water is rapidly increasing in many countries worldwide. In the U.S., it rose to about 50 L per capita per year.[1] Part of this rise concerns 'mineral waters'. This term includes spring waters, natural or pumped waters, and processed tap water, all being carbonated (sparkling) or not, while the term "natural" indicates that the added $CO_2$ is derived from the same spring.[2] This chapter deals with spring waters, since they can contain minerals at concentrations high enough to be of nutritional importance when taken regularly. It also concentrates on those minerals that have a proven influence on bone health, independently of the fact that the water is carbonated or not. The observation that the consumption of 'carbonated beverages' in schoolchildren is associated with a higher incidence of bone fractures,[3] mainly concerns Cola drinks and not mineral waters. The bone fractures were explained by the 'displacement of milk' rather than by a specific effect of these drinks on bone metabolism, apart from the small calciuric effect of caffeine-containing drinks.[4]

The mineral content of mineral waters varies greatly.[1] For instance, in a series of mineral waters consumed in the U.S., the calcium content varied between 2 and 358 mg/L[2] and in a review of 226 German mineral waters it varied between 1.4 and about 500 mg/L[5] while one source indicated 801.6 mg/L, a concentration which comes close to that found in dairy products. However, the composition of the analyzed waters concerning the ions with a proven specific effect on bone metabolism is usually incompletely indicated in these reports, depending on the particular interest of each author. Maximal concentrations of such ions can be as high as 801.6 mg $L^{-1}$ Calcium, 12 830 mg $L^{-1}$ Potassium (Bad Mergentheimer Albertquelle, Germany), 1040 mg $L^{-1}$ Chloride (Roisdorfer Mineralwasser, Germany), 1708 Sodium, 9 mg $L^{-1}$ Fluoride (StYorre, France), 3486 mg $L^{-1}$ Bicarbonate (Vichy Célestins, France). If 1 L or more of such a water with a high content of a given mineral is regularly consumed over several years, the metabolic effect on bone can be of considerable importance. An example mentioned is the increased bone density observed in subjects used to drinking a fluoride-rich mineral water over many years.[6]

This chapter reviews the published studies on these effects, and attempts to derive from these data conclusive observations and some recommendations.

# Calcium

For obvious reasons, the calcium content of mineral waters has been given the most attention and often an exclusive one. Calcium from mineral water is as well absorbed, as that from milk or cheese, as demonstrated by four studies over 24–36 h using labeled calcium.[7] While in one study with lactase-deficient subjects, calcium from mineral water was even slightly better absorbed,[8] the other studies showed remarkably similar figures,[9,10] and one showed significantly higher absorption from water when taken together with a meal.[11] In these studies, the fractional absorption depended mainly on the amount taken, in the expected inverted function of the load,[12] and varied between 25 and 45%.[7] A comparison of three mineral waters with calcium concentrations between 10.4 and 467 mg $L^{-1}$ showed no significant difference in fractional absorption (33–39%) when the amount of calcium provided was kept at 200 mg.[13]

For this reason, calcium taken in the form of mineral water has the same biologic effects as that taken in oral calcium preparations. About 200 mg calcium from mineral water caused a sharp fall in plasma PTH by 34–43% within 1 h, and a decrease in serum telopeptides CTX by 17%, effects which could not be obtained with a low calcium water providing only 5 mg.[14,15]

It is only partially known whether this biologic effect on PTH and resorption markers improves bone density in the long term. A crossover study on 255 women drinking regularly calcium-rich water or not and so differing in their total calcium intake, showed a significantly higher lumbar BMD in the subgroup of postmenopausal women with the higher calcium intake.[16] This suggests that the regular intake of a calcium-rich mineral water represents an essential contribution to the total calcium intake with a measurable effect on bone density. However, conclusive interventional trials are still rare. In one controlled follow-up study over 13±1 months, the intake of a high-calcium water in postmenopausal women led to an increase of BMC at the distal radius, when compared with the intake of a low-calcium water.[17]

It therefore might be in the interest of public health to promote the consumption of calcium-rich mineral waters, especially in regions where the concentration in the tap water is low. There is no risk of promoting renal stones, since water reduces this risk by dilution, and because calcium-poor waters increase the urinary concentration of oxalate.[18] In North America, tap water in the major cities is relatively low in calcium ($\pm 34$ mg $L^{-1}$), and does not exceed 83 mg $L^{-1}$; bottled spring waters are even lower, while mineral waters vary between 1 and 310 mg $L^{-1}$.[19] In Germany, which includes both flatlands and alpine regions, 12% of the mineral waters contain more than 400 mg $L^{-1}$ calcium, while only 15% contain less than 40 mg $L^{-1}$.[7] Some mineral waters are marketed as rich in calcium, when the content is above 200 mg $L^{-1}$, but the use of this denomination is not regulated. Assuming a daily fluid intake of 1 L in the form of bottled water, and considering a calcium supplementation as significant when exceeding 300 mg $day^{-1}$, a mineral water could be considered as rich in calcium when it contains at least 300 mg $L^{-1}$. This would apply to about 20% of all the German mineral waters,[7] 12% of the major French mineral waters, and to about 10% of the European mineral waters that are commercially available in the U.S.[19] If 200 mg $L^{-1}$ was the limit, 30–40% of mineral water in European countries with alpine regions could be called 'calcium-rich'.

# Sodium

As discussed in the chapter by Flynn and Cashmam, a high sodium intake increases renal calcium excretion and bone resorption, while low salt intake has the opposite effect. Theoretically, mineral waters can contribute to the total sodium intake. But in fact, even a high sodium concentration of several hundred mg per liter is still a modest contribution to the dietary sodium intake, even in a patient on a low salt diet. Therefore, it is improbable that such a mineral water has a negative effect on bone health, or that the choice of a water with a low sodium content contributes effectively to the eventual benefit for bone health of a low sodium diet. It only has to be considered that the rare mineral waters with about 1 g of Na per liter, as *e.g.*, Vichy Springs California U.S., not only are a noticeable contribution to dietary sodium, but also might have a calciuric effect. This, however, has never been demonstrated.

The question remains, if a high content of sodium in a given mineral water counterbalances the calcium-sparing effect of potassium and bicarbonate. There is no data answering this question. It can be noted that only $KHCO_3$ but not $NaHCO_3$ reduce urinary calcium excretion in healthy men,[20] while they both increase renal base excretion by a similar amount.[21]

# Acid Load

The body normally produces 50–100 mEq of non carbonic protons by metabolizing the ingested food.[22] Each minor decrease of the pH stimulates bone resorption and inhibits tubular reabsorption of calcium, while each increase of the pH has the opposite effect. Therefore, acid load from food theoretically stimulates bone resorption. Indeed, an oral acid load, such as 2 mmol $NH_4Cl_2$ per kg body weight, increased urinary excretion of calcium (and magnesium) in correlation with the net acid excretion (NAE).[23] This leads to the conclusion that waters with an acid load can theoretically lead to increased renal loss of calcium, or at least diminish the profit derived from their calcium content. Therefore, mineral waters should not increase the acid load of food; it might even be advantageous for bone health, if they provide an alkaline load.

The potential renal acid load (PRAL) of a mineral water can be calculated with the formula proposed for nutrients by Remer and Manz:[24] NAE in mEq $= (Cl + P + SO_4 + \text{organic acid}) - (Na + K + Ca + Mg)$. For each ion, the fractional intestinal absorption is taken into account, *i.e.*, Na mg $\times 0.95$ (95% absorption) $\times 0.0413 = \text{mEq}$, Ca $\times 0.25$ (25% absorption) $\times 0.0125 = \text{mEq}$, *etc.* NAE is significantly correlated with urinary pH. According to this calculation, many mineral waters represent an acid load, *e.g.*, San Pellegrino (Italy) 6.75 mEq $L^{-1}$, Contrex (France) 13.2 mEq $L^{-1}$, Aproz (Switzerland) 15.3 mEq $L^{-1}$, while others represent an alkaline load, like Passugger (Switzerland) 4.64 mEq $L^{-1}$, Vichy (France) 42.7 mEq $L^{-1}$.[25] This is still a relatively small acid load, as already shown for Cola-drinks.[4] Moreover, the calculation is probably incorrect for waters rich in bicarbonate, since it ignores bicarbonates, assuming that they are totally metabolized, producing $CO_2$ with no product eliminated in the urine. But the intake of bicarbonate increases urinary excretion of bicarbonate. Therefore, when bicarbonates are included in the formula with an absorption of 95%, the renal acid load from mineral waters containing some bicarbonate, must be smaller: *e.g.*, for San Pellegrino (223 mg $L^{-1}$ bicarbonate) 3.3 mEq $L^{-1}$, for Contrex (386 mg $L^{-1}$ bicarbonate) 7.2 mEq $L^{-1}$, and for Aproz (153 mg $L^{-1}$ bicarbonate) 12.9 mEq $L^{-1}$.

Since these figures are low compared to the normal acid load induced by food, the acid load induced by some mineral waters can probably be ignored.

# Bicarbonate

Bicarbonate is present in most mineral waters, from a low 50–100 mg $L^{-1}$ up to several g $L^{-1}$. Bicarbonate has an anticalciuric effect.[26] In the form of tablets, K-bicarbonate (3.7–7.4 g $day^{-1}$ bicarbonate), given to postmenopausal women on a diet with a normal endogenous acid production (EAP) of 1.25 mEq $kg^{-1}$ body weight, decreased EAP nearly to zero,[27] improved calcium balance by +56 mg $day^{-1}$ and phosphorus balance by 47 mg $day^{-1}$, and lowered urinary excretion of hydroxyproline, a marker of bone resorption, slightly, while net renal acid excretion dropped by 82%.[28] Given in the form of mineral water, 3.4 g $day^{-1}$ bicarbonate decreased urinary calcium excretion,[29] and combined with a high potassium alkalinizing diet, 4.4 g $day^{-1}$ bicarbonate lowered bone resorption and urinary calcium excretion significantly, although calcium, sodium and protein intakes were kept constant.[30]

# Potassium

The anticalciuric effect of bicarbonate and the inhibition of bone resorption by an alkaline food load could lead to the proposition that for the prevention of osteoporosis, mineral waters rich in bicarbonate are to be preferred to high-calcium waters with an acid load. But the question remains open, as to whether the advantageous effect of bicarbonate-rich waters is not due to the potassium content rather to the bicarbonate. Indeed, and as already mentioned, $KHCO_3$ but not $NaHCO_3$ reduced urinary calcium excretion and improved calcium balance,[20] although both $KHCO_3$ and $NaHCO_3$ increase renal base excretion by a similar amount.[21] After several days of delay, K deprivation results in increased calcium excretion. The latter can be corrected when K is added to the diet, whether the accompanying ion is $Cl^-$ or $HCO_3^-$.[31] Potassium causes calcium retention in healthy adults, and reduces urinary calcium excretion even in the presence of a high protein load.[32] Therefore, potassium seems to exert a calcium sparing effect, independently of bicarbonate. Indeed, the reduction of calcium excretion as well as the improvement of the calcium balance, obtained by the intake of $KHCO_3$, both occur in relation to the urinary potassium excretion.[33]

The mechanism of this effect is unknown. Potassium and bicarbonate might have additional effects. It has to be considered that in combination with thiazides, $KHCO_3$ reduces calcium excretion more than KCl, since it neutralizes completely the EAP.[34] This allows us to conclude that mineral waters rich in both potassium and bicarbonate might have a positive potential for bone health.

There are mineral waters that are especially rich in potassium and in bicarbonate, but the literature on this eventual advantage is poor. As already mentioned, a short term study showed that such a mineral water (Vichy), together with a diet with an alkaline load, not only decreased urinary calcium excretion, but also markers of bone resorption.[30] In the median term too, bicarbonate-rich mineral waters seem to fulfil the expectations, according to one study. Given over a month, a mineral water rich in bicarbonate (1762 mg $L^{-1}$) and potassium (52 mg $L^{-1}$), although less than the extreme example of Vichy (3245 and 72 mg $L^{-1}$) decreased urinary markers of bone resorption.[35] It remains open whether the positive effect of these waters can be confirmed on the long term. Since a modern diet leads to low grade metabolic acidosis and K deficiency,[36] the recommendation to increase K and bicarbonate intake could be achieved, at least partially, with mineral waters.

# Fluoride

Fluoride stimulates bone formation, and for this reason the fluoride content of water has been examined in respect to its effect on bone. One French mineral water (St.Yorre) contains 8.5 mg $L^{-1}$, an exceptionally high concentration. Regular consumers of this water with a minimal daily intake of $\frac{3}{4}$ L over at least 5 years, *i.e.*, 6.38 mg $day^{-1}$, showed elevated bone mineral density in the lumbar spine, and a high plasma level of osteocalcine.[6] This is an exceptional finding, since most mineral waters have less than 1 mg $L^{-1}$, and only one out of 220 German mineral waters contains more than 4 mg $L^{-1}$. But 6% contain more than 2 mg $L^{-1}$,[5] as well as 16% of mineral waters consumed in the U.S.[2] At this concentration, fluoride might have an effect on bone when consumed over many years. But the question if this involuntary long-term intake of low doses of fluoride has a protective effect against osteoporosis, as shown in the rat,[37] or has no antifracture effect as the treatment with 15–20 mg $day^{-1}$, remains open.

# Sulphate

Sulphate is an almost regular component of mineral waters, and might exceed the standard of 250 mg $L^{-1}$ of drinking water in the U.S. and the E.E.C. It has no proven effect on bone. Although an increase of urinary calcium excretion could be shown in sheep on a sulphate-rich diet containing 4 mg $kg^{-1}$ dry food,[38] no such effect was shown in humans. Sulphate in mineral water neither inhibits the absorption of its calcium, nor does it increase urinary calcium excretion,[10] and in rats it has no effect on growth.[39] Even 1 L of a water that is particularly rich in sulphate as, *e.g.*, Contrexeville (France) with 550 mg $L^{-1}$, provides less sulphate than a normal diet.

# Conclusions

Although many 'mineral waters' contain only low concentrations of minerals, others are rich in ions that have a known influence on bone metabolism, especially calcium and fluoride. Mineral waters have shown to decrease plasma

PTH and bone resorption markers due to their calcium content. Regular intake of such a mineral-rich water over several years sums up to relevant amounts with noticeable effects on bone health. However, mineral waters are not only vehicles of nutritional calcium. Although the eventual acid load of a given mineral water is insignificant, the alkaline load might be of crucial importance. In addition, human and animal data point to a strong effect of potassium on renal handling of calcium. Some waters are rich in bicarbonate, and/or in potassium, and have been shown to decrease urinary calcium excretion and bone resorption. When also characterized by a relatively high concentration of calcium, these waters might have a positive long-term effect in the prevention of osteoporosis, and should be preferred to high-calcium waters, which are low in potassium and bicarbonate, and rich in sodium. However, this has first to be proven.

# References

1. Ph. Garzon and M.J. Eisenberg, Variation in the mineral content of commercially available bottled waters: implications for health and disease, *Am. J. Med.*, 1998, **105**, 125–130.
2. H.E. Allen, M.A. Halley-Henderson and C.N. Hass, Chemical composition of bottled mineral water, *Arch. Environ. Health*, 1989, **44**(2), 102–116.
3. G. Wyshak, Teenaged girls, carbonated beverage consumption, and bone fractures, *Arch. Pediatr. Adolesc. Med.*, 2000, **154**, 610–613.
4. R.P. Heaney and K. Rafferty, Carbonated beverages and urinary calcium excretion, *Am. J. Clin. Nutr.*, 2001, **74**, 343–347.
5. Willershausen, H. Kroes and M. Brandenbusch, Evaluation of the contents of mineral water, spring water, table water and SPA water, *Eur. J. Med. Res.*, 2000, **5**, 251–262.
6. P.J. Meunier, M. Femenias, F. Duboeuf, M.C. Chapuy and P.D. Delmas, Increased vertebral bone density in heavy drinkers of mineral water rich in fluoride, *Lancet*, 1989, **1**(8630), 152.
7. H. Böhmer, H. Müller and K.L. Resch, Calcium supplementation with calcium-rich mineral waters: a systematic review and meta-analysis of its bioavailability, *Osteoporosis Int.*, 2000, **11**, 938–943.
8. G.M. Halpern, J. Van de Water, A.M. Delabroise, C.L. Keen and M.E. Gershwin, Comparative uptake of calcium from milk and a calcium-rich mineral water in lactose intolerant adults: implications for treatment of osteoporosis, *Am. J. Prev. Med.*, 1991, **7**(6), 379–383.
9. R.P. Heaney and M.S. Dowell, Absorbability of the calcium in a high-calcium mineral water, *Osteoporosis Int.*, 1994, **4**, 323–324.
10. F. Couzy, P. Kastenmayer, M. Vigo, J. Clough, R. Munoz-Box and D.V. Barclay, Calcium bioavailability from a calcium- and sulfate-rich mineral water, compared with milk, in young adult women, *Am. J. Clin. Nutr.*, 1995, **62**, 1239–1244.
11. W. Van Dokkum, V. de La Guéronnière, G. Schaafsma, C. Bouley, J. Luten and C. Latgé, Bioavailability of calcium of fresh cheeses, enteral food and mineral water. A study with stable calcium isotopes in young adult women, *Br. J. Nutr.*, 1996, 893–903.

12. R.P. Heaney, C.M. Weaver and M.L. Fitzsimmons, The influence of calcium load on absorption fraction, *J. Bone Mineral Res.*, 1990, **11**, 1135–1138.

13. A. Wynckel, C. Hanrotel, A. Wuillai and J. Chanard, Intestinal calcium absorption from mineral water, *Miner. Electrolyte Metab.*, 1997, **23**, 88–92.

14. J. Guillemant, H.T. Le, S. Guillemant, A.M. Delabroise and M.J. Arnaud, Acute effects induced by a calcium-rich mineral water on calcium metabolism and on parathyroid function, *Osteoporosis Int.*, 1997, **7**, 85–86.

15. J. Guillemant, H.T. Le, C. Accarie, S. Tézenas du Montcel, A.M. Delabroise, M.J. Arnaud and S. Guillemant, Mineral water as a source of dietary calcium acute effects on parathyroid function and bone resorption in young men, *Am. J. Clin. Nutr.*, 2000, **71**, 999–1002.

16. D. Costi, P.G. Calcaterra, N. Iori, S. Vourna, G. Nappi and M. Passeri, Importance of bioavailable calcium drinking water for the maintenance of bone mass in post-menopausal women, *J. Endocrinol. Invest.*, 1999, **22**, 852–856.

17. C. Cepollaro, G. Orlandi, S. Gonnelli, G. Ferrucci, J.C. Arditti, D. Borracelli, E. Toti and C. Gennari, Effect of calcium supplementation as a high-calcium mineral water on bone loss in early postmenopausal women, *Calcif. Tissue Int.*, 1996, **59**(4), 238–239.

18. P. Jaeger, L. Portmann, A.F. Jacquet and P. Burckhardt, Drinking water for stone formers: is the calcium content relevant? *Eur. Urol.*, 1984, **10**, 53–54.

19. A. Azoulay, P. Garzon and M.J. Eisenberg, Comparison of the mineral content of tap water and bottled waters, *J. Gen. Intern. Med.*, 2001, **16**, 168–175.

20. J. Lemann, R.W. Gray and J.A. Pleuss, Potassium bicarbonate, but not sodium bicarbonate, reduces urinary calcium excretion and improves calcium balance in healthy men, *Kidney Int.*, 1989, **35**, 688–695.

21. M.I. Lindinger, T.W. Franklin, L.C. Lands, P.K. Pedersen, D.G. Welsh and G.J.F. Heigenhauser, $NaHCO_3$ and $KHCO_3$ ingestion rapidly increases renal electrolyte excretion in humans, *J. Appl. Physiol.*, 2000, **88**, 540–550.

22. B.D. Rose, Clinical Physiology of Acid–Base and Electrolyte Disorders, 3rd edn, McGraw Hill, 1977.

23. P. Houillier, M. Normand, M. Froissart, A. Blanchard, P. Jungers and M. Paillard, Calciuric response to an acute acid load in healthy subjects and hypercalciuric calcium stone formers, *Kidney Int.*, 1996, **50**, 987–997.

24. T. Remer and F. Manz, Potential renal acid load of foods and its influence on urine pH, *J. Am. Dietetic Assoc.*, 1995, **95**(7), 791–797.

25. F. Kieffer, Neubeurteilung des Säure- und Basenpotentials von Lebensmitteln. Unpublished data, Berne, November 2001.

26. J. Lemann, J.A. Pleuss, L. Hornick and R.G. Hoffman, Dietary NaCl-restriction prevents the calciuria of KCl-deprivation and blunts the calciuria of KHCO3-deprivation in healthy adults, *Kidney Int.*, 1995, **47**(3), 899–906.

27. L. Frassetto, R.C. Morris and A. Sebastian, Potassium bicarbonate reduces urinary nitrogen excretion in postmenopausal women, *J. Clin. Endocrinol. Metab.*, 1997, **82**(1), 254–259.

28. A. Sebastian, S.T. Harris, J.H. Ottaway, K.M. Todd and R.C. Morris, Improved mineral balance and skeletal metabolism in postmenopausal women treated with potassium bicarbonate, *N. Engl. J. Med.*, 1994, **330**(25), 1776–1781.

29. T. Kessler and A. Hesse, Cross-over study of the influence of bicarbonate-rich mineral water on urinary composition in comparison with sodium potassium citrate in healthy male subjects, *Br. J. Nutr.*, 2000, **84**, 865–871.

30. T. Buclin, M. Cosma, M. Appenzeller, A.F. Jacquet, L.A. Décosterd, J. Biollaz and P. Burckhardt, Diet acids and alkalis influence calcium retention in bone, *Osteoporosis Int.*, 2001, **12**, 493–499.

31. J. Lemann, J.A. Pleuss, R.W. Gray and R.G. Hoffmann, Potassium administration reduces and potassium deprivation increases urinary calcium excretion in healthy adults, *Kidney Int.*, 1991, **39**, 973–983.

32. S.J. Whiting, D.J. Anderson and S.J. Weeks, Calciuric effects of protein and potassium bicarbonate but not of sodium chloride or phosphate can be detected acutely in adult women and men, *Am. J. Clin. Nutr.*, 1997, **65**(5), 1465–1472.

33. J. Lemann, J.A. Pleuss and R.W. Gray, Potassium causes calcium retention in healthy adults, *J. Nutr.*, 1993, **123**, 1623–1626.

34. L.A. Frassetto, E. Nash, R.C. Morris and A. Sebastian, Comparative effects of potassium chloride and bicarbonate on thiazide-induced reduction in urinary calcium excretion, *Kidney Int.*, 2000, **58**, 748–752.

35. P. Burckhardt, S. Waldvogel Abramovski, J.M. Aeschlimann and M.J. Arnaud, Bicarbonate in mineral water inhibits bone resorption, *J. Bone Miner. Res.*, 2002, **17** (Suppl. 1), S476–M360.

36. R.C. Morris, O. Schmidlin, M. Tanaka, A. Forman, L. Frassetto and A. Sebastian, Differing effects of supplemental KCl and KCHCO3: pathophysiological and clinical implications, *Semin. Nephrol.*, 1999, **19**(5), 487–493.

37. C.H. Turner, M.P. Akhter and R.P. Heaney, The effects of fluoridated water on bone strength, *J. Orthop. Res.*, 1992, **10**(4), 581–587.

38. L. Guéguen and P. Besançon, Influence des sulfates sur le métabolisme phospho-calcique, *Ann. Biol. Animale, Biochim. Biophys.*, 1972, **12**(4), 589–598.

39. H.P. Würzner, Exposure of rats during 90 days to mineral water containing various amounts of sulphate, *Z. Ernährungswiss.*, 1979, **18**, 119–127.

CHAPTER 14

# Dietary Vitamin K and Skeletal Health*

SARAH L. BOOTH

Jean Mayer USDA Human Nutrition Research Center on Aging, Tufts University, Boston, MA, 02111, Email: sarah.booth@tufts.edu

**Key Points**

- The presence of vitamin K dependent proteins in the skeleton has been used to support the hypothesis that vitamin K has a role in bone biology.
- Average self-reported dietary intakes of vitamin K are below recommended intakes among adults in the US and UK. Low dietary intakes of vitamin K are potentially a modifiable risk factor for osteoporosis.
- In humans, much of the supporting evidence of a role for dietary vitamin K in age-related bone loss is based on reported associations between dietary intakes or biological markers of vitamin K status and bone mineral density (BMD) or hip fracture risk.
- One criticism of the epidemiologic data is the potential confounding effect of overall poor nutrition.
- There is some evidence that different forms of vitamin K may have variable influence on bone turnover.
- The proposed dietary role of vitamin K as a risk factor for osteoporosis would be strengthened if controlled changes in vitamin K intakes, obtained from the diet or supplements, were consistently shown to influence bone metabolism.

## Structure of Vitamin K

Vitamin K refers to a family of compounds with a common chemical structure, 2-methyl-1,4-napthoquinone (Figure 14.1). This parent compound, also known

*This material is based upon work supported by the US Department of Agriculture, under agreement No. 58-1950-9-001. Any opinions, findings, conclusion, or recommendations expressed in this publication are those of the authors and do not necessarily reflect the view of the US Department of Agriculture.

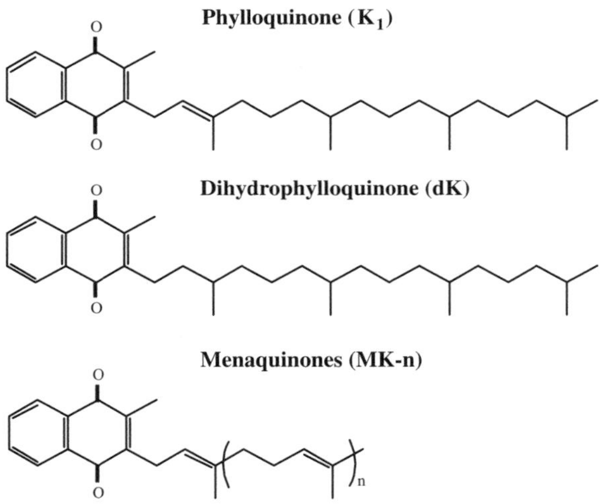

**Figure 14.1**   *Structural formulas for different forms of vitamin K*

as menadione or vitamin $K_3$, is not occurring naturally. Phylloquinone, or vitamin $K_1$, is a component of the Photosystem I of plants,[1] and is present in foods of plant origin. More recently, phylloquinone has been added in varying amounts to some dietary supplements. During the process of hydrogenation of certain phylloquinone-rich vegetable oils, phylloquinone is converted to $2',3'$-dihydrophylloquinone.[2] Dihydrophylloquinone differs from the parent form phylloquinone, by saturation of the $2',3'$ double bond of the phytyl side chain (Figure 14.1). Bacterial forms of vitamin K, referred to as the menaquinones or vitamin $K_2$, differ in structure from phylloquinone in their 3-substituted lipophilic side chain. The major menaquinones contain 4–10 repeating isoprenoid units, indicated by MK-4 to MK-10, although forms up to 13 isoprenoid groups have been identified (Figure 14.1). Menaquinone-4 (MK-4) is not a major constituent of bacterial production; instead it is alkylated from menadione present in animal feeds or is the product of tissue-specific conversion directly from dietary phylloquinone.[3]

Different forms of vitamin K have a tissue-specific distribution. The liver, which is the major storage site for vitamin K, contains long-chain menaquinones (MK-7 through to MK-13) with limited capacity for storing phylloquinone.[4] In plasma and bone, the major forms of vitamin K are phylloquinone, followed by short-chain menaquinones, such as MK-4 through to MK-8.[5] Animal studies confirm that phylloquinone and the short-chain menaquinones are the predominant form in extrahepatic tissues, but the proportion of phylloquinone, MK-4 and MK-6 varies with age, gender,[6] tissue[7] as well as diet.[8] Dietary phylloquinone is converted to MK-4 in extrahepatic tissues, such as brain and kidney, but not in the liver.[3]

The physiological relevance of this tissue-specific distribution of different forms of vitamin K is not well understood.

## Potential Mechanisms

The only established biochemical role for vitamin K is as a cofactor specific to the formation of γ-carboxyglutamyl (Gla) residues from specific glutamate residues in certain proteins.[9] Several hepatic and extrahepatic proteins have now been identified that contain Gla, and are termed vitamin K dependent or Gla-containing proteins.[10] The observation that vitamin K dependent proteins are present in the skeleton has been used to support the hypothesis that vitamin K has a role in bone biology. The Gla residues in these vitamin K dependent proteins confer mineral binding properties. It has been shown that Gla is essential for the biological activity of the coagulation proteins.[11] It is currently assumed that Gla is also essential for the biological activity of those vitamin K dependent proteins in bone.

There are at least three vitamin K dependent proteins present in bone and cartilage: osteocalcin, matrix Gla protein (MGP) and protein S.[10] Osteocalcin is a small 49-amino acid protein produced by osteoblasts during bone matrix formation, and is one of the most abundant non-collagenous proteins in bone.[12] Osteocalcin synthesis is regulated by, but not absolutely dependent on, 1,25-dihydroxvitamin D, and its hydroxyapatite-binding capacity is associated with the vitamin K dependent γ-carboxylation of its three glutamate residues (residues 17, 21 and 24). Osteocalcin is thought to act as a regulator of bone mineral, although its precise physiological function is yet to be elucidated.[13] The degree of carboxylation of Glu residues in osteocalcin obtained from human adult bone has been shown to be incomplete, particularly at the first potential Gla residue at position 17.[14] The Gla residue at position 17 is essential for the conformation that facilitates the selective binding of osteocalcin to hydroxyapatite,[15] such that partially carboxylated osteocalcin may have reduced binding to the hydroxyapatite.

MGP is present in cartilage and bone.[16] Although there is some sequence homology between MGP and osteocalcin, and the synthesis of both are stimulated by 1,25-dihydroxyvitamin D, these proteins differ by 80% of their amino acid sequence, and their solubility in water. MGP also appears earlier in calcification than osteocalcin and binds to both the organic and mineral components of bone, whereas osteocalcin is anchored to bone exclusively by its binding to hydroxyapatite. Whereas osteocalcin can only be detected in bone, other tissues, such as kidney, lung and spleen synthesize MGP. In a knockout mice model, MGP-deficient mice have abnormal calcification, leading to osteopenia, fractures and premature death due to arterial calcification.[17] Protein S, a protein cofactor for the anticoagulant activities of protein C, is also synthesized by bone. Protein S has significant homology to a vitamin K dependent growth arrest specific gene product (Gas6), which has been identified in chondrocytes and shown to regulate osteoclast activity.[18]

Although multiple vitamin K dependent proteins have been identified in the skeleton, analytical developments have been limited to osteocalcin.

As reviewed elsewhere, the measurement of the amount of osteocalcin that is undercarboxylated is a sensitive marker of vitamin K status, although inherent limitations of the assay exist.[19] In the presence of vitamin K antagonists, such as warfarin, the percentage of undercarboxylated osteocalcin (ucOC) increases rapidly.[20,21] The percentage ucOC has also been shown to increase within several days of dietary restriction of vitamin K, and decrease in response to dietary supplementation, even when coagulation times remain constant.[22,23] Thus, subclinical deficiency of vitamin K may have deleterious effects on bone without evidence of coagulopathy.

Alternatively, vitamin K may affect bone resorption through a mechanism associated with the geranylgeranyl side chain, as proposed by investigators based on animal studies using MK-4.[24,25] Whereas the active site for the carboxylation reaction is on the napthoquinone ring, which is identical in both compounds (Figure 14.1), MK-4 structurally differs from phylloquinone in its side chain configuration. The mechanisms underlying this putative effect of vitamin K on bone turnover are not understood, but warrant further investigation.

## Dietary Sources and Intakes

Since the publication of a revised provisional table for phylloquinone (vitamin $K_1$) content in some foods,[26] there has been a major expansion in the food composition data available for this fat-soluble vitamin. Several European countries have since incorporated phylloquinone into their national food tables,[27,28] in addition to one US database.[29] Although phylloquinone is the predominant dietary source of vitamin K, other forms of the vitamin are yet to be systematically determined, such that their overall contribution to total vitamin K intake is not currently known.

Green, leafy vegetables contain the highest content of phylloquinone (Table 14.1), and contribute up to 60% of total phylloquinone intake.[30,31] Plant oils, margarine, spreads and salad dressings, derived from plant oils, are higher in phylloquinone concentrations compared to animal fat sources, such as butter.[32,33] The phylloquinone content of oils varies with the plant species, such as soybean, canola and olive oils, which are rich sources whereas corn and peanut oils are not.[34] The phylloquinone content of oils can also vary with exposure to fluorescent light and storage,[34] the percentage fat in the margarine or spread and the degree of hydrogenation.[33] Dihydrophylloquinone is found exclusively in hydrogenated phylloquinone-rich vegetable oils, which are widely used by industry in food preparation because of their physical characteristics and oxidative stability.

Of the limited data available, MK-6 to MK-9 are found in significant amounts in cow livers, some animal meats, and in foods whose preparation involves a bacterial fermentation, such as cheese and fermented soybean products.[35–37]

**Table 14.1.** *Ranking of phylloquinone concentrations in select vegetables*

| Vegetable | Serving size |
|---|---|
| **Very high ( > 100 µg per serving size)** | |
| Spinach, cooked | 1/2 cup |
| Collard greens, cooked | 1/2 cup |
| Kale, raw | 1/2 cup |
| Brussel sprouts | 1/2 cup |
| Swiss chard, raw | 1/2 cup |
| Broccoli, cooked | 1/2 cup |
| Endive, raw | 1/2 cup |
| **High (51–100 µg per serving size)** | |
| Cabbage, cooked or raw | 1/2 cup |
| Okra, boiled | 1/2 cup |
| Asparagus, cooked | 1/2 cup |
| **Medium (11–50 µg per serving size)** | |
| Lettuce, romaine, raw | 1/2 cup |
| Lettuce, bib, raw | 1/2 cup |
| Pumpkin, canned | 1/2 cup |
| Peas, green, frozen, cooked | 1/2 cup |
| Mixed vegetables, frozen, cooked | 1/2 cup |
| Cabbage, red, raw | 1/2 cup |
| Celery, raw | 1 medium stalk |
| Avocado, raw | 1/2 medium |
| Cauliflower, cooked | 1/2 cup |
| **Low (0–10 µg per serving size)** | |
| Carrot, raw | 1 medium |
| Lettuce, iceberg, raw | 1/2 cup |
| Squash, summer, cooked | 1/2 cup |
| Green pepper, raw | 1/2 medium |
| Tomato, raw | 1/2 medium |
| Cucumber with peel, raw | 1/2 cup |
| Mushroom, raw | 1/2 cup |

Median values obtained from Booth *et al.*[26] Bolton-Smith *et al.*[27] Shearer *et al.*[35] and unpublished data.

Very little is known about the contribution of dietary menaquinones to overall vitamin K nutrition and although it is a generally held belief that approximately 50% of the daily requirement for vitamin K is supplied by the gut flora, there is insufficient experimental evidence to support this conviction.[38]

Based on representative dietary intake data, the adequate intake (AI) for vitamin K was recently set at 120 and 90 µg per day for men and women, respectively.[39] With the widespread availability of food composition data, there have been recent surveys of dietary phylloquinone intakes in North America, Europe and Asia, the findings of which indicate that dietary intakes of phylloquinone are lower than previously assumed.[40] Compared to younger adults, older adults report higher average phylloquinone intakes (80–210 µg per day), attributed to their greater vegetable consumption. However, in a study among a national sample of British elderly men and women using 4-d weighed diet records, mean phylloquinone intakes were 70 and 61 µg per day, respectively.[30]

# Evidence for a Role of Vitamin K in Bone Health

## Epidemiological Associations

Most of the evidence in support of a role for dietary vitamin K and age-related bone loss in humans has been based on associations between dietary intakes of biochemical markers of vitamin K status and BMD or fracture rate.

There are several reports of statistically significant associations between low dietary phylloquinone intake and hip fracture risk, notably among women. In the Nurses' Health Study ($n = 72\ 327$), there was a significantly higher risk ($R$:1.43; 95% CI: 1.08, 1.89) for hip fractures among women in the lowest quintile of phylloquinone intake compared to those in the other quintiles combined.[41] When stratified by estrogen replacement use, there was a suggestion that phylloquinone intake only had a protective effect on hip fracture risk among women reporting no estrogen use (Table 14.2). In the Study of Osteoporotic Fractures, there was also a trend towards higher rates of hip fracture among elderly women with low vitamin K intake.[42] Among the original cohort of elderly men and women (mean age: 75 years) participating in the Framingham Heart Study, low phylloquinone intakes (median: 56 µg per day) were associated with an increased risk of hip fracture compared to those with high phylloquinone intakes (median: 254 µg per day) (Table 14.3).[43] Interestingly, these low phylloquinone intakes are within the range of average intakes reported among a national sample of British elderly men and women (mean: 70 and 61 µg day, respectively).[30] Dietary intakes of another form of vitamin K, menaquinone-7 (MK-7) have also been associated with a lower risk for hip fracture among postmenopausal Japanese women.[44] Geographical differences in hip fracture have been attributed to corresponding regional differences in intake of natto, a fermented soybean food that is a rich source of MK-7.

**Table 14.2.** *Age-adjusted relative risks (RR) with 95% confidence intervals (CIs) for risk of hip fracture by quintiles 2–5 (Q2–Q5) compared to quintile 1 (Q1) of vitamin K intake in 72 327 postmenopausal women participating in the Nurses' Health Study as stratified by use of estrogen replacement medication*

| Estrogen replacement medication use | Person-years (thousands) Q1/Q2–Q5 | Number of cases | |
|---|---|---|---|
| | | Q1/Q2–Q5 | RR (95% CI) |
| Never | 37.9/157.0 | 31/86 | 0.65 (0.43, 0.98) |
| Past | 16.4/77.0 | 15/57 | 0.78 (0.44, 1.38) |
| Current | 23.1/109.3 | 8/41 | 1.01 (0.47, 2.14) |

From Ref. [41], reproduced with permission.

**Table 14.3.** *Relative risks for hip fracture in elderly men and women as stratified by quartile of phylloquinone intake*

| Quartile | Number of hip fractures (total n) | Median phylloquinone intake (µg per day) | Multivariate RR (95% CI)[a,b] |
|---|---|---|---|
| 1 | 16 (223) | 56 | 1.00 |
| 2 | 11 (222) | 105 | 0.53 (0.22, 1.28) |
| 3 | 10 (228) | 156 | 0.59 (0.25, 1.39) |
| 4 | 7 (227) | 254 | 0.35 (0.13, 0.94) |

[a]Adjusted for smoking status, calcium and vitamin D supplement use, alcohol consumption, BMI, age, energy intake, physical activity score, vitamin D, calcium and caffeine intake.
[b]$P = 0.047$ for linear trend across quartiles of phylloquinone intake.
Source: Ref. [43], reproduced with permission.

In contrast, the associations between dietary phylloquinone and BMD are equivocal. Among the elderly men and women in the Framingham Heart Study, dietary phylloquinone intakes were not associated with BMD (either cross-sectionally or prospectively), even though there was a positive association with hip fracture risk.[43] In the Framingham Offspring Study, which is younger (mean age: 59 years) than the original cohort, women in the lowest quartile of phylloquinone intake (mean: 70 µg per day) had significantly lower BMD at the femoral neck (mean±SEM: $0.854 \pm 0.006$ g cm$^{-2}$) and spine (mean±SEM: $1.140 \pm 0.010$ g cm$^{-2)}$ compared to those in the highest quartile of phylloquinone intake (mean: 309 µg per day) (mean±SEM: $0.888 \pm 0.006$ and $1.190 \pm 0.010$ g cm$^{-2}$, respectively).[45] These associations remained after controlling for potential confounders, such as dietary potassium, and stratification by age or supplement use. However, no significant association was found between dietary vitamin K intake and BMD among men. Sex-specific differences in the changes in biological markers of bone turnover and/or vitamin K status in response to vitamin K depletion, supplementation or antagonism have not been demonstrated in metabolic studies.[20,22,46] A caveat to metabolic studies is their short-term duration, whereas the food frequency questionnaire used in these studies captures long-term (one year) dietary phylloquinone intake. Therefore, it is not known with the limited data available if there is a sex-specific effect of vitamin K on bone.

While suggestive of a role of vitamin K status in age-related bone loss, one consistent criticism of the epidemiological data has been the potential confounding effect of overall poor nutrition. In the Nurses' Health Study, the primary dietary source of phylloquinone was lettuce.[41] In the Study of Osteoporotic Fractures, the primary dietary source of phylloquinone was broccoli.[42] In the Framingham Heart Study, high dietary vitamin K intakes were positively associated with high dietary intakes of green, leafy vegetables,[31] suggestive of an overall health lifestyle. Therefore, a high vitamin K intake may simply be a marker for an overall healthy diet that includes high fruit and vegetable consumption. In addition, high intakes of alkaline-producing foods, specifically fruits and vegetables, and their associated minerals, potassium

**Table 14.4.** *Bone marker concentrations in men and women (n = 15, mean age: 29.5 years) in response to phylloquinone depletion and repletion*

| | Residency period 1 | | | Residency period 2 | | | P-value[1] |
|---|---|---|---|---|---|---|---|
| | Control 100 µg K₁ | Depletion 10 µg K₁ | Repletion 200 µg K₁ | Control 100 µg K₁ | Depletion 10 µg K₁ | Repletion 200 µg DK | |
| *Serum markers* | | | | | | | |
| Total osteocalcin (µg per Liter) | 7.1 (0.5)[a] | 8.4 (0.6)[b] | 7.1 (0.5)[a] | 7.3 (0.5)[a] | 7.9 (0.6)[a] | 8.1 (0.6)[a] | 0.002 |
| UcOC (%) | 27.7 (3.3)[a] | 46.8 (4.6)[b] | 20.3 (2.0)[a] | 29.1 (3.0)[a] | 42.0 (4.0)[b] | 42.5 (3.9)[b] | <0.001 |
| BAP (U/L) | 18.5 (1.1)[a] | 18.7 (1.0)[a] | 18.5 (0.8)[a] | 19.7 (1.1)[a] | 18.8 (1.0)[b] | 18.5 (1.1)[b] | 0.01 |
| 25(OH)D(ng per mL) | 28.8 (2.3)[a] | 27.6 (2.3)[a] | 28.7 (2.1)[a] | 30.1 (2.3)[a] | 29.3 (2.1)[a] | 29.8 (2.3)[a] | 0.85 |
| PTH (pg per mL) | 26.2 (2.3)[a] | 28.2 (2.8)[a] | 28.8 (2.1)[a] | 27.3 (1.9)[a] | 29.0 (1.9)[a] | 29.8 (3.6)[a] | 0.99 |
| *Urinary markers* | | | | | | | |
| NTx (nM/mM Cr) | 31.5 (2.9)[a] | 35.5 (4.1)[a] | 29.6 (3.5) | 30.4 (2.9)[a] | 38.2 (4.0)[b] | 37.8 (4.7)[b] | 0.08 |
| DPD (nM/mM Cr) | 4.2 (0.3)[a] | 4.5 (0.3)[a] | 4.2 (0.3) | 4.4 (0.3)[a] | 4.3 (0.3)[a] | 4.0 (0.3)[a] | 0.56 |

Source: Ref. [23], reproduced with permission.
[1] *P*-value for the diet-by-residency period interaction. Means within rows and residency periods with different subscript letters are significantly different, P<0.05.

and magnesium have been associated with greater BMD.[47,48] Likewise, soybean, and its associated products, including natto, is a rich dietary source of MK-7; it is also a rich source of isoflavones, also associated with reduced bone loss.[49] While regional differences in intake of soybean products, other than natto, were not associated with differences in the incidence of hip fracture,[44] by nature of their study design, observational studies preclude the ability to isolate the effects of a single nutrient on bone health from those of the dietary patterns associated with high intakes of food(s) rich in that nutrient. In the absence of a mechanism to explain the collective observations from these epidemiological studies, the associations between low phylloquinone or MK-7 intakes and low BMD or increased risk of hip fracture may be suggestive of causality or simply consistent with an overall healthy diet.

Several small studies have reported lower plasma phylloquinone concentrations among adults with low BMD compared to those with normal BMD.[50,51] However, as a biological marker for vitamin K status, plasma phylloquinone measurements are limited by the dependence of this measure on dietary intake within the last 24 h.[40] Plasma and liver phylloquinone concentrations are rapidly reduced under conditions of poor dietary phylloquinone intake,[22] surgery and fasting.[52] The impact of low vitamin K intakes on phylloquinone concentrations in human bone is not known. Therefore, the findings of these studies may be spurious or indicative of a true association.

As previously described, the percentage of ucOC measured in serum is a more sensitive measure of vitamin K status.[53] ucOC has been used in several epidemiological studies to demonstrate a positive association between vitamin K status and BMD.[54–56] In prospective studies, Szulc *et al.*[57,58] and Vergnaud *et al.*[55] reported an inverse association between ucOC and risk of hip fracture in elderly institutionalized and free-living women, respectively. In the study by Szulc *et al.*[57] there was also an inverse association between ucOC and serum 25(OH)D concentrations, and a subsequent reduction of ucOC by vitamin D supplementation. These findings cannot be explained by our current understanding of the biochemical role of vitamin K, but suggest that vitamin D may influence the ucOC level. These data highlight the need to consider vitamin D status when examining the associations between vitamin K and age-related bone loss. Because ucOC is responsive to changes in phylloquinone intake[22] and possibly vitamin D status, a low ucOC may also be a marker of optimal nutritional status.

## Metabolic Studies

In metabolic studies of various age groups, dietary restriction of phylloquinone (vitamin $K_1$) to less than 35 µg per day caused rapid decreases in plasma phylloquinone[23,59,60] and urinary excretion of Gla residues, a measure of turnover of all vitamin K dependent proteins,[23,59,60] and increases in the under-γ-carboxylated forms of the vitamin K dependent proteins, osteocalcin[23] and prothrombin.[23,59,60] Collectively, these data are consistent with the development of a subclinical deficiency of vitamin K. However, the clinical significance of these changes in individual biological markers is not known because, coagulation times – the classic measure of vitamin K status – consistently remained stable. Plasma phylloquinone, undercarboxylated prothrombin (PIVKA-II) and under-carboxylated osteocalcin (ucOC) have also been demonstrated to respond to dietary supplementation.[23,46] However, there is a lack of dose response data using a single assay for each biological measure so the optimal amount of dietary phylloquinone intake is currently not known. As previously mentioned, the current AI for vitamin K is based on representative intake data.[39]

Few metabolic studies have examined the role of dietary phylloquinone depletion and repletion on markers of bone turnover. In a randomized crossover study in a metabolic unit, 15 young adults were fed a phylloquinone restriction diet (10 µg per day) for 15 days, followed by 10-day repletion (200 µg per day) of phylloquinone or its hydrogenated form, dihydrophylloquinone. PIVKA-II responded to supplementation of both phylloquinone and vitamin K, $2'$,$3'$-dihydrophylloquinone, whereas %ucOC and urinary Gla responded to supplementation of phylloquinone, but not $2'$,$3'$-dihydrophylloquinone. These results suggest that the carboxylation of prothrombin was mostly conserved in response to dietary depletion, whereas carboxylation of osteocalcin was not. It is also consistent with the suggestion that the liver has a high efficiency for utilization of vitamin K to conserve coagulation, hence that vitamin K requirements for extra-hepatic proteins are higher than those required for hepatic proteins.[63]

In this study,[23] there was an increase, and subsequent decrease, in measures of bone formation (serum total osteocalcin) and resorption (urinary NTx) following dietary phylloquinone restriction and repletion, respectively (Table 14.3). In comparison to phylloquinone, dihydrophylloquinone was less absorbed and had no measurable biological effect on measures of bone formation and resorption. This study, as supported by increases in both total osteocalcin and NTX, suggest that vitamin K may have a direct effect on bone turnover, consistent with data from animal studies.[24,25] However, two other human studies reported no effect of either short-term phylloquinone or long-term MK-4 supplementation on bone resorption markers in healthy adults or osteoporotic women, respectively.[46,61]

There is a current lack of consensus regarding vitamin K supplementation and its effect on total osteocalcin. There is one report of an observed increase in serum osteocalcin concentrations among post-menopausal women following two weeks of supplementation at 1 mg phylloquinone per day.[56] However, there was no significant change in serum osteocalcin concentrations among the premenopausal women on the same regimen.[56] Indeed, among younger and older adults, supplementation at 1 mg phylloquinone per day has been reported as having no effect[56,62] or an observed decrease[23,46] in serum osteocalcin. Because different antibodies are used for measurement of total osteocalcin among the different studies, the discrepancy in results may be an artifact of the variation in affinities of different antibodies for the carboxylated form of osteocalcin.[53]

Comparison of plasma concentrations for phylloquinone and dihydrophylloquinone following repletion suggest that dihydrophylloquinone is not as well absorbed as phylloquinone.[23] Alternatively, dihydrophylloquinone may be more rapidly metabolized and excreted compared to the parent form, phylloquinone. Therefore, it is probable that differences in relative biological activity between the two forms of vitamin K at least reflect differences in availability as a cofactor for carboxylation of (γ-carboxyglutamic acid (Gla) residues in vitamin K dependent proteins. By implication, the presence of dihydrophylloquinone in margarines and processed foods containing hydrogenated phylloquinone-rich plant oils may reduce the contribution of this food source to overall vitamin K status. Comparable metabolic studies evaluating the absorption and biological activity of other dietary forms of vitamin K, such as MK-4 and MK-7, would further contribute to our limited understanding of the role of the vitamin K side chain in bone metabolism.

## Clinical Trials

There is very little information available concerning the long-term effect of phylloquinone supplementation on age-related bone loss. Douglas[62] demonstrated in 20 postmenopausal women that 14 days of phylloquinone supplementation increased carboxylation of osteocalcin. These findings were later confirmed by Binkley et al.[46] Orimo[64] reported increases in bone mass among 272 postmenopausal women following supplementation with MK-4 (menatetrenone) for 24–48 weeks. Shiraki et al.[61] demonstrated that MK-4 supplementation for 24-months increased serum total osteocalcin and reduced occurrence of new fractures in osteoporotic women. However, these investigators used doses of 45 mg of MK-4 per day, in contrast to the current AI of 90–120 μg of phylloquinone per day.[39] It has been suggested that pharmacological doses of MK-4 have an anti-resorptive effect on bone, presumably mediated through a mechanism associated with the geranylgeranyl side chain.[65] Reports from long-term clinical trials of the effects of phylloquinone supplementation on BMD are beginning to emerge. Braam et al.[66] reported that 3-year intake of a dietary supplement containing calcium, vitamin D and phylloquinone significantly reduced age-related bone loss at the femoral in women, aged 50–60 years, compared to a placebo or a supplement containing calcium and vitamin D, but not phylloquinone. While encouraging, more clinical trials need to be completed before conclusions can be drawn regarding the efficacy of phylloquinone supplementation in reducing age-related bone loss.

## Conclusions

There is emerging evidence that vitamin K may have a protective role against age-related bone loss that is mediated through the carboxylation of the vitamin K dependent proteins in bone. Average self-reported dietary intakes of vitamin K are below recommended intakes among adults in the US and UK.

Low dietary intakes of vitamin K are potentially a modifiable risk factor for osteoporosis. However, the putative mechanisms of action are not yet understood, which limits the interpretation of existing studies. Much of the evidence in humans is limited to epidemiological studies, the results of which may be confounded by overall poor nutrition.

While epidemiological studies are useful in defining associations between nutrient intake and health outcomes, metabolic studies and clinical trials would be more definitive in isolating the putative effects of vitamin K on bone loss. To define the potential dietary role of phylloquinone in age-related bone loss, clinical trials are required that use doses of phylloquinone that are nutritionally optimal, yet attainable in the diet.

# References

1. A. Ortiz and F.J. Aranda, The influence of vitamin K1 on the structure and phase behaviour of model membrane systems, *Biochim. Biophys. Acta*, 1999, **1418**(1), 206–220.
2. K. Davidson, S. Booth, G. Dolnikowski and J. Sadowski, Conversion of vitamin K-1 to $2',3'$-dihydrovitamin K1 during the hydrogenation of vegetable oils, *J. Agri. Food Chem.*, 1996, **44**, 980–983.
3. R.T. Davidson, A.L. Foley, J.A. Engelke and J.W. Suttie, Conversion of dietary phylloquinone to tissue menaquinone-4 in rats is not dependent on gut bacteria, *J. Nutr.*, 1998, **128**(2), 220–223.
4. P. Newman and M.J. Shearer, Vitamin K metabolism, *Subcellular Biochem.*, 1998, **30**, 455–488.
5. M.J. Shearer, The roles of vitamins D and K in bone health and osteoporosis prevention, *Proc. Nutr. Soc.*, 1997, **56**(3), 915–937.
6. A.M. Huber, K.W. Davidson, M.E. O'Brien-Morse and J.A. Sadowski, Gender differences in hepatic phylloquinone and menaquinones in the vitamin K-deficient and -supplemented rat, *Biochim. Biophys. Acta*, 1999, **1426**(1), 43–52.
7. A.M. Huber, K.W. Davidson, M.E. O'Brien-Morse and J.A. Sadowski, Tissue phylloquinone and menaquinones in rats are affected by age and gender, *J. Nutr.*, 1999, **129**(5), 1039–1044.
8. J. Ronden, M.-J. Drittij-Reijnders, C. Vermeer and H. Thijssen, Intestinal flora is not an intermediate in the phylloquinone-menaquinone -4 conversion in the rat, *Biochim. Biophys. Acta*, 1998, **1379**, 69–75.
9. B. Furie, B.A. Bouchard and B.C. Furie, Vitamin K-dependent biosynthesis of gamma-carboxyglutamic acid, *Blood*, 1999, **93**(6), 1798–1808.
10. G. Ferland, The vitamin K-dependent proteins: an update, *Nutr. Rev.*, 1998, **56**(8), 223–230.
11. B.C. Furie and B. Furie, Structure and mechanism of action of the vitamin K-dependent gamma-glutamyl carboxylase: recent advances from mutagenesis studies, *Thrombos. Haemostas.*, 1997, **78**(1), 595–598.
12. P.V. Hauschka, J.B. Lian, D. Cole and C.M. Gundberg, Osteocalcin and matrix Gla protein: vitamin K-dependent proteins in bone, *Physiol. Rev.*, 1989, **69**(3), 990–1047.
13. P. Ducy, C. Desbois, B. Boyce, G. Pinero, B. Story, C. Dunstan, E. Smith, J. Bonadio, S. Goldstein, C. Gundberg, A. Bradley and G. Karsenty, Increased bone formation in osteocalcin-deficient mice, *Nature*, 1996, **382**(6590), 448–452.

14. J.R. Cairns and P.A. Price, Direct demonstration that the vitamin K-dependent bone Gla protein is incompletely gamma-carboxylated in humans, *J. Bone Miner. Res.*, 1994, **9**(12), 1989–1997.
15. M. Nakao, Y. Nishiuchi, M. Nakata, T. Kimura and S. Sakakibara, Synthesis of human osteocalcins: gamma-carboxyglutamic acid at position 17 is essential for a calcium-dependent conformational transition, *Pep. Res.*, 1994, **7**(4), 171–174.
16. P.A. Price, Gla-containing proteins of bone, *Connective Tissue Res.*, 1989, **21**(1–4), 51–57.
17. G. Luo, P. Ducy, M.D. McKee, G.J. Pinero, E. Loyer, R.R. Behringer, G. Karsenty, Spontaneous calcification of arteries and cartilage in mice lacking matrix GLA protein, *Nature*, 1997, **386**(6620), 78–81.
18. Y.S. Nakamura, Y. Hakeda, N. Takakura, T. Kameda, L. Hamaguchi, T. Miyamoto, S. Kakudo, T. Nakano, M. Kumegawa and T. Suda, Tyro 3 receptor tyrosine kinase and its ligand, Gas6, stimulate the function of osteoclasts, *Stem Cells*, 1998, **16**(3), 229–238.
19. C. Gundberg, Biology, physiology, and clinical chemistry of osteocalcin, *J. Clin. Ligand Assay*, 1998, **21**(2), 128–138.
20. A.U. Bach, S.A. Anderson, A.L. Foley, E.C. Williams and J.W. Suttie, Assessment of vitamin K status in human subjects administered "minidose" warfarin, *Am. J. Clin. Nutr.*, 1996, **64**(6), 894–902.
21. L.J. Sokoll, M.E. O'Brien, M.E. Camilo and J.A. Sadowski, Undercarboxylated osteocalcin and development of a method to determine vitamin K status, *Clin. Chem.*, 1995, **41**, 1121–1128.
22. S.L. Booth, M.E. O'Brien-Morse, G.E. Dallal, K.W. Davidson and C.M. Gundberg, Response of vitamin K status to different intakes and sources of phylloquinone-rich foods: comparison of younger and older adults, *Am. J. Clin. Nutr.*, 1999, **70**(3), 368–377.
23. S.L. Booth, A.H. Lichtenstein, M. O'Brien-Morse, N.M. McKeown, R.J. Wood, E. Saltzman and C. Gundberg, Effects of a hydrogenated form of vitamin K on bone formation and resorption, *Am. J. Clin. Nutr.*, 2001, **74**(6), 783–790.
24. K. Hara, Y. Akiyama, T. Nakamura, S. Murota and I. Morita, The inhibitory effect of vitamin K2 (menatetrenone) on bone resorption may be related to its side chain, *Bone*, 1995, **16**(2), 179–184.
25. Y. Akiyama, K. Hara, T. Tajima, S. Murota and I. Morita, Effect of vitamin K2 (menatetrenone) on osteoclast-like cell formation in mouse bone marrow cultures, *Eur. J. Pharmacol.*, 1994, **263**(1–2), 181–185.
26. S. Booth, J. Sadowski, J. Weihrauch and G. Ferland, Vitamin K-1 (phylloquinone) content of foods: a provisional table, *J. Food Compos. Anal.*, 1993, **6**, 109–120.
27. C. Bolton-Smith, R.J. Price, S.T. Fenton, D.J. Harrington and M.J. Shearer, Compilation of a provisional UK database for the phylloquinone (vitamin K1) content of foods, *Br. J. Nutr.*, 2000, **83**(4), 389–399.
28. G. Deharveng, U.R. Charrondiere, N. Slimani, D.A. Southgate and E. Riboli, Comparison of nutrients in the food composition tables available in the nine European countries participating in EPIC. European Prospective Investigation into Cancer and Nutrition, *Eur. J. Clin. Nutr.*, 1999, **53**(1), 60–79.
29. N.M. McKeown, H.M. Rasmussen, J.M. Charnley, R.J. Wood and S.L. Booth, Accuracy of phylloquinone (vitamin K-1) data in 2 nutrient databases as determined by direct laboratory analysis of diets, *J. Am. Diet. Assoc.*, 2000, **100**(10), 1201–1204.

30. C. Thane, A. Paul, C. Bates, C. Bolton-Smith, A. Prentice and M. Shearer, Intake and sources of phylloquinone (vitamin K-1): variation with socio-demographic and lifestyle factors in a national sample of British elderly people, *Br. J. Nutr.*, 2002, **87**, 605–613.

31. N.M. McKeown, P.F. Jacques, C.M. Gundberg, J.W. Peterson, K.L. Tucker, D.P. Kiel, P.W.F. Wilson and S.L. Booth, Dietary and non-dietary determinants of vitamin K biochemical measures in men and women, *J. Nutr.*, 2002, **32**, 1329–1334.

32. V. Piironen, T. Koivu, O. Tammisalo and P. Mattila, Determination of phylloquinone in oils, margarines and butter by high-performance liquid chromatography with electrochemical detection, *Food Chem.*, 1997, **59**, 473–480.

33. J.W. Peterson, K.L. Muzzey, D. Haytowitz, J. Exler, L. Lemar and S.L. Booth, Phylloquinone (vitamin K-1) and dihdyrophylloquinone content of fats and oils, *J. Am. Oil Chem. Soc.*, 2002, **79**, 641–646.

34. G. Ferland and J. Sadowski, Vitamin K1 (phylloquinone) content of edible oils: effects of heating and light exposure, *J. Agri. Food Chem.*, 1992, **40**, 1869.

35. M.J. Shearer, A. Bach and M. Kohlmeier, Chemistry, nutritional sources, tissue distribution and metabolism of vitamin K with special reference to bone health, *J. Nutr.*, 1996, **126**(4 Suppl), 1181S–1186S.

36. L. Schurgers, J. Geleijnse, D. Grobbee, H. Pols, A. Hofman J.C.M. Witteman and C. Vermeer, Nutritional intake of vitamins K1 (phylloquinone) and K2 (menaquinone) in the Netherlands, *J. Nutr. Environ. Med.*, 1999, **9**, 115–122.

37. T.S.N. Sakano, T. Nagaoka, A. Morimoto, K. Fujimoto, S. Masuda, Y. Suzuki and K. Hirauchi, Measurement of K vitamins in food by high-performance liquid chromatography with fluorometric detection, *Vitamins (Japan)*, 1988, **62**, 393–398.

38. J.W. Suttie, The importance of menaquinones in human nutrition, *Annu. Rev. Nutr.*, 1995, **15**, 399–417.

39. Institute of Medicine, In *Dietary Reference Intakes for Vitamin A, Vitamin K, Arsenic Boron, Chromium, Copper, Iodine, Iron, Manganese, Molybdenum, Nickel, Silicon, Vanadium and Zinc.*, National Academy Press, Washington, DC, 2001.

40. S.L. Booth and J.W. Suttie, Dietary intake and adequacy of vitamin K, *J. Nutr.*, 1998, **128**(5), 785–788.

41. D. Feskanich, P. Weber, W.C. Willett, H. Rockett, S.L. Booth and G.A. Colditz, Vitamin K intake and hip fractures in women: a prospective study, *Am. J. Clin. Nutr.*, 1999, **69**(1), 74–79.

42. K. Stone, T. Duong, D. Sellmeyer, J. Cauley, R. Wolfe and S. Cummings, Broccoli may be good for bones: dietary vitamin K-1, rates of bone loss and risk of hip fracture in a prospective study of elderly women, *J. Bone Miner. Res.*, 1999, **14**(Suppl. 1), S263.

43. S.L. Booth, K.L. Tucker, H. Chen, M.T. Hannan, D.R. Gagnon, L.A. Cupples, P.W.F. Wilson, J. Ordovas, E.J. Schaefer, B. Dawson-Hughes and D.P. Kiel, Dietary vitamin K intakes are associated with hip fracture but not with bone mineral density in elderly men and women, *Am. J. Clin. Nutr.*, 2000, **71**(5), 1201–1208.

44. M. Kaneki, S. Hedges, T. Hosoi, S. Fujiwara, A. Lyons, S. Crean, N. Ishida, M. Nakagawa, M. Takechi, Y. Sano, Y. Mizuno, S. Hoshino, M. Miyao, S. Inoue, K. Horiki, M. Shiraki, Y. Ouchi and H. Orimo, Japanese fermented soybean food as the major determinant of the large geographic difference in circulating levels of vitamin K2 - possible implications for hip-fracture risk, *Nutrition*, 2001, **17**, 315–321.

45. S. Booth, K. Broe, D. Gagnon, K. Tucker, M. Hannan, R. McLean, B. Dawson-Hughes, P.W.F. Wilson, L.A. Cupples and D.P. Kiel, Vitamin K intakes and bone mineral density in women and men, *Am. J. Clin. Nutr.*, 2003, **77**, 512–516.
46. N.C. Binkley, D.C. Krueger, J.A. Engelke, A.L. Foley and J.W. Suttie, Vitamin K supplementation reduces serum concentrations of under-gamma- carboxylated osteocalcin in healthy young and elderly adults, *Am. J. Clin. Nutr.*, 2000, **72**(6), 1523–1528.
47. S.A. New, C. Bolton-Smith, D.A. Grubb and D.M. Reid, Nutritional influences on bone mineral density: a cross-sectional study in premenopausal women, *Am. J. Clin. Nutr.*,1997, **65**(6), 1831–1839.
48. K.L. Tucker, M.T. Hannan, H. Chen, L.A. Cupples, P.W. Wilson and D.P. Kiel, Potassium, magnesium, and fruit and vegetable intakes are associated with greater bone mineral density in elderly men and women, *Am. J. Clin. Nutr.*, 1999, **69**(4), 727–736.
49. B.H. Arjmandi, M.J. Getlinger, N.V. Goyal, L. Alekel, C.M. Hasler, S. Juma, M.L. Drum, B.W. Hollis and S.C. Kukreja, Role of soy protein with normal or reduced isoflavone content in reversing bone loss induced by ovarian hormone deficiency in rats, *Am. J. Clin. Nutr.*, 1998, **68**(6 Suppl), 1358S–1363S.
50. J.P. Hart, M.J. Shearer, L. Klenerman, A. Catterall, J. Reeve, P.N. Sambrook, R.A. Dodds, L. Bitensky and J. Chayen, Electrochemical detection of depressed circulating levels of vitamin K1 in osteoporosis, *J. Clin. Endocrinol. Metab.*, 1985, **60**(6), 1268–1269.
51. T. Kanai, T. Takagi, K. Masuhiro, M. Nakamura, M. Iwata and F. Saji, Serum vitamin K level and bone mineral density in post-menopausal women, *Int. J. Gynaecol. Obstet.*, 1997, **56**(1), 25–30.
52. Y. Usui, H. Tanimura, N. Nishimura, N. Kobayashi, T. Okanoue and K. Ozawa, Vitamin K concentrations in the plasma and liver of surgical patients, *Am. J. Clin. Nutr.*, 1990, **51**(5), 846–852.
53. C.M. Gundberg, S.D. Nieman, S. Abrams and H. Rosen, Vitamin K status and bone health: an analysis of methods for determination of undercarboxylated osteocalcin, *J. Clin. Endocrinol. Metab.*, 1998, **83**(9), 3258–3266.
54. P. Szulc, M. Arlot, M.C. Chapuy, F. Duboeuf, P.J. Meunier and P.D. Delmas, Serum undercarboxylated osteocalcin correlates with hip bone mineral density in elderly women, *J. Bone Miner. Res.*, 1994, **9**(10), 1591–1595.
55. P. Vergnaud, P. Garnero, P.J. Meunier, G. Breart, K. Kamihagi and P.D. Delmas, Undercarboxylated osteocalcin measured with a specific immunoassay predicts hip fracture in elderly women: the EPIDOS Study, *J. Clin. Endocrinol. Metab.*, 1997, **82**(3), 719–724.
56. M.H. Knapen, A.C. Nieuwenhuijzen-Kruseman, R.S.M.E. Wouters and C. Vermeer, Correlation of serum osteocalcin fractions with bone mineral density in women during the first 10 years after menopause, *Calcif. Tissue Int.*, 1998, **63**, 375–379.
57. P. Szulc, M.C. Chapuy, P.J. Meunier and P.D. Delmas, Serum undercarboxylated osteocalcin is a marker of the risk of hip fracture in elderly women, *J. Clin. Invest.*, 1993, **91**(4), 1769–1774.
58. P. Szulc, M.C. Chapuy, P.J. Meunier and P.D. Delmas, Serum undercarboxylated osteocalcin is a marker of the risk of hip fracture: a three year follow-up study, *Bone*, 1996, **18**(5), 487–488.
59. G. Ferland, J.A. Sadowski and M.E. O'Brien, Dietary induced subclinical vitamin K deficiency in normal human subjects, *J. Clin. Invest.*, 1993, **91**(4), 1761–1768.

60. J.W. Suttie, L.L. Mummah-Schendel, D.V. Shah, B.J. Lyle and J.L. Greger, Vitamin K deficiency from dietary vitamin K restriction in humans, *Am. J. Clin. Nutr.*, 1988, **47**(3), 475–480.

61. M. Shiraki, Y. Shiraki, C. Aoki and M. Miura, Vitamin K2 (menatetrenone) effectively prevents fractures and sustains lumbar bone mineral density in osteoporosis, *J. Bone Miner. Res.*, 2000, **15**(3), 515–521.

62. A.S. Douglas, S.P. Robins, J.D. Hutchison, R.W. Porter, A. Stewart and D.M. Reid, Carboxylation of osteocalcin in post-menopausal osteoporotic women following vitamin K and D supplementation, *Bone*, 1995, **17**(1), 15–20.

63. C. Vermeer and L.J. Schurgers, A comprehensive review of vitamin K and vitamin K antagonists, *Hematol. Oncol. Clin. N. Am.*, 2000, **14**(2), 339–353.

64. H. Orimo, M. Shiraki, T. Fujita, T. Onomura, T. Inoue and K. Kushida, Clinical evaluation of menatetrenone in the treatment of involutional osteoporosis – a double-blind multicenter comparative study with 1 alpha hydroxy vitamin D3, *J. Bone Miner. Res.*, 1992, **7**, S122.

65. N.C. Binkley and J.W. Suttie, Vitamin K nutrition and osteoporosis, *J. Nutr.*, 1995, **125**(7), 1812–1821.

66. L. Braam, M. Knapen, P. Geusens, F. Brouns, K. Hamulyak, M. Gerichhausen and C. Vermeer, Vitamin K1 supplementation retards bone loss in postmenopausal women between 50 and 60 years of age, *Calcif. Tissue Int.*, 2003, Apr 3, [Epub ahead of print].

CHAPTER 15

# The Role of Dietary Magnesium Deficiency in Osteoporosis

GARRISON TONG and ROBERT K. RUDE

University of Southern California, School of Medicine, 1975 Zonal Avenue, Los Angeles, California 90089-9317, USA, E-mail: rrude60075@aol.com

**Key Points**

- Dietary magnesium (Mg) deficiency may be a risk factor for osteoporosis.
- Animal studies demonstrate a link between Mg depletion and skeletal integrity.
- Human epidemiologic studies suggest a relationship between dietary Mg and bone mass, except in early post menopausal women.
- Studies examining bone turnover markers, Mg status and Mg therapy have shown conflicting results.
- Potential mechanisms of Mg deficiency-induced osteoporosis are discussed.

## Introduction

Magnesium (Mg) is the second most abundant intracellular cation in the body. Its principal functions include regulation of enzyme activity, control of various calcium and potassium channels, and promotion of membrane stabilization. Magnesium depletion has been associated with a number of clinical conditions including hypocalcaemia, hypokalaemia, hypertension, cardiac arrhythmias, and myocardial infarction.[1] It is known that Mg deficiency, when severe, will markedly disturb calcium homeostasis resulting in impaired parathyroid (PTH) secretion and PTH end-organ resistance leading to hypocalcaemia and neuromuscular hyperexcitability.[1] Mg exists in macronutrient quantities in bone and dietary Mg deficiency has been implicated as a risk factor for osteoporosis. The U.S. Food Nutrition Board established the Recommended Daily Allowance for Mg for adult males at approximately 420 mg per day and for adult females at 320 mg per day.[2] The usual dietary Mg intake, however, falls below this recommendation in a large proportion of the population. According to the USDA, the mean Mg intake for males is 323 mg per day and females is 228 mg per day.[3] Mg intake for individuals over age 70 years is the lowest for any adult group. This chapter will review human and animal studies designed to assess the role of dietary Mg deficiency in the development of osteoporosis.

# Experimental Animal Models of Mg Deficiency

The effect of dietary Mg depletion on bone and mineral homeostasis in animals (primarily the rat) has been studied since the 1940s. Dietary restriction of Mg is usually severe, ranging from 0.2 to 8 mg per 100 g chow (normal = 50–70 mg per 100 g). A universal observation has been a decrease in growth of both the whole body as well as the skeleton.[4–7]

Studies of magnesium depletion in animals suggest alterations in chondrocyte organization, osteoblast and osteoclast activity, as well as overall skeletal integrity. The epiphyseal and diaphyseal growth plate is characterized by thinning and a decrease in the number and organization of chrondrocytes.[6] There is a reduction of osteoblastic bone formation as observed by quantitative histomorphometry.[7,8] Moreover, serum and bone alkaline phosphatase,[6,9,10] serum and bone osteocalcin[5,7,11] and bone osteocalcin mRNA[7,11] are reduced suggesting a decrease in osteoblastic function. This is supported by an observed decrease in collagen formation and sulfation of glycosaminoglycans.[12] A decrease in tetracycline labeling also suggests impaired mineralization.[7,13] Data on osteoclast function has been conflicting. A decrease in urinary hydroxyproline[14] and deoxypyridinoline[11] in a rat model has suggested decrease in bone resorption, however, Rude *et al.* recently reported an increase in the number and activity of osteoclasts in the Mg deficient rat.[8] Bone from Mg deficient rats is described as brittle and fragile.[10,15] Biomechanical testing demonstrates increased skeletal fragility in both rat and pig.[5,16–19] Reduction in bone mass occurs with dietary Mg depletion by 6 weeks or longer.[5,7,8,18,19] Bone implants into Mg deficient rats also show osteoporosis in the implanted bone.[20,21] The effect of higher than the recommended dietary Mg intake on mineral metabolism in the rat was reported by Toba *et al.*[22] In this study, increasing dietary Mg from 48 mg per 100 g chow to 118 mg per 100 g chow resulted in a decrease in bone resorption and an increase in bone strength in ovariectomized rats. No loss of bone mineral density (BMD) was observed suggesting a beneficial effect of Mg in acute sex steroid deficiency.

# Epidemiological Studies

In humans, epidemiologic studies provide a major link associating dietary Mg inadequacy to osteoporosis. The majority of studies are cross-sectional in design and assess the effect of dietary nutrients on appendicular (*i.e.*, distal radius, ulna, heel) or axial (*i.e.*, hip and spine) BMD. Several studies have

**Table 15.1.** *Epidemiologic studies examining the correlation between dietary magnesium intake and bone mineral density (BMD)*

| Author/(reference) | (N) | Population | Mean Mg intake | Correlation | Site investigated |
|---|---|---|---|---|---|
| New[23] | 994 | Premenopausal women | 311 | Yes | Lumbar spine |
| Angus[24] | 89 | Premenopausal women | 243 | Yes | Distal forearm |
| | 71 | Recent menopausal | 253 | No | Distal forearm |
| Houtkooper[25] | 66 | Premenopausal women | 289 | Yes | Lumbar spine |
| Wang[26] | 61 | Girls age 9–11 | 240 | Yes | Calcaneous |
| Freudenheim[27] | 17 | Premenopausal women | 243 | Yes | Distal forearm |
| | 67 | Postmenopausal women | 249 | Yes | Distal forearm |
| Tranquilli[28] | 194 | Postmenopausal women | 288 | Yes | Distal forearm |
| Tucker[29] | 345 | Men ages 69–97 | 300 | Yes | Hip/distal forearm |
| | 562 | Postmenopausal women | 288 | Yes | Hip/distal forearm |
| Yano[30] | 1208 | Men ages 61–81 | 238 | No | Appendicular |
| | 259 | Men ages 61–81 taking Mg supplements | 381 | Yes | Appendicular |
| | 912 | Women ages 43–80 | 191 | Yes | Appendicular |
| | 217 | Subset of women taking Mg supplements | 321 | No | Appendicular |
| New[31] | 65 | Pre and postmenopausal | 326 | Yes | Distal forearm |
| Michaelsson[32] | 175 | Women ages 28–74 | 262 | No | Lumbar spine |

found a correlation between dietary Mg and BMD in premenopausal and post menopausal women (Table 15.1).

New and colleagues, in a study of 994 premenopausal women found a significant correlation of BMD of the lumbar spine with Mg intake.[23] Another study by Angus and colleagues involving 89 premenopausal women found a positive correlation of BMD of the forearm, but not femur or spine with Mg intake.[24] The reason for this difference is unclear; although, in the study by New and colleagues, the mean dietary Mg was higher and the number of subjects studied was greater (see Table 15.1). In 66 premenopausal women, Houtkooper and colleagues showed a significant relationship between dietary Mg intake as well as exercise and body composition, and rate of change of BMD over a one year period of the lumbar spine and total body calcium.[25] In younger individuals, Wang evaluated the effect of dietary Mg intake of preadolescent girls on bone mass/quality, using ultrasound determination of bone mass of the calcaneus.[26] The group showed that Mg intake was positively related to quantitative ultrasound properties of bone suggesting that this nutrient was important in skeletal growth and development. A small study of 17 premenopausal and 67 postmenopausal women by Freudenheim *et al.* did not find a cross-sectional correlation in either group between Mg intake and BMD measured in the distal forearm.[27] However, a longitudinal observation over 4 years demonstrated that loss of bone mass was inversely related to Mg intake in both premenopausal and postmenopausal women.

A study by Tranquilli and colleagues involving 194 postmenopausal women showed a significant positive correlation between BMD of the forearm and Mg intake.[28] Moreover, Tucker *et al.* assessed Mg intake in older females in a cross-sectional study and a two year longitudinal study of a subset of these subjects.[29] In the cross-sectional analysis, there was a positive correlation of BMD of the hip and Mg intake, but not in the longitudinal assessment.

While studies that have examined premenopausal and postmenopausal women as discreet groups have found a correlation between Mg intake and BMD, studies that have grouped a wide range of ages, or included perimenopausal women have not shown such a correlation as consistently. Yano and colleagues did show a positive correlation between Mg intake and appendicular BMD in 912 women aged 43–80.[30] In a further study by New and colleagues of women 45–55 years of age, higher intakes of Mg were associated with higher bone mass of the forearm (but not femoral neck or hip).[31] However, Angus and colleagues did not find a correlation in 71 recently menopausal women.[24] In addition, a cross-sectional study that combined 175 premenopausal and postmenopausal women age 28–74 found no correlation with BMD at the lumbar spine, femoral neck, or total body calcium.[32]

Few epidemiologic studies have been done in men. In a cross sectional study by Tucker *et al.* on 345 elderly men ages 69–97, there was a significant correlation between Mg intake and BMD at the hip. A subsequent longitudinal analysis over a 4 year period showed less bone loss in men with higher intakes of Mg, potassium, fruit and vegetables.[33] Yano and colleagues also investigated Mg intake in males.[30] In 1208 men aged 61–81 years there was no correlation between mean Mg intake and appendicular BMD; however, in a subset of 259 men taking Mg supplements, there was a significant correlation between Mg intake and appendicular BMD.

In summary, these epidemiological studies link dietary Mg intake to bone mass. Exceptions appear to include women in the early postmenopausal period in which the effect of acute sex steroid deficiency may mask the effect of dietary factors such as Mg. In addition, diets deficient in Mg are usually deficient in other nutrients which also affect bone mass. Further investigations are needed to provide a firm relationship of dietary Mg inadequacy with osteoporosis.

# Bone Turnover

Bone markers are the biochemical products released into the circulation during the process of bone formation (*e.g.*, serum osteocalcin and bone specific alkaline phosphatase) or resorption (*e.g.*, urine deoxypyridinoline and *N*-telopeptide of collagen cross-links). Elevations in these markers reflect an increase in bone turnover, and have been used in various studies to predict bone loss.

In two of the epidemiological studies cited above, markers of bone turnover were determined. In one, where no correlation was found between BMD and dietary Mg intake, serum osteocalcin did not correlate with Mg (or any other nutrient) intake.[32] New and colleagues also found that serum osteocalcin was not associated with dietary intake of Mg or other nutrients.[31] Mg intake, however, was significantly negatively correlated with urinary excretion of pyridinoline and deoxypyridinoline suggesting that a low Mg diet was associated with increased bone resorption.[31]

The effect of short-term administration of Mg on bone turnover in young normal subjects has been conflicting. Dimai and colleagues administered 360 mg Mg per day for 30 days to 12 normal males age 27–36 (mean dietary intake prior to supplementation was 312 mg day$^{-1}$) and compared markers of bone formation and bone resorption with 12 age-matched controls.[34] Both markers of formation and resorption were significantly suppressed, but only during first 5–10 days of the study. In a similar trial[35] with a double blind, placebo-controlled randomized crossover design, 26 females age 20–28 years received Mg, 240 mg day$^{-1}$, or placebo for 28 days (mean dietary Mg intake was 271 mg day$^{-1}$ prior to and during study). No effect of Mg supplementation was observed on serum osteocalcin, bone-specific alkaline phosphatase, or urinary pyridinoline and deoxypyridinoline excretion.

# Mg Status

Several investigators have found a significant correlation between serum Mg and bone metabolism. Gur and colleagues demonstrated lower serum values of Mg, copper and zinc in 70 osteoporotic subjects compared to 30 non-osteoporotic postmenopausal women.[36] In a similar study involving 20 perimenopausal and 53 postmenopausal women, women with severe osteoporosis had significantly lower ionized magnesium levels.[37] In a study of 168 patients with Crohn's disease, 95 of whom were females with a mean age of 33 years, higher serum Mg predicted higher BMD at the femur, however, a negative correlation was observed in the spine.[38]

Recently, Kantorovich and colleagues have assessed the magnesium status and BMD in a family with an autosomal dominant form of primary hypomagnesemia due to renal Mg wasting.[39] Affected family members demonstrated significant reductions in serum and lymphocyte Mg concentrations. Moreover, affected members had significantly reduced BMD at the lumbar spine and proximal femur (Figure 15.1). The data from this study suggests that hypomagnaesemia may be associated with low bone mass.

Measurement of the serum Mg concentration is the most commonly employed test to assess Mg status. However, Mg is principally an intracellular cation and less than 1% of the body Mg content is in the extracellular fluid

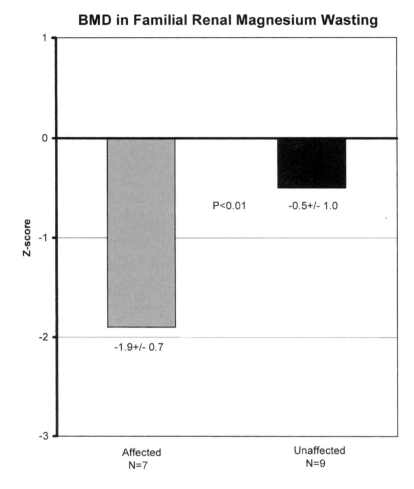

**Figure 15.1** *Comparison of bone mineral density between family members affected by renal magnesium wasting (Z score −1.9±0.7) and unaffected family members (Z score −0.5±1.0). A Z score is the number of standard deviations above or below the mean bone mineral density compared to the conrol population adjusted for sex, race and age. (P<0.01)* Data extrapolated from Kantorovich, *et al.*[39]

compartments. Therefore, serum Mg concentration may not reflect dietary Mg intake or the intracellular Mg content. In order to more accurately determine intracellular Mg content, researchers have employed magnesium tolerance testing, as well as measuring red blood cell Mg and skeletal Mg content.

Few studies have been done assessing Mg status in patients with osteoporosis despite interest in the possible role that dietary Mg insufficiency may play as a risk factor for osteoporosis. Cohen and Kitzes studied a small group of 15 osteoporotic subjects (10 female, 5 male) aged 70–85. The presence or absence of osteoporosis was determined by radiographic features and compared the results with those obtained in 10 control non-osteoporotic subjects.[40] Both groups had normal serum Mg concentrations which were not significantly different from each other. Results of the Mg tolerance test however revealed a significantly greater retention in the osteoporotic patients, 38%, as compared to 10% in the control subjects suggesting Mg deficiency. In a second study by Cohen *et al.* 12 younger women age 55–65 with osteoporosis (as determined by X-ray) had significantly lower serum Mg concentrations than 10 control subjects, but no difference in the Mg tolerance test was observed.[41] Reginster and colleagues found red blood cell Mg to be significantly lower in 10 postmenopausal women who had at least one vertebral fracture as compared to 10 subjects with degenerative osteoarthritis; no difference in plasma Mg was found.[42] In a second report by Reginster, Mg levels in 10 postmenopausal women age $68.9\pm9$ years with vertebral crush fracture were compared to 10 non-osteoporotic women age $67.2\pm6$ years.[43] The osteoporotic subjects had a significantly lower serum Mg but no difference in red blood cell Mg.

The majority of body Mg, 50–60%, resides in the skeleton and skeletal Mg reflects Mg status. In two studies cited above in which a Mg deficit was suggested by either Mg tolerance testing or low serum Mg concentration, the Mg content of iliac crest trabecular bone was significantly reduced in the osteoporotic patients.[40,41] Two additional studies also found a lower bone Mg concentration in elderly osteoporotic patients.[44,45] Reginster[43] however found no difference in bone Mg concentration between osteoporotic subjects and bone obtained from cadavers. Another study found no difference between patients with osteoporosis and control subjects in cortical bone[46] while two studies reported higher bone Mg concentration in osteoporotic bone.[46,47]

In summary, Mg status has been assessed in very few osteoporotic patients. Low serum and red blood cell Mg concentrations as well as high retention of parenterally administered Mg has suggested a Mg deficit, however these results are not consistent from one study to another. Similarly, while low skeletal Mg content has been observed in some studies, others have found normal or even high Mg content.

# Effect of Mg Therapy

The effect of dietary Mg supplementation on bone mass in patients with osteoporosis has not been extensively studied. Abraham administered 600 mg of Mg per day to 19 patients over 6–12 months.[48] BMD of the calcaneus increased by 11% compared to a 0.7% rise in seven control subjects. All subjects were postmenopausal (age 42–75) and on sex steroid replacement therapy. The subjects who received Mg also received 500 mg of calcium per day as well as many other dietary supplements however, making if difficult to conclude that Mg alone was the sole reason for the increase in bone mass. Steidl and Ditmar[49] administered 243 mg Mg per day to 61 females (age 40–83) and noted improvement of back pain and movement. No objective measurements such as bone density were performed. In a retrospective study, 200 mg per day of Mg, given to 6 postmenopausal women (mean age 59) was observed to produce a small non-significant 1.6% rise in bone density of the lumbar spine; no change was seen in the femur.[50] Stendig-Linberg *et al.* conducted a 2-year trial in which 31 postmenopausal osteoporotic women were given 250 mg Mg per day increasing to a maximum of 750 mg per day for 6 months depending on tolerance.[51] All subjects were given 250 mg Mg per day from months 6 to 24. Twenty-three age-matched subjects served as control. At 1 year there was a significant 2.8% increase in bone density of the distal radius. Twenty-two of the 31 subjects had an increase in bone density while five did not change. Three subjects that showed a decrease in bone density had primary hyperparathyroidism and one underwent a thyroidectomy. No significant effect of Mg supplementation was shown at 2 years, although only 10 subjects completed the trial. In a small uncontrolled trial, Rude and Olerich[52] reported a significant increase in bone density of the proximal femur and lumbar spine in celiac sprue patients who received approximately 575 mg Mg per day for 2 years. These subjects had shown evidence of reduced free Mg in red blood cells and peripheral lymphocytes. In summary, the effect of Mg supplements on bone mass has generally led to an increase in BMD, although study design limits useful information.

# Possible Mechanisms for Mg Deficiency-induced Osteoporosis

There are several potential mechanisms, which may account for a decrease in bone mass in Mg deficiency (Table 15.2). Mg is mitogenic for bone cell growth in cell culture and this may contribute to the decrease in osteoblast number observed in Mg deficient rat which may directly result in a decrease in bone formation.[53] Mg deficiency will result in impaired PTH secretion as well as end organ resistance to the action of PTH.[54] In addition, serum 1,25(OH)$_2$-vitamin D levels are also low in Mg deficient humans and rats.[55] Both hormones are trophic for bone and their deficiency may result in a decrease in osteoblast

**Table 15.2.** *Potential mechanisms of magnesium deficiency-induced osteoporosis*

Decreased bone formation
   Decreased mitogenic effect Mg on osteoblast
   Decreased PTH secretion
   Low serum 1,25(OH)$_2$ vitamin D
   Low serum IGF-1

Increased bone resorption
   Increased serum TNF-α, IL-1, IL-6, IL-11
   Increased bone TNF-α, IL-1, IL-6

Abnormal bone mineralization
   Larger, more perfect crystals

activity and hence, a decrease in bone formation. Serum IGF-1 levels have also been observed to be low in the Mg deficient rat which could affect skeletal growth.[56] While the above may explain low bone formation, it does not explain the observation of an increase in osteoclast number observed in the Mg deficient rat. Acute Mg depletion in the rat causes an immediate rise in serum concentrations of substance P followed by a rise in inflammatory cytokines (TNFα, IL-1, IL-6, and IL-11).[57] Immunohistochemical studies have demonstrated an increase in TNFα and IL-1 in trabecular bone in Mg deficient mice.[57] These cytokines stimulate the development and activity of osteoclastic bone formation and explain the uncoupling of bone formation and bone resorption observed in the rat.[8] Mg affects crystal formation; a lack of Mg results in a larger, more perfect crystal which may affect bone strength.[58] Whether these effects of Mg deficiency are operative in people with suboptimal chronic dietary Mg and osteoporosis is unknown and in need of further investigation.

## Conclusions

Dietary Mg intake falls below the RDA in a large proportion of the population; the clinical consequences of which are not fully understood. One consequence may be an increased risk of developing osteoporosis. Animal studies demonstrate a link between Mg depletion and compromised skeletal integrity; however, the same relationship in humans has been more difficult to establish. Epidemiologic studies do suggest a correlation between dietary intake and bone mass, except in early postmenopausal women, where acute estrogen deficiency may mask this relationship. Studies examining bone turnover markers and assessing Mg status show conflicting results and involve too few subjects. Likewise, the effect of Mg therapy on bone mass has not been adequately studied. Osteoporosis has a major impact on the quality of life of women, as well as a financial impact on society. Further research is necessary to clarify the relationship between dietary Mg intake and bone mass.

## References

1. R.K. Rude, Magnesium Deficiency: a heterogeneous cause of disease in humans, *J. Bone Miner. Res.*, 1998, **13**, 749–758.
2. *Dietary Reference Intakes for Calcium, Phosphorus, Magnesium, Vitamin D, and Fluoride*, National Academy Press, Washington, DC, 1997, pp. 190–249.
3. L.E. Cleveland, J.D. Goldman and L.G. Borrude, *Data Tables: Results From USDA's 1994 Continuing Survey of Food Intakes by Individuals and 1994 Diet and Health Knowledge Survey*, Agricultural Research Service, U.S. Department of Agriculture, Beltsville, MD.
4. M.A. Kenney, H. McCoy and L. Williams, Effects of magnesium deficiency on strength, mass and composition of rat femur, *Calcif. Tissue Int.*, 1994, **54**, 44–49.
5. A.L. Boskey, C.M. Rimnac, M. Bansal, M. Federman, J. Lian and B.D. Boyan, Effect of short-term hypomagnesemia on the chemical and mechanical properties of rat bone, *J. Orthpaedic Res.*, 1992, **10**, 774–783.

6. J.M. Mirra, N.W. Alcock, M.E. Shils and P. Tannenbaum, Effects of calcium and magnesium deficiencies on rat skeletal development and parathyroid gland area, *Magnesium*, 1982, **1**, 16–33.

7. T.O. Carpenter, S.J. Mackowiak, N. Troiano and C.M. Gundberg, Osteocalcin and its message: relationship to bone histology in magnesium-deprived rats, *Am. J. Physiol.*, 1992, **263**, E107–E114.

8. R.K. Rude, M.E. Kirchen, H.E. Gruber, M.H. Meyer, J.S. Luck and D.L. Crawford, Magnesium deficiency-induced osteoporosis in the rat: uncoupling of bone formation and bone resorption, *Magn. Res.*, 1999, **12**, 257–267.

9. B.W. Loveless and F.W. Heaton, Changes in the alkaline phosphatase (EC 3.1.3.1) and inorganic pyrophosphatase (EC 3.6.1.1) activities of rat tissues during magnesium deficiency. The importance of controlling feeding pattern, *Br. J. Nutr.*, 1976, **36**, 487–495.

10. C.C. Lai, L. Singer and W.D. Armstrong, Bone composition and phosphatase activity in magnesium deficiency in rats, *J. Bone Joint Surg.*, 1975, **57**, 516–522.

11. A. Creedon, A. Flynn and K. Cashman, The effect of moderately and severely restricted dietary magnesium intakes on bone composition and bone metabolism in the rat, *Br. J. Nutr.*, 1999, **82**, 63–71.

12. H.O. Trowbridge and J.L. Seltzer, Formation of dentin and bone matrix in magnesium-deficient rats, *J. Periodont. Res.*, 1967, **2**, 147–153.

13. J.E. Jones, R. Schwartz and L. Krook, Calcium homeostasis and bone pathology in magnesium deficient rats, *Calcif. Tissue*, 1980, **31**, 231–238.

14. J. MacManus and F.W. Heaton, The effect of magnesium deficiency on calcium homeostasis in the rat, *Clin. Sci.*, 1969, **36**, 297–306.

15. J. Duckworth, W. Godden and G.M. Warnock, The effect of acute magnesium deficiency on bone formation in rats, *Biochem. J.*, 1940, **34**, 97–108.

16. M.A. Kenney, H. McCoy and L. Williams, Effects of dietary magnesium and nickel on growth and bone characteristics in rats, *J. Am. Col. Nutr.*, 1992, **11**, 687–693.

17. E.R. Miller, D.E. Ullrey, C.L. Zutaut, B.V. Baltzer, D.A. Schmidt, J.A. Hoefer and R.W. Luecke, Magnesium requirement of the baby pig, *J. Nutr.*, 1965, **85**, 13–20.

18. O. Heroux, D. Peter and A. Tanner, Effect of a chronic suboptimal intake of magnesium on magnesium and calcium content of bone and on bone strength of the rat, *Can. J. Physiol. Pharmacol.*, 1974, **53**, 304–310.

19. B.S. Smith and D.I. Nisbet, Biochemical and pathological studies on magnesium deficiency in the rat. "I. Young animals", *J. Comp. Path.*, 1968, **78**, 149–159.

20. L.F. Belanger, J. Robichon and M.R. Urist, The effects of magnesium deficiency on the host response to intramuscular bone matrix implanted in the rat, *J. Bone Joint. Surg.*, 1975, **57**, 522–526.

21. R. Schwartz and A.H. Reddi, Influence of magnesium depletion on matrix-induced endochondral bone formation, *Calcif. Tissue Int.*, 1979, **29**, 15–20.

22. T. Toba, Y. Kajita R. Masuyama, Y. Takada, K. Suzuki and S. Aoe, Dietary magnesium supplementation affects bone metabolism and dynamic strength of bone in ovariectomized rats, *J. Nutr.*, 2000, **130**, 216–220.

23. S.A. New, C. Bolton-Smith, D.A. Grubb and D.M. Reid, Nutritional influences on bone mineral density: a cross-sectional study in premenopausal women, *Am. J. Clin. Nutr.*, 1997, **65**, 1831–1839.

24. R.M. Angus, P.N. Sambrook, N.A. Pocock and J.A. Eisman, Dietary intake and bone mineral density, *Bone Miner.*, 1988, **4**, 265–277.

25. L.B. Houtkooper, C. Ritenbaugh, M. Aickin, T.G. Lohman, S.B. Going, J.L. Weber, K.A. Greaves, T.W. Boyden, R.W. Pamenter and M.C. Hall,

Nutrients, body composition and exercise are related to change in bone mineral density in premenopausal women, *J. Nutr.*, 1995, **125**, 1229–1237.

26. M.C. Wang, E.C. Moore, P.B. Crawford, M. Hudes, Z.I. Sabry, R. Marcus and L.K. Bachrach, Influence of pre-adolescent diet on quantitative ultrasound measurements of the calcaneus in young adult women, *Osteoporosis Int.*, 1999, **9**, 532–535.

27. J.L. Freudenheim, N.E. Johnson and E.L. Smith, Relationships between usual nutrient intake an bone-mineral content of women 35–65 years of age: longitudinal and cross-sectional analysis, *Am. J. Clin. Nutr.*, 1986, **44**, 863–876.

28. A.L. Tranquilli, E. Lucino, G.G. Garzetti and C. Romanini, Calcium, phosphorus and magnesium intakes correlate with bone mineral content in postmenopausal women, *Gynecol. Endocrinol.*, 1994, **8**, 55–58.

29. K.L. Tucker, M.T. Hannan, H. Chen, L.A. Cupples, P.W.F. Wilson and D.P. Kiel, Potassium, magnesium and fruit and vegetable intakes are associated with greater bone mineral density in elderly men and women, *Am. J. Clin. Nutr.*, 1999, **69**, 727–736.

30. K. Yano, L.K. Heilbrun, R.D. Wasnich, J.H. Hankin and J.M. Vogel, The relationship between diet and bone mineral content of multiple skeletal sites in elderly Japanese-American men and women living in Hawaii, *Am. J. Clin. Nutr.*, 1985, **42**, 877–888.

31. S.A. New, S.P. Robins, M.K. Campbell, J.C. Martin, M.J. Garton, C. Bolton-Smith, D.A. Brubb, S.J. Lee and D.M. Reid, Dietary influences on bone mass and bone metabolism: further evidence of a positive link between fruit and vegetable consumption and bone health, *Am. J. Clin. Nutr.*, 2000, **71**, 142–151.

32. K. Michaelsson, L. Holmberg, H. Mallmin, A. Wolk, R. Bergstrom and S. Ljunghall, Diet, bone mass, and osteocalcin: a cross-sectional study, *Calcif. Tissue Int.*, 1995, **57**, 86–93.

33. K.L. Tucker, M.T. Hannan and D.P. Kiel, The acid-base hypothesis: diet and bone in the Framingham Osteoporosis Study, *Eur. J. Nutr.*, 2001, **40**, 231–237.

34. H.P. Dimai, S. Porta, G. Wirnsberger, M. Lindschinger, I. Pamperl, H. Dobnig, M. Wilders-Truschnig and K.H.W. Lau, Daily oral magnesium supplementation suppresses bone turnover in young adult males, *J. Clin. Endocrinol. Metab.*, 1998, **83**, 2742–2748.

35. L. Doyle, A. Flynn and K. Cashman, The effect of magnesium supplementation on biochemical markers of bone metabolism or blood pressure in healthy young adult females, *Eur. J. Clin. Nutr.*, 1999, **53**, 255–261.

36. A. Gur, L. Colpan, K. Nas, R. Cevik, J. Sarac, F. Erdogan and M. Duz, The role of trace minerals in the pathogenesis of postmenopausal osteoporosis and a new effect of calcitonin, *J. Bone Miner. Metab.*, 2002, **20**, 39–43.

37. J. Brodowski, Levels of ionized magnesium in women with various stages of postmenopausal osteoporosis progression evaluated on the densitometric examinations, *Przegl Lek.*, 2000, **57**, 714–716.

38. A. Habtezion, M.S. Silverberg, R. Parkes, S. Mikolainis and A. Steinhart, Risk factors for low bone density in Crohn's disease, *Inflammatory Bowel Dis.*, 2002, **8**, 87–92.

39. V. Kantorovich, J.S. Adams, J.E. Gaines, X. Guo, M.R. Pandian, D.H. Cohn and R.K. Rude, Genetic heterogeneity in familial renal magnesium wasting, *J. Clin. Endocrinol. Met.*, 2002, **87**, 612–617.

40. L. Cohen and A.L. Kitzes, Bone magnesium, crystallinity index and state of body magnesium in subjects with senile osteoporosis, maturity-onset diabetes and women treated with contraceptive preparations, *Magnesium*, 1983, **2**, 70–75.

41. L. Cohen, A. Laor and R. Kitzes, Magnesium malabsorption in postmenopausal osteoporosis, *Magnesium*, 1983, **2**, 139–143.

42. J.Y. Reginster, B. Maertens de Noordhout, A. Albert, A. Dupont-Onkelinx and P. Franchimont, Serum and erythrocyte magnesium in osteoporotic and osteoarthritic postmenopausal women, *Magnesium*, 1985, **4**, 208.

43. J.Y. Reginster, Serum magnesium in postmenopausal osteoporosis, *Magnesium*, 1989, **8**, 106.

44. D.H. Manicourt, S. Orloff, J. Brauman and A. Schoutens, Bone mineral content of the radius:good correlations with physicochemical determinations in iliac crest trabecular bone of normal and osteoporotic subjects, *Metabolism*, 1981, **30**, 57–62.

45. K. Milachowski, D. Moschinske and R.R. Jaeschock, Die bedeutung des magnesiums beider medialen schenkelhasfrakur des alter menschen, *Mag. Bull.*, 1981, **3**, 90–102.

46. M.F. Basle, Y. Mauras, M. Audran, P. Clochon, A. Rebel and P. Allain, Concentration of bone elements in osteoporosis, *J. Bone Miner. Res.*, 1990, **5**, 41–47.

47. J.M. Burnell, D.J. Baylink, C.H. Chestnut III, M.W. Mathews and E.J. Teubner, Bone matrix and mineral abnormalities in postmenopausal osteoporosis, *Metabolism*, 1982, **31**, 1113–1120

48. G.E. Abraham, The importance of magnesium in the management of primary postmenopausal osteoporosis, *J. Nutr. Med.*, 1991, **2**, 165–178.

49. L. Steidl and R. Ditmar, Osteoporosis treated with magnesium lactate, *Acta Univ. Palacki Olomuc Fac Med.*, 1991, **129**, 99–106.

50. J. Eisinger and D. Clairet, Effects of silicon, fluoride, etidronate and magnesium on bone mineral density: a retrospective study, *Magn. Res.*, 1993, **6**, 247–249.

51. G. Stendig-Lindberg, R. Tepper and I. Leichter, Trabecular bone density in a two year controlled trial of peroral magnesium in osteoporosis, *Magn. Res.*, 1993, **6**, 155–163.

52. R.K. Rude and M. Olerich, Magnesium deficiency: Possible role in osteoporosis associated with gluten-sensitive enteropathy, *Osteoporosis Int.*, 1996, **6**, 453–461.

53. C.C. Liu, J.K. Yeh and J.F. Aloia, Magnesium directly stimulates osteoblast proliferation, *J. Bone Miner. Res.*, 1988, 3(Suppl. 1), S104.

54. R.K. Rude, S.B. Oldham, C.F. Sharp and F.R. Singer, Parathyroid hormone secretion in magnesium deficiency, *J. Clin. Endocrinol. Metab.*, 1978, **47**, 800–806.

55. S. Fatemi, E. Ryzen, J. Flores, D.B. Endres and R.K. Rude, Effect of experimental human magnesium depletion on parathyroid hormone secretion and 1,25 dihydroxyvitamin D metabolism, *J. Clin. Endocrinol. Metab.*, 1991, **73**, 1067–1072.

56. I. Dorup, A. Flyvbjerg, M.E. Everts and T. Clausen, Role of insulin-like growth factor-1 and growth hormone in growth inhibition induced by magnesium and zinc deficiencies, *Br. J. Nutr.*, 1991, **66**, 505–521.

57. W.B. Weglicki, B.F. Dickens, T.L. Wagner, J.J. Chemielinska and T.M. Phillips, Immunoregulation by neuropeptides in magnesium deficiency: *ex vivo* effect of enhanced substance P production on circulation T lymphocytes from magnesium-deficient mice, *Magn. Res.*, 1996, **9**, 3–11.

58. R.K. Rude, H.E. Gruber, L.Y. Wei, A. Rausto and B.G. Mills, Magnesium Deficiency: Effect on bone and mineral metabolism in the mouse, *Calcif. Tissue Int.*, 2003, **72**, 32–41.

59. L. Cohen, A. Laor and R. Kitzes, Bone magnesium, crystallinity index and state of body magnesium in subjects with senile osteoporosis, maturity-onset diabetes and women treated with contraceptive preparations, *Magnesium*, 1983, **2**, 70–75.

CHAPTER 16

# The Role of Trace Elements in Bone Health

PAULA MICHELLE WALSH, JACQUELINE MARY
O'CONNOR and JOHN JOSEPH STRAIN

Northern Ireland Centre for Food and Health (NICHE), Faculty of Life and
Health Sciences, University of Ulster, Coleraine, BT52 1SA,
Email: p.walsh@ulster.ac.uk

**Key Points**
- Iron overload may be detrimental to bone health.
- Zinc is needed for adequate osteoblastic activity, and calcification.
- Copper deficiency, in various species, results in bone abnormalities.
- Manganese is required for synthesis of connective tissue in cartilage and bone.
- Fluoride accumulates in new bone formation sites and results in net gain in bone mass.
- Other trace elements including vanadium, silicon and boron may emerge as being essential for bone health.

## Introduction

Most studies investigating dietary minerals and bone health in the pathophysiology and prevention of osteoporosis have focused on the role of dietary calcium and calcium supplementation. Bone, however, is an active, living tissue, undergoing continual remodelling through osteoblastic and osteoclastic activity and, thereby, constantly participating in a wide range of biochemical reactions.[1] It is, therefore, not surprising, that other metabolically active minerals, including trace and ultra trace elements, have been implicated in bone health.

A 'trace element' is one that occurs in the human body, at a concentration of $<0.01\%$.[2] An 'ultra trace element' is one with an estimated or suspected requirement in terms of $\mu g$ day$^{-1}$ for humans.[3] Trace elements are needed for cellular functions ranging from structural components of body tissue to essential components of enzymes and other important molecules. They occur in the body as inorganic elements or as constituents of organic molecules, for example proteins, fats, carbohydrates and nucleic acids.[4] In order to carry out these vital processes, trace element concentrations and functional forms must be maintained within narrow limits.[5]

There are indications that some trace elements have a role in bone health and perhaps osteoporosis.[6-9] Although perceived as minor building components in teeth and bone, trace elements play important functional roles in bone metabolism and turnover.[9] In addition, some trace elements (copper, zinc, boron, manganese) can promote the formation of strong bones and connective tissues.[10] The adverse effects of trace element deprivation on bone metabolism and skeletal development in young animals and man has been recognised for many years.[11] In contrast, some trace elements at high concentrations can have a toxic effect on bone metabolism with harmful consequences. Although each trace element is treated separately in this chapter, it is important to recognise that interactions of some of the trace elements with each other and with other food constituents exist and are important to health.[4]

The UK Recommended Nutrient Intake (RNI) and USA Recommended Dietary Allowance (RDA)/ Adequate Intake (AI) for some of the trace elements are given in Table 16.1.[12-14]

**Table 16.1** *Adult UK Recommended Nutrient Intake (RNI) and USA Recommended Dietary Allowance (RDA)/ Adequate Intake (AI) for trace elements which have been implicated in the maintenance of bone health*

|  | UK (RNI) Mg day$^{-1}$ | USA (RDA) (AI*) Mg day$^{-1}$ |
| --- | --- | --- |
| *Iron* | | |
| Adult men 19–50 years | 8.7 | 8 |
| Adult women 15–50 years | 14.8 | 18 |
| Adults over 51 years | 8.7 | 8 |
| *Zinc* | | |
| Adult men | 9.5 | 11 |
| Adult women | 7 | 8 |
| *Copper* | | |
| Adult men and women | 1.2 | 0.9 |
| *Manganese* | | |
| Adult men | | 2.3* |
| Adult women | | 1.8* |
| *Fluoride* | | |
| Adult men | | 4* |
| Adult women | | 3* |

UK, RNI; Committee on Medical Aspects of Food Policy (COMA) (1991) and USA RDA; USA Food and Nutrition Board (2001).

# Iron

Iron may play a role in bone formation. Experimental work has established a critical role for iron in the synthesis of the collagen of connective tissue.[15] The classical work of Stetten and Schoenheimer, quoted by Prockop (1971), showed that free proline is a precursor of both the proline and the hydroxyproline in the collagen molecule. Furthermore, iron is required for the hydroxylation of both the proline and lysine in protocollagen.[15]

There are no reports of iron deficiency being associated with osteoporosis. However, a high intake of dietary or supplemental calcium reduces the incidence of osteoporosis and this may have an adverse effect on the absorption of iron and other trace elements.[16,17] This is not a consistent finding, however, as no effect of long-term calcium supplementation was demonstrated on iron status in infants,[18] adolescent girls,[19] adults[20] or lactating women.[21]

In contrast, iron overload may be detrimental to bone. Abnormal bone histology has been reported in patients with impaired iron metabolism and severe iron overload such as haemochromatosis[22] and Bantu siderosis.[23] In both these disorders, osteoporosis is a prominent finding but it is not clear whether the effect on bone is from iron itself, or iron induced hypervitaminosis, C or both.[24,25] Both histochemical and radionuclide studies have shown that ionised iron is taken up by cortical and trabecular bone, is localised at trabecular and endosteal surfaces and at the calcified bone boundaries, and is deposited within osteoclasts, osteoblasts and osteocytes.[26-30] In chronic renal disease, accumulation of iron in bone has been found in some patients undergoing dialysis. In a study of 12 dialysis patients, five patients showed that iron overload was associated with an increased frequency of dynamic bone disease.[31] In addition, McCarthy *et al.*[32] reported that deposition of iron in bone occurs in young patients who tended to have higher serum ferritin and lower serum immnunoreactive parathyroid (PTH) levels.

At present there is little evidence suggesting a role for iron sufficiency in bone formation or metabolism, but there are indications that iron overload may be detrimental to bone health.

# Zinc

Over the decades, the importance of zinc in bone metabolism and in the normal growth of humans and animals has been recognised in numerous studies.[33-36] Studies of the distribution of zinc in bone using histochemical and microradiographic methods[33] concluded that zinc was found in close proximity to sites of calcification and increased uptake of $^{65}$Zn has been found at the site of bone healing in rats.[37] The concentration of zinc in bone is higher than in most other tissues and is lowered quickly by zinc depletion.[38] Zinc has a structural role in the bone matrix where *inter alia*, it complexes with fluoride in the hydroxyapatite crystal of bone.[39] It is thought that the Zn–F complex may be important in the formation and structure of biological apatite.[40]

Many of the effects of zinc deficiency on bone metabolism may be related to a generalised impairment of nucleic acid and protein metabolism. Further to the well-known role of zinc deficiency in growth retardation and other skeletal abnormalities, a series of investigations (*in vivo* and *in vitro*) by Yamaguchi and colleagues[41–43] have demonstrated zinc related increases in protein synthesis, alkaline phosphatase activity and bone collagen content. A compound, β alanyl-L-histidinato zinc (AHZ), in which zinc has been chelated to β alanyl-L-histidine (L-carnosine) has been proposed to be of value therapeutically in osteoporosis.[44] *In vivo* studies in elderly rats[45] and weanling rats[46] have demonstrated that AHZ directly increases bone growth factors and bone matrix protein, which are involved in the stimulation of bone formation and cell proliferation in osteoblasts. Furthermore, zinc deficiency has been reported to induce bone growth retardation and cause a decrease in serum insulin like growth factor (IGF-I) levels in rats, suggesting an involvement for zinc in IGF-I production.[47] In addition, Yamaguchi *et al.*[48] demonstrated that zinc can directly stimulate the production of IGF-I from osteoblasts *in vitro* and Matsui *et al.*[49] showed that the anabolic effect of IGF-I in osteoblasts is enhanced by zinc supplementation. The peptide produced a significant increase in osteoblast number, protein concentration and deoxyribonucleic acid (DNA) content. It was also found that the presence of both IGF-I and zinc sulphate synergistically increased alkaline phosphatase activity.[49]

The teratogenic effects of severe zinc deficiency in mammals have been demonstrated in numerous studies.[50–54] Hurley and colleagues[50–53] reported that pregnant rats fed diets severely deficient in zinc (day 0–21) gained less weight and had significantly fewer live foetuses at term than did controls. In addition, surviving foetuses from the zinc deficient group were characterised by a variety of congenital malformations encompassing a wide variety of organs and systems including the skeleton. The skeletal defects were manifested as deformities of the skull, fusion of the ribs, missing vertebrae, shortened long bones, missing centres of ossification, club foot, cleft palate and micrognathia (undersized mandible).[50–53] Defects in skeletal development have also been reported in zinc deficient chicks, pigs, cows and rhesus monkeys.[55,56]

The role of zinc in bone metabolism and in the pathogenesis of osteoporosis was elucidated further in a study investigating the effects of zinc depletion on bone structure and turnover.[36] This study demonstrated that alimentary zinc deficiency induced cancellous bone osteopoenia in growing rats. The osteopoenia observed in the study was characterised as a low turnover osteopoenia *i.e.*, as an osteopoenia that is accompanied by a reduction both in osteoblastic bone formation and osteoclastic bone resorption. In patients with established osteoporosis it is thought that cancellous bone loss may be due to a decrease in osteoblast performance. Therefore, the study of Eberle *et al.*[36] may support a possible role of zinc deficiency as a risk factor in the development of osteoporosis. However, as Eberle *et al.* state "one should be cautious in extrapolating experimental findings in growing rats to adult humans".

Studies in humans have shown that zinc intake decreases with age, especially in women.[57,58] An inverse correlation between zinc intake and bone loss in postmenopausal women was observed by Freudenheim *et al.*[59] Zinc deficit is accompanied by hypogonadism, which causes osteopoenia, and zinc supplements enhance the effect of oestrogen on bone.[60] Atik *et al.*[61] found patients with senile osteoporosis had lower levels of skeletal zinc than a control group; although others have not confirmed this.[62] Excessive urinary zinc loss has also been found in patients with osteoporosis. Women with osteoporosis have been reported to excrete more than $800\ \mu g\ Zn\ g^{-1}$ creatinine in urine, and urinary zinc has been suggested as a marker of bone resorption.[63] Relea *et al.*[64] reported that the urinary clearance of zinc in postmenopausal women was higher than in premenopausal women. In addition, Angus *et al.*[65] found a significant correlation between forearm bone mineral content and zinc intake in pre-menopausal women. Low serum zinc levels have also been found in elderly people who are housebound, immobilised or institutionalised.[66,67] Work by Strause and Saltmam[8,68] has shown that bone loss in calcium supplemented postmenopausal women can be further arrested by concomitant increases in trace mineral intake including that of zinc, copper and manganese.

Zinc is essential for many biological functions in man and animals. Enough is now known about the clinical and public health importance of zinc function and metabolism, and zinc deficiency, to establish beyond doubt the relevance of this trace element to the skeleton. The role of zinc in bone appears to be two fold; it appears to diminish resorption and to stimulate formation and it has been shown to increase bone mass in osteopenic environments in animal models. These factors serve as a stimulus and cornerstone for expanded research, preferably in conventional clinical trials in humans investigating further the role of zinc in bone health.

# Copper

Copper is an essential component for several enzymes involved in bone formation. The important role for copper in collagen maturation can be linked primarily to its function as a cofactor for lysyl oxidase (EC 1.4.3.13). Lysyl oxidase is a cuproprotein that is present in high concentrations in dense connective tissue.[69] It catalyses the oxidative deamination of lysine residues in elastin and collagens as an initial step in their extracellular assembly into insoluble fibres.[70,71] Therefore, decreased activity of this copper dependent enzyme is the mechanism most probably responsible for the loss of bone architecture in copper deficiency.[72] *In vitro* studies have shown that copper reduces active bone resorption.[73] The trace element plays a role in the inhibition of bone resorption through its action as a cofactor for superoxide dismutase (EC 1.15.1.1). This is an antioxidant enzyme containing two atoms of zinc and copper, and acts as a free radical scavenger, neutralising the superoxide radicals produced by osteoclasts during bone resorption.[74] Furthermore, it has been suggested that caeruloplasmin (EC 1.16.3.1) may be mechanistically involved in bone changes as this protein is stimulated by oestrogen which declines at the menopause.[72]

Bone abnormalities have been found in various species given copper deficient diets, including sheep, cattle, rats, pigs, and dogs.[75–77] Rats[78] and lambs[79,80] fed a copper deficient diet exhibit bone loss, inhibition of bone growth and formation and promotion of pathological changes similar to those seen in osteoporosis. More recently, Rico et al.[81] investigated the effect of copper on the prevention of bone mass loss induced by ovariectomy in rats. The results from this study showed that a copper supplement of 15 mg kg$^{-1}$ of feed in rats, during a period of 30 days post ovariectomy, inhibited the loss of vertebral and femoral bone mass induced by ovariectomy. In addition, Yee et al.[82] demonstrated that osteopoenia was more acute in rats with ovariectomy combined with copper deficiency than with ovariectomy alone.

Cornatzer et al.[83] demonstrated that copper deficiency produced a significant decrease in the activity of choline phosphotransferease (CP). CP phosphorylates choline, the first step in the synthesis of the phospholipid phosphatidylcholine from choline. Although copper deficiency produced a significant decrease in the activity of CP, a significant decrease was also found in the concentration of total phospholipid-P, phosphatidylcholine-P, phosphatidylethanolamine-P, phosphatidylinositol-P, sphingomyelin-P and cardiolipin-P in rat heart microsomes.[83] Furthermore, recent data confirm the presence of choline-containing phospholipids in the peripheral membrane of osteoblasts, in the interfibrillary spaces of the osteoid border, in matrix vesicles, and in the peripheral zone of the calcification nodules during the early stages of calcification.[84] These phospholipids have an active role during vesicle-induced biological calcification and degradation of these phospholipids, apparently enables the egress of crystalline mineral from the vesicle lumen.[85] In particular, phosphatidylcholine, which has been demonstrated to be significantly decreased during copper deficiency in rats,[83] appears to be involved in signal transduction in osteoblasts.[86,87]

Severe copper deficiency caused by inadequate copper intake is known to occur only in infants.[88] Graham and Cordano[88] were among the first to observe that copper deficiency was accompanied by osteopoenia, among other abnormalities. There are also reports of infants developing osteoporosis-like symptoms because of nutritional copper deficiency.[77,89] One study, in particular, reported systemic bone disease which developed in small, premature infants with copper deficiency.[90] Symptoms included: epiphyseal separations, metaphyseal cupping, subperiosteal new bone growth, systemic perosis and rib fractures.[90]

The most severe form of copper deficiency in humans, owing to a defect in intestinal absorption of dietary copper, is the inherited sex linked recessive syndrome known as Menkes syndrome.[91] The symptoms reported in Menkes syndrome include osteopoenia, cupping and flaring of metaphyses of long bones, submetaphyseal fracture, and spontaneous fractures, most frequently in the ribs.[92,93] The disease, in contrast to dietary copper deficiency, does not respond to copper supplementation and is considered incurable at the present time.

It has been postulated that dietary copper deficiency may be an important factor in the onset and progression of osteoporosis[94] and there is some evidence to suggest a role of copper deficiency in age-related osteoporosis. It has been shown that there was no loss of lumbar bone mass in women ($n = 38$) (45–56 years) who were treated for two years with a supplement of 3 mg of copper but that there was loss in a group ($n = 35$) of comparable age receiving a placebo.[72] In a study with 46 elderly patients with fractures of the femoral neck, serum copper levels were found to be significantly lower than those of a group of controls matched for age and sex.[95] Furthermore, Howard et al.[96] reported that postmenopausal women with a high dietary calcium intake combined with a high serum copper level had a greater lumbar bone density than women with low calcium intake and low serum copper. These studies suggest that inadequate dietary copper intake may be a contributory factor to age-related bone loss in older women.

In addition, the implications for bone health in younger individuals with marginal copper intakes was demonstrated when increasing dietary copper intake in healthy adult males (aged 20–59 years, living in a residential metabolic research unit for 6 weeks) from a relatively low to a relatively high level significantly decreased the rate of urinary pyridinium crosslink excretion.[97] The urinary pyridinium crosslinks, pyridinoline and deoxypyridinoline, are sensitive and specific biochemical markers of bone resorption and the increased excretion of these biomarkers is indicative of increased rate of bone resorption.[97] In another placebo-controlled crossover study, however, there was no effect of increasing copper intakes above the usual dietary intake on biomarkers of bone metabolism in healthy adults over a 6 week period.[98]

Copper overload can also interfere with bone metabolism. In patients with Wilson's disease, a genetic disorder resulting in copper accumulation in the body, copper overload is associated with generalised loss of bone density, and rickets.[99] Kaji et al.[100] demonstrated that chick femurs incubated with copper (2.5 µM and above) showed a reduction in collagen content of both diaphysis and epiphysis mainly owing to inhibition of collagen synthesis. Furthermore, high levels of copper induced inhibition of mineralisation and matrix formation.

There is a lack of reliable biomarkers to permit the determination and interpretation of an individuals copper status.[101] Consequently assessing copper status makes it difficult to ascertain the benefits of increased copper intake for at risk groups, in the prevention of disease, or to determine threshold values for copper toxicity. Copper deficiency in man and animals results in bone abnormalities. Although copper deficiency is rare in healthy adults, it does not rule out the possibility that sub-optimal copper intakes over long periods of time may be involved in the precipitation of chronic diseases such as osteoporosis.

# Manganese

Manganese deficiency occurring during foetal and early postnatal life results in abnormal skeletal development in a number of animal species. One of the first structural abnormalities recognised as a potential consequence of manganese deficiency was perosis in chickens.[102] This disorder is characterised by shortened and thickened limbs, curvature of the spine, and swollen and enlarged joints. Because of this disorder, attention was turned to a possible involvement of manganese in the synthesis of the organic matrix of cartilage.[102] The role of manganese in normal epiphyseal cartilage metabolism appears to be with its involvement in the biosynthesis of proteoglycans. These are complex macromolecules, which are major constituents of the cartilage extracellular matrix. Manganese is the preferred cofactor of glycosyltransferases, enzymes involved in the synthesis of the glycoaminoglycan side chains and which are required for the synthesis of proteoglycans.[103]

In rats and mice fed manganese deficient diets during early development, bone defects similar to those seen in chicks were observed. Rats fed a manganese deficient diet had smaller, less dense bones with less resistance to fractures than those fed adequate amounts of manganese.[104] Also, Strause et al.[105] reported defects in chondrogenesis, osteogenesis and bone resorption in manganese deficient rats. The skeletal abnormalities observed in animals are characteristic of manganese deficiency and could be due to a reduction in proteoglycan synthesis secondary to a reduction in the activities of several manganese dependent glycosyltransferases. However, another mechanism has been suggested that could contribute to these skeletal abnormalities in manganese deficient animals. More recently, Clegg et al.[106] observed in male rats that chronic manganese deficiency altered metabolism of IGF. The deficient animals displayed lower circulating concentrations of IGF-1 and insulin compared to controls. These alterations could contribute to the growth and bone abnormalities observed in deficient animals.[106]

Although manganese deficiency has been observed in many species of animals, there is little evidence of its deficiency in humans. It has been reported that prolonged manganese deficiency is associated with osteoporosis in man[107] and that osteoporotic subjects have low serum manganese levels.[108] However, it is important to note, that these subjects were deficient in several other minerals and elements as well. In addition Saltman et al.[8] reported data that supported the possibility that sub-optimal manganese status could be related to osteoporosis. A supplement containing manganese, copper and zinc in combination with a calcium supplement was found to be more effective than the calcium supplement alone in preventing spinal bone loss over a period of two years. However, the presence of other trace elements in the supplement makes it impossible to determine whether manganese supplementation was the beneficial agent for maintaining bone mineral density.

Although manganese deficiency has been observed in many species of animals and various skeletal abnormalities have been reported, there is little evidence of such deficiency in humans. Evidence from animal studies suggests that manganese is critical for proper development of both bone and cartilage mainly through its role in proteoglycan biosynthesis. At present there is no direct evidence that manganese supplements can help to prevent osteoporosis. As the role of manganese in human nutrition becomes more apparent, research should be directed towards determining the optimal dietary intakes of this mineral.

# Fluoride

Fluoride appears to function in the crystalline structure of bone, where, after absorption, it becomes incorporated into the calcium phosphate crystal structure apatite, replacing the hydroxyl group and forming fluorapatite.[109] It is present in the hard and soft tissues of the body in extremely small amounts with the adult human containing some 3–7 mg. The wide distribution of fluoride in food and water, plants and animals as well as human tissues suggests that fluoride may have a definitive physiological effect. At very minute quantities, fluoride plays a role in controlling certain enzymatic reactions in various organs of the human body. Following oral ingestion of fluoride salts, almost 100% are completely and readily absorbed in the gastrointestinal tract by passive diffusion. The kidney excretes approximately 50% of the fluoride ion and the remainder crosses the cell membrane and becomes incorporated into the teeth, bone and calcified cartilage, where it is stored.[110–112]

For many years the ability of fluoride to affect the biological function of bone cells as well as the physiochemical properties of bone mineral has led to an extensive amount of research on fluoride (sodium fluoride), in relation to industrial and endemic fluorosis and the treatment of osteoporosis. Fluoride affects bone in at least two ways: (1) high serum levels of fluoride cause an increase in osteoblast activity. *In vitro* studies have demonstrated that osteoblastic cells from osteoporotic subjects had a positive histological response to fluoride therapy, and had greater maximal DNA synthesis compared to normal controls and a group of similarly treated osteoporotic patients in whom there was minimal histological response to therapy.[113] From this study, it was concluded that the net increase in bone formation in the responders resulted from enhancement of osteoblastic proliferation rather than increased osteoblastic differentiation. In addition, Lau *et al.*[114] suggested that fluoride may act as a mitogen by inhibiting acid phosphatase (EC 3.1.3.2)/ phosphotyrosyl protein phosphatase (EC 3.1.3.16) activity in osteoblasts, leading to increased cellular tyrosyl phosphorylation and bone cell proliferation. (2) A second rationale for using a fluoride supplement was that the fluoride ions replace hydroxyl ions in bone crystals to form fluorapatite. During fluoride exposure, fluoride incorporation in bone occurs in bone mineral substances and results in various modifications of the characteristics of apatite crystals. Fluoride replaces the hydroxyl group to form fluorapatite. The crystal lattice of fluorapatite is quite different from that of hydroxyapatite in that it is larger[115] and more stable.[116] On the other hand, if crystals are excessively large, as in the case of skeletal fluorosis, bone may become brittle and more fragile.[109]

Histological evidence suggests that fluoride is capable of directly inducing the differentiation of osteoclasts, thereby bypassing the normal remodelling sequence of activation–resorption–formation.[117,118] This ability to contribute to a positive skeletal balance makes fluoride a potential therapeutic agent in diseases such as osteoporosis.[119] Following administration of fluoride, the bone produced is defective in structure and in degree of mineralisation. The histological effects of sodium fluoride are dose dependent, with doses $<30$ mg day$^{-1}$ having no consistent effect and doses $>80$ mg day$^{-1}$ usually associated with marked abnormalities.[119] The bone matrix with excess fluoride (fluorosis) is irregularly fibrous and has a woven rather than lamellar appearance. It has a 'mottled' appearance consisting of irregularly distributed osteocytes lying in enlarged lacunae and surrounded by halos of low mineral density. Histologically, the picture resembles osteomalacia; this led to the concept of administration of vitamin D and calcium during fluoride therapy.[119-122] However, the addition of calcium is not completely effective in preventing fluoride induced osteomalacia and it has been suggested that vitamin D appears to blunt the overall effectiveness of fluoride promoting new bone formation.[120,123]

Rich and Ensinck[124] reported the first clinical trial of sodium fluoride for osteoporosis in the hope that the induction of subclinical fluorosis might strengthen the skeleton but not lead to other changes. Findings from this study showed that calcium retention increased and since then, many reports on the success and failures on the effect of fluoride therapy in osteoporosis have appeared in the literature.[125-128] Sodium fluoride has been shown to increase spinal bone mass in a dose dependent manner.[129,130] However, sodium fluoride has never been convincingly demonstrated to reduce the vertebral fracture rate in established spinal osteoporosis, even when bone mass has been increased to above the theoretical fracture threshold.[120] Based on results of four prospective placebo controlled trials[131-134] on the effect of sodium fluoride on vertebral fracture rate in osteoporosis, it was concluded that sodium fluoride has no significant therapeutic advantage over placebo in reducing vertebral fracture rates.[122] These reports are in sharp contrast to uncontrolled trials,[135,136] in which a dramatic effect of sodium fluoride in reducing vertebral fracture rate was found[137,138] while others reported no effect.[139-141] The most specific side-effect of sodium fluoride therapy is the painful lower extremity pain syndrome. This has been reported in 10–50% of patients treated with fluoride therapy.[112] In a French study,[131] ankle and foot pain syndrome was noted significantly more often in fluoride treated patients than in the non-supplemented group. Riggs et al.[133] reported that lower extremity pain syndrome occurred more frequently in the fluoride (75 mg fluoride day$^{-1}$) group compared to the calcium group.

The evidence regarding the safety of fluoride therapy in human studies remains inconclusive. More than any other currently available therapeutic agent, sodium fluoride therapy has clearly been shown to have pronounced effects on the skeleton. These effects appear to be both beneficial and potentially detrimental at the same time. Although slow release therapy has been used in many European countries as a therapy for osteoporosis, approval in the US is still pending.[142,143] At this time, treatment of humans with fluoride may continue only as an experimental approach in properly conducted clinical research studies rather than being accepted as evidence based standard practice.

# Conclusions

This review has presented evidence that nutritional deficiencies or excess of trace elements including iron, zinc, copper and manganese can affect bone health. In addition to the essential trace elements discussed above, other elements including vanadium, silicon and boron may emerge as being important for bone health. Although trace elements are minor building components in teeth and bone, they have important functional roles in bone metabolism and bone turnover. It is possible that chronic deficiencies of one or more of these elements may play an important role in the aetiology of osteoporosis.

# References

1. A.R. Gaby and M.D. Wright, Nutrients and Osteoporosis, *J. Nutr. Med.*, 1990, **1**, 63–72.
2. N.W. Solomons and M. Ruz, Trace element requirements in Humans: An update, *J. Trace Elements Exp. Med.*, 1998, **11**, 177–195.
3. F.H. Nielsen, Ultratrace elements in Nutrition:Current Knowledge and Speculation, *J. Trace Elements Exp. Med.*, 1998, **11**, 251–274.
4. J.J. Strain and K.D. Cashman, Mineral and trace elements, In *Introduction to Human Nutrition*, M.J. Gibney, H. Hester and F.J. Kok (eds.), Blackwell Science, Oxford, 2002, pp. 177–224.
5. E.J. Underwood and W. Mertz, Introduction, In *Trace Elements in Human and Animal Nutrition*, W. Mertz (ed.), Academic Press, San Diego, CA, 1987, vol. 1, pp. 1–20.
6. A. Gur, L. Colpan, K. Nas, R. Cevik, J. Sarac, F. Erdogan and M.Z. Duz, The role of trace minerals in the pathogenesis of postmenopausal osteoporosis and a new effect of calcitonin, *J. Bone Miner. Metab.*, 2002, **20**, 39–43.
7. L.G. Strause, J. Hegenauber, P. Saltman, R. Cone and D. Resnick, Effects of long term dietary manganese and copper deficiency on rat skeleton, *J. Nutr.*, 1985, **116**, 135–141.
8. P. Saltman and L. Strause, Trace elements in bone metabolism (Abstract), *J. Inorg. Biochem.*, 1991, **43**, 284.
9. T. Okano, Effects of essential trace elements on bone turnover-relation to the osteoporosis, *Nippon Rinsho. Jpn. J. Clin. Med.*, 1996, **54**, 148–154.
10. H. Rico, Minerals and Osteoporosis, *Osteoporosis Int.*, 1991, **2**, 20–25.
11. R.M. Leach, The role of trace elements in the development of cartilage matrix, In *Trace elements in Man and Animals*, L.S. Hurley, C.L. Keen, B. Lonnerdal and R.B. Rucker (eds.), Plenum Press, New York, 1988, pp. 267–271.
12. Institute of Medicine. Dietary Reference Intakes for Vitamin A, Vitamin K, Arsenic, Boron, Chromium, Copper, Iodine, Iron, Molybdenum, Nickel, Silicon, Vanadium and Zinc. Food and Nutrition Board, National Academy Press, Washington, DC, 2001.
13. P. Trumbo, A.A. Yates, S. Schlicker and M. Poos, Vitamin A, Vitamin K, Arsenic, Boron, Chromium, Copper, Iodine, Iron, Manganese, Molybdenum, Nickel, Silicon, Vanadium and Zinc, *J. Am. Diet. Assoc.*, 2001, **101**(3), 294–301.
14. Report of the Panel on Dietary Reference Values of the Committee on Medical Aspects of Food Policy (COMA). Dietary Reference Values of the Committee on Medical Aspects of Food Policy. HMSO, London, 1991.

15. D.J. Prockop, Role of Iron in the synthesis of collagen in connective tissue, *Federation Proc.*, 1971, **30**(3), pp. 984–990.
16. P.A. Seligman, J.H. Caskey, J.L. Fazier, R.M. Zucker, E.R. Podell and R.H. Allen, Measurements of iron absorption from prenatal multivitamin-mineral supplements, *Obstet. Gynecol.*, 1983, **61**, 356–362.
17. B. Dawson-Hughes, F.H. Seligson and V.A. Hughes, Effects of calcium carbonate and hydroxyapatite on zinc and iron retention in postmenopausal women, *Am. J. Clin. Nutr.*, 1986, **44**, 83–88.
18. M.A. Dalton, J.D. Sargent, G.T. O'Connor, E.M. Olmstead and R.Z. Klein, Calcium and phosphorus supplementation of iron fortified infant formula: no effect on iron status of healthy full-term infants, *Am. J. Clin. Nutr.*, 1997, **65**, 921–926.
19. J.Z. Ilich-Ernst, A.A. McKenna, N.E. Badenhop, A.C. Clairmont, M.B. Andon, R.W. Nahhas, P. Goel and V. Matkovic, Iron status, menarche, and calcium supplementation in adolescent girls, *Am. J. Clin. Nutr.*, 1998, **68**, 880–887.
20. A.M. Minihane and S.J. Fairweather-Tait, Effect f calcium supplementation on daily nonheme-iron absorption and long term iron status, *Am. J. Clin. Nutr.*, 1998, **68**, 96–102.
21. H.J. Kalkwarf and S.D. Harrast, Effects of calcium supplementation and lactation on iron status, *Am. J. Clin. Nutr.*, 1998, **67**, 1244–1249.
22. S. De Seze, J. Solnica, D. Mitrovic, L. Miravet and H. Dorfmann, Joint and bone disorders and hypoparathyroidism in hemochromatosis, *Semin. Arthritis Rheumat.*, 1972, **2**(1), 71–94.
23. H.C. Seftel, C. Malkin and A. Schmaman, Osteoporosis, scurvy and siderosis in Johannesburg Bantu, *Br. J. Med.*, 1966, **1**, 642–646.
24. S.R. Lynch, H.C. Seftel, A.A. Wapnick, R.W. Charlton and T.H. Bothwell, Some aspects of calcium metabolism in normal and osteoporotic Bantu subjects with special reference to the effects of iron overload and ascorbic acid depletion, *S. Afr. J. Med. Sci.*, 1970, **35**(2), 45–56.
25. A.A. Wapnick, S.R. Lynch, H.C. Seftel, R.W. Charlton, T.H. Bothwell and J. Jowsey, The effect of siderosis and ascorbic acid depletion on bone metabolism, with special reference to osteoporosis in the Bantu, *Br. J. Nutr.*, 1971, **25**(3), 367–376.
26. G.M. Gratwick, P.G. Bullough, W.H. Bohne, A.L. Markenson and C.M. Peterson, Thalassemic osteoarthropathy, *Ann. Intern. Med.,* 1978, **88**(4), 494–501.
27. M.C. de Vernejoul, R. Girot, J. Gueris, L. Cancela, S. Bang, J. Bielakoff, C. Mautalen, D. Goldberg and L. Miravet, Calcium phosphate metabolism and bone disease in patients with homozygous thalassemia, *J. Clin. Endocrinol. Metab.*, 1982, **54**(2), 276–281.
28. M.C. de Vernejoul, A. Pointillart, C.C. Golenzer, C. Morieux, J. Bielakoff, D. Modrowski and L. Miravet, Effects of iron overload on bone remodeling in pigs, *Am. J. Pathol.*, 1984, **116**(3), 377–384.
29. K.R. Phelps, V.J. Vigorita, M. Bansal and T.A. Einhorn, Histochemical demonstration of iron but not aluminum in a case of dialysis-associated osteomalacia, *Am. J. Med.*, 1988, **84**, 775–780.
30. H.J. Huser, P. Eichenberger and H. Cottier, Incorporation of iron into osteoid tissue and bone, *Schweiz. Med. Wochenschr.*, 1971, **101**(49), 1815.
31. F.L. Van de Vyver, W.J. Visser, P.C. D'Haese and M.E. De Broe, Iron overload and bone disease in chronic dialysis patients, *Nephrol. Dialysis Transplant.*, 1990, **5**(9), 781–787.

32. J.T. McCarthy, S.F. Hodgson, V.F. Fairbanks and T.P. Moyer, Clinical and histologic features of iron related bone disease in dialysis patients, *Am. J. Kidney Dis.*, 1991, **17**, 551–561.

33. S. Haumont, Distribution of zinc in bone tissue, *J. Histochem. Cytochem.*, 1961, **9**, 141–145.

34. H. Ronaghy, M.R.S. Fox, S.M. Garn, H. Israel, A. Harp, P.G. Moe and J.A. Halsted, Controlled zinc supplementation for malnourished school boys: a pilot experiment, *Am. J. Clin. Nutr.*, 1969, **22**, 1279.

35. N.R. Calhouen, J.C. Smith and K.L. Becker, The effects of zinc on ectopic bone formation, *Oral Surg. Oral Med. Oral Pathol. Oral Radiol. Endodont.*, 1975, **39**, 698–706.

36. J. Eberle, S. Schmidmayer, R.G. Erben, M. Stangassinger and H.P. Roth, Skeletal effects of zinc deficiency in growing rats, *J. Trace Elements Med. Biol.*, 1999, **13**, 21–26.

37. N.R. Calhoun and J.C. Smith, Uptake of 65Zn in fractured bones, *Lancet*, 1968, **2**, 682.

38. N.R. Calhoun, J.C. Smith and K.L. Becker, The role of zinc in bone metabolism, *Clin. Orthop. Relat. Res.*, 1974, **103**, 212–234.

39. E.J. Murray and H.H. Messer, Turnover of bone zinc during normal and accelerated bone loss in rats, *J. Nutr.*, 1981, **111**, 1641–1647.

40. R. Lappalainen, M. Knuuttila, S. Lammi and E.M. Alhava, Fluoride content related to the elemental composition, mineral density and strength of bone in healthy and chronically diseased persons, *J. Chronic Dis.*, 1983, **36**, 707–713.

41. M. Yamaguchi and R. Yamaguchi, Action of zinc on bone metabolism in rats. Increases in alkaline phosphatase activity and DNA content, *Biochem. Pharmacol.*, 1986, **35** (5), 773–777.

42. M. Yamaguchi, H. Oishi and Y. Suketa, Stimulatory effect of zinc on bone formation in tissue culture, *Biochem. Pharmacol.*, 1987, **36**(22),4007–4012.

43. M. Yamaguchi and R. Matsui, Effect of dipicolinate, a chelator of zinc, on bone protein synthesis in tissue culture. The essential role of zinc, *Biochem. Pharmacol.*, 1989, **38**(24), 4485–4489.

44. M. Yamaguchi and H. Miwa, Stimulatory effect of beta-alanyl-L-histidinato zinc on bone formation in tissue culture, *Pharmacology*, 1991, **42**(4), 230–240.

45. M. Yamaguchi and K. Ozaki, Effect of the new zinc compound beta-alanyl-L-histidinato zinc on bone metabolism in elderly rats, *Pharmacology*, 1990a, **41**(6), 345–349.

46. M. Yamaguchi and K. Ozaki, A new zinc compound, beta-alanyl-L-histidinato zinc, stimulates bone growth in weanling rats, *Res. Exp. Med. (Berl.)*, 1990b, **190**(2), 105–110.

47. G. Oner, B. Bhaumick and R.M. Bala, Effect of zinc deficiency on serum somatomedin levels and skeletal growth in young rats, *Endocrinology*, 194, **114**(5), 1860–1863.

48. M. Yamaguchi and M. Hashizume, Effect of beta-alanyl-L-histidinato zinc on protein components in osteoblastic MC3T3-El cells: increase in osteocalcin, insulin-like growth factor-I and transforming growth factor-beta, *Mol. Cell. Biochem.*, 1994, **136**(2), 163–169.

49. T. Matsui and M. Yamaguchi, Zinc modulation of insulin-like growth factor's effect in osteoblastic MC3T3-E1 cells, *Peptides*, 1995, **16**(6), 1063–1068.

50. L.S. Hurley, The consequences of fetal impoverishment. *Nutr. Today*, 1968, **3**, 2.

51. L.S. Hurley, J. Gowan and H. Swenerton, Teratogenic effects of short-term and transitory zinc deficiency in rats, *Teratology*, 1971, **4**, 199.

52. L.S. Hurley and H. Swenerton, Congenital malformations resulting from zinc deficiency in rats, *Proc. Soc. Exp. Biol. Med.*, 1971, **123**, 692.

53. L.S. Hurley and H. Swenerton, Lack of mobilization of bone and liver zinc under teratogenic conditions of zinc deficiency in rats, *J. Nutr.*, 1971, **101**(5), 597–603.

54. J. Warkany and H.J. Petering, Congenital malformations of the central nervous system in rats produced by maternal zinc deficiency, *Teratology*, 1972, **5**, 319.

55. L.S. Hurley, Teratogenic aspects of manganese, zinc, and copper nutrition, *Physiol. Rev.*, 1981, **61**(2), 249–295.

56. K.M. Hambidge, C.E. Casey and N.F. Krebs, Zinc, In *Trace elements in Human and Animal Nutrition*, W. Mertz, (ed.), Academic Press, San Diego, CA 1986, vol. 2, pp. 1–137.

57. J.L. Greger, Dietary intake and nutritional status in regard to zinc of institutionalized aged, *J. Gerontol. Nursing*, 1977, **32**(5), 549–553.

58. P.A. Wagner, M.L. Krista, L.B. Bailey, G.J. Christakis, J.A. Jernigan, P.E. Araujo, H. Appledorf, C.G. Davis and J.S. Dinning, Zinc status of elderly black Americans from urban low-income households, *Am. J. Clin. Nutr.*, 1980, **33**(8), 1771.

59. J.L. Freudenheim, N.E. Johnson and E.L. Smith, Relationship between usual nutrient intake and bone mineral content of women 35–65 years of age: longitudinal and cross sectional analysis, *Am. J. Clin. Nutr.*, 1986, **44**, 863–876.

60. C Seco, M. Revilla, E.R. Hernandez, J. Gervas, J. Gonzalez-Riola, L.F. Villa and H. Rico, Effects of zinc supplementation on vertebral and femoral bone mass in rats on strenuous treadmill training exercise, *J. Bone Miner. Res.*, 1998, **13**(3), 508–512.

61. O.S. Atik, Zinc and Senile osteoporosis, *J. Am. Geriatrics Soc.*, 1983, **31**, 790–791.

62. R. Lappalainen, M. Knuuttila, S. Lammi, E.M. Alhava and H. Olkkonen, Zn and Cu content in human cancellous bone, *Acta Orthopaed. Scand.*, 1982, **53**(1), 51–55.

63. M. Herzberg, A. Lusky, J. Blonder and Y. Frenkel, The effect of estrogen replacement therapy on zinc in serum and urine, *Obstet. Gynecol.*, 1996, **87**, 145–154.

64. P. Relea, M. Revilla, E. Ripoll, L. Arribas, L.F. Villa and H. Rico, Zinc, biochemical markers of nutrition, and type I osteoporosis, *Age Ageing*, 1995, **24**(4), 303–307.

65. R.M. Angus, P.N. Sambrook, N.A. Pocock and J.A. Eisman, Dietary intake and bone mineral density, *Bone Miner*, 1988, **4**(3), 265–277.

66. V.W. Bunker, L.J. Hinks, M.F. Stansfield, M.S. Lawson and B.E. Clayton, Metabolic balance studies for zinc and copper in housebound elderly people and the relationship between zinc balance and leukocyte zinc concentrations, *Am. J. Clin. Nutr.*, 1987, **46**(2), 353–359.

67. A.J. Thomas, V.W. Bunker, L.J. Hinks, N. Sodha, M.A. Mullee and B.E. Clayton, Energy, protein, zinc and copper status of twenty-one elderly inpatients: analysed dietary intake and biochemical indices, *Br. J. Nutr.*, 1988, **59**(2), 181–191.

68. L. Strause, P. Saltman, K.T. Smith, M. Bracker and M.B. Andon, Spinal bone loss in Postmenopausal Women supplemented with calcium and trace minerals, *J. Nutr.*, 1994, **124**, 1060–1064.

69. R.B. Rucker, N. Romero-Chapman, T. Wong, J. Lee, F.M. Steinberg, C. McGee, M.S. Clegg, K. Reiser, T. Kosonen, J.Y. Uriu-Hare, J. Murphy and C.L. Keen, Modulation of lysyl oxidase by dietary copper in rats, *J. Nutr.*, 1996, **126**(1), 51–60.

70. H.M. Kagan and P.C. Trackman, Properties and function of lysyl oxidase, *Am. J. Respir. Cell Mol. Biol.*, 1991, **5**(3), 206–210.

71. K. Reiser, R.J. McCormick and R.B. Rucker, Enzymatic and nonenzymatic cross-linking of collagen and elastin, *Federation Am. Soc. Exp. Med.*, 1992, **6**(7), 2439–2449.

72. J. Eaton-Evans, E.M. McIlrath, W.E. Jackson, H. McCartney and J.J. Strain, Copper supplementation and the maintenance of bone mineral density in middle aged women, *J. Trace Elements Exp. Med.*, 1996, **9**, 87–94.

73. T. Wilson, J.M. Katz and D.H. Gray, Inhibition of active bone resorption by copper, *Calcif. Tissue Int.*, 1981, **33**(1), 35–39.

74. L.L. Key, W.L. Ries, R.G. Taylor, B.D. Hays and B.L. Pitzer, Oxygen derived free radicals in osteoclasts: the specificity and location of the nitroblue tetrazolium reaction, *Bone*, 1990, **11**(2), 115–119.

75. G.K. Davis and W. Mertz, Copper, In *Trace Elements in Human and Animal Nutrition*, W. Mertz (ed.), Academic Press, San Diego, CA, 1987, vol. 1, pp. 301–364.

76. L.S. Hurley, Teratogenic aspects of manganese, zinc and copper nutrition, *Physiol. Rev.*, 1981, **61**, 249–295.

77. H.H.A. Dollwet and J.F.J. Sorenson, Roles of copper in bone maintenance and healing, *Biol. Trace Element Res.*, 1988, **18**, 39–48.

78. L.G. Strause, J. Hegenauer, P. Saltman, R. Cone and D. Resnick, Effects of long-term dietary manganese and copper deficiency on rat skeleton, *J. Nutr.*, 1986, **116**(1), 135–141.

79. N.F. Suttle, K.W. Angua, D.I. Nisbet and A.C. Field, Osteoporosis in copper depleted lambs, *J. Comparat. Pathol.*, 1972, **82**, 93–96.

80. A. Whitelaw, R.H. Armstrong, C.C. Evans and A.R. Fawcett, A study of the effects of copper deficiency in Scottish blackface lambs on improved hill pasture, *Vet. Record*, 1979, **104**(20), 455–460.

81. H. Rico, C. Roca-Botran, E.R. Hernandez, C. Seco, E. Paez, M.J. Valencia and L.F. Villa, The effect of supplemental copper on osteopenia induced by ovariectomy in rats, *Menopause*, 2000, **7**(6), 3–6.

82. C.D. Yee, K.S. Kubena, M. Walker, T.H. Champney and H.W. Sampson, The relationship of nutritional copper to the development of postmenopausal osteoporosis in rats, *Biol. Trace Element Res.*, 1995, **48**(1), 1–11.

83. W.E. Cornatzer, J.A. Haning and L.M. Klevay, The effect of copper deficiency on heart microsomal phosphatidylcholine biosynthesis and concentration, *Int. J. Biochem. Cell Biol.*, 1986, **18**(12), 1083–1087.

84. E. Bonucci, G. Silvestrini and P. Mocetti, MC22-33F monoclonal antibody shows unmasked polar head groups of choline-containing phospholipids in cartilage and bone, *Eur. J. Histochem.*, 1997, **41**, 177–190.

85. L.N.Y. Wu, B.R. Genge, M.W. Kang, L. Arsenault and R.E. Wuthier, Changes in phospholipids extractability and composition accompany mineralization of chicken growth plate cartilage matrix vesicles, *J. Biol. Chem.*, 2002, **277**(7), 5126–5133.

86. H. Kaneki, J. Yokozawa, M. Fujieda, S. Mizuochi, C. Ishikawa and H. Ide, Phorbol ester-induced production of prostaglandin E2 from phosphatidylcholine through the activation of phospholipase D in UMR-106 cells, *Bone*, 1998, **23**, 213–222.

87. A.T.K. Singh, J.G. Kunnel, P.J. Strieleman and P.H. Stern, Parathyroid hormone (PTH)-(1-34), [Nle$^{8,18}$, Tyr$^{34}$]PTH-(3-34) amide, PTH-(1-31) amide, and PTH-related peptide-(1-34) stimulate phosphatidylcholine hydrolysis in UMR-106 osteoblastic cells: comparison with effects of phorbol 12,13-dibutyrate, *Endocrinology*, 1999, **140**, 131–137.

88. G.G. Graham and A. Cordano, Copper depletion and deficiency in the malnourished infant, *Johns Hopkins Med. J.*, 1969, **124**(3), 139–150.

89. T.M. Allen, A. Manoli II and R.L. LaMont, Skeletal changes associated with copper deficiency, *Clini. Orthopaed. Relat. Res.*, 1982, (**168**), 206–210.

90. N.T. Griscom, J.N. Craig and E.B. Neuhauser, Systemic bone disease developing in small premature infants, *Pediatrics*, 1971, **48**(6), 883–895.

91. D.M. Danks, Copper deficiency in humans, *Annu. Rev. Nutr.*, 1988, **8**, 235–257.

92. R.A. Al-Rashid and J. Spangler, Neonatal copper deficiency, *N. Engl. J. Med.*, 1971, **285**(15), 841–843.

93. J.R. Seely, G.B. Humphrey and B.J. Matter, Copper deficiency in a premature infant fed on iron-fortified formula, *N. Engl. J. Med.*, 1972, **286**(2), 109–110.

94. J.J. Strain, A Reassessment of Diet and Osteoporosis – possible role for copper, *Med. Hypothesis*, 1988, **27**, 333–338.

95. D. Conlan, R. Korula and D. Tallentire, Serum copper levels in elderly patients with femoral-neck fractures, *Age Ageing*, 1990, **19**(3), 212–214.

96. G. Howard, M. Andon, P. Saltman and L. Strausse, Serum copper concentration, dietary calcium intake and bone density in postmenopausal women: Cross-sectional measurements, *J. Bone Miner. Res.*, 1990, **9**, 1515–1523.

97. A. Baker, L. Harvey, G. Majask-Newman, S. Fairweather-Tait, A. Flynn and K. Cashman, Effect of dietary copper intakes on biochemical markers of bone metabolism in healthy adult males, *Eur. J. Clin. Nutr.*, 1999, **53**(5), 408–412.

98. A. Baker, E. Turley, M.P. Bonham, J.M. O'Connor, J.J. Strain, A. Flynn and K.D. Cashman, No effect of copper supplementation on biochemical markers of bone metabolism in healthy adults, *Br. J. Nutr.*, 1999, **82**(4), 283–290.

99. C.A. Seymour, Copper toxicity in man, in *Copper in Animals and Man*. J. Mc C. Howell and J.M. Gawthorne (eds.), CRC Press, Boca Raton, FL, 1987, pp. 79–106.

100. T. Kaji, R. Kawatani, M. Takata, T. Hoshino, T. Miyahara, H. Kozuka and F. Koizumi, The effects of cadmium, copper or zinc on formation of embryonic chick bone in tissue culture, *Toxicology*, 1988, **50**(3), 303–316.

101. J.J. Strain, Defining Optimal copper status in humans: concepts and problems, In *Trace Elements in Man and Animals*. A.M. Roussel, R.A. Anderson and A.E. Favier (eds.), Kluwer Academic/Plenum Publishers, New York, 2000, pp. 923–928.

102. R.M. Leach and A.M. Muenster, Studies on the role of manganese in bone formation. Effect upon the mucopolysaccharide content of chick bone, *J. Nutr.*, 1962, **78**, 51–56.

103. R.M. Leach, Role of manganese in mucopolysaccharide metabolism, *Federation Proc.*, 1971, **30**(3), 991–994.

104. M.O. Amdur, L.C. Norris and G.F. Heuser, The need for manganese in bone development by the rat, *Proc. Soc. Exp. Biol. Med.*, 1945, **59**, 254–255.

105. L. Strause, P. Saltman and J. Glowacki, The effect of deficiencies of manganese and copper on osteoinduction and on resorption of bone particles in rats, *Calcif. Tissue Int.*, 1987, **41**(3), 145–150.

106. M.S. Clegg, S.M. Donovan, M.H. Monaco, D.L. Baly, J.L. Ensunsa and C.L. Keen, The influence of manganese deficiency on serum IGF-1 and IGF binding proteins in the male rat, *Proc. Soc. Exp. Biol Med*, 1998, **219**(1), 41–47.

107. C.W. Asling and L.S. Hurley, The influence of trace elements on the skeleton, *Clin. Orthopaed.*, 1963, **27**, 213–264.

108. J.Y. Reginster, L.G. Strause, P. Saltman and P. Franchimont, Trace elements and postmenopausal osteoporosis: a preliminary study of decreased serum manganese, *Med. Sci. Res.*, 1988, **16**, 337–338.

109. M.D. Grynpas, Fluoride effects on bone crystals, *J. Bone Miner. Res.*, 1990, **5**(Suppl. 1), S169–S175.

110. M. Kleerekoper and R. Balena, Fluorides and osteoporosis. *Annu. Rev. Nutr.*, 1991, **11**, 309–324.

111. ADA Reports. Position of the American Dietetic Association: The impact of fluoride on health, *J. Am. Diet. Assoc.*, 2000, **100**(10), 1208–1213.

112. G. Boivin, J. Dupuis and P.J. Meunier, Fluoride and osteoporosis, *World Rev. Nutr. Diet.*, 1993, **73**, 80–103.

113. P.J. Marie, M.C. De Vernejoul and A. Lomri, Stimulation of bone formation in osteoporosis patients treated with fluoride associated with increased DNA synthesis by osteoblastic cells in vitro, *J. Bone Miner. Res.*, 1992, **7**(1), 103–113.

114. W.K.H. Lau, J.R. Farley, T. Freeman and D. Baylink, A proposed mechanism of the mitogenic action of fluoride on bone cells: Inhibition of the activity of an osteoblastic acid phosphatase, *Metabolism*, 1989, **38**(9), 858–868.

115. P. Fratzl, P. Roschger, J. Eschberger, B. Abendroth and K. Klaushofer, Abnormal bone mineralization after fluoride treatment in osteoporosis: a all-angle X-ray-scattering study, *J. Bone Miner. Res.*, 1994, **9**(10), 1541–1549.

116. E.D. Eanes and A.H. Reddi, The effect of fluoride on bone mineral apatite, *Metab. Bone Dis. Relat. Res.*, 1979, **2**, 3–10.

117. D. Briancon and P.J. Meunuer, Treatment of Osteoporosis with Fluoride, Calcium and Vitamin D, *Orthopedic Clin. N. Am.*, 1981, **12**(3), 629–648.

118. A.M. Parfitt, The coupling of bone formation to bone resorption: a critical analysis of the concept and of its relevance to the pathogenesis of osteoporosis, *Metab. Bone Dis. Relat. Res*, 1982, **4**(1), 1–6.

119. M. Kleerekoper and R. Balena, Fluorides and osteoporosis, *Annu. Rev. Nutr.*, 1991, **11**, 309–324.

120. M. Kleerekoper and D.B. Mendlovic, Sodium Fluoride Therapy of Postmenopausal Osteoporosis, *Endocrine Rev.*, 1993, **14**(3), 312–323.

121. M. Kleerekoper and B. Raffaella, Fluorides and Osteoporosis, *South. Med. J.*, 1992, **85**(2), 2534–2542.

122. M. Kleerekoper, The role of fluoride in the prevention of osteoporosis, *Endocrinol. Metab. Clin. N. Am.*, 1998, **27**(2), 441–452.

123. M. Kleerekoper, I. Oliver, B. Frame, D.S. Rao, A.R. Villanueva and A.M. Parfitt, Adverse effect of vitamin D in fluoride treated osteoporosis, *Vet. Human Toxicol.*, 1982, **24**, 388–393.

124. C. Rich and F. Ensinck, Effect of sodium fluoride on calcium metabolism of human beings, *Nature*, 1961, **191**, 184–185.

125. D.S. Bernstein, C.D. Guri and P. Cohen, The use of sodium fluoride in metabolic bone disease, *J. Clin. Invest.*, 1963, **42**, 916.

126. P. Cohen and F.H. Gardener, Induction of subacute skeletal fluorosis in a case of multiple myeloma, *N. Engl. J. Med.*, 1964, **271**, 1129–1133.

127. J. Jowsey, B.L. Riggs and P.J. Kelly, Effects of combined therapy with sodium fluoride, vitamin D and calcium in osteoporosis, *Am. J. Med.*, 1972, **53**, 43–49.

128. B.L. Riggs, S.F. Hodgson and D.L. Hoffman, Treatment of primary osteoporosis with fluoride and calcium: Clinical tolerance and fracture occurrence, *J. Am. Med. Assoc.*, 1980, **243**, 446–449.

129. T. Hanson and B. Roos, The effect of fluoride and calcium on spinal bone mineral content: A controlled, prospective (3 year) study, *Calcif. Tissue Int.*, 1987, **40**, 315–317.

130. H. Resch, C. Libanati, S. Farley, P. Bettica, E. Schulz and D.J. Baylink, Evidence that fluoride therapy increases trabecular bone density in a peripheral skeletal site, *J. Clin. Endocrinol. Metab.*, 1993, **76**(6), 1622–1624.

131. M.A. Dambacker, J. Ittner and P. Ruegsegger, Fluoride therapy of post-menopausal osteoporosis, *Bone*, 1986, **7**, 199–205.

132. N. Mamelle, R. Dusan and J.L. Martin, Risk-benefit of sodium fluoride treatment in primary vertebral fracture osteoporosis, *Lancet*, 1988, **2**, 361–365.

133. M. Kleerekoper, E. Peterson, E. Phillips and D.A. Nelson, Continuous sodium fluoride therapy does not reduce vertebral fracture rate, *J. Bone Miner. Res.*, 1989, **7**(Suppl. 1), S376.

134. B.L. Riggs, S.F. Hodgson, W.M. O'Fallon, E.Y. Chao, H.W. Wahner, J.M. Muhs and S.L. Cedel, Effect of fluoride treatment on the fracture rate in postmenopausal women with osteoporosis, *N. Engl. J. Med.*, 1990, **322**(12), 802–809.

135. P.D. Delmas, J. Dupuis, F. Duboeuf, M.C. Chapuy and P.J. Meunier, Treatment of vertebral osteoporosis with disodium monofluorophosphate: comparison with sodium fluoride, *J. Bone Miner. Res.*, 1990, **5** (suppl. 1), S143–S147.

136. C.Y. Pak, K. Sakhaee, J.E. Zerwekh, C. Parcel, R. Peterson and K. Johnson, Safe and effective treatment of osteoporosis with intermittent slow release sodium fluoride: augmentation of vertebral bone mass and inhibition of fractures, *J. Clin. Endocrinol. Metab.*, 1989, **68**(1), 150–159.

137. J. Farrerons, A. Rodriguez de la Serna, N. Guanabens, L. Armadans, A. Lopez-Navidad, B. Yoldi, A. Renau and J. Vaque, Sodium fluoride treatment is a major protector against vertebral and nonvertebral fractures when compared with other common treatments of osteoporosis: a longitudinal, observational study, *Calcif. Tissue Int.*, 1997, **60**(3), 250–254.

138. J.Y. Reginster, L. Meurmans, B. Zegels, L.C. Rovati, H.W. Minne, G. Giacovelli, A.N. Taquet, I. Setnikar, J. Collette and C. Gosset, The effect of sodium monofluorophosphate plus calcium on vertebral fracture rate in postmenopausal women with moderate osteoporosis. A randomized, controlled trial, *Ann. Intern. Med.*, 1998, **129**(1), 1–8.

139. E. Seeman, Osteoporosis: trials and tribulations, *Am. J. Med.*, 1997, **103**(2A), 74S–87S; discussion 87S–89S.

140. R.D. Blank and R.S. Bockman, A review of clinical trials of therapies for osteoporosis using fracture as an end point, *J. Clin. Densitometry*, 1999, **2**(4), 435–452.

141. G. Boivin, J. Dupuis and P.J. Meunier, Fluoride and osteoporosis, *World Rev. Nutr. Diet.*, 1993, **73**, 80–103.

142. ADA Reports. Position of the American Dietetic Association: The impact of fluoride on health, *J. Am. Diet. Assoc.*, 2000, **100**(10), 1208–1213.

143. J.Z. Ilich and J.E. Kerstetter, Nutrition in bone health revisited: A story Beyond Calcium, *J. Am. Coll. Nutr.*, 2000, **19**(6), 715–737.

# CHAPTER 17

# Vitamin A and Fracture Risk

HÅKAN MELHUS

Department of Medical Sciences, University Hospital, SE-751 85 Uppsala, Sweden, Email: hakan.melhus@medsci.uu.se

## Key Points

- The most prominent features of high doses of retinoids in laboratory animals are accelerated bone resorption, a reduction of the bone diameter, and spontaneous fractures.
- *In vitro* retinoids induce bone resorption and regulate gene transcription *via* the nuclear receptors, which are found in both osteoclasts and osteoblasts.
- High doses of beta-carotene or other provitamin A carotenoids have no adverse effects on bone.
- The bioefficacy of provitamin A carotenoids in plant foods is much less than was previously thought and the conversion to retinol seems to be regulated by the body's needs.
- Several case reports suggest that the skeletal effects of hypervitaminosis A in man are similar to those found in animals.
- Epidemiological studies that distinguish between retinol and carotenoid intakes show that an excessive intake of retinol is associated with an increased risk of fractures in man.

## Introduction

The term *vitamin A* refers to a family of essential, fat-soluble dietary compounds.[1–3] Generally, their structure consists of a β-ionine ring, a conjucated isoprenoid side chain and a polar terminal group. In 1931, Karrer[4] proposed the structural formula for the alcohol form of vitamin A, which is also termed *all-trans-retinol* (Figure 17.1). This parent compound and its fatty acid ester derivatives, *retinyl esters* (most commonly retinyl palmitate), are referred to as *preformed vitamin A*. All-*trans*-retinol functions as a prehormone which is transported to target tissues where it is converted to the active forms of vitamin A. In the eye, the active metabolite is *11-cis-retinal*. In all other

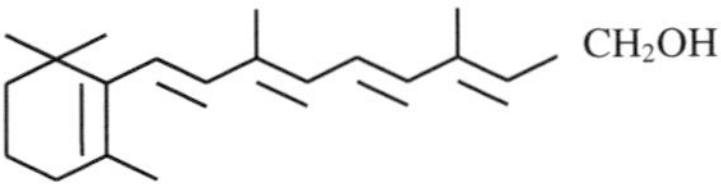

All-trans-retinol

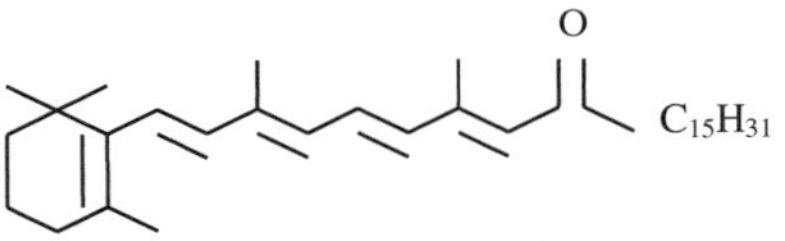

Retinyl palmitate

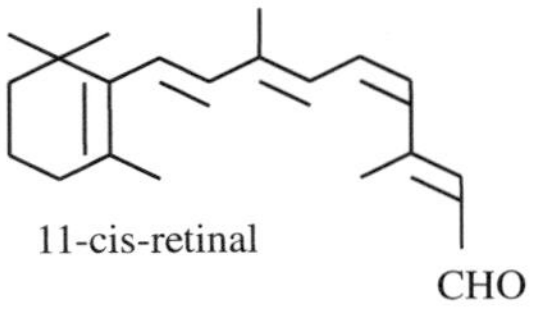

11-cis-retinal

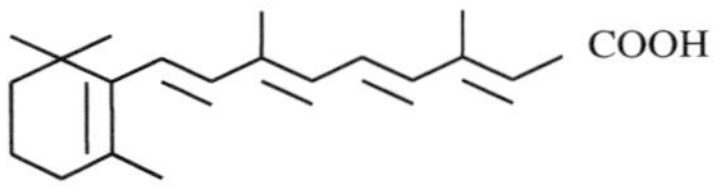

CHO

All-trans-retinoic acid

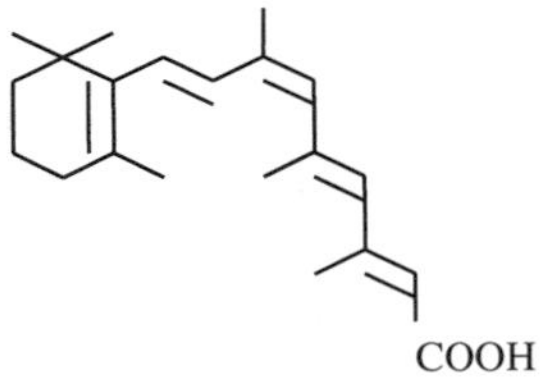

9-cis-retinoic acid

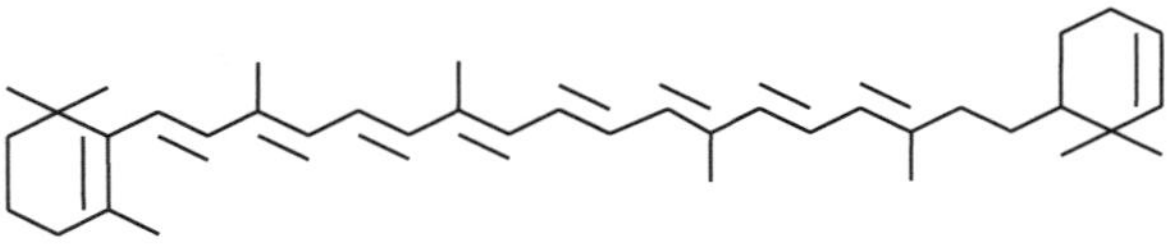

Beta-carotene

**Figure 17.1** *Chemical structures of different forms of vitamin A and of beta-carotene*

target organs all-*trans*-retinol is converted to *all-trans-retinoic acid* or to *9-cis retinoic acid*. The term *retinoids* refers to retinol, its metabolites, and synthetic analogues that have a similar structure.

Preformed vitamin A does not occur in plant products, but its precursor carotene occurs in several forms. *Beta-carotene* is actually a tail-joined retinal dimer and is more efficiently converted to retinol than other compounds of the class of plant pigments called *carotenoids*, owing to their relation to the carotenes. Beta-carotene and other carotenoids that the body can transform to active forms of vitamin A are commonly referred to as *provitamin A*.[2,3]

# Sources of Vitamin A

The major natural sources of vitamin A are the long-chain fatty acid esters, retinyl esters, provided by animal foods, such as liver, meats, milk products, eggs, and fat fish. Common sources today are also fortified foods (margarines; and in some countries also other foods, such as low-fat dairy products) and dietary supplements. Supplements usually contain relatively high levels of retinol, 0.5–1.5 mg, *i.e.*, levels close to the recommended dietary intake (RDI) per dose. Some supplements can contain even higher amounts.

More than 600 carotenoids have been isolated and characterized in nature and about 10% have provitamin A activity.[5] In practice, however, only a handful of these provitamins are commonly encountered in food, and food composition data are available for only three of these (beta-carotene, alpha-carotene, and beta-cryptoxanthin). Many orange fruits and green vegetables are sources of provitamin A carotenoids, *e.g.*, carrots, spinach, broccoli, squash, mango, apricots, peas, papaya.[1–3]

# Recommended Dietary Intake

There has been no consensus in different countries as to the amount of retinol and carotenoids that should be consumed in order to maintain optimal health. The dietary recommendations are generally based on the FAO/WHO definition 'retinol equivalent' (RE), where 1 µg RE = 1 µg retinol = 1.78 µg retinyl palmitate = 6 µg β-carotene = 12 µg other carotenoids with provitamin A activity = 3.33 international units (IU) vitamin A activity from retinol.[6] Most countries recommend between 500 and 1000 µg RE for adults per day. The FAO/WHO RDI is 600 µg RE per day for adult males and 500 µg RE for adult females, the United Kingdom reference nutrient intake (RNI) is 700 µg RE for adult males and 600 µg RE for adult females. In Nordic countries, the RDI is 900 and 800 µg RE, respectively. The recommended dietary allowance (RDA), as used in the United States and in a few other countries, was 1000 µg RE for men and 800 µg RE for women.

Based on data demonstrating that the vitamin A activity of dietary beta-carotene is one-sixth, rather than one-third, the vitamin activity of purified beta-carotene in oil, the term retinol activity equivalents (RAEs) has been introduced, where 1 µg = retinol = 2 µg of supplemental beta-carotene = 12 µg dietary beta-carotene = 24 µg of other dietary provitamin A carotenoids. Revised dietary recommendations based on RAE for the United States have recently been published[7] (see Table 17.1).

**Table 17.1.** *Adequate intakes for infants and recommended daily allowance of vitamin A for children and adults in the revised dietary recommendations based on retinol activity equivalents (RAE)*

| | |
|---|---|
| *Infants* | |
| 0–6 months | 400 μg RAE day$^{-1}$ |
| 7–12 months | 500 |
| *Children* | |
| 1–3 years | 300 |
| 4–8 | 400 |
| *Males* | |
| 9–13 | 600 |
| 14+ | 900 |
| *Females* | |
| 9–13 | 600 |
| 14+ | 700 |

However, rapid progress in the methodology to assess the bioavailability of carotenoids, specifically stable-isotope dilution methods, will lead to a better understanding of the actual contribution of provitamin A carotenoids to vitamin A status.[5] Thus, dietary recommendations are likely to change during the coming years.

# The Bioconversion of Provitamin A Carotenoids to Vitamin A

For about 30 years it was generally accepted that 6 μg of beta-carotene and 12 μg of other provitamin A carotenoids had the same vitamin activity as 1 μg of retinol.[6] These conversion factors were essentially based on the so called Sheffield study, in which the amount of beta-carotene in oil required to reverse and prevent abnormal dark adaptation was compared with the amount of retinol to carry out the same functions – in two men![8] In the 1988, FAO/WHO revised recommendation[9] it was concluded that further studies of these conversion factors were warranted because studies on animals had indicated that beta-carotene, and presumably other carotenoids, were absorbed much more effectively from oily solution than from plant material. The conversion factors remained the same, but for intakes of beta-carotene above 4 mg per meal a conversion factor of 0.10 instead of 0.167 was suggested.[9]

Many factors have the potential to reduce carotenoid bioavailability and bioefficacy. These are incorporated into the mnemonic SLAMENGHI:[10] Species of carotenoid, molecular Linkage, Amount of carotene consumed in a meal, Matrix in which the carotenoid is incorporated, Effectors of absorption and bioconversion, Nutrient status of the host, Genetic factors, Host-related factors, and Interactions between the other factors. The most important of these influencing bioavailability is the food matrix.[11]

The conversion factors were first challenged by de Pee *et al.*[10,12] based on their observations that a high proportion of pregnant Indonesian women who had intakes of provitamin A that would have been expected to supply about three times the recommended daily allowance of vitamin A, still had marginal vitamin A deficiency.[11] Several studies (see Ref. [11]) have now confirmed that the bioefficacy of beta-carotene in plant foods is much less than was previously thought. Therefore, the US Institute of Medicine, Food and Nutrition Board[13] recently introduced the term RAE. However, these lowered estimates have been considered as still too high, since they are not based on studies in developed countries.[11,14] Indeed, a recent study using a double-tracer design[15] shows that the vitamin A activity of beta-carotene can be zero in American men fed a mixed diet.

An important reason why vitamin A was not identified as a risk factor for osteoporosis in humans, was that preformed retinol and carotenoid sources of vitamin A were not distinguished. Melhus *et al.*[16] challenged the conversion factors, but from a different angle of approach than de Pee *et al.*[10] They focused on the fact that carotenoids do not cause hypervitaminosis A or teratogenicity even when ingested in large amounts, indicating that the conversion of provitamin A carotenoids must be determined by the body's needs for the vitamin. Several animal studies support this notion. In 1961, Olson[17] reported that injection of vitamin A together with [$^{14}$C]-beta-carotene into ligated loops of rat intestine inhibited the formation of [$^{14}$C]-retinyl ester, and formation of retinyl ester in intestinal mucosa increased linearly at low but not at high doses. The activity of the enzyme responsible for the conversion of beta-carotene to retinal in the intestine, the $\beta,\beta$-carotene 15,15'-monooxygenase ($\beta$CMOOX), is up-regulated in vitamin A-deficiency and down-regulated by an increased intake of retinyl esters or beta-carotene.[18] Recently, it was shown that retinoic acid reduces, and a retinoic acid receptor (RAR) alpha antagonist significantly increases the intestinal enzyme activity. Moreover, a retinoic acid responsive element was found in the promoter of the $\beta$CMOOX gene. These results are consistent with a transcriptional feedback regulation of the enzyme by RA *via* the specific nuclear receptors.[19] The bioconversion of plant carotenoids seems to be regulated by vitamin A status also in humans.[15,20]

# Transport and Metabolism of Preformed Vitamin A

Intestinal absorption of preformed vitamin A occurs following the processing of retinyl esters in the lumen of the small intestine. Within the water-miscible micelles formed from bile salts, solubilized retinyl esters are hydrolyzed to retinol (Figure 17.2). A specific cellular retinol-binding protein II[21,22] within the brush border of the enterocyte facilitates retinol uptake by the mucosal cells. Retinol is then re-esterified before incorporation into the chylomicrons, which are transported *via* the lymph into the general circulation. After entry into the vascular compartment, chylomicrons are metabolized in extrahepatic

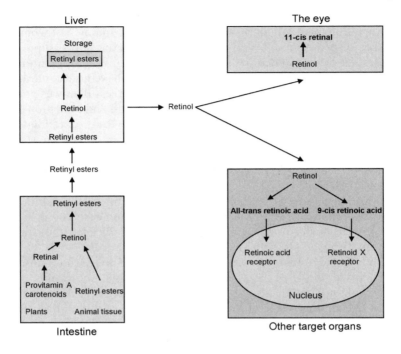

**Figure 17.2**  *An overview of retinoid metabolism (Active metabolites in bold)*

tissues by the lipolytic removal of much of the chylomicron triglyceride. The chylomicron remnants thus formed are mainly cleared by the liver. After uptake of the chylomicron retinyl esters, hydrolysis and reesterification occur in the liver. The resulting retinyl esters (predominantly retinyl palmitate) are stored in the liver. When vitamin A is mobilized from the liver, stored retinyl esters are hydrolyzed prior to its release into the bloodstream, and retinol is transported to target tissues bound to a specific plasma protein, retinol-binding protein.[23,24] Small amounts of retinoic acid derived from dietary retinoids is absorbed *via* the portal system bound to albumin.[25]

## Transport and Metabolism of Provitamin A Carotenoids

After consumption of carotenoid-containing foods, carotenoids are released from their food matrix before they are incorporated into micelles. Carotenoids appear to be absorbed by the mucosa of the small intestine (mainly in the duodenum) *via* passive diffusion to become packaged into chylomicrons. Therefore, the release of carotenoids from the food matrix and dissolution in the lipid phase are critical steps in the absorption process. Provitamin A carotenoids can partly be converted to vitamin A in the intestinal mucosa, and both carotenoids and retinyl esters are incorporated into chylomicrons and transported to the liver (Figure 17.2). The liver secretes carotenoids associated with hepatic very low density lipoproteins (VLDL), but in the fasting state most plasma carotenoids are associated with low density lipoproteins (LDL) and high density lipoproteins (HDL).[5]

# Vitamin A in Target Cells

The mechanism through which retinol is taken up from the circulation by peripheral cells has not been conclusively established. In the target cells, the active metabolites are usually synthesized by a complex metabolic system involving numerous enzymes and binding proteins. A simplified overview of the formation of these metabolites is shown in Figure 17.2.

Already in the 19th century, investigators reported that the color of dark-adapted frog retinas changed from purple to yellow when exposed to light.[26] In the 1930s, Wald[27] identified a vitamin A derivative, the aldehyde form retinal, in the visual pigment. Subsequent work has shown that in the eye all-*trans*-retinol is esterified to all-*trans*-retinyl esters and then isomerohydrolyzed to 11-*cis*-retinol[28] and then further oxidized by retinol dehydrogenases (RDH) to 11-*cis*-retinal, which is transferred from the retinal pigment epithelium to the photoreceptor cells, probably by the interphotoreceptor retinoid binding protein (IRBP). In the photoreceptor cells, 11-*cis*-retinal is bound covalently to opsin to form rhodopsin. Opsin-bound 11-*cis*-retinal can be activated by light, triggering the isomerization of 11-*cis*-retinal to all-*trans*-retinal, which eventually generates a visual response. All-*trans*-retinal is then reduced back to all-*trans*-retinol to complete the visual cycle.[29] It has recently been suggested that retinoids act in non-visual light sensing in the eye, thus regulating the circadian clock that controls our daily 24-h rhythm.[30]

Vitamin A also has important functions in vertebrate growth and development, supporting cell differentiation, embryonic development, the immune response, and reproduction. With the exception of the visual cycle, most processes are related to the control of gene expression. In nonocular tissues, all-*trans*-retinol is oxidized to the active metabolite RA in a two-step reaction. First, retinol is oxidized by retinol dehydrogenase to retinal and then further by retinal dehydrogenase to RA. RA is the active metabolite that controls the activation or inhibition of gene transcription *via* the RARs and retinoid X receptors (RXRs).[31] It is generally accepted that all-*trans*-RA serves as a physiologic ligand for the RARs and that 9-*cis*-RA acts as the preferred ligand for RXRs. RAR and RXR regulate transcription by binding DNA at sites called retinoid response elements (RARE and RXRE). RAR is only able to bind DNA if it is heterodimerized with RXR. In functional RAR/RXR heterodimers, RXR serves as a silent ligand-independent partner. RXR binds DNA also as a homodimer, which can be activated by 9-*cis*-RA. RXR can also form heterodimers with the vitamin D receptors (VDRs), the thyroid hormone receptors,[31] liver X receptors (LXRs)[32] and peroxisome proliferator-activated receptors (PPARs).[33]

# Skeletal Effects of Provitamin A Carotenoids

There is no evidence of bone toxicity from provitamin A carotenoids in laboratory animals or humans.[34-36] Beta-carotene produces no bone resorption or other significant change in bone *in vitro*.[35,37,38] Moreover, these compounds do not cause skeletal malformations.[36,39]

# Skeletal Effects of Vitamin A Concentrates and Purified Retinol in Animals

The first observations of hypervitaminosis A were published in 1925 by Takahashi *et al.*[40] When they reviewed the literature dealing with vitamins, they found that, "without exception, all of the previous investigators have made observations in regard to the untoward influences of the deficiency or the starvation of the allied substances, and to the marked efficacy of vitamins against the malnutrition exclusively caused by their deficiencies" and "none of them has dealt with the influences of the intake of an excessive amount of superphysiological amount of vitamins against the animal body". Between 1921 and 1924, they tested the effect of massive doses of their vitamin A-rich fish-oil concentrate 'Biosterin' in rats, mice and dogs. The symptoms of albino rats and mice, caused by the intake of daily doses that were about 10 000 times the amount required to sustain normal growth were "alopecia of the head, which develops first of all, and then paralysis of the hind legs and at last they will not be able to stand. All this while serious emaciation is gradually developing." Post mortem findings were: "fatty degeneration of the liver and hemorrhages in the digestive system and the lungs." Also in dogs "the hind legs became powerless and the animal could no more stand upright."

In 1933 two independent investigators reported skeletal lesions in hypervitaminotic rats. Collazo and Rodriguez[41] administered 20 000 Ratteneinheten ('rat units') in the form of a fish-oil concentrate (0.5 mL Vogan), corresponding to approximately 18 mg of vitamin A daily to three rats, 4-weeks of age (35 g). X-ray examination revealed that the animals had sustained spontaneous fractures of the long bones (Figure 17.3). They noted that this 'vitamin dystrophy' was reversible and that the fractures began to heal once the administration of overdoses of vitamin A ceased. Blood levels of calcium and phosphate were normal. Bomskov and Seeman[42] used 0.2 or 0.3 mL Vogan for 3–4 weeks in seven 3-weeks old and four 8-weeks old rats and spontaneous fractures were found in the young animals. The content of calcium and phosphate in bone was reduced, although the blood level of calcium and phosphate did not change significantly. Moreover, the vitamin A concentrate inhibited the curative effect of vitamin D in rats with rickets. Strauss[43] reported a thinning of the long bones, especially at the diaphysis, with a very thin cortex in young rats, which could explain the fractures, whereas no fractures were seen in older (200 g) rats. The major histological finding was a decreased density of osteoblasts. Davies and Moore[44] using a distillate containing less impurity than the previously used concentrates noted

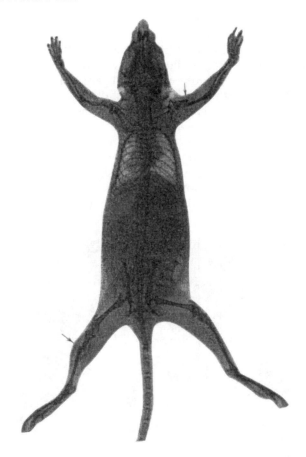

**Figure 17.3** *Spontaneous fractures (arrows) caused by hypervitaminosis A.* From Collazo and Rodriguez 1933[41]

that "the most striking feature was the extreme softness of the bones. X-ray examination revealed that several of the larger leg bones had sustained spontaneous fractures, the broken ends of the bones being healed together in irregular shapes with the formation of calluses". They also made the important observation that a toxic state could not be produced in animals with carotene.

However, the question arose as to whether the harmful effects should be ascribed to vitamin A or other components in cod liver oil. In a review of the experimental work on this subject written in 1939, Clausen came to the conclusion that "the literature was so contradictory at that time so as to afford no evidence that vitamin A would have a harmful effect on human beings".[45]

In 1945, the toxicity of overdoses of vitamin A was proved beyond doubt when it was possible to conduct the experiments with purified, crystalline retinyl acetate.[46] The most characteristic lesions in rats treated with doses of

8.5–13.6 mg were skeletal fractures, which occurred most consistently in young animals, and internal or external hemorrhage. These effects of hypervitaminosis are associated with drastic histological changes. Wolbach[47] described that the processes of bone remodeling during growth (epiphyseal cartilage sequences, resorption, apposition) were greatly accelerated. Vitamin A caused a premature closure of the epiphyses in a dose-dependent way and with sufficiently high doses osteoclastic resorption of the cortex of the long bones occurred, particularly in regions of active remodelling, and with a somewhat lower and more prolonged dosage the shaft diameters became reduced in relation to their length. Rodahl[48] used systematic X-ray examination of the long bones at various stages of hypervitaminosis A and found that there was a gradual thinning of the shaft with a reduction of the diameter of the femur (Figure 17.4) and an abnormal thinning of the cortex in hypervitaminotic rats.

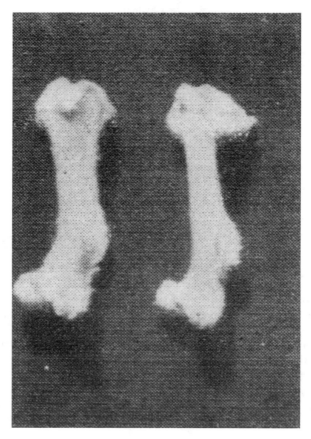

**Figure 17.4**  *Femur of adult hypervitaminotic rat (right) compared with the femur of a normal control rat of the same age (left) showing the difference in diameter* From Rodahl 1950[48]; reprinted with permission from the Norwegian Polar Institute

Microscopical examination demonstrated a marked hyperemia and subperiosteal and osseous hemorrhages, with destruction of bone spicules, an increased number of osteoclasts, and thinning of the compact bone. Chemical analyses showed no significant reduction in the calcium or phosphorus content of the ash. Contrary to the findings previously reported, there was a similar frequency of spontaneous fractures in young and adult animals, although the fractures developed later in the adult rats.

Whether the changes resulting from hypervitaminosis A were due to a direct action of the vitamin on the bone or to an indirect action through the endocrine glands, was uncertain. Barnicot[37] approached this problem by grafting experiments. He attached fragments of crystalline retinyl acetate to small pieces of parietal bone cut from 10-day-old mice, and the combination was inserted into the cerebral hemisphere of litter-mates. When the grafts were removed after 7 or 14 days, a well-marked resorption, accompanied by numerous osteoclasts, and often leading to perforation of the bone was apparent after 14 days, thus suggesting that vitamin A is capable of producing active local bone resorption if it is brought into close contact with the surface of young bone.

Nieman and Obbink[34] provided an extensive review of the toxicity of vitamin A that appeared in the literature up to the early 1950s. Skeletal lesions were reported in 31 different studies, and laboratory animals included rat, mouse, guinea pig, rabbit, chicken, duck, and dog. The lowest dose that could produce negative bone effects was about 3 mg day$^{-1}$ for young and 7.5 mg day$^{-1}$ for adult rats.

More recent investigations have confirmed and expanded these early observations. Frankel et al.[49] studied 3-week-old rats given 15 mg RE of retinyl palmitate three times per week for 6 weeks. Histomorphometry of the tibia showed an approximate doubled number of osteoclasts and a similar reduction in osteoid surface compared with control rats. Serum PTH was not detectable and serum 25-hydroxyvitamin $D_3$ was significantly lower than in controls. At 7.5 mg RE, three times a week for 3 weeks, S-PTH was suppressed to undetectable levels but there was no effect on serum 25-hydroxyvitamin $D_3$. A single dose of 82 mg RE per kg given to adult rats had no effect on S-PTH.

Increased bone resorption (increased osteoclast size and number) and reduced bone formation was also found by Hough et al.[50] using tibial histomorphometry to evaluate the skeletal effects in young rats treated daily with 3 or 7.5 mg RE of retinyl palmitate for 21 days by stomach tube. There was also a paucity of trabecular surfaces covered with osteoid. Spontaneous limb fractures and increased skeletal turnover as measured by serum-ALP and urinary OH-proline excretion was demonstrated in the high dose group. Serum calcium, PTH, 1,25 and 25-hydroxyvitamin $D_3$ were comparable in the treated animals and the controls suggesting a direct effect on bone.

Similar findings were reported by Wu et al.[51] who used retinoic acid 70 mg kg$^{-1}$ per day for 14 days to establish a model of osteoporosis in 3-month-old rats.

The cancellous and compact bone volume were markedly reduced, the trabecular number were decreased, the thickness of the cortex was decreased, and osteoclasts were activated which promoted bone resorption.

In a recent review, Binkley and Kreuger[52] pointed out that most animals studies have used very high doses administered to young growing animals, and that results from such studies may not be directly relevant to human osteoporosis.

Interestingly, Li *et al.*[53] studied the effects of long-term ingestion of moderate excesses of retinol in aged rats. In rats over 18 months of age fed a diet containing only 90 µg day$^{-1}$ (5 × daily requirement, DR) for 14 months, the trabecular area decreased to about 81% compared to the control group. Furthermore, in the 5 × DR group the fractional resorption surface increased by almost 40% and the fractional formation surface decreased to about 41%.

That 'subclinical' hypervitaminosis A can produce adverse skeletal effects was confirmed by Johansson *et al.*[54] Mature rats fed approximately 2.7 mg retinyl palmitate per day for 3 months had the characteristic thinning of the cortex and reduction of the diameter of long bones as measured by peripheral quantitative computed tomography. In addition, the mechanical strength was reduced as determined by the three-point bending analysis.

A summary of the findings in laboratory animals is shown in Table 17.2.

# Skeletal Effects *In vitro*

Vitamin A was the first agent demonstrated to stimulate bone resorption *in vitro*. Fell and Mellanby[38] pointed out that in Barnicot's experiments (above) the conditions were not wholly comparable to those in a hypervitaminotic animal. In the grafts, the bone was in direct contact with a solid mass of vitamin, whereas in the hypervitaminotic animal it is only exposed to the relatively minute concentration present in the blood stream. They chose to

**Table 17.2.** *Summary of skeletal effects in laboratory animals with hypervitaminosis A*

Spontaneous fractures
Reduced diameter of long bones
Thinning of bone cortex
Periosteal thickening
Premature closure of epiphyses
Retardation of growth

Increased bone resorption
Decreased bone formation
Increased number of osteoclasts
Decreased density of osteoblasts

Elevated S-Ca
Elevated S-ALP
Elevated U-Hydroxyproline

study the effects of vitamin A on embryonic long bones cultivated *in vitro*, using the vitamin in concentrations similar to those in the blood of animals suffering from hypervitaminosis A. In limb-bone rudiments from 5- to 6-day chick embryos growth was arrested and the most striking histological effect of excess vitamin A was on the cartilage matrix, which shrank and softened. Similarly, the addition of retinyl acetate or retinol to the medium of late fetal mouse bones had drastic effects (Figure 17.5). The cartilage matrix rapidly dwindled, lost its metachromasia and finally disappeared, leaving the chondroblasts naked. The bone was quickly absorbed, so when the experiment was terminated after 7 days, the bone had nearly disappeared from the hypervitaminotic rudiments. The younger the fetuses from which the long bones were obtained, the greater was the effect of the vitamin A and the higher the concentration of added vitamin A in the medium, the more rapid was its effects on the explanted mouse bones.[38]

Several more recent *in vitro* studies have demonstrated direct effects of retinoids on bone cells. For example excess retinol decreases collagen synthesis in embryonic chick calvaria,[55] and it inhibits incorporation of [³H] proline into collagen and stimulates bone resorption in rat calvaria.[50] All-*trans*-RA has been reported to inhibit growth of osteoblast-like cells and to regulate several important genes in the osteoblast, such as alkaline phosphatase,[56,57] osteonectin,[56] and procollagen I.[58] Both retinol and all-*trans*-RA cause several modifications of the metabolic status of isolated osteoclasts that result in augmented rates of bone resorption.[59] Furthermore, the quantity and the topography of podosomes may be modulated by retinol, which increases bone resorbing activity of osteoclasts both *in vitro* and *in vivo*[60] and all-*trans*-RA stimulates osteoclast formation.[61] Both all-*trans*-RA and 9-*cis* RA stimulate bone resorption in mice calvaria[62] and all-*trans*-RA regulates the gene expression of cathepsin K/OC-2, a dominant cysteine proteinase,[63] and osteopontin[64] in osteoclasts. Vitamin A effects are mediated through gene regulation *via* the specific nuclear receptors that have been shown to be

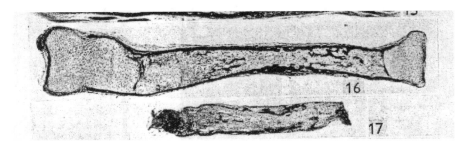

**Figure 17.5** *Radii from a late mouse foetus after 7 days' cultivation in normal medium ('16') and in medium with added retinol ('17'). Note the enormous shrinkage of the entire rudiment, the extensive absorption of bone and the density of the marrow reticulum*
From Fell and Mellanby 1952[38]; reprinted with permission from publisher

**Table 17.3.** *Summary of the retinoid effects on bone in vitro*

Growth arrest
Shrinkage and disappearance of cartilage matrix
Marked bone resorption
Decreased collagen synthesis
Regulation of gene expression in osteoblasts and osteoclasts

expressed in both osteoblasts[65] and in osteoclasts.[63] As in another target organ, skin, these RARs and binding proteins are regulated by RA.[66,67] The effects of retinoids on bone tissue and bone cells *in vitro* are summarized in Table 17.3.

## Skeletal Effects in Humans

It has long been known among Eskimos and arctic travelers that the ingestion of polar-bear liver by men and dogs causes severe illness. It was not until 1942, however, that the toxic substance was identified as being vitamin A. Rodahl and Moore[68] examined specimens of the livers of the polar-bear and found it to be very rich in retinol, 5.4–8.1 mg g$^{-1}$ liver. The ingestion of excessive amounts of bear liver by rats led in one instance to fatal hypervitaminosis A. They, therefore, concluded that the well-known poisonous action of bear liver in man probably is due to its high content of vitamin A.

One of the first descriptions of illness (somnolence, stomach pain, nausea, headache, diarrhea, vertigo, peeling of the skin) was given in the logbook of Gerrit de Veer written during the hibernation of the Dutch polar explorer Willem Barendsz on Nova Zembla in 1596 during an attempt to reach Indonesia by the northern passage. He states that he and his men became gravely ill after eating polar-bear liver. They feared for their lives but ultimately recovered.[69–71] Symptoms of hypervitaminosis A as a result of consuming natural vitamin A concentrates was reported in 1912 by Czerny,[72] who had administered large doses of cod liver oil to tuberculosis children and observed the development of seborrheic dermatitis of the face and scalp.

Clinical recognition of chronic vitamin A poisoning was first described by Josephs in 1944.[73] A 3-year-old boy, who from the age of about 3 months had been given one teaspoon of halibut-liver oil daily, corresponding to approximately 72 mg RE day$^{-1}$ for 3 months. Moreover, he had on occasion drunk the oil directly from the bottle in undetermined amounts. He presented with abnormal skeletal development. X-rays examinations of the chest and skull showed no abnormalities, but those of the limb revealed considerable irregularity of the cortical structure. In the phalanges and metacarpals the cortex was extremely thin and the epiphyses of the upper end of the humeri and tibiae were mottled in appearance, and the distances between epiphyses and diaphyses were greater than normal.

The first adult case was reported in 1951.[74] The patient had been taking about 180 mg vitamin A daily during 18 months. Although the patient had generalized joint and bone pains, the roentgenological features regularly and characteristically found in children, such as periosteal bone formation causing the characteristic hyperostoses in metacarpals, metatarsals, and other tubular bones such as the ulna, tibia, and fibula, were missing in this case. Three years later Gerber and associates[75] reported osteoporotic change accompanied by markedly increased urinary calcium in a 28-year-old female with an intoxication, which lasted eight and half years. She had multiple severe joint and bone pains, and X-ray and bone biopsy studies indicated calcification with or without true bone formation in the pericapsular, ligamentous, tendinous and subperiosteal tissues. Also noted were decalcification in the skull, scapulae and vertebral bodies.

Pease[76] reported seven pediatric cases with serious alteration of skeletal development. The findings of retardation of growth and early closure of epiphyses were similar to those reported by Wolbach[47] in guinea pigs and rats 15 years earlier.

An interesting article from Sweden was published in 1965.[77] Five infants with vitamin A intoxication during the first half-year of life were described. The periods of overdosing were short (1–3 months) and the doses considerably lower (5.6–18 mg vitamin A per day) in comparison with those of the cases reported earlier. During the 1950s, AD-vitamins replaced cod liver oil as the prophylaxis against rickets given to all infants and children in Sweden. After 1955, AD-vitamins soluble in water were introduced and increasingly used. Although it was known that an aqueous dispersion of vitamin A gave about four times higher blood concentration (see below), the vitamin A content had remained the same, corresponding to a daily dose of about 2.5–3.0 mg. All five infants had received water-soluble AD-preparations. The roentgenological findings were in the most advanced cases a general reduction of the calcium content of the skeleton. The bones of the cranium were thin. In all cases cup-shaped deformations of the widened metaphyses were observed, which were sharply demarcated from the epiphyseal cartilages in the skeleton of wrists and ankles. In Sweden, the vitamin A content in AD-vitamins has been reduced several times and today contain a daily dose of 0.33 mg.

Although no abnormalities were found on X-ray, a clear effect on bone turnover was demonstrated by Jowsey and Riggs.[78] The microradiographic appearance of bone taken during and after the period of high vitamin A intake (60–90 mg day$^{-1}$) showed that there was an increase in the size of the osteocyte lacunae, resorption surfaces were a factor of six above the normal, and there was an absence of a comparative increase in bone formation. Similar findings with numerous resorptive lacunae and no index of bone formation have been described in a French patient.[79]

Only one case of fracture due to hypervitaminosis A has been described in man. Ruby *et al.*[80] reported that a girl 1 year and 3 months old had been given a daily dose of 15 mg in a water-soluble preparation for 7 months. Roentgenograms showed a fracture of the right humerus and periostal new-bone formation on the left ulna and left tibia. Only remnants of the distal femoral epiphyses were visible and both were seemingly impressed into the metaphyses, which were irregular. The serum vitamin A was elevated by a factor of 5.

Although hypercalcemia had been reported by several investigators[81–83] most reports had involved children and in many of these reports a history of concomitant vitamin D ingestion was common. To exclude the possibility of vitamin D intoxication or hyperparathyroidism as the cause of hypercalcemia in their case, Ragavan *et al.*[84] measured serum vitamin A, S-PTH and 25-hydroxyvitamin D. Both the latter were normal and based on their findings and two other case reports,[85,86] they concluded that hypercalcemia can indeed be caused solely by vitamin A toxicity. Hypercalcemia leading to renal failure has been described[79] and it is well known that patients with chronic renal failure are at risk of vitamin A intoxication.[87,88] An extreme case of hypercalcemia and metastatic calcifications in hypervitaminosis A was reported in a neonate,[89] who died after having ingested more than 60 times the suggested dose of vitamin A for 11 days. The hospital course was marked by hypercalcemia, hyperphosphatemia, bleeding disorder, and pulmonary insufficiency. An autopsy showed extensive calcifications of the alveolar septa and bronchioles. Metastatic calcifications were also present in the kidney, stomach, soft tissue, and skin. The skeleton showed prominent alteration of the endochondral bone formation and there was also evidence of accelerated bone resorption, which was thought to have caused the hypercalcemia and calcifications. A summary of the skeletal effects described in case reports is presented in Table 17.4.

The lowest dose inducing chronic hypervitaminosis A has not been defined. In some cases doses as low as about 0.1–0.3 mg kg$^{-1}$ per day[90,91] have been suggested to cause toxicity in children.

# Water-soluble Preparations of Retinol are more Toxic

Lewis *et al.*[92,93] showed more than half a century ago, that aqueous emulsions of retinol and retinyl esters resulted in higher plasma peak values, higher liver

**Table 17.4.** *Summary of skeletal effects in human subjects with hypervitaminosis A*

Bone and joint pain
Periosteal bone formation causing characteristic hyperostoses in metacarpals, metatarsals, and other tubular bones such as the ulna, tibia, and fibula
Premature closure of epiphyses and retardation of growth
Calcification of tendons, ligaments, and other tissues
Thinning of bones or bone cortex
Decalcification of bones
Fracture (one single case)

Increased bone resorption
Hypercalcemia
Elevated S-ALP
Elevated U-Ca

values and lower faecal losses compared to oily retinol preparations. With comparable doses the symptoms of chronic hypervitaminosis A appear significantly earlier (by a factor of 6) after emulsified or equivalent preparations than after oily emulsions.[94] This is because the retinyl esters in the oily preparations must be hydrolyzed to retinol and solubilized by the bile salts before uptake by the enterocytes.[95] Supplements of water-soluble retinyl esters have higher bioavailibility compared to retinyl esters given as liver, and the peak plasma concentrations and the area under the concentration-time curve (AUC) of all-*trans*-RA were up to 20 times higher after supplements compared to the same dose as liver.[96] In contrast, the formation of several retinoic acid metabolites are higher after consumption of liver paste compared to a supplement containing retinyl esters in oil.[97] It is therefore not surprising that the most severe skeletal effects described in humans are seen in infants who have ingested large amounts of water-soluble retinol (see above).

# Bone Effects of Synthetic Retinoids in Laboratory Animals

The bone toxicity of different retinoids is qualitatively very similar. Initial studies of the subchronic toxicity of tretinoin (all-*trans*-RA) in rats and dogs.[98–101] showed elevation of serum alkaline phosphatase, but cartilage and bone appeared normal. Although no attempt was made to identify the isoenzymes, the elevated values were attributed to 'osteoblastic hyperactivity'. Evidence for bone remodeling was seen at 10 mg kg$^{-1}$ per day in rats.[99] Kurtz *et al.*[102] found that rats given 14 mg kg$^{-1}$ showed clear signs of retinoid intoxication including growth depression, serum alkaline phosphatase elevation, and bone fracture. Hixson *et al.*[103,104] found that isotretinoin (13-*cis*-RA) appeared to be somewhat less toxic than tretinoin, since it produced the same number and incidence of fractures at doses three to five times that of tretinoin during a 3-week study. Later, bone fractures in mice exposed to dietary administration of 13-*cis*-RA for 23 weeks was shown to be a more sensitive index of retinoid toxicity than body weight. Osteotoxicity was found at a total dose of 13-*cis*-RA which was only about twice the total dose used clinically.[105]

The subchronic toxicity of etretinate in rats and dogs has been summarized by Teelmann.[106] Alkaline phosphatase showed a dose-dependent elevation, which was slight in the case of dogs, and somewhat more pronounced in rats. The most prominent symptoms were reduction of body weight gain and skeletal alterations. The duration of onset and frequency of fractures were clearly dose-related. The occurrence of fractures was noted as early as 10–14 days at doses of 10–15 mg kg$^{-1}$ per day. In contrast, with 3 mg kg$^{-1}$ per day fractures were not observed earlier than 5–6 months after commencement of compound administration. The fractures were preceded by a markedly intensified remodeling of the bones with deformations, haemorrhages, increased periostal osteoclastic activity, thickening of the periosteum, and accelerated ossification of the epiphyseal line. Furthermore an augmented endosteal osteoblastic activity resulting in marked reduction of bone diameter and increasing porosity, particularly of the long bones, was noted. All side effects were reversible after cessation of treatment.

Studies in dogs given retinoids resulted in less prominent bone changes. No fractures were reported, but in pups the clinical signs were evident pain in limb joints and retarded growth, radiological changes were decreases in overall length and thickness of bones, development of osteophytes, periostal reaction, and premature closure of epiphyses. Pathologic changes were degenerative epiphyseal plate, haemorrhages and exostotic proliferation of periosteum.[107]

# Bone Effects of Synthetic Retinoids in Humans

From preclinical studies it was clear that synthetic retinoids have effects on bones that mimic effects of hypervitaminosis A. When synthetic retinoids were introduced to the clinic,[108,109] premature closure of epiphyses,[110] hypercalcemia,[111] hyperostosis,[112] thinning of long bones,[113] and calcification of ligaments and tendons[114] were all reported within a few years. Whereas short-term therapy with isotretinoin (13-*cis*-RA) has not been found to influence bone density in three studies[115–117] Leachmann *et al.*[118] reported that bone mineral density (BMD) at the Ward triangle decreased a mean of 4.4% after 6 months and four patients showed decreased density of more than 9%. Long-term isotretinoin caused spontaneous vertebral fracture in one individual[119] and long-term treatment with etretinate has been reported to lead to osteoporosis by several investigators.[116,120–122]

# Vitamin A and Vitamin D have Antagonistic Effects

Interactions between vitamin A and vitamin D have been suggested ever since they were first described in 1933 by Bomskov and Seeman[42] who noted that high doses of a vitamin A concentrate inhibited the healing effect of vitamin D in rats with rickets. Simultaneous administration of excess vitamin A and vitamin D reduced the effects of hypervitaminosis D in the rat[123] whereas extra vitamin D protected against vitamin A toxicosis.[124] Convincing evidence of (weak) antagonism have not been presented until more recently. In a series of experiments in broiler chicks, Aburto *et al.* demonstrated that high dietary levels of vitamin A (450 µg of retinyl acetate $kg^{-1}$) reduced bone ash and increased rickets only when $D_3$ was present at a marginal level (12.5 µg $kg^{-1}$) in the diet, but not when it was synthesized in the bird by exposure to UV light or supplemented at 62.5 µg $kg^{-1}$ in the diet, indicating that the antagonism is at the level of intestinal absorption.[125] Consistent with the initial findings by Moore and Wang,[46] who found no effects of hypervitaminosis A on bone ash, these experiments suggest that vitamin A has no or little direct effects on bone mineralization, instead the vitamin A:D ratio is important. Rohde *et al.*[126] reported that at low levels of vitamin D and with a normocalcemic diet, high levels of vitamin A can reduce the serum calcium response to vitamin D in rats.

In a randomized, double-blind, crossover clinical trial[127] it was found that a single large dose of retinyl palmitate (15 mg) decreased the serum calcium response to a single dose of the activated form of vitamin D (2 μg of 1,25 dihydroxyvitamin $D_3$) several hours after the administration of both vitamins together.

Thus, if the exposure to UV light or the intake of vitamin D is adequate, measurements of BMD may not be the optimal method to detect the effects of vitamin A on bone.

# Epidemiological Studies

Although animal, *in vitro*, and human data are consistent and all indicate that excess vitamin A may contribute to osteoporosis, surprisingly few studies have addressed this issue. There are only two BMD studies that have distinguished between retinol and carotenoid intake, and two that have used a biochemical marker (Table 17.5). The most important outcome, fractures, has only been investigated in four studies (Table 17.6). Several other studies on diet and bone mass, that have included total vitamin A in the analysis, are cited in reviews on this subject. Therefore, these studies will also be mentioned in this chronological review of the epidemiological studies.

Sowers *et al.*[128] noted that most community studies had focused on calcium intake and not considered other nutrients which substantially influence the role of calcium in bone mineralisation. They selected two study communities with greatly divergent calcium content in their municipal water supplies, and the study cross-sectionally explored the correlates of mid-radius bone density in 324 postmenopausal women. Dietary intake was estimated by the 24-h recall method, bone mass was measured by single photon absorptiometry, and a stepwise multiple regression procedure was used in the statistical analysis. They found that mid-radius bone density decreased with age while humeral muscle area, extended estrogen use, thiazide use, and vitamin D intake were independently and positively associated with bone density. "Vitamin A intake was negatively correlated with bone density values at a level which approached significance."

Using similar methods, Yano *et al.*[129] investigated the relationship between diet and bone mass in a population of elderly Japanese residents (1208 men and 912 women) living in Hawaii. After controlling for age, weight, height, strenuous exercise (men), history of nonviolent fracture, thiazide use, and estrogen use (women) three dietary variables were significantly and positively associated with bone mineral content: calcium, milk, and vitamin D.

An improved dietary measurement was used by Freudenheim *et al.*[130] In a 4-year clinical trial, the effect of energy and 14 nutrients on bone density was studied in 99 women 35–65 years of age randomly assigned to placebo or calcium supplements. Dietary intake was estimated by multiple 24-h records, which were obtained from different days of the week and in two cycles of 29 days during the final 3 years of the trial, and it was possible to collect up to 72 records for each subject. A positive relationship between vitamin A and rate of humeral bone loss was found among the 9 premenopausal women in the

**Table 17.5.** *Studies of retinol and bone mineral density*

| Year | Study design | Sample size | Age | Dietary assessment | Biomarker | BMD measurement | Findings | References |
|---|---|---|---|---|---|---|---|---|
| 1990 | Cross-sectional | 246 ♀ | 55–80 | 24-h recall + interview | Serum retinol | SPA | No association between vitamin A supplement use or serum retinol and radial bone mass | 131 |
| 1998 | Cross-sectional | 175 ♀ | 28–74 | Four 1-week records | – | DXA | For intake > 1.5 mg day$^{-1}$ *versus* ≤0.5 mg day$^{-1}$ BMD was reduced 10% at the femoral neck, 14% at lumbar spine, and 6% for total body | 16 |
| 2001 | Cross-Sectional | 2888 ♀ 2902 ♂ | 20–80+ | – | Serum retinylesters | DXA | No association between serum retinyl esters and BMD at any site | 140 |
| 2002 | 4-y longitudinal | 570 ♀ 388 ♂ | 55–92 | FFQ | – | DXA×2 | Inverse U-shaped association between retinol intake and BMD. BMD optimal at 600–840 μg day$^{-1}$ | 146 |

**Table 17.6.** *Studies of retinol and the risk of osteoporotic fracture*

| Year | Study design | Sample size | Vitamin A assessment | Identification of cases of fracture | Follow-up | Risk for fracture | References |
|---|---|---|---|---|---|---|---|
| 1990 | Retrospective | 246 ♀, 56 fractures | Serum retinol | Patient interview | – | No association | 131 |
| 1998 | Nested case-control | Cohort 66,651 ♀, 247 hip fractures, 873 controls | FFQ[a] | Hospital discharge records | 2–64 months | OR = 2.05 >1.5 mg day$^{-1}$ vs. ≤0.5 mg day$^{-1}$ | 16 |
| 2002 | Prospective cohort study | Cohort 72,337 ♀, 603 hip fractures | Repeated FFQs | Questionnaire | 18 years | RR = 1.89 ≥2 mg day$^{-1}$ vs. ≤0.5 mg day$^{-1}$ | 145 |
| 2003 | Prospective cohort study | Cohort 2322 ♂, 266 fractures, 84 hip fractures | Serum retinol Serum β-carotene | Hospital discharge register, Orthopedic records, Radiographic records | 30 years | OR = 2.5, 5th vs. 3rd quintile No assoc β-carotene | 144 |

[a]Food frequency questionnaire.

placebo-treated group. In contrast, in the postmenopausal women of the treatment group, there was an inverse correlation between vitamin A intake and the rate of change in ulna (but not in radius or humerus) bone mineral content. In a single patient receiving a high supplemental dose (average intake about 4.4 mg RE day$^{-1}$) bone loss was very rapid with no other reason apparent.

The first study that was designed to investigate the relationship between vitamin A intake and bone status was published in 1990 by Sowers and Wallace (Ref. [131], see Table 17.6). It was also the first to include a biochemical marker of vitamin A intake. The study was motivated by the increasing use of vitamin A supplements and several of the observations described in this chapter: the well-documented adverse skeletal effects of retinol in animals, the anecdotal reports of accidental vitamin A poisoning in humans and the skeletal effects of retinoids used in the treatment of dermatologic conditions suggesting impaired bone remodeling, and finally their previous finding of a negative association between vitamin A intake and bone density which approached significance. They specifically wanted to test the two hypotheses (a) that vitamin A supplementation, or elevated retinol as a biologic marker of vitamin A, may be associated with decreased bone mass in postmenopausal women and (b) that levels of serum retinol or vitamin A from supplement may be associated with an increase in the report of fracture in postmenopausal women.

Serum retinol and bone mass was measured in 246 postmenopausal women, and 56 women reported an atraumatic fracture of the hip, ribs, spine or wrist postmenopausally (21 of these occurred within the past 10 years). As in the three previous studies, single photon densitometry was used. Nutrient intake was quantified from a nutritional supplements interview and 24-h food recall. The correlation between vitamin A supplement intake estimate and serum retinol levels was 0.14 ($p<0.03$), whereas the correlation between dietary food intake estimate from 24-h food recall and serum retinol was only 0.02 ($p<0.70$). Serum retinol levels were within the normal range in these women; even in the 90th percentile serum retinol increased from about 2 μmol L$^{-1}$ at ages 40–44 to about 2.5 μmol L$^{-1}$ at ages 75–80 years. There was a statistically significant relationship in mean serum retinol levels between 89 vitamin A supplement users (1.69±0.52 μmol L$^{-1}$) and 157 non-users (1.55±0.52 μmol L$^{-1}$, $p<0.03$). However, after controlling for age, current estrogen replacement, and current thiazide use, they observed no significant relationship between vitamin A supplement use or serum retinol with radial bone mass or fractures. As pointed out by the authors, this study was not powered to test an association of bone mass among women ingesting more than 2000 μg RE day$^{-1}$ from supplements; only 8% of the study population had intakes exceeding this level. In addition, about 1/3 of the population was <60 years of age and therefore likely to be heterogeneous with regards to estrogen depletion bone loss, and other important con-founders such as body mass index, smoking, physical activity etc. were not included in the analysis.

Hernández-Avila[132] evaluated the influence of dietary, anthropomorphic, and hormonal factors on bone density in a cross-sectional sample of 281 pre- and perimenopausal women. Information was obtained using a previously validated semiquantitative food frequency questionnaire (FFQ) listing 116 foods.[133] Bone density was measure using single-photon absorptiometry in the midshaft and the ultradistal radius. The investigators observed no important associations between dietary variables and midshaft bone density, but retinol, vitamin D, and vitamin C were all positively associated with radius density. All of these associations were very weak unless supplements were included in the calculation of intake. Women who used multivitamins had higher values than those who did not. Because of the high correlation among the nutrients included in multivitamins, it was not possible to distinguish which nutrients were responsible for this association.

Houtkooper[134] studied relationships among total energy intake, nutrient intake, body composition, exercise group status, and annual rates of change in BMD in 66 premenopausal women taking calcium supplements. The results showed that both nutrient intake and body composition variables were independent factors that influenced the rates of change of BMD. The specific regression model for prediction of total body BMD slope that had the best fit included vitamin A, initial fat mass, and slope of fat mass and accounted for about one third of the variance. The regression coefficients for these variables were weakly positive ($0.007$ mg cm$^{-2}$ per year) and significant. The estimated mean intake of total vitamin A from the diet was $1220\pm472$ µg RE day$^{-1}$. Surprisingly, the model that included carotene (instead of total vitamin A) had an almost equivalent fit to the model.

Earnshaw[135] did not find any correlation between any dietary variable and BMD in 426 postmenopausal women. However, dietary assessment was performed only with a 3 day unweighed dietary record. The quality of this method must be questioned since, for example, there was no association between dietary intake of vitamin D and serum 25-hydroxycholecalciferol, which is in contrast to most other studies.[136–138]

An association between excessive dietary vitamin A intake and osteoporosis was first reported in 1998 (Ref. [16], Table 17.6). It was also the first to distinguish between dietary retinol and beta-carotene. The study was prompted by the consistent animal and *in vitro* data, reports on osteoporosis as a toxic effect of long-term therapy with synthetic retinoids, and by observations of high incidence of osteoporotic fractures and high retinol intake in northern Europe. It included a cross-sectional examination of BMD in 175 women 28–74 years of age, and a nested case-control study of 247 women 40–76 years of age who had a first hip fracture within 2–64 months after enrollment and 873 age-matched controls in a cohort of 66 651 women. In the former, dietary vitamin A intake was assessed by review of four 1-week dietary records and BMD was measured by dual energy X-ray absorptiometry. In the case-control component of the study dietary vitamin A intake was estimated from a semiquantitative FFQ, and hip fracture was identified by using hospital discharge records and confirmed by record review. In a multivariate analysis adjusted for the body-mass index, energy intake, level of physical activity, smoking status, estrogen status, and use of estrogen, retinol intake was negatively associated with BMD. For every 1-mg increase in daily intake of retinol, risk for hip fracture increased by 68%. For intake greater than 1.5 mg day$^{-1}$ compared with intake less than 0.5 mg day$^{-1}$, BMD was reduced by 10% at the femoral neck, 14% at the lumbar spine, and 6% for the total body and the risk for hip fracture was doubled. Dietary intake of beta-carotene was not associated with BMD or hip fracture risk.

Elmståhl[139] studied dietary risk factors for fracture in Swedish men 45–69 years of age. The diet was assessed using a combined 7-day menu book for hot meals, beverages and dietary supplements and a quantitative FFQ. The incident fractures that occurred during a four-year period were retrieved from the local hospital registry and registry of X-ray examinations. This study with relatively young men and short follow-up time was not designed to study osteoporotic fractures. Of the 160 cases with fractures, only 51 were defined as fragility fractures. There were no hip or vertebral fractures and only two fractures of the humerus. Interestingly, the mean retinol intake among the 160 cases was 1.96 mg day$^{-1}$ compared to 1.82 mg day$^{-1}$ among 6416 controls, but there was no association between retinol intake and fracture risk after adjustment for age, energy intake and previous fractures. Important confounders such as body mass index, smoking, physical activity, *etc* were not included in the multivariate analysis.

In a cross-sectional analysis of the Third National Health and Nutrition Examination Survey (NHANES III), the association between fasting serum retinyl esters and BMD was examined in 5790 adults.[140] About one third of the participants had fasting serum retinyl esters ⩾10% of total serum vitamin A and one quarter or more had osteopenia/osteoporosis at one or more sites. Data were also collected on age, body mass index, smoking, alcohol consumption, use of dietary supplements, diabetes, physical activity, and in women, use of oral contraceptives or hormone replacement therapy, menopausal status and parity. These covariates were controlled for using multiple linear regresseion. The study showed no significant association between fasting serum retinyl ester concentration, or fasting serum retinyl esters as percent of total vitamin A, and BMD as assessed at the femoral neck, trochanter, intertrochanter and total hip. However, serum retinyl esters are not the optimal biochemical marker of vitamin A status. Based on experiments in rats and three human cases, Smith and Goodman[141] presented the theory that vitamin A toxicity occurs when retinyl esters, which do not bind to RBP, are presented to cell membranes. A number of more recent studies have demonstrated that retinoic acid metabolites mediate most, if not all the toxic effect of retinol,[96,97,142] and that the targeted disruption of the mouse gene for RBP, does not cause any toxicity symptoms.[143] Serum retinyl esters, which increase after every vitamin A-containing meal, more likely reflect a temporary excess in vitamin A intake rather than long-term vitamin intake and storage (Ref. [144], Table 17.6).

In 2002 two studies that confirmed and extended the data by Melhus *et al.* were published. In the Nurses Health Study[145] 603 incident hip fractures resulting from low or moderate trauma were identified among more than 70 000 postmenopausal women during 18 years of follow-up. Importantly, this study considered total vitamin A intake, as well as retinol and beta-carotene separately. After controlling for confounding factors, women in the highest quintile of total vitamin A intake from food and supplements ($\geq$3 mg RE day$^{-1}$), had a significantly elevated RR of hip fracture (1.48 (95% CI 1.05–2.07)) compared with women in the lowest quintile of intake ($<$1.25 mg day$^{-1}$ of RE). This increased risk was attributable primarily to retinol (RR 1.89 (95% CI 1.33–2.68) comparing $\geq$2 mg day$^{-1}$ *versus* $<$0.5 mg day$^{-1}$). Beta-carotene did not contribute significantly to fracture risk. A retinol intake equal to or greater than 1.5 mg day$^{-1}$ as compared with an intake of less than 0.5 mg day$^{-1}$ was associated with a relative risk of 1.64 for hip fracture similar to the OR of 1.54 reported by Melhus *et al.* Using intake as a continuous variable, for every additional 0.5 mg day$^{-1}$ increase in retinol intake, hip fracture risk increased by 33% for retinol from food only, which should be compared with the 68% in increased risk per mg retinol in the Swedish study. Multivitamins were the primary contributors to total retinol (35–43% of intake), and carrots contributed the most to total beta-carotene intake (30–41% of intake). Liver was the primary food source of retinol (22% in 1980 and 15% in 1994). Multivitamins were used by 34% of the cohort in 1980 and 53% by 1996.

In the Rancho Bernardo Study,[146] the association between BMD and bone loss, and total and supplemental retinol intake was investigated in a cohort of 570 elderly women and 388 men, aged 55–92 years at baseline. Dietary intake of vitamin A was assessed by dietary questionnaire over a 4-year period. This ended four years before the start of a 4-year-period in which annual measurements of BMD and bone loss were made. Regression analyses, adjusted for standard osteoporosis covariates, showed an inverse U-shaped association of retinol intake with baseline BMD, BMD measured 4 years later, and BMD change. BMD was optimal when the retinol intake was 0.6–0.84 mg day$^{-1}$. The high retinol levels found in this study to be associated with BMD and bone maintenance were attained almost exclusively by those taking supplemental retinol.

The first prospective study that used a biological marker of vitamin A status to assess the risk of fractures in humans was published by Michaëlsson et al in 2003. In a long-term, population-based study of 2322 men, serum retinol and beta-carotene levels were measured at base line. Fractures were documented in 266 men during 30 years of follow-up. The relative risk was 1.64 for any fracture and 2.47 for hip fracture among men in the highest quintile for serum retinol ($>$2.64 µmol L$^{-1}$), as compared with the middle quintile (2.17–2.36 µmol L$^{-1}$). The normal level of retinol is poorly defined, but has been suggested to be 0.7–2.8 µmol L$^{-1}$.[147] There was an especially steep rise in the rate-ratio curve for men with serum levels above the 95th percentile (3.1 µmol L$^{-1}$) and men with retinol levels in the 99th percentile ($>$3.60 µmol L$^{-1}$) had an overall risk of fracture that exceeded the risk among men with lower levels by a factor of 7 (Figure 17.6). The serum

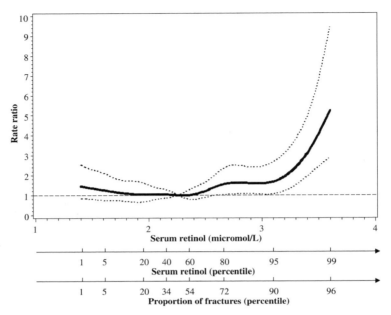

**Figure 17.6**   *Smoothed plot of rate ratios for any fracture according to the serum retinol level. The rate ratios (solid line) and 95 % confidence intervals (dotted lines) were estimated by restricted cubic-spline Cox regression analysis, with the median serum retinol level, 2.26 μmol L$^{-1}$, as the reference value*
From Michaëlsson *et al.* 2003 (144); reprinted with permission from the publisher

beta-carotene level was not associated with the risk of fracture. In addition, a dietary assessment had been performed 20 years after the serum samples had been obtained in a subgroup of 1221 men using a 7-day dietary record. The highest quintile for estimated retinol intake (>1.5 mg day$^{-1}$) was associated with an energy-adjusted rate ratio of 2.00 (95% CI 1.00–3.99). Michaëlsson and colleagues conclude that serum levels above 3 μmol L$^{-1}$ may increase the risk of fracture and should thus be avoided.

## Conclusions

The issue for vitamin A and bone health is not whether mechanisms exist, but rather at what levels retinol increases bone fragility.[52,148] The studies summarized in Tables 17.5 and 17.6 suggest that increasing retinol intake may have a detrimental effect on bone at retinol levels not far in excess of the recommended daily intake and considerably lower than today's upper limit of 3 mg day$^{-1}$. We cannot be certain that a retinol intake of approximately 1.5 mg day$^{-1}$ is without adverse consequences. Excessive amounts can easily be attained from supplements containing retinol. Therefore supplements should not be routinely used, especially not by the elderly. Hypervitaminosis A may explain, in part, the high incidence of hip fractures in Scandinavia and the United States, countries where vitamin supplements are commonly used.

A more judicious supplementation and fortification with retinol is recommended in Western countries, where the prevalence of osteoporosis is increasing. Retinol levels in multivitamins and processed foods can easily be lowered or partially replaced with beta-carotene without risk of deficiency.[149]

# References

1. M. Sporn, A. Roberts and D. Goodman, In *The Retinoids: Biology, Chemistry and Medicine*, Raven Press, New York, NY, 1994.
2. G. Combs Jr, In *The Vitamins. Fundamental Aspects in Nutrition and Health*, 2nd edn, Academic Press, San Diego, 1998.
3. L. McDowell, In *Vitamins in Animal and Human Nutrition*, 2nd edn, Ames, Iowa State University Press, Iowa, 2000.
4. P. Karrer, R. Morf and K. Schöpp, Zur kenntnis des vitamins-A aus fischtranen II, *Helv. Chim. Acta*, 1931, **14**, 1431–1436.
5. K.J. Yeum and R.M. Russell, Carotenoid bioavailability and bioconversion, *Annu. Rev. Nutr.*, 2002, **22**, 483–504.
6. FAO/WHO, Requirements of vitamin A, thiamine, riboflavin and niacin. In *Report of a Joint FAO/WHO Expert Group; 1967*, Rome, Italy: WHO Technical Report Series; 1967.
7. *Food and Nutrition Board, Institute of Medicine, Dietary Reference Intakes for Vitamin A, Vitamin K, Arsenic, Boron, Chromium, Copper, Iodine, Iron, Manganese, Molybdenum, Nickel, Silicon, Vanadium, and Zink*, National Academy Press, Washington, DC, 2001.
8. E.K. Hume and H.A. Krebs, Vitamin A requirements of human adults. *Medical Research Council Special Report*; 1949. Report No. 264.
9. FAO/WHO, Requirements of vitamin A, iron, folate, and vitamin B12, In FAO (ed.), *Report of a Joint FAO/WHO Expert Consultation, 1988*, FAO Press, Rome, Italy, 1988, pp. 85–89.
10. S. de Pee and C.E. West, Dietary carotenoids and their role in combating vitamin A deficiency: a review of the literature, *Eur. J. Clin. Nutr.*, 1996, **50**(Suppl. 3), S38–S53.
11. C.E. West, A. Eilander and M. van Lieshout, Consequences of revised estimates of carotenoid bioefficacy for dietary control of vitamin A deficiency in developing countries, *J. Nutr.*, 2002, **132**(Suppl. 9), 2920S–2926S.
12. S. de Pee, C.E. West, Muhilal, D. Karyadi and J.G. Hautvast, Lack of improvement in vitamin A status with increased consumption of dark-green leafy vegetables, *Lancet*, 1995, **346**(8967), 75–81.
13. *U.S. Institute of Medicine Food and Nutrition Board, Standing Committee on the Scientific Evaluation of Dietary Reference Intakes. Dietary Reference Intakes for Vitamin A, Vitamin K, Arsenic, Boron, Chromium, Copper, Iodine, Iron, Manganese, Molybdenum, Nickel, Silicon, Vanadium and Zinc*, National Academy Press, Washington, DC, 2000.
14. A.J. Edwards, C.H. Nguyen, C.S. You, J.E. Swanson, C. Emenhiser and R.S. Parker, Alpha- and beta-carotene from a commercial puree are more bioavailable to humans than from boiled-mashed carrots, as determined using an extrinsic stable isotope reference method, *J. Nutr.*, 2002, **132**(2), 159–167.

15. S.J. Hickenbottom, J.R. Follett, Y. Lin, S.R. Dueker, B.J. Burri, T.R. Neidlinger and A.J. Clifford, Variability in conversion of beta-carotene to vitamin A in men as measured by using a double-tracer study design, *Am. J. Clin. Nutr.*, 2002, **75**(5), 900–907.

16. H. Melhus, K. Michaëlsson, A. Kindmark, R. Bergström, L. Holmberg, H. Mallmin, A. Wolk and S. Ljunghall, Excessive dietary intake of vitamin A is associated with reduced bone mineral density and increased risk for hip fracture, *Ann. Intern. Med.*, 1998, **129**, 770–778.

17. J.A. Olson, The conversion of radioactive beta-carotene into vitamin A by the rat intestine in vivo, *J. Biol. Chem.*, 1961, **236**, 349–356.

18. T. van Vliet, M.F. van Vlissingen, F. van Schaik and H. van den Berg, beta-Carotene absorption and cleavage in rats is affected by the vitamin A concentration of the diet, *J. Nutr.*, 1996, **126**(2), 499–508.

19. H. Bachmann, A. Desbarats, P. Pattison, M. Sedgewick, G. Riss, A. Wyss, *et al.*, Feedback regulation of beta,beta-carotene 15,15′-monooxygenase by retinoic acid in rats and chickens, *J. Nutr.*, 2002, **132**(12), 3616–3622.

20. J.D. Ribaya-Mercado, F.S. Solon, M.A. Solon, M.A. Cabal-Barza, C.S. Perfecto, G. Tang, J.A. Solon, C.R. Fjeld and R.M. Russell, Bioconversion of plant carotenoids to vitamin A in Filipino school-aged children varies inversely with vitamin A status, *Am. J. Clin. Nutr.*, 2000, **72**(2), 455–465.

21. D. Ong, B. Kakkad and P. MacDonald, Acyl-CoA-independent esterification of retinol bound to cellular retinol-binding protein (type II) by microsomes from rat small intestine, *J. Biol. Chem.*, 1987, **262**(6), 2729–2736.

22. S.E. Dew and D.E. Ong, Specificity of the retinol transporter of the rat small intestine brush border, *Biochemistry*, 1994, **33**(40), 12340–12345.

23. D. Goodman and W. Blaner, Biosynthesis, absorption, and hepatic metabolism of retinol, In *The Retinoids*, Academic Press Inc, Orlando, 1984, pp. 1–39.

24. R. Blomhoff, M.H. Green, T. Berg and K.R. Norum, Transport and storage of vitamin A, *Science*, 1990, **250**(4979), 399–404.

25. W. Blaner and J.A. Olson, Retinol and retinoic acid metabolism, In *The Retinoids*, M. Sporn, A. Roberts and D. Goodman (eds.), Raven Press, New York, 1994, pp. 229–255.

26. K. Ewald and W. Kühne, Über kunstliche Bildung des Sehpurpurs, *Centralbl. Med. Wissensch.*, 1877, **15**, 753–754.

27. G. Wald, Molecular basis of visual excitation, *Science*, 1968, **162**, 230–239.

28. D.R. Gollapalli and R.R. Rando, All-*trans*-retinyl Esters Are the Substrates for Isomerization in the Vertebrate Visual Cycle, *Biochemistry*, 2003, **42**(19), 5809–5818.

29. J. Saari, Retinoids in the photosensitive systems, In *The Retinoids*, M. Sporn, A. Roberts and D. Goodman (eds.), Raven Press, New York, 1994, pp. 351–385.

30. R.J. Lucas, S. Hattar, M. Takao, D.M. Berson, R.G. Foster and K.W. Yau, Diminished pupillary light reflex at high irradiances in melanopsin-knockout mice, *Science*, 2003, **299**(5604), 245–247.

31. D. Mangelsdorf, K. Umesono and R. Evans, The retinoid receptors, In *The Retinoids*, 2nd edn, M. Sporn, A. Roberts and D. Goodman (eds.), Raven Press, New York, 1994, pp. 319–349.

32. P.J. Willy, K. Umesono, E.S. Ong, R.M. Evans, R.A. Heyman and D.J. Mangelsdorf, LXR, a nuclear receptor that defines a distinct retinoid response pathway, *Genes. Dev.*, 1995, **9**(9), 1033–1045.

33. K.L. Gearing, M. Gottlicher, M. Teboul, E. Widmark and J.A. Gustafsson, Interaction of the peroxisome-proliferator-activated receptor and retinoid X receptor, *Proc. Natl Acad. Sci.*, USA, 1993, **90**(4), 1440–1444.

34. C. Nieman and H. Obbink, The biochemistry and pathology of hypervitaminosis A, *Vitam Horm.*, 1954, **12**, 69–99.

35. J. Kamm, K. Ashenfelter and C. Ehmann, Preclinical and clinical toxicology of selected retinoids, In *The Retinoids*, 1st edn, M. Sporn, A. Roberts and D. Goodman, (eds.), Academic Press, Inc (London) Ltd, London, 1984, pp. 287–326.

36. R. Armstrong, K. Ashenfelter, C. Eckoff, A. Levin and S. Shapiro, General and reproductive toxicology of retinoids, In *The Retinoids*, 2nd edn, M. Sporn, A. Roberts and D. Goodman (eds.), Raven Press, New York, 1994, pp. 545–572.

37. N. Barnicot, The local action of vitamin A on bone, *J. Anat.*, 1950, **84**, 374–387.

38. H. Fell and E. Mellanby, The effect of hypervitaminosis A on embryonic limb-bones cultivated in vitro, *J. Physiol.*, 1952, **116**, 320–349.

39. R. Heywood, A.K. Palmer, R.L. Gregson and H. Hummler, The toxicity of beta-carotene, *Toxicology*, 1985, **36**(2-3), 91–100.

40. K. Takahashi, Z. Nakamiya, K. Kawakimi and T. Kitasato, On the physical and chemical properties of biosterin (a name given to fat-soluble A) and on its physiological significance, *Sci. Papers Inst. Phys. Chem. Res.* (Tokyo), 1925, **3**, 81–145.

41. J. Collazo, J. Rodriguez and A. Hypervitaminose II, Exophtalmus und spontan-frakturen, *Klin. Wochschr.*, 1933, **12**, 1768–1771.

42. C. Bomskov and G. Seeman, Über eine wirkung des vitamin A auf den mineralhaushalt, *Z. Ges. Exp. Med.*, 1933, **89**, 771–779.

43. K. Strauss, Beobachtungen bei hypervitaminose A, *Anat. U. Allgem. Pathol.*, 1934, **94**, 345–352.

44. A. Davies and T. Moore, Vitamin A and Carotene. The distribution of vitamin A in the organs of the normal and hypervitaminotic rat, *Biochemistry*, 1934, **28**, 289–295.

45. S. Clausen, Pharmacology and therapeutics of vitamin A, *JAMA*, 1939, **111**, 144.

46. T. Moore and Y. Wang, Hypervitaminosis A, *Biochem. J.*, 1945, **39**, 222–228.

47. B. Wolbach, Vitamin-A deficiency and excess in relation to skeletal growth, *J. Bone Joint Surg. Br.*, 1947, **29**, 171–192.

48. K. Rodahl, Hypervitaminosis A in the rat, *J. Nutr.*, 1950, **41**, 399–421.

49. T.L. Frankel, M.S. Seshadri, D.B. McDowall and C.J. Cornish, Hypervitaminosis A and calcium-regulating hormones in the rat, *J. Nutr.*, 1986, **116**, 578–587.

50. S. Hough, L.V. Avioli, H. Muir, D. Gelderblom, G. Jenkins, H. Kurasi, E. Slatopolsky, M.A. Bergfeld and S.L. Teitelbaum, Effects of hypervitaminosis A on the bone and mineral metabolism of the rat, *Endocrinology*, 1988, **122**(6), 2933–2939.

51. B. Wu, B. Xu, T.Y. Huang and J.R. Wang, A model of osteoporosis induced by retinoic acid in male Wistar rats, *Yao Hsueh Hsueh Pao*, 1996, **31**(4), 241–245.

52. N. Binkley and D. Krueger, Hypervitaminosis A and bone, *Nutr. Rev.*, 2000, **58**(5), 138–144.

53. X.F. Li, B. Dawson-Hughes, R. Hopkins, R.M. Russell, W.S. Jee, D. Bankson and X.J. Li, The effects of chronic vitamin A excess on bone remodeling in aged rats, *Proc. Soc. Exp. Biol. Med.*, 1989, **191**(2), 103–107.

54. S. Johansson, P. Lind, H. Håkansson, J. Örberg and H. Melhus, Subclinical hypervitaminosis causes fragile bones in rats, *Bone*, 2002, **31**, 685–689.

55. I. Dickson and J. Walls, Vitamin A and bone formation. Effect of an excess of retinol on bone collagen synthesis in vitro, *Biochem. J.*, 1985, **226**(3), 789–795.

56. K. Ng, P. Gummer, V. Michelangeli, J. Bateman, T. Mascare, W. Cole and T.J. Martin, Regulation of alkaline phosphatase expression in a neonatal rat clonal calvarial cell strain by retinoic acid, *J. Bone Miner. Res.*, 1988, **3**, 53–61.

57. J. Heath, L. Suva, K. Yoon, M. Kiledjian, T.J. Martin and G.A. Rodan, Retinoic acid stimulates transcriptional activith from the alkaline phosphatase promoter in the immortalized rat calvarial cell line, RCT-1, *Mol. Endocrinol.*, 1992, **6**, 636–646.

58. J. Heath, S. Rodan, K. Yoon and G. Rodan, Rat calvarial cell lines immortalized with SV-40 large T antigen: Constitutive and retinoic acid-inducible expression osteoblastic features, *Endocrinology*, 1989, **124**, 3060–3068.

59. R.O. Oreffo, A. Teti, J.T. Triffitt, M.J. Francis, A. Carano and A.Z. Zallone, Effect of vitamin A on bone resorption: evidence for direct stimulation of isolated chicken osteoclasts by retinol and retinoic acid, *J. Bone Miner. Res.*, 1988, **3**(2), 203–210.

60. A. Zambonin-Zallone, A. Teti, A. Carano and P.C. Marchisio, The distribution of podosomes in osteoclasts cultured on bone laminae: effect of retinol, *J. Bone Miner. Res.*, 1988, **3**(5), 517–523.

61. B.A. Scheven and N.J. Hamilton, Retinoic acid and 1,25-dihydroxyvitamin D3 stimulate osteoclast formation by different mechanisms, *Bone*, 1990, **11**(1), 53–59.

62. A. Kindmark, H. Melhus, S. Ljunghall and O. Ljunggren, Inhibitory effects of 9-cis and all-*trans* retinoic acid on 1,25(OH)2 vitamin D3-induced bone resorption, *Calcif. Tissue Int.*, 1995, **57**(3), 242–244.

63. S. Saneshige, H. Mano, K. Tezuka, S. Kakudo, Y. Mori, Y. Honda, A. Iitabashi, T. Yamada, K. Miyata, Y. Hakeda, J. Ishii and M. Kumegawa, Retinoic acid directly stimulates osteoclastic bone-resorption and gene-expression of cathepsin K/OC-2, *Biochem. J.*, 1995, **309**, 721–724.

64. H. Kaji, T. Sugimoto, M. Kanatani, M. Fukase, M. Kumegawa and K. Chihara, Retinoic acid induces osteoclast-like cell formation by directly acting on hemopoietic blast cells and stimulates osteopontin mRNA expression in isolated osteoclasts, *Life Sci.*, 1995, **56**(22), 1903–1913.

65. A. Kindmark, H. Torma, A. Johansson, S. Ljunghall and H. Melhus, Reverse transcription-polymerase chain reaction assay demonstrates that the 9-cis retinoic acid receptor alpha is expressed in human osteoblasts, *Biochem. Biophys. Res. Commun.*, 1993, **192**(3), 1367–1372.

66. H. Melhus, A. Gobl and S. Ljunghall, Competitive PCR demonstrates that 9-*cis* retinoic acid induces cellular retinoic acid-binding protein-II more efficiently than all-*trans* retinoic acid in human osteosarcoma cells, *Biochem. Biophys. Res. Commun.*, 1994, **200**(2), 1125–1129.

67. H. Harada, R. Miki, S. Masushige and S. Kato, Gene expression of retinoic acid receptors, retinoid-X receptors, and cellular retinol-binding protein I in bone and its regulation by vitamin A, *Endocrinology*, 1995, **136**(12), 5329–5335.

68. K. Rodahl and T. Moore, The vitamin A content and toxicity of bear and seal liver, *Biochem. J.*, 1943, **37**, 166–168.

69. J. Bauernfeind, *The Safe Use of Vitamin A*. A report of the international vitamin A consultative group, Nutrition Foundation, Washington, DC, 1980.

70. P. Lips, Hypervitaminosis A and fractures, *N. Engl. J. Med.*, 2003, **348**(4), 347–349.

71. V. Roeper and D. Wildeman, *Om de Noord: De tochten van Willem Barentsz en Jacob van Heemskerck en de overwintering op Nova Zembla zoals opgetekend door Gerrit de Veer*, Nijmegen, the Netherlands: Uitgeverij SUN, 1996.

72. A. Czerny, Beitrag zur lebertrantherapie, *Therapie der Gegenwart*, 1912, **53**, 49–50.

73. H. Josephs, Hypervitaminosis A and carotenemia, *Am. J. Dis. Child.*, 1944, **67**, 33–43.

74. M. Sulzberger and M. Lazar, Hypervitaminosis A. Report of a case in an adult, *JAMA*, 1951, **146**, 788–793.
75. A. Gerber, A. Raab and A. Sobel, Vitamin A poisoning in adults, *Am. J. Med.*, 1954, **16**, 729–745.
76. C. Pease, Focal retardation and arrestment of growth of bones due to vitamin A intoxicaton, *JAMA*, 1962, **182**, 980–985.
77. B. Persson, R. Tunell and K. Ekengren, Chronic vitamin A intoxication during the first half year of life, *Acta Paediat. Scand.*, 1965, **54**, 49–60.
78. J. Jowsey and B. Riggs, Bone changes in a patient with hypervitaminosis A, *J. Clin. Endocrinol. Metab.*, 1968, **28**, 1833–1835.
79. A. Baglin, C. Hagege, B. Franc, M. Richaud and J. Prinseau, A systemic-like disease: chronic vitamin A poisoning, *Ann. Med. Intern. (Paris)*,1986, **137**(2), 142–146.
80. L.K. Ruby and M.A. Mital, Skeletal deformities following chronic hypervitaminosis A; a case report, *J. Bone Joint Surg. Am.*, 1974, **56**(6), 1283–1287.
81. E. Shaw and J. Niccoli, Hypervitaminosis A: Report of a case in adult male, *Ann. Intern. Med.*, 1953, **39**, 131–134.
82. C. Katz and M. Tzagournis, Chronic adult hypervitaminosis A with hypercalcemia, *Metabolism*, 1972, **21**, 1171–1176.
83. R.G. Wieland, F.H. Hendricks, Y. Amat, F. Leon, L. Gutierrez and J.C. Jones, Hypervitaminosis A with hypercalcemia, *Lancet*, 1971, **1**, 698.
84. V. Ragavan, J. Smith and J. Bilezikian, Vitamin A toxicity and hypercalcemia, *Am. J. Med. Sci.*, 1982, **283**, 161–164.
85. B. Frame, C.E. Jackson, W.A. Reynolds and J.E. Umphrey, Hypercalcemia and skeletal effects in chronic hypervitaminosis A, *Ann. Intern. Med.*, 1974, **80**(1), 44–48.
86. K. Hofman, F. Milne and C. Schmidt, Acne, hypervitaminosis A and hypercalcemia, *S. A. Med. J.*, 1978, **54**, 579–580.
87. K. Farrington, P. Miller, Z. Varghese, R.A. Baillod and J.F. Moorhead, Vitamin A toxicity and hypercalcaemia in chronic renal failure, *Br. Med. J. (Clin. Res. Ed.)*, 1981, **282**(6281), 1999–2002.
88. E.E Gleghorn, L.D. Eisenberg, S. Hack, P. Parton and R.J. Merritt, Observations of vitamin A toxicity in three patients with renal failure receiving parenteral alimentation, *Am. J. Clin. Nutr.*, 1986, **44**(1), 107–112.
89. M.E. Bush and B.B. Dahms, Fatal hypervitaminosis A in a neonate, *Arch. Pathol. Lab. Med.*, 1984, **108**(10), 838–842.
90. D. Schurr, J. Herbert, E. Habibi and A. Abrahamov, Unusual presentation of vitamin A intoxication, *J. Pediatr. Gastroenterol. Nutr.*, 1983, **2**(4), 705–707.
91. T.O. Carpenter, J.M. Pettifor, R.M. Russell, J. Pitha, S. Mobarhan, M.S. Ossip, S. Wainer and C.S. Anast, Severe hypervitaminosis A in siblings: evidence of variable tolerance to retinol intake, *J. Pediatr.*, 1987, **111**(4), 507–512.
92. J. Lewis, S. Cohlan, O. Bodansky and J. Birmingham, Vitamin A – comparative absorption, excretion, and storage of oily and aqueous preparations, *J. Pediatr.*, 1947, **31**, 496–508.
93. J. Lewis and S. Cohlan, Comparative absorption of various types of vitamin A preparations, *Med. Clin. N. Am.*, 1950, **34**, 413–424.
94. W.F. Korner and J. Vollm, New aspects of the tolerance of retinol in humans, *Int. J. Vitam. Nutr. Res.*, 1975, **45**(4), 363–372.
95. R. Blomhoff, M.H. Green, J.B. Green, T. Berg and K.R. Norum, Vitamin A metabolism: new perspectives on absorption, transport, and storage, *Physiol. Rev.*, 1991, **71**(4), 951–990.

96. N.E. Buss, E.A. Tembe, B.D. Prendergast, A.G. Renwick and C.F. George, The teratogenic metabolites of vitamin A in women following supplements and liver, *Hum. Exp. Toxicol.*, 1994, **13**(1), 33–43.

97. T. van Vliet, E. Boelsma, A.J. de Vries and H. van den Berg, Retinoic acid metabolites in plasma are higher after intake of liver paste compared with a vitamin A supplement in women, *J. Nutr.*, 2001, **131**(12), 3197–3203.

98. R. Kretzschmar and F. Leuschner, Biosynthesis and metabolism of retinoic acid, *Acta Derm. Venereol. Suppl. (Stockh.)*, 1975, **74**, 25–28.

99. J. Cahn, P. Bordier, M. Herold and M.T. Meunier, Pharmacological studies on retinoic acid in Wistar and in atrichos rats, *Acta. Derm. Venereol. Suppl. (Stockh)*, 1975, **74**, 33–35.

100. M. Herold, J. Cahn and P. Gomont, Toxicology of vitamin A acid, *Acta. Derm. Venereol. Suppl. (Stockh.)*, 1975, **74**, 29–32.

101. G. Zbinden, Pharmacology of vitamin A acid (beta-all transretinoic acid), *Acta. Derm. Venereol. Suppl. (Stockh.)*, 1975, **74**, 21–24.

102. P.J. Kurtz, D.C. Emmerling and D.J. Donofrio, Subchronic toxicity of all-*trans*-retinoic acid and retinylidene dimedone in Sprague-Dawley rats, *Toxicology*, 1984, **30**(2), 115–124.

103. E.J. Hixson and E.P. Denine, Comparative subacute toxicity of all-*trans*- and 13-*cis*-retinoic acid in Swiss mice, *Toxicol. Appl. Pharmacol.*, 1978, **44**(1), 29–40.

104. E.J. Hixson, J.A. Burdeshaw, E.P. Denine and S.D. Harrison Jr., Comparative subchronic toxicity of all-*trans*- and 13-*cis*- retinoic acid in Sprague–Dawley rats, *Toxicol. Appl. Pharmacol.*, 1979, **47**(2), 359–365.

105. K.S. Forsyth, R.R. Watson and H.L. Gensler, Osteotoxicity after chronic dietary administration of 13-*cis*-retinoic acid, retinyl palmitate or selenium in mice exposed to tumor initiation and promotion, *Life Sci.*, 1989, **45**(22), 2149–2156.

106. K. Teelmann, Experimental toxicology of the aromatic retinoid Ro 10-9359 (Etretinate), In *Retinoids: Advances in Basic Research and Therapy*, C. Orfanos, O. Braun-Falco, E. Farber, C. Grupper, M. Polano and R. Schuppli (eds.), Springer-Verlag, Berlin and Heidelberg, 1981, pp. 41–47.

107. D.Y. Cho, R.A. Frey, M.M. Guffy and H.W. Leipold, Hypervitaminosis A in the dog, *Am. J. Vet. Res.*, 1975, **36**(11), 1597–1603.

108. G.L. Peck and F.W. Yoder, Treatment of lamellar ichthyosis and other keratinising dermatoses with an oral synthetic retinoid, *Lancet*, 1976, **2**(7996), 1172–1174.

109. G.L. Peck, T.G. Olsen, F.W. Yoder, J.S. Strauss, D.T. Downing, M. Pandya, D. Butkus and J. Arnaud-Battandier, Prolonged remissions of cystic and conglobate acne with 13-*cis*-retinoic acid, *N. Engl. J. Med.*, 1979, **300**(7), 329–333.

110. L.M. Milstone, J. McGuire and R.C. Ablow, Premature epiphyseal closure in a child receiving oral 13-*cis*-retinoic acid, *J. Am. Acad. Dermatol.*, 1982, **7**(5), 663–666.

111. J. Cassidy, M. Lippman, A. Lacroix and G. Peck, Phase II trial of 13-*cis*-retinoic acid in metastatic breast cancer, *Eur. J. Cancer Clin. Oncol.*, 1982, **18**(10), 925–928.

112. R.A. Pittsley and F.W. Yoder, Retinoid hyperostosis. Skeletal toxicity associated with long-term administration of 13-*cis*-retinoic acid for refractory ichthyosis, *N. Engl. J. Med.*, 1983, **308**(17), 1012–1014.

113. J. Prendiville, E.A. Bingham and D. Burrows, Premature epiphyseal closure – a complication of etretinate therapy in children, *J. Am. Acad. Dermatol.*, 1986, **15**(6), 1259–1262.

114. J.J. DiGiovanna, R.K. Helfgott, L.H. Gerber and G.L. Peck, Extraspinal tendon and ligament calcification associated with long-term therapy with etretinate, *N. Engl. J. Med.*, 1986, **315**(19), 1177–1182.

115. M. Kocijancic, 13-*cis*-retinoic acid and bone density, *Int. J. Dermatol.*, 1995, **34**(10), 733–734.

116. J.J. DiGiovanna, R.B. Sollitto, D.L. Abangan, S.M. Steinberg and J.C. Reynolds, Osteoporosis is a toxic effect of long-term etretinate therapy, *Arch. Dermatol.*, 1995, **131**(11), 1263–1267.

117. D.J. Margolis, M. Attie and J.J. Leyden, Effects of isotretinoin on bone mineralization during routine therapy with isotretinoin for acne vulgaris, *Arch. Dermatol.*, 1996, **132**(7), 769–774.

118. S.A. Leachman, K.L. Insogna, L. Katz, A. Ellison and L.M. Milstone, Bone densities in patients receiving isotretinoin for cystic acne, *Arch Dermatol.*, 1999, **135**(8), 961–965.

119. L.M. Milstone, A.F. Ellison and K.L. Insogna, Serum parathyroid hormone level is elevated in some patients with disorders of keratinization, *Arch. Dermatol.*, 1992, **128**(7), 926–930.

120. L. Halkier-Sørensen, G. Leurberg and J. Andersen, Bone changes in children on long-term treatment with etretinate, *J. Am. Acad. Dermatol.*, 1987, **16**, 999–1006.

121. R. Ruiz-Maldonado and L. Tamayo, Retinoids in disorders of keratinization: their use in children, *Dermatologica*, 1987, **175**(Suppl 1), 125–132.

122. N. Okada, M. Nomura, S. Morimoto, T. Ogihara and K. Yoshikawa, Bone mineral density of the lumbar spine in psoriatic patients with long term etretinate therapy, *J. Dermatol.*, 1994, **21**(5), 308–311.

123. I. Clark and C. Bassett, The amelioration of hypervitaminosis D in the rat with vitamin A, *Exp. Med.*, 1962, **115**, 147–155.

124. E. Vedder and C. Rosenberg, Concerning the toxicity of vitamin A, *J. Nutr.*, 1938, **16**, 57–68.

125. A. Aburto and W.M. Britton, Effects of different levels of vitamins A and E on the utilization of cholecalciferol by broiler chickens, *Poult. Sci.*, 1998, **77**(4), 570–577.

126. C.M. Rohde, M. Manatt, M. Clagett-Dame and H.F. DeLuca, Vitamin A antagonizes the action of vitamin D in rats, *J. Nutr.*, 1999, **129**(12), 2246–2250.

127. S. Johansson and H. Melhus, Vitamin A antagonizes calcium response to vitamin D in man, *J. Bone Miner. Res.*, 2001, **16**(10), 1899–1905.

128. M.R. Sowers, R.B. Wallace and J.H. Lemke, Correlates of mid-radius bone density among postmenopausal women: a community study, *Am. J. Clin. Nutr.*, 1985, **41**(5), 1045–1053.

129. K. Yano, L.K. Heilbrun, R.D. Wasnich, J.H. Hankin and J.M. Vogel, The relationship between diet and bone mineral content of multiple skeletal sites in elderly Japanese-American men and women living in Hawaii, *Am. J. Clin. Nutr.*, 1985, **42**(5), 877–888.

130. J.L. Freudenheim, N.E. Johnson and E.L. Smith, Relationships between usual nutrient intake and bone-mineral content of women 35–65 years of age: longitudinal and cross-sectional analysis, *Am. J. Clin. Nutr.*, 1986, **44**(6), 863–876.

131. M.F. Sowers and R.B. Wallace, Retinol, supplemental vitamin A and bone status, *J. Clin. Epidemiol.*, 1990, **43**(7), 693–699.

132. M. Hernandez-Avila, M.J. Stampfer, V.A. Ravnikar, W.C. Willett, I. Schiff, M. Francis, C. Longcope, S.M. McKinlay and C. Longscope, Caffeine and other predictors of bone density among pre- and perimenopausal women, *Epidemiology*, 1993, **4**(2), 128–134.

133. W.C. Willett, R.D. Reynolds, S. Cottrell-Hoehner, L. Sampson and M.L. Browne, Validation of a semi-quantitative food frequency questionnaire: comparison with a 1-year diet record, *J. Am. Diet. Assoc.*, 1987, **87**(1), 43–47.

134. L.B. Houtkooper, C. Ritenbaugh, M. Aickin, T.G. Lohman, S.B. Going, J.L. Weber, K.A. Greaves, T.W. Boyden, R.W. Pamenter and M.C. Hall, Nutrients, body composition and exercise are related to change in bone mineral density in premenopausal women, *J Nutr.*, 1995, **125**(5), 1229–1237.

135. S.A. Earnshaw, A. Worley and D.J. Hosking, Current diet does not relate to bone mineral density after the menopause. The Nottingham Early Postmenopausal Intervention Cohort (EPIC) Study Group, *Br. J. Nutr.*, 1997, **78**(1), 65–72.

136. H.K. Kinyamu, J.C. Gallagher, K.E. Balhorn, K.M. Petranick and K.A. Rafferty, Serum vitamin D metabolites and calcium absorption in normal young and elderly free-living women and in women living in nursing homes, *Am. J. Clin. Nutr.*, 1997, **65**(3), 790–797.

137. C.J. Lamberg-Allardt, T.A. Outila, M.U. Karkkainen, H.J. Rita and L.M. Valsta, Vitamin D deficiency and bone health in healthy adults in finland: could this be a concern in other parts of Europe? *J. Bone Miner. Res.*, 2001, **16**(11), 2066–2073.

138. C. Brot, P. Vestergaard, N. Kolthoff, J. Gram, A.P. Hermann and O.H. Sorensen, Vitamin D status and its adequacy in healthy Danish perimenopausal women: relationships to dietary intake, sun exposure and serum parathyroid hormone, *Br. J. Nutr.*, 2001, **86**(Suppl 1), S97–S103.

139. S. Elmstahl, B. Gullberg, L. Janzon, O. Johnell and B. Elmstahl, Increased incidence of fractures in middle-aged and elderly men with low intakes of phosphorus and zinc, *Osteoporos. Int.*, 1998, **8**(4), 333–340.

140. C. Ballew, D. Galuska and C. Gillespie, High serum retinyl esters are not associated with reduced bone mineral density in the Third National Health And Nutrition Examination Survey, 1988–1994, *J. Bone Miner. Res.*, 2001, **16**(12), 2306–12.

141. F. Smith and D. Goodman, Vitamin A transport in human vitamin A toxicity, *N. Engl. J. Med.*, 1976, **294**, 805–808.

142. C. Eckhoff, J.R. Bailey, M.D. Collins, W. Slikker and H. Nau, Influence of Dose and Pharmaceutical Formulation of Vitamin-A on Plasma Levels of Retinyl Esters and Retinol and Metabolic Generation of Retinoic Acid Compounds and beta-Glucuronides in the Cynomolgus Monkey, *Toxicol. Appl. Pharmacol.*, 1991, **111**(1), 116–127.

143. L. Quadro, W.S. Blaner, D.J. Salchow, S. Vogel, R. Piantedosi, P. Gouras, S. Freeman, M.P. Cosma, V. Colantuoni and M.E. Gottesman, Impaired retinal function and vitamin A availability in mice lacking retinol-binding protein, *EMBO J.*, 1999, **18**(17), 4633–4644.

144. K. Michaelsson, H. Lithell, B. Vessby and H. Melhus, Serum retinol levels and the risk of fracture, *N. Engl. J. Med.*, 2003, **348**(4), 287–294.

145. D. Feskanich, V. Singh, W.C. Willett and G.A. Colditz, Vitamin A intake and hip fractures among postmenopausal women, *JAMA*, 2002, **287**(1), 47–54.

146. J.H. Promislow, D. Goodman-Gruen, D.J. Slymen and E. Barrett-Connor, Retinol intake and bone mineral density in the elderly: the Rancho Bernardo Study, *J. Bone Miner. Res.*, 2002, **17**(8), 1349–1358.

147. B. Underwood, Vitamin A in animal and human nutrition, In *The Retinoids*, 1st edn, M. Sporn, A. Roberts and D. Goodman (eds.), Academic Press, New York: 1984, pp. 282–392.

148. J.N. Hathcock, Does high intake of vitamin A pose a risk for osteoporotic fracture? *JAMA*, 2002, **287**(11), 1396–1397.

149. D. Feskanich, W.C. Willett and G.A. Colditz, Does high intake of vitamin A pose a risk for osteoporotic fracture? *JAMA*, 2002, **287**, 1396–1397.

CHAPTER 18

# The Influence of Food Groups upon Bone Health

## MARIAN T. HANNAN[1] and KATHERINE L. TUCKER[2]

[1]Hebrew Rehabilitation Center for Aged, Research and Training Institute and Harvard Medical School Division on Aging, Boston, MA, USA 02131-1097 Email: hannan@mail.hrca.harvard.edu
[2]Jean Mayer Human Nutrition Research Center on Aging, Tufts University, Boston, MA, USA 02111 Email: katherine.tucker@tufts.edu

**Key Points**

- Dietary food groups have not been well studied for their impact on bone health, yet food groups represent an important modifiable risk factor.
- Six distinct dietary patterns, as they exist in the population, are reported in this chapter for how they relate to bone density results.
- Greater intakes of fruits and vegetables are associated with less bone loss and higher bone mineral density (BMD).
- Greater intakes of candy are associated with significantly lower BMD (8–16% less than the fruit, vegetable and cereal groups).
- For women, greater intakes of alcohol tend to be linked to higher BMDs than other dietary groups.
- A good quality diet with high intake of fruits, vegetables and breakfast cereals, and limited in less nutrient dense foods, may contribute to better bone density. The use of dietary patterns is helpful in placing observed individual food and nutrient associations with bone into the context of actual dietary choices.

## Introduction

Osteoporosis and related fractures represent major public health problems that will only increase in importance as the population ages and as other chronic disease treatments lead to increased life expectancy. An estimated 26 million women in the US have low bone mass[1] and the lifetime risk of osteoporosis-related fractures in America exceeds 40% in women and 13% in men.[2] Osteoporosis causes significant morbidity, loss of independence and mortality risk, and in the US, it accounts for over \$13 billion in healthcare expenses annually.[1,3,4] Fractures of the hip and spine are the most serious and frequent outcomes of osteoporosis and the most disabling of its consequences. BMD is considered to be one of the best predictors of the risk for osteoporotic fracture. The identification of those factors that contribute to a reduction in bone mass should lead to effective interventions to reduce fracture.

Preventing osteoporotic fractures is a critical health priority. A renewed focus on treatments and prevention for osteoporosis is likely to increase in importance as the population ages.[3,5] Recent findings from longitudinal studies, such as the Framingham Study and the Rotterdam Study, have suggested that bone loss in the elderly continues at a rate comparable to that of younger individuals or may actually increase with age.[6-9] Better understanding of the factors that may contribute to bone loss may help to refine the types of interventions to preserve bone mass for older persons. In particular, developments in nutritional analysis offer an opportunity to examine risk factors for bone loss that have not been well studied. Examination of dietary factors and their effects upon bone loss may aid in defining interventions that preserve bone mass in populations. In this chapter, we examine dietary risk factors in the form of food groups for the impact on bone loss.

As seen in prior research and in other chapters of this book, precise research on single nutrients has yielded important information for bone health. However, food choices, as they exist in the population, most certainly also affect bone status, since the interactions of simultaneous nutrients in the diet may be important for bone health. The impact of food groups upon bone health has not been well studied, and more importantly, these dietary groupings represent risk factors that are *modifiable*. Specifically, we will present information on whether food patterns, as they exist in the population, affect bone status. The basic premise in this chapter is that low bone mass is associated with certain dietary patterns as designated by food intake patterns.

# Background

## Age Related Bone Loss

The patterns of bone loss that lead to osteoporotic fractures have been studied intensively since the advent of technologies such as single and dual photon absorptiometry, quantitative computed tomography, and most recently, dual energy X-ray absorptiometry (DXA). Each of these methods is characterized by high precision and accuracy.[10] Several investigators have demonstrated that the rate of bone loss in old age continues at the same rate,[9,11,12] or may even increase[6,8] relative to bone loss rates seen in younger individuals. Thus, risk factors that are modifiable are even more important as the population ages.

## Dietary Risk Factors for Osteoporosis

The identification of dietary risk factors for osteoporosis is particularly important from the prevention standpoint because they are modifiable. The extensive data available on calcium and vitamin D nutrition as it relates to bone has led to successful interventions to prevent bone loss and fractures[13–18]. Based largely on experimental studies, most agree that calcium and vitamin D are important nutrients to bone health. Indeed, both calcium and vitamin D are widely prescribed to prevent osteoporosis and hip fracture.[15,19,20] In fact, the bulk of attention to dietary risk factors for osteoporosis has focused on these two nutrients. A number of papers have contributed to our understanding of the influence of caffeine and alcohol consumption to bone loss and fracture,[21–25] and osteoporosis.[26–28] Much less is known about the effects of other nutrients on bone, although effects have been evaluated for protein, magnesium, potassium, and vitamin K.[2,13,29–33] Effects have also been hypothesized for saturated fat, phosphorus, vitamin C, sodium, and several trace minerals, including manganese, zinc, copper, and silicon.[13,34–37] Nutritional factors are important to bone health, yet most of these factors are not well studied in human populations. Yano *et al.* noted the need for longitudinal studies to examine bone loss with dietary nutrients,[38] however, very little has been published on patterns of food consumption as they relate to bone.

## Dietary Insufficiency

The relation between dietary intake and bone health may be even further complicated in the elderly. Massey[39] notes that dietary factors in some elders may be further affected by reduced activity in aged kidneys and intestines, influencing absorption and calcium imbalance. In addition, Heaney suggests that total dietary insufficiency, often evidenced as low protein intake, is associated with frailty and fracture in the elderly.[13,40] Additional studies of protein supplementation in elderly women post-hip fracture clearly showed benefits in terms of bone mass and muscle strength,[41,42] implying that dietary insufficiency, particularly in the oldest old, contributes to osteoporosis. In addition, two studies reported that hip fracture patients had diets particularly deficient in overall energy and protein.[43,44] Eaton-Evans also reported that low protein intake probably reflected low intake of other nutrients and as such was associated with low bone mass[45]. There is no doubt, as Massey notes that "our understanding of dietary factors on bone is rudimentary and critical questions remain to be answered".[39]

## Fruit and Vegetable Intakes

In 1999, we investigated associations of both cross-sectional and longitudinal bone loss with dietary potassium, magnesium, and fruits and vegetables upon BMD in elderly men and women.[29] Greater potassium intake was significantly associated with greater BMD for men and for women ($p < 0.05$), as was magnesium intake. Fruit and vegetable intake was associated with BMD for men and for women. In addition, greater intakes of potassium, magnesium and greater fruit and vegetable intake were associated with less bone loss over the longitudinal follow-up. The finding that fruit and vegetable intake is positively associated with BMD is also supported by New *et al.*[46] who found significant associations between past reported fruit intake and BMD among premenopausal women. In 2000, New *et al.* reported further evidence of higher bone density in women with high fruit and vegetable intakes[47]. We published a paper in 2001 discussing an acid–base mechanism to help explain the impact of diet upon bone.[48] The overall results suggest that a good quality diet, with high intake of magnesium, potassium, fruit, and vegetables, with adequate protein, and limited in less nutrient dense foods, may contribute to better bone health in older age. The effects of proteins appear to be complex, with beneficial effects perhaps exceeding negative effects within normal intake ranges. Perhaps, the effect of dietary protein is modified by other components in the protein foods themselves or in the mixed diet. Further studies on food interactions, and on the interactions of other metabolic responses to dietary intake, are needed to understand their cumulative effects.

All these studies suggest that adequate dietary intakes of a wide spectrum of nutrients contribute to bone health. Evidence suggests that under-nutrition, as well as malnutrition, is associated with osteoporosis and fracture. Both dietary excesses and deficient intakes are clearly problems in human populations, however, it is unclear what effect normal variations in dietary food groups will have upon bone health.

## Food Intake Patterns

Few epidemiological studies have examined the relation of food choices to bone health. Studies of single nutrients with an outcome like bone are likely to yield important information for possible preventive interventions; however, this information may be limited by the complex interactions of simultaneous mixtures of nutrients and food components in the diet. For example, as noted in the section above, intake of protein may interact with other nutrients that commonly are consumed with protein, and in addition, the single nutrient (protein) actually may be masking the effect from the other nutrient(s). Interventions that alter single nutrients may have unforeseen effects because the intervention leads to changes in the intake of certain foods containing the single nutrient. Due to the complex relation between bone and diet, it may be misleading to examine nutrients individually. Nutrients are packed together in foods and therefore associations seen with a single nutrient may, in fact, be due to a more complex constellation of other nutrients consumed at the same time. Conversely, adjusting for other nutrient contributors to bone density may make it difficult to see a true association due to the close association with one another in a healthy diet. For these reasons we chose to evaluate whether food choices, as they exist in the population, affect bone status. One approach is to classify people by diet pattern, for example, assigning persons to quintile levels of milk and dairy product intake. Another promising method is to characterize persons by the types of foods they eat most, basing these patterns on cluster analysis of intake, expressed as the percent of energy consumed for a set of food groups. Variations of this method have been used to examine nutritional status as an outcome,[49–52] but it has only recently been used to examine diet and bone.

Our early work with this method in elderly persons has shown it to assign individuals to clear and logical existing food patterns associated with nutritional status.[53] Repeated use of the method across three data sets has also shown it to be reliable in that very similar patterns emerge across data sets, even when dietary data are collected by different methods.[53,54] If a variety of individual nutrients are in fact important, then overall dietary patterns that maximize those nutrients should also be associated with bone mass. This approach allows us to consider overall dietary recommendations for improving bone health.

# Dietary Pattern Studies

Since nutrients occur together in foods and diet patterns, and the overall effects of dietary choices are not well understood, we evaluated the associations between overall dietary patterns and bone density in elderly men and women.[55] The following section describes our experience with the Framingham Osteoporosis Study.

## Framingham Osteoporosis Studies

We have evaluated food groups and bone density data in the Framingham Osteoporosis Studies using the original Framingham Cohort – a population-based random sample of the town of Framingham selected in 1948 and examined biennially.[56] Many important risk factors for osteoporosis have been measured in this group with comprehensive longitudinal follow-up. The Framingham Osteoporosis Study was initiated in 1988 to evaluate bone density and associated risk factors in older men and women.[57] The Boston University Institutional Review Board approved our study, and written informed consent has been obtained for all study subjects. BMD data of multiple skeletal sites was collected during 1988–1989 with repeat measures of BMD in 1992–1993 and 1996–1997. The Framingham Osteoporosis Studies have been the subject of biologic and behavioral risk factor analyses. Over the past several years, we have examined the association between dietary intake components and patterns in relation to BMD and fracture risk in the Framingham Osteoporosis Study, including the effect of dietary patterns with high intakes of fruit and vegetables, potassium and magnesium, and of protein.

## The Original Framingham Cohort

The Framingham Study began in 1948 taking a population-based sample of 5,209 adults, aged 28–62 residing in Framingham, a small town to the west of

Boston, Massachusetts, USA. The Framingham Heart Study is a longitudinal cohort study that originated to examine risk factors for heart disease and has continued for more than 50 years. The age and sex distribution of the Framingham cohort survivors is close to the population distribution of elders for the town of Framingham.[57] Furthermore, the age specific incidences of hip fracture mirror national figures,[58] suggesting that this sample is representative of a national sample. Fewer than 4% of the members are lost to follow-up. The Framingham cohort is examined on a continuous cycle that is repeated every two years. The evaluation for each subject consists of a history, physical examination, questionnaires, and laboratory and cardiovascular tests.

At the baseline osteoporosis examination in 1988–89, we measured spine and femoral BMD by dual and single photon absorptiometry. BMDs of the proximal right femur (femoral neck, trochanter and Ward's area) were measured in g cm$^{-2}$, using a Lunar dual photon absorptiometer (DP3). Bone density at the 33% radial shaft was measured in g cm$^{-2}$ using a Lunar SP2 single photon absorptiometer (Lunar Radiation Corporation, Madison, WI). Coefficients of variation in normals measured twice with repositioning were 2.6% at the femoral neck, 2.8% at the trochanter, 4% at Wards area, and 2% at the radial shaft bone site.[57]

At the follow-up examinations, BMDs of the hip, spine, and forearm scans were done in the cohort for the purpose of evaluating longitudinal change in BMD using dual X-ray absorptiometry (DXA). We have compared the information obtained using both DPA and DXA absorptiometer technologies and, with correct techniques used, found very close values between these two technologies.[59] Thus, our longitudinal measures of change in BMD are based on absolute replication of technique and current state-of-the-art precision. Bone density scans at baseline and follow-up examinations were analyzed using the exact same analysis method, since the validity of longitudinal findings are predicated on absolute replication of analytic technique.

## Dietary Intake Measures

Usual dietary intake was assessed at the 1988–1989 examination using a semi-quantitative 126-item food frequency questionnaire.[60,61] Questionnaires were mailed to the subjects prior to the examination, and they were asked to complete them based on their intake over the previous year. This food frequency questionnaire has been validated for many nutrients, and in several populations, against multiple diet records and blood measures.[60–64] Individuals who reported energy intakes below 600 and above 4000 kilocalories (2.51–16.74 MJ) per day, or with over 12 food items left blank were considered invalid (8%) and excluded from analysis.

## Measurement of Confounders

Many important risk factors for osteoporosis have been collected or measured in this cohort with comprehensive longitudinal follow-up. In addition to diet, factors reported to affect BMD include body weight, height, and body mass index (BMI, weight in kg/height in m$^2$),[65] physical activity,[66] smoking,[67] estrogen use by women,[58,68] and use of calcium and/or vitamin D supplements.[19–20] Because BMI is a measure of relative weight, designed to be independent of height, we often include both BMI and height in our study model equations to capture the total effect of total body mass on BMD.[69]

Smoking status was assessed *via* questionnaire as current cigarette smoker (smoked regularly in the past year), former smoker or never smoked. Physical activity was measured with the Framingham physical activity index, which queries the number of hours spent in heavy, moderate, light, or sedentary activity and hours spent sleeping during a typical day. Each component was then multiplied by a validated weighting factor of associated energy expenditure and summed to derive a physical activity score.[70,71] For women, estrogen use was defined as current use (continuous use for at least one year), *vs.* never or past users, based on evidence that past use does not sustain bone benefits.[72] Alcohol and dietary calcium, vitamin D, magnesium, potassium, and vitamin C intakes were assessed from the food frequency questionnaire, while use of calcium or vitamin D supplements were ascertained from the supplement section of the food frequency questionnaire.

## Analytical Approach

In our analyses we employed a different approach to the definition of dietary influences, in that we did not evaluate a specific nutrient, but rather we examined the role of dietary food patterns on bone. Since we were interested in the effect of food pattern on bone density of elderly persons, we hypothesized that different food intake patterns, as they exist in this population, relate to bone density. As a first step, food items from the 126-item food frequency questionnaire were collapsed into 34 food groups, based on similarity in food and nutrient composition. Percent of daily energy intake contribution from each food group was calculated. This standardization by energy contribution helps to remove dietary variation due to differences in sex, age, body size, and physical activity, and to retain the proportionally based food intake patterns.

These food pattern groups were then entered into cluster analyses to classify individuals into maximally different existing food patterns. This procedure compares Euclidean distances between each subject and each cluster center in an interactive, reiterative process. We evaluated several cluster analyses, specifying the number of clusters to vary between 3 and 10 groups. The final 6-cluster set was selected by comparing between-cluster variance and within-cluster variance ratios, and by examining the content of the clusters for nutritionally meaningful separation.[73] This methodology has been described and examined for reliability.[53,54] The rationale for using the food intake pattern analysis method is to examine total diets as they currently exist in the population to get a sense of how overall food choice can affect bone status. The associations were then estimated with and without control for potential confounding variables.

Each of the femoral and radius measures of bone density was regressed onto the categorical dietary pattern variable separately for men and women. Potential confounders included age, BMI, height, physical activity index, smoking status, total energy intake, use of calcium supplement, use of vitamin D supplement, and for women, current estrogen use. The least squares means for each BMD measure were compared across all pairwise combinations of dietary pattern groups adjusted for confounding effects.

## Dietary Food Groups in the Framingham Osteoporosis Study

Six dietary food groups were formed from the cluster analysis, with relatively greater proportions of intake from: (1) meat, dairy products and bread; (2) meat and baked products; (3) baked products; (4) alcohol; (5) candy; and (6) fruits, vegetables and cereals. Relative to most other groups, the meat, dairy and bread group derived more of their total energy intake from meat, poultry, and fish (18% *vs.* 12–17% in other groups), milk and dairy products (13% *vs.* 8–11%), and bread (11% *vs.* 7–9%), and less from sweet-baked products. The Meat and baked products group also had high meat intake (17% of total energy intake), including the highest intake of processed meats, and had a moderately high intake of sweet baked products (14%). The Baked products group had high intake of sweet-baked products, including cakes, pies, doughnuts, and cookies (29% *vs.* 5–14%), and low intake of fruit, vegetables, bread, and cereal. The Alcohol group had high intake of alcohol (19% *vs.* 2–3%). The Candy group had high candy intake (20% *vs.* 3–5%). Finally, the Fruits, vegetables and cereals group had the highest intake of fruit and vegetables (30% *vs.* 11–16%) and of breakfast cereal (7% *vs.* 3–5%) and the lowest intake of red meat, processed meats, candy, and soft drinks.

The dietary pattern groups are shown in Table 18.1, along with several descriptive characteristics of the Framingham Study subjects in this analysis. Body mass index, physical activity score and use of either calcium or vitamin D supplements did not differ across diet pattern groups. Age ranged from an average of 74 years in the Alcohol group to 76 years in the Fruits, vegetables and cereals group and differed significantly across the groups ($p < 0.05$). Although men comprised 38% of the total sample, they made up a larger proportion of the Alcohol group (53%), and a smaller proportion of the Fruit, vegetables and cereal group (30%). The Alcohol group also contained the most current and former smokers (28 and 56%, respectively). The fewest smokers were in the Fruits, vegetables and cereals group (5% current and 39% former smokers).

These dietary patterns, defined by food intake, also differed considerably in nutrient intake as seen in Table 18.2. The Candy group had the highest energy intake; while the Fruits, vegetables and cereals and the Meat, dairy and bread groups had the lowest energy intake. The highest protein intakes were in the Meat, dairy and bread group. The lowest protein intakes were in the Candy and Baked products groups. After adjusting for total energy intake, the Fruit, vegetables and cereal, and Meat, dairy and bread groups had the highest

**Table 18.1.** *Descriptive characteristics of Framingham Osteoporosis Study participants across dietary pattern groups*

| | Meat, dairy and bread (n = 313) | Meat and baked products (n = 260) | Baked products (n = 69) | Alcohol (n = 81) | Candy (n = 75) | Fruits, vegetables and cereals (n = 109) | P-value |
|---|---|---|---|---|---|---|---|
| Age (years) | 74.8±4.7 | 75.7±4.9 | 75.5±5.2 | 73.6±4.1 | 75.1±4.8 | 76.2±5.2 | 0.003 |
| Men (%) | 33.23 | 41.15 | 43.48 | 53.09 | 37.3 | 30.28 | 0.008 |
| Body Mass Index (kg m$^{-2}$) | 26.7±4.7 | 26.8±4.8 | 26.7±5.4 | 25.7±3.4 | 26.3±4.5 | 26.7±4.4 | 0.50 |
| Physical activity score | 33.3±5.0 | 33.9±6.3 | 33.8±6.1 | 34.0±5.6 | 33.4±5.3 | 32.4±5.3 | 0.27 |
| Cigarette Smoking (%) | | | | | | | |
| Past smoker | 50.48 | 45.0 | 39.1 | 55.56 | 45.3 | 39.4 | 0.0001 |
| Current Smoker | 7.35 | 10.0 | 14.49 | 28.4 | 14.67 | 4.59 | |
| Vitamin D supplements (%) | 31.3 | 23.08 | 21.7 | 27.16 | 24.0 | 29.36 | 0.25 |
| Calcium supplements (%) | 21.09 | 16.15 | 14.49 | 12.35 | 13.33 | 19.27 | 0.29 |

Adapted from reference 55, with permission.
P-value indicates the measure of statistical variability across the dietary pattern groups.

**Table 18.2.** Total energy intake (mean±SD) and energy adjusted nutrient intakes (mean±SE) across the six dietary patterns among Framingham Osteoporosis Study participants[1]

| | Meat, dairy and bread (n = 313) | Meat and baked products (n = 260) | Baked products (n = 69) | Alcohol (n = 81) | Candy (n = 75) | Fruits, vegetables and cereals (n = 109) | P-value |
|---|---|---|---|---|---|---|---|
| Energy (MJ) | 6.8±2.3 | 7.3±2.3 | 8.1±2.7 | 7.4±2.6 | 8.4±3.0[a] | (6.8±2.4) | 0.0001 |
| Protein (g) | 76.6±0.7[a] | 67.1±0.8 | (57.6±1.6) | 61.0±1.4 | (58.0±1.5) | 64.2±1.2 | 0.0001 |
| Calcium (mg) | 933±20[a] | 731±21 | (633±42) | 696±38 | 705±40 | 873±33 | 0.0001 |
| Vitamin D (IU) | 384±14[a] | 296±16 | (246±31) | 297±28 | (261±30) | 375±24[a] | 0.0001 |
| Phosphorus (mg) | 1237±13[a] | 1041±14 | (898±27) | 966±25 | 942±26 | 1139±21 | 0.0001 |
| Magnesium (mg) | 319±3.7 | 276±4.1 | (229±8.0) | 268±7.3 | (250±7.7) | 341±6.3[a] | 0.0001 |
| Potassium (mg) | 3210±31 | 2807±34 | (2381±67) | 2598±61 | 2494±64 | 3493±53[a] | 0.0001 |
| Vitamin K (µg) | 165±6.0 | 156±6.5 | (115±13) | 138±12 | 122±12 | 198±10[a] | 0.0001 |

Adapted from reference 55, with permission.

P-value indicates the measure of statistical variability across the dietary pattern groups.

[a]Intakes are greater than those in most other groups; the numbers in parentheses indicate lower intakes than other groups.

micronutrient intakes, and the Candy and Baked products groups tended to have the lowest micronutrient intakes.

## Associations between Dietary Pattern and BMD

Figure 18.1 presents the chart comparing bone density at the femoral neck across the six food cluster groups for men and for women. The pattern of results was generally similar for all the bone density sites we evaluated at the hip and radius, although statistically significant differences differed across bone sites. For men, these inter-pattern differences were statistically significant at the femoral neck site ($p = 0.001$) and approached significance at the trochanter and radius sites ($p < 0.06$). For women, the overall comparison across groups was significant only at the radial shaft ($p = 0.004$).

For men, the Fruits, vegetables and cereals dietary pattern group had the highest mean BMD (Figure 18.1). For all of the hip sites, this group's BMD was significantly greater than that of any of the other groups ($p < 0.05$). At the radial shaft, this group showed significantly greater BMD than the Candy group (data not shown). The Candy group had the lowest BMD at most bone

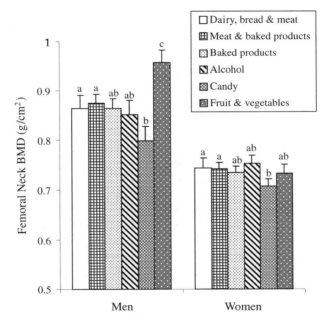

**Figure 18.1** *Adjusted mean ($\pm SE$) femoral neck BMD by dietary pattern, after adjusting for BMI, height, age, energy intake, physical activity score, smoking, vitamin D supplement use, calcium supplement use, season and (in women), current estrogen use. Overall significance of pattern at the femoral neck in men (p = 0.001) and in women, (p = 0.3). Bars with different letters are statistically significantly different (p < 0.05)*[55]
Reprinted from reference 55, with permission

sites. At the femoral neck and Ward's area, the Candy group's mean BMD was significantly lower than that of the Meat, dairy and bread and Meat and baked products groups. At the radial shaft, the Candy group had lower BMD than all groups, except the Meat and baked products group. The mean BMD among men in the Candy group was significantly lower than that of the Fruits, vegetables & cereals group at all bone sites, with differences of 16.5% at the femoral neck, 8.0% at the radius, 10.9% at the trochanter, and 23.8% at Ward's area (data not shown).

For women, (Figure 18.1), the Candy group also had the lowest average BMD at most sites–significantly lower than all other groups at the radius, and than at least one other diet group at each of the hip sites. The advantage of the Fruits, vegetables and cereals group was not as clear for the women as for the men, but their BMD tended to be higher than that of several other groups. At the radial shaft, the Candy group had 11.5% significantly lower BMD compared to the Fruits, vegetables and cereals group. In contrast to men, the Alcohol group tended to have higher BMDs than other dietary groups among the women. Compared with the Alcohol group, women in the Candy group had significantly lower BMD at the radius (by 12.8%), and lower BMDs in both the Candy and Baked products groups (by 9.4 and 8.7%, respectively) at the trochanter site.

Thus, from the six distinct dietary patterns in the population-based cohort of Framingham elders, we found that these dietary patterns were associated with BMD. Men with a diet high in fruits, vegetables and cereals had significantly greater BMD than other diet pattern groups. Although women in the Fruits, vegetables and cereals group tended to have higher BMD than other groups, the Alcohol group also had high BMD, compared to the Fruits, vegetables and cereals group at most sites. This apparent protective effect of alcohol was not seen among men, and in women it may be due to the effects of alcohol on adrenal androgens or estrogen levels.[74] Thus, the association may have been seen in women and not in men due to the estrogen boost from alcohol in postmenopausal women, or because of the lower levels of alcohol consumption among women. The Meat, dairy and bread, and Meat and baked products groups tended to have intermediate BMD. Both men and women consuming the most candy had significantly lower BMDs than most other groups. While any single comparison must be viewed with caution, there is a consistency of results across bone sites and also for men and women.

Findings from two population groups have previously reported that greater fruit and vegetable intakes were associated with greater BMD,[29,46,47] and these dietary pattern results are consistent with those results. Not surprisingly, compared with other diet pattern groups, our Fruits, vegetables and cereals group had the highest intakes of magnesium, potassium, vitamin C and vitamin K. Intakes of the nutrients most associated with bone, calcium and vitamin D were greatest in the dairy group, but were also high in the Fruits, vegetables and cereals group.

Protein has been associated with calcium loss,[75] however, epidemiologic studies of protein and bone have shown conflicting results. In the Framingham population, total protein and animal protein intake were protective against bone loss,[31] and others have also shown better bone status with higher protein intake.[32,76] Conversely, another study reported greater bone loss at the femoral neck and more hip fracture with higher animal/vegetable protein ratio.[33] As Heaney notes,[13] the actual effect of protein intake on bone is complicated and dependent on other components in the diet. He suggests that the calciuric effect of protein may be offset by increased intestinal calcium absorption, unless calcium intakes are quite low. Other nutrients in the diet are also likely to affect the association between protein intake and calcium loss. This is one of the reasons that the use of dietary pattern analysis is particularly useful. In this analysis, those with the highest protein intake were in the Meat, dairy and bread, and Meat and baked products groups, both with average BMDs that were higher than the Candy group, but lower than the Fruits, vegetables and cereals group. These groups also had higher intakes of calcium, vitamin D, phosphorous, magnesium, potassium and vitamin C than did the Alcohol, Candy or Baked products groups. Consistent with these overall nutrient profiles, the Candy group, followed by the Baked products group, had the lowest BMD, suggesting that the displacement of nutrient dense foods in the diet may explain why high intake of these foods is detrimental to bone status. These results suggest that a good quality diet with high intake of fruits, vegetables and breakfast cereals, and limited in less nutrient dense foods, may contribute to better bone density in older people. The use of dietary patterns is helpful in clarifying associations of actual dietary choices on bone and in placing observed individual food and nutrient associations into context.

In addition to cluster analysis, common methods of defining dietary patterns include principal component analysis and scores related to dietary recommendations. In these methods, dietary patterns have been associated with different nutrient profiles, socioeconomic factors and health outcomes.[53,77,78] As with many methods, cluster analysis is data dependent and different dietary patterns will result in population groups with differing dietary intakes. Despite this, investigations in US and European populations have generated similar dietary pattern groups that usually include groups high in bread or sweets, in alcohol, in fruits and vegetables, and in meat.[53,78–81] The examination of dietary patterns allows a more inclusive and comprehensive evaluation and thus adds insight to the more common analysis of individual nutrients. These results may both confirm the findings of individual nutrients and foods when embedded in actual intake patterns and tell us that the constellation of food choices is associated with health outcomes. Due to the complexity of simultaneous effects of multiple nutrients on bone, the examination of total dietary pattern is helpful.

# Conclusions

The rate of osteoporosis rises dramatically with age, and has enormous personal and economic consequences, including increased mortality. As populations age, osteoporosis and related fractures will only escalate in importance as major public health problems. Dietary factors represent an important understudied area in osteoporosis research. Research in this area could identify important, potentially modifiable factors that could have enormous ramifications for the risk of osteoporosis and diets of millions of men and women. Studies suggest a role for food groups in bone health. Further, food choices can be evaluated, as they exist in the population, for their effect upon bone. Studies of single nutrients with bone outcome are likely to yield important information for possible preventive interventions; however, this information may be limited by the complex simultaneous interactions of nutrients and food components in the diet. Since interventions altering single nutrients may have unforeseen effects due to changing the intake of certain foods containing the single nutrient, it is essential to examine dietary intake patterns. Studies of dietary patterns of intake on bone density will advance three major components of osteoporosis research. First, certain intake patterns may be identified as risk factors for osteoporosis. Second, feasible changes in diet pattern may be suggested that are protective for bone health. Finally, these studies may add further context to understanding the findings of individual nutrients to bone status. Certain food groupings could be more widely recommended as means of preserving bone mass, with major public health implications.

# References

1. L.J. Melton, How many women have osteoporosis now? *J. Bone Miner. Res.*, 1995, **10**, 175–177.
2. J.A. Kanis and the WHO Study Group, Assessment of fracture risk and its application to screening for postmenopausal osteoporosis: synopsis of a WHO report, *Osteoporosis Int.*, 1994, **4**, 368–381.
3. H.K. Genant, C. Cooper, G. Poor, I. Reid, G. Ehrlich *et al.*, Interim report and recommendations of the World Health Organization task-force for osteoporosis, *Osteoporosis Int.*, 1999, **10**, 259–264.
4. N.F. Ray, J.K. Chan, M. Thamer and L.J. Melton, Medical expenditures for the treatment of osteoporotic fractures in the United States in 1995: report from the National Osteoporosis Foundation, *J. Bone Miner. Res.*, 1997, **12**, 24–35.
5. Osteoporosis prevention, diagnosis, and therapy, *J Am. Med. Assoc.*, 2001, **285**, 785–795.
6. K.E. Ensrud, L. Palermo, D.M. Black, J. Cauley, M. Jergas, E.S. Orwoll, M.C. Nevitt, K.M. Fox and S.R. Cummings, Hip and calcaneal bone loss increases with advancing age. Longitudinal results from the Study of Osteoporotic Fractures, *J. Bone Miner. Res.*, 1995, **10**, 1778–1787.
7. H. Burger, C. de Laet, P. van Daele, A. Weel, J. Witteman, A. Hofman and H. Pols, Risk factors of increased bone loss in an elderly population: the Rotterdam Study, *Am. J. Epidemiol.*, 1998, **147**, 871–879.
8. G. Jones, T. Nguyen, P. Sambrook, P.J. Kelly and J.A. Eisman, Progressive loss of bone in the femoral neck in elderly people: longitudinal findings from the Dubbo osteoporosis epidemiology study, *Br. Med. J.*, 1994, **309**, 691–695.

9. M.T. Hannan, D.T. Felson, B. Dawson-Hughes, K.L. Tucker, L.A. Cupples, P. Wilson and D.P. Kiel, Risk factors for longitudinal bone loss in elderly men and women: the Framingham Osteoporosis Study, *J. Bone Miner. Res.*, 2000, **15**, 710–720.

10. C.C. Johnston, L.J. Melton, R. Lindsay and D.M. Eddy, Clinical indications for bone mass measurements, *J. Bone Miner. Res.*, 1989, **4**(Suppl. 2),1–28.

11. K.E. Broe, M.T. Hannan, C.E. Stewart and D.P. Kiel, Elderly long-term care residents continue to lose bone, but not uniformly, *J. Bone Miner. Res.*, 2002, **17**(Suppl. 1), S146.

12. S.L. Greenspan, L.A. Maitland, E.R. Myers, M.B. Krasnow and T.H. Kido, Femoral bone loss progresses with age: a longitudinal study in women over age 65, *J. Bone Miner. Res.*, 1994, **9**, 1959–1965.

13. R.P. Heaney, Nutritional factors in osteoporosis, *Ann. Rev. Nutr.*, 1993, **13**, 287–316.

14. I.R. Reid, R.W. Ames, M.C. Evans, G.D. Gamble and S.J. Sharpe, Effect of calcium supplementation on bone loss in postmenopausal women, *N. Engl. J. Med.*, 1993, **328**, 460–464.

15. B. Dawson-Hughes, G.E. Dallal, E.A. Krall, L. Sadowski, N. Sahyoun and S. Tannenbaum, A controlled trial of the effect of calcium supplementation on bone density in postmenopausal women. *N. Engl. J. Med.*, 1990, **323**, 878–883.

16. B. Dawson-Hughes, S.S. Harris, E.A. Krall, G.E. Dallal, G. Falconer and C.L. Green, Rates of bone loss in postmenopausal women randomly assigned to one of two dosages of vitamin D, *Am. J. Clin. Nutr.*, 1995, **61**, 1140–1145.

17. M.C. Chapuy, M.E. Arlot, P.D. Delmas, P.J. Meunier, Effect of calcium and cholecalciferol treatment for three years on hip fractures in elderly women, *Br. Med. J.*, 1994, **308**, 1081–1082.

18. M.C. Chapuy, M.E. Arlot, F, Duboeuf, J. Brun, B. Crouzet, S. Arnaud, P.D. Delma and P.J. Meunier, Vitamin D3 and calcium to prevent hip fractures in elderly women, *N. Engl. J. Med.*, 1992, **327**, 1637–1642.

19. B. Dawson-Hughes, Calcium supplementation and bone loss: a review of clinical trials, *Am. J. Clin. Nutr.*, 1991, **54**(Suppl. 1), 274S–280S.

20. B. Dawson-Hughes, G.E. Dallal, E.A. Krall, S. Harris, L.J. Sokoll and G. Falconer, Effect of vitamin D supplementation on wintertime and overall bone loss in health postmeno-pausal women, *Ann. Internal Med.*, 1991, **115**, 505–512.

21. D.P. Kiel, D.T. Felson, M.T. Hannan, J.J. Anderson, P.W. Wilson, Caffeine and the risk of hip fracture: the Framingham Study, *Am. J. Epidemiol.*, 1990, **132**, 675–684.

22. E. Barrett-Connor, J.C. Chang and S.L. Edelstein, Coffeeassociated osteoporosis offset by daily milk consumption: the Rancho Bernardo Study, *J. Am. Med. Assoc.*, 1994, **271**, 280–283.

23. L.K. Massey, Caffeine and bone: directions for research, *J. Bone Miner. Res.*, 1991, **6**, 1149–1151.

24. M.J. Barger-Lux and R.P. Heaney, Caffeine and the calcium economy revisited, *Osteoporosis Int.*, 1995, **5**, 97–102.

25. S.S. Harris and B. Dawson-Hughes, Caffeine and bone loss in healthy postmenopausal women, *Am. J. Clin. Nutr.*, 1994, **60**, 573–578.

26. D.T. Felson, Y. Zhang, M.T. Hannan and D.P. Kiel, Alcohol intake and bone mineral density in elderly men and women: the Framingham study, *Am. J. Epidemiol.*, 1995, **142**, 485–492.

27. T.L. Holbrook, E. Barrett-Connor, A prospective study of alcohol consumption and bone mineral density, *Br. Med. J.*, 1993, **306**, 1506–1509.

28. D. Scharpira, Alcohol abuse and osteoporosis, *Semin. in Arthritis Rheum.*, 1990, **19**, 371–376.

29. K.L. Tucker, M. Hannan, H. Chen, L.A. Cupples, P. Wilson and D.P. Kiel, Potassium, magnesium, and fruit and vegetable intakes are associated with greater bone mineral density in elderly men and women, *Am. J. Clin. Nutr.*, 1999, **69**, 727–736.

30. S.L. Booth, K.L. Tucker, H. Chen, M.T. Hannan, P. Wilson, L.A. Cupples, B. Dawson-Hughes and D.P. Kiel, Vitamin K intake, bone mineral density and hip fractures in elderly women and men, *Am. J. Clin. Nutr.*, 2000, **71**, 1201–1208.

31. M.T. Hannan, K.L. Tucker, B. Dawson-Hughes, L.A. Cupples, D.T. Felson and D.P. Kiel, Effect of dietary protein on bone loss in elderly men and women: the Framingham Osteoporosis Study, *J. Bone Miner. Res.*, 2000, **15**, 2504–2512.

32. R.G. Munger, J.R. Cerhan and B.C.-H. Chiu, Prospective study of dietary protein intake and risk of hip fracture in postmenopausal women, *Am. J. Clin. Nutr.*, 1999, **69**, 147–152.

33. D.E. Sellmeyer, K.L. Stone, A. Sebastian and S.R. Cummings, A high ratio of dietary animal to vegetable protein increases the rate of bone loss and the risk of fracture in postmenopausal women, *Am. J. Clin. Nutr.*, 2001, **73**, 118–122.

34. G.R. Wohl, L. Loehrke, B.A. Watkins and R.F. Zernicke, Effects of high-fat diet on mature bone mineral content, structure, and mechanical properties, *Calcif. Tissue Int.*, 1998, **63**, 74–79.

35. S.P. Robins, S.A. New, Markers of bone turnover in relation to bone health, *Proc. Nutr. Soc.*, 1997, **56**, 903–914.

36. J. Eisinger, D. Clairet, Effects of silicon, fluoride, etidronate and magnesium on bone mineral density: a retrospective study, *Magnesium Res.*, 1993, **6**, 247–249.

37. S.J. Whiting, J.L. Boyle, A. Thompson, R.L. Mirwald and R.A. Faulkner, Dietary protein, phosphorus and potassium are beneficial to bone mineral density in adult men consuming adequate dietary calcium, *J. Am. College Nutr.*, 2002, **21**, 402–409.

38. K. Yano, L.K. Heilbrun, R.D. Wasnich, J.H. Hankin and J.M. Vogel, The relationship between diet and bone mineral content of multiple skeletal sites in elderly Japanese-American men and women living in Hawaii, *Am. J. Clin. Nutr.*, 1985, **42**, 877–888.

39. L. Massey, Dietary factors influencing calcium and bone metabolism: introduction, *J. Nutr.*, 1993, **123**, 1609–1610.

40. G. Geinoz, C.H. Rapin, R. Rizzoli, R. Kraemer, B. Buchs, D. Slosman, J.P. Michel, and J.P. Bonjour, Relationship between bone mineral density and dietary intakes in the elderly, *Osteoporosis Int.*, 1993, **3**, 242–248.

41. M. Delmi, C.H. Rapin, J.M. Bengoa, P.D. Delmas, H. Vassey and J.P. Bonjour, Dietary supplementation in elderly patients with fractured neck of the femur, *Lancet*, 1990, **335**, 1013–1016.

42. L. Tkatch, C.H. Rapin, R. Rizzoli *et al.*, Benefits of oral protein supplement in elderly patients with hip fracture. *J. Am. College Nutr.*, 1992, **11**, 519–525.

43. J.E. Jensen, T.G. Jensen, T.K. Smith, D.A. Johnston and S.J. Dudrick, Nutrition in orthopedic surgery, *J. Bone Joint Surg. [Am.]*, 1982, **64**, 1263–1272.

44. M.D. Bastow, J. Rawlings and S.P. Allison, Undernutrition, hypothermia and injury in elderly women with fractured femur: an injury response to altered metabolism? *Lancet*, 1983, **1**, 143–146.

45. J. Eaton-Evans, Osteoporosis and the role of diet, *Br. J. Biomed. Sci.*, 1994, **51**, 358–370.

46. S.A. New, C. Bolton-Smith, D.A. Grubb and D.M. Reid, Nutritional influences on bone mineral density: a cross-sectional study in premenopausal women, *Am. J. Clin. Nutr.*, 1997, **65**, 1831–1839.

47. S.A. New, S.P. Robins, M.K. Campbell, J.C. Martin, M.J. Garton, C. Bolton-Smith, D.A. Grubb, S.J. Lee and D.M. Reid, Dietary influences on bone mass and bone metabolism: further evidence of a positive link between fruit and vegetable consumption and bone health? *Am. J. Clin. Nutr.*, 2000, **71**, 142–151.
48. K.L. Tucker, M.T. Hannan and D.P. Kiel, The acid-base hypothesis: diet and bone in the Framingham Osteoporosis Study, *Eur. J. Nutr.*, 2001, **40**,(5), 231–237.
49. H.S. Schwerin, J.L. Stanton, J.L. Smith, A.M. Riley Jr. and B.E. Brett, Food, eating habits and health: a further examination of the relationship between food eating patterns and nutritional health, *Am. J. Clin. Nutr.*, 1982, **35**, 1219–1325.
50. H. Boeing, U. Klein, A. Hendricks *et al.*, Strategies for analyzing nutritional data for epidemiological purposes–food patterns by means of cluster–analytical procedures, *Nutr. Reports Int.*, 1989, **49**, 189–198.
51. J.S. Akin, D.K. Guilkey, B.M. Popkin and M.T. Fanelli, Cluster analysis of food consumption patterns of older Americans, *J. Am. Dietetic Assoc.*, 1986, **86**, 616–624.
52. B.E. Millen, P.A. Quatromoni, D.R. Gagnon, L.A. Cupples, M.M. Franz and R.B. D'Agostino, Dietary patterns of men and women suggest targets for health promotion: the Framingham Nutrition Studies, *Am. J. Health Promotion*, 1996, **11**, 42–52.
53. K. Tucker, G. Dallal and D. Rush, Dietary patterns of elderly Bostonarea residents defined by cluster analysis, *J. Am. Dietetic Assoc.*, 1992, **92**,(12), 1487–1491.
54. K. Newby, D. Muller, J. Hallfrisch, N. Qiao, R. Andres and K.L. Tucker, Dietary patterns and changes in body mass index and waist circumference in adults, *Am. J. Clin. Nutr.*, 2003, **77**, 1417–1425.
55. K.L. Tucker, H. Chen, M.T. Hannan, L.A. Cupples, P. Wilson, D. Felson and D.P. Kiel, Bone mineral density and dietary patterns in older adults: the Framingham Osteoporosis Study, *Am. J. Clin. Nutr.*, 2002, **76**, 245–252.
56. T. Dawber, G. Meadors and F.J. Moore, Epidemiological approaches to heart disease: the Framingham Study, *Am. J. Public Health*, 1951, **41**, 279–286.
57. M.T. Hannan, D.T. Felson and J.J. Anderson, Bone mineral density in elderly men and women: results from the Framingham Osteoporosis Study, *J. Bone Miner. Res.*, 1992, 7, 547–553.
58. D.P. Kiel, D.T. Felson, J.J. Anderson and M.A. Moskowitz, Hip fracture and the use of estrogens in postmenopausal women: the Framingham Study, *N. Engl. J. Med.*, 1987, **317**, 1169–1174.
59. D.P. Kiel, C.A. Mercier, B. Dawson-Hughes, C. Cali and M.T. Hannan, Effects of scan analysis technique and analytic software on the comparison of dual-x-ray absorptiometry with dual photon absorptiometry of the hip in the elderly, *J. Bone Miner. Res.*,1995, **10**, 1130–1136.
60. W.C. Willett, L. Sampson, M.J. Stampfer, *et al.*, Reproducibility and validity of a semi-quantitative food frequency questionnaire, *Am. J. Epidemiol.*, 1985, **122**, 51–65.
61. W. Willett and M.J. Stampfer, Total energy intake: implications for epidemiologic analyses, *Am. J. Epidemiol.*, 1986, **124**, 17–27.
62. E.B. Rimm, E.L. Giovanncci, M.J. Stampfer, G.A. Colditz, L.B. Litin and W.C. Willett, Reproducibility and validity of an expanded self-administered semiquantitative food frequency questionnaire among male health professionals, *Am. J. Epidemiol.*, 1992, **135**, 1114–11126.
63. A. Ascherio, M. Stampfer, G. Coldtiz, E. Rim, L. Litin and W. Willett, Correlation of vitamin A and E intake with plasma concentrations of carotenoids and tocopherols among American men and women, *J. Nutr.*, 1992, **122**, 1792–1801.

64. P.F. Jacques, S.I. Sulsky, J.A. Sadowski *et al.*, Comparison of micronutrient intake measure by a dietary questionnaire and biochemical indicators of micronutrient status, *Am. J. Clin. Nutr.*, 1993, **57**, 182–189.

65. D.T. Felson, Y. Zhang, M.T. Hannan and J.J. Anderson, Effects of weight and body mass index on bone mineral density in men and women: the Framingham Study, *J. Bone Miner. Res.*, 1993, **8**, 567–573.

66. G.A. Greendale, E. Barrett-Connor, S. Edelstein, S. Ingles and R. Haile, Lifetime leisure exercise and osteoporosis: the Rancho Bernardo Study, *Am. J. Epidemiol.*, 1995, **141**, 951–959.

67. D.P. Kiel, Y. Zhang, M.T. Hannan, J.J. Anderson, J.A. Baron and D.T. Felson, The effect of smoking at different life stages on bone mineral density in elderly men and women, *Osteoporosis Int.*, 1996, **6**, 240–248.

68. D.T. Felson, Y. Zhang, M.T. Hannan and J.J. Anderson, The effect of postmenopausal and later life estrogens on bone mineral density in elderly women, *N. Engl. J. Med.*, 1993, **329**, 1141–1146.

69. K.B. Michels, S. Greenland and B.A. Rosner, Does body mass index adequately capture the relation of body composition and body size to health outcomes? *Am. J. Epidemiol.*, 1998, **147**, 167–172.

70. W. Kannel and P. Sorlie, Some health benefit of physical activity: the Framingham Study, *Arch. Intern. Med.*, 1979, **139**, 857–861.

71. M.T. Hannan, D.T. Felson, J.J. Anderson, A. Naimark, Habitual physical activity and knee steoarthritis in the Framingham Study, *J. Rheumatol.*, 1993, **20**,(4), 704–709.

72. J.A. Cauley, D.G. Seeley, K. Ensrud, B. Ettinger, D. Black and S.R. Cummings, Estrogen replacement therapy and fractures in older women. Study of Osteoporotic Fractures Research Group, *Ann. Intern. Med.*, 1995, **122**, 9–16.

73. J. Hartigan, In *Clustering Algorithms*, John Wiley and Sons, New York, NY, 1975.

74. R.A. Wild, J.R. Buchanan, C. Myers and L.M. Demers, Declining adrenal androgens: an association with bone loss in aging women, *Proc. Soc. Exp. Biol. Med.*, 1987, **186**, 355–360.

75. R.P. Heaney and R.R. Recker, Effects of nitrogen, phosphorus, and caffeine on calcium balance in women. *J. Lab. Clin. Med.*, 1982, **99**, 46–55.

76. J.L. Freudenheim, N.E. Johnson and E.L. Smith, Relationships between usual nutrient intake and bonemineral content of women 35–65 years of age: longitudinal and crosssectional analysis, *Am. J. Clin. Nutr.*, 1986, **44**, 863–876.

77. E. Randall, J. Marshall, S. Graham and J. Brasure, High-risk health behaviors associated with various dietary patterns, *Nutr. Cancer*, 1991, **16**, 135–151.

78. M.A. Bernstein, K.L. Tucker, N.D. Ryan, E.F. O'Neill, K.M. Clements, M.E. Nelson, W.J. Evans and M.A. Fiatarone Singh, Higher dietary variety is associated with better nutritional status in frail elderly people, *J. Am. Dietetic Assoc.*, 2002, **102**(8), 1096–104.

79. A.K. Wirfalt and R.W. Jeffery, Using cluster analysis to examine dietary patterns: nutrient intakes, gender, and weight status differ across food pattern clusters, *J. Am. Dietetic Assoc.*, 1997, **97**, 272–279.

80. P.P. Huijbregts, E.J. Feskens and D. Kromhout, Dietary patterns and cardiovascular risk factors in elderly men: the Zutphen Elderly Study, *Int. J. Epidemiol.*, 1995, **24**, 313–320.

81. SENECA Investigators, Food Patterns of Elderly Europeans, *Eur. J. Clin. Nutr.*, 1996, **50**, S86–S100.

CHAPTER 19

# *Soy Isoflavones and Bone Health*

DAVID J. CAI[1], LISA A. SPENCE[2] and
CONNIE M. WEAVER[2]

[1]William Wrigley Jr. Company, Chicago, IL 60609
[2]Department of Foods and Nutrition, Purdue University, West Lafayette,
IN 47906, Email: weavercm@cfs.purdue.edu

**Key Points**

- Concern over HRT has prompted the search for alternative therapies. Phytoestrogens, specifically soy isoflavones, have received attention as a possible alternative to provide benefits of oestrogen without side effects.
- Soy isoflavones have a similar chemical structure to estradiol and possess weak oestrogenic activity.
- The evidence of the efficacy and safety of soy isoflavones in preventing bone loss is equivocal.
- Epidemiological data of an association between the consumption of high levels of soy isoflavones from soy products and lower prevalence of hip fractures and greater spine BMD are inconclusive.
- Various studies using the ovariectomized rat model have shown comparable bone-sparing effects of 17β-estradiol and soy protein isolate, genistein or daidzein, or their respective 6'-O-succinylated products obtained after soybean fermentation. However, the maturity of the animal influences the effects. Studies in young rats have demonstrated that soy isoflavones reduce bone loss; whereas, studies in adult rats have not shown favorable effects from soy isoflavones or soy protein on bone loss.
- While promising effects of soy protein isolates containing isoflavones and isolated isoflavones have been demonstrated on BMD in peri- and post-menopausal women, more randomized, controlled trials of longer duration and sufficient sample size are needed to determine the role of soy protein and soy isoflavones on BMD and fracture risk.
- In order to ascertain the mode of action of isoflavones on the skeleton, the effects of soy isoflavones on calcium metabolism need to be studied.

Oestrogen receptor-mediated modulation of intestinal calcium absorption or calcium reabsorption by the kidney, where oestrogen receptors also are located, might be a target in postmenopausal women for estradiol or SERMs, such as phytoestrogens or isoflavones.
- The role of soy isoflavones in bone health is far from determined and much remains to be learned about the bioactivity of compounds, dose effectiveness, soy protein *vs.* isoflavone contribution, and mode of action.

## Introduction

Oestrogen therapy (ET) has been prescribed to postmenopausal women for decades to treat menopausal symptoms. For women with an intact uterus, oestrogen plus progestin have been used as hormone replacement therapy (HRT) in recent years to lower the risk of uterine cancer associated with unopposed oestrogen. However, compliance has been poor because of undesirable side effects including periodic bleeding, increased risk of thrombosis and gallstones, and the fear of breast cancer. Approximately 38% of women in the US are prescribed HRT (58.7% of those who have undergone hysterectomy and 19.6% of those who have not had a hysterectomy).[1] However, most reports indicate that typically ~25% of these women stop HRT within 6 months and few remain on HRT for more than 1 or 2 years.[2] At the time of this writing, concern over HRT is at an all time high with the recent report of a large study of HRT in over 16 000 postmenopausal women which showed a decreased risk of hip fracture and colon cancer, but an increased risk of heart attacks, stroke, breast cancer, and blood clots.[3] The increased risk out of 10 000 was eight for breast cancer, seven for heart attacks, eight for strokes, and 18 for blood clots, which caused the Data and Safety Monitoring Board to stop the HRT arm of the Women's Health Initiative (WHI) 3 years early. The National Institutes of Health (NIH) released a statement that postmenopausal women should not take HRT to prevent cardiovascular disease or osteoporosis. In fact, NIH is no longer using the term HRT, but has removed the term 'replacement' and shortened it to HT because 'replacement' implies risk-free. Furthermore, the American Heart Association is recommending that women not be prescribed oestrogen solely to prevent stroke and heart disease.

This series of events leave many clinicians in doubt when prescribing HT and many women afraid to take it. Thus, the interest in alternatives to HT is expected to grow appreciably. Newer drugs, the bisphosphonates in particular, can be used instead of HT to prevent bone loss and in preventing vertebral and nonvertebral fractures.[4] Women who turn to phytoestrogens are hoping that they will provide the benefits of oestrogen without the side effects. But are they effective? This chapter reviews the evidence on the efficacy and safety of soy isoflavones in preventing bone loss.

# Abundance and Structure

Plants contain a wide variety of phytochemicals. Many of these phytochemicals have biological activities that may be effective in the treatment or prevention of chronic diseases such as cardiovascular disease, cancer, osteoporosis, and cognitive function degeneration. Phytoestrogens have chemical structures similar to 17β-estradiol with phenolic rings and hydroxyl groups. These compounds exert oestrogenic effects on the central nervous system, induce estrus, and stimulate growth of the genital tract in female animals.[5] Phytoestrogens are divided into three groups: isoflavones, coumestans, and ligans. Isoflavones and coumestans have been identified as the most common oestrogenic compounds in plants.[6] Soybeans and foods derived from soybeans contain large amounts of these isoflavones.[7] Isoflavones are found in legumes predominately as glycoside conjugates (genistin, daidzin, and glycitin), their storage forms (Figure 19.1). The concentration of isoflavones in soybeans and soy foods varies depending on the variety of the soybean, growth location, harvesting season, moisture conditions, bean maturity, and processing.[8] Traditional soy foods, such as tofu, soybeverage, tempeh, and miso are rich sources of isoflavones, providing about 30–40 mg (as aglycones) per serving size. Soy sauce and soybean oil do not contain isoflavones.[7]

# Physiological Metabolism

Although isoflavones are naturally present in foods mainly as glycones, they are biotransformed by intestinal microflora (Figure 19.2). The glycoside conjugated isoflavones, genistin, daidzin, and glycitin, are hydrolyzed by glucosidases in the gut to their respective aglycone forms: genistein, daidzein, and glycitein. The aglycones undergo further changes within the gut lumen by intestinal microflora. Lactase has recently been shown to be responsible for hydrolysis of isoflavone

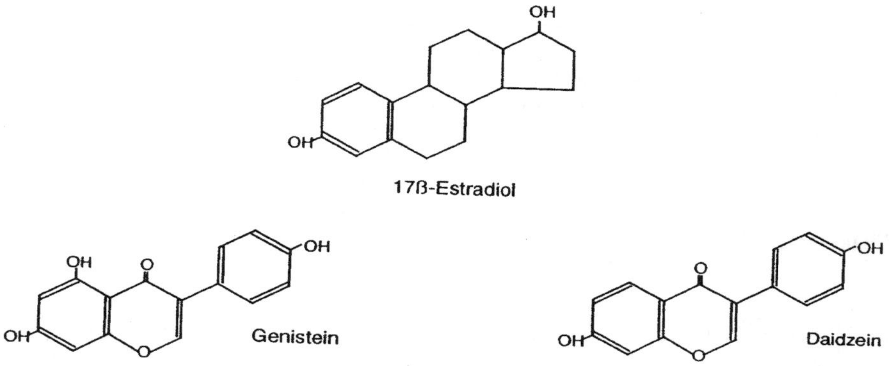

**Figure 19.1** *Chemical structures of estradiol vs. soy phytoestrogens*

*Diet*

Soy Isoflavones - Malonyl and acetyl glycosides, β-glycosides

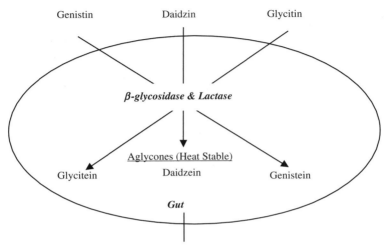

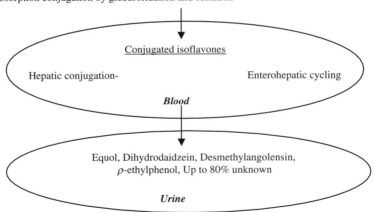

**Figure 19.2** *Metabolism of soy isoflavones*

β-glycosides,[9] suggesting that those with intestinal lactase deficiency may be less able to metabolize isoflavones. Isoflavones are mainly absorbed through the intestinal wall as aglycones and their derivatives. Isoflavones follow the same pathway as other lipid-soluble molecules.[10] Blood concentrations of isoflavones can reach $10^{-9}$ g $L^{-1}$ in individuals consuming soy foods with 50 mg day$^{-1}$

intakes of isoflavones, compared to normal oestrogen concentrations at $10^{-12}$ g $L^{-1}$.[11] Japanese consumption of isoflavones is estimated to be between 20 and 50 mg day$^{-1}$ from a daily intake of 30 g of soy.[12] Plasma concentration of aglycone isoflavones peaks at 4–6.6 h after ingestion compared to about 9 h for β-glycosides, as the hydrolytic cleavage takes several hours.[13] Individual variation in blood isoflavone concentration can vary 10-fold from the same administered dose.[14] The volume of distribution is great, suggesting extensive tissue distribution after absorption.[13] Daidzein distributes more widely than genistein; thus, plasma concentrations are lower. The half-life for elimination was 6.78 h for genistein and 9.34 h for daidzein.[13] Not only do the glycosides peak later, but they clear more quickly than their aglycone forms by about 2 h.

Daidzein is eliminated mainly *via* urine as equol. All rats typically produce equol, but only one-third of humans are equol producers. There is much current interest in the differential responsiveness to soy isoflavone consumption by equol *vs.* nonequol producers, possibly related to relatively higher degrees of oestrogenicity of equol or to equol being a marker of bioavailability. There are many metabolites in addition to equol that are not fully characterized, which makes it difficult to quantify urinary excretion for recovery analysis.[15] Equol and 2-(4-hydroxyphenyl) propionic acid as metabolites are more oestrogenic than their precursors[16] and may be the bioactive forms that actually reach the tissues.

Isoflavones possess weak oestrogenic activity, having between 1/1000 and 1/100 000 the activity of 17β-estradiol in terms of affinity to oestrogen receptors (ER).[17] Genistein has a higher affinity for ERβ than ERα. ERβ is present in bone, brain, bladder, reproductive tissues, and vascular epithelia.[18] Soy isoflavones have a 10-fold lower affinity for sex hormone binding globulin than do oestrogens,[19] and thus, are more likely to be free to bind with receptors. The isoflavones are believed to have anti-oestrogenic effects on reproductive tissue, which may reduce the risk of breast and uterine cancer with isoflavone use compared to ET.[20] However, one group of researchers has shown that isoflavones stimulate oestrogen receptor-positive human breast cancer (MCF-7) cells in mice causing an increase in breast tumors.[21] Isoflavones have also been shown to stimulate breast proliferation in women with benign or malignant breast disease.[22] The oestrogenic effect of isoflavones depends on dose and the profile of oestrogen-responsive genes. Even in the same tissue, physiological oestrogens, phytoestrogens, and synthetic oestrogens differentially regulate gene expression.[23] It will require microassay technologies to unravel the interactions of isoflavones with the many oestrogen-responsive genes that are responsible for various phenotypes. Isoflavones directly or indirectly have been shown to lower total cholesterol,[24] possess anticancer activity,[17] relieve menopause symptoms,[25] and inhibit bone resorption.[26]

# Phytoestrogens and Osteoporosis

Vulnerability to osteoporosis relates to both genetic and lifestyle factors. Aging, diet, exercise, and hormonal status not only affect the rate of bone loss but also peak bone mass. Hormonal status is influenced by (1) natural or surgical menopause resulting in ovarian hormone deficiency, (2) oophorectomy or ovarian dysfunction due to chemotherapy, (3) use of HT, (4) excessive exercise leading to amenorrhea, as well as other underlying causes of ovarian failure.[27] Oestrogen deficiency leads to a rapid and progressive loss of bone, followed by sustained bone loss that typically accompanies aging. Hence, women, especially postmenopausal women over the age of 50, constitute the population most vulnerable to osteoporosis.

ET and HT have been used in the prevention and treatment of osteoporosis. Although ET and HT are thought to primarily reduce bone resorption, HT has been shown to increase bone mineral density (BMD)[28] and combined oestrogen and androgen therapy has been shown to increase biochemical markers of bone formation.[29] Thus, ET has been used effectively to stabilize skeletal BMD in postmenopausal women,[30] with epidemiological evidence suggesting that ET prevents fractures.[31,32] Long-term administration of ET or HT is required to produce the maximal reduction in fracture risk, because bone loss ensues when treatment is stopped.[32,33] Epidemiological data also suggest that ET may reduce the risk of ischemic heart disease, may improve cognitive function,[34] and relieves symptoms of menopause, such as hot flushes and vaginal dryness.[35] However, evidence from the HERS trial (The Heart and Oestrogen/Progestin Replacement Study), the first prospective report on a large sample of postmenopausal women, cast doubt on the effectiveness of HT in reducing coronary heart disease (CHD) risk, at least in women with documented CHD.[36] This 4-year, randomized, double-blind, placebo-controlled trial of daily use of conjugated equine oestrogens (0.625 mg) plus medroxyprogesterone acetate (progestin) (2.5 mg) investigated the combined rate of nonfatal myocardial infarction and CHD death, among 2800 postmenopausal women with coronary disease. The primary outcome measurements, occurrence of nonfatal myocardial infarction or CHD death, along with the secondary outcome measurements of other cardiovascular outcomes, were not reduced as a result of HT intervention. Indeed, despite improvements in LDL-C and HDL-C, cardiovascular events were greater in years 1 and 2, but then declined in years 3 through 5, compared with the placebo group. Additionally, in the HERS II follow-up trial, after 6.8 years, HT did not reduce the risk of cardiovascular events in women with CHD.[37] Furthermore, there were no significant differences in several other end points, since the power was limited, including fracture, cancer, or total mortality, whereas there was an increase in deep vein thrombosis, pulmonary emboli, and incidence of gallbladder disease. Subsequently, the WHI was the first large randomized primary prevention trial, with mostly healthy non-hysterectomized women (only 7.7% of whom had prior cardiovascular disease) aged 50–79 years. Women in the HT group had slight but significant increases in risk of invasive breast cancer (hazard ratio [HR] = 1.26), CHD (HR = 1.29), stroke (HR = 1.41), and pulmonary embolism (HR = 2.13).[3] These additional risks were considered to outweigh any benefits in reducing hip fractures (HR = 0.66) or colorectal cancer (HR = 0.63). The trial was terminated after an average of 5.2 years, more than three years early.

Due to perceived undesirable side effects and potential health risks, compliance with HT regimens is low.[31] With the HERS I and II trials and the recent WHI release, it may be expected that compliance to HT will further decrease and adherence to alternatives to HT increase as women search for relief of menopausal symptoms. However, evidence varies on the efficacy and side effects of these alternatives. Among the alternatives are bisphosphonates, selective oestrogen receptor modulators (SERMs), calcitonin, and phytoestrogens or isoflavones. Of the alternatives to HT, the least is known about the efficacy of the phytoestrogens or isoflavones.

## Epidemiological Evidence

The interest in soy isoflavones as an alternative to HT for maintaining skeletal health began with epidemiological studies of elderly women in Asian countries who consume high levels of isoflavones from soy products, but had lower prevalence of hip fractures.[38] However, this type of association is inconclusive, as Asian women are thought to have a different hip geometry than Caucasian women. In a subsample from the Study of Osteoporotic Fractures in four cities in the US where Asians were recruited from Hawaii, shorter hip axis lengths of Asians were observed and contributed to the lower prevalence of hip fracture compared to Caucasians, despite similar or lower bone density values.[39] Several epidemiological studies have been conducted in Hong Kong and Japan to directly examine the linkage between dietary isoflavone intake and BMD in Asian women. A greater lumbar spine BMD was observed in postmenopausal Chinese women ($n = 650$, age $= 19$–$86$)[40] and postmenopausal Japanese women ($n = 478$)[41] who consumed the highest levels of dietary isoflavones compared to those who consumed the least. Another study from Hong Kong showed that high soy isoflavone intake contributed to greater spinal BMD in premenopausal women ($n = 132$, age $= 30$–$40$) who were followed for three years.[42] The results demonstrated that greater BMD was positively linked to dietary isoflavone intake in Asian women. It is not clear if the skeletal protective effects of soy and/or their isoflavones are causal, and if so, whether length of exposure is important. For the majority of Asians, length of exposure is usually over the entire life span, since soy products are consumed over a lifetime. However, this consumption pattern is changing with increasing adoption of a Western Lifestyle.

## Animal Models

Various studies using the ovariectomized rat model have shown comparable bone-sparing effects of 17β-estradiol and soy protein isolate,[43] genistein or daidzein,[41–46] or their respective 6'-*O*-succinylated products obtained after soybean fermentation.[47] In a study by Arjmandi *et al.*[43] ovariectomized rats that consumed isoflavone-containing soy protein isolate had significantly greater femur BMD than rats on a casein-based diet. However, there was no effect of isoflavone-rich soy on reducing bone resorption markers, indicating that soy isoflavones did not suppress bone turnover in a pattern comparable to oestrogen. Similar results were found in a study by Fanti *et al.*[45] Purified genistein at 5 μg g$^{-1}$ body weight in young ovariectomized rats prevented trabecular bone loss, but not bone resorption, compared to ovariectomized control rats. Anderson *et al.*[48] demonstrated a biphasic response to genistein in ameliorating bone loss in lactating, ovariectomized rats, suggesting oestrogenic agonist action only at low doses. However, all these studies were conducted in young rats, with some of these animals provided calcium-deficient diets.[44,47] Germane to this issue, oestrogen has been shown to be more effective in postmenopausal women when dietary calcium is adequate.[49] Because human postmenopausal bone loss starts after skeletal maturity, an appropriate rat model should be based on skeletally mature rats. Ovariectomized rats younger than 3 months are not suitable models for studying postmenopausal bone loss because ovariectomy-induced bone remodeling occurs simultaneously with bone modeling and skeletal growth, confounding the interpretation of treatment effects.[50] Hence, an appropriate animal model is essential for studying hormone deficiency-induced osteopenia.

In contrast, studies using adult or aged rats have not shown favorable effects from soy isoflavones or soy protein on bone loss.[51-54] An investigation of the effects of the phytoestrogens, coumestrol and zearalanol, and a mixture of isoflavones on oestrogen-dependent bone loss in six-month old ovariectomized rats showed different effects from injected isolated phytoestrogens *vs.* oral mixtures of isoflavones.[52] Coumestrol injected intramuscularly significantly reduced urinary calcium excretion and bone resorption markers after one week. Animals treated with coumestrol and zearalanol had significantly greater bone density than controls. However, animals treated with oral isoflavones showed no changes in BMD, urinary calcium, or bone resorption markers.[52] In another study, bone-sparing effects of environmental oestrogens, including coumesterol and genistein, were compared to 17β-estradiol in virgin 6-month-old ovariectomized Sprague-Dawley rats.[51] Doses were tested at 0.1, 1.0, 10, and 30 mg kg$^{-1}$ of body weight. Bone effects were determined by measuring tibial BMD at the proximal metaphysis (excised), a region rich in cancellous bone. Of these environmental oestrogens, coumesterol, but not genistein prevented ovariectomy-induced bone loss, at the doses tested.[51] When purified daidzein or genistein were given daily to 12-month-old rats at 10 μg g$^{-1}$ body weight for 3 months, daidzein was more effective than genistein in preventing ovariectomy-induced bone loss from cancellous bone and reducing ovariectomy-induced bone turnover.[53] Genistein failed to show any effect on BMD in trabecular-rich bone. When isoflavone (IF) concentrates (1 g contained 159 mg genistin and 156 mg daidzin) were given to 7 month-old rats for 12 weeks at 0, 20, 40, or 80 mg IF kg$^{-1}$ body weight per day, none of the doses were effective in preventing trabecular bone loss as indicated by BMD or bone volume, trabecular number, thickness, and separation.[54] None of the IF doses had an effect on mechanical testing or BMD of cortical bone.

## Clinical Studies

Clinical trials on the effects of soy isoflavones on bone are limited and have shown mixed results. Studies have used both soy protein isolate (SPI) with isoflavones and isolated isoflavones. In a study by Potter *et al.*[55] designed to examine the lipid-related effects of soy protein, 66 postmenopausal women were randomized to one of three treatments: (1) casein + nonfat dry milk protein, (2) SPI (40 g) with 56 mg day$^{-1}$ of isoflavones, or (3) SPI (40 g) with 90 mg day$^{-1}$ of isoflavones. These hypercholesterolemic women were *heterogeneous* with respect to time since menopause and age (49–83 years). After 6 months of treatment, women in the high isoflavone group experienced an increase in bone mineral content (BMC) and BMD at the lumbar spine ($\sim$2%; $p < 0.05$), whereas those in the casein + milk-based protein had slight decreases. However, women in the high isoflavone group started the study with lower BMD and BMC than the other two groups, and baseline values were not taken into account in assessing the effects. The effect of treatment on bone is typically greater in those with lower bone mass[56] and thus baseline BMD should be considered. Alekel *et al.*[57] carried out a randomized (double-blind) trial with 69 perimenopausal women treated for 24 weeks, with the dose expressed as aglycone units: isoflavone-rich soy (SPI$^+$, 80.4 mg day$^{-1}$; $n = 24$), isoflavone-poor soy (SPI$^-$, 4.4 mg day$^{-1}$; $n = 24$), or whey (control; $n = 21$) protein. While the subjects consuming the whey protein ($-1.3$%, $p = 0.004$) lost BMD compared to the baseline, subjects receiving either the low ($-0.7$%, $p = 0.1$) or high ($-0.2$%, $p = 0.7$) isoflavone soy protein had no significant decline in spine BMD. The response of spine BMD to SPI in which isoflavones had been removed by alcohol extraction was intermediate, but not statistically significantly different from the other groups. There was no treatment benefit to the femur in the Alekel study. Both the Potter and Alekel studies reported that 80–90 mg isoflavones per day had a positive effect on spine BMD of 1–2% relative to a milk-based protein control group.

In a randomized double-blind placebo-controlled 12 month trial, Morabito *et al.*[58] tested the effects of HRT and genistein (54 mg day$^{-1}$) in 90 healthy, postmenopausal women. Bone resorption decreased evident by a significant decrease in urinary pyridinium cross-links in the women treated with genistein. The women receiving HRT experienced similar but not significant decreases in bone resorption while the placebo group did not experience any modifications in bone resorption. Bone alkaline phosphatase and osteocalcin increased in the women treated with genistein indicative of increased bone formation. In contrast, women receiving HRT experienced decreases in bone formation while bone formation in the placebo group remained unchanged. Significant increases in BMD of the femur and lumbar spine occurred in the genistein and HRT treated women. This trial provides the strongest clinical evidence to date of a positive effect of a soy isoflavone on bone.

A few short-term studies have evaluated the effects of soy isoflavones on biochemical markers of bone turnover. In a non-placebo controlled study,[59]

42 postmenopausal women consumed whole soy foods containing 60 mg isoflavones per day for 12 consecutive weeks. A bone resorption marker in urine was negatively correlated with urinary excretion of isoflavones. Bone formation markers in plasma were positively correlated with plasma levels of isoflavones. However, serum bone-specific alkaline phosphatase did not change, whereas osteocalcin increased ($p<0.03$) and N-telopeptides decreased ($p<0.02$) from baseline to week 12. Nonetheless, the lack of placebo control in this study makes interpretation very difficult, given the biologic variability of bone turnover markers. Wangen *et al.*[60] conducted a randomized, crossover 3-month SPI (63 g day$^{-1}$) study to examine the dose-response effect of isoflavones on bone markers in 14 premenopausal and 17 postmenopausal women. They found that markers of bone turnover in both pre- and postmenopausal women were mildly affected by approximately 65 and 130 mg day$^{-1}$ soy isoflavones consumed for 3 months. There were no effects of treatment on osteocalcin in the premenopausal women, but during the early follicular phase, deoxypyridinoline crosslinks increased in response to both isoflavone diets; during the periovulatory phase, serum insulin-like growth factor-I (IGF-I) increased in the low (64 mg day$^{-1}$) compared with high (128 mg day$^{-1}$) isoflavone diets. In the postmenopausal women, the low (65 mg day$^{-1}$) and high (132 mg day$^{-1}$) isoflavone diets decreased bone-specific alkaline phosphatase. The high isoflavone group experienced a trend toward a decrease in osteocalcin and in IGF-I. Compared with baseline, all three diets increased bone-specific alkaline phosphatase, osteocalcin, and IGF-I, although the increases in the latter two markers were not statistically significant at the highest isoflavone dose. The authors suggest that although isoflavones modestly affected markers of bone turnover, the changes were small and not likely to be clinically relevant. Hsu *et al.*[61] found similar results in 37 postmenopausal women who were given undefined isoflavone supplements (150 mg day$^{-1}$) twice daily for 6 months. No significant changes were found in calcaneus (heel) BMD after the 6 months of isoflavone supplementation. This study is difficult to interpret because there was no control group. Alternatively, because the calcaneus is weight-bearing and has greater trabecular content[62] than vertebral bone, it may respond differently to isoflavone treatment than the lumbar spine.

In summary, published clinical studies with isoflavone-containing soy protein or phytoestrogens have been of very short duration with only one 12-month trial. Although promising effects of soy protein isolates containing isoflavones were demonstrated on spine BMD in peri- and postmenopausal women, these 6 month studies are too short to adequately evaluate the impact of an intervention on BMD, while there is only one 12-month trial providing evidence of a positive effect of genistein on bone. Moreover, there were inconsistent changes in biochemical markers of bone turnover in some of the studies. Longer term randomized controlled clinical trials with sufficient sample sizes, powered to determine the effect of soy isoflavones on BMD in early postmenopausal women are currently underway.

Along with the investigations of the effect of phytoestrogens or isoflavones on bone, studies have been conducted using ipriflavone, a synthetic isoflavone derivative. Ipriflavone is similar in chemical structure to isoflavones. Ipriflavone is maximally effective only when it is metabolized; one of its metabolites is daidzein.[63] Ipriflavone has been found to reduce bone loss and increase bone mass in ovariectomized rats.[63] Studies of ipriflavone in human subjects have shown that it prevents both axial and peripheral skeletal bone loss in postmenopausal women with low bone mass and that it is well tolerated.[64.] In contrast, one large landmark study raises doubt about the effectiveness of ipriflavone on bone mass.[65] Ipriflavone was tested in a double-blind, placebo-controlled, 4-year study carried out across Europe in four centres. Postmenopausal women ($n = 475$) were randomized to oral ipriflavone (600 mg day$^{-1}$) or placebo. Neither bone loss nor biochemical markers of bone were affected by ipriflavone, whereas lymphocytopenia was documented in some women.

## Effect on Calcium Economy

None of the studies in animals or humans have assessed the effect of soy isoflavones on calcium metabolism. This information is essential in ascertaining the mode of action of isoflavones on the skeleton. Calcium is the most abundant mineral in the body and bone is its major storage site. The skeleton serves as a calcium reservoir for maintaining calcium homeostasis. An inadequate calcium intake leads to calcium deficiency, eventually manifesting itself in bone loss. Any changes in bone mass are reflected in changes in calcium balance.

In postmenopausal women, ovarian hormone deficiency, independent of vitamin D status, is associated with decreased intestinal absorption of calcium[66,67] and increased calcium excretion by the kidney.[68] Impaired calcium absorption and impaired reabsorption by the kidney may exacerbate negative calcium balance, contributing to postmenopausal bone loss. Evidence in rats demonstrates that oestrogen increases intestinal absorption by direct regulation through oestrogen receptors in the intestine without increasing plasma 1,25 dihydroxyvitamin D.[69] Oestrogen receptor-mediated modulation of intestinal calcium absorption or calcium reabsorption by the kidney, where oestrogen receptors also are located, might be a target in postmenopausal women for estradiol or SERMs, such as phytoestrogens or isoflavones.

In addition to the possible role of isoflavones in calcium metabolism, many soy foods are a good source of calcium and soy protein is less calciuric than animal protein. Although soybeans contain both oxalate and phytate, which are inhibitors of calcium absorption, calcium absorption from soy beans has been shown to be equivalent to that from milk.[70] An exception is calcium-fortified soy beverage from which calcium absorption was reported to be 75% of that from cows milk.[71] Urinary calcium loss plays a significant role in maintaining calcium homeostasis. One factor that increases urinary calcium loss is protein intake. For every gram of protein consumed, urinary calcium increases by about 1.75 mg.[72,73] The sulfur-containing amino acids appear to be responsible for this effect by increasing the acid load due to the production of hydrogen,[67,72,74] which in turn increases calcium loss. Calcium is used to buffer hydrogen ions, as both are excreted in the urine. In comparison to animal protein, soy protein contains less sulfur-containing amino acids. One study found that subjects consuming animal protein-based diets excreted 150 mg of urinary calcium per day, whereas those consuming a soy protein-based diet only excreted 103 mg of urinary calcium per day.[75] However, unpublished data from our laboratory suggests that the urinary calcium-conserving effect of soy protein does not translate into an overall advantage in calcium balance.

# Safety

The efficacy of soy isoflavones to reduce bone loss has not been resolved, nor has the upper level for safety been established. The early concern about infertility observed in sheep consuming a subterranean clover species has not been demonstrated in humans with typical dietary intakes of soy protein. In a cohort of infants, fed soy or milk protein-based formulas, followed into adulthood to 30–40 years of age, anthropometric measurements were similar, but women who had been fed soy formula reported slightly longer duration of menstrual bleeding, with no difference in severity of menstrual flow.[76] For more than 30 additional outcomes of interest, no statistically significant differences were observed between groups in either women or men. The implications of these findings for chronic disease outcomes later in life are unknown.

The evidence of increased lymphocytopenia in some women in the ipriflavone (600 mg day$^{-1}$) trial mentioned earlier[65] is another potential concern for isoflavones. Iprilfavone is given in much larger doses than have been studied for phytoestrogens. Acute doses of genistein (1, 2, 4, 8, or 16 mg kg$^{-1}$ body weight) administered to 30 healthy men 40–69 years of age were higher than previously reported doses. At the highest dose, 24 h after administration, 1 case of grade 2 (intermediate) leukopenia was reported,[77] but this side effect has yet to be associated with chronic or low dose exposure.

Of all the concerns, risk of stimulating oestrogen-sensitive cancers due to soy isoflavone consumption is the greatest. Asian women who consume soy isoflavones lifelong have a lower incidence of breast cancer, but this protection is lost within a generation after migrating to the US.[78] However, postmenopausal women who initiate consumption of soy isoflavones to prevent bone loss may not have similar protection as those consuming isoflavones early in life. The worse case scenario would be that consumption of isoflavones by postmenopausal women might increase their risk of breast cancer, cardiovascular disease, or other events, as was observed for those on HT in the WHI trial.[3] However, we have no evidence to date to suggest that the intake of isoflavone-rich soy foods confers an increase in risk for any chronic disease. On the contrary, we have good evidence to suggest that soy foods may favorably modulate lipid profiles and may decrease cancer risk. In reality, we are not likely to determine elevated disease risk associated with soy isoflavone intake based on relatively small clinical trials, given that 16 000 subjects were needed to determine the slight, but significantly greater risk of some diseases associated with HT use. Decisions on the most appropriate therapy should be made by the individual in consultation with her primary care physician, after considering her own risk profile based on family history and health records. Germane to this discussion, one recent review concluded that, based on current knowledge, oral doses of isoflavones below $2 \text{ mg kg}^{-1}$ body weight per day should be considered safe for most populations.[79]

# Conclusions

Soybeans and soy foods have attracted much attention because they are unique in not only providing an abundant source of dietary isoflavones, but also other important nutrients, such as protein and minerals. More evidence is needed to confirm the beneficial attributes of isoflavones that could further highlight the value of soy-derived foods. The current understanding of the role of soy isoflavones in bone health is still in its infancy. There is much to learn about which compounds are bioactive, what doses are effective, the relative contribution of the soy protein *vs.* isoflavones, and the nature of action on bone. Further studies must be conducted to corroborate or refute the skeletal effects of isoflavones and to determine how they spare bone tissue in the face of oestrogen deficiency. Long-term randomized controlled trials will ultimately be required to determine the role of the soy protein and soy isoflavones on BMD and fracture risk.

# Acknowledgement

Funded by NIH grant P50 AT00477. The authors greatefully ackowledge the help of Lee Alekel and Joanna Cadogan in carefully reading this manuscript.

# References

1. N. Keating, P. Clear, A. Aossi, A. Zaslavsky and J. Ayanlan, Use of hormone replacement therapy by postmenopausal women in the United States, *Ann. Intern. Med.*, 1999, **130**, 545–553.
2. D.W. Sturdee, The importance of patient education in improving compliance, *Climacteric*, 2000, **3**, 9–13.
3. Writing Group for the Women's Health Initiative Investigators. Risk and benefits of estrogen plus progestin in healthy postmenopausal women. Principal results from the Women's Health Initiative randomized controlled trial. *JAMA*, 2002, **288**, 321–333.
4. I. Fogelman, C. Ribot, R. Smith, D. Ethgen, E. Sod and J.-Y. Reginster, Risedronate reverses bone loss in postmenopausal women with low bone mass: results from a multinational, double-blind, placebo-controlled trial. *J. Clin. Endocrinol. Metab.*, 2000, **85**, 1895–1900.

5. S. Lieberman, Are the differences between estradiol and other estrogens, naturally occurring or synthetic, merely semantical? *J. Clin. Endocrinol. Metab.*, 1996, **81**, 850.

6. M.S. Kurzer and X. Xu, Dietray phytoestrogens, *Ann. Rev. Nutr.*, 1997, **17**, 353–381.

7. L. Coward, N.C. Barnes, E.D.R. Setchell and S. Barnes, Genistein, daidzein and their β-glucoside conjugates: antitumor isoflavones in soybean foods from American diets and Asian diets, *J. Agri. Food Chem.*, 1993, **41**, 1961–1967.

8. P.A. Murphy, Phytoestrogen content of processed soybean foods, *Food Technol.*, 1982, **34**, 60–64.

9. A.J. Day, F.J. Canada, J.C. Diaz, P.A. Kroon, R. Mclauchlan, C.B. Faulds, G.W. Plumb, M.R.A. Morgan and G. Williamson, Dietary flavonoid and isoflavone glycosides are hydrolyzed by the lactase site of lactase phlorizin hydrolase. *FEBS Lett.*, 2000, **468**, 166–170.

10. T. Isumi, M.K. Piskula, S. Osawa, A. Obata, K. Tobe, M. Saito, S. Kataoka, Y. Kubota and M. Kikuchi, Soy isoflavone aglycones are absorbed faster and in high amounts than their glucosides in humans, *J. Nutr.*, 2000, **130**, 1695–1699.

11. X. Xu, K.S. Harris, H.-J. Wang, P.A. Murphy and S. Hendrich, Bioavailability of soybean isoflavones depends upon gut microflora in women, *J. Nutr.*,1995, **125**, 2307.

12. C. Nagata, N. Takatsuka, Y. Kunsu and H. Shimizu, Decreased serum total cholesterol is associated with high intake of soy products in Japanese men and women, *J. Nutr.*, 1998, **128**, 209–218.

13. K.D.R. Setchell, N.M. Brown, P. Desai, L. Zimmer-Nechemias, B.E. Wolfe, W.T. Brashear, A.S. Kirschner, A. Cassidy and J.E. Heubi, Bioavailability of pure isoflavones in healthy humans and analysis of consumed isoflavones supplements, *J. Nutr.*, 2001, **131**, 1362S–1375S.

14. D. Urban, W. Irwin, M. Kirk, M.A. Markiewicz, R. Myers, M. Smith, H. Weiss, W.E. Grizzle and S. Barnes, The effect of isolated soy protein in plasma biomarkers in elderly men with elevated serum prostate specific antigen, *J. Urol*, 2001, **165**, 294–300.

15. S. Barnes, L. Coward, M. Kirk and J. Sfakianos, HPLC-mass spectrometry analysis of isoflavones, *Exp. Biol. Med.*, 1998, **217**, 254–262.

16. N.G. Coldham, L.C. Howells, A. Santi, C. Montesissa, C. Langlais, L.J. King, D.D. Macpherson and M.J. Sauer, Biotransformation of genistein in the rat: elucidation of metabolite structure by product ion mass fragmentology, *J. Steroid Biochem. Mol. Biol.*, 1999, **70**, 169–184.

17. M.J. Messina, V. Persky, K.D.R. Setchell and S. Barnes, Soy intake and cancer risk: a review of the in vitro and in vivo data, *Nutr. Cancer*, 1994, **21**, 113.

18. G.G. Kuiper, E. Enmarch, M. Pelto-Huikko, S. Nielsson and J.A. Gustafson, Cloning of the novel receptor expressed in rat prostate and ovary, *Proc. Natl Acad. Sci.*, 1996, **93**, 5925–5930.

19. S.C. Nagel, S. Vorn and W.V. Welshons, The effective free fraction of estradiol and xenoestrogens in human serum measured by whole cell uptake assays: Physiology of delivery modifies estrogenic activity, *Proc. Soc. Exp. Biol. Med.*, 1998, **2176**, 300–309.

20. D.C. Knight and J.A. Eden, A review of the clinical effects of phytoestrogens, *Obstet. Gynecol.*, 1996, **87**, 897–904.

21. C.-Y. Hsieh, R.C. Santell, S.Z. Haslam and W.G. Helferich, Estrogenic effects of genistein on the growth of estrogen receptor-positive human breast cancer (MCF-7) cells in vitro and in vivo, *Cancer Res.*, 1998, **58**, 3833–3838.

22. D.F. McMichael-Phillips, C. Harding, M. Morton, S.A. Roberts, A. Howell, S. Potter and N.J. Bundred, Effects of soy-protein supplementation on epithelial proliferation in the historically normal human breast, *Am. J. Clin. Nutr.*, 1998, **68**, 1431S–1435S.

23. P. Diel, K. Smolnikar, T. Schulz, L. Laudenbach, U. Leschowski, H. Michna and G. Vollmer, Phytoestrogens and carcinogenesis – differential effects of genestein in experimental models of normal and malignant rat endometriosis. *Human Reprod.*, 2001, **16**, 977–1006.

24. M.S. Anthony, T.B. Clarkson, C.L. Hughes, T.M. Morgan and G.L. Burke, Soybean isoflavones improve cardiovascular risk factors without affecting the reproductive system of peripubertal rhesus monkeys, *J. Nutr.*, 1996, **126**, 43–50.

25. D.H. Upmalis, R. Lobo, L. Bradley, M. Warren, F.L. Cone and C.A. Lamia, Vasomotor symptom relief by soy isoflavone extract tablets in postmenopausal women. A multicenter, double-blind, randomized, placebo-controlled study, *Menopause*, 2000, **7**, 236–242.

26. J.J.B. Anderson, M. Anthony, M. Messina and S.C. Garner, Effects of phyto-estrogens on tissues. *Nutr. Res. Rev.*, 1999, **12**, 75–116.

27. L.V. Avioli, The future of ipriflavone in the management of osteoporosis syndromes, *Calcif. Tissue Int.*, 1997, **61**, S33–S35.

28. R.R. Recker, K.M. Davies, R.M. David and R.P. Heaney, The effect of low-dose continuous estrogen and progesterone therapy with calcium and vitamin D on bone in elderly women, *Ann. Intern. Med.*, 1999, **130**, 897–904.

29. L.G. Raisz, B. Wiita, A. Artis, A. Bowen, S. Schwartz, M. Trahiotis, K. Shoukri and J. Smith, Comparison of the effects of estrogen alone and estrogen plus androgen on biochemical markers of bone formation and resorption in postmenopausal women, *J. Clin. Endocrinol. Metab.*, 1996, **81**, 37–43.

30. R. Lindsay, Sex steroids in the pathogenesis and prevention of osteoporosis, In *Osteoporosis: Etiology, Diagnosis and Management*, B.L. Riggs (ed.), Raven Press, New York, 1988, pp. 333–358.

31. B. Ettinger, Overview of estrogen replacement therapy: a historical perspective, *PSEBM*, 1998, **217**, 2–5.

32. J.A. Cauley, D.G. Seeley, K. Ensurd, B. Ettinger, D. Black and S.R. Cummings, Estrogen replacement therapy and fractures in older women. Study of Osteoporotic Fractures Research Group, *Ann. Intern. Med.*, 1995, **122**, 9–16.

33. R. Lindsay, D.M. Hart, A. MacLean, A.C. Clark, A. Kraszewski and J. Garwood, Bone response to termination of estrogen treatment, *Lancet.*, 1978, **1**, 1325–1327.

34. B.S. Hulka, Hormone-replacement therapy and the risk of breast cancer, *Cancer*, 1990, **40**, 289–296.

35. T.L. Bush, H.B. Wells, M.K. James *et al.*, Effects of hormone therapy on bone mineral density: results from the postmenopausal estrogen/progestin interventions (PEPI) trial, *JAMA*, 1996, **276**, 1389–1396.

36. S. Hulley, D. Grady, T. Bush, C. Furberg, D. Herrington, B. Riggs and E. Vittinghoff, Randomized trial of estrogen plus progestin for secondary prevention of coronary heart disease in postmenopausal, *JAMA*, 1998, **280**, 605–613.

37. D. Grady, D. Herrington, V. Bittner *et al.*, Cardiovascular disease outcomes during 6.8 years of hormone therapy: heart and estrogen/progestin replacement study follow-up (HERS II), *JAMA*, 2002, **288**, 49–57.

38. J.M. Lacey and J.J.B. Anderson, Older women in Japan and the United States: physical and nutritional comparisons, In *Bone Morphometry: Proceedings of the*

*Fifth International Congress*, H.E. Takahashi (ed.), Nishimura, Japan 1991, pp. 562–565.

39. S.R. Cummings, J.A. Cauley, L. Palermo, P.D. Ross, R.D. Wasnick, D. Black and K.G. Faulkner, Racial differences in hip axis lengths might explain racial differences in rates of hip fracture, *Osteoporosis. Int.*, 1994, **4**, 226–229.

40. J. Mei, S.S. Yeung and A.W. Kung, High dietary phytoestrogen intake is associated with higher bone mineral density in postmenopausal but not premenopausal women, *J. Clin. Endocrinol. Metab.*, 2001, **11**, 5217–5221.

41. Y. Somekawa, M. Chiguchi, T. Ishbashi and T. Aso, Soy intake related to menopausal symptoms, serum lipids, and bone mineral density in postmenopausal Japanese women, *Obstet. Gynecol.*, 2001, **97**, 109–115.

42. S.C. Ho, S.G. Chan, Q. Yi, E. Wong and P.C. Leung, Soy intake and the maintenance of peak bone mass in Hong Kong Chinese women, *J. Bone Miner. Res.*, 2001, **16**, 1363–1369.

43. B.H. Ajrmandi, L. Alekel, B.W. Hollis, D. Amin, M. Stacewicz-Sapuntzakis, P. Guo and S.C. Kukreja, Dietary soybean protein prevents bone loss in an ovariectomized rat model of osteoporosis, *J. Nutr.*, 1996, **126**, 161–167.

44. H. Ishida, T. Uesugi, K. Hirai, T. Toda, H. Nukaya, K. Yokotsuka and K. Tsuji, Preventative effects of the plant isoflavones, daidzein and genistein, on bone loss in ovariectomized rats fed a calcium-deficient diet, *Biol. Pharm. Bull.*, 1998, **21**, 62–66.

45. P. Fanti, M.C. Monier-Faugere, Z. Geng, J. Schmidt, P.E. Morris, D. Cohen and H.H. Malluche, The phytoestrogen genistein reduces bone loss in short term ovariectomized rats, *Osteoporosis. Int.*, 1998, **8**, 274–281.

46. T. Uesugi, T. Toda, K. Tsuji and H. Ishida, Comparative study on reduction of bone loss and lipid metabolism abnormality in ovariectomized rats by soy isoflavones, daidzin, genistin and glycitin, *Biol. Pharm. Bull.*, 2001, **24**, 368–372.

47. T. Toda, T. Uesugi, K. Hirai, H. Nukaya, K. Tsuji and H. Ishida, New 6-*O*-acyl isoflavones glycosides from soybeans fermented with Bacillus subtilis (Natlo). I. 6-*O*-Succinylated isoflavone glycosides and their preventative effect on bone loss in ovariectomized rats fed a calcium-deficient diet, *Biol. Pharm. Bull.*, 1991, **22**, 1193–1201.

48. J.J.B. Anderson, W.W. Ambrose and S.C. Garner, Biphasic effects of genistein on bone tissue in the ovariectomized, lactating rat model, *PSEBM*, 1998, **218**, 345–350.

49. J.W. Nieves, L. Koniar, F. Cosman and R. Lindsay, Calcium potentiates the effect of estrogen and calcitonin on bone mass: Review and analysis, *Am. J. Clin. Nutr.*, 1998, **67**, 18–24.

50. D.N. Kalu, C.C. Liu, R.R. Hardin and B.W. Hollis, The aged rat model of ovarian hormone deficiency bone loss, *J. Endocrinol.*, 1989, **124**, 7–16.

51. J.A. Dodge, A.L. Glasebrook, D.E. Magee, D.L. Philips, M. Sato, L.L. Short and H.U. Bryant, Environmental estrogens: effects on cholesterol lowering and bone in the ovariectomized rat, *J. Steroid Biochem. Mol. Biol.*, 1996, **59**, 155–161.

52. C.R. Draper, M.J. Edel, I.M. Dick, A.G. Randall, G.B. Martin and R.L. Prince, Phytoestrogens reduce bone loss and bone resorption in oophorectomized rats, *J. Nutr.*, 1997, **127**, 1795–1799.

53. C. Picherit, V. Coxam, C. Bennetau-Pelissero, S. Kati-Coulibaly, M.J. Davicio, P. Lebecque and J.P. Barlet, Daidzein is more efficient than genistein on preventing ovariectomy-induced bone loss in rats, *J. Nutr.*, 2000, **130**, 1675–1681.

54. C. Picherit, C. Bennetau-Pelissero, B. Chanteranne, P. Lebecque, M.J. Davicio, J.P. Barlet and V. Coxam, Soybean isoflavones dose-dependently reduce bone

turnover but do not reverse established osteopenia in adult ovariectomized rats, *J Nutr.*, 2001, **131**, 723–281.

55. S.M. Potter, J.A. Baum, H. Teng, R.J. Stillman, N.F. Shay and J.W. Erdman, Soy protein and isoflavones: their effects on blood lipids and bone density in postmenopausal women, *Am. J. Clin. Nutr.*, 1998, **68**, 1375S–1379S.

56. A. Pines, H. Katchman, Y. Villa, V. Mijatovic, I. Dotan, Y. Levo and D. Ayalon, The effect of various hormonal preparations and calcium supplementation on bone mass in early menopause. Is there a predictive value for the initial bone density and body weight? *J. Intern. Med.*, 1999, **246**, 357–361.

57. L. Alekel, A. St. Germain, C.T. Peterson, K. Hanson, J.W. Stewart and T. Toda, Isoflavone-rich soy protein isolate exerts significant bone-sparing in the lumbar spine of perimenopausal women, *Am. J. Clin. Nutr.*, 2000, **729**, 844–852.

58. N. Morabito, A. Crisafulli, C. Vergara, A. Gaudio, A. Lasco, N. Frisina, R. D'Anna, F. Corrado, M.A. Pizzoleo, M. Cincotta, D. Altavilla, R. Ientile and F. Squadrito, Effects of genistein and hormone-replacement therapy on bone loss in early postmenopausal women: a randomized double-blind placebo-controlled study, *J. Bone Miner. Res.*, 2002, **17**, 1904–1912.

59. M.D. Scheiber, J.H. Liu, M.T. Subbiah, R.W. Rebar and K.D. Setchell, Dietary inclusion of whole soy foods results on significant reductions in clinical risk factors for osteoporosis and cardiovascular disease in normal postmenopausal women, *Menopause*, 2001, **8**, 384–392.

60. K.E. Wangen, A.M. Duncan, B.E. Merz-Demlow, X. Xu, R. Marcus, W.R. Phipps and M.S. Kurzer, Effects of soy isoflavones on markers of bone turnover in premenopausal and postmenopausal women, *J. Clin. Endocrinol. Metab.*, 2000, **85**, 3043–3048.

61. C.S. Hsu, W.W. Shen, Y.M. Hsueh and S.L. Yeh, Soy isoflavoine supplementation in postmenopausal women. Effects on plasma lipids, antioxidant enzyme activities and bone density, *J. Reprod. Med.*, 2001, **46**, 221–226.

62. J.W. Davis, R. Novotny, R.D. Wasnich and P.D. Ross, Ethnic, anthropometric, and lifestyle associations with regional variations in peak bone mass, *Calcif. Tissue Int.*, 1999, **55**, 100–105.

63. M.G. Cecchini, H. Fleisch and R.C. Muhlbauer, Ipriflavone inhibits bone resorption in intact and ovariectomized rats, *Calcif. Tissue Int.*, 1997, **61**, S9–S11.

64. C. Gennari, S. Adami, D. Agnusdei, L. Bufalino, R. Cervetti, G. Crepaldi, C. Di Marco, O. Di Munno, L. Fantasia, G.C. Isaia and G.F. Mazzuoli, Effect of chronic treatment with ipriflavone in postmenopausal women with low bone mass, *Calcif. Tissue Int.*, 1997, **61**, S19–S22.

65. P. Alexandersen, A. Toussaint, C. Christiansen, J.P. Devogelaer, C. Roux, J. Fechtenbaum, C. Gennari and J.Y. Reginster, Ipriflavone in the treatment of postmenopausal osteoporosis, *JAMA*, 2001, **285**, 1482–1488.

66. J.C. Gallagher, B.L. Riggs, J. Eisman, A. Hamstra, S.B. Arnaud and H.F. DeLuca, Intestinal calcium absorption and serum vitamin D metabolites in normal subjects and osteoporotic patients, *J. Clin. Invest.*, 1979, **64**, 729–736.

67. R.P. Heaney, R.R. Recker and P.D. Saville, Menopausal changes in bone remodeling, *J. Lab. Clin. Med.*, 1978, **92**, 964–970.

68. R.L. Prince, I. Dick, A. Devine, R.I. Price, D.H. Gutteridge, D. Kerr, A. Criddle, P. Garcia-Webb and A. St John, The effects of menopause and age on calcitropic hormones: a cross-sectional study of 655 healthy women aged 35 to 90, *J. Bone Miner. Res.*, 1995, **10**, 835–842.

69. B.H. Arjmandi, B.W. Hollis and D.N. Kalu, In vivo effect of 17β-estradiol on intestinal calcium absorption in rats, *Bone Miner.*, 1994, **26**, 181–189.
70. A.G. Poneros and J.W. Erdman Jr., Bioavailability of calcium from tofu, tortillas, nonfat dry milk and mozzarella cheese in rats: effect of supplemental ascorbic acid, *J. Food Sci.*, 1993, **58**, 382–384.
71. R.P. Heaney, M.S. Dowell, K. Rafferty and J. Bierman, Bioavailability of the calcium in fortified soy imitation milk with some observations on method, *Am. J. Clin. Nutr.*, 2000, **71**, 1166–1169.
72. R. Heaney, Protein intake and the calcium economy, *J. Am. Diet. Assoc.*, 1993, **93**, 1259–1260.
73. J.E. Kerstetter and L.H. Allen, Dietary protein increases urinary calcium. *J. Nutr.*, 1990, **120**, 134–136.
74. G. Trilok and H.H. Draper, Effect of a high protein intake on acid-base balance in adult rats, *Calcif. Tissue Int.*, 1989, **44**, 339–342.
75. N.A. Breslau, L. Brinkley, K.D. Hill and C.Y. Pak, Relationship of animal protein-rich diet to kidney stone formation and calcium metabolism, *J. Clin. Endocrin. Meter.*, 1988, **66**, 140–146.
76. B.L. Strom, R. Schinnar, E.E. Ziegler, K.T. Barnhart, M.D. Sammel, G.A. Macones, V.A. Stallings, J.M. Drulis, S.E. Nelson and S.A. Hanson, Exposure to soy-based formula in infancy and endocrinological and reproductive outcomes in young adulthood, *JAMA*, 2001, **286**, 807–814.
77. M.G. Busby, A.R. Jeffcoat, L.T. Bloedon, M.A. Kock, K.J. Dix, W.D. Heizer, B.F. Thomas, J.M. Hill, J.A. Crowell and S.H. Zeisel, Clinical characteristics and pharmacokinetics of purified soy isoflavones: single-dose administration to healthy men, *Am. J. Clin. Nutr.*, 2002, **75**, 126–136.
78. H. Shimizu, R.K. Ross, L. Bernstein, R. Yatani, B.E. Henderson and T.M. Mack, Cancers of the prostate and breast among Japanese and white immigrants in Los Angeles County, *Br. J. Cancer*, 1991, **63**, 963–966.
79. S. Barnes, Phytoestrogens and osteoporosis – What is a safe dose? *Br. J. Nutr.*, 2003, **39**, S101–S108.

CHAPTER 20

# Alcohol and Caffeine: Effects on Osteoporosis Risk

JOHN W. DICKERSON[1], SUSAN A. NEW[1]
and LINDA K. MASSEY[2]

[1]Centre for Nutrition and Food Safety, School of Biomedical and Molecular
Sciences, University of Surrey, Guildford, Surrey, GU2 7XH,
Email: s.new@surrey.ac.uk
[2]Department of Human Nutrition, Washington State University Spokane,
Spokane, WA 99210-1495, USA, Email: massey@wsu.edu

## Alcohol and Bone Health

**Key Points**

- Alcohol consumption is a strong feature of many societies, with wide variations in drinking habits. Both rates of absorption and metabolism of alcohol vary between individuals due to endogenous factors and exogenous factors.
- Alcohol provides 29 kJ ml$^{-1}$ (*i.e.*, 7 kcal) of energy, as well as moderate amounts of protein, carbohydrates and fat. Beers contain small quantities of the vitamin B complex. Red/white wines contain several trace elements as well as flavonols, flavones, falvan-3-ols, flavanones and anthocyanins.
- Excessive alcohol consumption has also been shown to decrease osteoblast number, osteoid formation and osteoblast proliferation.
- Chronic alcohol abuse is an important risk factor for risk of osteoporosis and fracture. Alcohol in large amounts is directly toxic to osteoblasts, thus reducing bone formation and may also directly affect bone mass by impairing liver function and altering both vitamin D and calcium metabolism.
- The trauma associated with alcohol-related falls is important to fracture risk.
- There is evidence, particularly for postmenopausal women, of a positive correlation between bone mineral density and moderate alcohol consumption.

- Mechanisms of action for a beneficial effect of moderate alcohol intake on bone health include a reduction in bone remodeling *via* decreases in PTH concentrations as well as increases in serum calcitonin.
- Alcohol consumption also affects endogenous hormone levels and is involved in inducing adrenal production of androstenedrone and its adrenal conversion to estrone. Higher estrogen concentrations have been found in postmenopausal women who consume alcohol.

## Introduction

The major component of most alcohol beverages is water, with the ethanol content depending on both the type of beverage consumed and the particular production process used.[1] Alcohols are compounds with the general formula $C_nH_{2n+1}OH$. The type of alcohol that is present in alcoholic beverages is ethanol ($C_2H_5OH$). Pure alcohol is known as absolute alcohol.[2]

It is a requirement by law that ethanol for human consumption can only come from the fermentation of agricultural crops and not industrial feed stock. The average percentage of alcohol by volume of alcoholic beverages is shown in Table 20.1 (see Ref. [3]). One unit of alcohol is equivalent to:

- one single measure of spirit (25 mL);
- one small glass of sherry/fortified wine (50 mL);
- one small glass of wine (125 mL);
- ¼ pint of strong lager, beer or cider (142 mL);
- ½ pint of ordinary strength lager, beer or cider (284 mL);
- 2 pints of low alcohol lager, beer or cider (1136 mL).

Alcohol is, in part, a nutritional source providing 29 kJ mL$^{-1}$ (*i.e.*, 7 kcal) of energy, as well as moderate amounts of protein, carbohydrates and fat. Beers contain small amounts of a variety of the B vitamins and both red/white wines contain several trace elements including iron, potassium, copper and sodium, although this varies (predominantly for iron), depending on year of harvest, wine preparation method, grape type, *etc.*[4] Other important components are

**Table 20.1.** *Typical percentage of alcohol by volume of alcoholic beverages (ABV)*

| Alcoholic beverage | Percentage ABV |
|---|---|
| Beer, Cider, Sherry | 4–6 |
| Wines | 9–13 |
| Spirits | 37–45 |
| Liqueurs | 20–40 |
| Fortified wines | 18–25 |

From Ref. 3.

also found in alcoholic beverages including flavonols, flavones, falvan-3-ols, flavanones and anthocyanins.[5]

The Department of Health in 1995 reported UK guidelines on alcohol consumption providing guidelines on a daily and weekly upper limits. It is specifically noted that there are no significant health risks at levels of consumption of between three and four units a day or less for men and between two and three units a day or less for women.[2] However, trends in alcohol consumption are a cause for concern. For example, the Health Survey for England found that 38% (2/5ths) of men had drink more than four units of alcohol on at least one day in the previous week and approximately 20% of women (1/5th) had drink more than three units of alcohol on at least one day in the previous week. Mean alcohol consumption per week was 16.4 units for men and 6.4 units for women. Similar findings have also been noted in a number of European countries. In 1999, in the UK alone, a total of 16 830 casualties in traffic accidents involved illegal alcohol levels which was reflected as 5% of all traffic accident categories.[6] The lifetime prevalence of alcohol abuse and dependence in women has been estimated to be as high as 10% and even higher in men.[7]

## Alcohol and Metabolic Bone Disease

Alcohol consumption has been shown to diminish osteoblastic activity in both the human and animal model.[8] Guerri and Sanchis (1985) found that the offspring of alcoholic animals had a deficient corporal development, an effect that was mediated either by hormonal or nutritional deficiencies with the net result being cellular impairment with decreased activity.[9] Turner *et al.* (1987) observed a mineralisation defect and a decrease in the synthesis of bone matrix.[10] Reduced osteocalcin levels have been reported after sporadic acute alcoholic intoxication in subjects with prolonged alcohol intake.[11] Excessive alcohol consumption has also been shown to decrease osteoblast number, osteoid formation and osteoblast proliferation.[12]

## Alcoholism and Osteoporosis Risk

There is good evidence in the literature that chronic alcohol abuse is an important risk factor for risk of osteoporosis and fracture.[13-15] The mechanisms remain to be clearly defined but there is evidence to suggest that in the human model, alcohol in large amounts is directly toxic to osteoblasts, thus reducing bone formation and may also directly affect bone mineral density by impairing liver function and altering both vitamin D and calcium metabolism.[16] Furthermore, the trauma associated with alcohol-related falls is clearly important to fracture risk.[17] Recent work by Clark and Sowers suggests that women treated for alcohol abuse had a lower femoral neck and lumbar spine BMD when compared with non-alcohol abusing women. Women in treatment and recovery reported more fractures during childhood and early adolescence than non-alcohol-dependent women, suggesting that factors other than acute toxification contributed to the greater fracture prevalence.[18]

## Moderate Alcohol Consumption and Bone Health

Evidence is conflicting for a positive effect of moderate alcohol consumption on bone. A number of published studies have found no association between moderate alcohol intake and bone mass, including data on American,[19] British,[20] Finnish,[21] Australian[22] and Danish[23] women. However, data predominantly for postmenopausal women do indicate a positive correlation between bone mineral density and alcohol consumption. In a study of elderly women by Felson *et al.* (1995), an alcohol intake of greater than 210 mL (7 oz) per week was found to be associated with a higher bone mineral density compared with those women consuming less than this amount[24] and in a prospective study by Holbrook and Barrett-Connor, social drinking was associated with higher bone mass, a finding that was consistent across both genders.[25] In a recent study of 489 elderly women aged 65–77 years by Rapuri *et al.* (2000), moderate alcohol intake was found to be associated with higher bone mass.[26]

Few studies have undertaken thorough investigations into the mechanisms of action of the proposed beneficial effect of moderate alcohol intake on bone health. In the study by Rapuri *et al.* a reduction in bone remodeling was observed, as evidenced by the decrease in bone resorption markers as well as a decrease in serum parathyroid concentrations. Other possible reasons for the increase seen in bone mass with moderate alcohol consumption include: (i) the stimulation of calcitonin production[8] which has been shown to increase spine bone mineral density, with little effect on femoral sites;[27] (ii) the effect of alcohol on endogenous hormone levels and the influence of alcohol upon inducing adrenal production of androstenedrone and its adrenal conversion to estrone.[28] Higher estrogen concentrations have been found in postmenopausal women who consume alcohol.[29–30]

## Conclusions

Alcohol consumption has been shown to diminish osteoblastic activity in both the human and animal model. Excessive alcohol consumption has also been shown to decrease osteoblast number, osteoid formation and osteoblast. Chronic alcohol abuse is an important risk factor for risk of osteoporosis and fracture, with the mechanisms involved including a direct toxic to osteoblasts, impairment of liver function and altering both vitamin D and calcium metabolism. Moderate alcohol consumption has been shown to be beneficial to bone health, particularly in postmenopausal women. Whilst further research is required concerning the mechanisms of action for a beneficial effect, these include a reduction in osteoclastic activity, enhanced by a decrease in PTH, stimulation of calcitonin production and the effect of alcohol on endogenous hormone levels.

# Caffeine: Caffeinated Beverages and Bone Health

**Key Points**

- Caffeine is found in many beverages, including coffee, tea, soft drinks and hot cocoa.
- Caffeine increases urinary calcium for several hours after its consumption.
- Caffeine and caffeinated beverages have small negative effects on calcium balance, bone mineral density and fracture, most notably in postmenopausal women consuming low amounts of dietary calcium.
- Risk of bone loss due to caffeine-induced hypercalciuria may be limited to certain genotypes.
- Consumption of recommended amounts of dietary calcium appears to offset the small calcium losses caused by caffeine.
- Individuals who consume diets rich in milk, fruits and vegetables are unlikely to have detrimental effects on bone health from drinking moderate amounts of caffeinated beverages.

## Introduction

Caffeine is consumed by populations worldwide, mostly in the form of caffeinated beverages. Coffee has the highest caffeine concentration per ounce of the beverages (Table 20.2). Although the caffeine concentration in soft drinks is relatively low, the amount of caffeine from soft drinks can be substantial because these beverages are widely consumed, and in large volumes.[31] When considering the role of caffeine in bone health, consideration must be given to the other components of the beverages, which may also affect calcium metabolism and/or bone. Cocoa contains only very small amounts of caffeine (Table 20.2), and large amounts of milk. Because of its composition, cocoa is unlikely to affect calcium and bone negatively, and will not be discussed in this chapter.

After ingestion, caffeine is rapidly absorbed and distributed through the lean body mass. Plasma levels peak about 30 min after consumption. Caffeine has a half-life of 3–5 h in healthy adults; smoking decreases half-life by increasing its rate of catabolism. In humans 80% of caffeine is metabolized by demethylation by hepatic P450 enzyme CYP1A2 to paraxanthine; 16% is converted to theobromine and theophylline. These metabolites are further catabolized by demethylation and oxidation to uric acid derivatives, which are excreted in the urine.[32]

Caffeine affects metabolism by its antagonism of adenosine receptors. Caffeine does not directly affect bone;[33] it affects calcium metabolism by its hypercalciuric action on the kidney and possibly decreased intestinal calcium absorption. Although tolerance to caffeine's neural stimulation develops within a few days, full hypercalciuric action persists after at least a week of daily caffeine consumption compared to complete abstinence.[34] The hypercalciuric effect is biphasic with a small nocturnal reduction mostly compensating for the acute response within 6 h after consumption.[35] The 24 h net loss of urinary calcium due to 6 mg caffeine kg$^{-1}$ body weight dose is about 3 mg.

**Table 20.2.** *Caffeine content of foods and beverages*

| Food/beverage | Type | Caffeine content per unit (mg) | Size of unit |
|---|---|---|---|
| Coffee | Ground roasted[a] | 85 | 150 mL (5 oz) cup |
| | Instant[a] | 60 | 150 mL (5 oz) cup |
| | Decaffeinated[a] | 3 | 150 mL (5 oz) cup |
| Tea | Leaf or bag[a] | 31 | 150 mL (5 oz) cup |
| | Instant[a] | 20 | 150 mL (5 oz) cup |
| | Bottled[b] | 6–21 | 360 mL (12 ox) |
| Soft drinks with caffeine[b,c] | | 15–47 | 360 mL (12 oz) can |
| High energy soft drinks[d] | | 80 | 360 mL (12 oz) can |
| Hot chocolate, cocoa, chocolate milk[a] | | 4 | 150 mL (5 oz) cup |
| Chocolate candy[a] | | 5–20 | 100 g |
| | | 1.5–6 | 1 oz |

[a]Ref. [31].
[b]www.usda.org (accessed January 29, 2003).
[c]Not all colas contain caffeine, while some non-cola drinks do. In the US, if caffeine has been added, it must be listed as an ingredient on the label.
[d]www.red-bull.org (accessed January 29, 2003).

A series of studies has shown that caffeine effects calcium balance, although the amounts are small (Table 20.3). In all these studies, subjects were on their usual self-selected diets and typical calcium intakes were less than currently recommended. Even though the negative balances were small, if continued over many years, they would be clinically significant in increasing bone loss. The three studies on postmenopausal women showed a negative effect on calcium balance, while the study on premenopausal women did not.

**Table 20.3.** *Effect of caffeine on calcium balance in controlled metabolic studies*

| First author and date | Experimental Design | Change in calcium balance |
|---|---|---|
| Heaney and Recker (1982)[36] | Pre- and post-menopausal women, 8 day balance | −4 mg per 175 mg caffeine |
| Barger-Lux et al. (1990)[37] | 1.4 L cola day$^{-1}$ (400 mg caffeine), 19 day crossover, 16 premenopausal women | n.s. |
| Hasling et al. (1992)[38] | 4 day balance postmenopausal women with osteoporosis | −6 mg per 100 mL coffee |
| Barger-Lux and Heaney (1995)[39] | Pre- and post-menopausal women, 8 day balance | −4 mg per 180 mL coffee |

Several dozen epidemiological studies have looked for a relationship between caffeine, coffee or tea consumption and bone loss or fractures. Because prospective studies have greater power than cross-sectional and case-controls to detect such an association, only studies of this design will be discussed (Table 20.4). The cross-sectional and case control studies are discussed by Heaney.[53] The important interaction of dietary calcium with caffeine consumption was seen in the study by Harris and Dawson-Hughes.[36] These investigators found that bone loss from the spine and total-body bone mineral density increased only in postmenopausal women who had both low calcium intakes (440–744 mg day$^{-1}$) and higher caffeine intakes (450–1120 mg day$^{-1}$).

The results of the metabolic and prospective epidemiological studies suggest that older persons appear to be more susceptible to caffeine-induced bone loss. As humans age, they are less able to adapt to the losses of calcium with appropriate changes in calcium absorption. This failure of adaptation was reported by Yeh *et al.*[54] in rats consuming moderately low calcium diets. Young rats (4 week) increased their 1,25 dihydroxyvitamin D, and therefore absorbed more calcium in response to caffeine; older rats (12–13 months) did not show this compensating increase. Thus, calcium balance in young rats was unaltered, but in older rats it became significantly negative. More of the larger and longer prospective studies on the elderly show a modest effect of caffeine consumption on bone density or fracture (Table 20.4).

Genotype appears to determine the bone response to caffeinated beverages. Rapuri *et al.*[51] found that caffeine intakes of > 300 mg day$^{-1}$ were associated with bone loss in a 3 year prospective study of postmenopausal women. However caffeine was only associated with increased bone loss in women with the *tt* VDR genotype. Both the BB VDR genotype and the *tt* VDR genotype may be associated with a lower ability to increase calcitriol-mediated calcium absorption. Because only 11 women had this genotype, the authors were not able to determine if dietary calcium intake affected the response to higher intakes of caffeine.

Caffeinated beverages usually contain other ingredients that will affect calcium metabolism and possibly bone. The most important of these are potassium and sugar. Potassium has a hypocalciuric effect; Lemann found that each 100 mg of dietary potassium reduced urinary calcium by 1.5 mg.[55] Brewed coffee contains 128 mg of potassium per 240 mL,[56] so coffee would reduce urinary calcium by about 2 mg less than an equivalent dose of pure caffeine. Sugars also have a hypercalciuric effect.[57] Heaney and Rafferty[58] calculated that the sugar in a 12 oz can of carbonated beverage increased urinary calcium by 4.5 mg. In the same study Heaney and Rafferty[58] showed that the amounts of citric and phosphoric acids used as acidulants in carbonated beverages did not increase urinary calcium in the 5 h postconsumption. Antioxidants have been hypothesized to have beneficial effects on bone, and both coffee and tea are rich in antioxidant compounds. However only cross-sectional studies of tea and bone health have been published at the time of writing.[59]

**Table 20.4.** *Prospective epidemiological studies on caffeine and bone health*

| Author, year | Population | Length of study (years) | Diet measure | Outcome | Significance |
|---|---|---|---|---|---|
| Lloyd et al. (1998)[40] | 81 females, 12–18 years | 6 | Caffeine mg day$^{-1}$ | Total body bone gain | NO |
| Packard and Recker (1996)[41] | 145 females, 19–26 years | 1.6–4.0 | Caffeine mg day$^{-1}$ | Change in lumbar BMC<br>Lumbar BMD<br>Total body mineral | NO<br>NO<br>YES |
| Holbrook et al. (1988)[42] | 426 men, 531 women, 50–79 years | 12 | RR per 352 mg caffeine day$^{-1}$ | RR 1.1 for hip fracture | NO |
| Kiel et al. 1990[43] | 3170 white males and females, 50+ years, 12 years | 12 | Unit = 1 cup coffee or 2 cups tea | RR 1.69 for hip fracture for 2.5 or more units | YES |
| Hernandez-Avila et al. (1991)[44] | 84 484 females, Nurses' study 6 years | 6 | Quintiles of caffeine mg day$^{-1}$ | Hip and forearm fracture | Hip YES<br>Forearm NO |
| Hansen et al. (1991)[45] | 121 postmenopausal women | 12 | Caffeine mg day$^{-1}$ | Bone loss | NO |
| Harris and Dawson-Hughes (1994)[46] | 205 postmenopausal women | 2 | Caffeine mg day$^{-1}$ | Higher bone loss if caffeine <300 mg day$^{-1}$ and dietary calcium low | YES |
| Cummings et al. (1995)[47] | 9516 women 65+ years | 4 | Caffeine mg day$^{-1}$ | Hip fractures | YES |
| Reid et al. (1994)[48] | 122 postmenopausal women | 2 | Caffeine mg day$^{-1}$ | Bone loss at several sites | NO |
| Meyer et al. (1997)[49] | 39 787 men and women 37–58 years | 11 | Cups of coffee | Hip fracture | YES if over 9 cups per day |
| Lloyd et al. (2000)[50] | 138 Postmenopausal women | 2 | Caffeine mg day$^{-1}$ | Total body or femoral neck bone | NO |
| Hannan et al. (2000)[51] | 800 men and women 67–90 years | 4 | Caffeine mg day$^{-1}$ | Bone loss at six sites | NO |
| Rapuri et al. (2001)[52] | 489 women 65–77 years | 3 | Caffeine mg day$^{-1}$ | Spinal bone loss | YES |

# Conclusions

Caffeine and caffeinated beverages have small negative effects on calcium balance, bone mineral density and fracture, most notably in postmenopausal women consuming low amounts of dietary calcium. Risk may be limited to certain genotypes. Consumption of recommended amounts of dietary calcium appears to offset the small calcium losses caused by caffeine. Individuals who consume diets rich in milk, fruits and vegetables are unlikely to have detrimental effects on bone health from drinking moderate amounts of caffeinated beverages.

# References

1. W.P.T. James and A. Ralph, Alcohol; its metabolism and effects, In *Human Nutrition and Dietetics*, 10th edn, J.S. Garrow, W.P.T. James and A. Ralph (eds.), Churchill Livingstone, Chapter 8, 2000, pp. 121–133.
2. British Nutrition Foundation. Briefing Paper on Alcoholic Beverages and Health. 2001.
3. McCance and Widdowson, *The Composition of Foods*, 5th edn, Royal Society of Chemistry, Cambridge, UK, 1995.
4. W.V. Rumpler, D.G. Rhodes and D.J. Baer, Energy value of moderate alcohol consumption by humans, *Am. J. Clin. Nutr.*, 1996, **64**, 108–114.
5. J.S. Gavaler, A.F. Imhoff, C.R. Pohl, E.R. Rosenblum and D.H. Van Thiel, Alcoholic beverages: a source of estrogenic substances, *Alcohol Alcohol. Suppl.*, 1987, **1**, 545–549.
6. Royal College of Physicians, A great and growing evil. The medical consequences of alcohol abuse, Tavistock, London, 1987 Report.
7. P. Peris, A. Pares and N. Guanabens, Reduced spinal and femoral bone mass and deranged bone mineral metabolism in chronic alcoholics, *Alcohol Alcohol*, 1992, **27**, 619–625.
8. H. Rico, Alcohol and bone disease, *Alcohol Alcohol.*, 1990, **25**, 345–352.
9. C. Guerri and R. Sanchis, Acetaldehyde and alcohol levels in pregnant rats and their fetuses, *Alcohol*, 1985, **2**, 267–270.
10. R.T. Turner, V.S. Greene and N.H. Bell, Demonstration that ethanol inhibits bone matrix synthesis and mineralisation in the rat, *J. Bone Miner. Res.*, 1987, **2**, 61–66.
11. H. Rico, H.A. Cabranes, J. Cabello, F. Gomez-Castresana and E.R. Hernandez, Low serum osteocalcin in acute alcohol intoxication: a direct toxic effect of alcohol on the osteoblast, *Bone Miner.*, 1987, **2**, 221–225.
12. D. Bickle, H. Genant, C. Cann, R. Recker, B. Haloran and G. Stewler, Bone disease in alcohol abuse, *Ann. Intern. Med.*, 1985, **103**, 42–48.
13. R. Crilly, C. Anderson, D. Hogan and L. Delaquerriere-Richardson, Bone histomorphometry, bone mass and related parameters in alcoholic males, *Calcif. Tissue Int.*, 1988, **43**, 269–276.
14. B. Lalor, M. France, P. Powell, P.H. Adams and R.B. Counihan, Bone and mineral metabolism and chronic alcohol abuse, *Q. J. Med.*, 1986, **59**, 497–511.
15. T. Pepersack, M. Fuss, J. Otero, P. Bergmann, J. Valsamis and J. Corvilain, Longitudinal study of bone metabolism after ethanol withdrawal in alcoholic patients, *J. Bone Miner. Metab.*, 1992, **7**, 383–387.

16. G. Wilkinson, T. Cundy, V. Parsons and P. Lawson-Mathew, Metabolic bone disease and fractures in male alcoholics: a pilot study, *Br. J. Addict.*, **80**, 65–68.
17. F.R. Klein, K.A. Fausti and A.S. Carlos, Ethanol inhibits human osteoblastic cell proliferation, *Alcohol Clin. Exp. Res.*, 1996, **20**, 572–578.
18. M.K. Clark and M.F.R. Sowers, Bone mineral density and fractures among alcohol-dependent women in treatment and in recovery, *Osteoporosis Int.*, 2003, **14**, 396–403.
19. D.C. Bauer, W.S. Browner and J.A. Cauley, Factors associated with appendicular bone mass in older women. The Study of Osteoporotic Fracture Research Group, *Ann. Int. Med.*, 1993, **118**, 657–665.
20. J.C. Stevenson, B. Lees, M. Devenport, M.P. Cust and K.F. Ganger, Determinants of bone density in normal women: risk factors for future osteoporosis.? *Br. Med. J.*, 1989, **298**, 924–928.
21. M. Tuppurainen, H. Kroger and R. Honkanen, Risks of perimenopausal fractures – a prospective population-based study, *Acta Obstet. Gynaecol. Scand.*, 1995, **74**, 624–628.
22. R.M. Angus, P.N. Sambrook and N.A. Pocock, Dietary intake and bone mineral density, *Bone Miner.*, 1988, **4**, 265–277.
23. M.A. Hansen, K. Overgaard, B.J. Riis and C. Christiansen, Potential risk factors for development of postmenopausal osteoporosis – examined over a 12-year period, *Osteoporosis Int.*, 1991, **1**, 95–102.
24. D.R. Felson, Y. Zhang, M.T. Hannan, W.B. Kannel and D.P. Kiel, Alcohol intake and bone mineral density in elderly men and women. The Framingham Study, *Am. J. Epidemiol.*, 1995, **142**, 485–492.
25. T.L. Holbrook and E. Barrett-Connor, A prospective study of alcohol consumption and bone mineral density, *Br. Med. J.*, 1993, **306**, 1506–1509.
26. P.B. Rapuri, J.C. Gallagher, K.E. Balhorn and K.L. Ryschon, Alcohol intake and bone metabolism in elderly women, *Am. J. Clin. Nutr.*, 2000, **72**, 1206–1213.
27. R. Civitelli, S. Gonnelli and F. Zacchei, Bone turnover in postmenopausal osteoporosis. Effect of calcitonin treatment, *J. Clin. Invest.*, 1988, **82**, 1268–1274.
28. J.S. Gavaler and D.H. Van Thiel, The association between moderate alcoholic beverage consumption and serum oestradiol and testosterone levels in normal postmenopausal women: relationship to the literature, *Alcohol Clin. Exp. Res.*, 1991, **16**, 87–92.
29. K. Katsouyanni, P. Boyle and D. Trichopoulos, Diet and urine oestrogens among postmenopausal women, *Oncology*, 1991, **48**, 490–494.
30. S.E. Hankinson, W.C. Willett and J.E. Manson, Alcohol, height and adiposity in relation to oestrogen and prolactin levels in postmenopausal women, *J. Natl Cancer Ins.*, 1995, **87**, 1297–1302.
31. J.J. Barone and H.R. Roberts, Caffeine consumption, *Food Chem. Toxicol.*, 1996, **34**, 119–129.
32. H.G. Mandel, Update on caffeine consumption, disposition and action, *Food Chem. Toxicol.*, 2002, **40**, 1231–1234.
33. E. Bergman, L.K. Massey, W. Wise and D. Sherrard, Effect of oral caffeine on renal handling of calcium and magnesium in adult women, *Life Sci.*, 1990, **47**, 557–564.
34. L.K. Massey and A. Opryszek, Effects of adaptation to caffeine on calcium excretion in young adult human females, *Nutr. Res.*, 1990, **10**, 741–747.
35. S.A. Kynast-Gales and L.K. Massey, Effect of caffeine on circadian excretion of urinary calcium and magnesium, *J. Am. Coll. Nutr.*, 1994, **13**, 467–472.

36. R.P. Heaney and R.R. Recker, Effects of nitrogen, phosphorus and caffeine on calcium balance in women, *J. Lab. Clin. Med.*, 1982, **99**, 46–55.

37. M.J. Barger-Lux, R.P. Heaney and M.R. Stegman, Effect of moderate caffeine intake on the calcium economy of premenopausal women, *Am. J. Clin. Nutr.*, 1990, **52**, 722–725.

38. C. Hasling, K. Sondergaard, P. Charles and L. Mosekilde, Calcium metabolism in postmenopausal women is determined by dietary calcium and coffee intake, *J. Nutr.*, 1992, **122**, 1119–1126.

39. M.J. Barger-Lux and R.P. Heaney, Caffeine and the calcium economy revisited, *Osteoporosis Int.*, 1995, **5**, 97–102.

40. T. Lloyd, N.J. Rollings, K. Kieselhorst, D.F. Eggli and E. Mauger, Dietary caffeine intake is not correlated with adolescent bone gain, *J. Am. Coll. Nutr.*, 1998, **17**, 454–457.

41. P.T. Packard and R.R. Recker, Caffeine does not affect the rate of gain in spine bone in young women, *Osteoporosis Int.*, 1996, **6**, 149–152.

42. T.L. Holbrook, E. Barrett-Connor and D.L. Wingard, Dietary calcium and risk of hp fracture: 14-Year prospective population study, *Lancet*, 1988 (Nov 5), 1046–1049.

43. D.P. Kiel, D.T. Felson, M.T. Hannan, J.J. Anderson and P.W.F. Wilson, Caffeine and the risk of hip fracture: The Framingham study, *Am. J. Epidemiol.*, 1990, **132**, 675–684.

44. M. Hernandez-Avila, G.A. Colditz, M.J. Stampfer, B. Rosner, F.E. Speizer and W.C. Willett, Caffeine, moderate alcohol intake, and risk of fractures of the hip and forearm in middle-aged women, *Am. J. Clin. Nutr.*, 1991, **54**, 157–163.

45. M.A. Hansen, K. Overgaard, B.J. Riis and C. Christiansen, Potential risk factors for development of postmenopausal osteoporosis, *Osteoporosis Int.*, 1991, **1**, 95–102.

46. S.S. Harris and B. Dawson-Hughes, Caffeine and bone loss in healthy postmenopausal women, *Am. J. Clin. Nutr.*, 1994, **60**, 573–578.

47. S.R. Cummings, M.C. Nevitt, W.S. Browner, K. Stone, K.M. Fox, K.E. Ensrud, J. Cauley, D. Black and T.M. Vogt, Risk factors for hip fracture in white women, *N. Engl. J. Med.*, 1995, **332**, 767–773.

48. I.R. Reid, R.W. Ames, M.C. Evans, S.J. Sharpe and G.D. Gamble, Determinants of the rate of bone loss in normal postmenopausal women, *J. Clin. Endocrinol. Metab.*, 1994, **79**, 950–954.

49. H.E. Meyer, J.I. Petersen, E.B. Loken and A. Tverdal, Dietary factors and the incidence of hip fracture in middle-aged Norwegians: A prospective study, *Am. J. Epidemiol.*, 1997, **145**, 117–123.

50. T. Lloyd, N. Johnson-Rollings, D.F. Eggli, K. Kieselhorst, E.A. Mauger and D. Cardomone-Cusatis, Bone status among postmenopausal women with different habitual caffeine intakes: A longitudinal investigation, *J. Am. Coll. Nutr.*, 2000, **19**, 256–261.

51. M.T. Hannan, D.T. Felson, B. Dawson-Hughes, K.L. Tucker, L.A. Cupples, P.W.F. Wilson and D.P. Kiel, Risk factors for longitudinal bone loss in elderly men and women: The Framingham Osteoporosis study, *J. Bone Miner. Res.*, 2000, **15**, 710–720.

52. P. Rapuri, J.C. Gallagher, H.K. Kinyamu and K.L. Ryschon, Caffeine intake increases the rate of bone loss in elderly women and interacts with vitamin D receptor genotypes, *Am. J. Clin. Nutr.*, 2001, **74**, 694–700.

53. R.P. Heaney, Effects of caffeine on bone and the calcium economy, *Food Chem. Toxicol.*, 2002, **40**, 1263–1270.

54. J.K. Yeh and J.F. Aloia, Differential effect of caffeine administration on calcium and vitamin D metabolism in young and old rats, *J. Bone Min. Res.*, 1986, **1**, 251–258.

55. J. Lemann, R.W. Gray and J.A. Pleuss, Potassium bicarbonate, but not sodium bicarbonate, reduces urinary calcium excretion and improves calcium balance in healthy men, *Kidney Int.*, 1989, **35**, 688–695.

56. E.S. Hands, *Nutrients in Food*, Lippincott Williams & Wilkins, 2000.

57. L.K. Massey and P.W. Hollingbery, Acute effects of dietary caffeine and sucrose on urinary mineral excretion of healthy adolescents, *Nutr. Res.*, 1988, **8**, 1005–1012.

58. R.P. Heaney and K. Rafferty, Carbonated beverages and urinary calcium excretion, *Am. J. Clin. Nutr.*, 2001, **74**, 343–347.

59. D.L. McKay and J.B. Blumberg, The role of tea in human health: An update, *J. Am. Coll. Nutr.*, 2002, **21**, 1–13.

CHAPTER 21

# Influence of Nutrition on Bone Health: The Twin Model Approach

JOHN D. WARK[1] and CARYL NOWSON[2]

[1]Department of Medicine, The University of Melbourne, Bone and Mineral Service, The Royal Melbourne Hospital, Victoria 3050, Australia,
Email: jdwark@unimelb.edu.au
[2]School of Health Sciences, Deakin University, Burwood 3125, Australia

**Key Points**

- Studies of nutrition in twin and non-twin population samples are highly complementary.
- Certain characteristics of twins make them valuable for nutrition and other biomedical research, particularly that twins within a pair are matched for age, for all or on-average half their genes and for many environmental exposures.
- Sources of variation in a trait such as bone mineral density (BMD) can be decomposed into additive genetic, common environmental (*i.e.*, shared within twin pairs) and individual environmental components using classical analytical approaches with twin data.
- Co-twin association studies have demonstrated the associations of within-pair differences in factors related to nutrition (dietary calcium intake, alcohol consumption, cigarette smoking, physical activity, soft tissue composition) with within-pair differences in BMD.
- A refinement of the co-twin study design involves the selection of twin pairs who are discordant for an exposure of interest: this approach was used to characterise the adverse effects of smoking on BMD in smoking-discordant twin pairs, and the lack of any long-term deleterious effect of pregnancy and lactation on BMD.
- Co-twin controlled intervention trials offer major advantages in statistical power compared with similar studies in unrelated individuals: several

randomised, co-twin, placebo-controlled trials of calcium supplementation performed in young twins have all demonstrated positive effects of calcium on BMD.
- Dizygotic twins are a special, age-matched example of sibling pairs, where the degree of environmental sharing is enhanced compared with non-twin siblings. Twin data are therefore finding valuable use in the identification of specific or individual genes that determine complex traits such as BMD.
- There is great potential for the application of the twin model approach to examine the interplay between osteoporosis-related genes and environmental factors including nutrition.

## Introduction

The scientific value of twins began to be appreciated from the 1870s when Galton first explored questions of 'nature *vs.* nurture' by observing the growth and development of twins.[1] Since that time, many study designs, analytical methods and applications of information derived from twins have been developed. It is clear now that twin research offers a number of unique advantages and that studies involving twin and non-twin population samples, respectively, are highly complementary in addressing many scientific questions. Some of the characteristics of twins that make them attractive for biomedical research are (1) twins within a pair are matched for age, for all or on-average half their genetic makeup, and for many environmental exposures, especially in the early stages of life; (2) reduced collinearity, compared with studies of unrelated individuals, for example, when studying traits that are influenced by body size; (3) enhanced statistical power in co-twin controlled intervention studies[2] – for example, in a trial using BMD as an endpoint, one may require one quarter the number of monozygotic (MZ) twins as of unrelated individuals; (4) accessibility and approachability for research, particularly through twin registries – our ability to recruit substantial numbers of twins for studies of bone health, body composition and blood pressure has been underpinned by the resources of the Australian Twin Registry and there are a number of other organisations providing similar opportunities. These points are summarised in Table 21.1.

## Approaches to Research Using Twins

What types of approach might be readily applied in nutritional research utilising twins? One might seek evidence of the heritability of traits and of disease risk influenced by nutrition, by comparing the within-pair correlations in MZ and dizygotic (DZ) twins. Under the assumption that within-pair sharing of environmental exposures is the same for MZ and DZ twins, a greater within-pair correlation in MZ than DZ twins is due to genetic factors.

**Table 21.1.** *Why use twins?*

---

- Enhances statistical power
- Adjusts for confounders
- Matches for age and height
- Similar environment particularly during childhood/adolescence
- Controls all or, on average, half of additive genetic effects
- Controls a number of shared environmental effects
- Availability, for example, through twin registries

---

MZ twins share 100% of their genes and DZ twins share, on average, 50% of their genes. Therefore, for a trait whose variation is determined purely by genetic factors, the within-pair correlation would be 1 in MZ twins and 0.5 in DZ twins.

Using more sophisticated analytical approaches such as multivariate normal modelling,[3] variance in a continuously distributed trait such as BMD can be decomposed into additive genetic (A), common environmental (C) (*i.e.*, shared within a twin pair) and individual environmental (E) sources of variation. E includes variation due to measurement error. For traits showing moderate-to-high levels of heritability, either AE or ACE models give the best fit. This modelling approach can indicate what proportion of variation in a trait within a particular population may be due to additive genetics and to environmental influences such as nutritional factors.

Studies that captured much interest were those where twins who were separated at birth and brought up in different environments were assessed in later life. A striking number of similarities were found in these adult twins. For example, body weight appeared to be very similar indicating the strong influence of genetic factors on body weight within a similar background of adequate nutrition.[4]

An important, and challenging application of twin data is the study of gene–environment interactions. These effects are best understood as 'different genotypes responding differently to the same environment; or viewed from the other end, some genotypes being more sensitive to changes in the environment than others (different reaction ranges)'.[5]

The above study designs generally require observations in both MZ and DZ twins. There are also a number of powerful co-twin study designs that can use data from MZ, DZ or a pooled sample of MZ and DZ twins. The within-pair relationship between traits such as BMD and their putative determinants can be examined in co-twin association studies. For example, we have explored the associations of within-pair differences in dietary calcium intake, alcohol consumption, physical activity and cigarette use with within-pair differences in BMD in several twin studies.[6–8] The strength of association is given by the regression coefficients expressed as the percentage difference in BMD per unit difference in the independent variable. This approach is powerful, particularly because it controls inherently for age and for genetic and environmental factors shared by twins. The shared characteristics include various measures of body size, which are highly correlated within pairs of twins.[6] On the other hand, the co-twin observational approach loses strength where environmental exposures of interest tend to be shared, for example, dietary and lifestyle factors during growth and development in twins reared in the same household. Multivariate modelling of variation in a trait that is subject to environmental influences in childhood and adolescence may reveal a 'common environment effect' that wanes in early adult life. This phenomenon has been demonstrated for BMD.[9]

A co-twin study design of particular interest involves the selection of twins who show within-pair discordance for a continuous or categorical variable under investigation, for example, cigarette smoking.[10] We have used this study design to examine the associations of pregnancy and lactation, and of hormone replacement therapy use with BMD.[11,12] Again, this design controls the many factors shared within twin pairs, and data can be adjusted for the effects of covariates.

Perhaps the most powerful co-twin design is the co-twin controlled intervention study. The within-pair matching gives this approach a major advantage in statistical power compared with the use of unrelated individuals (see above). In the field of nutrition and bone health, several co-twin, placebo-controlled, calcium intervention trials in young twins have been reported.[13–15] One should be particularly alert to the possibility of cross-contamination with the intervention when this design is used. Ensuring blinding of the intervention may reduce the risk of cross-contamination.

A potentially vast field in which twins are being utilised increasingly is the identification of specific or individual genes responsible for complex traits such as BMD and blood pressure. DZ twins are a special, age-matched example of sibling pairs where the degree of environmental sharing is enhanced compared with non-twin siblings.[16] Thus, twin data have found notable use in candidate gene association studies for traits such as BMD.[17] The identification of genotypes of interest enables gene-environment interactions to be sought, for example, the genotype-dependent BMD response to varying calcium intake.[18] Major research efforts are now ongoing to map genes for quantitative traits using large twin databases.[19] Success in these efforts will open the way to understanding the interplay between osteoporosis-related genes and environmental factors including nutrition.

# Twin Studies Related to Bone Mineral Density

Early twin studies demonstrated that BMD was a highly heritable trait.[20,21] In the study of older female twins by Flicker *et al.* the best-fitting model indicated that 75% of variation in adjusted BMD at the spine and hip was due to additive genetic factors.[7] The residual variation was attributed to environmental factors unique to the individual. There was no evidence for a shared environment effect in these older female twins, except at the forearm where, intriguingly, the best-fitting model included both common and unique environmental effects, but not additive genetics. Based on a large body of evidence, up to 80% of population variability in bone mass at non-forearm sites may be genetically determined, leaving 20% of variability in a given population to be influenced by environmental factors such as nutrition and physical activity.

Peak bone mass is the maximal lifetime amount of bone tissue accrued in the skeleton during growth. Low peak bone mass is now considered to be an important determinant of low BMD (osteoporosis) in old age. Maximising peak bone mass during the first few decades of life is currently seen as a primary component of osteoporosis prevention. The period of most rapid skeletal development occurs over several years in childhood and adolescence, accounting for about 50% of the total accrual of skeletal mass. This period is therefore seen as the best opportunity to maximise peak bone mass. Calcium is the major mineral in bone and increasing dietary calcium intake has been proposed to be an effective way of increasing peak bone mass. During periods of most rapid growth and skeletal consolidation (particularly during infancy and adolescence, followed by childhood and young adulthood), a significant increase in calcium requirements has been demonstrated.[22] A number of studies have shown an increase in bone density ranging from 1.6 to 5.1% compared to controls, when dietary calcium is increased to between 1200 and 1600 mg per day.

The interactive effect of calcium supplementation and pubertal status is controversial, and the effects of augmented calcium intake during bone development are not well understood. A recent review of the calcium supplementation trials on BMD in younger and older children indicated inconsistent results, with trials observing significant effects at different pubertal ages and at some bone sites but not others.[23]

## Co-twin Intervention Studies

The co-twin approach is likely to be particularly useful in intervention studies measuring bone-related outcomes during growth. This is because genetic variance in BMD has been shown to surge strongly in adolescence and the co-twin design controls for this powerful source of variation.[9]

The first intervention studies making use of twins in the area of bone health were conducted by Johnston *et al.* who recruited young MZ twins, provided one twin a calcium supplement and one twin a placebo and assessed the effect on bone density after 2 years of intervention.[13] This study clearly demonstrated a positive effect of calcium supplementation on bone density utilising only 45 twin pairs, where bone density increased by 5.1% more at the distal radius in the twin on the calcium supplement. There was no effect of increased calcium on bone density in female twins who were either post-menarcheal or who passed through puberty during the study. However, those twins who were pre-menarcheal showed significant increases in the lumbar spine and distal radius. It should be noted that the numbers in each group (pre-pubertal, $n = 22$; post-pubertal, $n = 23$) were relatively low and that this study was conducted in both boys and girls. It may be relevant that the peak annual accretion of bone occurs earlier in girls (12.5 years) than boys (14.2 years) making it problematic to combine data for girls and boys in this type of study.[24] The above results contrast to our studies in twin girls. In 42 pairs of twin girls with a mean age of 14 years at baseline, we found that calcium supplementation increased both hip (1.3%) and spine (1.5%) BMD significantly compared with co-twin controls over 6 months, despite 74% of our subjects having achieved menarche.[14] These results were not different from a further study in 52 pairs of pre-menarcheal female twins with a mean age of 10.3 years, where calcium supplementation increased BMD in the hip by 1.6% at 6 months, and by a similar amount at the spine at 12 months compared with the placebo group.[15] The increased bone mass was maintained for at least 12 months with calcium supplementation in both age groups, but there was no incremental effect after the initial 6-month gain. Bonjour and co-workers found that supplementation (719 mg calcium per day in foods) of unrelated pre-pubertal females for 48 weeks resulted in a 1.6–2.4% increase in bone density compared with controls at the radial diaphysis, femoral trochanter and femoral diaphysis.[25] In contrast to our study, Bonjour and colleagues did not observe a significant difference in lumbar spine bone density between the two groups after 48 weeks of intervention. After 6 months of intervention, there was no difference between groups at the lumbar spine in our 52 pairs of pre-menarcheal female twins, but by 12 months the within-pair difference in lumbar spine BMD was 1.5% ($p < 0.05$). Therefore, an augmented calcium intake (by 1000–1200 mg per day) for a period of 1 year or more in a population with a mean dietary intake of approximately 700 mg per day (estimated using 4-day food records) increased bone density in children/adolescents. However, we are yet to assess whether this effect is maintained over time.

It is thought that positive results shown in calcium supplementation studies may be explained by remodelling suppression and that any positive effect on bone may be lost once supplementation has ceased. We will be assessing the long-term effect on bone in our twins in their twenties, as there have been conflicting reports regarding the maintenance of increased bone density after calcium supplementation has ceased. Follow-up studies in twins[26] and in unrelated teenagers have indicated that the effect of calcium supplementation on bone density was only temporary and disappeared post-supplementation.[27,28] On the other hand, another childhood calcium intervention study, which utilised milk calcium-fortified foods, reported maintenance of benefit several years post-intervention.[29]

## Co-twin Association Studies: Within-twin Pair Difference Analysis

In addition to utilising the co-twin control design in intervention studies, we have utilised a similar approach by assessing the association of the within-twin pair differences in two parameters. For example, the correlation of the within-twin pair difference in lean mass with the difference in bone density can be assessed.

Measuring the difference in one variable within a twin pair and comparing this to the within-twin pair difference in another variable provides evidence of an association. To assess the association between two variables, regression was performed through the origin as previously described.[10] This co-twin difference method of analysis to evaluate determinants of bone density matches for age, sex and height, and has been shown to reduce the problem of collinearity affecting parameters related to body size, for example, lean mass and fat mass. We initially utilised this approach in 122 MZ and 93 DZ female twin pairs aged 10–26 years and found that the within-twin pair difference in lean mass, not fat mass, was the major independent determinant of the within-pair difference in BMD at the hip, both pre- and post-menarche.[6] We then assessed the longitudinal changes in bone mineral measures in 60 MZ, 44 DZ and 78 unpaired twins in the same age range measured 2.4 times over an average of 1.8 years. This study confirmed the strong association with lean mass such that changes in bone mineral measures were strongly associated with changes in lean mass during linear growth. Changes in fat mass, however, were the predominant although weaker predictor once linear growth had ceased, that is, by 4 years post-menarche.[30] This study clearly demonstrated that a strong association between lean mass and bone mineral measures is established during growth and development and that after the pubertal growth spurt, fat mass emerges as a prominent determinant of bone change in healthy young adult females.

We have also utilised the within-twin pair approach to assess constitutional, dietary and lifestyle determinants of BMD in 146 female twin pairs aged 30–65 years (83 pairs pre-menopausal, 63 pairs post-menopausal).[8] Using this approach we measured BMD at multiple sites, together with lean and fat mass by dual energy X-ray absorptiometry. Height and weight were measured and menopausal status, current dietary calcium intake, current physical activity, lifetime tobacco use and lifetime alcohol consumption were determined by questionnaire. We found that the within-twin pair differences in lean mass and fat mass were positively associated with the within-twin pair difference in bone mass at all sites. In terms of environmental factors, we found that lifetime smoking was negatively related to bone density at all sites except the forearm, with stronger effects in post-menopausal women. In all women, sporting activity was found to have a positive relationship with total hip BMD, with a consistent positive relationship across all bone sites in pre-menopausal women. Results of our within-pair difference analysis indicated that for each 2 h per week of discordance for playing sport, the more active twin had a 1.6% greater total hip BMD. A subgroup analysis in pre-menopausal twins indicated a positive, independent association between current sporting activity and bone mineral measures at all sites. For example, 2 h of sporting activity per week was associated with 1.7% higher total hip BMD. An intervention study of high impact exercise in pre-menopausal women reported an increase in femoral neck bone density of 1.6% compared with the control group suggesting that the estimate of the effect of exercise made in our study is realistic.[31]

There was a suggestion of an adverse effect of alcohol intake but, in the best-fitting model, the association with alcohol consumption was significant only for total body BMC in post-menopausal twins. For the group as a whole, best-fitting models identified significant associations with calcium intake for forearm BMD and for total body BMC (+1.4% per 1000 mg calcium daily for each). In the subgroup analysis of pre- and post-menopausal twins, the only significant association with calcium intake was at the forearm in the post-menopausal subgroup (+2.7% per 1000 mg calcium daily). It is of interest to compare these findings with earlier observations in 215 pairs of 10–26-year-old female twins.[6] In post-menarchial pairs (*n* = 158), there was a univariate association between within-pair differences in calcium intake and total body BMC, but this relationship was not seen at regional bone sites. Moreover, the coefficient for calcium intake became non-significant across all pairs once adjustments were made for menarchial status and anthropometric factors. In post-menarchial twins, a marginal positive association was seen between lifetime alcohol consumption and lumbar spine BMD. There were no significant within-pair associations of lifetime cigarette smoking or of physical activity with any of the bone mineral measures. As stated above, young twins cohabitating tend to share environmental exposures, so the ability of this co-twin observational approach to quantify lifestyle effects will, in general, be limited in this age group. Concerning the assessment of calcium intake and its potential consequences, this limitation is compounded by the problems in dietary calcium assessment[32] and the inherent variability in calcium intake over time in younger subjects. Figure 21.1 shows the variation in recorded intake of calcium in a pair of MZ twins aged 12 years at baseline. Four-day food records were completed at six-month intervals over 18 months. The mean daily intakes of calcium for each twin were 541 and 672 mg.

## Studies Using Discordant Twins

In a study using this approach we investigated the influence of pregnancy and breast-feeding on bone density by assessing the relationship between discordance within twin pairs for number of children and length of breastfeeding and bone

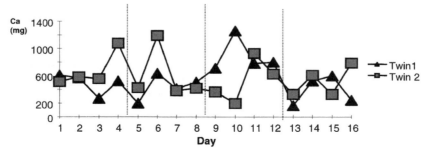

**Figure 21.1** *Daily dietary calcium intake for an MZ twin pair (four 4-day food records over 18 months)*

density at different sites.[11] Pregnancy and lactation put significant stress on maternal calcium homeostasis, potentially resulting in substantial changes in bone mineral. While bone loss in the first 6 months of lactation has been documented, with an estimated 4–6% loss, there is no clear consensus regarding the recovery of bone mineral from the effects of either pregnancy or lactation. If the lost bone is not completely restored, pregnancy and/or lactation may increase the risk of osteoporosis in later life. Alternatively, the temporary reduction in bone density induced by pregnancy and lactation could act as a stimulus to build up additional bone. We measured BMD in female twins, sisters and female relatives aged 18 years and over. In 83 female twin pairs (21 MZ and 62 DZ) with a mean age 42.2 years, who were discordant for ever being pregnant beyond 20 weeks, there were no significant within-pair differences in unadjusted BMD or BMD adjusted for age, height and fat mass at any skeletal site. Similarly, in a larger group with a differing number of pregnancies within a twin pair (498 twin pairs, mean age 42.3 (15.0) years), there was no significant within-pair difference in measures of bone mineral or body composition related to the within-pair difference in number of pregnancies. Cross-sectionally, we found that individuals with one or two ($n = 455$), and three or more pregnancies ($n = 473$) had higher lumbar spine BMD (adjusted for age, height and fat mass) by 2.9% and 3.8%, respectively, ($P = 0.001$), and total body BMC by 2.2% and 3.1%, respectively, ($P < 0.001$) compared with non-parous individuals ($n = 426$). Our results are similar to those of Laskey and colleagues who found that the changes in BMD were reversible and 'do not persist after a subsequent pregnancy, even when conception occurs during lactation at a time when bone loss is still evident'.[33] Therefore, we found no long-term detrimental effect of pregnancy or breast-feeding on bone mineral measures using the powerful within-pair analysis and some evidence from cross-sectional analysis after adjustment for body size to suggest that pregnancy may be associated with increased bone density.

# Conclusions

Twins provide the opportunity to apply several interesting and powerful study designs. Twin studies allow variation in continuously distributed traits such as measures of BMD to be apportioned into genetic and environmental components. Co-twin difference analyses provide valuable evidence of associations between bone density determinants and various bone mineral measures. Data from twin pairs who are discordant for nutritional and other factors are particularly useful. Co-twin controlled intervention trials give greatly enhanced statistical power and have an obvious role in studies of nutritional factors. A number of these approaches have been used successfully to investigate the impact of nutritional factors on bone health. Novel information has been obtained regarding the effects of bone density determinants including soft tissue composition, calcium intake, physical activity, cigarette use and alcohol consumption. The genetics revolution will open the way for wide-ranging studies of gene–environment interactions in relation to nutrition and bone health.

# References

1. F. Galton, The history of twins, as criterion of the relative powers of nature and nurture, *Fraser's Mag.*, 1875 Nov., 566–576.
2. J.C. Christian and K.W. Kang, Efficiency of human monozygotic twins in studies of blood lipids, *Metabol.: Clin. Exp.*, 1972, **21**, 691–699.
3. J.L. Hopper and J.D. Mathews, A multivariate normal model for pedigree and longitudinal data, and the software package "FISHER", *Aust. J. Stat.*, 1994, **36**, 153–176.
4. D.B. Allison, J. Kaprio, M. Korkeila, M. Koskenvou, M.C. Neale and K. Hayakawa, The heritability of body mass index among an international sample of monozygotic twin reared apart, *Int. J. Obes. Related Metabol. Disord.*, 1996, **20**(6), 501–506.
5. N. Martin, Gene-environment interaction and twin studies, In *Advances in Twin and Sib-pair Analysis*, T.D. Spector, H. Snieder and A.J. MacGregor (eds.), Greenwich Medical Media, London, 2000, 143–150.
6. D. Young, J.L. Hopper, C.A. Nowson, R.M. Green, J. Sherwin, B. Kaymakci, M. Smid, C. Guest, R.G. Larkins and J.D. Wark, Determinants of bone mass in 10 to 26 year old females: a twin study, *J. Bone. Miner. Res.*, 1995, **10**(4), 558–567.
7. L. Flicker, J.L. Hopper, L. Rodgers, B. Kaymakci, R.M. Green and J.D. Wark, Bone density determinants in elderly women: a twin study, *J. Bone Miner. Res.*, 1995, **10**, 1607–1613.
8. R.J. MacInnis, C. Cassar, C.A. Nowson, L.M. Paton, L. Flicker, J.L. Hopper, R.G. Larkins and J.D. Wark, Determinants of bone density in 30- to 65-year-old women: a co-twin study, *J. Bone Miner. Res.*, 2003, **18**(9), 1650–1656.
9. J.L. Hopper, R. Green, C. Nowson, D. Young, R.G. Larkins and J.D. Wark, Genetic common environment and individual specific components of variance for bone mineral density in 10-26 year-old females: a twin study, *Am. J. Epidemiol.*, 1998, **147**, 17–29.
10. J.L. Hopper and E. Seeman, The bone density of female twins discordant for tobacco use, *N. Engl. J. Med.*, 1994, **330**, 387–392.
11. L.M. Paton, J.L. Alexander, C.A. Nowson, C. Margerison, M.G. Frame, B. Kaymakci and J.D. Wark, Pregnancy and lactation have no long-term deleterious effect on measures of bone mineral in healthy women: a twin study, *Am. J. Clin. Nutr.*, 2003, **77**(3), 707–714.
12. C. Margerison, L.M. Paton, C. Nowson, S.F. Hossain, B. Kaymakci and J.D. Wark, Hormone replacement therapy and bone mineral density: a co-twin approach, *Menopause*, 2002, **9**, 436–442.
13. C.C. Johnston Jr., J.Z. Miller, C.W. Slemenda, T.K. Reister, S. Hui, J.C. Christian and M. Peacock, Calcium supplementation and increases in bone mineral density in children, *N. Engl. J. Med.*, 1992, **327**, 82–87.
14. C.A. Nowson, R.M. Green, J.L. Hopper, A.J. Sherwin, D. Young, B. Kaymakci, *et al.*, A co-twin study of the effects of calcium supplementation on bone density during adolescence, *Osteoporosis Int.*, 1997, **7**, 219–225.
15. M. Cameron, C. Nowson, R. MacInnis, J. Alexander, A. Sherwin, J. Hopper, B. Kaymakci and J.D. Wark, A co-twin controlled study of the effect of calcium supplementation on bone density during adolescence. Proceedings of the first international conference on children's bone health, *Osteoporosis Int.*, 2000, **11**, S19–S20.
16. N.J. Schork and X. Xu, The use of twins in quantitative trait locus mapping, In *Advances in Twin and Sib-pair Analysis*, T.D. Spector, H. Snieder and A.J. MacGregor (eds.), Greenwich Medical Media, London, 2000, 189–202.

17. M. Peacock, C.H. Turner, M.J. Econs and T. Foroud, Genetics of osteoporosis, *Endocr. Rev.*, 2002, **23**, 303–326.
18. E.A. Krall, P. Parry, J.B. Lichter and B. Dawson-Hughes, Vitamin D receptor alleles and rates of bone loss: influences of years since menopause and calcium intake, *J. Bone Miner. Res.*, 1995, **10**(6), 978–984.
19. S.G. Wilson, P.W. Reed, A. Bansal, M. Chiano, M. Lindersson, M. Langdown, R.L. Prince, D. Thompson, E. Thompson, M. Bailey, P.W. Kleyn, P. Sambrook, M.M. Shi and T.D. Spector, Comparison of genome screens for two independent cohorts provides replication of suggestive linkage of bone mineral density to 3p21 and 1p36, *Am. J. Human Genetics*, 2003, **72**(1), 144–155.
20. D.N. Smith, W.E. Nance, K.W. Kang, J.C. Christian and C.C. Johnston Jr., Genetic factors in determining bone mass, *J. Clin. Invest.*, 1973, **52**, 2800–2808.
21. N.A. Pocock, J.A. Eisman, J.L. Hopper, M.G. Yeates, P.N. Sambrook and S. Eberl, Genetic determinants of bone mass in adults, *J. Clin. Invest.*, 1987, **80**, 706–710.
22. V. Matkovic, Calcium metabolism and calcium requirements during skeletal modelling and consolidation on bone mass, *Am. Clin. Nutr.*, 1991, **54**, S245–S260.
23. K. Wosje and B. Specker, Role of calcium in bone during childhood, *Nutr. Rev.*, 2000, **58**(9), 253–268.
24. C. Molgaard, B.L. Thomsen and K.F. Michaelson, Whole body bone mineral accretion in healthy children and adolescents, *Arch. Dis. in Childhood*, 1999, **81**, 10–15.
25. J.P. Bonjour, A.L. Carrie, S. Ferrari, H. Clavien, D. Slosman, G. Theintz and R. Rizzoli, Calcium enriched foods and bone mass growth in prepubertal girls: a randomised, double-blind, placebo-controlled trial, *J. Clin. Invest.*, 1997, **99**, 1287–1294.
26. C.W. Slemenda, M. Peacock, W. Charles, S. Hui, L. Zhou and C.C. Johnston, Reduced rates of skeletal remodelling are associated with increased bone mineral density during the development of peak skeletal mass, *J. Bone Miner. Res.*, 1997, **12**, 676–682
27. W.T. Lee, S.S. Leung, D.M. Leung and J.C. Cheng, A follow-up study on the effects of calcium supplement withdrawal and puberty on bone acquisition of children, *Am. J. Clin. Nutr.*, 1996, **64**, 71–77.
28. M.J. Merrilees, E.J. Smart, N.L. Gilchrist, C. Frampton, J.G. Turner, E. Hooke and R.L. March, Effects of dairy food supplements on bone mineral density in teenage girls, *Eur. J. Nutr.*, 2000, **39**, 256–262.
29. J.P. Bonjour, T. Chevalley, P. Ammann, D. Slosman and R. Rizzoli, Gain in bone mineral mass in prepubertal girls 3.5 years after discontinuation of calcium supplementation: a follow-up study, *Lancet*, 2001, **358**(9289), 1208–1212.
30. D. Young, J. Hopper, R.J. Macinnins, C.A. Nowson, N.H. Hoang and J.D. Wark, Changes in body composition as determinants of longitudinal changes in bone mineral measures in 8–26 year-old female twins, *Osteoporosis Int.*, 2001, **12**, 506–515.
31. A. Heinonen, P. Kannus, H. Sievanen, P. Oja, M. Pasanen, M. Rinne, K. Ususi-Rasi and I. Vuori, Randomised controlled trial of effect of high-impact exercise on selected risk factors for osteoporotic fractures, *Lancet*, 1996, **16**(348), 1343–1347.
32. C.A. Nowson, A.J. Sherwin, R.M. Green and J.D. Wark, Limitations of dietary calcium assessment in female twins of different ages, In *Challenges of Modern Medicine*, P. Burkhardt and R.P. Heaney (eds.), Ares-Serono Symposia Publications, Rome, Italy, 1995, vol. 7, 97–104.
33. M.W. Laskey and A. Prentice, Effect of pregnancy on recovery of lactational bone loss, *Lancet*, 1997, **349**, 1518–1519.

# CHAPTER 22

# Nutrient–Gene Interactions Influencing the Skeleton

## HELEN M. MACDONALD

Osteoporosis Research Unit, Victoria Pavilion, Woolmanhill Hospital, Aberdeen, AB25 1LD, Email: h.macdonald@abdn.ac.uk

**Key Points**

- Although there is general agreement that the genetic contribution to osteoporosis is high (accounting for up to 80% bone mineral density (BMD)), there is no single gene or handful of genes that dominates BMD or fracture risk.
- Vitamin D receptor (VDR) site polymorphism is the most extensively studied gene in the field of bone health and yet results are inconsistent. Although interaction with dietary Ca could be partly responsible, it is only at low Ca intakes that subjects with BB genotype appear to be slightly disadvantaged. For the normal population with adequate Ca, the role of VDR polymorphism on BMD or fracture risk appears to be negligible.
- The effect of VDR may be influenced through body size or age; and there are many other potential interactions involving other nutrients (vitamin D status, caffeine, fat intake) or other lifestyle factors that could affect bone health outcome in addition to gene–gene interactions.
- The focus of research so far has been on VDR and Ca, but there are many other potential nutrient–gene interactions for which there are plausible reasons for their involvement in bone metabolism (vitamin K or dietary fatty acids with APOE; B vitamin complex with MTHFR; nutrients affecting acid–base balance with genes such as carbonic anhydrase and chloride channel pump).
- Other bone diseases are less well studied but nutrient–gene interaction could influence disease progression (*e.g.*, rheumatoid arthritis) and more research is required in this area.

- Complex statistical techniques and large numbers of subjects are required to adequately study gene-interaction at a population level. Interpretation of studies involving interaction between nutrients or dietary patterns will be difficult and concomitant studies at a cellular level are needed to further progress understanding in this field.

# Introduction

Depending on your viewpoint, the term 'nutrient–gene interaction' can be interpreted in different ways; but a 2-fold definition adapted from Ottman[1] will serve as a starting point to this chapter. Gene–nutrient interaction (or nutrient–gene interaction) is either a different effect of diet on disease risk in persons with different genotypes, or a different effect of a genotype on disease risk in persons with different diets. However, in many cases, the strict statistical definition is not what is implied by the expression 'nutrient–gene interaction' and the term is used to simply describe the interplay between genes and diet. At the cellular level, key nutrients (*e.g.*, fatty acids; vitamin A; 1,25 dihydroxy vitamin D [1,25(OH)2D]) affect gene transcription[2] and can also influence post-transcriptional events.[3] Also, nutrients, in addition to their role as signalling molecules and controlling factors, are required for the basic structure and maintenance of the body. We are literally at least part of what we eat. However, what determines how nutrients came into contact with cellular structures, and what we utilise to grow, develop and live our lives on a day-to-day basis depends on a plethora of other factors. Depending on the nutrient, this could include where the food is obtained; how the food is processed (at farm, factory and how it is prepared at home), whether it is available, at an appropriate price; whether the subject can or wants to eat it; and, depending on the food matrix or if eaten with other foods, how much is absorbed. Different genes can affect how much of a nutrient is absorbed and how much is transported to the target organ (*e.g.*, fatty acids and fat soluble vitamins such as vitamin K). Psychological factors have a role to play, in that we can override our food likes or dislikes. However, our genetic makeup can also influence our food choices. In a twin study, it was found that the amount eaten at a meal was influenced by the time of day meals were consumed, how hungry the subjects felt, their estimation of stomach contents before the meal, and the number of people present at the meal; and that these influences appeared to be heritable.[4] Also, there appear to be genetic differences between tasters and non-tasters of bitter compounds,[5] which suggests that genetic makeup may determine food likes and dislikes.

This introduction illustrates the complexity of interactions and potential confounders involving nutrients and genes. The rest of the chapter focuses on our current understanding of nutrient–gene interactions on bone health from population studies. First, the concept of heritability is covered, followed by a detailed review of the interaction of VDR genotype and dietary calcium, in relation to osteoporosis. Other potential gene–nutrient interactions are discussed, and a brief overview of other bone diseases is included. The final section deals with the problems of statistical analysis of gene–nutrient interaction in population-based studies.

# Heritability

People cannot be studied in a controlled environment (at least not for long periods of time) and it is difficult to work out how much of a trait is due to genes, and how much is due to the environment in normal populations. Comparisons between monzygotic and dizygotic twins who share similar environments but differ in the similarity of their genetic makeup have been used to estimate the heritability of osteoporosis. (A previous chapter is devoted to the study of genetics in twin studies.) From twin studies it was suggested that genes control 70–80% of BMD. This oft-quoted remark has led to the assumption that environmental factors can make only a minor contribution to what is already programmed in our genes. However, as explained by Seeman *et al.* heritability is not a constant and there have been increases in mean height and age of menarche in many populations over the last 50 years, despite these characteristics being under strong genetic control.[6] Furthermore, there are examples of traits that would be considered to be solely 'genetic' in origin (*e.g.*, phenylketonuria) but modifying the environment can influence the outcome (in the case of phenylketonuria, brain damage can be avoided by minimising consumption of dietary amines soon after birth). Also, the heritability obtained from twin studies may be an overestimate given that monozygotic twins may share more environmental factors compared to dizygotic twins.[7] Our genes have not changed in the last 10 000 years but there have been major changes in the diet with increases in saturated fat, and sodium chloride, and a reduction in the ratio of *n*-3 to *n*-6 fatty acids, calcium, potassium, protein and vitamin C.[8] We are also considerably less physically active in the Western world, and our energy requirements are less. Change in lifestyle/diet has been blamed for a number of diseases: the rising obesity amongst Pima Indians;[9] and increase in asthma prevalence as a result of decreased dietary antioxidants[10,11] or increased salt intake.[12] The Western diet has been linked to cancer[13,14] with diet accounting for 35% of all cancer deaths.[15]

Other inherited factors, besides BMD, may affect bone strength, including bone-related factors such as bone architecture, bone size; and also muscle strength and adiposity.[16] In fact muscle strength appeared to explain the small association observed between VDR *Bsm*I genotype and BMD in a group of elderly women with a body mass index of less than 30 kg m$^{-2}$.[17]

# Nutrient–Gene Interactions Affecting Bone Mineral Density

## Calcium and Vitamin D Receptor Site (VDR) Genotype

The VDR genotype present in the q12–14 region of chromosome 12 is currently the most extensively studied gene in the field of osteoporosis.

Yet, whether it has a role to play in the aetiology of osteoporosis is still uncertain.

It has been shown that dietary phosphate deprivation in rats induces gene expression for α1hydroxlase (the renal enzyme that converts 25 hydroxy vitamin D to the active form of the vitamin, $1,25(OH)2D)^{18}$ and dietary phosphate appears to regulate the expression of the gene for 24-hydroxylase (the renal enzyme involved in the catabolism of vitamin D compounds).[19] However, in epidemiological studies, it is dietary calcium in relation to the VDR genotype that has been the focus of attention for nutrient–gene interaction in the field of bone health. Whether the VDR genotype is important in the development of osteoporosis has been a topic of some controversy ever since a link with osteocalcin was reported[20] and it was first suggested that BMD could be predicted from VDR alleles.[21] There are four endonuclease enzymes that have been used in the study of VDR polymorphisms. The restriction enzymes *Bsm*I, *Taq*I, *Apa*I have closely linked sites of action in or around exon 8 and 9 whereas the *Foq*1 site is located in exon 2 towards the 5' end of the gene. The former three sites are in strong linkage disequilibrium and the polymorphisms do not alter the protein sequence of VDR. However, the *Fok*I translation initiation polymorphism produces a shorter (by three amino acids) form of VDR. Although *in vitro* studies suggest that the shortened form has greater transcriptional activity,[22] it is not known whether this mutation has any effect *in vivo* that would result in altered BMD. Recently, the half maximal growth inhibition by 1,25 dihyroxyvitamin D of cultured human blood cells stimulated by phytohemagglutinin was shown to be dependent on VDR *Fok*I genotype and not *Bsm*I genotype,[23] which suggests that there is a potential mechanism for its involvement in bone health.

Most studies have used *Bsm*I restriction enzyme that cuts at intron 8 on chromosome 12.[16] This produces two DNA fragments: for example, depending on the primers used, 650 base pairs and 175 base pairs in length. These can be separated by gel electrophoresis, allowing carriers of this allele to be distinguished from those without the polymorphism, which produces a single 825-base pair fragment.[24] For heterozygotes, all three fragments would be detected. The common consensus is to use the upper case letter 'B' to denote the absence of the *Bsm*I restricted site and lower case 'b' the presence of the restricted site. There appears to be under-representation of the B allele in certain ethnic groups so that the BB homozygotes are almost absent in Chinese populations.

Although Morrison *et al.* concluded that VDR genotype accounted for 75% of the genetic influence on bone mass,[21] reanalysis of the data showed a more modest effect.[25] Other studies have confirmed that BB homozygotes have lowest BMD or greatest bone loss[26,27] but some studies have found no relationship between VDR and bone mass.[28–30] There have also been reports of the opposite effect, albeit weak that lower BMD was associated with the b allele.[31,32] In this instance, it was concluded that the link with BMD was perhaps not due to VDR gene but to another gene that was in strong linkage disequilibrium. Also, it has been found that the upward trend in BMC on moving from BB to Bb to bb genotypes disappeared after adjusting for weight, suggesting that the VDR polymorphism may affect bone mass through its influence on size.[33]

Although, a meta analysis in 1996 found overall that BB had 2% lower BMD at the hip, spine and radius compared to bb, there were wide differences in the extent of the effect observed between different studies.[34] Environmental influences such as dietary calcium intake, which can vary from less than 300 to over 1000 mg a day, could perhaps account for some of the differences. Eisman suggested that low exposures to a particular environmental factor could confer an advantage on one genotype over the other two resulting in greater bone density; but with increased exposure, the order of genotypes with respect to bone density could be reversed, as the environmental change causes a shift in advantage from one genotype to another.[35]

## Confounding with Other Risk Factors: Age and Body Size

It is not certain at what stage in life the influence of VDR genotype is most critical: in-utero, infancy, childhood or adolescence for peak bone mass optimisation; or during post-menopausal or elderly bone loss. There is evidence to suggest that what happens very early on in development affects our future bone mass.[36] A retrospective study found that women with tt genotype (equivalent to BB homozygotes) were 7% heavier than the other genotypes after 1 year (not at birth) but this did not track into adulthood.[37] Australian prepubertal girls with TT (equivalent to bb) had greater BMD simply because TT homozygotes were heavier and taller than their tt counterparts. Again, the differences in height and weight were not there at birth suggesting that VDR may play a role in controlling growth after birth.[38] Interestingly, gender differences in body size have been observed in infants with Bb and bb genotypes but not in BB homozygotes.[39] Also, prepubertal girls of Mexican descent had greater BMD if they were bb homozygotes compared to BB genotype; but in this case weight, height or BMI did not differ between the groups.[40] In prepubertal European children, dietary calcium was found to increase bone size and BMD[41] and this was dependent on VDR genotype.[42]

A more recent examination of papers published on VDR polymorphism found that a positive relationship with BMD was more likely to be observed in studies of women before the menopause.[43] As people age, the accumulation of environmental influences may alter the genetic effect. For example, differences in BMD seen with *Bsm*I genotypes in childhood were not observed in adults.[44] However, there is also a suggestion of age interaction resulting in a stronger gene effect, with tt genotype (equivalent to BB) showing greater bone loss in subjects older than 70 years.[24] A similar effect has been noted in the Col1A1 sp1 binding site polymorphism where the gene effect on BMD increased with age.[45] In addition, there may be interaction with body size with high BMI protecting against the lower BMD seen with the f and b alleles.[46]

## Recent Studies on VDR and BMD

There have been more publications on the subject of VDR and bone since the 1996 meta-analysis, some of which have been discussed above. Again, some studies have reported a weak effect (with B allele associated with low BMD)[47–51] and others a limited or no effect[24,52–57] and one where the b allele was associated with low BMD.[46] A few studies have measured dietary calcium intake (Table 22.1), but most studies have limited power because of small sample size, and other factors such as physical activity level have not been taken into account. Also, the larger studies tend to be heterogeneous in nature, especially with respect to age.

More recent work has tended to use the *Foq*1 enzyme where presence of ATG codon is represented by f and absence by F, whereas in Japanese studies M is used for ATG and m for ACG. Unlike the *Bsm*, *Apa*, *Taq* polymorphisms, this polymorphism appears to have an effect on the function of 1,25(OH)2D at the cellular level.[23] In studies that found a relationship, it appeared that it was the FF (or mm) genotype that was associated with greater BMD, and ff (MM) with lower BMD (or increased bone loss) across a range of ethnic types[22,57–64] However, there were still differences in the site affected (hip or LS) and whether or not there was an association with bone resorption markers. Also, there were still some studies that did not find an association between *Foq*1 polymorphism and BMD[53] (see Table 22.1).

A recent study involving 53 families (630 subjects), in which the proband (the family member chosen for a particular characteristic from whom the rest of the family are recruited) was selected for low BMD. Linkage and association tests were used to explore the relationship between a number of candidate genes and BMD. Interestingly, they found the *Apa*1 and *Fok*1 polymorphisms of the VDR to be associated with lower LS BMD but not FN BMD.[65] In contrast, a similar study with subjects from the Framingham study (332 extended families containing 1062 individuals and a second sample of 169 sibships comprising 284 full-sib pairs) did not provide any evidence for linkage of BMD to chromosome 12q12–14.[66] Again these discrepancies suggest that environmental factors must be involved.

## VDR and Fracture Risk

There have been fewer studies examining VDR genotype and fracture risk. A study comparing 44 vertebral compression fractures with controls showed no difference in genotype distribution.[31] No information on dietary calcium was given. However, the Nurses Health Study found that women with the BB genotype had more than a 2-fold increased risk of hip fracture compared with the bb genotype (although this failed to reach statistical significance). Mean calcium intake was over 1100 mg a day by FFQ and risk was greater for women who were older ($>66$ years), leaner ($BMI < 24.2$ kg m$^{-2}$), less physically active or who had a lower calcium intake ($<1078$ mg day$^{-1}$).[67]

**Table 22.1.** *A selection of recent studies involving the vitamin D receptor site polymorphisms Bsm1 and FokI and its effect on BMD*

| Study | Size (n) | Dietary calcium (approximate) | Summary of findings |
|---|---|---|---|
| Vandevyver et al. 1997[47] | 748 (190 obese) | 600 mg day$^{-1}$ Ca by questionnaire | Non-obese Belgian bb women 5% higher FN BMD compared to BB |
| McClure et al. 1997[48] | 103 | Ca 1.2–1.3 g day$^{-1}$ by FFQ | Mexican American women Trend for BB lower BMD but not significant |
| Sigurdsson et al. 1997[49] | 83 | Icelandic population Ca > 1000 mg | Icelandic women with b allele high bone mass |
| Kung et al. 1998[50] | 144 | Ca intake 500 mg by FFQ | Southern Chinese women bb slightly higher BMD compared to Bb |
| Zmuda et al. 1997[24] | 101 | No details on normal diet | African American women no effect on BMD but BB allele had 14% lower Ca fractional absorption than bb |
| Francis et al. 1997[52] | 48 | Usual Ca not measured | English men – no difference in BMD or fractional Ca absorption between genotypes |
| Laskey et al. 1998[105] | 47 | 25–37 mmol (1000–1480 mg) by FFQ 17–31 mmol (680–1240 mg) by 7-day food diary | Bone loss in breast feeding mothers independent of VDR *BsmI* genotype |
| Kikuchi et al. 1999[51] | 191 | No details on diet | Japanese postmenopausal women with Bb lost BMD faster than bb |
| Lau et al. 1999[54] | 509 | No details on diet | Elderly Chinese men ($n = 237$) and women ($n = 272$) from Hong Kong – no effect of VDR (*BsmI, ApaI, TaqII*) on BMD |
| Bagger et al. 2000[55] | 499 | No details on diet | Danish women – no effect of VDR (*BsmI, ApaI, TaqII*) on bone markers |
| Holmberg-Marttila et al. 2000[56] | 43 | Ca intake > 1000 mg day$^{-1}$ (up to 1.8 g day$^{-1}$) by 1 week Ca intake diary | Postpartum Finnish women – BMD decreased but no effect of *Bsm* I genotype |
| Kubota et al. 2001[57] | 126 | 24 h recall 500 mg day$^{-1}$ | Japanese premeopausal women – no effect of BsmI polymorphism But mm (= FF) greater bone mass and ultrasound stiffness index compared to MM. Low milk intake in childhood more detrimental for MM |

**Table 22.1.** cont.

| Study | Size (n) | Dietary calcium (approximate) | Summary of findings |
|-------|----------|-------------------------------|---------------------|
| Eccleshall et al. 1998[53] | 174 | Ca intake 800 mg day$^{-1}$ by questionnaire | French premenopausal women no assoc BMD and most bone markers but NTx resorption higher in ff compared to FF |
| Gross et al. 1996[59] | 100 | Ca 1100 mg day$^{-1}$ but no details on method | Mexican American ff lower LS BMD and greater BMD loss at hip but no bone markers |
| Harris et al. 1997[60] | 154 | Dietary Ca by FFQ: 500 mg day$^{-1}$ for black women 800–900 mg day$^{-1}$ for white women | Premenopausal American women (72 black and 82 white) – ff genotype in white women had 12.1% lower FN BMD and 4.3% lower whole body BMD compared to FF but no difference in LS. No difference in BMD at any site for black women |
| Arai et al. 1997[22] | 110 | No details on diet | Japanese premenospausal women 12% greater LS BMD for mm compared to MM |
| Ferrari et al. 1998[61] | 761 | 800 mg day$^{-1}$ by FFQ | Swiss girls and premenopausal women. Trend for FF greater BMD at higher Ca intake (Daily mean 1–1.2 g cf 600 mg) |
| Gennari et al. 1999[62] | 400 | Calcium questionnaire: 550–600 mg day$^{-1}$ | Postmenopausal Italian women – higher prevalence of ff in osteoporotic and in vertebral Fx cases. Weak assoc with LS BMD (not FN BMD) – ff lower BMD and more pronounced in women < 60 year |
| Cheng and Tsai 1999[58] | 101 | Dietary Ca by FFQ 500 mg day$^{-1}$ | Late preomenopausal Taiwanese women – no difference in bone turnover markers or BMD at LS, FN or whole body |
| Kanan et al.[63] | 96 | No details on diet | English men (median age 50 year). LS BMD greatest for FF cf ff genotypes. No effect on FN BMD |
| Tofteng et al. 2002[46] | 429 | 4–7 day diet records Ca 800 mg day$^{-1}$ | Danish women (BMI 25 kg m$^{-1}$ b allele low LS BMD and f allele low hip (FN and total) BMD |
| Lau et al. 2002[64] | 684 | 500 mg day$^{-1}$ for early postmenopausal women and 300 mg day$^{-1}$ in elderly but no details on method | Chinese men and women. No effect on BMD in men and younger women but weak effect seen in elderly women (70–79 year). No effect on vertebral Fx. No interaction with Ca |

A smaller study showed a trend towards a lower prevalence and incidence of vertebral fractures with women bb homozygotes although this just failed to reach statistical significance ($p = 0.07$) and, in this study, calcium intake data were collected in the form of low or medium to high intake of dairy products.[68] However, no effect of VDR genotype on fracture was seen for the Study of Osteoporotic Fractures, in which fracture subjects ($n = 531$) and controls were selected from a cohort of 9704 elderly women, and mean calcium intake was 740 mg day$^{-1}$ by FFQ.[69] Also, no differences in VDR genotype distribution were found between hip fracture patients ($n = 153$) and controls ($n = 239$) [70] In this study, the patients were significantly taller with low BMI and although no details on dietary calcium were given their mean 25-hydroxyvitamin D (25(OH)D) level (a marker of vitamin D status) was markedly lower compared to the controls. Although the *Fok*I polymorphism had no effect on fracture risk in a comparison of 192 fractures with 207 controls, the B allele was associated with increased fracture risk. However, it was the inverse 'baT' VDR haplotype that was over-represented among fracture cases ($p = 0.009$), corresponding to an increased risk of 1.8 for heterozygous carriers and 2.6 for homozygous carriers of the risk haplotype.[71] The effect was similar for vertebral and non-vertebral fractures and was independent of BMD. Dietary calcium in this cohort was around 1100 mg a day by FFQ. As with the BMD findings, the discrepancies could be explained by dietary confounders or linkage disequilibrium.

## Calcium–VDR Gene Interaction Studies

A few studies have specifically examined whether there is an interaction between VDR genotype and a particular dietary factor (mainly calcium, but also vitamin D[72] and caffeine[73]) and these are summarized in Table 22.2. The results of the intervention studies show similar trends with BB having lower BMD or greater bone loss. In terms of calcium (Ca) interaction, Ca supplementation protected against BMD loss in elderly women with BB genotype,[74] elderly men and women with Bb genotype[75] and both BB and Bb genotypes for prepubertal girls.[42] It was found, on changing from a low to a high Ca intake, that women with BB had a smaller increase in fractional Ca absorption compared to bb.[76] Similarly, men with BB had greater serum PTH levels compared to the Bb or bb genotypes whether on a low or high Ca/phosphate diet, but on the low Ca/phosphate diet they also had reduced phosphate reabsorption and lower plasma phosphate levels.[77]

Two observational studies on dietary Ca shown in Table 22.2 show markedly different results. For late premenopausal women, BMD was greater for BB and Bb genotypes when daily Ca intake was greater than 1036 mg, whereas the BMD of bb homozygotes was unaffected by dietary Ca.[78] However, for elderly Framingham subjects, the opposite was true, with greater BMD observed for bb genotypes when Ca was more than 800 mg a day (compared to less than 500 mg day$^{-1}$) and the BB and Bb genotypes showed no relationship between BMD and Ca.[79] A more recent study found that elderly post-menopausal Australian women homozygous for T (equivalent to bb) lost bone less rapidly

**Table 22.2.** Studies with nutrient–gene interaction involving the vitamin D receptor site polymorphism Bsm1 (BB, Bb, bb) or Taq1 (TT, Tt, tt)

| Study | Subjects | Intervention | Genotype | Findings |
|---|---|---|---|---|
| Ferrari et al. 1995[75] | 72 Elderly patients (mainly women) Mean age±SD: 73±7 years Switzerland | Once only 300 000 IU vitamin D Ca supplements (800 mg day⁻¹) Normal Ca 550 mg day⁻¹ Time: 18 months | BB 9 Bb 37 bb 26 | BB had greatest LS bone loss in spite of Ca supplement. Subjects with bb had least bone loss and Ca gave no added benefit. For Bb, bone loss was inversely related to Ca intake |
| Krall et al. 1995[74] | 229 Postmenopausal women. Mean age±SD: 59±6 years US | Two thirds took Ca supplement (500 mg day⁻¹) to make total Ca 892 mg day⁻¹ Remainder (placebo) Ca from diet 376 mg day⁻¹ Time: 2 years | BB 44 Bb 102 bb 83 | For women > 10 years postmenopause BB had lowest FN BMD and greatest bone loss. Bone loss appeared to be protected by Ca |
| Dawson et al. 1995[76] | 60 postmenopausal women. Mean age±SD: 68±6 years US | High Ca diet (1500 mg) 2 weeks Low Ca diet (<300 mg day⁻¹) 2 weeks | BB 26 bb 34 | Absorption of Ca was similar for BB and bb at low Ca intakes. At high Ca intake, BB had smaller increase in fractional Ca absorption compared to bb |
| Salamone et al. 1996[78] | 470 Premenopausal women aged 44–50 years US | Observational | BB 97 Bb 184 bb 189 | Greater BMD for BB and Bb when Ca intake > 1036 mg day⁻¹ Women with bb unaffected by Ca |
| Graafmans et al. 1997[72] | 81 women aged > 70 years Netherlands | Vitamin D supplementation 400 IU for at least 2 years | BB 23 Bb 39 bb 19 | Vitamin D supplements showed greater response in Bb and BB women compared to bb |
| Ferrari et al. 1998[42] | 108 Prepubertal girls Mean age±SD: 7.9±0.7 years Switzerland | Ca supplements (added to foods) 850 mg day⁻¹ Time: 1 year | BB 14 Bb 47 bb 40 | 108 out of 144 completed study. If Ca intake > median (880 mg day⁻¹) Ca supplement showed increase in BMD (50–70%), bone area and height. FN BMD was greater for bb and Ca had no effect, but Ca supplement increased BMD for BB and Bb. No effect on LS BMD for any genotype |

cont.

| Reference | Subjects | Method | Genotype counts | Results |
|---|---|---|---|---|
| Kiel 1997[79] | 328 Elderly Framingham subjects aged 69–90 years US | Observational Ca estimated from Food frequency questionnaire | BB 46 Bb 161 bb 121 | Greater BMD for bb with Ca > 800 mg day$^{-1}$ compared to Ca < 500 mg day$^{-1}$ but no relationship between BMD and Ca for BB or Bb |
| Ferrari et al. 1999[77] | 25 Men aged 24.3±3.1 years Switzerland | Low Ca phosphorous diet for 5 days and high Ca phosphorous diet for 5 days | BB 10 bb 15 | BB subjects had greater serum PTH levels on either diet, compared to Bb or bb. On low Ca/Pi diet BB had reduced Pi reabsorption and plasma Pi levels |
| Rubin et al. 1999[81] | 677 young women aged 18–35 years Canada | Observational Ca estimated from FFQ (mean 460 mg day$^{-1}$) | tt (= BB) 108 Tt 320 TT (= bb) 249 | Significant interaction between Ca and genotype (P < 0.01) was suggested but difficult to assess this from the 3-D figure. Quartile group > 684 mg day$^{-1}$ had greatest FN BMD compared to other Ca quartiles for all genotypes, and tt greatest FN BMD compared to other genotypes across the Ca range |
| Brown et al. 2001[80] | 193 elderly postmenopausal women mean age 69 years Australia | Observational Ca estimated by FFQ: Lowest third: 100–456 mg day$^{-1}$ Middle third: 461–592 mg day$^{-1}$ Highest third: 705–2237 mg day$^{-1}$ | tt (= BB) 38 Tt 77 TT (= bb) 74 | TT homozygotes (i.e., bb) lost bone less rapidly from FN and LS. Interaction with Ca at FN (not LS). Also noted *Taq1* VDR association with BMD more marked at low Ca intakes (both sites affected) compared to middle Ca intake (FN only affected) and highest Ca (neither site affected by VDR) |
| Rapuri et al. 2001[73] | 96 Elderly women aged 65–77 years US | Observational 7 day food diary Mean daily Ca intake 800 mg Time 3 years | tt 11 (= BB) Tt 38 TT 38 (= bb) | Caffeine increased bone loss of spine. Women with tt (BB) had higher rates of bone loss than women with TT (bb) when caffeine intake greater than 300 mg day$^{-1}$ |

over 6 years compared to the other genotypes and this was more marked at low Ca intakes ($<460$ mg day$^{-1}$).[80] Although, a Canadian study also claimed to observe interaction between genotype and Ca intake, it was not clear which genotype was selectively influenced by Ca intake.[81]

Eisman's model[35] might explain some differences, but it does not explain why there are differences when calcium intakes are similarly high for different studies, especially if one accepts the argument that there is a threshold for dietary calcium intake, above which there is no additional benefit.

## Other Potential Dietary Confounders in Nutrient–VDR Gene Interaction Studies

Calcium is not the only nutrient that could influence VDR phenotype. Vitamin D supplementation showed greater response in BB and Bb genotypes compared to bb homozygotes, as shown in Table 22.2.[72] Vitamin D status can be affected by dietary intake but it is through sunlight exposure that the normal population obtain vitamin D. In the US, where milk is supplemented with vitamin D (but not in the UK) there is a high degree of correlation between calcium intake and vitamin D intake, as noted in a study finding lower incidence of colorectal cancer in BB homozygotes at low dietary vitamin D intake.[82] Also, there is controversy about how important meat is in relation to vitamin D status. Improved techniques for measuring vitamin D and its metabolites (25(OH)D; 1,25(OH)2D and 24,25(OH)2D) have led to the introduction of vitamin D data for meat into the UK National Food Survey in 1995/6.[83] Using a 5 times potency factor for 25(OH)D resulted in an apparent increase in dietary vitamin D intake in 1995 and 1996; and it now appears that meat may be the most important natural source of vitamin D in the UK diet,[84] with meat and meat products providing 10% of the dietary vitamin D. However, there is still debate regarding what value should be used for the potency factor. Vitamin A may play a role by antagonizing the action of vitamin D as shown in rats.[85] One explanation for this antagonism, at a molecular level, is that both vitamins use retinoid X receptor proteins for heterodimer formation during transcription. Vitamin A has been reported to be detrimental to bone health in population-based studies, with elevated fracture risk associated with intakes greater than 1.5 mg retinol or 3 mg total vitamin A.[86,87] However, it is possible that these findings might be caused by other components in the diet.

Since dietary phosphate has been shown to influence gene expression of the renal enzymes required for vitamin D activation and catabolism,[18,19] it is likely that phosphate will also affect calcium absorption. A low ratio of calcium to phosphorous (e.g., 0.25) can result in persistently elevated parathyroid hormone (PTH) levels that are detrimental to bone health of young, and possibly older, women.[88] Calcium absorption is influenced by whether it is given in a single dose or throughout the day.[89] Bone metabolism follows a circadian rhythm[90,91] and in post-menopausal women, calcium intake in the evening markedly suppressed the nocturnal rise in bone turnover markers whereas calcium taken in the morning had no effect.[92] Fasting was also found to diminish this circadian rhythm.[90] In subjects with established osteoporosis it was noted that although calcium intake in the evening increased urinary calcium the following morning, it had only a marginal influence on bone resorption markers.[93]

It is important to remember than not all Ca absorption is regulated by vitamin D; but that there is a passive pathway that is non-saturable and depends on quantity and availability of Ca in the diet. It has been suggested that intakes above 3 mmol (120 mg) Ca in a meal will be absorbed in this way.[94] Also, there are a number of other dietary factors that can affect how much calcium is taken up and retained. Calcium is poorly absorbed and what food is eaten influences the amount absorbed. Data collected by Weaver *et al.* show that the Ca in certain foods thought to be rich in Ca such as spinach and rhubarb is poorly absorbed due to the presence of oxalates.[95] A number of studies have found that dietary fat is associated with reduced BMD and we have suggested that dietary fat may reduce absorption of Ca intake through formation of divalent ion soaps in the intestine (Macdonald *et al.* unpublished). It has been noted in the rat that increasing dietary Ca intake increased the concentration of long chain fatty acids (both saturated and unsaturated), chelated with Ca, in the faeces.[96] Ca absorption is also known to decrease with age,[97] which may be a result of reduced 1,25(OH)2D levels[98] caused by impaired cutaneous synthesis of vitamin D[99] and reduced renal $\alpha$1 hydroxylase activity.[100]

There are a number of dietary components that can increase Ca urinary losses. Caffeine is known to increase calcium excretion and it was found that the effect of VDR on BMD in elderly women was dependent on caffeine intake.[73] Urinary calcium increases with increased dietary sodium intake[101] and with increased protein intake.[102] However, a low protein diet reduces calcium absorption.[103]

The number of dietary factors that can affect how much calcium is absorbed and lost highlights one aspect of the problem of diet measurement. Even if the amount of nutrient retained from the type of food eaten could be accounted for, along with any other confounding influences, such as the time of day the food is eaten, we still assume that people eat what they tell us they eat. However, we know that this is not the case, with under-reporting a common occurrence, especially in the overweight.[104] There may also be differences in calcium intake according to the dietary measurement technique employed (7-day weighed intake *vs* food frequency questionnaire) as can be seen in the study of Laskey *et al.* as shown in Table 22.1.[105] Dietary energy intake is also a possible confounder, with high calcium diets likely to contain more energy compared to low calcium diets. For subjects of similar weight, a difference in energy intake suggests differences in physical activity levels, which will confound studies of bone health.

If one considers the number of confounding factors purely from diet alone, it is perhaps not surprising that there is so much inconsistency in the findings between different studies examining the same gene. In conclusion, at low Ca intakes subjects with BB genotype appear to be slightly disadvantaged, but for the normal population with adequate Ca intake the issue of VDR genotype appears to be largely irrelevant.

## Other Potential Gene–Nutrient Interactions

There are many other potential gene–nutrient interactions that could influence bone health besides VDR polymorphism with Ca and vitamin D. Hormone replacement therapy has been considered as an environmental factor that can interact with certain gene polymorphisms. For example, HRT helps protect against the accelerated bone loss seen in the rare 'ss' homozygotes of the COL1A1 polymorphism around the time of menopause[106] and the bisphosphonate etidronate selectively improved FN BMD of SS but not those carrying the s allele.[107] HRT was shown to influence the effect of estrogen receptor site polymorphism[108] and it was found that muscle strength and BMD response to HRT was dependent on angiotensin-I converting enzyme (ACE) genotype (an insertion allele being associated with lower serum ACE activity and improved muscle efficiency in response to physical training).[109] It is, therefore, possible that dietary phytoestrogens may contribute a weak environmental influence on the effect of certain polymorphisms on markers of bone health, through their actions on estrogen receptor sites.[110] Dietary contaminants such as bisphenol have been shown to influence the effect of the aromatase gene in mice,[111] which suggests that there is the potential that dietary components may interact with similar genes in humans with resultant changes in hormone levels causing altered bone metabolism. Also the heterocyclic amine, 2-amino-1-methyl-6-phenylimidazo[4,5-b]pyridine (PhIP), which originates in the diet from cooked meats or meat products, and is structurally similar to estradiol, may interfere with $\alpha$ estrogen receptor, progesterone receptors and has been shown to affect prolactin secretion in rats,[112] which could impact on bone metabolism.[113] Response to HRT was found to be greater in the pp homozygotes of the PvuII polymorphism of the $\alpha$ estrogen receptor site[108] and our group found that the beneficial effect of alcohol intake was more pronounced in those carrying the p allele.[114] One theory put forward to explain the beneficial effects of modest alcohol intake on bone health was that alcohol stimulates the adrenal production of androstenedione and conversion to estrone and it has been observed that alcohol intake was positively associated with estrone and estradiol levels.[115] However, there is likely to be confounding with other genes for example those that control alcohol metabolism.[116]

There are examples of candidate genes that influence BMD for which there are environmental interactions that affect other disease outcomes, for example methylenetetrahydrofolate reductase (MTHFR) genotype and dietary folic acid in relation to plasma homocysteine levels (an independent risk factor for coronary heart disease).[117] Other examples of interactions that could influence bone health include dietary components that may affect leptin expression. An association between serum leptin levels and BMD was observed in non-obese

**Table 22.3.** *Plausible potential gene–nutrient interactions*

| Candidate gene | Nutrient |
|---|---|
| Vitamin D Receptor (VDR) | Calcium |
| | Phosphate |
| | Dietary vitamin D (and sunlight exposure) |
| Apolipoprotein E (APOE) | Vitamin K |
| | Dietary fat |
| Methylenetetrahydrofolate reductase (MTHFR) | B vitamins: Vitamin B12, Folate, riboflavin |
| | Copper |
| | Alcohol |
| Collagen 1 alpha 1 Sp1 binding site (COL1A1) | Vitamin C, Micronutrients affecting collagen synthesis (Zn, Cu) |
| | Phytoestrogens |
| Estrogen Receptor | Alcohol |
| | Phytoestrogens |
| Metallothionein Chloride channel (ClC) | Dietary zinc |
| | Acid–base balance (protein versus fruit and vegetable intake) |
| | Phytoestrogens |
| Carbonic anhydrase | Acid–base balance |
| Calcium sensing receptor (CaSR) | Calcium |
| Aromatase gene | Dietary contaminants such as bisphenol |
| | Phytoestrogens |
| Tumour necrosis factor (TNF)–alpha/ Interleukin (IL-6)/Tumour necrosis factor receptor (TNFR) | Dietary fatty acids, Phytoestrogens |

women[118] and leptin has been proposed as the link between fat and bone mass.[119] At a functional level, metallothionein expression was shown to be dependent on dietary zinc in rats and may be important in the development of bone marrow cells.[120] Transferrin polymorphism affects metabolism of vitamin C, but only in black individuals, as the common allele is found exclusively in Caucasians.[121] Vitamin C is important for cross-linking collagen and can act as a specific inducer of collagen synthesis in cells producing type I collagen.[122]

Table 22.3 lists some possible nutrient–gene interactions. The list is not exhaustive and merely highlights a few areas where nutrient–gene interactions are plausible. Three potential interactions are discussed in more detail: APOE with vitamin K or dietary fat; MTHFR with B vitamins, carbonic anhydrase and/or chloride channels with diet in relation to acid–base balance.

## Vitamin K or Dietary Fat Interaction with Apoliprotein E (APOE) polymorphism

The apolipoprotein E gene (APOE) is located on chromosome 19q13.2. There are 3 alleles: numbered epsilon 2, 3 and 4, each producing a protein that differs in amino acid sequence at positions 112 and 158. APOE2 has a cysteine at both positions; APOE4 an arginine at both positions; and APOE3 a cysteine at 112 and an arginine at 158. Also, the amount of protein produced by each genotype differs, with greater concentrations of lipoprotein being produced for APOE2 compared to APOE4. Since the APOE3 allele is usually present in 60% or more of the population, the genotype 3/3 tends to be the most common followed by 4/3,3/2, 4/4, 4/2, 2/2. There are differences in the frequencies of APOE2 and APOE4 between different populations with Northern Europeans having twice as many APOE4 compared to Southern Europeans (14–19% compared to 7–12%). In Native Americans, the APOE2 allele appears to be rare (2–4% of the population). APOE has also been reported to be associated with differences in risk of premature atherosclerosis and development of Alzheimer's disease.

There is recent evidence to show that the presence of the apolipoprotein APOE4 genotype is associated with reduced BMD;[123] with any fracture;[124] with hip fracture risk independent of BMD;[125] and with rates of bone loss in perimenopausal women not receiving HRT[126] (although no effect on BMD was seen in a Finnish study of early post-menopausal women[127]). Older women (>65 years) with APOE4 had greater risk of severe vertebral deformities but in men, the association between APOE4 and BMD was only seen in the group aged 65–69 years and not in the older age group.[128] Furthermore, for women not taking HRT, APOE2 was associated with lower rates of bone loss at the hip and increased serum alkaline phosphatase activity.[129] It was suggested that APOE alleles may alter vitamin K transport in blood[130] and that this in turn may affect bone health, because vitamin K is necessary in the post-translational modification of the osteoblast-specific protein, osteocalcin where glutamic acid is converted to γ-carboxyglutamic acid. Levels of undercarboxylated osteocalcin have been associated with increased risk of hip fractures[131,132] and lower levels have been found in post-menopausal women with osteoporosis.[133]

Although no interactions of this gene with vitamin K have yet been reported, this may in part be due to lack of food composition data for vitamin K. (In the UK, preliminary food composition tables for vitamin K have been published in an effort to address this shortfall.[134]) It has also been suggested, however, that APOE may have a direct effect on bone metabolism or its influence may be through other well-known effects of APOE genotype such as lipoprotein metabolism and its role in atherosclerosis.[129] It is interesting that an atherogenic high-fat diet reduced bone mineralization in mice, which appeared to be the result of blocking differentiation of osteoblast precursors.[135] It was suggested that oxidised lipids could be responsible for osteoporosis in addition to coronary heart disease since the two diseases are linked epidemiologically.[136] Also, alcohol appears to decrease LDL cholesterol in men with the APOE2 allele but increase LDL cholesterol in men with the APOE4 allele.[137]

## B Vitamin Interaction with Methylenetetrahydrofolate Reductase Genotype

Methelenetetrahdyrofolate reductase (MTHFR) polymorphism C to T produces an enzyme, which is heat labile and less active than the wild type.

This enzyme removes homocysteine and requires folate and vitamin B12 as cofactors. The MTHFR genotype has been studied in relation to risk of coronary heart disease and plasma homocysteine levels since raised homocysteine is an independent risk factor for coronary disease.[138,139]

As there is a high prevalence of osteoporosis in homocystinuria, a condition characterized by very high levels of homocysteine,[140] it is possible that slightly elevated levels may be involved in the aetiology of osteoporosis in the normal population. Homocysteine competitively inhibits lysyl oxidase, an enzyme that is involved in the synthesis of cross-links that stabilise the collagen fibrils[141] and there is evidence that deficient collagen cross-linking is to blame in patients with homocystinuria.[142] Homocysteine levels increase with age,[143] after the menopause[144] whereas HRT reduces homocysteine levels.[145]

MTHFR polymorphism was associated with low bone mass in a Japanese population.[146] A diet high in folate was shown to affect serum folate and homocysteine dependent on MTHFR genotype with greatest reduction of homocysteine (18%) seen in the TT homozygotes.[147] A similar reduction of 20% was also observed for TT homozygotes when folic acid supplements were given and these appeared to be more effective compared to 5 methyltetrahydrofolate.[117] McNulty *et al.* found lower red blood cell folate and higher homocysteine in the TT homozygotes but further analysis of the data showed that this was dependent on riboflavin status (riboflavin being the precursor for FAD the cofactor for MTHFR).[148] Also, as copper depletion is known to raise homocysteine levels and copper 'insufficiency' has been implicated in osteoporosis,[149] the picture becomes more complicated. However, if the population under investigation is replete in certain nutrients but not others, a suitable hypothesis can be generated and tested to examine whether the effect of the MTHFR in bone health is mediated by diet.

## Acid–Base Balance Interaction with Carbonic Anhydrase and Chloride Channel Pump

The acidity of the diet has been implicated in bone health[102] and in the late 1960s Wachman & Berstein proposed a model of osteoporosis based on the chronic degradation of bone to balance this acidity.[150] At the cellular level, using mouse calveriae, it was noted that by lowering the bicarbonate concentration of the medium, calcium efflux increased.[151] In vitro experiments using rat calverial cultures found that bone resorption, as measured by calcium release into the media, was stimulated in the presence of 1,25(OH)2D and prostaglandin E2. An increase in carbonic anhydrase activity concomitant with the Ca release was also observed within 2–3 days.[152,153] Carbonic anhydrases are zinc metalloenzymes that catalyse the reversible hydration of $CO_2$ to form $HCO_3^-$ and protons according to the reaction: $CO_2 + H_2O \leftrightarrow H_2CO_3 \leftrightarrow HCO_3^- + H^+$. The first reaction is catalysed by carbonic anhydrase and the second reaction occurs instantaneously.[154] It was found that 1,25(OH)2D regulated the expression of carbonic anhydrase in bone marrow[155] and acid pH stimulated the production of carbonic anhydrase in mature osteoclasts.[156] It was suggested that $HCO_3^-$ acidosis stimulates resorption by activating mature osteoclasts already present in calvarial bones, rather than by inducing formation of new osteoclasts.[157] It was recently found that the carbonic anhydrase gene is the target for a transcription factor that is required for osteoclast differentiation.[158]

Bone resorption by osteoclasts requires massive transcellular acid transport, which is accomplished by a proton pump coupled to a chloride channel in the ruffled border of the osteoclast.[159,160] Mice deficient for the chloride channel ClC-7 exhibit severe osteopetrosis, a condition characterised by increased BMD. This appeared to be a result of defective osteoclasts, unable to resorb bone, and it was suggested that the ClC-7 channel provides the chloride conductance required for an efficient proton pumping by the vacuolar H(+)-ATPase of the osteoclast ruffled membrane.[161] Furthermore, mutations in the ClC-7 chloride channel gene were associated with Albers–Schonberg disease in humans, a type of autosomal dominant osteopetrosis.[162]

One of the explanations for the beneficial effect of fruit and vegetables in bone health is in restoring acid–base balance. If this hypothesis is correct, the effect of carbonic anhydrase or chloride channel genotypes may depend on the acidity of the diet. Even small changes in pH that are still within physiological range have been shown to affect bone resorption.[163] Furthermore it has been found that genistein a dietary phytoestrogen, which is an inhibitor of tyrosine kinase, inhibits bone resorption.[164] It was suggested that tyrosine kinase inhibition directly inhibits osteoclast membrane hydrochloric acid transport. Therefore, it is possible that dietary phytoestrogens may also modify the influence of the chloride channel genotype.

# Nutrient–Gene Interactions and Other Diseases of the Skeleton

There has been very little work done in relation to gene–nutrient interaction on other major bone diseases such as osteoarthritis and rheumatoid arthritis.

The *Taq*I polymorphism of the VDR appears to be associated with osteoarthritis of the knee, with the T allele associated with increased risk compared to the t allele.[165] No association was observed in a Japanese population but tt (or BB) genotypes are extremely rare in this population.[166] The COL2A1 gene, which is also present on chromosome 12q, has also been implicated in osteoarthritis at the knee.[167] However, other studies have not found an association between COL2A1 and osteoarthritis.[168,169]

The *Bsm*I VDR polymorphism has been investigated in relation to rheumatoid arthritis and the BB/tt genotype was found to be weakly associated with early onset rheumatoid arthritis[170] with some evidence of VDR fos-c gene interaction at the molecular level.[171] Again, other studies have not confirmed this.[172–174]

As with osteoporosis, results are mixed and dietary factors tend to have been disregarded; and there is potential for gene–nutrient interaction to affect other diseases of bone.

# Statistical Methods for Examining Nutrient–Gene Interaction in Epidemiological Studies

## Statistical Models

There are different models of gene–nutrient interaction depending on the subject and area of expertise of the investigator. There is ambiguity in use of the term 'interaction' between the statistical definition and biological/causal interpretation[175] with confusion in definitions and difficulties in biological interpretation arising in studies examining gene–gene interaction in complex disease.[176]

Ottman suggested five models for use in studying disease risk in relation to genotype and environmental exposure, generated from a purely statistical viewpoint, but with plausible biological examples.[1] The relative risks or odds ratios can be displayed in a 'two by four table' with the three levels of genetic exposure possible with a bi-allelic gene (absence, one or two alleles) reduced to presence and absence of allele; and diet categorized into presence or absence of dietary component or more usually into low and high intakes. Suggested values of relative risk are shown in Table 22.4 to illustrate the differences that might be seen between Ottman's five models.

**Table 22.4.** *Possible values of relative risk (RR) for different models of nutrient–gene interaction*

| Model | RR | | | | |
|---|---|---|---|---|---|
| | Gene present | | Gene absent | | |
| | High intake | Low intake | High intake | Low intake | |
| A | 2 | 1 | 2 | 1 | |
| B | 6 | 1 | 2 | 1 | |
| C | 6 | 2 | 1 | 1 | |
| D | 2 | 1 | 1 | 1 | |
| E | 6 | 2 | 2 | 1 | |

The unexposed normal population (without the high risk gene) is the reference. Model A: Increased risk due to environmental exposure (EE) is the same with or without the susceptible gene (SG) (*i.e.*, no interaction). Model B: EE causes increased risk in the normal population but greatly increased risk in SG. Model C: Normal population is unaffected by EE, but SG show increased risk in absence of EE and greatly increased risk with EE. Model D: Normal population (with or without EE) and unexposed SG show no symptoms, but there is increased risk for SG with EE. Model E: Increased risk for both exposed normal population and unexposed SG but there is greatly increased risk for the SG with EE.
Adapted from Ottman 1996.[1]

There has been much disagreement as to whether additive or multiplicative models are more appropriate in measuring gene environment interaction.[1] Some have suggested that if the disease involves a multistage process, then two factors acting at the same point would fit an additive model, but if they acted at different stages then they would fit a multiplicative model. However, the scale can depend on whether the main goal is to understand disease aetiology (in which case it is suggested that a multiplicative scale should be used) or whether it is to predict the number of cases in the population (in which case an additive scale is more appropriate).[177] In a recent review, it was suggested the multiplicative model is appropriate when measuring ratio of disease incidence between exposed and unexposed subjects, as in a case control design, but that the additive model could be used when measuring a rate difference between two factors.[175]

## Power and Sample Size

There is little doubt that power and sample size estimations are critical for the design and statistical evaluation of epidemiological studies investigating gene–environment interaction. Hwang *et al.* suggested that sample sizes of 600 (200 cases, 400 controls) were required to detect a 4-fold difference in odds ratio for exposure, without and with the susceptible genotype.[178] This approach corresponded to model B described by Ottman above and was limited by exposure being reduced to a binary variable. Dietary intake is not a dichotomous variable and although subjects can be categorized according to low or high intakes, there may be difficulties in determining a suitable cut-off point. Foppa and Spiegelman extended their method based on Ottman's model E above to include continuous exposure variables.[179] They found that sample sizes of less than 1000 were adequate only if the interaction was >6 and the gene relatively common (>10% for 80% power and >25% for 90% power). Sample sizes of tens of thousands would be needed to detect an interaction of 1.5, with greater numbers being required if the gene was rare (5–10% of the population). Both these approaches were criticised for overestimating power (and underestimating sample size) but they provide approximate estimates when the interaction is small, or the odds ratios for the genetic and exposure effects are small.[180] Case only designs[181] and counter-matching[182] have been suggested as ways for improving efficiency in detecting gene–environmental interactions, particularly when the gene is rare. Clayton and McKeigue reviewed gene–environment interaction in relation to complex diseases and concluded that for modest risk ratios, the case control design is more feasible than the cohort design because of the issue of statistical power.[175]

Most theoretical epidemiological studies have concentrated on disease risk and generally assumed that the environmental exposure is dichotomous. However, in the case of bone health, a disease outcome such as fracture risk is not feasible for many studies. The surrogate BMD (usually areal BMD measured by DXA) or BMC is most often used. Recently, Luan *et al.* proposed sample sizes based on a linear regression model, relating a continuous outcome (which could be BMD) to a continuously distributed exposure variable (which could be dietary calcium intake).[183] The slope of each regression line depends on genotype and interaction term is given by the ratio of the slopes for each genotype. However, the sample sizes required for adequate power are still prohibitive for many studies and there may be other ways in which knowledge of genotype can help further understanding.

## The Use of Genetic Information in Population Studies

One way of using our knowledge of genetics is to be able to design intervention studies involving nutrients/diet and remove potential confounding by genotype by imposing stratified randomisation of genotype prior to intervention. Alternatively, rare genotypes that are known to affect bone health (such as COL1A1 ss genotype) could be excluded. 'Mendelian randomisation' has been suggested to help remove bias in epidemiological studies, and certain genotypes could be used as a marker of nutrient status.[175] An example of this approach is a case control study involving the MTHFR gene with the C and T variants, where the reduced activity of the T variant can partly be reversed by dietary folate. The interpretation of the finding that TT genotype had increased risk of neural tube defects could be translated to dietary folate being required to reduce risk of neural tube defects.[184] However, there will still be confounders, as dietary status of other nutrients such as riboflavin and copper influences the activity of the enzyme.

## Conclusions

The huge advances in genetics as a result of the human genome project and improved techniques for genotyping have resulted in an explosion in the number of candidate genes for osteoporosis (see Chapter 5), which will undoubtedly continue and accelerate with the increased availability of microarray (DNA chip) technology. However, how this avalanche of molecular information translates to our understanding of osteoporosis at an individual level or population level is uncertain.

Although it is generally agreed that genetics accounts for 50–80% of the variation in BMD, the few genes examined so far have accounted for only a small percentage of BMD variation; and there is little agreement between different studies investigating the same gene (*e.g.*, VDR). Possible explanations for the discrepancies include linkage disequilibrium and interaction with environmental influences (*e.g.*, diet) and influence of other genes. There are a number of potential dietary confounders that make interpretation difficult.

The study of nutrient–gene interaction in populations is complex, requiring specialised statistical techniques and large study numbers for adequate statistical power. However, with appropriately designed studies and suitable statistical techniques, backed up by investigations at the cellular level, it is possible that the pattern of this complex disease may be unravelled.

There are many potential nutrient–gene interactions that warrant attention. In the future it may be possible to give people, with a genetic predisposition to osteoporotic fracture, appropriate dietary advice, tailored to the individual. What influences willingness to change lifestyle is also likely to be partly hereditary (genetically determined) and partly dependent on past experience (environmental exposure).

# References

1. R. Ottman, Gene–environment interaction: definitions and study designs, *Prev. Med.*, 1996, **25**, 764–770.
2. J.-A. Gustafsson, Fatty acids in control of gene expression, *Nutr. Rev.*, 1998, **56**, S20–S21.
3. J.E. Hesketh, M.H. Vasconcelos and G. Bermano, Regulatory signals in messenger RNA: determinants of nutrient–gene interaction and metabolic compartmentation, *Br. J. Nutr.*, 1998, **80**, 307–321.
4. J.M. de Castro, Heritability of diurnal changes in food intake in free-living humans, *Nutrition*, 2001, **17**, 713–720.
5. D.R. Reed, E. Nanthakumar, M. North, C. Bell, L.M. Bartoshuk and R.A. Price, Localization of a gene for bitter-taste perception to human chromosome 5p15, *Am. J. Hum. Genet.*, 1999, **64**, 1478–1480.
6. E. Seeman, J.L. Hopper, G. Pearce, A. Tabensky and M. Bradney, Interaction between genetic and nutritional factors, In *Nutritional Aspects of Osteoporosis. 3rd International Symposium on Nutritional Aspects of Osteoporosis, Switzerland, 1997*, P. Burckhardt, B. Dawson-Hughes and R.P. Heaney (eds.), Springer, Massachusetts, 1998, pp. 85–98.
7. T.V. Nguyen, G.M. Howard, P.J. Kelly and J.A. Eisman, Bone mass, lean mass, and fat mass: same genes or same environments? *Am. J. Epidemiol.*, 1998, **147**, 3–16.
8. S.B. Eaton and M. Konner, Paleolithic nutrition. A consideration of its nature and current implications, *N. Engl. J. Med.*, 1985, **312**, 283–289.
9. E. Ravussin, M.E. Valencia, J. Esparza, P.H. Bennett and L.O. Schulz, Effects of a traditional lifestyle on obesity in Pima Indians, *Diab. Care*, 1994, **17**, 1067–1074.
10. N. Hijazi, B. Abalkhail and A. Seaton, Diet and childhood asthma in a society in transition: a study in urban and rural Saudi Arabia, *Thorax*, 2000, **55**, 775–779.
11. S.O. Shaheen, J.A. Sterne, R.L. Thompson, C.E. Songhurst, B.M. Margetts and P.G. Burney, Dietary antioxidants and asthma in adults: population-based case-control study, *Am. J. Respir. Crit. Care Med.*, 2001, **164**, 1823–1828.
12. P. Burney, A diet rich in sodium may potentiate asthma. Epidemiologic evidence for a new hypothesis, *Chest*, 1987, **91**, 143S–148S.
13. B.A. Stoll, Breast cancer and the western diet: role of fatty acids and antioxidant vitamins, *Eur. J. Cancer*, 1998, **34**, 1852–1856.
14. M.L. Slattery, J.D. Potter, K.N. Ma, B.J. Caan, M. Leppert and W. Samowitz, Western diet, family history of colorectal cancer, NAT2, GSTM-1 and risk of colon cancer, *Cancer Causes Control*, 2000, **11**, 1–8.

15. World Cancer Research Fund. *Diet and Cancer Report. Diet and the cancer process.* 1997 [online], Available: http://www.wcrf-uk.org/report/chapter3 [2002].

16. P.J. Kelly, P.N. Sambrook, N.A. Morrison, T. Nguyen and J.A. Eisman, Genetics of osteoporosis, *World Rev. Nutr. Dietetics*, 1997, **80**, 126–144.

17. P. Geusens, C. Vandevyver, J. Vanhoof, J.J. Cassiman, S. Boonen and J. Raus, Quadriceps and grip strength are related to vitamin D receptor genotype in elderly nonobese women, *J. Bone Miner. Res.*, 1997, **12**, 2082–2088.

18. T. Yoshida, N. Yoshida, T. Monkawa, M. Hayashi and T. Saruta, Dietary phosphorus deprivation induces 25-hydroxyvitamin D(3) 1alpha- hydroxylase gene expression, *Endocrinology*, 2001, **142**, 1720–1726.

19. S. Wu, J. Finch, M. Zhong, E. Slatopolsky, M. Grieff and A.J. Brown, Expression of the renal 25-hydroxyvitamin D-24-hydroxylase gene: regulation by dietary phosphate, *Am. J. Physiol.*, 1996, **271**, F203–F208.

20. N.A. Morrison, R. Yeoman, P.J. Kelly and J.A. Eisman, Contribution of trans-acting factor alleles to normal physiological variability: vitamin D receptor gene polymorphisms and circulating osteocalcin, *Proc. Natl Acad. Sci. USA*, 1992, **89**, 6665–6669.

21. N.A. Morrison, J.C. Qi, A. Tokita, P.J. Kelly, L. Crofts, T.V. Nguyen, P.N. Sambrook and J.A. Eisman, Prediction of bone density from vitamin D receptor alleles, *Nature*, 1994, **367**, 284–287.

22. H. Arai, K. Miyamoto, Y. Taketani, H. Yamamoto, Y. Iemori, K. Morita, T. Tonai, T. Nishisho, S. Mori and E. Takeda, A vitamin D receptor gene polymorphism in the translation initiation codon: effect on protein activity and relation to bone mineral density in Japanese women, *J. Bone Miner. Res.*, 1997, **12**, 915–921.

23. E.M. Colin, A.E. Weel, A.G. Uitterlinden, C.J. Buurman, J.C. Birkenhager, H.A. Pols and J.P. van Leeuwen, Consequences of vitamin D receptor gene polymorphisms for growth inhibition of cultured human peripheral blood mononuclear cells by 1,25-dihydroxyvitamin D3, *Clin. Endocrinol.*, 2000, **52**, 211–216.

24. J.M. Zmuda, J.A. Cauley, M.E. Danielson, R.L. Wolf and R.E. Ferrell, Vitamin D receptor gene polymorphisms, bone turnover, and rates of bone loss in older African-American women, *J. Bone Miner. Res.*, 1997, **12**, 1446–1452.

25. N.A. Morrison, J.C. Qi, A. Tokita, P.J. Kelly, L. Crofts, T.V. Nguyen, P.N. Sambrook and J.A. Eisman, Prediction of bone density from vitamin D receptor alleles, *Nature*, 1997, **387**, 106.

26. B.L. Riggs, T.V. Nguyen, L.J. Melton III, N.A. Morrison, W.M. O'Fallon, P.J. Kelly, K.S. Egan, P.N. Sambrook, J.M. Muhs and J.A. Eisman, The contribution of vitamin D receptor gene alleles to the determination of bone mineral density in normal and osteoporotic women, *J. Bone Miner. Res.*, 1995, **10**, 991–996.

27. J.C. Fleet, S.S. Harris, R.J. Wood and B. Dawson-Hughes, The BsmI vitamin D receptor restriction fragment length polymorphism (BB) predicts low bone density in premenopausal black and white women, *J. Bone Miner. Res.*, 1995, **10**, 985–990.

28. M. Peacock, F.G. Hustmyer, S. Hui, C.C. Johnston and J. Christian, Vitamin D receptor genotype and bone mineral density. Evidence conflicts on link, *Br. Med. J.*, 1995, **311**, 874–875.

29. P. Garnero, O. Borel, E. Sornay-Rendu, M.E. Arlot and P.D. Delmas, Vitamin D receptor gene polymorphisms are not related to bone turnove, rate of bone loss,

and bone mass in postmenopausal women: the OFELY Study, *J. Bone Miner. Res.*, 1996, **11**, 827–834.

30. K.S. Tsai, S.H. Hsu, W.C. Cheng, C.K. Chen, P.U. Chieng and W.H. Pan, Bone mineral density and bone markers in relation to vitamin D receptor gene polymorphisms in Chinese men and women, *Bone*, 1996, **19**, 513–518.

31. L.A. Houston, S.F.A. Grant, D.M. Reid and S.H. Ralston, Vitamin D receptor polymorphism, bone mineral density, and osteoporotic vertebral fracture: studies in a UK population, *Bone*, 1996, **18**, 249–252.

32. A.G. Uitterlinden, H.A. Pols, H. Burger, Q. Huang, P.L. Van Daele, C.M. Van Duijn, A. Hofman, J.C. Birkenhager and J.P. Van Leeuwen, A large-scale population-based study of the association of vitamin D receptor gene polymorphisms with bone mineral density, *J. Bone Miner. Res.*, 1996, **11**, 1241–1248.

33. M.J. Barger-Lux, R.P. Heaney, J. Hayes, H.F. DeLuca, M.L. Johnston and G. Gong, Vitamin D receptor gene polymorphism, bone mass, body size, and vitamin D receptor density, *Calcif. Tissue Int.*, 1995, **57**, 161–162.

34. G.S. Cooper and D.M. Umbach. Are vitamin D receptor polymorphisms associated with bone mineral density? A meta-analysis, *J. Bone Miner. Res.*, 1996, **11**, 1841–1849.

35. J.A. Eisman, Vitamin D polymorphisms and calcium homeostasis: a new concept of normal gene variants and physiologic variation, *Nutr. Rev.*, 1998, **56**, s22-s29; discussion s54–s75.

36. N.K. Arden, P. Major, J.R. Poole, R.W. Keen, S. Vaja, R. Swaminathan, C. Cooper and T.D. Spector, Size at birth, adult intestinal calcium absorption and 1,25(OH)(2) vitamin D, *Quart. J. Med.*, 2002, **95**, 15–21.

37. R.W. Keen, P. Egger, C. Fall, P.J. Major, J.S. Lanchbury, T.M. Spector and C. Cooper, Polymorphisms of the vitamin D receptor, infant growth and adult bone mass, *Calcif. Tissue Int.*, 1997, **60**, 233–235.

38. C. Tao, T. Yu, S. Garnett, J. Briody, J. Knight, H. Woodhead and C.T. Cowell, Vitamin D receptor alleles predict growth and bone density in girls, *Arch. Dis. Childhood*, 1998, **79**, 488–493; discussion 493–494.

39. F. Suarez, F. Zeghoud, C. Rossignol, O. Walrant and M. Garabedian, Association between vitamin D receptor gene polymorphism and sex- dependent growth during the first two years of life, *J. Clin. Endocrinol. Metab.*, 1997, **82**, 2966–2970.

40. J. Sainz, J.M. Van Tornout, M.L. Loro, J. Sayre, T.F. Roe and V. Gilsanz, Vitamin D-receptor gene polymorphisms and bone density in prepubertal American girls of Mexican descent, *N. Engl. J. Med.*, 1997, **337**, 77–82.

41. J.P. Bonjour, A.L. Carrie, S. Ferrari, H. Clavien, D. Slosman, G. Theintz and R. Rizzoli, Calcium-enriched foods and bone mass growth in prepubertal girls: a randomized, double-blind, placebo-controlled trial, *J. Clin. Invest.*, 1997, **99**, 1287–1294.

42. S. Ferrari, R. Rizzoli and J.-P. Bonjour, Genetics–dietary calcium interaction and bone mass, In *Nutritional Aspects of Osteoporosis. Fourth International Symposium on Nutritional Aspects of Osteoporosis, Switzerland, 1997.* P. Burckhardt, B. Dawson-Hughes and R.P. Heaney (eds.), Academic Press, 1998.

43. G. Gong, H.S. Stern, S.C. Cheng, N. Fong, J. Mordeson, H.W. Deng and R.R. Recker, The association of bone mineral density with vitamin D receptor gene polymorphisms, *Osteoporosis Int.*, 1999, **9**, 55–64.

44. S.L. Ferrari, R. Rizzoli, D.O. Slosman and J.-P. Bonjour, Do dietary calcium and age explain the controversy surrounding the relationsjiop between bone mieral density and vitamin D receptor gene polymorphisms? *J. Bone Miner. Res.*, 1998, **13**, 363–370.

45. A.G. Uitterlinden, H. Burger, Q. Huang, F. Yue, F.E. McGuigan, S.F. Grant, A. Hofman, J.P. van Leeuwen, H.A. Pols and S.H. Ralston, Relation of alleles of the collagen type Ialpha1 gene to bone density and the risk of osteoporotic fractures in postmenopausal women, *N. Engl. J. Med.*, 1998, **338**, 1016–1021.

46. C.L. Tofteng, J.E. Jensen, B. Abrahamsen, L. Odum and C. Brot, Two polymorphisms in the vitamin D receptor gene–association with bone mass and 5-year change in bone mass with or without hormone-replacement therapy in postmenopausal women: the Danish Osteoporosis Prevention Study, *J. Bone Miner. Res.*, 2002, **17**, 1535–1544.

47. C. Vandevyver, T. Wylin, J.J. Cassiman, J. Raus and P. Geusens, Influence of the vitamin D receptor gene alleles on bone mineral density in postmenopausal and osteoporotic women, *J. Bone Miner. Res.*, 1997, **12**, 241–247.

48. L. McClure, T.R. Eccleshall, C. Gross, M.L. Villa, N. Lin, V. Ramaswamy, L. Kohlmeier, J.L. Kelsey, R. Marcus and D. Feldman, Vitamin D receptor polymorphisms, bone mineral density, and bone metabolism in postmenopausal Mexican-American women, *J. Bone Miner. Res.*, 1997, **12**, 234–240.

49. G. Sigurdsson, D.N. Magnusdottir, J.O. Kristinsson, K. Kristjansson and I. Olafsson, Association of BsmI vitamin-D receptor gene polymorphism with combined bone mass in spine and proximal femur in Icelandic women, *J. Intern. Med.*, 1997, **241**, 501–505.

50. A.W. Kung, S.S. Yeung and K.S. Lau, Vitamin D receptor gene polymorphisms and peak bone mass in southern Chinese women, *Bone*, 1998, **22**, 389–393.

51. R. Kikuchi, T. Uemura, I. Gorai, S. Ohno and H. Minaguchi, Early and late postmenopausal loss is associated with BsmI vitamin D receptor gene polymorphisms in Japanese women, *Calcif. Tissue Int.*, 1999, **64**, 102–106.

52. R.M. Francis, F. Harrington, E. Turner, S.S. Papiha and H.K. Datta, Vitamin D receptor gene polymorphism in men and its effect on bone density and calcium absorption, *Clin. Endocrinol.*, 1997, **46**, 83–86.

53. T.R. Eccleshall, P. Garnero, C. Gross, P.D. Delmas and D. Feldman, Lack of correlation between start codon polymorphism of the vitamin D receptor gene and bone mineral density in premenopausal French women: the OFELY study, *J. Bone Miner. Res.*, 1998, **13**, 31–35.

54. E.M. Lau, R.P. Young, S.C. Ho, J. Woo, J.L. Kwok, Z. Birjandi, G.N. Thomas, A. Sham and J.A. Critchley, Vitamin D receptor gene polymorphisms and bone mineral density in elderly Chinese men and women in Hong Kong, *Osteoporosis Int.*, 1999, **10**, 226–230.

55. Y.Z. Bagger, H.L. Jorgensen, A.M. Heegaard, L. Bayer, L. Hansen and C. Hassager, No major effect of estrogen receptor gene polymorphisms on bone mineral density or bone loss on Danish women, *Bone*, 2000, **26**, 111–116.

56. D. Holmberg-Marttila, H. Sievanen, T.L. Jarvinen and T.A. Jarvinen, Vitamin D and estrogen receptor polymorphisms and bone mineral changes in postpartum women, *Calcif. Tissue Int.*, 2000, **66**, 184–189.

57. M. Kubota, S. Yoshida, M. Ikeda, Y. Okada, H. Arai, K. Miyamoto and E. Takeda, Association between two types of vitamin D receptor gene polymorphism and bone status in premenopausal Japanese women, *Calcif. Tissue Int.*, 2001, **68**, 16–22.

58. W.C. Cheng and K.S. Tsai, The vitamin D receptor start codon polymorphism (Fok1) and bone mineral density in premenopausal women in Taiwan, *Osteoporosis Int.*, 1999, **9**, 545–549.

59. C. Gross, T.R. Eccleshall, P.J. Malloy, M.L. Villa, R. Marcus and D. Feldman, The presence of a polymorphism at the translation initiation site of the vitamin D receptor gene is associated with low bone mineral density in postmenopausal Mexican-American women, *J. Bone Miner. Res.*, 1996, **11**, 1850–1855.

60. S.S. Harris, T.R. Eccleshall, C. Gross, B. Dawson-Hughes and D. Feldman, The vitamin D receptor start codon polymorphism (FokI) and bone mineral density in premenopausal American black and white women, *J. Bone Miner. Res.*, 1997, **12**, 1043–1048.

61. S.L. Ferrari, R. Rizzoli, D. Manen, D. Slosman and J.-P. Bonjour, Vitamin D receptor gene start codon polymorphisms (Fok1) and bone mineral density: interaction with age, dietary calcium and 3'-end region polymorphisms, *J. Bone Miner. Res.*, 1998, **13**, 925–930.

62. L. Gennari, L. Becherini, R. Mansani, L. Masi, A. Falchetti, A. Morelli, E. Colli, S. Gonnelli, C. Cepollaro and M.L. Brandi, FokI polymorphism at translation initiation site of the vitamin D receptor gene predicts bone mineral density and vertebral fractures in postmenopausal Italian women, *J. Bone Miner. Res.*, 1999, **14**, 1379–1386.

63. R.M. Kanan, S.S. Varanasi, R.M. Francis, L. Parker and H.K. Datta, Vitamin D receptor gene start codon polymorphism (FokI) and bone mineral density in healthy male subjects, *Clin. Endocrinol.*, 2000, **53**, 93–98.

64. E.M. Lau, V. Lam, M. Li, K. Ho and J. Woo, Vitamin D receptor start codon polymorphism (Fok I) and bone mineral density in Chinese men and women, *Osteoporosis Int.*, 2002, **13**, 218–221.

65. H.W. Deng, H. Shen, F.H. Xu, H.Y. Deng, T. Conway, H.T. Zhang and R.R. Recker, Tests of linkage and/or association of genes for vitamin D receptor, osteocalcin, and parathyroid hormone with bone mineral density, *J. Bone Miner. Res.*, 2002, **17**, 678–686.

66. R.Y. Zee, R.H. Myers, M.T. Hannan, P.W. Wilson, J.M. Ordovas, E.J. Schaefer, K. Lindpaintner and D.P. Kiel, Absence of linkage for bone mineral density to chromosome 12q12-14 in the region of the vitamin D receptor gene, *Calcif. Tissue Int.*, 2000, **67**, 434–439.

67. D. Feskanich, D.J. Hunter, W.C. Willett, S.E. Hankinson, B.W. Hollis, H.L. Hough, K.T. Kelsey and G.A. Colditz, Vitamin D receptor site genotype and the risk of bone fractures in women, *Epidemiology*, 1998, **9**, 535–539.

68 C. Gomez, M.L. Naves, Y. Barrios, J.B. Diaz, J.L. Fernandez, E. Salido, A. Torres and J.B. Cannata, Vitamin D receptor gene polymorphisms, bone mass, bone loss and prevalence of vertebral fracture: differences in postmenopausal women and men, *Osteoporosis Int.*, 1999, **10**, 175–182.

69. K.E. Ensrud, K. Stone, J.A. Cauley, C. White, J.M. Zmuda, T.V. Nguyen, J.A. Eisman and S.R. Cummings, Vitamin D receptor gene polymorphisms and the risk of fractures in older women. For the Study of Osteoporotic Fractures Research Group, *J. Bone Miner. Res.*, 1999, **14**, 1637–1645.

70. J. Aerssens, J. Dequeker, J. Peeters, S. Breemans, P. Broos and S. Boonen, Polymorphisms of the VDR, ER and COLIA1 genes and osteoporotic hip fracture in elderly postmenopausal women, *Osteoporosis Int.*, 2000, **11**, 583–591.

71. A.G. Uitterlinden, A.E. Weel, H. Burger, Y. Fang, C.M. van Duijn, A. Hofman, J.P. van Leeuwen and H.A. Pols, Interaction between the vitamin D receptor gene and collagen type Ialpha1 gene in susceptibility for fracture, *J. Bone Miner. Res.*, 2001, **16**, 379–385.

72. W.C. Graafmans, P. Lips, M.E. Ooms, J.P. van Leeuwen, H.A. Pols and A.G. Uitterlinden, The effect of vitamin D supplementation on the bone mineral density of the femoral neck is associated with vitamin D receptor genotype, *J. Bone Miner. Res.*, 1997, **12**, 1241–1245.

73. P.B. Rapuri, J.C. Gallagher, H.K. Kinyamu and K.L. Ryschon, Caffeine intake increases the rate of bone loss in elderly women and interacts with vitamin D receptor genotypes, *Am. J. Clin. Nutr.*, 2001, **74**, 694–700.

74. E.A. Krall, P. Parry, J.B. Lichter and B. Dawson-Hughes, Vitamin D receptor alleles and rates of bone loss: influences of years since menopause and calcium intake, *J. Bone Miner. Res.*, 1995, **10**, 978–984.

75. S. Ferrari, R. Rizzoli, T. Chevalley, D. Slosman, J.A. Eisman and J.-P. Bonjour, Vitamin D receptor gene polymorphisms and change in lumbar spine bone mineral density, *Lancet*, 1995, **345**, 423–424.

76. B. Dawson-Hughes, S.S. Harris and S. Finnerman, Calcium absorption on high and low calcium intakes in relation to vitamin D receptor genotype, *J. Clin. Endocrinol. Metab.*, 1995, **80**, 3657–3661.

77. S. Ferrari, D. Manen, J.P. Bonjour, D. Slosman and R. Rizzoli, Bone mineral mass and calcium and phosphate metabolism in young men: relationships with vitamin D receptor allelic polymorphisms, *J. Clin. Endocrinol. Metab.*, 1999, **84**, 2043–2048.

78. L.M. Salamone, N.W. Glynn, D.M. Black, R.E. Ferrell, L. Palermo, R.S. Epstein, L.H. Kuller and J.A. Cauley, Determinants of premenopausal bone mineral density: the interplay of genetic and lifestyle factors, *J. Bone Miner. Res.*, 1996, **11**, 1557–1565.

79. D.P. Kiel, R.H. Myers, L.A. Cupples, X.F. Kong, X.H. Zhu, J. Ordovas, E.J. Schaefer, D.T. Felson, D. Rush, P.W. Wilson, J.A. Eisman and M.F. Holick, The BsmI vitamin D receptor restriction fragment length polymorphism (bb) influences the effect of calcium intake on bone mineral density, *J. Bone Miner. Res.*, 1997, **12**, 1049–1057.

80. M.A. Brown, M.A. Haughton, S.F. Grant, A.S. Gunnell, N.K. Henderson and J.A. Eisman, Genetic control of bone density and turnover: role of the collagen 1alpha1, estrogen receptor, and vitamin D receptor genes, *J. Bone Miner. Res.*, 2001, **16**, 758–764.

81. L.A. Rubin, G.A. Hawker, V.D. Peltekova, L.J. Fielding, R. Ridout and D.E. Cole, Determinants of peak bone mass: clinical and genetic analyses in a young female Canadian cohort, *J. Bone Miner. Res.*, 1999, **14**, 633–643.

82. H.S. Kim, P.A. Newcomb, C.M. Ulrich, C.L. Keener, J. Bigler, F.M. Farin, R.M. Bostick and J.D. Potter, Vitamin D receptor polymorphism and the risk of colorectal adenomas: evidence of interaction with dietary vitamin D and calcium, *Cancer Epidemiol., Biomarkers Prev.*, 2001, **10**, 869–874.

83. Department of Health, *Nutrition and Bone Health: With Particular Reference to Calcium and Vitamin D*, HMSO, London, 1998.

84. S.A. Gibson and M. Ashwell, New vitamin D values for meat and their implication for vitamin D intake in adults, *Proc. Nutr. Soc.*, 1997, **56**, 116A.

85. C.M. Rohde, M. Manatt, M. Clagett-Dame and H.F. DeLuca, Vitamin A antagonizes the action of vitamin D in rats, *J. Nutr.*, 1999, **129**, 2246–2250.

86. H. Melhus, K. Michaelsson, A. Kindmark, R. Bergstrom, L. Holmberg, H. Mallmin, A. Wolk and S. Ljunghall, Excessive dietary intake of vitamin A is associated with reduced bone mineral density and increased risk for hip fracture, *Ann. Intern. Med.*, 1998, **129**, 770–778.

87. D. Feskanich, V. Singh, W.C. Willett and G.A. Colditz, Vitamin A intake and hip fractures among postmenopausal women, *JAMA*, 2002, **287**, 47–54.

88. M.S. Calvo and Y.K. Park, Changing phosphorus content of the U.S. diet: potential for adverse effects on bone, *J. Nutr.*, 1996, **126**, 1168S–1180S.

89. R.P. Heaney, B. Berner and J. Louie-Helm, Dosing regimen for calcium supplementation, *J. Bone Miner. Res.*, 2000, **15**, 2291.

90. A. Schlemmer, C. Hassager, S.B. Jensen and C. Christiansen, Marked diurnal variation in urinary excretion of pyridinium cross-links in premenopausal women, *J. Clin. Endocrinol. Metab.*, 1992, **74**, 476–480.

91. W.D. Fraser, M. Anderson, C. Chesters, B. Durham, A. Ahmad, P. Chattington, J. Vora, C. Squire and M. Diver, Circadian rhythm studies of serum bone resorption markers: implications for optimal sample timing and clinical utility, In *Bone Markers: Biochemical and Clinical Perspectives,* R. Eastell, M. Baumann, N.R. Hoyle and L. Wieczorek (eds.), Martin Dunitz, London, 2001, pp. 107–118.

92. A. Blumsohn, K. Herrington, R.A. Hannon, P. Shao, D.R. Eyre and R. Eastell, The effect of calcium supplementation on the circadian rhythm of bone resorption, *J. Clin. Endocrinol. Metab.*, 1994, **79**, 730–735.

93. J. Aerssens, K. Declerck, B. Maeyaert, S. Boonen and J. Dequeker, The effect of modifying dietary calcium intake pattern on the circadian rhythm of bone resportion, *Calcif. Tissue Int.*, 1999, **65**, 34–40.

94. M.S. Sheikh, A. Ramirez, M. Emmett, C. Santa Ana, L.R. Schiller and J.S. Fordtran, Role of vitamin D-dependent and vitamin D-independent mechanisms in absorption of food calcium, *J. Clin. Invest.*, 1988, **81**, 126–132.

95. C.M. Weaver, W.R. Proulx and R. Heaney, Choices for achieving adequate dietary calcium with a vegetarian diet, *Am. J. Clin. Nutr.*, 1999, **70**, 543s–548s.

96. G.V. Appleton, R.W. Owen and R.C. Williamson, The effect of dietary calcium supplementation on intestinal lipid metabolism, *J. Steroid Biochem. Mol. Biol.*, 1992, **42**, 383–387.

97. B.E. Nordin, A.G. Need, Morris H.A., M. Horowitz and W.G. Robertson, Evidence for a renal calcium leak in postmenopausal women, *J. Clin. Endocrinol. Metab.*, 1991, **72**, 401–417.

98. J.C. Gallagher, B.L. Riggs, J. Eisman, A. Hamstra, S.B. Arnaud and H.F. DeLuca, Intestinal calcium absorption and serum vitamin D metabolites in normal subjects and osteoporotic patients: effect of age and dietary calcium, *J. Clin. Invest.*, 1979, **64**, 729–736.

99. J. MacLaughlin and M.F. Holick, Aging decreases the capacity of human skin to produce vitamin D3, *J. Clin. Invest.*, 1985, **76**, 1536–1538.

100. K.-S. Tsai, H. Heath, R. Kumar and B.L. Riggs, Impaired vitamin D metabolism with aging in women, *J. Clin. Invest.*, 1984, **73**, 1668–1672.

101. D.E. Sellmeyer, M. Schloetter and A. Sebastian, Potassium citrate prevents increased urine calcium excretion and bone resorption induced by a high sodium chloride diet, *J. Clin. Endocrinol. Metab.*, 2002, **87**, 2008–2012.

102. Barzel U.S. and Massey L.K. Excess dietary protein can adversely affect bone, *J. Nutr.*, 1998, **128**, 1051–1053.

103. J.E. Kerstetter, K. O'Brien and K. Insogna, Dietary protein and intestinal calcium absorption, *Am. J. Clin. Nutr.*, 2001, **73**, 990–992.

104. A.M. Prentice, A.E. Black, W.A. Coward and T.J. Cole, Energy expenditure in overweight and obese adults in affluent societies: an analysis of 319 doubly-labelled water measurements, *Eur. J. Clin. Nutr.*, 1996, **50**, 93–97.

105. M.A. Laskey, A. Prentice, L.A. Hanratty, L.M. Jarjou, B. Dibba, S.R. Beavan and T.J. Cole, Bone changes after 3 mo of lactation: influence of calcium intake, breast-milk output, and vitamin D-receptor genotype, *Am. J. Clin. Nutr.*, 1998, **67**, 685–692.

106. H.M. MacDonald, F.A. McGuigan, S.A. New, M.K. Campbell, M.H. Golden, S.H.Ralston and D.M. Reid, COL1A1 Sp1 polymorphism predicts perimenopausal and early postmenopausal spinal bone loss, *J. Bone Miner. Res.*, 2001, **16**, 1634–1641.

107. A.M. Qureshi, R.J. Herd, G.N. Blake, I. Fogelman and S.H. Ralston, COLIA1 Sp1 polymorphism predicts response of femoral neck bone density to cyclical etidronate therapy, *Calcif. Tissue Int.*, 2002, **70**, 158–163.

108. N. Kobayashi, T. Fujino, T. Shirogane, I. Furuta, Y. Kobamatsu, M. Yaegashi, N. Sakuragi and S. Fujimoto, Estrogen receptor alpha polymorphism as a genetic marker for bone loss, vertebral fractures and susceptibility to estrogen, *Maturitas*, 2002, **41**, 193–201.

109. D. Woods, G. Onambele, R. Woledge, D. Skelton, S. Bruce, S.E. Humphries and H. Montgomery, Angiotensin-I converting enzyme genotype-dependent benefit from hormone replacement therapy in isometric muscle strength and bone mineral density, *J. Clin. Endocrinol. Metab.*, 2001, **86**, 2200–2204.

110. A. Cassidy and M. Faughnan, Phyto-ostrogens through the life cycle, *Proc. Nutr. Soc.*, 2000, **59**, 489–496.

111 K. Toda, C. Miyaura, T. Okada and Y. Shizuta, Dietary bisphenol A prevents ovarian degeneration and bone loss in female mice lacking the aromatase gene (Cyp19), *Eur. J. Biochem.*, 2002, **269**, 2214–2222.

112. M. Venugopal, A. Callaway and E.G. Snyderwine, 2-Amino-1-methyl-6-phenyl-imidazo[4,5-b]pyridine (PhIP) retards mammary gland involution in lactating Sprague-Dawley rats, *Carcinogenesis*, 1999, **20**, 1309–1314.

113. A. Klibanski, R.M. Neer, I.Z. Beitins, E.C. Ridgway, N.T. Zervas and J.W. McArthur, Decreased bone density in hyperprolactinemic women, *N. Engl. J. Med.*, 1980, **303**, 1511–1514.

114. H.M. Macdonald, S.A. New, F.E. McGuigan, M.H.N. Golden, S.H. Ralston, D.A. Grubb and D.M. Reid, Modest alcohol intake reduces bone loss in peri and early postmenopausal Scottish women: an effect on estrogen receptor genotype? *Bone*, 2001, **28**, s95.

115. K. Katsouyanni, P. Boyle and D. Trichopoulos, Diet and urine estrogens among postmenopausal women, *Oncology*, 1991, **48**, 490–494.

116. L.M. Hines, M.J. Stampfer, J. Ma, J.M. Gaziano, P.M. Ridker, S.E. Hankinson, F. Sacks, E.B. Rimm and D.J. Hunter, Genetic variation in alcohol dehydrogenase and the beneficial effect of moderate alcohol consumption on myocardial infarction, *N. Engl. J. Med.*, 2001, **344**, 549–555.

117. I.P. Fohr, R. Prinz-Langenohl, A. Bronstrup, A.M. Bohlmann, H. Nau, H.K. Berthold and K. Pietrzik, 5,10-Methylenetetrahydrofolate reductase genotype determines the plasma homocysteine-lowering effect of supplementation with 5-methyltetrahydrofolate or folic acid in healthy young women, *Am. J. Clin. Nutr.*, 2002, **75**, 275–282.

118. J.A. Pasco, M.J. Henry, M.A. Kotowicz, G.R. Collier, M.J. Ball, A.M. Ugoni and G.C. Nicholson, Serum leptin levels are associated with bone mass in nonobese women, *J. Clin. Endocrinol. Metab.*, 2001, **86**, 1884–1887.

119. T. Thomas and B. Burguera, Is leptin the link between fat and bone mass? *J. Bone Miner. Res.*, 2002, **17**, 1563–1569.

120. K.L. Huber and R.J. Cousins, Metallothionein expression in rat bone marrow is dependent on dietary zinc but not dependent on interleukin-1 or interleukin-6, *J. Nutr.*, 1993, **123**, 642–648.

121. I. Kasvosve, J.R. Delanghe, Z.A. Gomo, I.T. Gangaidzo, H. Khumalo, M.R. Langlois, V.M. Moyo, T. Saungweme, E. Mvundura, J.R. Boelaert and V.R. Gordeuk, Effect of transferrin polymorphism on the metabolism of vitamin C in Zimbabwean adults, *Am. J. Clin. Nutr.*, 2002, **75**, 321–325.

122. R.I. Schwarz, Role of ascorbate in regulating the collagen pathway, In *Nutrition and Gene Expression* and C.D. Berdanier, J.L. Hargrove (eds.), CRC Press, Florida, 1993, 483–506.

123. M. Shiraki, Y. Shiraki, C. Aoki, T. Hosoi, S. Inoue, M. Kaneki and Y. Ouchi, Association of bone mineral density with apolipoprotein E phenotype, *J. Bone Miner. Res.*, 1997, **12**, 1438–1445.

124. M. Kohlmeier, J. Saupe, K. Schaefer and G. Asmus, Bone fracture history and prospective bone fracture risk of hemodialysis patients are related to apolipoprotein E genotype, *Calcif. Tissue Int.*, 1998, **62**, 278–281.

125. J.A. Cauley, J.M. Zmuda, K. Yaffe, L.H. Kuller, R.E. Ferrell, S.R. Wisniewski and S.R. Cummings, Apolipoprotein E polymorphism: a new genetic marker of hip fracture risk–the Study of Osteoporotic Fractures, *J. Bone Miner. Res.*, 1999, **14**, 1175–1181.

126. L.M. Salamone, J.A. Cauley, J. Zmuda, A. Pasagian-Macaulay, R.S. Epstein, R.E. Ferrell, D.M. Black and L.H. Kuller, Apolipoprotein E gene polymorphism and bone loss: estrogen status modifies the influence of apolipoprotein E on bone loss, *J. Bone Miner. Res.*, 2000, **15**, 308–314.

127. A.M. Heikkinen, H. Kroger, L. Niskanen, M.H. Komulainen, M. Ryynanen, M.T. Parviainen, M.T. Tuppurainen, R. Honkanen and S. Saarikoski, Does apolipoprotein E genotype relate to BMD and bone markers in postmenopausal women? *Maturitas*, 2000, **34**, 33–41.

128. S.M. Pluijm, M.G. Dik, C. Jonker, D.J. Deeg, D.J. Deeg, G.J. Kamp, P. Lips and P. Lips, Effects of gender and age on the association of apolipoprotein E epsilon4 with bone mineral density, bone turnover and the risk of fractures in older people, *Osteoporosis Int.*, 2002, **13**, 701–709.

129. L.U. Gerdes, P. Vestergaard, A.P. Hermann and L. Mosekilde, Regional and hormone-dependent effects of apolipoprotein E genotype on changes in bone mineral in perimenopausal women, *J. Bone Miner. Res.*, 2001, **16**, 1906–1916.

130. J. Saupe, M.J. Shearer and M. Kohlmeier, Phylloquinone transport and its influence on gamma-carboxyglutamate residues of osteocalcin in patients on maintenance hemodialysis, *Am. J. Clin. Nutr.*, 1993, **58**, 204–208.

131. P. Szulc, M.C. Chapuy, P.J. Meunier and P.D. Delmas, Serum undercarboxylated osteocalcin is a marker of the risk of hip fracture in elderly women, *J. Clin. Invest.*, 1993, **91**, 1769–1774.

132. P. Szulc, M.C. Chapuy, P.J. Meunier and P.D. Delmas, Serum undercarboxylated osteocalcin is a marker of the risk of hip fracture: a three year follow-up study, *Bone*, 1996, **18**, 487–488.

133. S.J. Hodges, M.J. Pilkinton, T.C.B. Stamp, A. Catterall, M.J. Shearer, L. Bitensky and J. Chayen, Depressed levels of circulating menaquinones in patients with osteoporotic fractures of the spine and femoral neck, *Bone*, 1991, **12**, 387–389.
134. C. Bolton-Smith, R.A.G. Price, S.T. Fenton, D.J. Harrington and M.J. Shearer, Compilation of a provisional UK database for the phylloquinone (vitamin K1) content of foods, *Br. J. Nutr.*, 2000, **83**, 389–399.
135. F. Parhami, Y. Tintut, W.G. Beamer, N. Gharavi, W. Goodman and L.L. Demer, Atherogenic high-fat diet reduces bone mineralization in mice, *J. Bone Miner. Res.*, 2001, **16**, 182–188.
136. R. Boukhris and K.L. Becker, Calcification of the aorta and osteoporosis. A roentgenographic study, *JAMA*, 1972, **219**, 1307–1311.
137. D. Corella, K. Tucker, C. Lahoz, O. Coltell, L.A. Cupples, P.W. Wilson, E.J. Schaefer and J.M. Ordovas, Alcohol drinking determines the effect of the APOE locus on LDL- cholesterol concentrations in men: the Framingham Offspring Study, *Am. J. Clin. Nutr.*, 2001, **73**, 736–745.
138. J. Ma, M.J. Stampfer, C.H. Hennekens, P. Frosst, J. Selhub, J. Horsford, M.R. Malinow, W.C. Willett and R. Rozen, Methylenetetrahydrofolate reductase polymorphism, plasma folate, homocysteine, and risk of myocardial infarction in US physicians, *Circulation*, 1996, **94**, 2410–2416.
139. M. Roest, Y.T. van der Schouw, D.E. Grobbee, M.J. Tempelman, P.G. de Groot, J.J. Sixma and J.D. Banga, Methylenetetrahydrofolate reductase 677 C/T genotype and cardiovascular disease mortality in postmenopausal women, *Am. J. Epidemiol.*, 2001, **153**, 673–679.
140. S.H. Mudd, F. Skovby, H.L. Levy, K.D. Pettigrew, B. Wilcken, R.E. Pyeritz, G. Andma, G.H.J. Boers, I.L. Brombeig, R. Cerone, B. Fowler, H. Gröbe, H. Schmidt and L. Swhweitzer, The natural history of homocystinuria due to cystathione beta-synthase deficiency, *Am. J. Hum. Genet.*, 1985, **37**, 1–31.
141. G. Liu, K. Nellaiappan and H.M. Kagan, Irreversible inhibition of lysyl oxidase by homocysteine thiolactone and its selenium and oxygen analogues, *J. Biol. Chem.*, 1997, **51**, 32370–32377.
142. B. Lubec, S. Fang-Kircher, T. Lubec, H.J. Blom and G.H.J. Boers, Evidence for McKuisik's hypothesis of deficient collagen cross-linking in patients with homocystinuria, *Biochim. Biophys. Acta*, 1996, **1315**, 159–162.
143. J. Selhub, P.F. Jacques, P.W. Wilson, D. Rush and I.H. Rosenberg, Vitamin status and intake as primary determinants of homocysteinemia in an elderly population, *JAMA*, 1993, **270**, 2693–2698.
144. A.E. Hak, A.A. Bak, J. Lindemans, J. Planellas, H.J. Coelingh Bennink, A. Hofman, D.E. Grobbee and J.C. Witteman, The effect of hormone replacement therapy on serum homocysteine levels in perimenopausal women: a randomized controlled trial, *Atherosclerosis*, 2001, **158**, 437–443.
145. M.R. Malinow, Homocyst(e)ine and arterial occlusive diseases, *J. Intern. Med.*, 1994, **236**, 603–617.
146. M. Miyao, H. Morita, T. Hosoi, H. Kurihara, S. Inoue, S. Hoshino, M. Shiraki, Y. Yazaki and Y. Ouchi, Association of methylenetetrahydrofolate reductase (MHTFR) polymorphism with bone mineral density in postmenopausal Japanese women, *Calcif. Tissue Int.*, 2000, **66**, 190–194.
147. M.L. Silaste, M. Rantala, M. Sampi, G. Alfthan, A. Aro and Y.A. Kesaniemi, Polymorphisms of key enzymes in homocysteine metabolism affect diet responsiveness of plasma homocysteine in healthy women, *J. Nutr.*, 2001, **131**, 2643–2647.

148. H. McNulty, M.C. McKinley, B. Wilson, J. McPartlin, J.J. Strain, D.G. Weir and J.M. Scott, Impaired functioning of thermolabile methylenetetrahydrofolate reductase is dependent on riboflavin status: implications for riboflavin requirements, *Am. J. Clin. Nutr.*, 2002, **76**, 436–441.

149. J. Eaton-Evans, E.M. McIlrath, W.E. Jackson, H. McCartney and J.J. Strain, Copper supplementation and the maintenance of bone mineral density in middle-aged women, *J. Trace Elements Exp. Med.*, 1996, **9**, 87–94.

150. A. Wachman and D.S. Bernstein, Diet and osteoporosis, *Lancet*, 1968, **i**, 958–959.

151. D.A. Bushinsky, Net calcium efflux from live bone during chronic metabolic, but not respiratory, acidosis, *Am. J. Physiol.*, 1989, **256**, F836–F842.

152. G.E. Hall and A.D. Kenny, Role of carbonic anhydrase in bone resorption induced by prostaglandin E2 in vitro, *Pharmacology*, 1985, **30**, 339–347.

153. G.E. Hall and A.D. Kenny, Role of carbonic anhydrase in bone resorption induced by 1,25 dihydroxyvitamin D3 in vitro, *Calcif. Tissue Int.*, 1985, **37**, 134–142.

154. S. Breton, The cellular physiology of carbonic anhydrases, *J. Pancreas*, 2001, **2**, 159–164.

155. A. Billecocq, J.R. Emanuel, R. Levenson and R. Baron, 1 alpha,25-dihydroxyvitamin D3 regulates the expression of carbonic anhydrase II in nonerythroid avian bone marrow cells, *Proc. Natl Acad. Sci. USA*, 1990, **87**, 6470–6474.

156. D.M. Biskobing and D. Fan, Acid pH increases carbonic anhydrase II and calcitonin receptor mRNA expression in mature osteoclasts, *Calcif. Tissue Int.*, 2000, **67**, 178–183.

157. S. Meghji, M.S. Morrison, B. Henderson and T.R. Arnett, pH Dependence of bone resorption: mouse calvarial osteoclasts are activated by acidosis, *Am. J. Physiol. – Endocrinol. Metab.*, 2001, **280**, E112–E119.

158. J.P. David, M. Rincon, L. Neff, W.C. Horne and R. Baron, Carbonic anhydrase II is an AP-1 target gene in osteoclasts, *J. Cell. Physiol.*, 2001, **188**, 89–97.

159. H.C. Blair, S.L. Teitelbaum, H.L. Tan, C.M. Koziol and P.H. Schlesinger, Passive chloride permeability charge coupled to H(+)-ATPase of avian osteoclast ruffled membrane, *Am. J. Physiol.*, 1991, **260**, C1315–C1324.

160. P.H. Schlesinger, H.C. Blair, S.L. Teitelbaum and J.C. Edwards, Characterization of the osteoclast ruffled border chloride channel and its role in bone resorption, *J. Biol. Chem.*, 1997, **272**, 18636–18643.

161. U. Kornak, D. Kasper, M.R. Bosl, E. Kaiser, M. Schweizer, A. Schulz, W. Friedrich, G. Delling and T.J. Jentsch, Loss of the ClC-7 chloride channel leads to osteopetrosis in mice and man, *Cell*, 2001, **104**, 205–215.

162. E. Cleiren, O. Benichou, E. Van Hul, J. Gram, J. Bollerslev, F.R. Singer, K. Beaverson, A. Aledo, M.P. Whyte, T. Yoneyama, M.C. deVernejoul and W. Van Hul, Albers-Schonberg disease (autosomal dominant osteopetrosis, type II) results from mutations in the ClCN7 chloride channel gene, *Hum. Mol. Genet.*, 2001, **10**, 2861–2867.

163. R.C. Morris, L.A. Frassetto, O. Schmidlin, A. Forman and A. Sebastian, Expression of osteoporosis as determined by diet-disordered electrolyte and acid–base metabolism, In *Nutritional Aspects of Osteoporosis. Fourth International Symposium on Nutritional Aspects of Osteoporosis, Switzerland, 2000.* P. Burckhardt, B. Dawson-Hughes and R.P. Heaney (eds.), Academic Press, 2001, pp. 357–378.

164. J.P. Williams, S.E. Jordan, S. Barnes and H.C. Blair, Tyrosine kinase inhibitor effects on avian osteoclastic acid transport, *Am. J. Clin. Nutr.*, 1998, **68**, 1369S–1374S.

165. A.G. Uitterlinden, H. Burger, Q. Huang, C.M. van Duijn, A. Hofman, J.C. Birkenhager, J.P.T.M. van Leeuwen and H.A.P. Pols, Vitamin D receptor genotype is associated with osteoarthritis, In *Osteoporosis*. S.E. Papapoulos, P. Lips, H.A.P. Pols, C.C. Johnston and P.D. Delmas (eds.), 1996, pp. 395–403.

166. J. Huang, T. Ushiyama, K. Inoue, T. Kawasaki and S. Hukuda, Vitamin D receptor gene polymorphisms and osteoarthritis of the hand, hip, and knee: a case-control study in Japan, *Rheumatology*, 2000, **39**, 79–84.

167. A.G. Uitterlinden, H. Burger, C.M. van Duijn, Q. Huang, A. Hofman, J.C. Birkenhager, J.P. van Leeuwen and H.A. Pols, Adjacent genes, for COL2A1 and the vitamin D receptor, are associated with separate features of radiographic osteoarthritis of the knee, *Arthritis Rheum.*, 2000, **43**, 1456–1464.

168. C.T. Baldwin, L.A. Cupples, O. Joost, S. Demissie, C. Chaisson, T. Mcalindon, R.H. Myers and D. Felson, Absence of linkage or association for osteoarthritis with the vitamin D receptor/type II collagen locus: the Framingham Osteoarthritis Study, *J. Rheumatol.*, 2002, **29**, 161–165.

169. J. Aerssens, J. Dequeker, J. Peeters, S. Breemans and S. Boonen, Lack of association between osteoarthritis of the hip and gene polymorphisms of VDR, COL1A1, and COL2A1 in postmenopausal women, *Arthritis Rheum.*, 1998, **41**, 1946–1950.

170. J.R. Garcia-Lozano, M.F. Gonzalez-Escribano, A. Valenzuela, A. Garcia and A. Nunez-Roldan, Association of vitamin D receptor genotypes with early onset rheumatoid arthritis, *Eur. J. Immunogenet.*, 2001, **28**, 89–93.

171. Y. Kuroki, S. Shiozawa, J. Kano and K. Chihara, Competition between c-fos and 1,25(OH)2 vitamin D3 in the transcriptional control of type I collagen synthesis in MC3T3-E1 osteoblastic cells, *J. Cell. Physiol.*, 1995, **164**, 459–464.

172. A. Gough, P. Sambrook, J. Devlin, J. Lilley, A. Huisoon, J. Betteridge, J. Franklyn, T. Nguyen, N. Morrison, J. Eisman and P. Emery, Effect of vitamin D receptor gene alleles on bone loss in early rheumatoid arthritis, *J. Rheumatol.*, 1998, **25**, 864–868.

173. Y.V. Ho, E.M. Briganti, Y. Duan, R. Buchanan, S. Hall and E. Seeman, Polymorphism of the vitamin D receptor gene and corticosteroid-related osteoporosis, *Osteoporosis Int.*, 1999, **9**, 134–138.

174. C.K. Lee, J.S. Hong, Y.S. Cho, B. Yoo, G.S. Kim and H.B. Moon, Lack of relationship between vitamin D receptor polymorphism and bone erosion in rheumatoid arthritis, *J. Korean Med. Sci.*, 2001, **16**, 188–192.

175. D. Clayton and P.M. McKeigue, Epidemiological methods for studying genes and environmental factors in complex diseases, *Lancet*, 2001, **358**, 1356–1360.

176. H.J. Cordell, J.A. Todd, N.J. Hill, C.J. Lord, P.A. Lyons, L.B. Peterson, L.S. Wicker and D.G. Clayton, Statistical modeling of interlocus interactions in a complex disease. Rejection of the multiplicative model of epistasis in type 1 diabetes, *Genetics*, 2001, **158**, 357–367.

177. K.J. Rothman, S. Greenland and A.M. Walker, Concepts of interaction, *Am. J. Epidemiol.*, 1980, **112**, 467–470.

178. S.J. Hwang, T.H. Beaty, K.Y. Liang, J. Coresh and M.J. Khoury, Minimum sample size estimation to detect gene–environment interaction in case-control designs, *Am. J. Epidemiol.*, 1994, **140**, 1029–1037.

179. I. Foppa and D. Spiegelman, Power and sample size calculations for case-control studies of gene–environment interactions with a polytomous exposure variable, *Am. J. Epidemiol.*, 1997, **146**, 596–604.

180. M. Garcia-Closas and J.H. Lubin, Power and sample size calculations in case-control studies of gene–environment interactions: comments on different approaches, *Am. J. Epidemiol.*, 1999, **149**, 689–692.

181. Q. Yang, M.J. Khoury and W.D. Flanders, Sample size requirements in case-only designs to detect gene environment interaction, *Am. J. Epidemiol.*, 1997, **146**, 713–720.

182. N. Andrieu, A.M. Goldstein, D.C. Thomas and B. Langholz, Counter-matching in studies of gene–environment interaction: efficiency and feasibility, *Am. J. Epidemiol.*, 2001, **153**, 265–274.

183. J.A. Luan, M.Y. Wong, N.E. Day and N.J. Wareham, Sample size determination for studies of gene–environment interaction, *Int. J. Epidemiol.*, 2001, **30**, 1035–1040.

184. D.C. Shields, P.N. Kirke, J.L. Mills, D. Ramsbottom, A.M. Molloy, H. Burke, D.G. Weir, J.M. Scott and A.S. Whitehead, The "thermolabile" variant of methylenetetrahydrofolate reductase and neural tube defects: an evaluation of genetic risk and the relative importance of the genotypes of the embryo and the mother, *Am. J. Hum. Genet.*, 1999, **64**, 1045–1055.

CHAPTER 23

# Nutrition and Bone Health in the Asian Population

SUZANNE C HO and YUMING CHEN

Department of Community and Family Medicine, The Chinese University of
Hong Kong, 4F School of Public Health, Prince of Wales Hospital, Shatin,
Hong Kong SAR, People's Republic of China,
Email: suzanneho@cuhk.edu.hk

**Key Points**

- Asian populations might need a lower calcium requirement to achieve zero balance than Caucasian due to a higher calcium absorption efficiency.
- 800–1000 mg calcium per day might be appropriate for Asian early postmenopausal women to maintain a zero balance and/or bone health.
- Vitamin D deficiency or subclinical deficiency is widespread in the northern regions of Asia, in which its supplementation tends to be of benefit to bone health.
- Preservation of the Asian traditional plant-based diet with the provision of soy and a fair amount of green vegetables would have a beneficial effect on bone health.
- Among Asian populations with only modest protein intake, an increase in protein intake might be promotive to bone health.
- Though salt–bone health association is inconsistent, a generally lower salt intake is recommended, particularly in regions with high salt consumption.
- The maintenance of body weight, particularly in the older populations, is important for bone mass maintenance and the prevention of fractures.

## Introduction

Osteoporosis is an important public health problem affecting both Western and Asian populations. It is predicted that by the year 2050, 50% of the hip fractures in the world will occur in Asia.[1] The incidence of hip fracture in Hong Kong has doubled over the past two decades,[2,3] and a similar rise has also been observed in other Asian cities.[3] The disease affects one-third of postmenopausal women and its prevalence increases exponentially with age.[4]

Nutritional factors are important modifiable environmental determinants of bone health. Many reports have discussed the role of nutrition in bone health in the Western populations, but there are few studies focusing on Asian populations. As Asian and Western populations' dietary intakes are quite different, bone health in the context of dietary patterns unique to the Asian populations will be dealt with in this chapter. Nutrition for women is highlighted, as osteoporosis is twice as prevalent in women as in men.

# Calcium and Bone Health

### Dietary Calcium Intake and Sources in Asian Populations

Asian populations have habitually low dietary intake of calcium. The mean dietary calcium intake among Asian population groups is in the range of 350–550 mg per day.[5–7] In Mainland China, the mean calcium intakes range from 362 mg per day in rural areas, to 453 mg day in urban regions.[8] Similar ranges of intake are also noted in other Asian regions including Hong Kong,[7] Japan,[9] Korea,[10] Malaysia,[11] Thailand,[12] and India.[13] The relatively low intake of calcium is mainly due to the low consumption of dairy products in the Asians, among whom about two-thirds do not consume milk or milk products regularly.[14]

While about 70–75% of dietary calcium in the Western populations is derived from dairy products, and less than 10% is from plant sources,[202] only about one-fifth of the total dietary Ca among Asians comes from dairy, and about half from vegetable sources, mainly green leafy vegetables and soybean products.[6] The proportion of dairy sources of calcium intake is even lower in the older populations.[6] A recent study in 685 early postmenopausal Chinese women showed that milk and milk products accounted for 23.5% of the total calcium intake, vegetables, 37%; soy foods, 9.6%; cereals, 9.6%; fish and shellfish, 8.1%; fruits, 6.2%; and the remaining food groups, such as meats and poultry, non-soy beans, Chinese Dim Sum, tea, *etc.*, 5.7% (Ho, unpublished data).

# Calcium Balance and Requirements in the Asian Populations

By definition, the calcium requirement is the amount required to achieve calcium balance in the body, where calcium balance is affected by calcium intake, absorption and retention.

### Calcium Absorption

Currently, there is very limited data on calcium absorption in the Asian population. The few published data reported that true fractional calcium absorption (TFCA) in Chinese ranged from 54–64%, and is significantly higher in the Chinese[15–17] than in the Western populations (mean ~25%).[18,19] Lee et al.[20] have measured TFCA in Chinese children aged seven years from Hong Kong ($n = 22$), and from Jiangmen southern China ($n = 12$), using a dual-label stable isotope technique. On 120 mg of testing calcium load (tracer + usual calcium) given in single dose, the values of TFCA in the combined samples with calcium intakes below and over 500 mg per day were 63.1 (SD 10.7) and 54.8% (SD7.3), respectively. With a 300 mg calcium per day supplementation over six months, the TFCA was significantly lower in the supplementation (55.6%) than the control group (64.3%).[15] In another study in which the oral tracer was administrated with the main three meals among 12 girls and four boys aged 9–16 years in Beijing, Lee et al.[16] observed a very similar TFCA (57.4±15.4%) to that obtained in southern Chinese children.

Two studies have examined TCFA in Chinese women. Chan et al.[21] reported a very high TFCA value of 91% in young women aged 19–40 years, and 65% in older women aged over 60 years with a single testing load of 20 mg calcium. As TFCA is a logarithmical function of the testing calcium load, the very low testing load might result in a substantial overestimation of the absorption efficiency. In another study, Kung et al.[17] have observed a TFCA of 61% among three groups of women (young, normal postmenopausal, and osteoporotic postmenopausal) with a single testing load of 100 mg calcium per day and unmodified diet. Our recently completed study,[122] in which 21 early postmenopausal southern Chinese women (aged 54 ± 4 years) were recruited for a three-phase study with their usual diet, their usual diet plus 500 mg calcium supplementation per day, and their usual diet plus 1000 mg calcium supplementation per day. TFCA was measured at the end of each phase using a dual tracer stable isotope technique. Our findings showed a TFCA value of 57 ± 12% for the usual diet, and a reduction of TFCA to 52 ± 12%, and 43 ± 13% with mid- and high-calcium supplementations.

In general, previous studies show that Chinese populations have almost doubled TFCA as compared with that of 15–45% observed in Western populations with habitual diets.[19,22] Reasons for the observed differences in TFCA are unclear, but racial differences, differences in the tested calcium loads, habitual dietary calcium intakes and calcium sources, and life-long adaptation to calcium intake levels, may contribute to the observed variations. As calcium absorption efficiency is the function of inverse logarithm of the size of the calcium load,[23] the absorption efficiency from the studies using a single-tested dose would be overestimated if the tested loads were less than one-third of the daily habitual intake. However, even when the oral tracer was given with the three main meals in Lee et al.'s[16] and our recently completed studies, very high levels of TFCA were also observed.

## Calcium Retention

To date, there are a few published studies available in Asia on calcium retention (on limited number of subjects). In 1973, Tandon et al.[24] tested calcium balance in seven young women over three tested doses of 123, 440, and 825 mg per day and concluded that 800 mg calcium per day would be required for maintaining calcium balance in healthy adult Indian women. Two studies conducted in Japan have measured calcium balance using a mass balance technique.[25,26] In the first study, deSouza et al.[25] studied calcium retention in nine osteoporotic (73.9 ± 2.7 years) and nine normal elderly women (67 ± 2.5 years) over two testing loads of 700 and 1474 mg per day. They concluded that 550 ± 50 and 648 ± 45 mg day are required in the normal and osteoporotic individuals to maintain zero balance, respectively. Another mass balance study by Uenishi et al.[26] in nine normal elderly women (aged 62–77 years) found a daily calcium requirement of 788 mg. Our recently completed balance study also showed that, assuming a dermal loss of 50 mg day, the net calcium retention values were −174 ± 108, 55 ± 145 and 114 ± 173 mg day, respectively, for the three testing doses of usual diet, usual diet plus 500, and 1000 mg calcium supplementation,[122]. The associations between retention and intake were well fitted using a logistic non-linear mixed-effects model,[27] which predicted that an intake of 735 mg day (95%CI 620–951) was required to achieve zero balance (unpublished data).

# Calcium Intake and Bone Health

## In Children and Adolescents

Both epidemiological and intervention studies have shown positive effects of calcium on bone mineral accretion in children before puberty. Beneficial effects are observed whether the calcium was in the form of calcium tablets, fortified foods, or dairy products.[28–30] Lee et al.[31] have found a significant positive association between cumulative calcium intake throughout 0–5 years and bone mineral content (BMC) in children aged 5 years. Another study comparing the BMC of two populations of Chinese 5-year-old children with very different calcium intakes[32] observed a significantly lower BMC values among children with lower calcium intake. The differences in BMC values remained even after adjusting for potential confounders. Two 18-month calcium supplementation studies conducted in children aged 7 years, from Jiangmen and Hong Kong,[33] showed the calcium effect was more pronounced in the Jiangmen children with habitually lower calcium intake than in the children from Hong Kong. Dawson-Hughes et al.[34] have also shown, in their double-blind, placebo-controlled, randomized trial that subjects with habitual low calcium intake would benefit most from a high calcium diet or supplement.

The beneficial effect of calcium after discontinuation of supplementation is still controversial.[35,36] Follow-up studies in Chinese children have attempted to examine whether calcium supplementation conveys a persistent benefit to bone mineral accretion.[37] The results revealed that the difference in bone mineral gain between the supplementation and control groups almost disappeared one year after the cessation of supplementation.[37,38] Long-term (>3 years) intervention studies in children are needed to clarify whether a sustainable high calcium intake is of benefit to bone mineral gains and peak bone mass.

Studies on the effects of calcium on bone mass in Asian adolescents have yielded inconsistent findings. A recent study conducted in Beijing among 649 pubertal girls aged 12–14 years showed that milk consumption had a beneficial effect on bone mass, even when adjusting for weight, bone age, Tanner age, and physical activity score.[14] A study in 197 healthy Japanese adolescent girls also observed an association between frequency of intakes of calcium-rich food and increased BMD[39]. However, a study in 87 Hong Kong girls aged 12–13 years with a mean habitual calcium intake of 550 mg per day did not observe a positive association between calcium intake and BMD in both the baseline cross-sectional and the subsequent 3-year follow-up studies.[40,41]

## Calcium Intake and Peak Bone Mass in Young Women

Peak bone mass is the maximum amount of bone an individual acquires before bone loss begins. Studies in the West as well as in Asia have suggested that peak bone mass is attained in the twenties or early thirties.[42,43] A small increase in peak bone mass may reduce subsequent fracture risks.[43,44] It has been reported that a 5% difference in bone mass may mean a 40% difference in fracture risk, and an increase of one standard deviation in bone mass may reduce the fracture risk by almost 100%. It is generally accepted that adequate calcium nutrition during childhood as well as early adulthood up to the age of 30–35 years is essential for optimizing peak bone mass.[45,46] Several population-based epidemiological studies in Caucasian populations have found a positive association between calcium intake and bone mass or density in children,[31,47] and between calcium intake in childhood and adult bone mass.[48] A number of studies have also observed a positive relation between the BMD of adult women and their milk consumption in childhood and adolescence.[48,49]

A cross-sectional study in young Chinese women has indicated an optimizing effect of high calcium intake on bone mass by 4–7% in women in their third decade of life, and high calcium density by up to 8% in women aged 31–40.[6] The difference in BMD between the high and low calcium intake groups was around 0.4 SD. Other researchers have also observed similar findings in studies in young women.[44,50] Hirota *et al.*[51] also found a significant contribution of dietary habitual calcium intake from infancy to young adults, and milk consumption in childhood to bone density in a study of 161 healthy Japanese female college students aged 19–25 years.

## Calcium Intake and Bone Health in Peri- and Postmenopausal Women

Oestrogen plays a key role in affecting bone loss in the early menopausal years, but other factors such as body weight,[52] and dietary intake[53–55] also play a role in influencing the rate of bone loss. Many observational studies have reported either a weak[54,55] or no[56,57] relationship between dietary calcium and bone mass and bone loss in postmenopausal Caucasian women. A few studies have also examined the association between habitual calcium intake or calcium supplementation and bone health in postmenopausal Asian women. Though results are inconsistent, findings have generally pointed toward a positive role of calcium intake on bone health.

Several observational studies in Asian women, including studies on 258 postmenopausal Taiwanese vegetarian women,[58] 137 postmenopausal Japanese women aged 39–60 years,[59] as well as 188 elderly Chinese men and women aged 70 years and above,[60] have observed no significant correlation between calcium intake and bone mass. However, another study of 843 women aged 35–75 years selected from five rural counties in China found that dietary calcium, especially from dairy sources, was associated with increased bone mass in middle-aged and elderly women, even after adjustment for age and body weight.[61] Our recent study examined the cross-sectional relation between habitual calcium intake and bone mass in 685 early postmenopausal Chinese women aged 48–63 years, and also with bone loss over an 18-month follow-up period. We found that the subjects in the top quartile of calcium intake had significantly higher BMD at the hip than those in the other three quartiles. Compared with women in the bottom calcium quartile, subjects of the top intake quartile (mean intake 882 mg) had 2–4% higher BMD at the whole body, lumbar spine, and the total hip, respectively. Further analyses showed that the beneficial effect of calcium was more obvious in women after the first four menopausal years, with lower body weight, or higher levels of protein intake (unpublished data). Case-control studies on risk factors of hip fractures in Asian women[62] revealed that women with low dietary calcium intake ( <498 mg per day) had a 2-fold increased risk as compared with women belonging to the highest quartile of calcium intake. Chan *et al.*[63] have observed a similar risk for vertebral fracture in a study among 481 Chinese women aged 70–79 years.

Haines *et al.*[64] have shown the addition of supplemental calcium may improve the bone mass of early postmenopausal Chinese women aged $43\pm5$ years using oestrogen replacement therapy. A randomized controlled trial in postmenopausal Chinese women aged 55–59 years found that subjects who had received 50 g milk powder per day containing 800 mg calcium over 24 months (with mean intakes ranging from 997 to 1046 mg per day) had less loss in both height and BMD than the controls.[65] Fujita *et al.*[66,67] have also observed a positive effect of calcium supplementation of 900 mg daily on bone density in 58 hospitalized elderly women with a mean age of 80 years. The study results in Asians seem to indicate that calcium intake at about 900–1000 mg per day would be beneficial to achieve better bone mass or a lower bone loss rate in postmenopausal women with habitually low calcium intake.

## Calcium Requirement for Postmenopausal Women

A review[68] of over 500 published studies has illustrated three main basic approaches in the assessment of human calcium requirements. These include epidemiological methods to evaluate health outcomes of the population as a function of dietary calcium intake; 'factorial' methods to calculate calcium intake requirements on the basis of balancing obligatory losses *via* urine, gastrointestinal tract, and skin; and balance studies with or without isotope label in normal persons for the measurement of balance performance on manipulated or habitual intakes.

Our calcium balance study (as previously described) in postmenopausal Chinese women within the first 10 years of natural menopause demonstrated that a daily calcium intake of 735 mg per day (95%CI, 620–951) was required to achieve zero balance, and 1200 mg to produce 95% (95%CI, 77–113%) of maximal calcium retention when assuming the dermal calcium loss is 50 mg day. Similar results of around 788 mg of calcium requirement were obtained in elderly Japanese women,[25] and 800 mg in young Indian women.[24] Allowing for individual variation of 20%, and taking into consideration the epidemiological association and bone mass or fracture outcomes, an intake of 900–1000 mg per day would most likely meet the calcium requirements of the majority of postmenopausal Chinese women, and possibly for similar populations in Asia. This estimation is very similar to the 1000 mg per day as 'adequate intake' for calcium developed by the Chinese Nutrition Society in 2000.[69] However, the value is lower than the recommended adequate intake of 1200 mg per day for North American women aged over 50 years.[70]

In conclusion, adequate dietary calcium intake is a major precondition for optimal bone health. However, bone health is also affected by many factors, including genetics, age, hormones, level and type of physical activity, life style, and other dietary factors. For an increase in calcium intake to optimize bone health, the other necessary dietary factors must also be met. In addition, due to threshold behavior of dietary calcium on calcium retention,[71,72] extra calcium intake beyond the plateau may not result in additional improvement in bone mass.

# Vitamin D and Bone Health in Asian

## Vitamin D Status in Asians

Vitamin D is essential for bone health. Vitamin D status in populations depends on latitude,[73] exposure to sunlight and dietary intake. Its deficiency is more common in northern than central and southern regions of Asia. Surveys have shown normal values in humans, including the elderly in sunny regions, such as Hong Kong[74] and Taiwan[75,76]. However, skin pigmentation may retard sunlight-mediated vitamin D synthesis, and low $25\text{-(OH)-D}_3$ concentrations have been observed, among healthy subjects in India[77] and Pakistan.[78] In areas of northern Asia, such as north China, Korea, and Japan, vitamin D deficiency or subclinical deficiency is widespread. In northern China, it was reported that ~40% infants and young children suffered from vitamin D deficiency rickets in the 1970s.[79] In a recent survey in Beijing, Du *et al.*[80] observed that 45% of adolescent girls had sub-clinical vitamin D deficiency in winter. Yan *et al.*[81]

reported similar results in adults in Shenyang, a city in northeast China, in which 48, 29, 15 and 13% of old men, young men, old women and young women were each respectively regarded as vitamin D-deficient in early spring. Similar results were also observed in Japan[82] and Korea.[83]

## Vitamin D and Bone Health

There seems to be little association between the circulating vitamin D and bone mass or bone markers in individuals with normal values of vitamin D,[76,84] whereas, subjects with hypovitaminosis D tend to have a decreased bone mass, especially in elderly subjects.[85,86] Lee *et al.*[16] has also reported a positive association between the circulating vitamin D and TFCA in Beijing adolescents with low vitamin D status. Intervention studies have generally established a positive effect of vitamin D treatment in the prevention of osteoporosis in postmenopausal women. A recent meta-analysis based on 25 eligible relevant trials[87] revealed that vitamin D reduced the incidence of vertebral fractures and non-vertebral fractures with RR 0.63 (95%CI:0.45–0.88) and RR:0.77 (95%CI:0.57–1.04) respectively. Hydroxylated vitamin D also has a consistently larger impact on bone density than standard vitamin D. However, the risk of early discontinuation of Vitamin D is high, due to either symptomatic adverse effects or abnormal laboratory results, as compared with controls.[87]

In Asia, few studies have examined the effects of supplementation of standard vitamin D on osteoporosis.[88] However, 1 alpha-OH-D3, an analog of the standard vitamin D3, has been extensively studied and used for treatment of osteoporosis in Japan.[89] A number of studies have demonstrated the beneficial effects of 1 alpha-OH-D3 on vertebral fracture, bone density, and cortical thickness, although most studies have small sample sizes or are of short durations.[88,89] A 6-month, placebo-controlled trial using 1.0 μg of 1 alpha-OH-D3 among 234 Japanese patients reported significantly higher radial and spinal bone densities and decreased vertebral fractures in the treated group than in the control group.[5] Another multicenter study by Fujita *et al.*[90] in a randomized group of 300 osteoporotic patients receiving alphacalcidol 0.75 μg per day or placebo for seven months, observed the favorable effects of alphacalcidol on BMD and cortical thickness of the second metacarpal bone. Similar benefits of alphacalcidol were also found in a number of other studies.[89,91]

Further studies have compared the effects of alphacalcidol with oestrogen[92] or alendronate[93] and have found that alphacalcidol is less effective on BMD and relevant bone markers than oestrogen and alendronate. In addition, Owada *et al.*[94] reported that the impact of alphacalcidol on spine BMD was more substantial in Japanese women having Bb of vitamin D receptor gene than the bb-gene group.

In conclusion, vitamin D deficiency is not a common problem in regions with lower latitude. However, vitamin D deficiency or subclinical deficiency is widespread particularly in the elderly populations of the northern regions of Asia. In these populations, vitamin D status is positively associated with BMD and the maintenance of bone mass. Supplementation of vitamin D or its analogs seems to be of benefit in retarding bone loss and reducing fracture rates.

# Soy Intake and Bone Health

Soy intake is part of the regular diet of the Asian populations. Observations that populations consuming soy have lower hip fracture incidences have given rise to the hypothesis that the intake of soy protein and/or soy-derived isoflavonoid phytoestrogens may be protective for bone health.[95] Soy protein consumption may range from 55 g per day in Japan[96] to <5 g in the US.[97] Soy consumption in Asia also varies from moderate intake in Southern Chinese[98] to high intake in the Japanese populations.[96] Variations, as reflected by urinary excretion of isoflavonoid phytoestrogens, have also been observed to be much higher in Asian than the Caucasian populations.[95,99]

The main forms of soy isoflavones are genistein and daidzein. They exist in the conjugated forms and are converted to equol by microflora, among the metabolites of isoflavones, which is structurally similar to estradiol 17β.[100] Soy isoflavones, in particular genistein, have particular affinity with oestrogen receptor (ER) beta. As bone cells are rich in ER-β, isoflavones may act as weak agonists and enhance osteoblast-like cell functions *via* ERs. Isoflavones may also reduce osteoclast-mediated bone resorption.[101,102] Animal studies have shown that soy-rich diets enhance bone formation in ovariectomized rat models.[103] Omi *et al.*[104] reported that rats which fed soymilk had greater BMD and mechanical bone strength than control rats which fed casein. A recent 6-month study showed that a soy-rich diet increased osteocalcin in postmenopausal women.[105] Two short-term studies in Japanese postmenopausal women have demonstrated a significant reduction of the bone resorption markers with soy isoflavones supplementation.[106,107]

Animal studies are also supportive of the positive role of soy isoflavonoid phytoestrogens in maintaining or modestly improving bone mass.[103] Studies to investigate the association between soy intake and bone mass in premenopausal as well as postmenopausal women have yielded inconsistent findings. Two short-term trials have observed a bone conserving effect of soy protein containing isoflavones on spinal BMD of Caucasian perimenopausal and postmenopausal women.[108,109] A recent 1-year randomized double-blind placebo-controlled study[110] on the effects of a daily intake of 54 mg of the phytoestrogen genistein on BMD in postmenopausal Caucasian women, revealed a 3% increase in the spine and femoral neck BMD compared to a −1.6% and −0.65% decrease in the respective bone sites in the placebo group. However, a 9-month randomized double-blind study on early postmenopausal women showed no effect for either 52 or 96 mg of isoflavones (contained in soy protein) on the spine and hip BMD when compared with the control group consuming alcohol-washed soy protein[111]. A small 1-year controlled double-blind intervention trial of isoflavones enriched soy protein containing 90 mg isoflavones in young women observed no effect in BMD or BMC.[112]

Limited studies on the effect of soy intake on bone mass have been conducted in the Asian populations. A few observational studies have revealed a positive association between habitual soy intake and bone health in premenopausal as well as postmenopausal women.[113-115] A cross-sectional observational study by Tsuchida et al.[115] in 995 healthy women aged 40–49 years reported a positive association between the frequency of soybean intake and BMD. A 3-year longitudinal study conducted by Ho et al.[113] in a cohort of 132 Chinese premenopausal women aged 30–40 years revealed that women belonging to the highest quartile of baseline soy protein consumption had significantly less reduction in spinal bone mass compared with women belonging to the lowest intake quartile. The positive effects of soy isoflavones on spinal BMD remained after adjusting for age and body size (height, weight, bone area). Results from these observational studies in Asian premenopausal women seem to contradict findings by Anderson et al.[112] on a lack of effect of soy isoflavones on bone mass in young premenopausal women aged 21–25 years. Reasons for this inconsistency of findings may include differences in study design, differences in circulating concentrations of oestrogens. Adaptation to phytoestrogens intake in Asian vs. Caucasian women may partly explain the inconsistent findings. It has been noted that Caucasian women have relatively higher circulating concentrations of oestrogens compared with Asian women,[116] and as such, isoflavones may have a stronger agonistic effect in Asian than in Caucasian premenopausal women.

Two recent cross-sectional studies in postmenopausal Japanese women[117,118] reported a significant positive association between soy protein or isoflavones intake and spinal BMD. We also conducted a population-based study, aimed at investigating the relation between soy protein consumption and bone mass in Chinese early postmenopausal women. The baseline cross-sectional analyses revealed a positive association between quartiles of soy protein intake and BMD at the various hip bone sites among women beyond the first few menopausal years.[119] The association remained significant after controlling for other confounding factors such as dietary calcium, protein intake, body weight and years since menopause. However, a non-significant association in women within the first few years of menopause was observed. Menopause plays a predominant role in bone loss due to oestrogen deprivation during the first few years of menopause,[120] and menopause-related bone loss would be less marked approximately 3–4 years after the last menses when the body had been accustomed to a low level of oestrogen.[121] Moderate habitual intake of soy during early menopause may be inadequate to counteract the drastic effect of oestrogen decline on bone loss. However, long-term soy intake seems to exert a protective effect on bone mass after those dramatic years. Another observational study in older Chinese postmenopausal women (mean age 63 years, and about 13 years postmenopausal) reported a positive association between phytoestrogens intake and spine and hip bone mass.[114] These studies in Asian women suggested that habitual soy protein intake may be beneficial in maintaining bone mass in menopausal women, and the effect is more obvious after the initial period of rapid bone loss.

We also conducted a one-year placebo-controlled randomized trial to examine the effects of two doses of isoflavones enriched soy extracts – 40 mg and 80 mg – on bone loss in postmenopausal Chinese women within the first 10 years since menopause.[122] We found a mild but statistically significant effect of high-dose soy-derived isoflavones in attenuating BMC loss at the trochanter, intertrochanter and the total hip after controlling for the years since menopause, body weight and height, baseline BMC at the relevant sites, as well as total dietary calcium, and protein intakes over the intervention period.

Though both observational and clinical trials have indicated a generally positive effect of soy protein or soy isoflavones on bone health. The effective or optimal dosage is still unclear. Results from the various short-term studies[108,109,122] seem to suggest a dosage of 80 mg per day or more of isoflavones would be required for the prevention of bone loss in postmenopausal women. However, bone remodeling is a relatively slow process and may take 6–18 months or even longer among an older population to reach a new equilibrium.[123] Animal experimental results have also suggested that a threshold dose of isoflavones needs to be consumed for a lengthy period of time for a measurable effect on bone to be observed.[124,125] Thus, longer-term (at least 2 years or more) trial(s) would be required to test the optimal dosages of isoflavones and their effects on bone mass. The risks and benefits of the use of suggested doses of isoflavones also need to be carefully evaluated in the longer-term studies.

## Mechanisms of Soy Intake on Bone

The effects of soy products on bone metabolism are still unclear. Researchers have proposed several mechanisms by which soy isoflavones help conserve bone mass. Soy isoflavones, particularly genistein, are very similar in structure to estradiol, which is known to decrease bone loss in postmenopausal women. Soy isoflavones may thus act as agonists in the ERs of bone tissues for bone mass preservation. Another mechanism by which soy isoflavones may help conserve bone mass could be mediated through the effects of calcium. Anderson *et al.* reported a highly significant calcium conserving effect of soy protein (containing isoflavones) in animal studies.[125] Moreover, soy protein induces less urinary calcium loss than animal protein. Anderson *et al.*[126] reported that, in contrast to the animal protein lactalbumin, soy protein (containing isoflavones) exerts a modest decrease in urinary calcium excretion. Anderson's findings were supported by Register *et al.*[127] who reported that serum calcium was significantly increased in the soy protein-fed compared with casein/lactalbumin-fed ovariectomized macaques. At present, a crossover trial in postmenopausal Chinese women is being conducted to test the effect of soy-derived isoflavones on calcium metabolism. The consumption of soy also involves the intake of high quality protein and a higher intake of other nutrients such as vitamin K. Vitamin $K_2$ (menakinone) is present in fermented soybeans and is known to stimulate bone formation and prevent bone loss.[128]

In conclusion, the habitual intake of soy, a traditional Asian food appears to be beneficial to bone health. The soy–bone action could be mediated through its oestrogenic or calcium conserving effects. In addition, soy intake also contributes to an increase in high quality protein intake. Some soy products are also important sources of calcium and vitamin K intakes. Therefore, this traditional dietary practice should be encouraged and preserved in Asian populations.

# Vitamin K and Bone Health

Two types of vitamin K occur in nature: vitamin $K_1$ (phylloquinone), and vitamin $K_2$ (menaquinones) (which is a series of vitamers with multi-isoprene units at the 3-position). Green vegetables (*e.g.*, broccoli) and certain plant oils such as soybean are good sources of vitamin $K_1$. The long chain vitamin $K_2$ (menaquinones, MK-7 through MK10) are synthesized by bacterial flora[129,130]. Fermented soybean (natto) is a rich source of vitamin $K_2$ and cheese also gives some vitamin $K_2$.

An increasing number of studies has shown a link between vitamin K and bone health,[131,132] and it has generally been assumed that the effect of vitamin K on bone is mediated through osteocalcin or matrix-Gla protein.[133] Other possible mechanisms include the enhancement of mineralization by osteoblasts, preservation of bone connectivity and reduction of urinary calcium excretion.[134,135]

Studies on the relation between vitamin K and bone health have also been conducted in Asia. Kaneki *et al.*[136] have observed a statistically significant inverse relation between regions with natto (fermented soybean) intake and hip fracture incidences in Japan. Our cross-sectional and longitudinal studies observed an association between vitamin $K_1$ intake and higher bone mass in perimenopausal Chinese women.[119] Orimo *et al.*[137] have reported that vitamin $K_2$ (mentatetrenone) supplementation improved bone density. A 24-month study of supplementation with 45 mg of mentatetrenone, a vitamin $K_2$ with four isoprene units, has also been shown to lower the incidence of fractures and to reduce spinal bone loss.[138]

Although 1 µg per kg per day of vitamin K intake is recommended, based on the classical function of vitamin K on blood coagulation,[139] insufficient data are available on the requirement of vitamin K for bone health. There is some indication that a level higher than the previous recommendation may be required.[133] As the hepatic storage capacity for vitamin K is low, and 60–70% of the daily dietary intake of vitamin $K_1$ is lost by excretion,[140] a daily intake of vitamin K appears necessary. The habitual daily consumption of green vegetables and soy products may help to meet the daily vitamin K requirements in the Asian population. Further studies are required to clarify the effectiveness and optimal level of vitamin K intake on bone mass maintenance.

# Dietary Protein and Bone Health

## Protein Intake and Calcium Metabolism

Most studies in both animals[141] and Western populations[142] have established that high intakes of dietary protein increase urinary calcium excretion due to increased urinary acidity[143,144], increased glomerular filtration rate (GFR)[141] and decreased renal calcium reabsorption.[145,146] In a review, Kerstetter *et al.*[142] reported that each 50 g increment of protein intake was associated with an extra 60 mg calcium excretion in urine.

Some studies among Asian populations have also observed a positive association between protein intake and urinary calcium excretion. In a survey of 764 middle-aged and elderly women in rural China, Hu *et al.* found that urinary calcium was positively associated with dietary animal protein and negatively associated with plant protein intakes.[143] Itoh *et al.* also reported that habitual high protein diets, rich in sulfur-containing amino acids might augment urinary calcium excretion, particularly in the elderly population[147]. However, our population-based study in Hong Kong Chinese consisting of 500 female and 510 male adults aged 24–74 years did not observe a positive association between dietary protein intake and urinary calcium, even in the those aged 50 years and over.[148]

Dietary protein may also influence calcium metabolism through other mechanisms, but the effects of dietary protein on intestinal calcium absorption seem to be inconsistent. A number of studies have reported a negative effect of protein on calcium absorption.[149,150] But, other studies have observed an increase in calcium absorption[151] and a reduction of intact PTH and 1,25-dihydroxyvitamin D.[152]

## Protein Intake and Bone Mass

Epidemiological and clinical data on the effect of protein on bone mass are still controversial. Many,[153,154] though not all,[155] studies reported higher fracture rates in populations with higher dietary protein intakes. Again, many[156,157] but not all[158] epidemiological studies have found a positive association between protein intake and bone mass, and the prevention of bone loss. Our studies in both young and early postmenopausal Chinese women have shown a weak but positive association between higher protein intake and higher bone mass.[2,119] The positive association could be due to the comparatively low level of protein intake in the Asian populations – with a mean intake of 60 g per day or less. This association between higher protein intake and better BMD among populations with only moderate levels of protein intake (68 ± 23.6 g per day) has also been observed in Caucasian elderly populations.[159,160] The Framingham Osteoporosis study[161] reveals that the lowest quartile of baseline protein intake was associated with higher bone loss over a 4-year follow-up in the Framingham elderly cohort aged 69–97 years.

Therefore, high protein intake, particularly of animal protein, can have an adverse effect on calcium metabolism and bone health; but could be beneficial in populations with moderate or low protein intake. Adequate protein intake in Asians with moderate or even low level of protein intake would be promotive of bone health.

# Sodium and Calcium Metabolism and Bone Health

## Sodium and Urinary Calcium Excretion

Both loading studies and surveys of free-living individuals in various countries showed a strong positive correlation between urinary calcium and sodium excretion in Western populations.[162] In Asia, several studies have also consistently shown a relationship between urinary calcium and sodium excretions. In a study of 334 young Chinese women aged 17–40 years, Chan et al.[163] observed a strong correlation ($r = 0.573$) between urinary and sodium excretions. About 33% of the variation of urinary calcium/creatinine (UCa/Cr) could be explained by urinary sodium/creatinine (UNa/Cr). In another population-based dietary survey conducted in the Hong Kong Chinese aged 24–74 years,[148] we estimated that urinary calcium excretion increased by about 1.4 mmol per 100 mmol (range 1.37–1.43) increase in urinary sodium (or 25 mg calcium per 1.0 g sodium). In a study of Japanese adults, Itoh et al.[164] estimated that each 100 mmol increase in urinary sodium was associated with an increase of 0.55 mmol urinary calcium in young adults aged 20–49 years, and of 0.97 mmol in older adults. Similar results regarding sodium and calcium excretions were observed in other studies in elderly Japanese men and women aged 50–79 years.[165,166] Assuming a calcium absorption rate of 50% as reported above, and not taking into consideration any adaptation of calcium retention to varied sodium intake, an increase of about 2.8 mmol of calcium dietary intake per 100 mmol increase in dietary sodium (~50 mg calcium per g sodium) would probably be required to maintain a stable calcium retention.[124] This estimated ratio is lower than that reported by Devine et al.[167] Based on a longitudinal study in elderly women, the authors suggested that the equivalent protective effect on bone loss could be achieved by either halving the current sodium intake of 150 mmol per day or by increasing dietary Ca by 22 mmol per day. However, differences in study methods, population difference in calcium absorption, and adaptation of calcium absorption to different levels of sodium intake may partly explain the observed variations in calcium/sodium ratio.

## Sodium Intake and Bone Health

Both animal and human studies have indicated an association between high sodium intake and high bone resorption. Chan et al.[163] and other researchers[168] found that urinary hydroxyproline (OHP) increased with greater salt intake in rats. The effect was independent of menopause status. In human studies, a cross-sectional study of 334 healthy Chinese women showed a significantly positive association between fasting urinary Na/Cr and urinary OHP/Cr ratios.[163] Such a positive association was also observed in many surveys in both Asians[164,166] and Caucasians[169,170] regardless of age and gender. A 10% increase in urinary sodium is associated with a 4% increase in hydroxyproline in Caucasian women aged 16–82,[169] or a 2% increase in bone resorption in Japanese adults.[164]

Several studies have shown inconsistent findings on the salt-BMD association. Studies in rats have shown that high sodium intake reduces bone mass.[171,172] Devine *et al.*[167] observed a positive association between salt intake and bone loss in a longitudinal study of postmenopausal women. However, other studies have found no significant association,[170,173] or even a small but statistically significant protective effect of sodium on bone mass in a 16-year longitudinal study of men.[174] As single point urinary sodium may not reflect the level of long-term exposure to sodium, further multiple-point assessments of urinary sodium excretion in longitudinal studies would be required to clarify the effects of high salt intake on bone health.

In conclusion, studies have shown consistent findings of a sodium calciuretic effect, but the salt-BMD association is less consistent. However, the relatively high salt consumption in parts of Asia, *e.g.*, Northern China, Japan, and Korea from preserved meats, vegetables, and sauces gives rise to concerns for their effect on bone health. As high salt intake is also associated with hypertension, which is prevalent in these parts of Asia, lower salt intake would generally be recommended.

## Lactose Intolerance and Bone Health

The prevalence of lactose intolerance is relatively high in Asian populations.[175] Nonetheless, data from studies have supported the idea that calcium absorption is not altered due to lactose maldigestion,[176] but lactose maldigesters may avoid calcium-rich dairy foods resulting in an overall low intake of calcium for bone mass development and maintenance.[177,178]

Our limited observational data in about 300 young Chinese women have found little difference in the mean BMD and total calcium intake between subjects with and without symptoms of lactose intolerance.[6] In these populations, dairy products accounted for less than one-quarter of their source of calcium intake. These study subjects obtained three-quarters of their calcium intake from non-dairy and mainly plant sources. The absorption fractions of calcium from plant sources (except spinach) have been found to be quite similar to that from dairy products.[179]

## Other Lifestyle Factors and Implications

A healthy, balanced diet should provide enough energy for the achievement of bone development. Body weight, lean mass and fat mass have been found to be among the strongest determinants of BMD.[42,67] Being underweight is a strong risk factor for low bone mass and osteoporosis.[180] Though over-nutrition is a cause of concern in developing countries, under-nutrition is still prevalent in many Asian populations and may have important implications on bone health. Both insufficient caloric and protein intake have been shown to be associated with hip fractures[181]; and low BMD[159] in elderly subjects and BMC in premenopausal women[182] The achievement of optimal weight, and maintenance of body weight, particularly in older populations, are important for the preservation of bone mass. Holbrook *et al.*[183] have reported dieting and weight loss are associated with lower BMD at all sites though these trends became non-significant after adjusting for current weight.

Associations between other nutritional factors with bone mass or fracture have also been noted. Our study in the young Chinese women aged 21–30 years has also observed a positive effect of high phosphorus and iron on BMD.[6] Other micronutrients like potassium and magnesium,[184] and vitamin C have also been found to be associated with bone health in women aged 45–55.[184] Our study of young women also observed an association between vitamin C intake and bone health.[6] These findings support the suggestions of a positive link between fruits and vegetables and bone health.[185]

There have been inconsistent findings on caffeine intake and BMD.[186,187] Among the Asian women, Yano *et al.*[5] found an inverse association between caffeine intake and BMC in Japanese–American women, but our study in young Chinese women failed to find such an association.[6] Some studies have identified tea drinking as a protective factor for osteoporosis or fractures.[157,188] However, increased consumption of carbonated cola drinks have been associated with increased urinary calcium excretion in premenopausal women and fractures in girls (mean age 14.34).[189,190]

Smoking has consistently been shown to be associated with low bone mass and hip fractures in both Caucasian and Asian populations.[62,191] The mechanism of smoking on bone mass is unclear, but an antiestrogenic effect of smoking has been suggested.[192] The prevalence of smoking has traditionally been low in Asian women. Recent increases in smoking in adolescents and young women[193] have been noted, and such trends may contribute to an increase in osteoporosis in later years.

Though moderate drinking has been associated with higher BMD in the middle-aged population as well as postmenopausal women,[185,194] excessive alcohol intake is associated with lower BMD. However, studies in Asian populations have generally identified alcohol drinking to be among the risk factors for low BMD or hip fractures.[62,180]

Weight bearing activities have been noted to exert a protective effect on bone mass or its maintenance.[195] Exercises with high impact loading also seem to exert a more noticeable effect than that with low impact loading.[195,196] Studies have observed that lean body mass, an indication of muscle strength and history of physical activity, is associated with bone mass.[195,197] Consistent with other studies, a study of the older Chinese population has shown body weight, and time on completing a 16-feet walk were the major determinants of femoral neck BMD.[60] As such body weight maintenance, particularly in the older population whose weight tends to decline with age, would play an important role in bone health and the prevention of hip fracture. Walking exercise, which is a major form of leisure time activity in the Asian populations, should be further developed for the prevention of falls and bone health.[198]

Westernization and urbanization have been associated with osteoporosis in the Asian populations.[199] The changing lifestyle and dietary pattern towards less intake of complex carbohydrates, particularly from fruits and vegetables, increased smoking (particularly in younger women), increased intake of carbonated drinks, dieting and reduced physical activity may all contribute to the increasing prevalence of osteoporosis in the Asian populations.

# Recommendations

Adequate and appropriate nutrition is important for overall health, including bone health. Among the nutrients, calcium is the specific nutrient considered most important for attaining peak bone mass, for preventing bone loss and for treating osteoporosis. The US National Institute of Medicine[70] has recommended a calcium intake of 1,300 mg per day for age 9–18, 1000 mg for 19–50 years, and 1200 for 50 years and above. Taking into account the smaller body size and probably better efficiency in calcium absorption in the Asian populations[17,20](Chen unpublished data), an intake of around 800–1100 mg per day would be recommended. With a mean intake in the range of 500 mg per day or even lower, a significant proportion of the Asian population has an inadequate intake of calcium. The challenge remains on how we might encourage people to meet their calcium requirements. This task should be approached, taking into account the habitual dietary practices of the local population. In addition to dairy products, which give the highest concentration of calcium density, non-dairy sources of calcium, such as bean products, tofu, fish with bones, Chinese cabbage, mustard greens, bok choy, kale, nuts, and seeds should be emphasized. The increasing availability of calcium fortified foods, including milk, soymilk, biscuits, and orange juice in the market, and increasing awareness of dietary risk of osteoporosis, are favorable towards an overall increase of calcium intake in Asia. An increasing trend of higher calcium intake has been observed in the younger age groups.[6]

Besides calcium, a number of dietary factors have also been shown to be relevant to bone health among the Asian populations. These include adequate protein intake, preservation of the cultural habits of soy intake and tea drinking, high intake of fruits and vegetables, in particular green leafy vegetables which are good sources of calcium as well as vitamin K, and reduction of sodium intake, and a sufficient dietary source of vitamin D for those who are housebound.

Caution has been advised in adding milk to the adolescent diet in populations that already have a high level of dietary protein, as this may not be the best strategy for preventing osteoporosis in later life.[200] With increasing westernization, a very high protein intake may be an upcoming issue in some urbanized affluent populations. Therefore, besides increasing the intake of milk and dairy products, Asian populations should be encouraged to increase the non-dairy sources of calcium intake as well.

At present, little is known about the nutritional needs of bone tissue for the other micronutrients with essential functions on bone metabolism. These include magnesium, zinc, manganese, iron, folate, vitamin C, and the B vitamins.[201] The role of these nutrients in bone health for the Asian populations should also be emphasized. As such, a balanced diet for overall health is also basic for the promotion and preservation of bone health.

## Conclusions

The mean calcium intake in the Asian population is low, and an increase of intake to around 800–1000 mg per day might be required, particularly in early postmenopausal women. Vitamin D supplementation would be of benefit to bone health in the regions where its deficiency or subclinical deficiency is widespread. An increase of protein intake in population groups with low level of protein intake might be beneficial to bone health. Soy products and green vegetable, which are usually part of the Asian traditional plant-based diet but may decline with urbanization, should be promoted. The maintenance of body weight, particularly in the older population, is important for bone mass maintenance and the prevention of fractures.

## References

1. C. Cooper, G. Campion and L.J. Melton III, Hip fractures in the elderly: a worldwide projection, *Osteoporosis Int.*, 1992, **2**, 285–289.
2. S.C. Ho, W.E. Bacon, T. Harris, A. Looker and S. Maggi, Hip fracture rates in Hong Kong and the United States, 1988 through 1989, *Am. J. Public Health*, 1993, **83**, 694–697.
3. E.M. Lau, Epidemiology of osteoporosis in urbanized Asian populations, *Osteoporosis Int.*, 1997, **7**(Suppl. 3), S91–S95.
4. S.C. Ho, E.M. Lau, J. Woo, A. Sham, K.M. Chan, S. Lee and P.C. Leung, The prevalence of osteoporosis in the Hong Kong Chinese female population, *Maturitas*, 1999, **32**, 171–178.
5. K. Yano, L.K. Heilbrun, R.D. Wasnich, J.H. Hankin and J.M. Vogel, The relationship between diet and bone mineral content of multiple skeletal sites in elderly Japanese-American men and women living in Hawaii, *Am. J. Clin. Nutr.*, 1985, **42**, 877–888.
6. S.C. Ho, P.C. Leung, R. Swaminathan, C. Chan, S.S. Chan, Y.K. Fan and R. Lindsay, Determinants of bone mass in Chinese women aged 21-40 years. II. Pattern of dietary calcium intake and association with bone mineral density, *Osteoporosis Int.*, 1994, **4**, 167–175.
7. S.S. Leung, S.C. Ho, J. Woo, T.H. Lam and E.D. Janus, *Hong Kong Adult Dietary Survey*, The Chinese University of Hong Kong, Hong Kong, 1997.
8. K.Y. Ge, F.Y. Zhai and H.C. Yan, *The Dietary and Nutritional Status of Chinese Population,* People's Medical Publishing House, Beijing, 1996.
9. N. Yoshiike, Y. Matsumura, M. Iwaya, M. Sugiyama and M. Yamaguchi, National Nutrition Survey in Japan, *J. Epidemiol.*, 1996, **6**, S189–S200.
10. South Korean Ministry of Health and Welfare, 1995 *National Nutrition Survey Report*, Seoul, South Korea, 1997.
11. W.S. Chee, A.R. Suriah, Y. Zaitun, S.P. Chan, S.L. Yap and Y.M. Chan, Dietary calcium intake in postmenopausal Malaysian women: comparison between the

food frequency questionnaire and three-day food records, *Asia Pac. J. Clin. Nutr.*, 2002, **11**, 142–146.

12. S. Nititham, S. Srianujata and T. Rujirawat, Dietary intake of phytate, zinc and calcium of self-selected diets of Ubon Ratchathani and Bangkok subjects, Thailand, *J. Med. Assoc. Thai.*, 1999, **82**, 855–861.

13. A. Wadhwa, M. Sabharwal and S. Sharma, Nutritional status of the elderly, *Indian J. Med. Res.*, 1997, **106**, 340–348.

14. X.Q. Du, H. Greenfield, D.R. Fraser, K.Y. Ge, Z.H. Liu and W. He, Milk consumption and bone mineral content in Chinese adolescent girls, *Bone*, 2002, **30**, 521–528.

15. W.T. Lee, S.S. Leung, Y.C. Xu, S.H. Wang, W.P. Zeng and J. Lau, Fairweather-Tait, Effects of double-blind controlled calcium supplementation on calcium absorption in Chinese children measured with stable isotopes (42Ca and 44Ca), *Br. J. Nutr.*, 1995, **73**, 311–321.

16. W.T. Lee, J. Jiang, P. Hu, X. Hu, D.C. Roberts and J.C. Cheng, Use of stable calcium isotopes (42Ca & 44Ca) in evaluation of calcium absorption in Beijing adolescents with low vitamin D status, *Food Nutr. Bull.*, 2002, **23**, 42–47.

17. A.W. Kung, K.D. Luk, L.W. Chu and P.K. Chiu, Age-related osteoporosis in Chinese: an evaluation of the response of intestinal calcium absorption and calcitropic hormones to dietary calcium deprivation, *Am. J. Clin. Nutr.*, 1998, **68**, 1291–1297.

18. S.A. Abrams, K.O. O'Brien, L.K. Liang and J.E. Stuff, Differences in calcium absorption and kinetics between black and white girls aged 5–16 years, *J. Bone Miner. Res.*, 1995, **10**, 829–833.

19. R.P. Heaney and R.R. Recker, Distribution of calcium absorption in middle-aged women, *Am. J. Clin. Nutr.*, 1986, **43**, 299–305.

20. W.T. Lee, S.S. Leung, S.J. Fairweather-Tait, D.M. Leung, H.S. Tsang, J. Eagles, T. Fox, S.H. Wang, Y.C. Xu and W.P. Zeng, True fractional calcium absorption in Chinese children measured with stable isotopes (42Ca and 44Ca), *Br. J. Nutr.*, 1994, **72**, 883–897.

21. E.L. Chan, E. Lau, C.C. Shek, D. MacDonald, J. Woo, P.C. Leung and R. Swaminathan, Age-related changes in bone density, serum parathyroid hormone, calcium absorption and other indices of bone metabolism in Chinese women, *Clin. Endocrinol. (Oxf.)*, 1992, **36**, 375–381.

22. S.A. Abrams, Pubertal changes in calcium kinetics in girls assessed using 42Ca, *Pediatr. Res.*, 1993, **34**, 455–459.

23. R.P. Heaney, C.M. Weaver and M.L. Fitzsimmons, Influence of calcium load on absorption fraction, *J. Bone Miner. Res.*, 1990, **5**, 1135–1138.

24. G.S. Tandon, S.P. Teotia, S.C. Yadav and S.K. Garg, A study of calcium balance in normal individuals and postmenopausal state, *J. Indian Med. Assoc.*, 1973, **61**, 214–217.

25. A.C. deSouza, T. Nakamura, K. Stergiopoulos, M. Shiraki, Y. Ouchi and H. Orimo, Calcium requirement in elderly Japanese women, *Gerontology*, 1991, **37**(Suppl. 1), 43–47.

26. K. Uenishi, H. Ishida, A. Kamei, M. Shiraki, I. Ezawa, S. Goto, H. Fukuoka, T. Hosoi and H. Orimo, Calcium requirement estimated by balance study in elderly Japanese people, *Osteoporosis Int.*, 2001, **12**, 858–863.

27. J.C. Pinheiro and D.M. Bates, Approximations to the Log-likelihood Function in the Nonlinear Mixed-effects Model, *J. Comput. Graph. Stat.*, 1995, **4**, 12–35.

28. J.P. Bonjour, A.L. Carrie, S. Ferrari, H. Clavien, D. Slosman, G. Theintz and R. Rizzoli, Calcium-enriched foods and bone mass growth in prepubertal girls: a randomized, double-blind, placebo-controlled trial, *J. Clin. Invest.*, 1997, **99**, 1287–1294.

29. G.M. Chan, K. Hoffman and M. McMurry, Effects of dairy products on bone and body composition in pubertal girls, *J. Pediatr.*, 1995, **126**, 551–556.

30. B. Dibba, A. Prentice, M. Ceesay, D.M. Stirling, T.J. Cole and E.M. Poskitt, Effect of calcium supplementation on bone mineral accretion in gambian children accustomed to a low-calcium diet, *Am. J. Clin. Nutr.*, 2000, **71**, 544–549.

31. W.T. Lee, S.S. Leung, S.S. Lui and J. Lau, Relationship between long-term calcium intake and bone mineral content of children aged from birth to 5 years, *Br. J. Nutr.*, 1993, **70**, 235–248.

32. W.T. Lee, S.S. Leung, M.Y. Ng, S.F. Wang, Y.C. Xu, W.P. Zeng and J. Lau, Bone mineral content of two populations of Chinese children with different calcium intakes, *Bone Miner.*, 1993, **23**, 195–206.

33. W.T. Lee, S.S. Leung, D.M. Leung, H.S. Tsang, J. Lau and J.C. Cheng, A randomized double-blind controlled calcium supplementation trial, and bone and height acquisition in children, *Br. J. Nutr.*, 1995, **74**, 125–139.

34. B. Dawson-Hughes, S. Harris, C. Kramich, G. Dallal and H.M. Rasmussen, Calcium retention and hormone levels in black and white women on high- and low-calcium diets, *J. Bone Miner. Res.*, 1993, **8**, 779–787.

35. J.P. Bonjour, T. Chevalley, P. Ammann, D. Slosman and R. Rizzoli, Gain in bone mineral mass in prepubertal girls 3.5 years after discontinuation of calcium supplementation: a follow-up study, *Lancet*, 2001, **358**, 1208–1212.

36. C.W. Slemenda, M. Peacock, S. Hui, L. Zhou and C.C. Johnston, Reduced rates of skeletal remodeling are associated with increased bone mineral density during the development of peak skeletal mass, *J. Bone Miner. Res.*, 1997, **12**, 676–682.

37. W.T. Lee, S.S. Leung, D.M. Leung and J.C. Cheng, A follow-up study on the effects of calcium-supplement withdrawal and puberty on bone acquisition of children, *Am. J. Clin. Nutr.*, 1996, **64**, 71–77.

38. W.T. Lee, S.S. Leung, D.M. Leung, S.H. Wang, Y.C. Xu, W.P. Zeng and J.C. Cheng, Bone mineral acquisition in low calcium intake children following the withdrawal of calcium supplement, *Acta Paediatr.*, 1997, **86**, 570–576.

39. N. Tsukahara, K. Sato and I. Ezawa, Effects of physical characteristics and dietary habits on bone mineral density in adolescent girls, *J. Nutr. Sci. Vitaminol. (Tokyo)*, 1997, **43**, 643–655.

40. J.C. Cheng, S.S. Leung, W.T. Lee, J.T. Lau, N. Maffulli, A.Y. Cheung and K.M. Chan, Determinants of axial and peripheral bone mass in Chinese adolescents, *Arch. Dis. Child*, 1998, **78**, 524–530.

41. J.C. Cheng, N. Maffulli, S.S. Leung, W.T. Lee, J.T. Lau and K.M. Chan, Axial and peripheral bone mineral acquisition: a 3-year longitudinal study in Chinese adolescents, *Eur. J. Pediatr.*, 1999, **158**, 506–512.

42. S.C. Ho, S.Y. Hsu, P.C. Leung, C. Chan, R. Swaminathan, Y.K. Fan and S.S. Chan, A longitudinal study of the determinants of bone mass in Chinese women aged 21 to 40. I. Baseline association of anthropometric measurements with bone mineral density, *Ann. Epidemiol.*, 1993, **3**, 256–263.

43. R.R. Recker, K.M. Davies, S.M. Hinders, R.P. Heaney, M.R. Stegman and D.B. Kimmel, Bone gain in young adult women, *JAMA*, 1992, **268**, 2403–2408.

44. C.C. Johnston Jr., J.Z. Miller, C.W. Slemenda, T.K. Reister, S. Hui, J.C. Christian and M. Peacock, Calcium supplementation and increases in bone mineral density in children, *N. Engl. J. Med.*, 1992, **327**, 82–87.

45. J.J. Anderson, Calcium, phosphorus and human bone development, *J. Nutr.*, 1996, **126**, 1153S–1158S.

46. V. Matkovic, Calcium intake and peak bone mass, *N. Engl. J. Med.*, 1992, **327**, 119–120.

47. K.S. Wosje and B.L. Specker, Role of calcium in bone health during childhood, *Nutr. Rev.*, 2000, **58**, 253–268.

48. R.B. Sandler, C.W. Slemenda, R.E. LaPorte, J.A. Cauley, M.M. Schramm, M.L. Barresi and A.M. Kriska, Postmenopausal bone density and milk consumption in childhood and adolescence, *Am. J. Clin. Nutr.*, 1985, **42**, 270–274.

49. S. Soroko, T.L. Holbrook, S. Edelstein and E. Barrett-Connor, Lifetime milk consumption and bone mineral density in older women, *Am. J. Public Health*, 1994, **84**, 1319–1322.

50. D. Picard, L.G. Ste-Marie, D. Coutu, L. Carrier, R. Chartrand, R. Lepage, P. Fugere and P. D'Amour, Premenopausal bone mineral content relates to height, weight and calcium intake during early adulthood, *Bone Miner.*, 1988, **4**, 299–309.

51. T. Hirota, M. Nara, M. Ohguri, E. Manago and K. Hirota, Effect of diet and lifestyle on bone mass in Asian young women, *Am. J. Clin. Nutr.*, 1992, **55**, 1168–1173.

52. T.L. Holbrook and E. Barrett-Connor, The association of lifetime weight and weight control patterns with bone mineral density in an adult community, *Bone Miner.*, 1993, **20**, 141–149.

53. J.F. Aloia, A.N. Vaswani, J.K. Yeh, P. Ross, K. Ellis and S.H. Cohn, Determinants of bone mass in postmenopausal women, *Arch. Intern. Med.*, 1983, **143**, 1700–1704.

54. B. Dawson-Hughes, P. Jacques and C. Shipp, Dietary calcium intake and bone loss from the spine in healthy postmenopausal women, *Am. J. Clin. Nutr.*, 1987, **46**, 685–687.

55. S. Murphy, K.T. Khaw, H. May and J.E. Compston, Milk consumption and bone mineral density in middle aged and elderly women, *BMJ*, 1994, **308**, 939–941.

56. S.A. Earnshaw, A. Worley and D.J. Hosking, Current diet does not relate to bone mineral density after the menopause. The Nottingham Early Postmenopausal Intervention Cohort (EPIC) Study Group, *Br. J. Nutr.*, 1997, **78**, 65–72.

57. E.C. van Beresteijn, M.A. 't Hof, G. Schaafsma, H. de Waard and S.A. Duursma, Habitual dietary calcium intake and cortical bone loss in perimenopausal women: a longitudinal study, *Calcif. Tissue Int.*, 1990, **47**, 338–344.

58. J.F. Chiu, S.J. Lan, C.Y. Yang, P.W. Wang, W.J. Yao, L.H. Su and C.C. Hsieh, Long-term vegetarian diet and bone mineral density in postmenopausal Taiwanese women, *Calcif. Tissue Int.*, 1997, **60**, 245–249.

59. S. Sasaki and R. Yanagibori, Association between current nutrient intakes and bone mineral density at calcaneus in pre- and postmenopausal Japanese women, *J. Nutr. Sci. Vitaminol. (Tokyo)*, 2001, **47**, 289–294.

60. S.C. Ho, S.S. Chan, J. Woo, P.C. Leung and J. Lau, Determinants of bone mass in the Chinese old-old population, *Osteoporosis Int.*, 1995, **5**, 161–166.

61. J.F. Hu, X.H. Zhao, J.S. Chen, J. Fitzpatrick, B. Parpia and T.C. Campbell, Bone density and lifestyle characteristics in premenopausal and postmenopausal Chinese women, *Osteoporosis Int.*, 1994, **4**, 288–297.

62. E.M. Lau, J.K. Lee, P. Suriwongpaisal, S.M. Saw, D.S. Das, A. Khir and P. Sambrook, The incidence of hip fracture in four Asian countries: the Asian Osteoporosis Study (AOS), *Osteoporosis Int.*, 2001, **12**, 239–243.

63. H.H. Chan, E.M. Lau, J. Woo, F. Lin, A. Sham and P.C. Leung, Dietary calcium intake, physical activity and the risk of vertebral fracture in Chinese, *Osteoporosis Int.*, 1996, **6**, 228–232.

64. C.J. Haines, T.K. Chung, P.C. Leung, S.Y. Hsu and D.H. Leung, Calcium supplementation and bone mineral density in postmenopausal women using estrogen replacement therapy, *Bone*, 1995, **16**, 529–531.

65. E.M. Lau, J. Woo, V. Lam and A. Hong, Milk supplementation of the diet of postmenopausal Chinese women on a low calcium intake retards bone loss, *J. Bone Miner. Res.*, 2001, **16**, 1704–1709.

66. T. Fujita, T. Ohue, Y. Fujii, A. Miyauchi and Y. Takagi, Effect of calcium supplementation on bone density and parathyroid function in elderly subjects, *Miner. Electrolyte Metab.*, 1995, **21**, 229–231.

67. T. Fujita, T. Ohue, Y. Fujii, A. Miyauchi and Y. Takagi, Heated oyster shell-seaweed calcium (AAA Ca) on osteoporosis, *Calcif. Tissue Int.*, 1996, **58**, 226–230.

68. M.I. Irwin and E.W. Kienholz, A conspectus of research on calcium requirements of man, *J. Nutr.*, 1973, **103**, 1019–1095.

69. Chinese Nutrition Society 2000, In *Chinese Dietary Reference Intakes*, Light Industry Press, Beijing, China, 458–458.

70. Institutes of Medicine. *Dietary reference intakes for calcium, phosphorus, magnesium, vitamin D and fluoride.* Food and nutrition board. Washington DC. National Academy Press, 1997, 71–145.

71. R.M. Forbes, K.E. Weingartner, H.M. Parker, R.R. Bell, and J.W. Erdman Jr., Bioavailability to rats of zinc, magnesium and calcium in casein-, egg- and soy protein-containing diets, *J. Nutr.*, 1979, **109**, 1652–1660.

72. V. Matkovic and R.P. Heaney, Calcium balance during human growth: evidence for threshold behavior, *Am. J. Clin. Nutr.*, 1992, **55**, 992–996.

73. P. Lips, T. Duong, A. Oleksik, D. Black, S. Cummings, D. Cox and T. Nickelsen, A global study of vitamin D status and parathyroid function in postmenopausal women with osteoporosis: baseline data from the multiple outcomes of raloxifene evaluation clinical trial, *J. Clin. Endocrinol. Metab.*, 2001, 1212–1221.

74. S.S. Leung, S. Lui and R. Swaminathan, Vitamin D status of Hong Kong Chinese infants, *Acta Paediatr. Scand*, 1989, **78**, 303–306.

75. W.P. Lee, L.W. Lin, S.H. Yeh, R.H. Liu and C.F. Tseng, Correlations among serum calcium, vitamin D and parathyroid hormone levels in the elderly in southern Taiwan, *J. Nurs. Res.*, 2002, **10**, 65–72.

76. K.S. Tsai, S.H. Hsu, J.P. Cheng and R.S. Yang, Vitamin D stores of urban women in Taipei: effect on bone density and bone turnover, and seasonal variation, *Bone*, 1997, **20**, 371–374.

77. R. Goswami, N. Gupta, D. Goswami, R.K. Marwaha, N. Tandon and N. Kochupillai, Prevalence and significance of low 25-hydroxyvitamin D concentrations in healthy subjects in Delhi, *Am. J. Clin. Nutr.*, 2000, **72**, 472–475.

78. L. Brunvand, S.S. Shah, S. Bergstrom and E. Haug, Vitamin D deficiency in pregnancy is not associated with obstructed labor. A study among Pakistani women in Karachi, *Acta. Obstet. Gynecol. Scand*, 1998, 77, 303–306.

79. Q.R. Guan, C. Ying and X.C. Ma, National survey on prevalence of rickets in infants and young children, In *Prevention and Treatment of Rickets*, Q.R. Guan, (ed.), vol.1, Hei Long Jiang Publishing Bureau, Harbin, China, 1984, 25–26.

80. X. Du, H. Greenfield, D.R. Fraser, K. Ge, A. Trube and Y. Wang, Vitamin D deficiency and associated factors in adolescent girls in Beijing, *Am. J. Clin. Nutr.*, 2001, **74**, 494–500.

81. L. Yan, A. Prentice, H. Zhang, X. Wang, D.M. Stirling and M.M. Golden, Vitamin D status and parathyroid hormone concentrations in Chinese women and men from north-east of the People's Republic of China, *Eur. J. Clin. Nutr.*, 2000, **54**, 68–72.

82. K. Nakamura, M. Nashimoto, Y. Tsuchiya, A. Obata, K. Miyanishi and M. Yamamoto, Vitamin D insufficiency in Japanese female college students: a preliminary report, *Int. J. Vitam. Nutr. Res.*, 2001, **71**, 302–305.

83. J.H. Kim and S.J. Moon, Time spent outdoors and seasonal variation in serum concentrations of 25-hydroxyvitamin D in Korean women, *Int. J. Food Sci. Nutr.*, 2000, **51**, 439–451.

84. K. Nakamura, M. Nashimoto and M. Yamamoto, Are the serum 25-hydroxy-vitamin D concentrations in winter associated with forearm bone mineral density in healthy elderly Japanese women? *Int. J. Vitam. Nutr. Res.*, 2001, **71**, 25–29.

85. K. Nakamura, M. Nashimoto, S. Matsuyama and M. Yamamoto, Low serum concentrations of 25-hydroxyvitamin D in young adult Japanese women: a cross sectional study, *Nutrition*, 2001, **17**, 921–925.

86. D.T. Villareal, R. Civitelli, A. Chines and L.V. Avioli, Subclinical vitamin D deficiency in postmenopausal women with low vertebral bone mass, *J. Clin. Endocrinol. Metab.*, 1991, **72**, 628–634.

87. E. Papadimitropoulos, G. Wells, B. Shea, W. Gillespie, B. Weaver, N. Zytaruk, A. Cranney, J. Adachi, P. Tugwell, R. Josse, C. Greenwood and G. Guyatt, Meta-analyses of therapies for postmenopausal osteoporosis. VIII: Meta- analysis of the efficacy of vitamin D treatment in preventing osteoporosis in postmenopausal women, *Endocr. Rev.*, 2002, **23**, 560–569.

88. T. Fujita, T. Inoue and H. Orimo, Clinical evaluation of the effect of calcitriol on osteoporosis. Multicenter double-blind study using alphacalcidol as the control drug, *J. Clin. Exp. Med.*, 1989, **148**, 833–857.

89. Y. Nishii, Active vitamin D and its analogs as drugs for the treatment of osteoporosis: advantages and problems, *J. Bone Miner. Metab.*, 2002, **20**, 57–65.

90. T. Fujita, Studies of osteoporosis in Japan, *Metabolism*, 1990, **39**, 39–42.

91. I.R. Reid, Vitamin D and its metabolites in the management of osteoporosis. In *Osteoporosis*, 2nd edn, R. Marcus, D. Feldman and J. Kelsey (ed.), Academic Press, San Diego, 2001, 553–575.

92. I. Gorai, O. Chaki, Y. Taguchi, M. Nakayama, H. Osada, N. Suzuki, N. Katagiri, Y. Misu and H. Minaguchi, Early postmenopausal bone loss is prevented by estrogen and partially by 1alpha-OH-vitamin D3: therapeutic effects of estrogen and/or 1alpha- OH-vitamin D3, *Calcif. Tissue Int.*, 1999, 65, 16–22.

93. M. Shiraki, K. Kushida, M. Fukunaga, H. Kishimoto, M. Taga, T. Nakamura, K. Kaneda, H. Minaguchi, T. Inoue, H. Morii, A. Tomita, K. Yamamoto, Y. Nagata, M. Nakashima and H. Orimo, A double-masked multicenter comparative study between alendronate and alfacalcidol in Japanese patients with osteoporosis. The Alendronate Phase III Osteoporosis Treatment Research Group, *Osteoporosis Int.*, 1999, 183–192.

94. M. Owada, K. Suzuki, T. Honda, H. Yamada, S. Tsukikawa, K. Hoshi and A. Sato, Effect of vitamin D receptor gene polymorphism on vitamin D therapy for postmenopausal bone loss, *Nippon Sanka Fujinka Gakkai Zasshi*, 1996, **48**, 799–805.

95. M.J. Messina, Legumes and soybeans: overview of their nutritional profiles and health effects, *Am. J. Clin. Nutr.*, 1999, **70**, 439S–450S.
96. C. Nagata, N. Takatsuka, Y. Kurisu and H. Shimizu, Decreased serum total cholesterol concentration is associated with high intake of soy products in Japanese men and women, *J. Nutr.*, 1998, **128**, 209–213.
97. M.J. Messina, V. Persky, K.D. Setchell and S. Barnes, Soy intake and cancer risk: a review of the in vitro and in vivo data, *Nutr. Cancer*, 1994, **21**, 113–131.
98. S.C. Ho, J.L. Woo, S.S. Leung, A.L. Sham, T.H. Lam and E.D. Janus, Intake of soy products is associated with better plasma lipid profiles in the Hong Kong Chinese population, *J. Nutr.*, 2000, **130**, 2590–2593.
99. V.J. Roach, T.F. Cheung, T.K. Chung, N.M. Hjelm, M.A. Waring, E.P. Loong and C.J. Haines, Phytoestrogens: dietary intake and excretion in postmenopausal Chinese women, *Climacteric*, 1998, **1**, 290–295.
100. S.J. Anderson, Anthony, M.J. Messina and S.C. Garner, Effects of phyto-oestrogens on tissues, *Nutr. Res. Rev.*, 1999, **12**, 75–116.
101. J.J. Anderson, and C.P. Miller, Lower lifetime estrogen exposure among vegetarian as a possible risk factor for osteoporosis: a hypothesis, *Veg. Nutr.*, 1998, 4–12.
102. E. Sugimoto and M. Yamaguchi, Anabolic effect of genistein in osteoblastic MC3T3-E1 cells, *Int. J. Mol. Med.*, 2000, **5**, 515–520.
103. B.H. Arjmandi and B.J. Smith, Soy isoflavones' osteoprotective role in postmenopausal women: mechanism of action, *J. Nutr. Biochem.*, 2002, **13**, 130–137.
104. N. Omi, S. Aoi, K. Murata and I. Ezawa, Evaluation of the effect of soybean milk and soybean milk peptide on bone metabolism in the rat model with ovariectomized osteoporosis, *J. Nutr. Sci. Vitaminol. (Tokyo)*, 1994, **40**, 201–211.
105. L.M. Chiechi, G. Secreto, M. D'Amore, M. Fanelli, E. Venturelli, F. Cantatore, T. Valerio, G. Laselva and P. Loizzi, Efficacy of a soy rich diet in preventing postmenopausal osteoporosis: the Menfis randomized trial, *Maturitas*, 2002, **42**, 295–300.
106. Y. Yamori, E.H. Moriguchi, T. Teramoto, A. Miura, Y. Fukui, K.I. Honda, M. Fukui, Y. Nara, K. Taira and Y. Moriguchi, Soybean isoflavones reduce postmenopausal bone resorption in female Japanese immigrants in Brazil: a ten-week study, *J. Am. Coll. Nutr.*, 2002, 21, 560–563.
107. T. Uesugi, Y. Fukui and Y. Yamori, Beneficial effects of soybean isoflavone supplementation on bone metabolism and serum lipids in postmenopausal Japanese women: a four- week study, *J. Am. Coll. Nutr.*, 2002, **21**, 97–102.
108. D.L. Alekel, A.S. Germain, C.T. Peterson, K.B. Hanson, J.W. Stewart and T. Toda, Isoflavone-rich soy protein isolate attenuates bone loss in the lumbar spine of perimenopausal women, *Am. J. Clin. Nutr.*, 2000, **72**, 844–852.
109. S.M. Potter, J.A. Baum, H. Teng, R.J. Stillman, N.F. Shay and J.W. Erdman Jr., Soy protein and isoflavones: their effects on blood lipids and bone density in postmenopausal women, *Am. J. Clin. Nutr.*, 1998, **68**, 1375S–1379S.
110. N. Morabito, A. Crisafulli, C. Vergara, A. Gaudio, A. Lasco, N. Frisina, R. D'Anna, F. Corrado, M.A. Pizzoleo, M. Cincotta, D. Altavilla, R. Ientile and F. Squadrito, Effects of genistein and hormone-replacement therapy on bone loss in early postmenopausal women: a randomized double-blind placebo- controlled study, *J. Bone Miner. Res.*, 2002, **17**, 1904–1912.
111. J.C. Gallagher, K. Rafferty, V. Haynatzka and M. Wilson, Effect of soy protein on bone metabolism, *J. Nutr.*, 1999, **130**, 667s (abstract).

112. J.J. Anderson, X. Chen, A. Boass, M. Symons, M. Kohlmeier, J.B. Renner and S.C. Garner, Soy isoflavones: no effects on bone mineral content and bone mineral density in healthy, menstruating young adult women after one year, *J. Am. Coll. Nutr.*, 2002, **21**, 388–393.

113. S.C. Ho, S.G. Chan, Q. Yi, E. Wong and P.C. Leung, Soy intake and the maintenance of peak bone mass in Hong Kong Chinese women, *J. Bone Miner. Res.*, 2001, **16**, 1363–1369.

114. J. Mei, S.S. Yeung and A.W. Kung, High dietary phytoestrogen intake is associated with higher bone mineral density in postmenopausal but not premenopausal women, *J. Clin. Endocrinol. Metab.*, 2001, **86**, 5217–5221.

115. K. Tsuchida, S. Mizushima, M. Toba and K. Soda, Dietary soybeans intake and bone mineral density among 995 middle-aged women in Yokohama, *J. Epidemiol.*, 1999, **9**, 14–19.

116. B.R. Goldin, H. Adlercreutz, S.L. Gorbach, M.N. Woods, J.T. Dwyer, T. Conlon, E. Bohn and S.N. Gershoff, The relationship between estrogen levels and diets of Caucasian American and Oriental immigrant women, *Am. J. Clin. Nutr.*, 1986, **44**, 945–953.

117. T. Horiuchi, T. Onouchi, M. Takahashi, H. Ito and H. Orimo, Effect of soy protein on bone metabolism in postmenopausal Japanese women, *Osteoporosis Int.*, 2000, **11**, 721–724.

118. Y. Somekawa, M. Chiguchi, T. Ishibashi and T. Aso, Soy intake related to menopausal symptoms, serum lipids, and bone mineral density in postmenopausal Japanese women, *Obstet. Gynecol.*, 2001, **97**, 109–115.

119. S.C. Ho, J. Woo, Y. Chen, A. Sham and J. Lau, Soy protein consumption and bone mineral density in early postmenopausal Chinese women, *Osteoporosis Int.*, 2003, in press.

120. S. Meema and H.E. Meema, Menopausal bone loss and estrogen replacement, *Isr. J. Med. Sci.*, 1976, **12**, 601–606.

121. R. Recker, J. Lappe, K. Davies and R. Heaney, Characterization of perimenopausal bone loss: a prospective study, *J. Bone Miner. Res.*, 2000, **15**, 1965–1973.

122. Y. Chen, S.C. Ho, S. Lam, S. Ho and J. Woo, Soy isoflavones have a favorable effect on bone loss in Chinese postmenopausal women with lower bone mass: a double-blind randomized-controlled trial, *J. Clin. Endoc. Metab.*, 2003, in press.

123. R.P. Heaney, J.C. Gallagher, C.C. Johnston, R. Neer, A.M. Parfitt and G.D. Whedon, Calcium nutrition and bone health in the elderly, *Am. J. Clin. Nutr.*, 1982, **36**, 986–1013.

124. J.J. Anderson, Plant-based diets and bone health: nutritional implications, *Am. J. Clin. Nutr.*, 1999, **70**, 539S–542S.

125. J.J. Anderson and S.C. Garner, The effects of phytoestrogens on bone, *Nutr. Res.*, 1997, **17**, 1617–1632.

126. J.J.B. Anderson, K. Thomsen and C. Christiansen, High protein meals, insular hormones, and urinary calcium excretion in human subjects, In *Osteoporosis*, C. Christiansen, J.S. Johnson and B. Riis (eds.), *Osteopress, ApS,* Copenhagen, Denmark, 1987, pp. 240–245.

127. T. Register, M. Anthony, M.J. Jayo and J.K. Williams, Effects of 17 β-oestradiol and consumption of phytoestrogen-rich soy or casein/lactalbumin-based diets on serum chemistries in ovariectomized female macaques, *J. Bone Miner. Res.*, 1997, **12**, S350–S350.

128. Y. Somekawa, M. Chigughi, M. Harada and T. Ishibashi, Use of vitamin K2 (menatetrenone) and 1,25-dihydroxyvitamin D3 in the prevention of bone loss induced by leuprolide, *J. Clin. Endocrinol. Metab.*, 1999, **84**, 2700–2704.

129. C. Bolton-Smith, R.J. Price, S.T. Fenton, D.J. Harrington and M.J. Shearer, Compilation of a provisional UK database for the phylloquinone (vitamin K1) content of foods, *Br. J. Nutr.*, 2000, **83**, 389–399

130. S.L. Booth, K.L. Tucker, H. Chen, M.T. Hannan, D.R. Gagnon, L.A. Cupples, P.W. Wilson, J. Ordovas, E.J. Schaefer, B. Dawson-Hughes and D.P. Kiel, Dietary vitamin K intakes are associated with hip fracture but not with bone mineral density in elderly men and women, *Am. J. Clin. Nutr.*, 2000, **71**, 1201–1208.

131. T. Kanai, T. Takagi, K. Masuhiro, M. Nakamura, M. Iwata and F. Saji, Serum vitamin K level and bone mineral density in post-menopausal women, *Int. J. Gynaecol. Obstet.*, 1997, **56**, 25–30.

132. T. Sugimoto, K. Nishiyama, F. Kuribayashi, and K. Chihara, Serum levels of insulin-like growth factor (IGF) I, IGF-binding protein (IGFBP)-2, and IGFBP-3 in osteoporotic patients with and without spinal fractures, *J. Bone Miner. Res.*, 1997, **12**, 1272–1279.

133. N.C. Binkley, J.W. Suttie, Vitamin K nutrition and osteoporosis, *J. Nutr.*, 1995, **125**, 1812–1821.

134. Y. Koshihara, K. Hoshi and M. Shiraki, Enhancement of mineralization in human osteoblast-like cells by vitamin K2 (menatetrenone), *J. Bone Miner. Res.*, 1992, **8**, s209 (abstract).

135. T. Mawatari, H. Miura, H. Higaki, T. Moro-Oka, K. Kurata, T. Murakami and Y. Iwamoto, Effect of vitamin K2 on three-dimensional trabecular microarchitecture in ovariectomized rats, *J. Bone Miner. Res.,* 2000, **15**, 1810–1817.

136. M. Kaneki, S.J. Hedges, T. Hosoi, S. Fujiwara, A. Lyons, S.J. Crean, N. Ishida, M. Nakagawa, M. Takechi, Y. Sano, Y. Mizuno, S. Hoshino, M. Miyao, S. Inoue, K. Horiki, M. Shiraki, Y. Ouchi and H. Orimo, Japanese fermented soybean food as the major determinant of the large geographic difference in circulating levels of vitamin K2: possible implications for hip-fracture risk, *Nutrition*, 2001, **17**, 315–321.

137. H. Orimo, M. Shiraki, Y. Hayashi, T. Hoshino, T. Onaya, S. Miyazaki, H. Kurosawa, T. Nakamura and N. Ogawa, Effects of 1 alpha-hydroxyvitamin D3 on lumbar bone mineral density and vertebral fractures in patients with postmenopausal osteoporosis, *Calcif. Tissue Int.*, 1994, **54**, 370–376.

138. M. Shiraki, Y. Shiraki, C. Aoki and M. Miura, Vitamin K2 (menatetrenone) effectively prevents fractures and sustains lumbar bone mineral density in osteoporosis, *J. Bone Miner. Res.*, 2000, **15**, 515–521.

139. J.W. Suttie, Vitamin K and human nutrition, *J. Am. Diet. Assoc.*, 1992, **92**, 585–590.

140. M.J. Shearer, A. Bach and M. Kohlmeier, Chemistry, nutritional sources, tissue distribution and metabolism of vitamin K with special reference to bone health, *J. Nutr.*, 1996, **126**, 1181S–1186S.

141. E.L. Chan and R. Swaminathan, The effect of high protein and high salt intake for 4 months on calcium and hydroxyproline excretion in normal and oophorectomized rats, *J. Lab. Clin. Med.*, 1994, **124**, 37–41.

142. J.E. Kerstetter and L.H. Allen, Protein intake and calcium homeostasis, *Adv. Nutr. Res.*, 1994, **9**, 167–181.

143. J.F. Hu, X.H. Zhao, B. Parpia and T.C. Campbell, Dietary intakes and urinary excretion of calcium and acids: a cross-sectional study of women in China, *Am. J. Clin. Nutr.*, 1993, **58**, 398–406.

144. J. Lemann Jr., Relationship between urinary calcium and net acid excretion as determined by dietary protein and potassium: a review, *Nephron*, 1999, **81**(Suppl. 1), 18–25.

145. Y. Kim and H.M. Linkswiler, Effect of level of protein intake on calcium metabolism and on parathyroid and renal function in the adult human male, *J. Nutr.*, 1979, **109**, 1399–1404.

146. S.A. Schuette, M.B. Zemel and H.M. Linkswiler, Studies on the mechanism of protein-induced hypercalciuria in older men and women, *J. Nutr.*, 1980, **110**, 305–315.

147. R. Itoh, N. Nishiyama and Y. Suyama, Dietary protein intake and urinary excretion of calcium: a cross-sectional study in a healthy Japanese population, *Am. J. Clin. Nutr.*, 1998, **67**, 438–444.

148. S.C. Ho, Y.M. Chen, J.L. Woo, S.S. Leung, T.H. Lam and E.D. Janus, Sodium is the leading dietary factor associated with urinary calcium excretion in Hong Kong Chinese adults, *Osteoporosis Int.*, 2001, **12**, 723–731.

149. H.H. Daper, L.A. Piche and R.S. Gibson, Effects of a high protein intake from common foods on calcium metabolism in a cohort of postmenopausal women, *Nutr. Res.*, 1991, **11**, 273–81.

150. R.P. Heaney, Dietary protein and phosphorus do not affect calcium absorption, *Am. J. Clin. Nutr.*, 2000, **72**, 758–761.

151. J.E. Kerstetter, K.O. O'Brien and K.L Insogna, Dietary protein affects intestinal calcium absorption, *Am. J. Clin. Nutr.*, 1998, **68**, 859–865.

152. J.E. Kerstetter, D.M. Caseria, M.E. Mitnick, A.F. Ellison, L.F. Gay, T.A. Liskov, T.O. Carpenter and K.L. Insogna, Increased circulating concentrations of parathyroid hormone in healthy, young women consuming a protein-restricted diet, *Am. J. Clin. Nutr.*, 1997, **66**, 1188–1196.

153. D. Feskanich,, W.C. Willett, M.J. Stampfer and G.A. Colditz, Protein consumption and bone fractures in women, *Am. J. Epidemiol.*, 1996, **143**, 472–479

154. H.E. Meyer, J.I. Pedersen, E.B. Loken and A. Tverdal, Dietary factors and the incidence of hip fracture in middle-aged Norwegians. A prospective study, *Am. J. Epidemiol.*, 1997, **145**, 117–123.

155. R.G. Munger, J.R. Cerhan and B.C. Chiu, Prospective study of dietary protein intake and risk of hip fracture in postmenopausal women, *Am. J. Clin. Nutr.*, 1999, **69**, 147–152.

156. J. Bell and S.J. Whiting, Elderly women need dietary protein to maintain bone mass, *Nutr. Rev.*, 2002, **60**, 337–341.

157. S.J. Whiting, J.L. Boyle, A. Thompson, R.L. Mirwald and R.A. Faulkner, Dietary protein, phosphorus and potassium are beneficial to bone mineral density in adult men consuming adequate dietary calcium, *J. Am. Coll. Nutr.*, 2002, **21**, 402–409.

158. R.D. Starling, P.A. Ades and E.T. Poehlman, Physical activity, protein intake, and appendicular skeletal muscle mass in older men, *Am. J. Clin. Nutr.*, 1999, **70**, 91–96.

159. G. Geinoz, C.H. Rapin, R. Rizzoli, R. Kraemer, B. Buchs, D. Slosman, J.P. Michel and J.P. Bonjour, Relationship between bone mineral density and dietary intakes in the elderly, *Osteoporosis Int.*, 1993, **3**, 242–248.

160. K.L. Tucker, M.T. Hannan and D.P. Kiel, The acid-base hypothesis: diet and bone in the Framingham Osteoporosis Study, *Eur. J. Nutr.*, 2001, **40**, 231–237.

161. M.T. Hannan, K.L. Tucker, B. Dawson-Hughes, L.A. Cupples, D.T. Felson and D.P. Kiel, Effect of dietary protein on bone loss in elderly men and women: the Framingham Osteoporosis Study, *J. Bone Miner. Res.*, 2000, **15**, 2504–2512.

162. L.K. Massey and S.J. Whiting, Dietary salt, urinary calcium, and bone loss, *J. Bone Miner. Res.*, 1996, **11**, 731–736.

163. E.L. Chan, C.S. Ho, D. MacDonald, S.C. Ho, T.Y. Chan and R. Swaminathan, Interrelationships between urinary sodium, calcium, hydroxyproline and serum PTH in healthy subjects, *Acta. Endocrinol. (Copenh.)*, 1992, **127**, 242–245.

164. R. Itoh and Y. Suyama, Sodium excretion in relation to calcium and hydroxyproline excretion in a healthy Japanese population, *Am. J. Clin. Nutr.*, 1996, **63**, 735–740.

165. R. Itoh, J. Oka, H. Echizen, K. Yamada, Y. Suyama and K. Murakami, The interrelation of urinary calcium and sodium intake in healthy elderly Japanese, *Int. J. Vitam. Nutr. Res.*, 1991, **61**, 159–165.

166. R. Itoh, Y. Suyama, Y. Oguma and F. Yokota, Dietary sodium, an independent determinant for urinary deoxypyridinoline in elderly women. A cross-sectional study on the effect of dietary factors on deoxypyridinoline excretion in 24-h urine specimens from 763 free-living healthy Japanese, *Eur. J. Clin. Nutr.*, 1999, 886–890.

167. A. Devine, R.A. Criddle, I.M. Dick, D.A. Kerr and R.L. Prince, A longitudinal study of the effect of sodium and calcium intakes on regional bone density in postmenopausal women, *Am. J. Clin. Nutr.*, 1995, **62**, 740–745.

168. A. Goulding and E. Gold, Effects of dietary NaCl supplementation on bone synthesis of hydroxyproline, urinary hydroxyproline excretion and bone 45Ca uptake in the rat, *Horm. Metab. Res.*, 1988, **20**, 743–745.

169. A. Goulding, Fasting urinary sodium/creatinine in relation to calcium/creatinine and hydroxyproline/creatinine in a general population of women, *NZ. Med. J.*, 1981, **93**, 294–297.

170. G. Jones, T. Beard, V. Parameswaran, T. Greenaway and R. von Witt, A population-based study of the relationship between salt intake, bone resorption and bone mass, *Eur. J. Clin. Nutr.*, 1997, **51**, 561–565.

171. A.Y. Chan, P. Poon, E.L. Chan, S.L. Fung and R. Swaminathan, The effect of high sodium intake on bone mineral content in rats fed a normal calcium or a low calcium diet, *Osteoporosis Int.*, 1993, **3**, 341–344.

172. E. Gold and A. Goulding, High dietary salt intakes lower bone mineral density in ovariectomised rats: a dual x-ray absorptiometry study, *Bone*, 1995, **16**, 115s (abstract).

173. B. Dawson-Hughes, S.E. Fowler, G. Dalsky and C. Gallagher, Sodium excretion influences calcium homeostasis in elderly men and women, *J. Nutr.*, 1996, **126**, 2107–2112.

174. G.A. Greendale, E. Barrett-Connor, S. Edelstein, S. Ingles and R. Haile, Dietary sodium and bone mineral density: results of a 16-year follow-up study, *J. Am. Geriatr. Soc.*, 1994, **42**, 1050–1055.

175. K.A. Jackson and D.A. Savaiano, Lactose maldigestion, calcium intake and osteoporosis in African-, Asian-, and Hispanic-Americans, *J. Am. Coll. Nutr.*, 2001, **20**, 198S–207S.

176. M. Horowitz, J. Wishart, L. Mundy and B.E. Nordin, Lactose and calcium absorption in postmenopausal osteoporosis, *Arch. Intern. Med.*, 1987, **147**, 534–536.

177. S.J. Birge Jr., H.T. Keutmann, P. Cuatrecasas and G.D. Whedon, Osteoporosis, intestinal lactase deficiency and low dietary calcium intake, *N. Engl. J. Med.*, 1967, **276**, 445–448.

178. G.R. Corazza, G. Benati, A. Di Sario, C. Tarozzi, A. Strocchi, M. Passeri and G. Gasbarrini, Lactose intolerance and bone mass in postmenopausal Italian women, *Br. J. Nutr.*, 1995, **73**, 479–487.

179. R.P. Heaney, Calcium in the prevention and treatment of osteoporosis, *J. Intern. Med.*, 1992, **231**, 169–180.

180. S.C. Ho and P.C. Leung, Determinants of peak bone mass in Chinese and Caucasian populations, *HKMJ*, 1995, **1**, 38–42.

181. A.M. Parfitt, Dietary risk factors for age-related bone loss and fractures, *Lancet*, 1983, **2**, 1181–1185.

182. C. Cooper, E.J. Atkinson, D.D. Hensrud, H.W. Wahner, W.M. O' Fallon, B.L. Riggs and L.J. Melton III, Dietary protein intake and bone mass in women, *Calcif. Tissue. Int.*, 1996, **58**, 320–325.

183. T.L. Holbrook and E. Barrett-Connor, The association of lifetime weight and weight control patterns with bone mineral density in an adult community, *Bone Miner.*, 1993, **20**, 141–149.

184. S.A. New, C. Bolton-Smith, D.A. Grubb and D.M. Reid, Nutritional influences on bone mineral density: a cross-sectional study in premenopausal women, *Am. J. Clin. Nutr.*, 1997, **65**, 1831–1839.

185. S.A. New, S.P. Robins, M.K. Campbell, J.C. Martin, M.J. Garton, C. Bolton-Smith, D.A. Grubb, S.J. Lee and D.M. Reid, Dietary influences on bone mass and bone metabolism: further evidence of a positive link between fruit and vegetable consumption and bone health? *Am. J. Clin. Nutr.*, 2000, **71**, 142–151.

186. E. Barrett-Connor, J.C. Chang and S.L. Edelstein, Coffee-associated osteoporosis offset by daily milk consumption. The Rancho Bernardo Study, *JAMA*, 1994, **271**, 280–283.

187. M.J. Grainge, C.A. Coupland, S.J. Cliffe, C.E. Chilvers and D.J. Hosking, Cigarette smoking, alcohol and caffeine consumption, and bone mineral density in postmenopausal women. The Nottingham EPIC Study Group, *Osteoporosis Int.*, 1998, **8**, 355–363.

188. O. Johnell, B. Gullberg, J.A. Kanis, E. Allander, L. Elffors, J. Dequeker, G. Dilsen, C. Gennari, V.A. Lopes and G. Lyritis, Risk factors for hip fracture in European women: the MEDOS Study. Mediterranean Osteoporosis Study, *J. Bone. Miner. Res.*, 1995, **10**, 1802–1815.

189. G. Wyshak and R.E. Frisch, Carbonated beverages, dietary calcium, the dietary calcium/phosphorus ratio, and bone fractures in girls and boys, *J. Adolesc. Health*, 1994, **15**, 210–215.

190. M.J. Barger-Lux, R.P. Heaney and M.R. Stegman, Effects of moderate caffeine intake on the calcium economy of premenopausal women, *Am. J. Clin. Nutr.*, 1990, **52**, 722–725.

191. M.R. Law and A.K. Hackshaw, A meta-analysis of cigarette smoking, bone mineral density and risk of hip fracture: recognition of a major effect, *BMJ*, 1997, **315**, 841–846.

192. T.J. Key, M.C. Pike, J.A. Baron, J.W. Moore, D.Y. Wang, B.S. Thomas and R.D. Bulbrook, Cigarette smoking and steroid hormones in women, *J. Steroid. Biochem. Mol. Biol.*, 1991, **39**, 529–534.

193. T.H. Lam, S.M. Stewart and L.M. Ho, An analysis of smoking and sexuality related data in the youth sexuality study. HSRC Report #832015: Health Services Research Committee of Hong Kong; 2002, pp. 1–4.

194. T.L. Holbrook and E. Barrett-Connor, A prospective study of alcohol consumption and bone mineral density, *BMJ*, 1993, **306**, 1506–1509.

195. S.C. Ho, E. Wong, S.G. Chan, J. Lau, C. Chan and P.C. Leung, Determinants of peak bone mass in Chinese women aged 21–40 years. III. Physical activity and bone mineral density, *J. Bone Miner. Res.*, 1997, **12**, 1262–1271.

196. T.L. Robinson, C. Snow-Harter, D.R. Taaffe, D. Gillis, J. Shaw and R. Marcus, Gymnasts exhibit higher bone mass than runners despite similar prevalence of amenorrhea and oligomenorrhea, *J. Bone Miner. Res.*, 1995, **10**, 26–35

197. C. Snow-Harter, M.L. Bouxsein, B.T. Lewis, D.R. Carter and R. Marcus, Effects of resistance and endurance exercise on bone mineral status of young women: a randomized exercise intervention trial, *J. Bone Miner. Res.*, 1992, **7**, 761–769

198. S.C. Ho, J. Woo, S.S. Chan, Y.K. Yuen and A. Sham, Risk factors for falls in the Chinese elderly population, *J. Gerontol. A. Biol. Sci. Med. Sci.*, 1996, **51**, M195–M198

199. W.T. Lee, J.C. Cheng, J. Jiang, P. Hu, X. Hu and D.C. Roberts, Calcium absorption measured by stable calcium isotopes ((42)Ca & (44)Ca) among Northern Chinese adolescents with low vitamin D status, *J. Orthop. Surg. (Hong Kong)*, 2002, **10**, 61–66

200. P. Appleby, Milk intake and bone mineral acquisition in adolescent girls. Adding milk to adolescent diet may not be best means of preventing osteoporosis, *BMJ*, 1998, **316**, 1747–1748

201. J.J. Anderson, Nutritional advances in human bone metabolism. Introduction, *J. Nutr.*, 1996, **126**, 1150S–1152S

202. K.H. Fleming, J.T. Heimbach, Consumption of calcium in the U.S.: food sources and intake levels, *J. Nutr.*, 1994, **124**, 1426S–30S.

CHAPTER 24

# Nutrition and Bone Health in Middle Eastern Women

JALALUDDIN A. KHAN and SAWSAN O. KHOJA

Department of Biochemistry, Faculty of Science, King Abdulaziz University,
PO Box 80174, Jeddah 21589, Saudi Arabia, Email: jahalkahn_k@hotmail.com

**Key Points**

- The population of Saudi Arabia is approximately 14.87 million and 38% of populations are above the age of 40 years.
- Throughout, last decade, there was not much attention on bone disease in Arab countries especially Saudi Arabia.
- The high prevalence of osteopenia and osteoporosis in Saudi population is due to the urbanization and westernization process.
- The percentage of obese women in Saudi Arabia and Gulf states is ranged from 68.3 to 83% and in children less than 6 years it is 14%.
- A high relationship between osteoporosis and eating habits was found during adolescence and early adulthood.
- Women in the Middle East wearing three different dress styles, covering whole body known as Nigab, sparing hands and face known as Hijab, and western type dress style.
- Dr. Maalouf and his colleagues (Lebanon) established the first Arabian reference population.
- The decline in stiffness index (SI) for the Lebanese women between 20 and 75 years is about 30.3% compared to 32% for the American or European reference curve.

## Introduction

Osteoporosis is a debilitating disease characterized by low bone mass and microarchitectural deterioration of bone tissue, leading to an enhanced bone fragility and risk of fracture.[1] Bone is continuously removed by osteoclasts (bone resorption) and replaced by osteoblasts (bone formation) in a process of remodeling. Although bone formation outstrips resorption in youth, accumulated small deficits from remodeling result in a gradual loss of bone mass after the third decade.[2] Primary osteoporosis, therefore, is usually an age-related disease. It can affect both sexes, though women are at higher risk due to accelerated bone loss to a variable degree, after the menopause. The most common cause of osteoporosis is postmenopausal lack of estrogen. It affects 22% of all women over 50 years of age and almost 50% of all women over 70 years of age.[3] Osteoporosis can also be secondary to a number of diseases, such as hypoparathyroidism, rheumatoid arthritis and hypogonadism, or to drugs such as corticosteroids and heparin.[3,4] Other risk factors contributing to the development of osteoporosis include inadequate nutrition, cigarette smoking, physical inactivity, high alcohol intake, a diet low in calcium, prolonged use of certain medication, early menopause, thin body type, low lifetime estrogen exposure, Caucasian and Asian ancestry, number of pregnancies and a family history of osteoporosis.[3-5] Recent studies indicate that White and Asian women have osteoporosis more often than Black women, largely due to differences in bone mass and density.[6] Skeletal health consists principally in having bones strong enough to support everyday activity without sustaining fractures. Bone strength is multifactorial and depends upon bone size, bone mass and micro- and macro-architecture.[7] Bones become more porous and fragile with advanced age. The most common sites of fracture are the vertebrae, femoral neck, forearm[8] (Figure 24.1).

**Figure 24.1** *The most common sites of fractures are the vertebrae, femoral neck and forearm*

**Table 24.1.** *Percentage of normal, osteopenia and osteoporosis in postmeno-pausal Saudi women*

| Age group (years) | Normal | Osteopenia | Osteoporosis |
|---|---|---|---|
| 50–59 | 42 | 34 | 24 |
| 60–69 | 11 | 27 | 62 |
| 70–79 | 4.62 | 21.5 | 73.8 |

Reproduced with permission from Desouki, MI. Abstracts book. PAOS meeting Beirut, Lebanon (2000).

# Osteoporosis in Saudi Arabia

The population of Saudi Arabia is estimated to be 19.9 million people with approximately 14.87 million Saudis. It is estimated that 38% of population are above the age of 40 years.[9] Throughout the last decade there was not much attention on bone diseases in Saudi Arabia with few studies published on the bone mineral density (BMD) of healthy Saudi females in Riyadh, the Capital city.[10–12] Attention has been largely focused in the present decade on osteoporosis for two reasons. Firstly, due to the increasing numbers of older people in populations of many nations of the world including Saudi Arabia, and secondly due to the high incidence of fractures, which is considered a significant worldwide problem.[13–29] Several studies have been published relating to the prevalence of osteoporosis and osteopenia in Saudi Arabia. The prevalence of osteopenia and osteoporosis in 321 healthy Saudi females 31 years old was found to be 18–41 and 0–7%, respectively; prevalence of osteopenia fall from 34 to 21.5% in the youngest to the oldest groups respectively, and that of osteoporosis increased from 24 to 73.8% over the same age range (Table 24.1) depending on the site examined.[10] The mean age $\pm$ SD of the subjects was 35.4 $\pm$ 11.3 years and peak BMD values were observed around age 35 years in the spine and earlier in the femur (Table 24.2).[10] The lumbar spine bone density in 802 postmenopausal Saudi women 50–80 years of age (average 59 years) is shown in Table 24.3.[30] This high prevalence of osteopenia and osteoporosis in the Saudi population is due to an acute urbanization and westernization process over the last 30 years, with adoption

**Table 24.2.** *BMD[a] of Lumbar spine and femur of healthy Saudi females*

| Age (year) | L2–L4 | Femoral neck | Ward's triangle | Trochanter |
|---|---|---|---|---|
| 10–20 | 1.130$\pm$0.113 | 0.981$\pm$0.104 | 0.931$\pm$0.145 | 0.808$\pm$0.115 |
| 21–30 | 1.145$\pm$1.161 | 0.930$\pm$0.12 | 0.884$\pm$0.149 | 0.765$\pm$0.112 |
| 31–40 | 1.181$\pm$0.131 | 0.937$\pm$0.119 | 0.847$\pm$0.137 | 0.784$\pm$0.114 |
| 41–50 | 1.173$\pm$0.162 | 0.947$\pm$0.112 | 0.832$\pm$0.125 | 0.800$\pm$0.117 |
| 50 | 0.997$\pm$0.243 | 0.857$\pm$0.117 | 0.742$\pm$0.164 | 0.761$\pm$0.129 |

Reproduced with permission from Ghannam N.N *et al.* (1999), Calif Tissue International 65:23–28.
[a]Mean$\pm$SD (g cm$^{-2}$).

**Table 24.3.** *Lumbar spine BMD in 802 postmenopausal Saudi women age (50–80 years) using DEX*

| Number of subjects | % of total subjects | Mean BMD | t-score | |
| --- | --- | --- | --- | --- |
| 234 | 29 | 1.117±0.11 | −0.66 | Normal |
| 247 | 31 | 0.983±0.11 | −2.4 | Osteopenia |
| 321 | 40 | 0.767±0.11 | −3.4 | Osteoporosis |

Reproduced with permission from Desouki, MI. Abstracts book. PAOS meeting. Beirut 2000.

of lifestyle patterns with their accompanying chronic disease problem common in nations like the United States.[31]

## Nutrition Factor

Nutrition plays a vital role in the pathogenesis, prevention and treatment of osteoporosis.[32] It is one of many factors that influence bone mass. Nutritional status can influence the tendency to fall and is a factor in the maintenance of adequate soft tissue mass to protect the bones from a fall related fracture. The diet in the Middle East has changed dramatically over the last two decades and is fast becoming more 'westernized'.[33] Diets are becoming higher in fat, animal products and refined foods and lower in fiber, vegetables, fruit and essential micronutrient content such as vitamins and some minerals.[34] In parallel with dietary changes there is a rapidly increasing prevalence of obesity, diabetes, hypertension, vascular disease and chronic diseases in Asia and Arab countries.[35–39] The percentage of obese women in the Middle East and North Africa is 17.2%.[40] However, the 1997 Demographic Health Services of maternal nutritional status reported that the prevalence of BMI $> 30$ kg m$^{-2}$ among women aged 15–49 years is 25% in Egypt and in Tunisia and Morocco, respectively, were 23 and 18% for women and 7 and 6% for men.[40] In Saudi Arabia and the Gulf States it ranged from 14% in children less than six years to 68.3–83% in women.[41,42] In Palestinian population it was high at 49 and 30% in women and men, respectively. Body weight is an important determinant of bone density, and body weight is positively correlated with BMD where weight loss is associated with bone loss.[43] Deficiency in both micronutrients and macronutrients in the traditional foods of Middle East countries appears to be strongly related with declines in BMD and increases in fractures. The epidemiologic projection of the rise in the number of fractures from 1.66 million in 1999 to 6.26 million by the year 2050 due to the huge increase in the elderly population of the world will be a major public health problem.[44] A few food intake studies have been conducted in Middle East countries in relation to osteoporosis.[45–47] Al-Masri *et al.* (2002)[45] presented results of the relationship between osteoporosis and eating habits during adolescence and early adulthood in a group of postmenopausal Jordanian women. The study was concerned with the likes and dislikes of the five food groups during adolescence and early adulthood. The five food groups were those that comprise the food guide pyramid, starches, fruits, vegetables, milk and meats.[48] The interesting point in the study was that the authors found the subjects disliked Labneh (concentrated yogurt, very popular in Lebanon, Jordan and Syria) and yogurt, which are highly implicated in the development of osteoporosis and osteopenia in the later life (Table 24.4). They concluded that women who disliked milk were about three times more prone to develop osteoporosis as compared with women who liked milk, and women who disliked yogurt during their adolescence and early adulthood were about six times more prone to osteoporosis as compared with women who liked yogurt, while the women who disliked Labneh and white cheese were about 10 times more liable to develop osteoporosis as women who liked Labneh. These results are consistent with several previous studies which assessed the relationship between early lifetime intake of calcium-rich foods (mainly milk and its derivatives) and both BMD and the risk of osteoporosis and fractures.[49-55] Nieves and colleagues[49] presented results of an extensive investigation into the relationship between self-reported lifetime calcium intake, current calcium and vitamin D intake, and peripheral bone mass and fracture incidence in white postmenopausal women. Consumption of diary food was evaluated during four periods – childhood, teenage, adulthood and present time. The results emphasized that an adequate lifetime calcium intake and current calcium and vitamin D intakes were associated with a reduction in the risk of osteoporosis, but not fracture risk. However, in the older age group > 65 years there was no association between calcium and vitamin D intake

**Table 24.4.** *The dairy food intake likes and dislikes during adolescence and early adulthood in postmenopausal Jordanian women*

| Independent variable Dislike vs. Like | Osteoporotic subjects odds ratio (95% CI) | P-value | Osteopentic subjects odds ratio (95% CI) | P-value |
|---|---|---|---|---|
| Milk | 3.27 | 0.003 | 1.13 | 0.76 |
| Milk intake as a beverage *vs.* alone | 2.17 | 0.16 | 0.88 | 0.82 |
| Yogurt | 6.62 | 6.03 | 0.56 | 0.23 |
| Cooked yogurt | 0.65 | 0.36 | 0.71 | 0.41 |
| Labneh | 10.35 | 0.01 | 0.80 | 1.00 |
| White cheese | 10.35 | 0.01 | 0.80 | 1.00 |
| Jamid | 2.27 | 0.04 | 1.52 | 0.27 |
| Processed cheese | 0.73 | 0.48 | 1.35 | 0.42 |
| Pudding | 0.59 | 0.41 | 0.69 | 0.45 |
| Milk with rice | 1.34 | 0.59 | 2.19 | 0.74 |
| Custard | 1.22 | 0.59 | 2.02 | 0.04 |
| Cream caramel | 1.07 | 0.89 | 2.13 | 0.06 |
| Ice cream | 2.12 | 0.35 | 1.89 | 0.30 |

With permission Dr. Al-Masri *et al.* (2002). Second Pan-Arab Osteoporosis Congress. Third International congress of the Egyptian Osteoporosis Prevention Society. Sharm El-Sheikh, Egypt.

and risk of osteoporosis.[50] In another study, Heaney et al. (2001) found that adding three servings of yogurt to the daily diet of older women with habitually low calcium intakes resulted in a significant reduction in urinary excretion of N-telepeptide, a marker for bone resorption.[55] Therefore, an adequate calcium intake during all stages of life, particularly throughout childhood, adolescence and early adulthood, is considered to be the main nutritional risk factor related to the development of osteoporosis.[56]

## Caffeine Factor

The drinking of coffee is highly prevalent in the Middle Eastern countries. Three different methods of preparation of coffee are common in Arab countries. For instance, in Saudi Arabia and the Gulf States the Arabic coffee (which is light and without sugar) is very popular and highly consumed (about 10–15 small cups per day). In Lebanon, Jordan, Syria and Egypt the very popular coffee is the 'Turkish coffee', which is heavy black with sugar and about 6–10 cups per day is consumed individually. In the other parts of the Middle East a third type of coffee known as American coffee is highly consumed. This is heavy black without sugar and consumed at about 4–5 mugs per day. The consumption of caffeine via coffee, tea, colas and chocolate products is relatively high. Although several studies have attempted to examine the effect of caffeine as a risk factor for osteoporosis, it remains unconfirmed. Conlisk and Galuska (2000)[57] have demonstrated that caffeine is not a risk factor for osteoporosis in young women, while Hansen et al. (2000)[58] found that caffeine slightly increases the risk of bone fracture in females, and the relative risk for combined fracture sites in postmenopausal women is 1.09% (95% CI = 0.99–1.21). In contrast, Hegarty et al. (2000)[59] reported that caffeine is a risk factor for osteoporosis. Al-Masri (2000)[45] also reported a significant association between a liking for coffee during adolescence and early adulthood and the incidence of osteoporosis.[45] However, it has been proved that the amount of caffeine in a cup of coffee can reduce calcium absorption by a few milligrams, but adding a tablespoon or two of milk can easily offset this loss. The negative effects of the caffeine in coffee on calcium balance are due to increasing urinary calcium, decreasing the intestinal calcium absorption and leading to delay in bone mineralization. Daily consumption of caffeine in amounts equal to or greater than that obtained from two to three servings of coffee by women who have inadequate calcium intake was associated with accelerated bone loss.[60] Much of the apparently harmful effect of caffeine appears to be due not to the caffeine itself, but to the fact that caffeine-containing beverages are often substituted for milk in the diet.[61] Tea drinking is also highly prevalent in the Middle East countries and is consumed daily at more than 10 cups per day. Nutrients found in tea may protect against osteoporosis in older women.[59] These conclusions are based on the fact that tea has naturally occurring nutrients called flavonoids, which may combat the onset of bone loss in older women.

# Role of Vitamin D in Bone Health

Ergocalciferol (Vitamin $D_2$) is a synthetic compound formed from plant ergosterol by ultraviolet light, and is assumed to have the same biological activity in the body as the natural cholecolciferol (Vitamin $D_3$) of animal origin.[62] The active form of vitamin D is synthesized in three sequential steps in the skin, liver and kidney. Most of the vitamin D is produced endogenously in the skin epidermis by the action of ultraviolet light on 7-dehydrocholesterol.[63–65] The product of this reaction, cholecalciferol or vitamin D becomes biologically active only after a first hydroxylation in the liver (at the 25-position) to produce 25-hydroxyvitamin D [25(OH) D], the predominant form in the body, which circulates in concentrations of 20–50 $\mu$g $L^{-1}$ This compound, 25(OH) D, is then bound to a specific vitamin D binding protein and transported to the kidney and other organs, where it is either hydroxylated at the 1-position to produce 1,25-dihydroxycholecalciferol (calcitriol or 1,25[OH]$_2$ D), the most biologically active form of vitamin D, or in the other positions to produce a variety of other steroids.[65] Vitamin D acting with parathyroid hormone (PTH), regulates calcium and phosphorus metabolism as well as promoting calcium absorption from the gut and kidney tubules.[66] The three vitamin D compounds have been used in the management of osteoporosis. Numerous studies in the last two decades have been published concerning the effect of vitamin D and calcium supplementation on bone loss in adolescent, premenopausal, postmenopausal and elderly populations. Vitamin D and calcium supplementation have been shown to significantly reduce fracture rates in both institutionalized[67] and free-living elderly,[68] whereas the studies of Lips *et al.* (1996)[69] and Mayer *et al.* (2000),[70] have shown that vitamin D supplementation alone is not effective. This conclusion was presented at the recent World Congress (2000). A supplementation trial using cod liver oil containing 10 $\mu$g (400 IU) vitamin D did not prevent fracture in 1144 nursing home residents compared with control.[70] The controversy may be explained by the differences in the level of vitamin D supplementation. In the Chapuy and Dawson–Hughes studies, the levels were 20 and 17.5 $\mu$g per day, respectively, whereas in the studies by Lips (1996)[69] and Meyer (2000)[70] only 10 $\mu$g per day. More studies were presented confirming the important role that vitamin D and calcium play in the prevention of fracture.[71] Strong evidence indicated that vitamin D supplementation, as a single agent does not prevent osteoporosis in healthy postmenopausal women > 70 years of age with normal vitamin D level.[72] A controversy among scientists on vitamin D levels throughout life remains a considerable debate. Mckenna and co-workers (1997)[73] have shown that vitamin D levels fall with age, whereas the study of Chailurkit and co-workers[74] reported no changes in the levels of serum 25-hydroxyvitamin D status with age of 20–84 years old women and men in Khonkown province in Thailand, but after the fourth decade, women displayed an increase in all biochemical markers of bone turnover and a sharp increase after sixth decade, whereas in men, all biochemical markers of bone turnover have a tendency to decrease with age, except for serum total alkaline phosphatase. In older women, a high plasma level of vitamin D enhances calcium absorption, whereas a high sodium, protein, alcohol and caffeine intake will cause increased urinary losses and negative calcium balance.[75] Many published studies have demonstrated that low calcium intake is associated with low bone mass and increased fracture risk.[76–80] The vitamin D level is subjected to seasonal variation[81–83] and by high latitudes. Information about vitamin D status in Middle East population is scarce. A few studies have been published about vitamin D levels in the populations of Lebanon,[84,85] Jordan[86] and Saudi Arabia[87,88] and UAE.[89]

It has been confirmed that at all ages, from neonates to older people, the lower plasma levels of 25(OH)vitamin D are associated with higher levels of PTH, and the concentration of ionized calcium will be lower, which leads to the liberation of calcium from bone by increasing the number of resorption pits. PTH also stimulates the renal synthesis of $1,25(OH)_2$ vitamin D, which enhances the active phase of absorption by stimulating the synthesis of calcium binding proteins. At the same time, the renal calcium re-absorption increases. If the level of ionized calcium is raised, PTH secretion is inhibited, calcitonin may be secreted by the thyroid gland and plasma calcium falls. A few studies have examined the correlation between vitamin D deficiency and lower bone mineral densities. The extent of vitamin D insufficiency and lifestyle characteristics on bone indices in premenopausal and postmenopausal Saudi women were investigated. The results (Tables 24.5 and 24.6) have shown that

**Table 24.5.** *Impact of the extent of vitamin D insufficiency and lifestyle characteristics on bone indices in premenopausal Saudi women*

|  | *Mean* | *SD* | *Range* |
|---|---|---|---|
| Age (years) | 23.1 | 3.6 | 20.0–33.0 |
| Weight (kg) | 61 | 15 | 38.70–112.50 |
| Height (m) | 1.59 | 0.06 | 1.45–1.73 |
| BMI (kg m$^{-2}$) | 24 | 5.7 | 16.0–50.0 |
| Lumbar spine *t*-score | −0.6 | 1.0 | −3.20–1.70 |
| Femoral neck *t*-score | −0.2 | 1.1 | −3.20–2.90 |
| BUA *t*-score | −0.7 | 0.8 | −2.62–1.50 |
| VOS *t*-score | −1.1 | 0.7 | −3.98–0.28 |
| 25OHD(nmol L$^{-1}$) | 22.1 | 18.47 | 2.40–108.02 |
| 1,25OHD (pg mL$^{-1}$) | 32.64 | 10.07 | 15.8–93.41 |
| PTH (pg mL$^{-1}$) | 70.14 | 54.76 | 16.83–437.0 |
| Ca (mmol L$^{-1}$) | 2.38 | 0.09 | 2.05–2.57 |
| Milk intake (glass per day) | 1.0 | 0.8 | 0.0–4.0 |
| Sunlight exposure (min per day) | 16.2 | 17 | 0.0–60.0 |
| Walking (time per week) | 3.2 | 2.8 | 0.0–10.0 |
| Coffee intake (cup per day) | 1.0 | 1.2 | 0.0–6.0 |
| Tea intake (cup per day) | 1.5 | 1.5 | 0.0–8.0 |
| Dairy product (per per week) | 2.2 | 1.5 | 0.0–8.50 |
| Fatty fish (intake per week) | 1.0 | 1.3 | 0.0–5.0 |

Reproduce with permission Khoja, S *et al.* (2002). Second Pan-Arab Osteoporosis Congress. Third International Congress of the Egyptian Osteoporosis Prevention Society. Sharm El-Sheikh, Egypt.

**Table 24.6.** *Impact of the extent of vitamin D insufficiency and lifestyle characteristics on bone indices in postmenopausal Saudi women*

|  | Mean | SD | Range |
|---|---|---|---|
| Age (years) | 49.6 | 5.01 | 43.0–60.0 |
| Weight (kg) | 75.7 | 14.6 | 51.5–115.0 |
| Height (m) | 1.56 | 0.06 | 1.42–1.74 |
| BMI (kg m$^{-2}$) | 30.9 | 5.5 | 21.2–47.3 |
| Lumbar spine *t*-score | −1.0 | 1.3 | −4.4–3.1 |
| Femoral neck *t*-score | −0.2 | 1.3 | −2.8–4.0 |
| BUA *t*-score | −1.2 | 0.9 | −2.93–1.8 |
| VOS *t*-score | −2.1 | 0.7 | −3.45–0.6 |
| 25OHD (nmol L$^{-1}$) | 22.54 | 18.21 | 1.30–99.71 |
| 1,25OHD (pg mL$^{-1}$) | 24.71 | 10.68 | 5.10–61.75 |
| PTH (pg mL$^{-1}$) | 97.56 | 53.45 | 34.88–386.4 |
| Ca (mmol L$^{-1}$) | 2.35 | 0.12 | 1.82–2.71 |
| Milk intake (glass per day) | 1.0 | 0.8 | 0.0–5.0 |
| Sunlight exposure (min per day) | 11.2 | 16.8 | 0.0–90.0 |
| Walking (time per week) | 1.7 | 3.1 | 0.0–21.0 |
| Coffee intake (cup per day) | 1.0 | 1.5 | 0.0–10.0 |
| Tea intake (cup per day) | 2.6 | 1.7 | 0.0–8.0 |
| Dairy product (per per week) | 2.1 | 1.4 | 0.0–7.83 |
| Fatty fish (intake per week) | 0.9 | 1.1 | 0.0–6.0 |

Reproduce with permission Khoja, S *et al.* (2002). Second Pan-Arab Osteoporosis Congress. Third International Congress of the Egyptian Osteoporosis Prevention Society. Sharm El-Sheikh, Egypt.

the level of 25(OH)vitamin D in premenopausal and postmenopausal Saudi women are lower than normal, they are at threshold of severe hypovitaminosis D[25(OH)D <20 nmol L$^{-1}$]. Limitation of data about vitamin D status whether in young or/and older people in the Middle East countries requires more investigation. A few studies have been published from Lebanon, UAE, Jordan and Saudi Arabia. A comparison of our findings with one from Lebanon presented by Gannage *et al.* (2000),[84] indicated that hypovitaminosis D was prevalent among pre- and postmenopausal women and is related mostly to low vitamin D intake, despite an adequate sunshine exposure.

# Effects of Dress Styles on Vitamin D Metabolism

The Middle East countries have always a Mediterranean claim of ample sunshine both in the summer and winter seasons, and where the women wear dress that cover the body to a variable extent for cultural and religious reasons. It is well known that vitamin D is predominantly derived from exposure of the human skin to solar ultraviolet B radiation.[90–94] However, there is a considerable debate concerning the effects of sun exposure and different dress styles on vitamin D metabolism in women. Women in Arab countries covering their bodies to a variable extent, with women wearing full covering dress styles, including the face and hand known as 'Nigab', women wearing covering dress styles sparing the hands and face known as 'Hijab', and women wearing western-type dress styles. The study by Mishal (2001),[86] concluded that there is no significant difference between the groups of women wearing the different dress styles. However, the 25(OH)vit D level in women covering whole body 'Nigab' and women covering part of body 'Hijab' were significantly lower than those found in male subjects. In contrast, a study by Gannage *et al.* (2000)[84] observed that the level of 25(OH)D is higher in nonveiled women (western style) compared with veiled women (covering) in healthy young Lebanese women. These data suggest that the dress styles are a likely contributory factor and further research in other Middle Eastern countries is clearly warranted.

## Assessment of a Lebanese Reference Population

Dr. Maalouf and his colleagues (Lebanon) established the first Arabian reference population. A cross-sectional study conducted involving 4320 women with a mean age of 52.5 years (age range 20–79 years) using three identical Achilles Express (GE/Lunar) and one Achilles Plus (GE/Lunar) ultrasonometry devices. Women were randomly selected and asked to participate in a nation wide screening program using media, conferences, telephone calls, *etc.* Broadband ultrasound attenuation (BUA) and speed of sound (SOS) and the SI of the os calcaneus was measured. Results showed that there was an overall decline of 19.2% for BUA, 3.1% for SOS and 30.3% for SI between late adolescence and old age. The authors found in premenopausal women that BUA decreased only slightly by 3%, while postmenopausal women showed a significant decrease of 16.2%. In contrast, SOS continuously decreased from the age of 42; there was a decline of 0.8% from adolescence to the menopause; postmenopausal women showed a larger decline of 2.4%. The SI of premenopausal women decreased by 6%, while postmenopausal women showed a significantly larger decline of 24.3%. The SI value for the female Lebanese young adult reference is 8% lower than that of the American and European women (92 SI units compared to 100) (Tables 24.7 and 24.8). At the age of 42, the SI value for the Lebanese women is 10.4% lower than that of the American women and 7.5% lower than that of the European women (86 SI units compared to 96 and 93 respectively). At the age of 75, SI values for the Lebanese women are 4.4% lower than that of American women and European women (65 SI units compared to 68). The decline in SI for the Lebanese women between age 20 and 75 years is about 30.3% compared to 32% for the American or European reference curve. The rate of decrease for the Lebanese women was 0.2 SI units per year for the premenopausal period, and 0.7 SI units per year for the postmenopausal period. They concluded that the age related female Lebanese reference curve was significantly different from the American and the European reference curves used by the manufacturer. Therefore, the use of their standardized reference data instead of the proposed US or European database reduced the risk of overestimating osteoporosis in the Lebanese population.

**Table 24.7.** *SI as a function of age*

| Age bracket | N | Mean age | SI | SD |
|---|---|---|---|---|
| 20–24 | 77 | 22.00 | 91.50 | 14.21 |
| 25–29 | 144 | 27.09 | 90.50 | 14.02 |
| 30–34 | 185 | 32.30 | 89.40 | 14.29 |
| 35–39 | 312 | 37.06 | 88.00 | 15.97 |
| 40–44 | 436 | 41.99 | 86.00 | 14.91 |
| 45–49 | 500 | 47.18 | 83.11 | 16.01 |
| 50–54 | 669 | 51.88 | 80.15 | 16.23 |
| 55–59 | 604 | 56.89 | 77.08 | 15.65 |
| 60–64 | 586 | 61.90 | 73.70 | 15.96 |
| 65–69 | 408 | 66.74 | 70.30 | 15.05 |
| 70–74 | 245 | 71.52 | 67.30 | 13.95 |
| 75–79 | 154 | 76.55 | 63.80 | 15.05 |

Reproduced with permission Maalouf *et al.* (2002).

**Table 24.8.** *Percentage of osteopenia and osteoporosis in the Lebanese population*

| Age bracket | Number n | Average age | Average stiffness | Average t-score | % Normal | % Osteo-penic | % Osteo-porotic |
|---|---|---|---|---|---|---|---|
| 20–24 | 77 | 22.00 | 94.31 | 0.17 | 90.91 | 9.09 | 0.00 |
| 25–29 | 144 | 27.09 | 98.35 | −0.13 | 85.42 | 14.58 | 0.00 |
| 30–34 | 185 | 32.30 | 90.38 | −0.07 | 88.11 | 10.81 | 1.08 |
| 35–39 | 312 | 37.06 | 88.12 | −0.20 | 78.53 | 20.19 | 1.28 |
| 40–44 | 436 | 41.99 | 83.28 | −0.48 | 71.56 | 26.61 | 1.83 |
| 45–49 | 500 | 47.18 | 83.11 | −0.49 | 68.20 | 30.60 | 1.20 |
| 50–54 | 669 | 51.88 | 80.15 | −0.67 | 62.18 | 35.87 | 1.94 |
| 55–59 | 604 | 56.89 | 77.08 | −0.85 | 53.48 | 42.38 | 4.14 |
| 60–64 | 586 | 61.90 | 73.70 | −1.05 | 43.00 | 51.02 | 5.97 |
| 65–69 | 408 | 66.74 | 69.72 | −1.28 | 33.33 | 58.82 | 7.84 |
| 70–74 | 245 | 71.52 | 65.83 | −1.51 | 23.27 | 65.31 | 11.43 |
| 75–79 | 154 | 76.55 | 63.68 | −1.64 | 24.03 | 58.44 | 17.53 |

Reproduced with permission Maalouf *et al.* (2002).

# Equipments of Measurements

The traditional technique for examining bones is by X-ray. This can show extremes of loss of bone mass with loss of opacity, but this methodology does not discriminate adequately for measuring smaller changes because approximately 30% of skeletal mass has to be lost before X-ray changes are apparent. Several other techniques are available for the assessment of bone mass. Dual energy X-ray absorptiometry (DXA) assesses bone mineral at both the axial and appendicular sites, has high reproducibility, and uses low doses of radiation. Single energy X-ray absorptiometry enables measurements only at appendicular sites, such as the forearm. Quantitative computer tomography enables differential measurement of cortical and trabecular bone in the spine or peripheral skeleton, but the equipment is expensive and the radiation dose is relatively high. Finally, ultrasonic measurements of bone at heel and knee are being evaluated for their reliability in assessing bone. This technique, which uses a non-ionizing radiation, is portable and relatively cheap. All these equipments may be available in most Middle Eastern countries, but they are not common. Currently, in Saudi Arabia, the number of DXA machines is inadequate to detect the entire population at risk, even if there were a sufficient number of DXA machines, the expensive cost would limit people from using it.

# The Requirements of Awareness for Osteoporosis in the Middle East

The newly gained affluence of many populations since the end of World War II has given rise to westernization, or adoption of lifestyle patterns common in nations like the United States, such as the increased consumption of animal protein, increased use of mechanical aids and a general decline in physical activities. The Middle East countries are a part of this world and are dramatically influenced to westernization. Therefore, the prevalence of osteopenia, osteoporosis and fractures are relatively high among adolescent, adulthood and older population in both genders. In most of these countries, enormous work is required to propose a methodology of health assessment that can contribute to the formulation of an effective reproductive health policy and identify the key health problem of Arab women. The reproductive health needs of Arab women opens by sketching the context of the modern Arab world in terms of female education, women's economic situation, patterns of marriage and levels of fertility, consanguineous marriage, contraceptive usage, maternal health, gynecological and related morbidity. The osteoporosis fractures are increasing at a dramatic rate and the disease is still poorly recognized especially among women. Therefore, they need numerous effort from specialists, physicians, societies and communities to create an awareness among people on this 'silent disease' to provide more information about this disease to population, to use all media available, TV advertisement, newspapers, magazines, internet and campaign like the First Global Osteoporosis Awareness Campaign, which was launched on 23, January 2003 in London encouraged women worldwide to gain control of their health by taking 'the 1-min osteoporosis risk test', which will help all to be aware of these risk factors and to take action against osteoporosis. The presence of DXA, the highly technical, highly precise, safe and noninvasive technology (Kulak and Bilezikian) to measure bone mass density, is the single most important predictor of fracture risk; it is a critically important tool to apply to the population at risk, which includes women who have definable risk factors for osteoporosis, such as menopause and family history of osteoporosis, life-long low calcium intake, smoking, extreme thinness, anorexia, certain diseases and medication. This device, which is now known as 'gold standard' test of the lumbar spine and hip, is not universally available, and on the other hand no large studies have evaluated the ability of the new, less expensive peripheral technologies to predict fractures and the subsequent fracture risk at the hip, wrist/forearm, spine and rib, which are increased in all populations.

## Conclusions

Osteoporosis is one of the most common metabolic bone diseases in Western countries, and is subjected to have a similar impact in the Middle East countries. It has been suggested that peak BMD, a major determinant of osteoporosis fractures, is lower in this part of the world compared with the western world.[85] However, the dramatically westernization change associated with social, economic and diet changes beside lack of awareness contribute to the rapidly increasing prevalence of osteopenia and osteoporosis. Accumulating evidence support the associations between early poor nutrition and later prone to osteoporosis, which is very common in Middle East. Vitamin D and calcium, which are very important nutrients for bone health are also inadequate, despite the greater availability of sunlight in our countries.

At this time, tremendous effort is put to introduce people to this 'silent disease', to increase the awareness with priority given to educate people that osteoporosis is treatable; much progress has been reflected to the knowledge of people to prevent and defeat this disease.

## References

1. World Health Organization (1994). Assessment of fracture risk and its application to screening for postmenopausal osteoporosis. WHO Technical Series Report 843. Geneva.
2. K. Parfitt, Bone modulating drugs, Martindale In *The Complete Drug Reference,* 32nd edn, Pharmaceutical Press, USA, 1999, 730.
3. D.W. Dempster and R. Lindsay, Pathogenesis of osteoporosis, *Lancet,* 1993, **341**, 791–801.
4. N. Peel and R. Eastell, Osteoporosis, *Br. Med. J.,* 1995, **310**, 989–992.
5. Jacob's Institute of Women's Health, The women's health data book, In *A Profile of Women's Health in the United States*, Washington DC, 1995.
6. S.C. Ho, Determination of bone mass in Chinese women aged 21–40 years, II. Pattern of dietary calcium intake and association with bone mineral density, *Osteoporosis Int.,*1994, **4**, 167–75.
7. R.P. Heaney, The importance of calcium intake for life long skeletal health, *Calcif. Tissue Int.,* 2002, **70,** 70–73.
8. J.A. Kanis, *Osteoporosis*, Black well science Ltd pub., UK, 1994, 28.
9. Economic Report. Saudi American bank. Central Department of Statistics (CDS) of the Ministry of planning release. SES 30 Oct. 2002.
10. N.N. Ghannam, M.M. Hammami, S.M. Bakheet and B.A. Khan, Bone mineral density of the spine and femur in healthy Saudi females: relation to Vitamin D status, pregnancy, and lactation, *Calcif. Tissue Int.,* 1999, **65**, 23–28.
11. M.I. Desouki, Osteoporosis in postmenopausal Saudi women using dual X-ray bone densitometry, In *Abstract Book PAOS Meeting*, Beirut, Lebanon, 1994.

12. A.R. Al-Nuaim, M. Kremli, M. Al-Nuaim and S. Sandkgi, Incidence of proximal femur fracture in an urbanized community in Saudi Arabia, *Calcif. Tissue Int.*, 1995, **56**, 536–538.

13. W.A. Wallace, The increasing incidence of fractures of the proximal femur, and orthopaedic epidemic, *Lancet*, 1983, **1**, 1413–1414.

14. W.J. Boyce and M.P. Vessey, Rising incidence of fractures of the proximal femur, *Lancet*, 1985, **1**, 150–151.

15. B.L. Riggs and L. Melton III., Involutional osteoporosis, *N. Engl. J. Med.*, 1985, **314**, 1676–1686.

16. H. Norimatsu, S. Mori, T. Uesats, T. Yoshikawa and N. Katsuyama, Bone mineral density of the spine and proximal femur in normal and osteoporotic subjects in Japan, *Bone Miner.*, 1989, **5**, 213–222.

17. L.J. Melton III., Hip fractures: a worldwide problem today and tomorrow, *Bone*, 1993, **14**, S1–S8.

18. J.C. Stevonson, B. Lees, M. Devenport, M.P. Cust and K.F. Ganger, Determinants of bone density in normal women: risk factors for future osteoporosis, *Br. Med. J.*, 1989, **298**, 924–928.

19. I.K. Han, W.K. Park, W.W. Choi, H.H. Skin and S.W. Kim, A study of hormonal changes and bone densities in Korean menopausal women, *J. Kor. Soc. Endocrinol.*, 1989, **4**, 21–28.

20. O. Lofman, L. Larsson, I. Ross, G. Toss and K. Berglund, Bone mineral density in normal Swedish women, *Bone*, 1997, **20**, 167–174.

21. T.W. O'Neill, D. Felsenberg, J. Varlow, C. Cooper, J.A. Kanis and A.J. Silman, The prevalence of vertebral deformity in European men and women: the European vertebral osteoporosis study, *J. Bone Miner. Res.*, 1996, **11**, 1010–1018.

22. T. Sugimoto, M. Tsutsumi, Y. Fujii, M. Kawakatsu, H. Negishi, M.C. Less, K.S. Tsai, M. Fukase and T. Fujita, Comparison of bone mineral content among Japanese, Koreans, and Taiwanese assessed by dual-photon absorptiometry, *J. Bone Miner. Res.*, 1992, **7**, 153–159.

23. S.M.R. Pluijm, M. Visser, J.H. Smit, C. Popp-Snijders, J.C. Roos and P. Lips, Determinants of bone mineral density in older men and women: body composition as mediator, *J. Bone Miner. Res.*, 2001, **16**(11), 2142–2151.

24. J.M. Nolla, C. Gomez-Vaquero, J. Fiter, D. Roig Vilaseca, L. Mateo, A. Rozadilla, M. Romera, J. Valverde and D. Roig Escofet, Usefulness of bone densitometry in postmenopausal women with clinically diagnosed vertebral fractures, *Ann. Rheum. Dis.*, 2002, **61**, 73–75.

25. B.L. Riggs and L.J. Melton, The prevention and treatment of osteoporosis, *N. Engl. Med.*, 1992, **327**, 620–627.

26. E.M.C. Lau and C. Cooper, The epidemiology and prevention of osteoporosis in urbanized Asian populations, *Osteoporosis Int.*, 1993, **3**(Suppl. 1), S23–S26.

27. J.M. Lacey, J.J.B. Anderson, T. Fujita, Y. Yoshimoto, M. Fukase, S. Tsuchie and G.G. Koch, Correlates of cortical bone mass among premenopausal and postmenopausal Japanese women, *J. Bone Miner. Res.*, 1991, **6**, 651–659.

28. C. Bergot, A.M. Laval-Jeantet, K. Hutchinson, I. Dautraix, F. Caulin and H.K. Genant, A comparison of spinal quantitative computed tomography with dual energy X-ray absorptiometry in European women with vertebral and nonvertebral fractures, *Calcif. Tissue Int.*, 2201, **68**, 74–82.

29. R.D. Ross, H. Norimatsu, J. Davis, K. Yano, R. Wasnich, S. Fujiwara, Y. Hosoda and L.J. Melton 3rd, A comparison of hip fractures incidence among Native

Japanese, Japanese Americans, and America Caucasians, *Am. J. Epidemiol.*, 1991, **133**(8), 801–809.

30. M.I. Desouki, (2000). Osteoporosis in postmenopausal Saudi women using dual X-ray bone densitometry. Pan Arab Osteoporosis Society Conference, Beirut, Lebanon. P 403 SU, Abstract.

31. M. Nevitt, S. Bent, L. Lui, Y. Zhang, W. Wu, S. Cummings and L. Xu, (2002). Vertebral deformities in urban Chinese men: the Beijing osteoarthritis study. *Osteoporosis* Int. **13**(Suppl. 1): S66. Program and abstract of the IOF World Congress on Osteoporosis, Lisbon, Portugal, Abstract P 210.

32. R.P. Heaney (1999). Nutrition and Osteoporosis. Ch. 50, *Primary on the Metabolic Bone Disease and Disorder of Mineral Metabolism*, 4th edn., Pub. Lippincott William and Wilkins, Official publication of the American Society for bone and mineral research.

33. S.S. Miladi, *Int. J. Food Sci. Nutr.*, 1998, **49**, S23–S30.

34. L.K. Tucker and S. Buranapin, Nutrition and aging in developing countries, *J. Nutr.*, 2001, **131**(9), S417–S2423.

35. R. Gupta and S. Singhal, Coronary heart disease in India, *Circulation*, 1997, **96**, 3785.

36. X.R. Pan, W.Y. Yang, G.W. Li and J. Liu, Prevalence of diabetes and its risk factors in China, *Diabetes Care*, 1997, **20**, 1664–1669.

37. Z.T. Bloomgarden, International Diabetes Federation of meeting, 1997, type 2 diabetes: its prevalence, causes, and treatment, *Diabetes Care*, 1998, **21**, 860–865.

38. Asian Acute Stroke Advisory Panel, Stroke epidemiological data of nine Asian countries, *J. Med. Assoc. Thai.*, 2000, **83**, 1–7.

39. R.B. Singh, I.L. Suh, V.P. Singh, S. Chaithirophan, R. Laothavorn, R.G. Sy, N.A. Babilonia, A.R. Rahman, S. Sheikh, B. Tomlinson and N. Sarraf-Zadlgan, Hypertension and stroke in Asia: prevalence, control and strategies in developing countries for prevention, *J. Hum. Hypertens.*, 2000, **14**, 749–763.

40. N. Mokhtar, J. Elati, R. Chabir, A. Bour, K. Elkari, N.P. Schlossman, B. Caballero and H. Agvenaou, Diet culture and obesity in Northern Africa, *J. Nutr.*, 2001, **131**, 8875–8925.

41. R. Martorell, L.K. Khan M.L. Hughes and L.M. Grummer-strawn, Obesity in women from developing countries, *Eur. J. Clin. Nutr.*, 2000, **54**(3), 247–252.

42. A.M. Khalid, Obesity in Saudi Arabia: a review, *Bahrain Med. Bull*, 2000, **22**(3), 113–118.

43. Nutrition and Skeleton, (2001), National Institution of Health Osteoporosis and Related Bone Disease ~ National Research Center. Nutrition and the Skeleton. The role of Calcium and other Nutrients. (MidLine).

44. C. Cooper, G. Campion and L.J. Melton III, Hip fracture in the elderly: a worldwide projection, *Osteoporosis Int.*, 1992, **2**, 285–289.

45. B.K. Al-Masri, A.M. Fagih, R.A. Al-Abbasi and F.M. Shakater, (2002). Second Pan-Arab Osteoporosis Congress. Third International Congress of the Egyptian Osteoporosis Prevention Socity, Sharm El Sheikh, Egypt 22–25 Oct. (Abstract book).

46. A. AL-Romaihi, M. Moloney and E.P. McNamara, (2002). Second Pan-Arab Osteoporosis Congress. Third International Congress of the Egyptian Osteoporosis Prevention Society, Sharm El Sheikh, Egypt 22–25 Oct. (Abstract book).

47. S.O. Khoja, J.A. Khan, R.A. Maimani and S.A. New, (2002). Second Pan-Arab Osteoporosis Congress. Third International Congress of the Egyptian

Osteoporosis Prevention Society, Sharm El Sheikh, Egypt 22–25 Oct. (Abstract book).

48. S.R. Williams and S.L. Anderson, *Nutrition and Diet Therapy*, St. Louis, Mosby.
49. J.W. Nieves, L. Komar, F. Consman and R. Lindsay, Calcium potentiates the effect of estrogen and calcitonin on bone mass, *Am. J. Clin. Nutr.*, 1998, **67**, 18–24.
50. R. Lindsay, Prevention and treatment of osteoporosis, *Lancet,* 1993, **341**, 801–805.
51. C.M. Weaver, M. Peacock and C.C. Johnston, Adolescent nutrition in the prevention of postmenopausal osteoporosis, *J. Clin. Endocr. Metab.*, 1999, **84**(6), 1839–1843.
52. J. Lodogan, R. Eastell, N. Jones and M.E. Berker, Milk intake and bone mineral acquisition in adolescent girls, *Br. J. Med.*, 1997, **315**, 1255–1260.
53. S.A. New, (2002). Nutrititional factors that influence bone health throughout the life cycle, 24th Annual Meeting of the American Society for Bone and Mineral Research. (Med Scape).
54. F. Ginty, S.J. Stear and S.C. Johnes, Impact of calcium supplementation on markers of bone and calcium metabolism in 16–18 years old boys, *J. Bone Miner. Res.*, 2002, **17**(suppl 1), S178 Abstract F003.
55. R.P. Heaney, K. Rafferty and M.S. Dowell, Effect of yogurt on a urinary marker of bone resorption in postmenopausal women, *J. Am. Diet. Assoc.*, 2002, **102**(11), 1672–1674.
56. S.A. New, C. Bolton-Smith, D.A. Grubb and D.M. Reid, Nutritional influence on bone mineral density, a cross sectional study in premenopausal women, *Am. J. Clin. Nutr.,* 1997, **65**,1831–1839.
57. A.J. Conlisk and D.A. Galuska, Is caffeine associated with bone mineral density in young adult women.? *Prev. Med.,* 2000, **31**, 562–568.
58. S.A. Hansen, A.R. Folsom, L.H. Kushi and T.A. Sellers, Association of fractures with caffeine and alcohol in postmenopausal women: the Iowa Women's Health Study, *Public Health Nutr.*, 2000, **3**, 253–261.
59. V.M. Hegarty, H.M. May and K.T. Khaw, Tea drinking and bone mineral density in older women, *Am. J. Clin. Nutr.*, 2000, **71**, 1003–1007.
60. NIH ORBD ~ NRC (2001), *Nutrition and Skeleton: The Role of Calcium and other Nutrients (Medline).*
61. H. Van den Berg, Bioavailability of vitamin D, *Euro. J. Clin. Nutr.*, 1997, **51**(Suppl. 1), 576–579.
62. M.W. Davie, D.E. Lawson, C. Emberson, J.L. Barnes, G.E. Roberts and N.D. Barnes, Vitamin D from skin: contribution to vitamin D status compared with oral vitamin D in normal and anticonvulsant – treated subjects, *Clin. Soc.*, 1982, **63**, 461–471.
63. D.R. Fraser, The physiological economy of vitamin D. *Lancet,* 1983, **1**, 969–972.
64. M.R. Holick, McCollum award lecture: vitamin D new horizons for the 21st century, *Am. J. Clin. Nutr.*, 1994, **60**, 619–630.
65. T.E. Andreoli, J.C. Bennett, C.J. Carpenter and F. Plum, (1997). Cecil Essentials of Medicine, 4th edn. W. B Saunder Company. A division of Harcourt Brace and Company, USA.
66. S.A. New, Nutrition, Bone mass and Preventing fracture: a review, *J. Natt. Osteo Soc.*, 2000, **8**(3).
67. M.C. Chapuy, M.E. Arlot, F. Duboruf, J. Brun, B. Crouzet, S. Arnaud, P.D. Delmas and P.J. Meunier, Vitamin D and calcium to prevent hip fractures in elderly women, *N. Engl. J. Med.*, 1992, **327**, 1637–1642.

68. B. Dawson-Hughes, S.S. Harris, E.A. Krall and G.E. Dallal, Effect of Ca and vitamin D supplementation on bone density in men and women 65 years of age or older, *N. Engl. J. Med.*, 1997, **339**, 670–676.

69. P. Lips, W.C. Graafmans, M.E. Ooms, D. Bezemer and L.M. Bouter, Vitamin D supplementation and fracture incidence in elderly person. A randomized placebo-controlled clinical trial, *Ann. Intern. Med.*, 1996, **124**, 400–406.

70. H.E. Meyer, J.A. Falch, E. Kvaavik, G.B. Smedshaug, A. Tverdal and J.I. Pederson, Can vitamin D supplementation reduce the risk of fracture in the elderly? A randomized controlled trial, *Osteoporosis Int.*, 2000, **11**, S114.

71. M.C. Chapuy, R. Pamphile, E. Paris, C. Kempf, M. Schlichting, S. Arnaud, P. Gamero and P.J. Meunier, Combined calcium and vitamin $D_s$ supplementation to prevent hip fracture in elderly women. A confirmatory study, Decalyos II, *Osteoporosis Int.*, 2002, **13**(Supp). Abstract P68.

72. D. Hunter, P. Major, N. Arden, R. Swaminathan, T. Andrew, A.J. MacGregor, R. Keen, H. Shieder and T.D. Spector, A randomized controlled trial of vit. D supplementation on preventing postmenopausal bone loss and modifying bone metabolism using identical twin pairs, *J. Bone Miner. Res.*, 2000, **15**, 2276–2283.

73. M.J. Mckenna, R. Freaney, A. Meade and F.P. Muldowney, Hypovitaminosis D and elevated serum alkaline phosphatase in elderly Irish people, *Am. J. Clin. Nutr.*, 1985, **41**, 101–109.

74. L.O. Chailurkit, C. Pongchaiyakul, S. Charonkiatkul, V. Kosulwat, N. Rojroongwasinkul and R. Rajatanavin, Different mechanism of bone loss in aging women and men in Khon Kaen Province, *J. Med. Assoc. Thai.*, 2001, **84**, 1175–1185.

75. M.P. Faine, Dietary factors related to preservation of oral and skeletal bone mass in women, *J. Prosthet. Dent. Jan.*, 1995, **73**(1), 65–72.

76. P.R. Ebeling, J.D. Wark, S. Yeung *et al.*, Effects of calcitriol or calcium on bone mineral density, bone turnover, and fractures in men with primary osteoporosis: a two-year randomized, double blind, double placebo study, *J. Clin. Endocrinol. Metab.*, 2001, **86**, 4098–4103.

77. C.C. Jonston, J.Z. Miller, C.W. Slemenda, T.K. Reister, S. Hui, J.C. Christian and M. Peacock, Calcium supplementation and increases in bone mineral density in children, *N. Engl. J. Med.*, 1992, **327**, 82–87.

78. T. Lioyd, M. Andon, N. Rollings, J.K. Martel, J.R. Landis, L.M. Demers, D.F. Eggli, K. Kieselhorst and H.F. Kulin, Calcium supplementation and bone mineral density in adolescent girls. *JAMA*, 1993, **270**, 841–844.

79. W.T.K. Lee, S.S.E. Leung, S.H. Wang, Y.C. Xu, W.P. Zeng, J. Lau, S.J. Oppenheimer and J.C. Cheng, Double blind, controlled calcium supplementation and bone mineral accretion in children accustomed to a low calcium diet, *Am. J. Clin. Nutr.*, 1994, **60**, 744–750.

80. C.A. Nowson, R.M. Green, J.L. Hopper, A.J. Sherwin, D. Young, E. Kaymakci, C.S. Guest, M. Smid, R.G. Larkins and J.D. Wark, A co-twin study of the effect of calcium supplementation on bone density during adolescence, *Osteoporosis Int.*, 1997, **7**, 219–225.

81. M. Davies, E.B. Mawer, J.T. Hann and J.L. Taylor, Seasonal changes in the biochemical indices of vitamin D deficiency in the elderly: a comparison of people in residential home, long stay words and attending a day hospital, *Age Ageing*, 1986, **15**, 77–83.

82. J. Giullemant, H.T. Le, A. Maria, A. Allemandou, G. Peres and S. Guillemant, Wintertime Vitamin D deficiency in male adolescents: effect on parathyroid

function and response to vitamin D supplementation, *Osteoporosis Int.*, 2001, **12**(10), 875–879.

83. E.A. Krall, N. Sahyoun, S. Tannenbaum, G.E. Dallal and B. Dawson-Hughes, Effect of vitamin intake on seasonal variations in parathydroid hormone secretion in postmenopausal women, *N. Engl. J. Med.*, 1989, **321**, 1777–1783.

84. M.H. Gannage-Yared, R. Chemali, N. Yaacoub and G. Halaby, Hypovitaminosis D in a sunny country: relation to lifestyle and bone markers, *J. Bone Miner. Res.*, 2000, **15**(9), 1856–1862.

85. G. El-hajjfuleihan, R. Baddoura, H. Awada, N. Salam, M. Salamoun and R. Rizk, Low people bone mineral density in healthy Lebanese subjects, *Bone*, 2002, **31**(4), 520–528.

86. A.A. Mishal, Effects of different dress styles on vitamin D levels in healthy young Jordanian women, *Osteoporosis Int.*, 2001, **12**(11), 931–935.

87. S.H. Sedrani, K.M. Al-Arabi, A. Abanmy and A. Elidrissy, In *Study of Vitamin D: Status and Factors Leading to its Deficiency in Saudi Arabia*, King Saud University Press, 1990.

88. N. Jurayyan and M.I. Desouki, (2000). Nutritional Rickets and Osteomalacia in school children and adolescents (6–18 years) in a major teaching hospital in Riyadh, Saudi Arabia. Pan Arab osteoporosis conference, Beirut, Lebanon P417.

89. H.F. Saadi, R.L. Reed, A.O. Carter, E.V. Dunn, H.S. Qazaq and A.R. Al-Suhaili, (2000). Bone density of the spine and femur in Arabian women: Relation to Quantitative ultrasound of the calcaneus and vitamin D status. Pan Arab Osteoporosis Conference Beirut, Lebanon. P409 SU. (Abstract).

90. M.F. Holick, Environmental factors that influence the cutaneous production of vitamin D, *Am. J. Clin. Nutr.*, 1995, **61**(suppl), S638–S645.

91. A.R. Webb and M.F. Holick, The role of sunlight in the cutaneous production of vitamin $D_s$, *Ann. Rev. Nutr.*, 1988, **8**, 75–99.

92. J.S. Adams, T.L. Clemens, J.A. Parrish and M.F. Holick, Vitamin D synthesis and metabolism after ultraviolet radiation of normal and vitamin D deficiency subjects, *N. Engl. J. Med.*, 1982, **306**, 722–725.

93. M.C. Chapuy, P. Preziosi, M. Maamer, S. Arnaud, P. Galen, S. Hercberg and P.J. Meunier, Prevalence of vitamin D deficiency in an adult normal population, *Osteoporosis Int.*, 1997, **7**, 439–443.

94. H. Reichel, H.P. Koeffler and A.W. Norman, The role of the vitamin D endocrine system in health and disease, *N. Engl. J. Med.*, 1989, **320**, 980–991.

95. G. Maalouf, S. Salem, A. Nehme, J. Wehbe, C. Cooper and P. Hadji, Age Associated Changes in Quantitative Ultrasonometry (QUS) of the Os calcis in Lebanese Women. Second Pan Arab Osteoporosis Congress. Third International Congress of the Egyptian Osteoporosis Prevention Society. Sharm El-Sheikh, Egypt. 2002, (Abstract).

CHAPTER 25

# Interaction Between Nutrition, Physical Activity and Skeletal Health

ADAM D.G. BAXTER-JONES[1], ROBERT A. FAULKNER[1] and SUSAN J. WHITING[2]

[1]College of Kinesiology, University of Saskatchewan, Saskatoon, Saskatchewan, S7N 5B2, Canada, Email: baxterjones@usask.ca
[2]College of Pharmacy and Nutrition, University of Saskatchewan, 110 Science Place, Saskatoon, Saskatchewan, S7N 5C9, Canada

**Key Points**

- Adequate calcium intake and regular physical activity are required for optimizing the genetic potential for skeletal health.
- Physical activity is quantitatively a more important factor in affecting bone than calcium intake.
- The greater importance of physical activity is particularly evident when calcium intakes approach recommended calcium requirements.
- Calcium has a permissive effect on bone mineral accrual and maintenance, while physical activity has a modifying effect.
- There appears to be a threshold of calcium intake that optimizes the effects of mechanical loading on bone, which may be age dependent
- The interaction effects of calcium intake and physical activity may vary depending on the relative amounts of trabecular or cortical bone at specific sites.

## Introduction

The independent roles of physical activity and calcium intake on skeletal health across the lifespan have been well documented[1–5]; however, it is not clear as to whether calcium intake and physical activity interactions can enhance these independent actions. In theory, calcium and exercise interventions should be synergistic if the mechanisms of action involve complementary pathways,[6] and if so, this interaction should be evident in children, adults and the elderly.

Cross-sectional studies comparing active or athletic populations, with their inactive counterparts, support claims that physical activity enhances[7] or maintains[8] bone health and then minimises[9] bone loss during the life span. The osteogenic effects of exercise appear to be region specific and load dependant. There is also evidence from bilateral studies that supports the idea that the estrogenic response has it greatest effect during growth in childhood compared to adulthood.[10] However, the changes to bone architecture and dimensions in response to loading are complex processes influenced by the characteristics of the strain applied and the loading history of the bone. Although athletic studies provide promising evidence that physical activity confers an osteogenic benefit to growing bone, results may also be influenced by selection bias that may contribute to apparent differences in bone mass between athletic and non-athletic populations.

During the growing years, it is logical that calcium intake and physical activity are both necessary to optimize bone gain, as the skeleton needs calcium as a major structural component and mechanical loading to stimulate mineral deposition.[11] Although the physiology is not fully understood, it seems plausible that calcium intake may have a potentiating effect in allowing physical activity to exert its effect on bone mineral accretion.[12,13] Several studies certainly suggest that increased consumption of calcium boosts bone mineral only when combined with adequate physical activity,[14,15] and in one case the interaction was region-specific.[16] Thus, if exercise is to be recommended as a major intervention for reducing bone loss, it is important to understand and consider the possible interactive effect of calcium intake with exercise on skeletal health.[17]

In spite of a theoretical basis for an interactive effect of physical activity and calcium intake, there have been relatively few studies specifically investigating an interaction. One of the reasons for this lack of information may be that physical activity as a contributor to bone development has been undervalued by nutrition scientists.[18] Specifically, Anderson[18] suggests the role of physical activity merits more careful attention as a mechanism to counter low calcium intakes in children and that physical activity may dominate as a determinant of bone mass in early life.

Our current knowledge is also hampered by the fact that no two studies have addressed the same hypothesis. Our inability to make comparisons between studies is related to the fact that sample sizes vary, as do age groups. In fact very few studies use the same measures of physical activity, calcium intake or bone mass. In addition, studies also use a verity of bone mass sites. This is problematic as peak bone mineral accrual is likely to occur at different times at different sites.

In this chapter, we review the data on physical activity and calcium intake interaction from childhood to the elderly. In addition, the mechanism for an interactive role of calcium and physical activity in bone mineral accrual and maintenance are discussed and research design issues are identified.

# Interaction Evidence

## Children and Adolescence

Bone mass accrued during childhood and adolescence is a major determinant of fracture risk in later life.[19] It is estimated that about 26% of adult total bone mineral is accrued during the 2 years surrounding peak growth.[20] Thus maximizing bone accrual during childhood and adolescence, when modeling is superimposed on growth, is critical in optimizing peak bone mass.[21] Both physical activity and calcium intake are thought to be important factors in optimizing bone accrual during the growing years; however evidence of an interactive effect of physical activity and calcium intake is sparse.[22,23]

Although there is evidence to suggest that individually, moderate exercise or calcium supplementation may enhance bone mass accrual in children,[24–27] there is limited evidence on the combined effects of exercise and calcium. There is one study on the effects of mechanical loading and nutritional intake on skeletal development in early life. Specker[28] followed 72 infants to investigate the effects of physical activity and calcium intakes on bone mass accrual during the rapid period of growth between 6 and 18 months of age. The infants were randomized into groups of gross motor (daily bone loading activities) and fine motor (activities designed to keep the child sitting quietly) activity groups, and calcium intakes were monitored. Evidence for an interactive effect between physical activity and calcium was found. Infants consuming a low-to-moderate calcium intake who were randomized to receive daily bone loading activities had a lower bone mass accretion than infants randomized to fine motor activities. There was no effect of activity on bone mass accretion with moderate to high dietary intakes of calcium. It was speculated that increased bone loading during periods of rapid skeletal growth may lead to an increased demand for calcium; that is, it may be that it is not possible for an infant with increased bone loading to maintain comparable bone mass accretion if the increased demand for calcium is not met.

In a randomized trial of physical activity and calcium supplementation on bone mineral content (BMC) in 3- to 5-year-old children, Specker and Binkley[29] reported that physical activity increased bone circumference, and when combined with higher calcium intake that there was a greater cortical thickness and area. The authors speculate that these results are due to less endosteal expansion. However, it was unclear whether increases in cortical thickness among children randomized to gross motor activities and a placebo supplementation occurred later in time compared to children randomized to gross motor activities and a calcium supplementation.

Results from studies investigating interactive effects in pubescent children are mixed. Gunnes and Lehmann,[30] investigated the effects of 1-year of weight bearing physical activity on forearm bone mineral density (BMD) in boys and girls 8–16 years of age. Weight bearing physical activity predicted bone gain, but in children under 11 years of age, the greatest gains were found in those with the highest calcium intakes – suggesting an interactive effect. In contrast, Slemenda et al.[31] reported a significant effect of calcium intake and physical activity on BMD at the lumbar spine (LS), proximal femur and distal radius in pre-pubescent children, but found no evidence of an interaction effect. A study by Iuliano-Burns et al.[16] revealed region-specific interaction of exercise (moderate impact) and additional calcium (~440 mg day$^{-1}$ to a diet containing ~700 mg day$^{-1}$) in pre-pubescent girls. In their randomized control trial, lasting 8.5 months, they found an exercise–calcium interaction at the femur but not at the tibia-fibula; in the latter site, there was a significant effect of exercise only.

There is evidence that physical activity and calcium intake during later childhood and adolescence affect adult bone status. Welten et al.[32] assessed the effects of calcium intake and physical activity at age 13 on LSBMD, measured at age 27, in 84 males and 98 females. Weight bearing physical activity at age 13 predicted adult BMD, while calcium intake was not a significant predictor when controlling for physical activity patterns. In boys, physical activity was related to a higher BMD in those with high calcium intakes suggesting an interaction effect; however, the calcium intake for the group as a whole was relatively high making it difficult to assess an interactive effect. Another study in 1359 Dutch children, 7–11 years of age, reported increased BMC only in those children with high levels of physical activity, while calcium intake was not associated with BMC. In this cross-sectional study, BMC was measured by quantitative roentgen micro-densitometry of the midphalanx and metaphyseal site. No interaction effect was reported; however, the calcium intakes in over 80% of the children were high – exceeding 1400 mg day$^{-1}$.[33] Results from these studies suggest that there may be a threshold of calcium necessary to optimize the effects of physical activity.

Tylavsky et al.[34] investigated the relationship of dietary calcium and physical activity patterns to radial BMD, in 705 female high school and college students (18–22 years). Those with both adequate calcium intakes and regular physical activity patterns during high school had greater bone mass than those with lower calcium intakes and less physical activity; however exercise exerted a greater overall beneficial effect than calcium intake. In addition, those with both high calcium and high physical activity levels had greater radial BMD compared to those with low/high combinations – suggesting an interactive effect of calcium and physical activity.

Data from a Hutterite population support an interactive effect of calcium and physical activity in late adolescence.[35] In this study 39 girls (age 11) and 192 women (age 35) were measured for total body (TB) and LSBMC and BMD. It was hypothesized that the Hutterite girls and women would have higher BMD values than normative comparative groups due to their higher physical activity patterns and calcium intake levels. Results showed the adults had higher BMD and BMC compared to US normative data, but there was no difference in the girls. It was suggested that the results in the girls might be related to lower physical activity levels at this age in Hutterite children (due to cultural differences) compared to the normal group. It is only after school age that physical activity levels increase substantially in the young women. Physical activity (as estimated from colony workload) of the women predicted BMD and bone area; however, calcium intake did not predict BMD in either the girls or adults. It was speculated that this result was likely due to the relatively high calcium intake of the colony; that is, the calcium intake was well above the threshold necessary for bone mineral accrual. It was concluded that the high levels of physical activity beginning in the post-school years, in the presence of adequate calcium intake over a lifetime may result in increased bone gain in Hutterite women compared to US normative data.

Results from other studies also support an interactive or threshold effect of calcium in combination with physical activity. Kanders *et al.*[36] reported a relationship between physical fitness and vertebral BMD in young women with higher calcium intakes. Exercise during childhood and young adulthood was a significant predictor of femoral neck (FN) BMD in both men and women at age 20–29 and in LSBMD in men.[37] In this study, calcium intake and physical activity were measured cross-sectionally when subjects were 9–18 years of age, and BMD was subsequently assessed when subjects were 20–29 years of age. Males were calcium replete with average intakes of over 1400 mg day$^{-1}$.

In a 2-year prospective study investigating the consolidation of bone mineral, after the cessation of linear growth (18–21 years), Parsons *et al.*[38] found that although there were significant increases in whole-body BMC in individuals, neither dietary calcium nor physical activity were significant independent predictors of BMC. This study also indicated that differences in BMC between individuals was largely a result of differences in body size. There is little pure longitudinal data on bone mineral accretion over the growing years; however our own laboratory has reported data on activity and nutrition in a cohort of children studied for seven years.[20] We have shown that at the time of maximum adolescent growth (age at peak height velocity (APHV)) stature, lean and fat mass accounts for 68.9, 26.2 and 3.2% of predicted wholebody BMC, respectively.[39] When aligned at APHV, and controlling for size differences, we reported group data that indicated active boys and girls had 9 and 17% greater total body bone mineral content (TBBMC) than their inactive peers.[21]

The advantage of using longitudinally gathered data is that the effects of normal growth and development on BMC accrual can be separated from the independent effects of physical activity and dietary intake. The site-specific BMC accrual data from our study, the Bone Mineral Accrual Study (BMAS) 1991–1997 are presented in Tables 25.1–25.3. We have used a multilevel random effects modeling procedure[40,41] to identify significant relationships ($\beta$ coefficients) between measure of growth (height and body mass), maturity (biological age, defined as the chronological age in years minus APHV), and environmental exposure (physical activity and calcium intake) on BMC accrual. Full details of the methodology are reported elsewhere.[20,39,41,42]

**Table 25.1.** *Multilevel regression analysis of TBBMC aligned on biological age, adjusted for height, body mass, physical activity, calcium intake and the interaction between physical activity and calcium intake*

| Fixed effect | Estimates ± SEM | | | |
|---|---|---|---|---|
| | Boys | | Girls | |
| Constant | −2336.0±209.1* | | −1716.9±207.3* | |
| Biological age | 61.4±10.5* | | 63.8±9.6* | |
| Biological age$^2$ | 6.54±0.74* | | 4.97±1.10* | |
| Biological age$^3$ | −1.18±0.21* | | −1.52±0.25* | |
| Height | 22.26±1.41* | | 17.64±1.47* | |
| Body mass | 9.45±1.02* | | 8.82±0.90* | |
| Physical activity | NS | | 15.2±6.1* | |
| Calcium intake | 0.017±0.007* | | NS | |
| Interaction | | | | |
| PA*Ca | NS | | NS | |
| Random effects | Level 1 (within individuals) | | | |
| Constant | 2273±193 | | 1650±138 | |
| | Level 2 (between individuals) | | | |
| | Constant | Biological age | Constant | Biological age |
| Constant | 30 368±4361* | 2740±663* | 18 255±4361* | 1290±376* |
| Biological age | 2740±663* | 886±167* | 1290±376* | 536±103* |

Fixed effect values are Estimated Mean Coefficients ± SEE (Standard Error Estimate) (TBBMC, g).
Random effects values Estimated Mean Variance ± SEE (Standard Error Estimate) (TBBMC, g$^2$).
Biological Age – years from APHV, Height (cm); Body mass (kg); Physical Activity Score (1 = low, 5 = high); Calcium Intake (mg day$^{-1}$); PA*Ca = physical activity calcium intake interaction (PA Score*mg day$^{-1}$).
*p < 0.05 if mean > 2*SEE; Not Significant (NS).

BMC increased with increasing biological age in the TB and also at the FN and LS, as indicated by the significant coefficients associated with the fixed effects of biological age (Tables 25.1–25.3). Height and body mass were also significant predictors of BMC at all sites in both genders, apart from body mass in boys at the LS. Once the confounders of growth and maturity are controlled the relationships between physical activity, calcium intake and their interactions are investigated. In girls, physical activity was a significant predictor of BMC at the LS and TB but not at the FN. In boys, physical activity was a significant predictor of BMC at the LS only. In girls calcium intake did not predict BMC at any site. In boys calcium intake was a significant predictor of TBBMC. Although, both physical activity and calcium intake had independent effects on BMC accrual, no interactive relationships were found at any BMC site (Tables 25.1–25.3).

Figure 25.1 illustrates the percentage contributions of each of these fixed effects to the prediction of TB, FN and LS BMC when individuals are aligned at age at peak height velocity (biological age = 0); and when height, body mass, physical activity and calcium intake take the average values at biological age zero. In the boys model of TBBMC height accounted for nearly 88% of the

**Table 25.2.** *Multilevel regression analysis of FNBMC aligned on biological age, adjusted for height, body mass, physical activity, calcium intake and the interaction between physical activity and calcium intake*

| Fixed Effect | Estimates ± SEM | | | |
|---|---|---|---|---|
| | Boys | | Girls | |
| Constant | −2.973±0.617* | | −2.175±0.641* | |
| Biological age | 0.101±0.030* | | 0.078±0.029* | |
| Biological age$^2$ | 0.015±0.004* | | 0.005±0.004 | |
| Biological age$^3$ | −0.003±0.001* | | −0.003±0.001* | |
| Height | 0.039±0.004* | | 0.030±0.004* | |
| Body mass | 0.014±0.003* | | 0.019±0.003* | |
| Physical activity | NS | | NS | |
| Calcium intake | NS | | NS | |
| *Interaction* | | | | |
| PA*CA | NS | | NS | |
| *Random effects* | Level 1 (within individuals) | | | |
| Constant | 0.031±0.003* | | 0.036±0.003* | |
| | Level 2 (between individuals) | | | |
| | Constant | Biological Age | Constant | Biological Age |
| Constant | 0.142±0.021* | 0.012±0.003* | 0.093±0.015* | 0.008±0.003* |
| Biological age | 0.012±0.003* | 0.003±0.001* | 0.008±0.003* | 0.003±0.001* |

Fixed effect values are Estimated Mean Coefficients ± SEE (Standard Error Estimate) (FNBMC, g). Random effects values Estimated Mean Variance ± SEE (Standard Error Estimate) (FNBMC, g$^2$). Biological Age – years from APHV (years); Height (cm); Body mass (kg); Physical activity Score (1 = low, 5 = high); calcium intake (mg day$^{-1}$); PA*CA = Physical activity calcium intake interaction (PA Score*mg day$^{-1}$).
*p < 0.05 if mean > 2*SEE; Not Significant (NS).

total TBBMC prediction (22.26 g BMC per 1 cm in height), whilst body mass accounted for 11.8% and calcium intake accounted for 0.5% of the predicted TBBMC (Figure 25.1 (a)). The large contributions of height and body mass on the prediction of BMC are illustrated in all the BMC sites in both genders. In boys at the LS, physical activity accounted for 2.2% of predicted BMC. In the girls models physical activity accounted for 1.4 and 1.3% of predicted BMC at the TB and LS respectively.

We concur with Parsons *et al.*[38] that bone and body size are the strongest predictors of BMC during the growing years, and after allowance for these factors it is difficult to demonstrate any age/size independent effects of calcium intake or physical activity. Our findings reflect the fact that the calcium intake for the group, average over 1000 mg day$^{-1}$,[43] was relatively high and likely above the threshold required to identify any independent beneficial effect on BMC accrual. Average physical activity was also fairly homogeneous between the children. Our results are in accord with others[38] who have shown that once BMC is adjusted for age, bone and body size independent effects of lifestyle factors are difficult to demonstrate. What are needed are prospective, longitudinal studies which follow children through adolescence into young

**Table 25.3.** *Multilevel regression analysis of LSBMC aligned on biological age, adjusted for height, body mass, physical activity, calcium intake and the interaction between physical activity and calcium intake*

| Fixed Effect | Estimates ± SEM | | | |
|---|---|---|---|---|
| | Boys | | Girls | |
| Constant | −84.60±7.79* | | −79.16±8.69* | |
| Biological age | 2.47±0.40* | | 1.22±0.40* | |
| Biological age$^2$ | 0.64±0.05* | | 0.38±0.05* | |
| Biological age$^3$ | −0.05±0.01* | | −0.04±0.01* | |
| Height | 0.74±0.05* | | 0.69±0.06* | |
| Body mass | NS | | 0.16±0.04* | |
| Physical activity | 0.90±0.32* | | 0.52±0.25* | |
| Calcium intake | NS | | NS | |
| *Interaction* | | | | |
| PA*CA | NS | | NS | |
| *Random effects* | Level 1 (within individuals) | | | |
| Constant | 2.23±0.23* | | 2.01±0.20* | |
| | Level 2 (between individuals) | | | |
| | Constant | Biological Age | Constant | Biological Age |
| Constant | 31.21±4.53* | 3.95±0.84* | 24.35±3.66* | 2.17±0.64* |
| Biological age | 3.95±0.84* | 1.22±0.25* | 2.17±0.64* | 1.15±0.22* |

Fixed effect values are Estimated Mean Coefficients±SEE (Standard Error Estimate) (LSBMC, g). Random effects values Estimated Mean Variance±SEE (Standard Error Estimate) (LSBMC, g$^2$). Biological Age – years from APHV (years); Height (cm); Body mass (kg); Physical Activity Score (1 = low, 5 = high); Calcium Intake (mg day$^{-1}$); PA*CA = physical activity calcium intake interaction (PA Score*mg day$^{-1}$).
*p < 0.05 if mean > 2*SEE; Not Significant (NS).

adulthood to determine whether calcium or physical activity or an interaction of the two play an important role in the determination of bone mineral mass.

Several other studies have investigated the relationship of physical activity or calcium to BMD or BMC in children and adolescents, but have not tested for interactive effects. Physical activity and calcium intake have been reported to be related to BMC at the radius in young women,[44] while physical activity, but not calcium intake predicted BMD, in 18-year-old females.[45] Calcium intake and physical activity were determinants of BMD at the spine and TB in 500 children and young adults 4–20 years of age, but were not significant predictors when BMD was corrected for size.[46] The main predictors of BMD during childhood were body weight in boys and pubertal development in girls. Both dietary calcium and physical activity were independent determinants of LS and femoral BMD in a group of children 7–15 years of age; however, physical activity had a greater effect.[47] There was no correlation between calcium intake and physical activity. Molgaard et al.[48] followed 192 girls and 140 boys, 5–19 years of age for 1 year. Bone area and BMC were adjusted for height and weight in order to account for the effects of size on the dependent variables. Size-adjusted BMC was positively associated with average calcium intake,

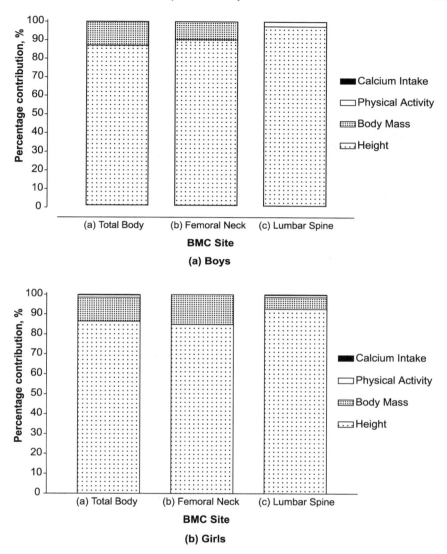

**Figure 25.1** *Site-specific percentage contributions of each fixed effect in the random effects models (Tables 25.1–25.3) on the predication of bone mineral content (BMC) at APHV. Note at PHV biological age = 0 and therefore makes no contribution to the predicted BMC value. Height, body mass, physical activity and calcium intake values were taken from the average values at biological age zero for (a) boys and (b) girls*

while size-adjusted accretion in BMC was positively associated with change in dietary calcium intake in boys only. In a group of 75 females, exercise and fitness levels, but not calcium intake (measured between ages 12–18 years) were significantly related to TBBMD and indices of bone strength (cross-sectional moment of inertia, section modulus) at the proximal femur at age 20.[49] In this study the average calcium intakes ranged from 486 to 1958 mg day$^{-1}$.

## Adults

Studies in adults have shown mixed results. Using a multiple regression approach, Recker *et al.*[50] found an interactive effect of calcium intake and physical activity on LSBMD in a sample of 156 women, 18–26 years of age. The effect of physical activity improved substantially with calcium intakes of greater than 1000 mg day$^{-1}$. In contrast, others found the greatest effects of physical activity on radial BMC at calcium intakes between 500 and 800 mg day$^{-1}$.[51] An 18-month resistance-training program in 56 women aged 28–29 years of age resulted in a significant increase in trochanter and LSBMD.[52] These subjects were supplemented at 500 mg day$^{-1}$ of calcium in addition to their average dietary intake of 1023 mg day$^{-1}$; that is, they were calcium replete.

Other studies have not reported an interaction effect of physical activity and calcium intake on bone parameters. Physical activity and calcium intake were associated with greater TBBMD in a cross-sectional study of 422 women representing an age range of 25–65 years.[53] Only exercise was associated with greater FNBMD. Neither physical activity nor calcium intake was related to higher radial BMD. Both physical activity and calcium intake were associated with indices of bone strength such as section modulus and cross-sectional moment of inertia. There was no interactive effect of calcium and physical activity on any of the bone parameters. The results of this study suggest that a relatively moderate level of physical activity or calcium intake maintained from childhood to adulthood improves mechanical competence of the skeleton in adulthood. Beshgetoor *et al.*[54] monitored the effects of an 18 month exercise program of either cycling or running on hip and LSBMD in 30 females aged 49 years. Compared to the control group, the exercise groups maintained BMD at the hip while BMD at the LS was maintained in the runners. There was no interaction effect of calcium intake and exercise.

The effects of calcium and physical activity on appendicular and axial BMD were evaluated in a 2- year study of 200–300 healthy young women aged 20–39 years.[55] No association of calcium intake with BMD and/or with BMD changes over the 2 years of study were observed. Daily activity had no effect on BMD and there was no apparent additive interaction of activity and calcium intake on BMD.

## Post-Menopausal Women and Elderly Men

Several studies in post-menopausal women suggest an interactive effect of calcium intake and physical activity on bone mass. Uusi-Rasi *et al.*[56] followed a group of 60–65 year old women over a 4-year period. BMC at the proximal femur and distal radius decreased in all subjects; however, those who maintained relatively higher calcium intakes and physical fitness had smaller decreases in BMC at the FN. Another study in post-menopausal women showed an increase in BMD at the proximal femur.[57] All the women in this study were calcium replete receiving a supplementation of 600 mg day$^{-1}$.

Prince *et al.*[58] found supplementing calcium intake from about 800 to 1800 mg day$^{-1}$ significantly reduced bone loss at the trochanter and intertrochanter sites; however, when supplementation was combined with physical activity there was also a significant reduction in rate of bone loss at the FN. This study showed that physical activity was ineffective unless calcium intake was sufficient supporting the theory that both factors are required to enhance bone health.

Epidemiological studies in elderly men and women support the theory of an interaction effect of calcium intake and physical activity. In a study of 709 men and 1080 women, BMD at the FN in those with higher calcium intakes and higher quadriceps muscle strength was significantly higher (about 5%) than in those with low calcium intakes and low strength.[17] In a subsequent study by this same group, osteoporosis prevalence was found to be lower in those in the highest tertiary for calcium intake, body mass index, and quadriceps strength compared to those in the lowest tertiles. The prevalence of osteoporosis was 12 and 1.5% for women and men, respectively compared to 64 and 40% in the subjects in the lowest tertiles for the three factors.[59]

In contrast to the previous studies, interactive effects have not been supported by others.[60,61] Nelson *et al.*[60] hypothesized that a combination of aerobic exercise with calcium supplementation would have a greater effect on BMD at the LS and FN than exercise alone. Results showed that exercise had a significant effect at the LS, while calcium supplementation significantly improved BMD at the FN. There was no interaction effect at either site. De Jong *et al.*[61] found no evidence of an interactive effect of a 17-week exercise program combined with a nutrient rich diet in frail elderly men and women.

## Summary of Interactive Evidence

Evidence for an interaction has been reported in infants[28] and in children less than 11 years of age[16,30]; however, others reported no evidence for an interaction in pre-pubertal children.[31] Evidence from several studies[33–36] supports the concept that during childhood and/or adolescence a threshold intake of calcium optimizes the effect of physical activity on adult bone status. Some studies in adults also support a threshold effect of calcium intake, although the level of the threshold required is not consistent.[50–52] Others, however have reported no evidence of an interaction.[53,54] Several studies in elderly populations support an interactive effect of physical activity or physical fitness and calcium[17,57–59]; however, others have reported no interaction effect.[60,61]

Two review papers have previously concluded that there is support for an interactive effect. Specker[62] summarized studies up until 1995. All studies were intervention trials involving randomized or non-randomized designs that reported calcium intake and physical activity. In analyzing the 17 studies that met the criteria, it was concluded that calcium and exercise might not act independently on bone; for example, a positive effect of physical activity was apparent only when calcium intake was greater than 1000 mg day$^{-1}$, and the benefits of high calcium intakes were only evident in the presence of physical activity.

Results of a meta-analyses of studies in post-menopausal women also suggest that a threshold of calcium intake is needed for optimizing the effects of physical activity.[13] Aerobic exercise had a site-specific effect on BMD at the proximal femur; however, the fact that the observed changes were greater in subjects consuming at least 1000 mg day$^{-1}$ of calcium, suggests that adequate calcium intake is necessary to maximize the effects of aerobic exercise.

In summary, although there is a paucity of data and results of some studies are inconsistent, the bulk of existing evidence supports the theory that calcium intake interacts (at a not yet well-defined level) with physical activity to affect bone mineral accrual in children and adolescence, bone maintenance in adults, and in reducing the rate of bone loss in the elderly.

## Mechanisms

It makes physiological sense that calcium and mechanical loading would have an interactive effect; that is, mechanical loading through physical activity is necessary to stimulate modeling and remodeling, while adequate calcium is necessary for bone mineralization.[2,12] Health status can also affect the response to calcium and physical activity; however we have not included clinical or pathological data in this review. It is also important to consider the specific skeletal site when assessing the effects of environmental factors on bone. About 85% of the total skeleton consists of cortical or compact bone and about 15% is trabecular bone; and, different bones and parts of bone have different proportions of these two types of bone. About 80–90% of the volume of cortical bone is calcified compared to only about 15–25% of trabecular bone. Cortical bone function is primarily mechanical and protective, while trabecular bone has a more metabolic function. Environmental factors such as nutrition and physical activity affect trabecular and cortical bone in different ways. For example, cortical bone responds to mechanical loading by potentially increasing in mineral content and/or changing in geometry, while trabecular bone may respond to mechanical loading by a change in orientation of individual trabeculae.[63]

The available evidence, from healthy subjects, supports the concept that nutritional factors such as calcium availability are permissive, while mechanical loading exerts the modifying effect for bone mineralization to reach its genetic potential.[64] Furthermore, calcium status is a major determinant of parathyroid hormone (PTH) and vitamin D activity, thus triggering metabolic events affecting bone. Low calcium intake is a signal for increased PTH release which itself is a trigger for 1,25-dihydroxyvitamin D production; higher intakes of calcium would attenuate action of these hormones.[65] Studies also support the concept that there is a threshold for calcium intake, below which accumulation of bone mass is a function of intake and above which it is not dependent on higher intakes.[66] However, the level of calcium intake to meet this theoretical threshold across age groups remains controversial.[12,67] It has been suggested that calcium and Vitamin D may modify the threshold of the intensity of exercise at which bone formation exceeds bone resorption; that is, a permissive affect of calcium may be that it makes bone more sensitive to the mechanical loading stimulus.[68] This conjecture is consistent with Frost's[69] mechanostat theory, which theorizes strain thresholds that control resorption and formation can be influenced by the hormonal and nutritional milieu. As proposed by Frost, muscle and bone form an operational unit; that is, what affects muscle (loading) also affects bone, and muscle contraction causes the largest loads (strain) on bone, affecting the biological mechanisms that determine bone strength.[70] Adequate calcium availability is unquestionably required for optimizing bone mineral accrual and/or reducing bone loss with age; however, as Frost eloquently states: "trying to increase significantly whole-bone strength in healthy subjects through nutrition (such as calcium supplements) or hormonal interventions (alone) is like trying to make a car go faster by adding petrol to its tank". In the study by Iuliano-Burns *et al.*[16] it was the loaded sites that experienced greater gains in bone mass with exercise and calcium.

There may also be other adaptations in bone mineral accrual during growth to compensate for low calcium intakes.[18] The average young adult skeleton contains about 928 g of calcium in males and 700 g in females; and about 26% of the adult skeleton is laid down during the 2 adolescent years surrounding peak skeletal growth.[71] The estimated daily accumulation rates during this time period are estimated at 359 mg day$^{-1}$ for boys and 284 mg day$^{-1}$ for girls.[71] The adult skeletal mass is accumulated in many children who have calcium intakes below recommended values; therefore there must be some adaptation occurring during growth to account for the accrual despite inadequate calcium intakes.[72] For example, no adverse affects on bone were apparent in children and women in Gambia despite calcium intakes of only 300–500 mg day$^{-1}$.[73] In addition, females who were active during their prepubertal and pubertal years achieved more than adequate bone mass despite sub-optimal calcium intakes.[74] Increased retention efficiency of dietary calcium is likely to be an important factor in this adaptation.[21] It also has been suggested that the adaptive mechanisms involve the muscle–bone linkage stimulated by exercise and vitamin D (sunlight exposure) and perhaps other yet unknown factors. For example, in animal models, in the absence of dietary calcium, exercise results in an increased skeletal accretion of calcium with transient hypocalcemia and hypophosphatemia, resulting in increased intestinal absorption of calcium and phosphorus.[75] Clearly, without assuming some adaptive mechanism it is difficult to explain the achievement of healthy skeletons in many children who have known nutritional inadequacies – but are otherwise healthy.[18] Low calcium intakes may be a marker for a diet that is inadequate in other nutrients as well.[76]

In summary, based on the current literature the following conclusions can be made on the interactive effects of physical activity and calcium intake on bone: adequate calcium intake and regular physical activity are required for optimizing the genetic potential for skeletal health; however, physical activity is quantitatively a more important factor in affecting bone than calcium intake.[77] The greater importance of physical activity is particularly evident when calcium intakes approach recommended calcium requirements.[78] Calcium has a permissive effect on bone mineral accrual and maintenance, while physical activity has a modifying effect. There appears to be a threshold of calcium intake that optimizes the effects of mechanical loading on bone. Results from several studies suggest that this threshold is over 1000 mg day$^{-1}$; however controversy remains as to the exact level.

# Research Design Issues

There are a number of research design issues that make interpretation of the results of many studies difficult. Many of the studies include relatively homogeneous or relatively high calcium intakes, and many have low statistical power. The optimal calcium intake to prevent osteoporosis is not known,[12,79] and the amount of calcium required for optimizing skeletal mass is difficult to determine.[72] The large differences in calcium intake recommendations in developed countries also make it difficult to interpret data from different areas of the world.[48]

Errors in self-reporting of nutritional intake, particularly in children and adolescents make interpretation of data difficult.[11,43] There is also a lack of consistency with the type of nutritional intervention used among studies with some intervention with supplements and others with general nutrient enrichment; these could have different effects and or different mechanisms for their effects.[23] Studies where calcium intake is provided[16] yield more accurate documentation of calcium intakes than studies using observational methods.

Although mechanical loading through muscle action primarily increases bone mass,[70] there is considerable variation in the definition and or type and intensity of physical activity intervention across studies. As with reporting of calcium intake, there are also considerable problems in the reporting of physical activity levels in observational studies – particularly in children.[80]

Many studies are not long enough to account for transient bone-remodeling. As proposed by Heaney[81] and summarized more recently by Barr and McKay,[22] the theory suggests that effects of interventions are not likely to be sustained over longer periods of time due to the transient bone-remodeling. The theory is particularly relevant to the interpretation of studies in older populations where remodeling predominates over bone formation. At any time, some remodeling units are in the resorption phase, while others are in the formation or mineralizing stage. Bone sites in the active remodeling stage will naturally have less bone compared to sites where remodeling is not active; thus, bone mineral assessment techniques (such as DXA) will underestimate the amount of bone tissue; and the degree of underestimation is directly related to the amount of remodeling occurring at the time of measurement. Calcium inhibits remodeling; thus, in the short term, calcium supplementation increases the amount of bone mineral measured; but with the completion of the remodeling cycle in sites already activated before the intervention, BMC would be expected to level off at a higher plateau and mineralization to be balanced. The resulting increase in BMC has been defined as a 'transient' because it returns to a lower level after the intervention is terminated. That is, when calcium intake returns to pre-supplementation levels, there is a subsequent increase in remodeling activity and a decrease in BMC measured. In order to take into account the bone-remodeling transient interventions need to be evaluated over sufficient time to include several modeling–remodeling cycles.

Finally, most of the evidence is based on cross-sectional data. Results from cross-sectional comparisons may be influenced by selection bias that may contribute to the apparent differences in bone mass between populations. In addition, many studies in children do not adequately adjust bone mineral measures for size differences or changes in growth over time.[48]

In summary, as supported by others,[15,18,22,23,72,78] in order to more fully understand the effects of calcium and physical activity on bone the following will be required:

more randomized-control studies

more accurate assessment of diet and physical activity

duration of interventions needs to be sufficient; that is, studies should span several remodeling cycles

initial status of calcium and physical activity needs to be factored into analysis in intervention studies

most importantly – is to refine the understanding of calcium requirements as a function of various levels of physical activity, and appropriate designs applied in order to investigate specifically interactive effects.

# Conclusions

Based on the current literature, the following conclusions can be made on the interactive effects of physical activity and calcium intake on bone: adequate calcium intake and regular physical activity are required for optimizing the genetic potential for skeletal health; however, physical activity is quantitatively a more important factor in affecting bone than calcium intake.[77] The greater importance of physical activity is particularly evident when calcium intakes approach recommended calcium requirements.[78] Calcium has a permissive effect on bone mineral accrual and maintenance, while physical activity has a modifying effect. There appears to be a threshold of calcium intake that optimizes the effects of mechanical loading on bone. Results from several studies suggest that this threshold is over 1000 mg day$^{-1}$; however controversy remains as to the exact level.

# Acknowledgements

The BMAS (1991–1998) was supported by the Canadian National Health and Research Development Program (NHRDP), grant no. 6608-1261. The BMAS (2002–2005) is supported by The Canadian Institute of Health Research, grant no. MOP 57671. BMAS Group members include D.A. Bailey, A.D.G. Baxter-Jones, P.E. Crocker, K.S. Davison, D.T. Drinkwater, E. Dudzic, R.A. Faulkner, K. Kowalski, H.A. McKay, R.L. Mirwald, W.M. Wallace and S.J. Whiting.

# References

1. D.A. Bailey, R.A. Faulkner and H.A. McKay, Growth, physical activity, and bone mineral acquisition, *Exerc. Sport Sci. Rev.*, 1996, **24**, 233–266.
2. S.I. Barr and H.A. McKay, Nutrition, exercise, and bone status in youth, *Int. J. Sport Nutr.*, 1998, **8**, 124–142.
3. R.P. Heaney, Calcium, dairy products and osteoporosis, *J. Am. Coll. Nutr.*, 2000, **19**, 825–842.
4. G.A. Kelley, Exercise and regional bone mineral density in postmenopausal women: a meta-analytic review of randomized trials, *Am. J. Phys. Med. Rehabil.*, 1998, **77**, 76–87.
5. T.L. Holbrook and E. Barrett-Connor, An 18-year prospective study of dietary calcium and bone mineral density in the hip, *Calcif. Tissue Int.*, 1995, **56**, 364–367.
6. M.A. Singh, Combined exercise and dietary intervention to optimize body composition in aging, *Ann. NY Acad. Sci.*, 1998, **854**, 378–393.
7. D.A. Bailey, The role of mechanical loading in the regulation of skeletal development during growth, In *New Horizons in Pediatric Exercise Science*, C.J.R. Blimkie and O. Bar-Or, (eds.), Human Kinetics, Champaign, IL, 1995, 97–108.
8. C. Snow-Harter and R. Marcus, Exercise and regulation of bone mass, *Bone*, 1989, **6**, 45–48.
9. E. Ernst, Exercise for female osteoporosis. A systematic review of randomised clinical trials, *Sports Med.*, 1998, **25**, 359–368.
10. H. Haapasalo, P. Kannus, H. Sievanen, M. Pasanen, K. Uusi-Rasi, A. Heinonen, P. Oja and I. Vuori, Effect of long-term unilateral activity on bone mineral density of female junior tennis players, *J Bone Miner. Res.*, 1998, **13**, 310–319.
11. T. Lloyd, V.M. Chinchilli, N. Johnson-Rollings, K. Kieselhorst, D.F. Eggli and R. Marcus, Adult female hip bone density reflects teenage sports-exercise patterns but not teenage calcium intake, *Pediatrics*, 2000, **106**, 40–44.
12. F. Branca, S. Valtuena and S. Vatuena, Calcium, physical activity and bone health – building bones for a stronger future, *Public Health Nutr.*, 2001, **4**, 117–123.
13. G.A. Kelley, Aerobic exercise and bone density at the hip in postmenopausal women: a meta-analysis, *Prev. Med.*, 1998, **27**, 798–807.
14. L.K. Bachrach, Making an impact on pediatric bone health, *J. Pediatr.*, 2000, **136**, 137–139.
15. R.D. Lewis and C.M. Modlesky, Nutrition, physical activity, and bone health in women, *Int. J. Sport Nutr.*, 1998, **8**, 250–284.
16. S. Iuliano-Burns, G. Naughton, K. Gibbons, S.L. Bass and L. Saxon, Regional specificity of exercise and calcium during skeletal growth in girls: a randomized control trial, *J. Bone Miner. Res.*, 2003, **18**, 156–162.

17. T.V. Nguyen, P.J. Kelly, P.N. Sambrook, C. Gilbert, N.A. Pocock and J.A. Eisman, Lifestyle factors and bone density in the elderly: implications for osteoporosis prevention, *J. Bone Miner. Res.*, 1994, **9**, 1339–1346.

18. J.J. Anderson, The important role of physical activity in skeletal development: how exercise may counter low calcium intake, *Am. J. Clin. Nutr.*, 2000, **71**, 1384–1386.

19. J.P. Bonjour, A.L. Carrie, S. Ferrari, H. Clavien, D. Slosman, G. Theintz and R. Rizzoli, Calcium-enriched foods and bone mass growth in prepubertal girls: a ranvdomized, double-blind, placebo-controlled trial, *J. Clin. Invest.*, 1997, **99**, 1287–1294.

20. D.A. Bailey, H.A. McKay, R.L. Mirwald, P.R. Crocker and R.A. Faulkner, A six-year longitudinal study of the relationship of physical activity to bone mineral accrual in growing children: the university of Saskatchewan bone mineral accrual study, *J. Bone Miner. Res.*, 1999, **14**, 1672–1679.

21. D.A. Bailey, Physical activity and bone mineral acquisition during adolescence, *Osteoporosis Int.*, 2000, **11**, S2–S3.

22. S.I. Barr and H.A. McKay, Nutrition, exercise, and bone status in youth, *Int. J. Sport Nutr.*, 1998, **8**, 124–142.

23. S.A. French, J.A. Fulkerson and M. Story, Increasing weight-bearing physical activity and calcium intake for bone mass growth in children and adolescents: a review of intervention trials, *Prev. Med.*, 2000, **31**, 722–731.

24. M. Bradney, G. Pearce, G. Naughton, C. Sullivan, S. Bass, T. Beck J. Carlson and E. Seeman, Moderate exercise during growth in prepubertal boys: changes in bone mass, size, volumetric density, and bone strength: a controlled prospective study, *J. Bone Miner. Res.*, 1998, **13**, 1814–1821.

25. J.P. Bonjour and R. Rizzoli, Bone acquisition in adolescence, In *Osteoporosis*, R. Marcus and D. Feldman, (eds.), Academic Press, San Diego, USA, 1996, 465–476.

26. C.C. Johnston, J.Z. Miller, C.W. Slemenda, T.K. Reister, S. Hui, J.C. Christian and M. Peacock, Calcium supplementation and increases in bone mineral density in children, *N. Engl. J. Med.*, 1992, **327**, 82–87.

27. F.L. Morris, G.A. Naughton, J.L. Gibbs, J.S. Carlson and J.D. Wark, Prospective ten-month exercise intervention, in premenarcheal girls: positive effects on bone and lean mass, *J. Bone Miner. Res.*, 1997, **12**, 1453–1462.

28. B.L. Specker, L. Mulligan and M. Ho, Longitudinal study of calcium intake, physical activity, and bone mineral content in infants 6–18 months of age, *J. Bone Miner. Res.*, 1999, **14**, 569–576.

29. B. Specker and T. Binkley, Randomized trial of physical activity and calcium supplementation on bone mineral content in 3- to 5-year-old children, *J. Bone Miner. Res.*, 2003, **18**, 885–892.

30. M. Gunnes and E.H. Lehmann, Physical activity and dietary constituents as predictors of forearm cortical and trabecular bone gain in healthy children and adolescents: a prospective study, *Acta Paediatr.*, 1996, **85**, 19–25.

31. C.W. Slemenda, T.K. Reister, S. Hui, J.Z. Miller, J.C. Christian and C.C. Johnston, Influences on skeletal mineralization in children and adolescents: evidence for varying effects of sexual maturation and physical activity, *J. Pediatr.*, 1994, **125**(2), 201–207.

32. D.C. Welten, H.C. Kemper, G.B. Post, W. Van Mechelen, J. Twisk, P. Lips and G.J. Teule, Weight-bearing activity during youth is a more important factor for peak bone mass than calcium intake, *J. Bone Miner. Res.*, 1994, **9**, 1089–1096.

33. M.F. VandenBergh, S.A. DeMan, J.C. Witteman, A. Hofman, W.T. Trouerbach and D.E. Grobbee, Physical activity, calcium intake, and bone mineral content in children in The Netherlands, *J. Epidemiol. Commun. Health*, 1995, **49**, 299–304.

34. F.A. Tylavsky, J.J. Anderson, V.R. Talmage and T.N. Taft, Are calcium intakes and physical activity patterns during adolescence related to radial bone mass of white college-age females? *Osteoporosis Int.*, 1992, **2**, 232–240.

35. K.S. Wosje, T.L. Binkley, N.L. Fahrenwald and B.L. Specker, High bone mass in a female Hutterite population, *J. Bone Miner. Res.*, 2000, **15**, 1429–1436.

36. B. Kanders, D.W. Dempster and R. Lindsay, Interaction of calcium nutrition and physical activity on bone mass in young women, *J. Bone Miner. Res.*, 1988, **3**, 145–149.

37. M.J. Valimaki, M. Karkkainen, C. Lamberg-Allardt, K. Laitinen, E. Alhava, J. Heikkinen, O. Impivaara, P. Makela, J. Palmgren and R. Seppanen, Exercise, smoking, and calcium intake during adolescence and early adulthood as determinants of peak bone mass. Cardiovascular Risk in Young Finns Study Group, *Br. Med. J.*, 1994, **309**, 230–235.

38. T.J. Parsons, A. Prentice, E.A. Smith, T.J. Cole and J.E. Compston, Bone mineral mass consolidation in young British adults, *J. Bone Miner. Res.*, 1996, **11**, 264–274.

39. A.D.G. Baxter-Jones, R.L. Mirwald, H.A. McKay and D.A. Bailey, A longitudinal analysis of sex differences in bone mineral accrual in healthy 8 to 19 year old boys and girls, *Ann. Hum. Biol.*, 2003, **30**, 160–175.

40. H. Goldstein, Multilevel Statistical Models, 2nd edn Arnold, London, 1995, 1–178.

41. A.D.G. Baxter-Jones and R.L. Mirwald, Multilevel modelling, In *Methods in Human Growth Research*. R.C. Hauspie, N. Cameron and L. Molinari (eds.), Cambridge University Press, Cambridge, 2003. (In Press).

42. L.M. Carter, S.J. Whiting, D.T. Drinkwater, G.A. Zello, R.A. Faulkner and D.A. Bailey, Self-reported calcium intake and bone mineral content in children and adolescents, *J. Am. Coll. Nutr.*, 2001, **20**, 502–509.

43. S.J. Whiting, C. Colleaux and T. Bacchetto, Dietary intakes of children age 8 to 15 years living in Saskatoon, *J. Can. Dietetic Assoc.*, 1995, **56**, 119–125.

44. J.A. Metz, J.J Anderson and P.N. Gallagher Jr., Intakes of calcium, phosphorus, and protein, and physical-activity level are related to radial bone mass in young adult women, *Am. J. Clin. Nutr.*, 1993, **58**, 537–542.

45. N.K. Henderson, R.I. Price, J.H. Cole, D.H. Gutteridge and C.I. Bhagat, Bone density in young women is associated with body weight and muscle strength but not dietary intakes, *J. Bone Miner. Res.*, 1995, **10**, 384–393.

46. A.M. Boot, M.A. de Ridder, H.A. Pols and E.P. Krenning, de Muinck Keizer-Schrama SM. Bone mineral density in children and adolescents: relation to puberty, calcium intake, and physical activity, *J. Clin. Endocrinol. Metab.*, 1997, **82**, 57–62.

47. J.C. Ruiz, C. Mandel and M. Garabedian, Influence of spontaneous calcium intake and physical exercise on the vertebral and femoral bone mineral density of children and adolescents, *J. Bone Miner. Res.*, 1995, **10**, 675–682.

48. C. Molgaard, B.L. Thomsen and K.F. Michaelsen, The influence of calcium intake and physical activity on bone mineral content and bone size in healthy children and adolescents, *Osteoporosis Int.*, 2001, **12**, 887–894.

49. T. Lloyd, T.J. Beck, H. Lin, M. Tulchinsky, D.F. Eggli, T.L. Oreskovic, P.R. Cavanagh and E. Seeman, Modifiable determinants of bone status in young women, *Bone*, 2002, **30**, 416–421.

50. R.R. Recker, K.M. Davies, S.M. Hinders, R.P. Heaney, M.R. Stegman and D.B. Kimmel, Bone gain in young adult women, *J. Am. Med. Assoc.*, 1992, **268**, 2403–2408.

51. L. Halioua and J.J. Anderson, Lifetime calcium intake and physical activity habits: independent and combined effects on the radial bone of health premenopausal caucasian women, *Am. J. Clin. Nutr.*, 1989, **49**, 534–541.

52. T. Lohman, S. Going, R. Pamenter, M. Hall, T. Boyden, L. Houtkooper, C. Ritenbaugh, L. Bare, A. Hill and M. Aickin, Effects of resistance training on regional and total bone mineral density in premenopausal women: a randomized prospective study, *J. Bone Miner. Res.*, 1995, **10**, 1015–1024.

53. K. Uusi-Rasi, H. Sievanen, I. Vuori, M. Pasanen, A. Heinonen and P. Oja, Associations of physical activity and calcium intake with bone mass and size in healthy women at different ages, *J. Bone Miner. Res.*, 1998, **13**, 133–142.

54. D. Beshgetoor, J.F. Nichols and I. Rego, Effect of training mode and calcium intake on bone mineral density in female master cyclist, runners, and non-athletes, *Int. J. Sports Nutr. Exerc. Metab.*, 2000, **10**, 290–301.

55. R.B. Mazess and H.S. Barden, Bone density in premenopausal women: effects of age, dietary intake, physical activity, smoking, and birth-control pills, *Am. J. Clin. Nutr.*, 1991, **53**, 132–142.

56. K. Uusi-Rasi, H. Sievanen, M. Pasanen, P. Oja and I. Vuori, Maintenance of body weight, physical activity and calcium intake helps preserve bone mass in elderly women, *Osteoporosis Int.*, 2001, **12**, 373–379.

57. D. Kerr, T. Ackland, B. Maslen, A. Morton and R. Prince, Resistance training over 2 years increases bone mass in calcium-replete postmenopausal women, *J. Bone Miner. Res.*, 2001, **16**, 175–181.

58. R. Prince, A. Devine, I. Dick, A. Criddle, D. Kerr, N. Kent, R. Price and A. Randell, The effects of calcium supplementation (milk powder or tablets) and exercise on bone density in postmenopausal women, *J. Bone Miner. Res.*, 1995, **10**, 1068–1075.

59. T.V. Nguyen, J.R. Center and J.A. Eisman, Osteoporosis in elderly men and women: effects of dietary calcium, physical activity, and body mass index, *J. Bone Miner. Res.*, 2000, **15**, 322–331.

60. M.E. Nelson, E.C. Fisher, F.A. Dilmanian, G.E. Dallal and W.J. Evans, A 1-y walking program and increased dietary calcium in postmenopausal women: effects on bone, *Am. J. Clin. Nutr.*, 1991, **53**, 1304–1311.

61. N. de Jong, M.J. Paw, L.C. de Groot, G.J. Hiddink and W.A. van Staveren, Dietary supplements and physical exercise affecting bone and body composition in frail elderly persons, *Am. J. Public Health*, 2000, **90**, 947–954.

62. B.L. Specker, Evidence for an interaction between calcium intake and physical activity on changes in bone mineral density, *J. Bone Miner. Res.*, 1996, **11**, 1539–1544.

63. M. Forwood, Physiology, In *Physical Activity and Bone Health*, K. Khan, H. McKay, P. Kannus, D. Bailey, J. Wark and K. Bennell (eds.), Human Kinetics, Champaign, IL, 2001, 11–19.

64. R.P. Heaney, Calcium, bone health, and osteoporosis. In *Bone and Mineral Research*, W.A. Peck (ed.) Elsevier, Amsterdam, 1986, 255–321.

65. H. Juppner, E.M. Brown and H.M. Kronenberg, Parathyroid hormone, In *Primer on the Metabolic Bone Diseases and Disorders of Mineral Metabolism*, 4th edn, M.J. Favus (ed.), Lipincott Williams & Wilkins, Philadelphia, 1999, pp. 80–87.

66. V. Matkovic, Calcium requirements for growth: are current recommendations adequate? *Nutr. Rev.*, 1993, **51**, 171–180.

67. A.F.M. Kardinaal, S. Ando, P. Charles, J. Charzewska, M. Rotily, K. Vaananen, A.M.J. van Erp-Baart, J. Heikkinen, J. Thomsen, M. Maggiolini, A. Deloraine, E. Chabros, R. Juvin and G. Schaafsma, Dietary calcium, and bone density in adolescent girls and young women in Europe, *J. Bone Miner. Res.*, 1999, **14**, 583–592.

68. J. Iwamoto, T. Takeda and S. Ichimura, Effect of exercise training and detraining on bone mineral density in postmenopausal women with osteoporosis, *J. Orthopaedic Sci.*, 2001, **6**, 128–132.

69. H.M. Frost, The role of changes in mechanical usage set points in the pathogenesis of osteoporosis, *J. Bone Miner. Res.*, 1992, **7**, 253–261.

70. H.M. Frost and E. Schonau, The "muscle–bone unit" in children and adolescents: a 2000 overview, *J. Pediatr.*, 2000, **13**, 571–590.

71. D.A. Bailey, A.D. Martin, H.A. McKay, S. Whiting and R. Mirwald, Calcium accretion in girls and boys during puberty: a longitudinal analysis, *J. Bone Miner. Res.*, 2000, **15**, 2245–2250.

72. J.J. Anderson, Calcium requirements during adolescence to maximize bone health, *J. Am. Coll. Nutr.*, 2001, **20**, 186S–191S.

73. A. Prentice, M.A. Laskey, J. Shaw, G. Hudson, K. Day, L.M.A. Jaru, B. Dibba and A.A. Paul, Calcium intakes of rural Gambian women, *Br. J. Nutr.*, 1993, **69**, 885–896.

74. S.M. Nickols-Richardson, P.J. O'Conner and S.A.L.R.D. Shapses, Longitudinal bone mineral density changes in female child artistic gymnasts, *J. Bone Miner. Res.*, 1999, **14**, 994–1002.

75. J.K. Yeh and J.F. Aloia, Effect of physical activity on calciotropic hormones and calcium balance in rats, *Am. J. Physiol.*, 1990, **258**, E263–E268.

76. K.H. Fleming and J.T. Heimbach, Consumption of calcium in the US: food sources and intake levels, *J. Nutr.*, 1994, **124**, 1426s–1430s.

77. J.J. Anderson, Exercise, dietary calcium, and bone gain in girls and young adult women, *J. Bone Miner. Res.*, 2000, **15**, 1437–1439.

78. C.M. Weaver, Calcium requirements of physically active people, *Am. J. Clin. Nutr.*, 2000, **72**, 579S–584S.

79. Standing Committee on the Scientific Evaluation of Dietary Reference Intakes. *Dietary Reference Intakes: Calcium, Magnesium, Phosphorus, Vitamin D, and Fluoride.* 1997. National Academy Press, Washington, D.C.

80. P. Crocker, D.A. Bailey, R.A. Faulkner and K.M.R. Kowalski, Measuring general levels of physical activity: Preliminary evidence for the Physical Activity Questionnaire for Older Children, *Med. Sci. Sports Exerc.*, 1997, **29**, 1344–1349.

81. R.P. Heaney, The bone-remodeling transient: implications for the interpretation of clinical studies of bone mass change, *J. Bone Miner. Res.*, 1994, **9**, 1515–1523.

CHAPTER 26

# The Treatment of Osteoporosis and Interaction of Medications with Nutrition

JERI W. NIEVES and FELICIA COSMAN

Columbia University and Helen Hayes Hospital, Route 9w, West Haverstraw, NY 10993, USA, Email: nievesj@helenhayshosp.org

**Key Points**

- Bone density at various skeletal sites can predict the risk of future fracture; Postmenopausal women at risk of osteoporosis and all women over age 65 should have a bone density test.
- It is important to encourage patients to reduce all possible risk factors, optimize nutrition, particularly calcium and vitamin D, ensure adequate protein intake and to encourage an exercise program.
- Pharmacologic therapies should be initiated in patients with osteoporosis and those at highest risk.
- There are numerous medications available to prevent the risk of future fracture including hormone therapy, tibolone, selective oestrogen receptor modulators, calcitonin, teriparatide and bisphosphonates. Each medication will have associated benefits and risks and the decision of which medication to use should be based on data presented here as well as the patient's age, and according to their personal and family medical history.
- All studies of the efficacy of treatments for osteoporosis have been studied in patients who were given calcium and vitamin D. Therefore, regardless of the pharmacologic therapy used, all women should be counseled to get sufficient calcium and vitamin D through their diet and supplementation if needed.

# Introduction

Osteoporosis compromises bone strength and predisposes a person to an increased risk of fracture. Bone strength primarily reflects the integration of bone quality and bone density. The goal of osteoporosis treatment is to prevent fractures. In order to determine the appropriate treatment strategy for an individual patient, it is important to first evaluate the patient through bone mass, bone turnover and clinical assessment.

# Patient Evaluation

## Clinical Evaluation of the Patient

All women who reach menopause should be interviewed and counseled about the risk factors for osteoporosis including their fracture history and family history of adulthood fracture. For patients with fractures it is important to determine whether the fractures are a result of either trauma or osteoporosis (determined by a bone density test), and not secondary to underlying malignancy. A careful history and physical examination is essential to evaluate the possible underlying causes of osteoporosis. In those who have had fractures, or in those who have bone mineral density (BMD) *t*-scores in the range of −2.0 or below, a laboratory evaluation should be considered to exclude diseases such as hyperthyroidism, hyperparathyroidism, celiac disease, vitamin D deficiency, hypercalciuria, Cushing's Syndrome, multiple myeloma and mastocytosis. The presence of height loss exceeding 1.5 inches, significant kyphosis or back pain (particularly with an onset after menopause) is an indication for radiography to rule out asymptomatic vertebral fractures. Newer procedures to measure the lateral spine by Dual X-ray Absorptiometry (DXA) may prove useful in the determination of vertebral deformity without radiography.

## Assessment of Bone Mass

The technique of bone mass measurement currently considered the gold standard is DXA. This X-ray based technique is highly accurate and precise. Attenuation of X-ray through the skeleton allows estimation of bone mineral for each region of interest within the spine, hip, forearm or total body. Results are reported as bone mineral content or bone mineral density (content in grams divided by area in cm$^2$). Since absolute bone density values vary depending on the manufacturer, it has become standard practice to compare the bone density results with values obtained in a reference population of the same race, age and sex. This comparison, results in a score commonly called the *Z*-score (Figure 26.1). In addition, the individual's results are compared with the normal values obtained in a young population (peak bone mass) of the same sex, the *T*-score (Figure 26.1). The World Health Organization defined osteoporosis as a *T*-score that is more than 2.5 standard deviations below the mean for bone mass of a young normal population (*i.e.*, compared with the

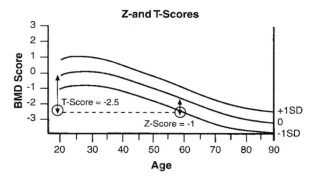

**Figure 26.1** *Illustration of a 59 year old patient with a Z-score of −1 and a t-score of −2.5*

normal range for young adults). When low Z-scores (below −2.0) are seen, consideration must be given to investigation of the patient for secondary causes of osteoporosis.

Other methods of bone mass measurement include central computed tomography (CT) that can be used to measure the spine or peripheral CT units that measure bone in the forearm or tibia. Ultrasound can also be used to measure bone mass. All of these techniques have been approved by the FDA based upon their capacity to predict the risk of fracture, and measurement of BMD is simply the estimation of a risk factor for fracture. It is clear that bone density at various skeletal sites can predict the risk of future fracture[1] and that many fractures are attributable to osteoporosis.[2]

Bone mass should be measured when the patient can be considered to be at risk of osteoporosis, or may be a candidate for treatment of osteoporosis. Clinical guidelines developed by the National Osteoporosis Foundation, based upon a cost effectiveness analysis of the available data, suggest the use of bone mass measurement in women after menopause, if they have risk factors for osteoporosis in addition to age, sex and oestrogen deficiency. The guidelines further suggest that measurement of bone mass should be performed in all women by age 65 years.[3] Major risk factors include personal history of fracture, family history of osteoporosis, low body weight, excessive alcohol consumption and cigarette smoking. A history of poor diet and little or no exercise are also risk factors for osteoporosis. Medication and disease related risk factors (Table 26.1) include all individuals who are on long-term glucocorticoids, or who are being started on glucocorticoid therapy that might be continued for more than three months.[4] There are currently no specific recommendations for the use of such tests in healthy men, but one suggestion might be to consider routine testing in men by age 70–75 years, or earlier if specific risk factors are present, such as history of adulthood fractures. A review of clinical evidence and endorsement by numerous expert organizations suggests that treatment be initiated in patients who have a *T*-score on bone density testing of −2.5 or lower. Those with additional risk factors for osteoporosis should be offered treatment when the *T*-score is higher (−2.0).

**Table 26.1-a.** *Diseases associated with an increased risk of generalized osteoporosis in adults*

| | |
|---|---|
| Acromegaly | |
| AIDS/HIV | Lymphoma and leukemia |
| Amyloidosis | Malabsorption syndromes |
| Ankylosing spondylitis | Mastocytosis |
| Celiac Disease | Multiple myeloma |
| Chronic obstructive pulmonary disease | Multiple sclerosis |
| Congenital porphyria | Organ transplantation |
| Cushing's syndrome | Osteogenesis imperfecta |
| Endometriosis | Pernicious anemia |
| Gastrectomy | Rheumatoid arthritis |
| Gaucher's Disease | Severe liver disease |
| Hemochromatosis | Spinal cord transection |
| Hyperparathyroidism | Stroke (CVA) |
| Hypogonadism | |
| Inflammatory Bowel Disease | Thalassemia |
| Immobilization | Thyrotoxicosis |
| Insulin-dependent diabetes mellitus | |

**Table 26.1-b.** *Drugs ssociated with an increased risk of generalized osteoporosis in adults*

| | |
|---|---|
| Anticonvulsants | Glucocorticoids and Adrenocorticotropin |
| Cigarette smoking | Gonadotropin-releasing hormone analogues |
| Coumadin | Long term use of heparin |
| Cytotoxic drugs | |
| Imunosuppressants | Suppressive thyroxine doses premenopasual tamoxifen use total parenteral nutrition |

# Biochemical Markers of Bone Turnover

Several biochemical tests are now available that provide an index of the overall rate of bone remodeling (at a single point in time). Biochemical markers are usually characterized as those related primarily to bone formation or bone resorption. Their clinical use has been limited by high biologic and analytic variability within and between individuals. Markers of bone resorption may help predict the risk of fracture, particularly in older individuals, adding to the predictive value of bone densitometry results.[5] In women 65 years of age or older when bone density results are above the treatment cut points noted above, a high level of bone resorption may indicate a need for treatment. Another use of biochemical markers at present is in monitoring response to treatment. If they are to be used, it is recommended that a level be obtained prior to starting therapy and then 3–4 months after initiating treatment.

# Treatment

## Fracture Treatment

Treatment of the patient with osteoporosis involves management of acute fractures, as well as treatment of the underlying disease. Hip fractures almost always require surgical repair if the patient is to become ambulatory again with procedures potentially including open reduction and internal fixation with pins and plates, hemiarthroplasties and total arthroplasties depending on the location and severity of the fracture, condition of the neighboring joint and general status of the patient. Long bone fractures often require either external or internal fixation. Some fractures such as vertebral, rib and pelvic fractures require only supportive care, with no specific orthopedic treatment.

Vertebral compression fractures only present with a sudden onset of back pain about 25–30% of the time. For those fractures that are acutely symptomatic, treatment with analgesics, including non-steroidal anti-inflammatory agents and/or acetaminophen, sometimes with the addition of a narcotic agent, is required. Short periods of bed rest may be helpful for pain management, although early mobilization is advantageous to help prevent any further bone loss. In some cases the use of a soft elastic style brace may help the patient become mobile sooner. Muscle spasms, which often occur with acute compression fractures, can be treated with muscle relaxants and moist or dry heat treatments. Constipation, which may be a secondary symptom of ileus related to acute compression fracture, should be treated with a mild cathartic agent to help prevent exacerbation of back pain. A recent technique for patients who have severe persistent pain from a vertebral compression fracture involves percutaneous injection of artificial cement (polymethymethacrylate) into the vertebral body (vertebroplasty) or through a balloon into the vertebral body (Kyphoplasty). These techniques have been reported to offer significant immediate pain relief in majority of patients but as of now all the data are observational, with no randomized controlled studies.[6] Some studies indicate that calcitonin, either injectable or nasal, might help abate the pain from vertebral compression fractures.[7–9]

## Treatment of the Underlying Disease

The management of the underlying disease, including risk factor reduction, is the next important step in caring for the osteoporotic patient. Patients should be thoroughly evaluated to reduce the likelihood of concomitant risks associated with bone loss and falls. Medications that might cause bone loss should be reviewed to make sure that the drug is truly indicated and is being given in minimally effective doses. For those on thyroid hormone replacement, TSH testing should be performed to ensure that an excessive dose is not being used, as this treatment can be associated with increased bone loss. In patients who are smokers, efforts should be made towards smoking cessation, and alcohol abuse should be evaluated and treated accordingly. Modifying the risk of falls also includes alcohol abuse treatment as well as a review of the medical regimen for any drugs that might be associated with orthostatic hypotension and/or sedation, including hypnotics and anxiolytics. Furthermore, patients should be instructed about environmental safety with regard to eliminating exposed wires, curtain strings, slippery rugs and mobile tables, and about providing good light in paths to bathrooms as well as just outside the home. Avoiding stocking feet on wood floors and checking carpet condition particularly on stairs are good preventive maintenances. Treatment for any underlying visual problem is recommended, particularly problems with depth perception, which are specifically associated with increased risk of falls.

High physical activity throughout the lifespan is likely to have a significant effect on bone mass and exercise is recommended for the prevention and treatment of osteoporosis. Postmenopausal women who initiate weight-bearing exercise (done while standing), can help prevent bone loss, but are unlikely to have substantial bone gain. A meta-analysis of the exercise programs indicates that exercisers can gain about 1% BMD *vs.* non-exercisers over 1 year.[10] If this difference continues each year, exercise could ultimately produce substantial effects on BMD. Specific muscle strengthening exercises can help prevent bone loss, and exercise is also recommended for effects on neuromuscular function, which can improve coordination, balance and strength, and thereby reduce the risk of falling. It is important that exercise is consistent over the long-term, optimally at least three times a week, but any exercise is better than none. Walking itself is associated with reduced mortality and may be enough particularly for people with other co-existing medical conditions.

A huge body of literature indicates that optimal calcium intakes can help reduce bone loss and suppress bone turnover (Chapter 8). All patients with postmenopausal osteoporosis should be consuming at least 1200 mg of calcium per day, including that from food sources, with dietary supplements as needed. The optimal calcium intake should be obtained in the context of a good diet, high in fruits and vegetables, low in saturated fat and salt, and moderate in protein. If a supplement is required, it should be taken in doses less than or equal to 600 mg at a time, as the calcium absorption fraction decreases with contents greater than that. Another important nutrient, Vitamin D, should be taken at doses of about 400 IU day$^{-1}$ in individuals below the age of 65 and at doses of 600–800 IU day$^{-1}$ in those 65 years of age and over (Chapter 9).

# Pharmocologic Therapies

## Oestrogens

There are numerous clinical trials indicating that oestrogens of various types (conjugated equine oestrogens, estradiol, estrone, esterified oestrogens, ethinyl-estradiol and mestranol) reduce bone turnover, prevent bone loss, and actually increase bone mass of the spine, hip and total body. This is true in women after either natural or surgical menopause, and is also true in late post-menopausal women and in women with established osteoporosis. Combined oestrogen/progestin preparations are now available in many countries. One large clinical trial, the Postmenopausal Oestrogen/Progestin Intervention trial (PEPI), indicated that C-21 progestins (micronized progesterone or MPA) did not augment the effect on bone mass of oestrogen alone.[11] In several small clinical trials, women receiving oestrogen had a 50% reduced vertebral fracture occurrence compared to those receiving placebo.[12,13] The effects of hormone therapy (HT) on non-vertebral fracture occurrence have recently been studied in the Women's Health Initiative (WHI), a randomized placebo controlled study of more than 16 000 menopausal women, which was prematurely halted in 2002 when preliminary results indicated that the risks associated with taking HT outweighed the benefits.[14] The specific HT that was compared to placebo was the drug Prempro, a combination of conjugated equine oestrogen and medoxyprogesterone acetate administered daily. For the first time in a randomized clinical trial (WHI) it was shown that HT as compared to placebo significantly reduced the risk of hip fractures by 34%, clinical vertebral fractures by 34% and other osteoporosis-related fractures by 24%. However, there are alternatives to HT for preventing bone loss and osteoporosis-related fractures. In many patients the risks of HT (discussed below) will outweigh the benefits and alternative treatments for osteoporosis should be considered.

In addition to the effects on the skeleton, oestrogens have complex effects throughout the body. They clearly decrease menopausal symptoms such as hot flashes and vaginal dryness. Unopposed oestrogen increases the risk of uterine cancer, so oestrogen must always be given with a progestin in women with intact uteri. Combined HT in the WHI did not increase the risk of endometrial cancer. Oestrogen use is associated with minor side effects including breast tenderness and engorgement and vaginal bleeding and/or spotting. However, with a daily continuous regimen of oestrogen/progestin, bleeding in most patients usually abates within six months.

Long-term risks of oestrogen therapy were recently reported in the WHI. One of the most important conclusions from the WHI is that the use of HT is not protective against heart disease as once believed; rather the use of HT actually increases the risk of heart attacks. The other major risks detected were an increased chance of developing invasive breast cancer, as well as an increased risk of strokes and venous blood clots (Figure 26.2). While the data also revealed that HT lowered the risk of osteoporosis and colon cancer, these benefits were not sufficient to outweigh the risks. It is important to put the WHI results in perspective by comparing absolute and relative risks. For example, if 10,000 women who took Prempro for one year were compared to 10,000 women taking a placebo, eight more women on Prempro would develop invasive breast cancer, seven more would have a heart attack or another coronary-related event, eight more would have a stroke and eighteen more would have venous thromboembolism. In addition, there would be six fewer cases of colon cancer and five fewer hip fractures overall in the Prempro group compared to the placebo group. When thinking about these numbers, however, remember that all should be multiplied by the number of years of use. For example, over five years there would be 40 excess cases of breast cancer in 10 000 women taking

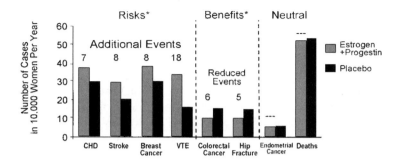

**Figure 26.2** *Effect of HRT on Event Rates in the Women's Health Initiative HRT Study[81]*

HT and 25 fewer hip fracture cases. Therefore, in most women HT/ET should not be used for prolonged periods of time.

The WHI study did not study other forms of HT and in addition, the portion of the WHI that is evaluating oestrogen therapy alone is ongoing. Therefore, it is unknown whether other forms of HT or ET will have the same benefits and risks as Prempro. However, the United States Food and Drug Administration has recommended general changes in the labeling of all HT and ET products since we cannot assume that other forms of HT will be safer. In addition, the WHI study did not include women under the age of 50 with natural or surgically induced premature menopause. HT and ET are highly effective in treating certain menopausal symptoms such as hot flashes, sleep disturbances due to nighttime flushes/sweats, and vaginal dryness. HT may still be appropriate for these patients depending on their individual benefit and risk profile. The healthcare provider must discuss with the patient their individual benefits to starting or continuing HT/ET with their individual risks for cardiovascular disease and breast cancer.

A review of the earlier clinical trials of oestrogen efficacy found that bone density response was greater in those women who had an adequate intake of calcium as compared to those women who took oestrogen with an insufficient amount of calcium (Figure 26.3).[15]

## Tibolone

Tibolone is a synthetic steroid that has oestrogenic, androgenic and progestagenic properties. Tibolone has been used in Europe for almost 20 years for the prevention of postmenopausal osteoporosis and the treatment of

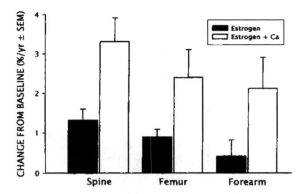

**Figure 26.3**   *Change in bone mineral density at three skeletal sites with oestrogen replacement in women with and without supplemental calcium*[15]

climacteric symptoms.[16] Tibolone acts on bone through stimulation of the oestrogen receptor[17–19] Tibolone has been compared to placebo in six randomized controlled studies. In early postmenopausal women there was a 3% higher bone mass by standard DXA and 15% higher trabecular bone mass by spinal computed tomography in women taking tibolone compared to women taking placebo after two years[18,20] Net increases in bone mass at the lumbar spine of 5% by DXA and 12% by DPA were found in later postmenopausal women with osteoporosis who took tibolone *vs.* placebo for two years.[21,22] Tibolone decreases bone turnover and significantly improves bone mass, especially trabecular bone mass. However, there are no long-term trials assessing fracture reduction with tibolone *vs.* placebo.

## Selective Oestrogens Receptor Modulators (SERMS)

Two SERMS are currently being used in postmenopausal women, raloxifene which is approved for prevention and treatment of osteoporosis, and tamoxifen, which is approved for the prevention and treatment of breast cancer. Tamoxifen has been shown to reduce bone turnover and bone loss in postmenopausal women when compared to placebo groups in patients with and without breast cancer. The Breast Cancer Prevention Trial showed a non-significant trend towards reduction in risk of non-spine fractures in healthy postmenopausal women over four years of Tamoxifen use.[23] The major benefit of tamoxifen is a 45% reduction in the incidence of new, invasive and non-invasive breast cancer in women who had been assessed to be at an elevated risk.

Raloxifene has oestrogenic effects on bone turnover and bone mass with potency similar to that of tamoxifen. In a large treatment study, over 7700 women with osteoporosis were randomly assigned to receive one of the two different doses of raloxifene *vs.* placebo in addition to calcium and Vitamin D.[24,25] Raloxifene reduced the occurrence of vertebral fracture by 30–50%, depending on the sub population (Figure 26.4). In this trial there was no

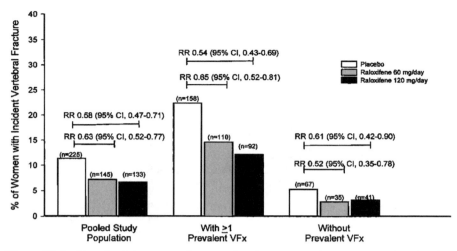

**Figure 26.4**   *Vertebral fracture reduction after four years of treatment with raloxifene (60 or 120 mg per day)*

suggestion of reduced risk of any other osteoporotic fracture other than that of the ankle. However, the study was neither designed, nor did it have adequate power, to evaluate risk of other fractures. A large randomized controlled trial called RUTH is currently underway which will evaluate non-spine fractures as one of its many outcomes. All randomized controlled trials of raloxifene have been in women with adequate calcium and vitamin D intakes; therefore these medications should be used in the context of adequate nutrition, particularly calcium and vitamin D.

Raloxifene, like tamoxifen and oestrogen, has effects throughout other organ systems.[25a] The most positive of these effects appears to be a reduction of about 65% in invasive breast cancer occurrence over 4 years in women taking raloxifene compared to placebo.[26] The persistence of this effect is still under study. The reduction is almost solely in oestrogen receptor positive tumors. In addition, raloxifene in multiple studies is not associated with an increase in the risk of uterine cancer or any benign uterine disease. Raloxifene, like tamoxifen, may increase the occurrence of hot flashes.[27] A cognitive function study in 143 women with osteoporosis revealed no differences in memory or mood between raloxifene and placebo groups[28] although long-term effects on the brain are not known.[29] Raloxifene reduces serum total and LDL cholesterol, lipoprotein (a) and fibrinogen. In one recent publication, where cardiovascular disease outcomes were evaluated in a secondary analysis of participants in the MORE trial, raloxifene did not appear to protect against cardiovascular events in the whole study population (Relative Risk = 0.86; 0.64–1.15). However, when analysis was restricted to women who had increased cardiovascular risk at baseline ($n = 1035$), those assigned to raloxifene had a significantly lower risk of cardiovascular events as compared with placebo (Relative Risk = 0.60; 0.38–0.95;).[30] This is being evaluated further in the RUTH study.

## Bisphosphonates

The bisphosphonate class of drugs is approved for treatment and prevention of osteoporosis as well as Paget's disease and hypercalcemia of malignancy. In addition, there is some evidence that some of these drugs may have an impact on progression of certain metastatic breast and prostate cancers as well. Two of these agents, alendronate and risedronate, are currently approved for osteoporosis prevention and treatment.

Alendronate decreased turnover and increased bone mass up to 8% *vs.* placebo in the spine and 6% *vs.* placebo in the hip over 3 years in patients with osteoporosis.[31] In addition, reduction in the occurrence of vertebral fracture, and reduced height loss were found in the alendronate treated women. The Fracture Intervention Trial (FIT) has subsequently provided evidences in over 2000 women with prevalent vertebral fractures (FIT-1;[32]) that daily alendronate treatment (5 mg per day for 2 years and 10 mg per day for 9 months afterwards), reduces the risk of vertebral fractures by about 50%, multiple vertebral fractures by up to 90%, and hip fractures by up to 50%. A similar study in more than 4000 women with low bone mass but no previous fracture (FIT-2)[33] suggested a 50% decrease in vertebral and hip fractures in those women with osteoporosis that were treated with alendronate. A later study in almost 2000 women with low bone density showed a 47% reduction in the incidence of all non-vertebral fractures.[34] A summary of the efficacy of alendronate in reducing vertebral fractures is given in Figure 26.5 and efficacy for non-vertebral fracture in Figure 26.6.

Alendronate must be given on an empty stomach, since it is poorly absorbed. It is contraindicated in patients who have stricture or inadequate emptying of the esophagus due to potential for esophageal irritation. Patients are recommended to remain upright after taking the medication for at least 30 min to avoid this possible esophageal irritation. Although cases of esophagitis, esophageal ulcer and esophageal stricture have been described, the true incidence of gastrointestinal toxicity from this agent is probably quite low. In clinical trials there was no increased risk from alendronate in overall gastrointestinal symptomatology.[35] Postmarketing experience, however, suggests that gastrointestinal effects are common and may be related to the underlying high risk of these symptoms in the elderly. Recently approved, flexible dosing for alendronate (70 mg once a week) is therapeutically equivalent and may avoid some of the apparent gastrointestinal toxicity.[36]

Risedronate is another potent bisphosphonate which also produces dramatic positive effects on bone mass, bone turnover and reduces the risk of vertebral fractures by 40–50% over 3 years.[37,38] These studies also report a reduction of about 33–40% in all non-spine fractures in risedronate treated patients. Preliminary reports from studies designed specifically to look at hip fracture suggest that risedronate is associated with a reduction in hip fracture of similar magnitude to that of alendronate in patients with proven osteoporosis.[39] A summary of the efficacy of risedronate in the prevention of vertebral fractures is shown in Figure 26.7 and for non-vertebral fractures is shown in Figure 26.8.

### Relative Risk with 95% CI for Vertebral Fractures for Doses of 5mg or Greater of Alendronate

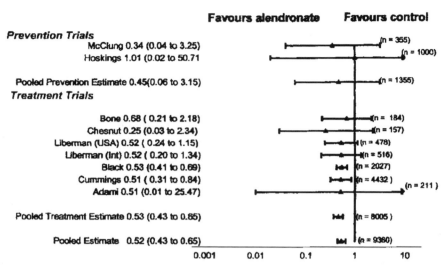

**Figure 26.5** *Relative risk with 95% CI for vertebral fractures for doses of 5 mg or greater of alendronate[79]*

### Risk Ratios and Summary Estimates with 95% CI for Non-Vertebral Fractures for Dose of 10mg or Greater of Alendronate

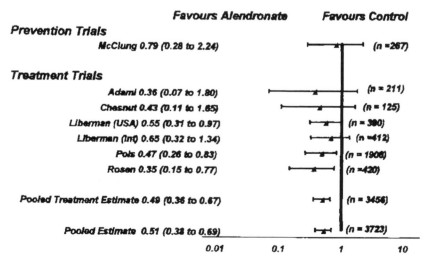

**Figure 26.6** *Risk ratios and summary estimates with 95% CI for non-vertebral fracture for dose of 10 mg or greater or alendronate[79]*

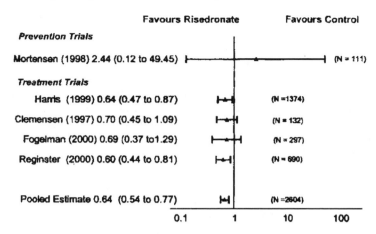

**Figure 26.7** *Relative Risk with 95% CI for vertebral fractures after treatment with risedronate*

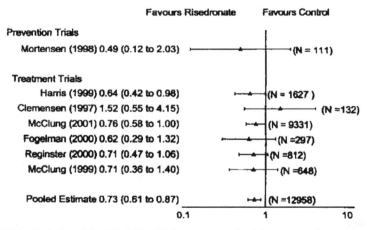

**Figure 26.8** *Relative risk with 95% CI for non-vertebral fractures after treatment with risedronate[80]*

Risedronate must be given in a similar fashion to alendronate to allow absorption. Bisphosphonates are retained over time in the skeleton, and may exert long-term effects. The optimal duration of therapy is unknown, although a 5-year course is probably safe and effective based on clinical trial data.[40]

Etidronate has not been approved in the US as a treatment for osteoporosis, but is used extensively to treat osteoporosis in Canada and Europe. Etidronate is given cyclically as 400 mg for 2 weeks every 3 months and must be taken on an empty stomach with only plain water consumed for 2 h before and two hours after the etidronate. The easiest time to take etidronate is before bed. There have been several randomized controlled trials to examine the efficacy of cyclical etidronate in women with osteoporosis in maintaining bone mass. These studies show that lumbar spine bone mass is 4–8% higher and femoral neck bone mass is 2–4% higher in women taking cyclical etidronate as compared to women taking placebo.[41–48] Some, but not all, studies also show a reduction in vertebral fracture of 50%.[43,44] There are no data on the efficacy of cyclical etidronate in preventing other fractures. Other bisphosphonates (for intravenous (IV) infusion pamidronate or zoledronic acid, or oral clodronate and tiludronate) that are approved in several countries for the treatment of Paget's disease of bone and/or hypercalcaemia of malignancy are available and can be used off-label for patients who cannot tolerate approved agents. For all of these agents the bone turnover and bone density data look good, however there are no fracture data available. A large clinical trial is currently underway to determine the fracture efficacy of an IV infusion of zoledronic acid.

All of the currently approved bisphosphonate treatments for osteoporosis have been tested with calcium[31–34,37–39] and in most studies with vitamin D;[32–34,37–39] therefore, any bone mass increments or fracture reductions were achieved in the context of adequate calcium and vitamin D.

## Calcitonin

Calcitonin is a polypeptide hormone normally produced by the thyroid gland. It can be either injected or administered intranasally. Its physiologic role is unclear, as no skeletal disease has been described in association with calcitonin deficiency or calcitonin excess. Calcitonins are FDA approved for Paget's disease, hypercalcemia, and osteoporosis in postmenopausal women who are more than five years from menopause. Injectable calcitonin has also been used with modest increments in spine BMD. In an observational study, The Mediterranean Osteoporosis Study there was only a modest decrease in the rate of hip fracture in those taking calcitonin.[49] There are no randomized placebo controlled trials evaluating fracture efficacy of this drug. Reported bone mass increments were found to be higher in individuals given calcitonin supplemented with calcium as compared to trials where calcitonin was given without additional calcium.[15]

A nasal spray containing calcitonin (200 IU day$^{-1}$) was approved in the United States in 1995 for treatment of osteoporosis in late postmenopausal women. Three trials have shown efficacy of nasal calcitonin against risk of vertebral fractures.[50–52] The largest study of nasal calcitonin (Proof) was a 5 year multicentre study of three different doses of calcitonin (100, 200 and 400 units) vs. placebo in patients with low bone mass and prevalent vertebral fractures at entry.[52] Of the original 1255 patients, whose mean age was 68 years, almost 70% dropped out of the study, thereby rendering conclusions difficult to interpret. All participants received 1000 mg of elemental calcium and 400 IU of Vitamin D daily. Overall, using intent to treat analysis, vertebral fracture reduction was 36% in the 200 unit dose of calcitonin only, with no statistically significant effect on non-spine fractures.

## Teriparatide (rDNA Origin) Injection

Endogenous parathyroid hormone (PTH) is an 84 amino acid peptide, which is largely responsible for calcium homeostasis. Observational studies in the past indicated that mild elevations in PTH could be associated with maintenance of cancellous bone mass, although elevated levels might be detrimental to cortical bone mass. On the basis of these observations, preclinical and early clinical studies were performed which indicated that exogenous PTH, (1–34 PTH) could produce dramatic increments in bone mass. The subsequent randomized controlled trials in postmenopausal women and men showed that PTH exerts dramatic effects on bone mass in the spine and positive effects throughout the skeleton.[53–58] Biopsy and bone turnover results indicate that the drug appears to work by stimulating new bone formation on trabecular and cortical bone surfaces by preferentially stimulating osteoblastic activity over osteoclastic activity.[59] A large multi-center clinical trial of Teriparatide (Forteo) in postmenopausal women found a significant increase in bone mass at the lumbar spine (9.7%) and total hip (2.6%) in teriapeptide treated women.[60] There was also a significant (65%) reduction in vertebral fractures (Figure 26.9) and 53% reduction in non-vertebral fractures in the teriapeptide treated *vs.* placebo group (Figure 26.10). PTH use may be limited by mode of administration, which currently must be by subcutaneous injection, although alternative modes of delivery are being investigated. The United States Food and Drug Administration approved the use of PTH or Teriparatide for up to 2 years of treatment of osteoporosis in postmenopausal women and men at high risk of fracture. Teriparatide should not be used in patients with Paget's disease or in patients with bone metastases or pre-existing hypercalcemia. The most common side effects associated with Teriparatide include dizziness and leg cramps. The fact that Teriparatide is in a class of drugs considered to be anabolic or bone forming would require that nutritional requirements, similar

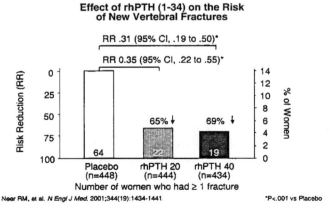

**Figure 26.9**   *Effect of rhPTH (1–34) on the risk of vertebral fractures*[60]

**Effect of rhPTH (1-34) on the Risk of
Nonvertebral Fragility Fractures**

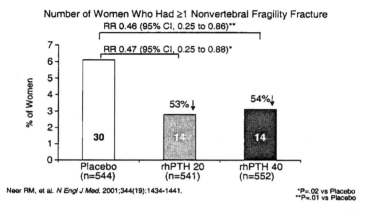

Figure 26.10    *Effect of rhPTH (1–34) on the risk of non-vertebral fractures*[60]

to those required during peak skeletal growth, be met to meet the skeletal demand of increasing bone mass.

## Experimental Agents

Fluoride has been available for many years and has been used in multiple osteoporosis studies with conflicting results.[61–67] Sodium fluoride consistently increases BMD but the bone quality and strength may be impaired.[67] Its use at this time must be considered experimental. Several small studies of growth hormone alone or in combination with other agents have not shown consistent or substantial positive effects on skeletal mass.[68–70] Anabolic steroids, mostly derivatives of testosterone, act primarily as anti-resorptive agents to reduce bone turnover, but may stimulate osteoblasts.[71,72] Effects on bone mass remain unclear, but, in general, are weak, and have been associated with frequent masculinizing side effects. Statin drugs currently used for hypercholesterolemia may also have effects on bone[73,74] but there are no clinical trials in humans yet published.

## Non-Pharmacological Approaches

Protective pads worn around the outer thigh, which cover the trochanteric region of the hip, were found to prevent hip fractures in elderly nursing home residents. In this study, no patients who fell while wearing a hip protector had a hip fracture. This was a dramatic improvement over those patients who fell without protective pads, where the fracture frequency was substantial.[75] The use of hip protectors is limited largely by compliance and comfort, but new devices are being developed which may resolve these problems, so that these agents might serve as adjunctive treatments to those pharmacologies mentioned above for prevention of hip fractures.

## Treatment Monitoring

There are currently no well-accepted guidelines for monitoring of treatment for osteoporosis. Since most of our osteoporosis treatments produce small or moderate bone mass increments on average, it is reasonable to use bone mineral density as a monitoring tool. As with any biological or assay determination, there is precision error with repeat measurements such that changes must exceed approximately 4% in the spine and 6% in the hip to be considered significant in any single individual. Since medication-induced increments may require several years to produce changes of this magnitude (if they do at all), it can be argued that BMD should not be repeated at intervals of less than 2 years. Moreover, since many individuals will not have BMD changes that equal or exceed the least significant change, only significant BMD reductions should prompt a change in medical regimen. Biochemical markers of bone turnover may be useful for treatment monitoring, particularly when either bisphosphonates or oestrogens are prescribed. A determination of turnover must be made prior to starting therapy and repeated four or more months after therapy is initiated. A change generally must exceed 30–40% from baseline to exceed the least significant change for any biochemical marker. The magnitude of the change required to achieve statistical significance is a product of the substantial intraindividual biologic and analytic variability. Biochemical markers are less useful for monitoring treatment with SERMS or calcitonin, since the average change in marker from treatment often does not exceed the least significant change.

# Interaction Between Nutrients and Treatments for Osteoporosis

Many of the available osteoporosis treatments are capable of improving bone mass by 1–10% at various skeletal sites. However, there must be an adequate nutritional supply in order to promote bone growth, similar to the needs that the skeleton has during growth in childhood. Obviously there is a need for calcium as a major substrate of bone. Vitamin D is also needed for bone health and to aid in calcium absorption during a period of increased need, for example medication related bone growth. There are also requirements for the nutrients needed to support osteoblast function. Evidence has been provided throughout this chapter to illustrate the relationship between each osteoporosis treatment and adequate calcium or vitamin D. A review of clinical trials of oestrogen efficacy found that bone density response was greater in those women who had an adequate intake of calcium as compared to those women who took oestrogen with an insufficient amount of calcium.[15] Tibolone, having a similar skeletal action to oestrogen, would also derive a greater skeletal benefit when given with an adequate intake of calcium. In a meta-analysis of published trials involving calcitonin, greater bone gain was found in individuals supplemented with calcium than in those in whom no additional calcium was provided.[15]

All recent large clinical trials of osteoporosis medications including calcitonin, raloxifene and the bisphosphonates have been performed in a setting of adequate calcium and vitamin D. All of the currently approved treatments for osteoporosis have been tested with calcium[31–34,37–39] and in most studies with vitamin D;[32–34,37–39] therefore, any bone mass increments or fracture reductions were achieved in the context of adequate calcium and vitamin D. Clinical trials are typically performed in a relatively healthy population where under nutrition is not a concern. However, the typical elderly osteoporotic patient may have inadequate nutrition with regard to calcium, vitamin D, phosphorus, magnesium and protein and the importance of adequate nutrition should be addressed with these patients. In fact, trials have shown that elderly hip fracture patients supplemented with protein had improved clinical outcomes as compared to those hip fracture patients who did not get protein supplementation.[82–84]

It was recently reported that in elderly patients with osteoporosis who do not respond to bisphosphonates, the addition of vitamin D (1000 IU per day) might improve BMD at the lumbar spine.[76] In another study it was shown that supplementation with calcium (1000 mg) and vitamin D (800 IU) led to more dramatic reductions in bone turnover in women taking alendronate compared to women taking alendronate alone.[77]

The fact that Teriparatide is in a class of drugs considered to be anabolic or bone forming would require that nutritional requirements, similar to those required during adolescent skeletal growth, be followed to meet the skeletal demands from these anabolic agents. This would include very large calcium intakes, as well as adequate vitamin D, protein and phosphorus.

In a recent review,[78] it was suggested, "optimal response to bone active agents demands good total nutrition," with a special emphasis on adequate calcium, vitamin D, protein, and phosphorus, as nutrients that are all required for normal bone metabolism. It is particularly important that more elderly, frail patients be counseled to get adequate nutrition, in addition to any osteoporosis medication being taken.

## Conclusions

In conclusion, there are a number of options for treatment of osteoporosis. The best approach is to reduce all possible risk factors, optimize nutrition, particularly calcium and vitamin D, ensure adequate protein intake and to encourage an exercise program, which has an overall health benefit as well as impact on the skeleton and muscle. Pharmacologic therapies should be initiated in patients with osteoporosis and those at highest risk. Different pharmacologic therapies might be appropriate for women at different ages and according to their personal and family medical history. There are numerous medications available to prevent the risk of future fracture including HT, tibolone, selective oestrogen receptor modulators, calcitonin, teriparatide and bisphosphonates. Each medication will have associated benefits and risks and the decision of which medication to use should be based on data presented here as well as the patient's age, and their personal and family medical history.

All studies of the efficacy of treatments for osteoporosis have been studied in patients who were given calcium and vitamin D. Recent smaller studies as well as reviews have shown an improved skeletal benefit for many of these medications if they are taken in the framework of adequate calcium and vitamin D.[15,76–78] Therefore, regardless of the pharmacologic therapy used, all women should be counseled to follow a healthy diet and get sufficient calcium and vitamin D through their diet and supplementation if needed.

# References

1. S.R. Cummings, D.M. Black, M.C. Nevitt, W. Browner, J. Cauley, K. Ensrud, H.K. Genant, L. Palermo, J. Scott and T.M. Vogt, Bone density at various sites for prediction of hip fractures, *Lancet*, 1993, **341**, 1273.
2. L.J. Melton III, M. Thamer, N.F. Ray, J.K. Chan, C.H. Chesnut III, T.A. Einhorn, C.C. Johnston, L.G. Raisz, S.L. Silverman and E.S. Siris, Gractures attributable to osteoporosis: the National Osteoporosis Foundation, *J. Bone Miner. Res.*, 1997, **12**, 16–23.
3. National Osteoporosis Foundation. Osteoporosis: Review of the evidence for prevention, diagnosis, and treatment and cost-effectiveness analysis, status report, *Osteoporosis Int.*, 1998, **4**(Suppl), S1–S88.
4. J.D. Adachi, W.P. Olsynski, D.A. Hanley, A.B. Hodsman, D.L. Kendler, K.G. Siminoski, J. Brown, E.A. Cowden, D. Goltzman, G. Ioannidis, R.G. Josse, L.G. Ste-Marie, A.M. Tenehouse, K.S. Davison, K.L. Blocka, A.P. Pollock and J. Sibley, Management of corticosteroid-induced osteoporosis, *Semin. Artihritis Rheum.*, 2000, **29**(4), 228–251.
5. A.C. Looker, D. Bauer, C.H. Chesnut III, C.M. Gundberg, M.C. Hochberg, G. Klee, M. Kleerekoper, N.B. Watts and N.H. Bell, Clinical use of biochemical markers of bone remodeling: current status and future directions, *Osteoporosis Int.*, 2000, **11**, 467–480.
6. M.E. Jensen, A.J. Evans, J.M. Mathis, D.F. Kallmes, H.J. Cloft and J.E. Dion, Percutaneous polymethylmethacrylate vertebroplasty in the treatment of osteoporotic vertebral body compression fractures: Technical aspects, *AJNR*, 1997, **18**, 1897–1904.
7. G.P. Lyritis, I. Paspati, T. Karachalios, D. Ioakimidis, G. Skarantavos and P.G. Lyritis, Pain relief from nasal salmon calcitonin in osteoporotic vertebral crush fractures. A double blind placebo controlled study, *Acta Ortho. Scand.* 1997, **68**(Suppl 275), 112–114.
8. A.E. Pontiroli, E. Pajetta, L. Scaglia, A. Rubinacci, G. Resmini, M. Arrigoni and G. Pozza, Analgesic effect of intranasal and intra-muscular salmon calcitonin in postmenopausal osteoporosis, *Aging Clin. Exp. Res.*, 1994, **6**, 459–463.
9. G.P. Lyritis, N. Tsakalakos, B. Magiasis, T. Karachalios, A. Yiatzides and M. Tsekoura Analgesic effect of salmon calcitonin in osteoporotic vertebral fractures. A double blind placebo controlled clinical study, *Calcif. Tissue Int.*, 1991, **49**, 369–372.
10. B.A. Wallace and R.G. Cumming, Systematic review of randomized trials of the effect of exercise on bone mass in pre- and postmenopausal women, *Calcif. Tissue Int.*, 2000, **67**, 10–18.
11. Writing group for the PEPI Trial. Effects of hormone therapy on bone mineral density: results from the Postmenopausal Estrogen/Pregestin Interventions (PEPI) Trial, *JAMA*, 1996, **275**, 1389–1396.
12. R. Lindsay, D.M. Hart, C. Forrest and C. Baird, Prevention of spinal osteoporosis in ooperectomized women, *Lancet*, 1980, **2**, 1151–1153.

13. E.G. Lufkin, H.W. Wahner, W.M. O'Fallon, S.F. Hodgson, M.A. Kotowicz, A.W. Lane, H.L. Judd, R.H. Caplan and B.L. Riggs, Treatment of postmenopausal osteoporosis with transderman estrogen, *Ann. Intern. Med.*, 1992, **117**, 1–9.

14. J.E. Rossouw, G.L. Anderson, R.L. Prentice, A.Z. LaCroix, C. Kooperberg, M.L. Stefanick, R.D. Jackson, S.A. Beresford, B.V. Howard, K.C. Johnson, J.M. Kotchen and J. Ockene (Writing group for the Women's Health Initiative Investigators), Risks and benefits of estrogen plus progestin in healthy postmenopausal women. Principal results from the Women's Health Initiative randomized controlled trial, *JAMA*, 2002, **288**(3), 321–333.

15. J.W. Nieves, L. Komar, F. Cosman and R. Lindsay, Calcium Potentiates the effect of estrogen and calcitonin on bone: review and analysis, *Am. J. Clin. Nutr.*, 1998, **67**, 18–24.

16. R. Lindsay and A. Hart, Prospective double-blind trial of synthetic steroid (org OD 14) for preventing postmenopausal osteoporosis, *Br. Med. J.*, 1980, **280**, 1207–1209.

17. H.J. Kloosterboer, Tibolone: a steroid with a tissue-specific mode of action, *Steroid Biochem. Mol. Biol.*, 2001, **76**, 231–238; J. Netelenbos, L.P. Siregar-Emck, F.C. vanGinkel, P. Lips and O.R. Leeuwendamp, Short-term effects of Org OD 14 and 17β-oestradiol on bone and lipid metabolism in early postmenopausal women, *Maturitas*, 1991, **13**, 137–149.

18. B. Berning, J.W. Kuijk, H.J. Bennink, P.M. Kicovic and B.C. Fauser, Effects of two doses of tibolone on trabecular and cortical bone loss in early postmenopausal women: a two-year randomized, placebo-controlled study, *Bone*, 1996, **19**, 395–399.

19. N. Bjarnason, J. Bjarnason, C. Rosenquist and C. Christiansen, Tibolone: prevention of bone loss in late postmenopausal women, *J. Clin. Endocrinol. Metab.*, 1996, **81**, 2419–2422.

20. P. Gallagher, D.J. Baylink and M. McClung, Prevention of bone loss with tibolone in postmenopausal women: results of two randomized, double-blind, placebo-controlled, dose-finding studies, *J. Clin. Endocrinol. Metab.*, 2001, **86**, 4717–4726.

21. P. Guesens, J. Dequeker and L.P. Schot, Non-linear increase in vertebral density induced by a synthetic steroid (Org OD 14) in women with established osteoporosis, *Maturitas*, 1991, **13**, 155–162.

22. N. Bjarnason, C. Bjarnason and C. Christiansen, The response in spinal bone mass to tibolone treatment is related to bone turnover in elderly women, *Bone*, 1997, **20**, 151–155.

23. B. Fisher, J.P. Costantino, D.L. Wickerham, C.K. Redmond, M. Kavanah, W.M. Cronin, V. Vogel, A. Robidoux, N. Dimitrov, J. Atkins, M. Daly, S. Wieand, E. Tan-Chiu, L. Ford and N. Wolmark, Tamoxifen for prevention of breast cancer. Report of the National Surgical Adjuvant Breast and Bowel Project P-1 Study, *J. Natl Cancer. Inst.*, 1998, **90**, 1371–1388.

24. B. Ettinger, D.M. Black, B.H. Mitlak, R.K. Knickerbocker, T. Nickelsen, H.K. Genant, C. Christiansen, P.D. Delmas, J.R. Zanchetta, J. Stakkestad, C.C. Gluer, K. Krueger, F.J. Cohen, S. Eckert, K.E. Ensrud, L.V. Avioli, P. Lips and S.R. Cummings, Reduction of vertebral fracture risk in postmenopausal women with osteoporosis treated with raloxifene. Results from a 3-year randomized clinical trial, *JAMA*, 1999, **282**, 637–645.

25. P.D. Delmas, K.E. Ensrud, J.D. Adachi, K.D. Harper, S. Sarkar, C. Gennari, J.Y. Reginster, H.A.P. Pols, R.R. Recker, S.T. Harris, W. Wu, H.K. Genant, D.M. Black and R. Eastell, For the Multiple Outcomes of Raloxifen Evaluation (MORE) Investigators. Efficacy of raloxifene on vertebral fracture risk reduction in postmenopausal women with osteoporosis: Four-Year results from a randomized clinical trial, *J. Clin. Endocrinol. Metab.*, 2002, **87**(8), 3609–3617.

26. J.A. Cauley, L. Norton, M.E. Lippman, S. Eckert, K.A. Krueger, D.W. Purdie, J. Farrerons, A. Karasik, D. Mellstrom, K.W. Ng, J.J. Stepan, T.J. Powles,

M. Morrow, A. Costa, S.L. Silfen, E.L. Walls, H. Schmitt, D.B. Muchmore, V.C. Jordan and L.G. Ste-Marie, Continued breast cancer risk reduction in postmenopausal women treated with raloxifene: 4-year results from the MORE trial, *Breast Cancer Rest. Treat.*, 2001, **65**, 124–134.

27. S.R. Cummings, S. Eckert, K.A. Krueger, D. Grady, T.J. Powles, J.A. Cauley, L. Norton, T. Nickelsen, N.H. Bjarnason, M. Morrow, M.E. Lippman, D. Black, J.E. Glusman, A. Costa and V.C. Jordan, The effect of raloxifene on risk of breast caner in postmenopausal women. Results from the MORE randomized trial, *JAMA*, 1999, **281**, 2189–2197.

28. T. Nickelsen, E.G. Lifkin, B.L. Riggs, D.A. Cox and T.H. Croo, Raloxifene hydrochloride, a selective estrogen receptor modulator: safety assessment of effects on cognitive function and mood in postmenopausal women, *Psychoneuroendocrinology*, 1999, **24**, 115–128.

29. K. Yaffe, K. Krueger, S. Sarkar, D. Grady, E. Barrett-Connor, D.A. Cox and T. Nickelsen, Multiple Outcomes of Raloxifere Evaluation Investigators, *N. Engl. J. Med.*, 2001, **344**, 1207–1213.

30. E. Barrett-Connor, D. Grady, A. Sashegyi, P.W. Anderson, D.A. Cox, K. Hoszowski, P. Pautaharju and K.D. Harper, For the MORE Investigators. Raloxifene and cardiovascular events in osteoporotic postmenopausal women, *JAMA*, 2002, **287**(7), 847–857.

31. U.A. Liberman, S.R. Weiss, J. Broll, H.W. Minne, H. Quan, N.H. Bell, J. Rodriguez-Portales, R.W. Downs Jr., J. Dequeker and M. Favus, Effect of alendronate on BMD and the incidence of fractures in postmenopausal osteoporosis, *N. Engl. J. Med.*, 1995, **36**, 612–613.

32. D. Black, S. Cummings, D. Karpf, J. Cauley, E. Thompson, M. Nevitt, D. Bauer, H. Genant, W. Haskell, R. Marcus, S. Ott, J. Torner, S. Quandt, T. Reiss and K. Ensrud, For the Fracture Intervention Trial Research Group. Randomised trial of effect of alendronate on risk of fracture in women with existing vertebral fractures, *Lancet*, 1996, **348**(9041), 1535–1541.

33. S.R. Cummings, D.M. Black, D.E. Thompson, W.B. Applegate, E. Barrett-Connor, T.A. Musliner, L. Palermo, R. Prineas, S.M. Rubin, J.C. Scott, T. Vogt, R. Wallace, A.J. Yates and A.Z. LaCroix, Effect of alendronate on risk of fracture in women with low bone density but without vertebral fractures, *JAMA*, 1998, **280**(24), 2077–2082.

34. H.A.P. Pols, D. Felsenberg, D.A. Hanley *et al.*, Multinational, placebo-controlled randomized trial of the effects of alendronate on bone density and fracture risk in postmenopausal women with low bone mass. Results of the FOSIT Study, *Osteoporosis Int.*, 1999, **9**, 461–468.

35. D.C. Bauer, D. Black, K. Ensrud, D. Thompson, M. Hochberg, M. Nevitt, T. Musliner and D. Freedholm, Upper gastrointestinal tract safety profile of alendronate. The Fracture Intervention Trial, *Arch. Intern. Med.*, 2000, **160**, 517–525.

36. T. Schnitzer, H.G. Bone, G. Crepaldi, S. Adami, M. McClung, D. Kiel, D. Felsenberg, R.R. Recker, R.P. Tonino, C. Roux, A. Pinchera, A.J. Foldes, S.L. Greenspan, M.A. Levine, R. Emkey, A.C. Santora 2nd, A. Kaur, D.E. Thompson, J. Yates and J.J. Orloff, Therapeutic equivalence of alendronate 70 mg once-weekly and alendronate 10 mg daily in the treatment of osteoporosis. Alendronate Once-Weekly Study Group, *Aging (Milano)*, 2000, **12**, 1–12.

37. S.T. Harris, N.B. Watts, H.K. Genant, C.D. McKeever, T. Hangartner, M. Keller, C.H. Chesnut III, J. Brown, E.F. Eriksen, M.S. Hoseyni, D.W. Axelrod and P.D. Miller, Effect of risedronate treatment on vertebral and non vertebral fractures in women with postmenopausal osteoporosis. A randomized controlled trial, *JAMA*, 1999, **282**(14), 1344–1352.

38. J.Y. Reginster, H. Minne, O. Sorensen *et al.*, Randomized trial of the effects of risedronate on vertebral fractures in women with established postmenopausal osteoporosis, *Osteoporosis Int.*, 2000, **11**, 83–91.

39. M.R. McClung, P. Geusens, P.D. Miller, H. Zippel, W.G. Bensen, C. Roux, S. Adami, I. Fogelman, T. Diamond, R. Eastell, P.J. Meunier and J.Y. Reginster; Hip Intervention Program Study Group. Effect of risedronate on the risk of hip fracture in elderly women, *N. Engl. J. Med.*, 2001, **344**(5),333–340.

40. F. Cosman, S.R. Cummings and R. Lindsay, How long should patients with osteoporosis be treated with Bisphosphonates? *J. Womens Health*, 2000, **9**, 81–84.

41. M.L.M. Montessori, W.H. Scheele, J.C. Netelenbos, J.F. Kerkhoff and K. Baker, The use of etidronate and calcium *versus* calcium alone in the treatment of postmenopausal osteopenia: results of three years of treatment, *Osteoporosis Int.*, 1997, **7**, 52–58.

42. S.T. Harris, N.B. Watts, R.D. Jackson, H.K. Genant, R.D. Wasnich, P. Ross, P.D. Miller, A.A. Licata and C.H. Chesnut 3rd, Effect of intermittent cyclical etidronate therapy on bone mass, fracture rate in women with postmenopausal osteoporosis: three years of blinded therapy followed by one year of open therapy, *Am. J. Med.*, 1993, **95**, 557–567.

43. T. Storm, G. Thamsborg, T. Steiniche, H.K. Genant and O.H. Sorensen, Effect of intermittent cyclical etedronate therapy on bone mass, fracture rate in women with postmenopausal osteoporosis, *N. Engl. J. Med.*, 1990, **322**, 1265–1271.

44. N.B. Watts, S.T. Harris, H.K. Genant, R.D. Wasnich, P.D. Miller, R.D. Jackson, A.A. Licata, P. Ross, G.C. Woodson 3rd, M.J. Yanover *et al.*, Intermittent cyclical etidronte treatment of postmenopausal osteoporosis, *N. Engl. J. Med.*, 1990, **323**, 73–79.

45. P.J. Meunier, E. Confavreux, I. Tupinon C. Hardouin, P.D. Delmas and R. Beléna, Prevention of early postmenopausal bone loss with cyclical etidronte therapy (a double-blind, placebo-controlled study and 1-year follow-up), *J. Clin. Endocrinol. Metab.*, 1997, **82**, 2784–2791.

46. R.J.M. Herd, R. Dalen, G.M. Blake, P.J. Ryan and I. Fogelman, The prevention of early postmenopausal bone loss by cyclical etidronate therapy: a 2-year, double-blind, placebo-controlled study, *Am. J. Med.*, 1997, **103**, 92–99.

47. J.H. Tobias, N. Dalzell, M. Pacianas and T.J. Chambers, Cyclical etidronate prevents spinal bone loss in early postmenopausal woment, *Br. J. Rheumatol.*, 1997, **36**, 612–613.

48. J.M. Pouilles, F. Tremollieres, C. Roux, J.L. Sebert, C. Alexandre, D. Goldberg, R. Treves, P. Khalifa, P. Duntze, S. Horlait, P. Delmas and D. Kuntz, Effects of cyclical etidronate therapy on bone loss in early postmenopausal women who are not undergoing hormonal replacement therapy, *Osteoporosis Int.*, 1997, **7**, 213–218.

49. J.A. Kanis, O. Johnell, B. Gulberg, E. Allander, G. Dilsen, C. Gennari, A.A. Lopez, G. P. Lyritis, G. Mazzuoli and L. Miravet, Evidence for efficacy of dugs affecting bone metabolism in preventing hip fracture, *BMJ*, 1992, **305**, 1124–1128.

50. H. Rico, M. Revilla, E.R. Hernandez, L.F. Villa and M. Alvarez de Buergo, Total and regional bone mineral content and fracture rate in postmenopausal osteoporosis treated with salmon calcitonin. A prospective study, *Calcif. Tissue Int.*, 1995, **56**, 181–185.

51. K. Overgaard, M.A. Hansen, S.B. Jensen and C. Christiansen, Effect of calcitonin given intranasally on bone mass and fracture rates in established osteoporosis. A dose response study, *BMJ*, 1992, **305**, 556–561.

52. C.H. Chesnut III, S. Silverman, K. Andriano, H. Genant, A. Gimona, S. Harris, D. Kiel, M. LeBoff, M. Maricic, P. Miller, C. Mouiz, M. Peacock, P. Richardson, N. Watts and D. Baylink, for the PROOF Study Group, Clinical studies. A randomized trial of nasal spray salmon calcitonin in postmenopausal women with established osteoporosis: the prevent recurrence of osteoporotic fractures study, *Am. J. Med.*, 2000, **109**, 267–276.

53. R. Lindsay, J. Nieves, C. Formica, E. Henneman, L. Woelfert, V. Shen, D. Dempster and F. Cosman, Randomized controlled study of effect of parathyroid hormone on vertebral-bone mass and fracture incidence among postmenopausal women on estrogen with osteoporosis, *Lancet*, 1997, **350**, 550–555.

54. E.S. Kurland, F. Cosman, D.J. McMahon, C.J. Rosen, R. Lindsay and J.P. Bilezikian, Parathyroid hormone as a therapy for idiopathic osteoporosis in men: effects on bone mineral density and bone marker, *J. Clin. Endrocrinol. Metab.*, 2000, **85**, 3069–3076.

55. N.E. Lane, S. Sanchez, G.W. Modin, H.K. Genant, E. Pierini and C.D. Arnaud, Parathyroid hormone treatment can reserve corticosteroid-induce osteoporosis. Results of a randomized controlled clinical, *J. Clin. Invest.*, 1998, **102**, 1627–1633.

56. E.B. Roe, S.D. Sanchez, A. DelPuerto, E. Pierini, P. Bacchetti, C.E. Cann and C.D. Arnaud, Parathyroid hormone 1–34 (hPTH 1–34) and estrogen produce dramatic bone density increases in postmenopausal osteoporosis. Results from a placebo-controlled randomized trial, *J. Bone. Miner. Res.*, 1999, **14**(Suppl. 1), S137.

57. R. Lindsay, A.B. Hodsman, H. Genant, A. Hanley, M.P. Ettinger, A. Bolognese, J. Fox and A.J. Metcalfe, for the PTH Working Group, A randomized controlled multi-center study of 1–84 hPTH for treatment of postmenopausal osteoporosis, *ASBMR-IBMS Second Joint Meeting*, 1998, 1109.

58. F. Cosman, J. Nieves, L. Woelfert, C. Formica, S. Gordon, V. Shen and R Lindsay, Parathyroid hormone added to establish hormone therapy:effects on vertebral fracture and maintenance of bone mass after PTH withdrawal, *J. Bone Miner. Res.*, 2001, **16**, 925–931.

59. D.W. Dempster, F. Cosman, E.S. Kurland, H. Zhou, J. Nieves, L. Woelfert, E. Shane, K. Plavetic, R. Muller, J. Bilezikian and R. Lindsay, Effects of Daily treatment with parathyroid hormone on bone microarchitecture and turnover in patients with osteoporosis: a paired biopsy study, *J. Bone. Miner. Res.*, 2001, **16**(10), 1846–1853.

60. R.M. Neer, C.D. Arnaud, J.R. Zanchetta, R. Prince, J.A. Gaich, J.Y. Reginster, A.B. Hodsman, E.F. Eriksen, S. Ish-Shalom, H.K. Genant, O. Wang and B.H. Mitlak, Effect of parathyroid hormone (1–34) on fractures and bone mineral density in postmenopausal women with osteoporosis, *N. Engl. J. Med.*, 2001, **344** (19): 1434–1441.

61. T. Hansson and B. Roos, The effect of fluoride and calcium on spinal bone mineral content. A controlled, prospective (3 years) study, *Calcif. Tiss. Int.*, 1987, **40**, 315–317.

62. M. Kleerekoper, E.J. Peterson, D.A. Nelson, E. Phillips, M.A. Schork, B.C. Tilley and A.M. Parfitt, A randomized trial of sodium fluoride as a treatment for postmenopausal osteoporosis, *Osteoporosis Int.*, 1991, **1**, 155–161.

63. C.Y. Pak, K. Sakhaee, B. Adams-Huet, V. Piziak, R.D. Peterson and J.R. Poindexter, Treatment of postmenopausal osteoporosis with slow-release sodium fluoride: Final report of a randomized controlled trial, *Ann. Inter. Med.*, 1995, **123**, 401–408.

64. J.Y. Reginster, L. Meurmans, B. Zegels, L.C. Rovati, H.W. Minne, G. Giacovelli, A.N. Taquet, I. Setnikar, J. Collette and C. Gosset, The effect of sodium monoflourophosphate plus calcium on vertebral fracture rate in postmenopausal women with moderate osteoporosis. A randomized controlled clinical trial, *Ann. Inter. Med.*, 1998, **129**, 1–8.

65. B.L. Riggs, Treatment of osteoporosis with sodium fluoride: an appraisal, *J. Bone Miner. Res.*, 1983, **8**, 366–393.

66. B.L. Riggs, P.F. Hodgson, W.M. O'Fallon, E.Y. Chao, H.W. Wahner, J.M. Muhs, S.L. Cedel and L.J. Melton 3rd, Effect of fluoride treatment on the fracture rate in postmenopausal women with osteoporosis, *N. Engl. J. Med.*, 1990, **322**, 802–809.

67. D. Haguenauer, V. Welch, B. Shea, P. Tugwell, J.D. Adachi and G. Wells, Flouride for the treatment of postmenopausal osteoporotic fractures: a meta-analysis, *Osteoporosis Int.*, 2000, **11**, 727–738.

68. C. Ohlsson, B.-A. Begisson, O.G.P. Isaksson, T.T. Andreassen and M.C. Slootweg, Growth hormone and bone, *Endo. Rev.*, 1998, **19**, 55–79.

69. R. Marcus, Recombinant human growth hormone as potential therapy for osteoporosis, *Baillieres Clin. Endocrinol. Metab.*, 1998, **12**, 251–260.

70. M. Saaf, A. Hilding, M. Thoren, S. Troell and K. Hall, Growth hormone treatment of osteoporotic postmenopausal women. A one-year placebo-controlled study, *Eur. J. Endocrinol.*, 1999, **140**, 390–399.

71. C. Chesnut, J.L. Ivey and H.E. Gruber, Stanozolol in postmenopausal osteoporosis therapeutic efficacy and possible mechanism of action, *Metabolism*, 1983, **32**, 571–580.

72. C.C. Christiansen, Androgens and androgenic progestins, In *Osteoporosis*, R. Marcus, D. Felman, J. Kelsey (eds.), Academic Press, San Diego, 1996, pp. 1279–1292.

73. Y.-S. Chung, M.-D. Lee, S.-K. Lee, H.M. Kim and L.A. Fitzpatrick, HMG-GA reductase inhibitors increase BMD in type 2 diabetes mellitus patients, *J. Clin. Endocrinol. Metab.*, 2000, **85**, 1137–1142.

74. C.J. Edwards, D.J. Hart and T.D. Spector, Oral statins and increased bone mineral density in postmenopausal women, *Lancet*, 2000, **555**, 2218.

75. P. Kannus, J. Parkkari, S. Niemi, M. Pasanen, M. Palvanen, M. Jarvinen and I. Vuori, Prevention of Hip Fracture in Elderly People with the Use of a Hip Protector, *N. Engl. J. Med.*, 2000, **343**, 1506–1513.

76. G.A. Heckman, A. Papaioannou, R.J. Sebaldt, G. Ioannidis, A. Petrie, C. Goldsmith and J.D. Adachi, Effect of vitamin D on bone mineral density of elderly patients with osteoporosis responding poorly to bisphosphonates, *BMC Musculoskeletal Disorders*, 2002, **3**, 6.

77. M. Brazier, S. Kamel, F. Lorget, Maamer, C. Tavera, N. Heurtebize, F. Grados, M. Mathiew, M. Garabedian, J.L. Sebert and P. Fardelle, Biological effects of supplementation with vitamin D and calcium in postmenopausal women with low bone mass receiving alendronate, *Clin. Drug Invest.*, 2002, **22**(12), 849–857.

78. R.P. Heaney, Constructive interactions among nutrients and bone active pharmacologic aagents with principal emphasis on calcium, phosphorus, vitamin D and protein, *J. Am. Coll. Nutri.*, 2001, **20**(5), 403S–409S.

79. A. Cranney, G. Wells, A. Willan, L. Griffith, N. Zytaruk, V. Robinson, D. Black, J. Adachi, B. Shea, P. Tugwell and G. Guyatt, The osteoporosis methodology group and the osteoporosis research advisory group. Meta-analysis of alendronate for the treatment of postmenopausal women, *Endocrinol. Rev.*, 2002, **23**(4), 508–516.

80. A. Cranney, P. Tugwell, J. Adachi, B. Weaver, N. Zytaruk, A. Papaioannou, V. Robinson, B. Shea, G. Wells and G. Guyatt, The osteoporosis methodology group and the osteoporosis research advisory group. Meta-analysis of risedronate for the treatment of postmenopausal osteoporosis, *Endocrinol. Rev.*, 2002, **23**(4), 517–523.

81. Women's Health Initiative. WHI HRT Update. http://www.nhlbi.nih.gov/whihrtupd/upd2002.htm.2002.

82. R. Rizzoli, P. Ammann, T. Chevalley and J.P. Bonjour, Protein Intake and Bone Disorders in the Elderly, *Joint Bone Spine*, 2001, **68**, 383–392.

83. M.-A. Schurch, R. Rizzoli, D. Slosman, L. Vadas, P. Vergnaud and J.-P. Bonjour, Protein supplements increase serum insulin-like growth factor-I levels and attenuate proximal femur bone loss in patients with recent hip fracture-a randomized, double-blind, placebo-controlled trial, *Ann. Intern. Med.*, 1998, **128**, 801–809.

84. L. Tkatch, C.H. Rapin, R. Rizzoli, D. Slosman, V. Nydegger, H. Vasey and J.P. Bonjour, Benefits of oral protein supplementation in elderly patients with fracture of the proximal femur, *J. Am. Coll. Nutr.*, 1992, **11**, 519–525.

CHAPTER 27

# Weight Reduction and Bone Health

SUE A. SHAPSES[1] and MARIANA CIFUENTES[2]

[1]Department of Nutritional Science, Rutgers University,
96 Lipman Drive, New Brunswick, NJ 08901-8525, USA,
E-mail: shapses@aesop.rutgers.edu
[2]Institute of Nutrition and Food Technology (INTA), University of Chile,
Santiago, Chile

**Key Points**

- Body weight and bone. There is a well-established direct relationship between body weight and bone mass. A leaner body weight is associated with lower bone mass and increased risk of fracture, whereas maintaining a relatively higher body weight protects against osteoporosis.
- Obesity and weight reduction. In an obesity epidemic, where weight reduction is widely encouraged to prevent co-morbid conditions associated with obesity, it is important to understand the means to prevent a negative impact on bone health.
- Weight reduction, bone mass and fracture. Involuntary weight reduction is associated with increased fracture risk. Voluntary weight reduction modulates bone metabolism and is associated with bone loss, yet the response of bone will vary and, is more detrimental in older individuals and in those with a history of weight cycling. Nutrition and lifestyle factors such as physical activity, as well as endocrine changes, are discussed.
- Measurement errors of dual energy X-ray absorptiometry (DXA) and body weight. DXA may have a reduced sensitivity when evaluating obese subjects or those who experience weight change. There should be caution when interpreting such data and the use of other methods to measure bone should be explored.

- Endocrine and other bone regulators during weight reduction. Multiple hormones and factors influence bone during weight change. Oestrogen is influenced by body fat mass and is an important regulator of bone during weight loss. Other potential regulators include parathyroid hormone, leptin, insulin-like growth factor-I and glucocorticoids.
- Conclusions. Further research is warranted on the mechanisms that influence bone mass and potential interventions (*i.e.*, dietary supplements or medications) which can address endocrine alterations that occur during weight reduction. The development of optimal diets and lifestyle recommendations for the maintenance of bone health is necessary.

## Introduction

This review discusses the relationship between bone and body weight, and describes nutritional and hormonal factors that regulate bone which emanate from changes in body weight.

## Body Weight and Bone

The importance of body weight in determining both bone mass and fracture risk has been emphasized by the National Osteoporosis Foundation. According to this group, low body weight is one of four major risk factors for osteoporosis, and the relationship between body weight and bone mass is well-established[1,2] (Figure 27.1). More recent studies consistently indicate that

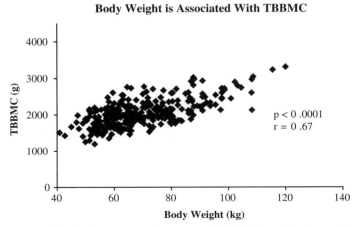

**Figure 27.1**   *Total body bone mineral content (TBBMC) and body weight in post-menopausal women[112]*

a low body weight in older individuals is a major risk factor for fracture,[3,4] and that maintenance of body weight can prevent bone loss.[5–8] Mechanisms that may be contributing to bone mineral density (BMD) in the obese include such factors as increased mechanical stimuli and an altered hormonal/metabolic environment. Although it is possible that measurement errors due to dual energy DXA scanning are affecting the sensitivity in obese individuals (see discussion below), there is sufficient evidence that bone mass is greater and regulated differently in obese than lean women.

# Obesity and Weight Reduction

Obesity is a major risk factor for many disease states, but is associated with a reduced risk of osteoporosis. The obese rarely suffer from low bone mass unless there are secondary causes (*i.e.*, medications including glucocorticoids and thyroid hormone). Nevertheless, compared with the severity of the co-morbid conditions associated with obesity (cardiovascular disease, hypertension, diabetes, cancer, and osteoarthritis), the benefits of a higher bone mass are small, and achieving of a normal body weight is encouraged. Overall, obese middle-aged individuals who lose weight and are no longer obese, have a lower mortality than persons who remain obese.[9]

Before 1998, co-morbid conditions were described only for obese individuals (BMI > 28 kg/m$^2$). In the new NIH guidelines established in 1998 the 'overweight' category was redefined (BMI 25–29.9 kg/m$^2$), a level which is also associated with the same co-morbid conditions.[10] In other words, a BMI between 25–28 kg/m$^2$ is no longer considered within the normal range and weight loss is recommended. Weight loss diets are therefore more popular than ever not only because of a new 'at risk' population, but also because obesity is increasing at an alarming rate.[11] We now know that weight loss increases bone turnover markers and may reduce bone mass.[12] Different populations based on age, gender, ethnicity and other factors may respond differently to weight loss with respect to bone, and this weight loss may not always be voluntary. Our intent in the next sections is to clarify the many studies of weight reduction and bone by classifying them according to study design and the population examined.

# Weight Reduction, Bone Mass and Fracture

### Voluntary *vs.* Involuntary Weight Reduction

Studies of involuntary weight reduction show a loss of bone,[5,6] and a greater rate of hip fracture in older women[13,14] and men,[15] and particularly in those who have a low body weight before weight loss.[6] It is possible that some of these low body weight individuals are also under-nourished, and this may be contributing to their increased risk of fracture. The major difference between voluntary and involuntary weight reduction is that the latter is usually associated with an underlying illness. Involuntary weight reduction typically is secondary to illnesses such as gastrointestinal disease or cancer, both of which may have detrimental effects on bone that are independent of weight loss.[16,17] Therefore, for a given amount of involuntary weight reduction, the bone loss may be greater than expected when it is compared to voluntary weight reduction. Depression is another cause of involuntary weight reduction that has also been independently associated with a reduced bone mass[18,19] and increased osteoporosis risk. Ageing (as well as a more sedentary lifestyle) may cause unintentional loss of muscle mass and body weight (sarcopenia). This is sometimes referred to as 'nutritional frailty' because the sarcopenia contributes to functional impairment.[20] The risk of fracture is greatly increased with aging due to both low bone mass and the high risk of falling. Overall, involuntary weight reduction is a risk factor for bone loss, especially in older populations. The next sections focus on voluntary weight reduction in healthy overweight and obese populations.

# Postmenopausal Women

It has been consistently observed in postmenoupausal women that they have about 1–2% bone loss with about a 10% weight loss over a 4–18 month period.[12] Since most studies do not include a control group of women who maintain their body weight, it is important to remember that most healthy women of this age group will lose about 1% bone per year. To our knowledge, there are three prospective studies[21–23] designed to examine the response of bone to weight reduction in postmenopausal women that have included a weight maintenance group (Table 27.1). Svendsen et al.[23] found that after 12 weeks of moderate energy restriction (~10 kg weight loss), there was a 2% decrease in lumbar spine BMD and a tendency for total body BMD to decrease. Avenell et al.[21] also found a significant BMD loss of ~5% at the lumbar spine compared to the 2.5% loss in the no intervention group. In our study,[22] we found a 1.4% total body BMD loss with a 10% weight loss. Although this is a small decrease, it was significantly greater compared to weight stable women. The data for the lumbar spine could not be reported[22] due to artifacts associated with osteoarthritis in this obese postmenopausal population. Other studies of postmenopausal women during weight loss also report the response of bone.[24,25] Chao[24] found no loss of bone (hip, spine or total body) with a small amount of weight loss compared to the no weight change intervention group (~4 kg vs. 1 kg, respectively), but found a significant correlation between weight loss and loss of total body bone density. A weakness in this study[24] was that its original aim was to examine the effects of diet on hypertension and therefore had no exclusion criteria typical of studies designed to examine bone. The study included postmenopausal women on hormone replacement therapy (HRT) (33%), diuretics (52%) and those adhering to a low sodium diet (44%), all of which could have independently

**Table 27.1.** *Controlled weight loss studies with weight maintenance group*[a]

| Author and Year | Time (months) | Pop | Groups | Wt. (BMI) kg (kg m$^{-2}$) | Wt. Δ (kg) | Calcium baseline (diet) (mg day$^{-1}$) | Diet information | BMD changes (percent change) Total | Fem. | Lumb. |
|---|---|---|---|---|---|---|---|---|---|---|
| Avenell (1994) | 12 | Post (60 years) | No Wt. loss (n16) | 63 (25) | 0.0 | 613 (No data) | No intervention | – | 2.5% | 2.5% |
| | | | Dieter (n = 43) | 81 (31) | –1.0 | 696 (578) | Hi fiber and counsel. | – | 2.1% | 4.8% |
| | | | | | (–3.3)[b] | | | NS | p = 0.03 | |
| Svendsen (1993) | 3 | Post (54 years) | No Wt. loss (n = 21) | 77 (30)[c] | 0.5 | 759 (750) | Regular diet | –1.2 | – | –0.4 |
| | | | Diet only (n = 51) | 78 (30)[c] | –9.5 | 761 (1190) | Partial liquid | –1.9 | – | –1.6 |
| | | | Diet + Ex. (n = 49) | 78 (30)[c] | –10.3 | 737 (1176) | Partial liquid | –1.9 | – | –2.4 |
| | | | | | | | | NS | | 0.02 |
| Ricci (2001) | 6 | Post (56 years) | No Wt. Loss (n = 13) | 87 (33) | 0.1 | 699 (636) | Nutr. Counsel. | +0.6% | – | – |
| | | | Dieter (n = 14) | 87 (33) | –8.9 | 666 (504) | Nutr. Counsel. (weekly) | –1.2% | – | – |
| | | | | | | | | p < 0.05 | | |
| Salamone (1999) | 18 | Pre (47 years) | No wt. loss (n = 121) | 67 (25) | 0.4 | 825 (771) | Ex. counsel. | – | –0.4% | –0.4% |
| | | | Dieter (n = 115) | 67 (25) | –3.4 | 804 (836) | Nutr. and ex. counsel. (weekly or monthly) | – | –0.8% | –0.7% |
| | | | | | | | | | p < 0.02 | p = 0.09 |
| Shapses (2001) | 6 | Pre (41 years) | No wt loss (n = 10) | 94 (35) | –0.5 | 1150 (795) | Nutr. counsel. | –0.1 | – | –0.8 |
| | | | Dieter (n = 14) | 94 (35) | –6.7 | 810 (450) | Nutr. counsel. | +0.4 | – | –0.2 |
| | | | | | | | | NS | | NS |
| Pritchard (1996) | 12 | Men (43 years) | No wt loss (n = 19) | 88 (27) | 0.0 | 786 (780) | Bimonthly | +0.4% | – | – |
| | | | Dieter (n = 18) | 88 (29) | 6.0 | 873 (732) | Bimonthly | –1.4% | – | – |
| | | | | | | | | p = 0.05 | | |

Abbreviations: BMI, Body mass index; BMD, Bone mineral density; Counsel, Counseling; Ex, Exercise; Fem, Femur; Lumb., Lumbar Spine; Nutr, Nutrition; Pop, Population; Post, Postmenopausal; Pre, Premenopausal; Time, Duration of study; Total, Total body; Wt, Weight.
[a]Only includes those studies with a homogeneous population (and for women, those without HRT or oral contraceptives).
[b]Women in this study lost –3.3 and –2.8 kg at 3 and 6 months, respectively.
[c]BMI of entire group was given, no data for individual groups.

affected bone mass.[24] In one additional weight reduction study that was compared to an exercising weight reduction group,[25] postmenopausal women lost 7% of their body weight and lost bone at the femoral neck (2.6%), with only a trend at the spine (1%) and none for the total body. Although regional differences in bone loss might be expected, with greater loss from more trabecular than cortical regions (or differences due to weight bearing), the data currently suggests that bone changes of about 1–2% occur at all sites. This may be due to systemic hormonal changes during weight reduction that affect all bone sites or imaging limitations, both of which are discussed later in this chapter.

# Pre- and Peri-menopausal Women and Men

In oestrogen replete women, studies of weight reduction with a control group are limited. We found that a 7.5% weight loss in obese premenopausal women was associated with no total body or lumbar spine bone loss compared to weight stable women.[26] This finding was surprising in light of other studies, including our own, showing that weight loss does decrease bone mass (Table 27.1). However, to our knowledge, there are no other controlled weight loss studies in obese premenopausal women who are not in their peri- or post-menopausal years. In a 1.5 year study of lean and overweight (BMI of ~25 kg m$^{-2}$) women beginning at 47 years of age, weight loss of 5% resulted in a small but significant loss of BMD at the hip (0.8% loss) compared to weight stable women[27] (Table 27.1). It is likely that a low initial body weight[6] and/or diminishing/irregular levels of oestrogen[28] in women older than ~45 years contributed to their weight reduction-induced bone loss. Although a control group is important, it could be argued that it is less necessary in populations like healthy young premenopausal women who generally do not lose bone. In weight reduction studies with a homogeneous group of premenopausal women without a control group, weight loss has been variable (5–17% loss) with a similar small amount of bone loss (<1% loss)[29,30] or bone loss was attributed to potential scanner artifacts.[31,32] Bone loss due to weight reduction in other studies including premenopausal women may be influenced by their mixed populations (*i.e.*, postmenopausal women and men) or a larger amount of weight loss.[33–39] Therefore, unlike the more consistent results in postmenopausal women, the results of studies in premenopausal women remain unclear, and it seems that age is an important consideration (Figure 27.2). The only bone study examining a homogeneous population of men after weight reduction shows a loss[40] (Table 27.1). Whether the distribution of adipose tissue influences the level of circulating hormones and affects bone mass is not clear.[41] Finally, differences in ethnic background are important determinants of bone mass,[42] yet previous studies have primarily included Caucasian populations. Ethnicity may be influencing the response of bone to weight reduction, and therefore should be addressed in future clinical trials.

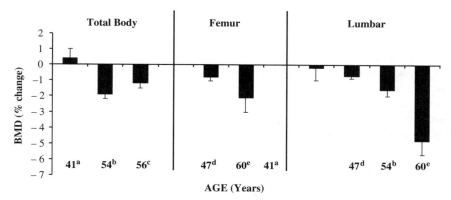

**Figure 27.2** *Increased loss of bone mineral density (BMD) with increasing age in controlled weight loss trials. Duration of study is listed below. (a) Shapses 2001 (6 months; 7% wt. loss); (b) Svendsen 1994 (3 months; 12% wt. loss); (c) Ricci 2001 (6 months; 10% wt loss); (d) Salamone 1999 (18 months; 5% wt loss); (e) Avenell 1994 (12 months; 4%)*

*"We're here for the weight loss class. Are any of us at risk of losing bone"?*

# Weight Regain (Weight Cycling)

The question as to whether bone mass is recovered upon weight regain, which is so common after voluntary weight reduction, is important. In humans, Avenell et al.[21] showed that the reduction in bone density of the lumbar spine due to weight reduction was not reversed by weight regain over a period of 6 months in postmenopausal women. Fogelholm[43] showed that premenopausal women with a weight cycling history had lower bone density of the spine and distal radius. In rodent studies, the reduction in bone quality and strength that has been shown to occur with energy restriction[44,45] is significantly compromised even after weight regain.[45] This is important because many overweight/obese individuals follow a pattern of repeated loss and regain of body weight, also known as weight cycling,[46] which may increase fracture risk.[47] It is possible that the stress associated with dietary restraint, which is common among weight cyclers, increases cortisol levels,[48] and this would be detrimental to bone. However, information about the effect of restrained eating on bone mass is controversial.[49,50] Weight cycling is likely to be highly detrimental to bone, yet a prospective trial is necessary to examine this appropriately.

## Physical Activity

The beneficial effects of physical activity on bone during weight stable conditions are well-established. A few studies have addressed its effects during weight loss.[23,25,27,31,33,40] In a controlled study by Ryan *et al.*[25] bone loss due to moderate weight loss was reduced with aerobic exercise in postmenopausal women. In a lifestyle intervention study,[27] it was found that those who included more physical activity (primarily aerobic) lost less bone than more sedentary women who lost weight (>8%). In contrast, other studies using either a walking intervention,[31] resistance exercise[33] or a combination of resistance and aerobic exercise[23] were not effective in preventing weight reduction-induced bone loss. In one other exercise study, the results were complicated by the weight loss differing in the intervention and control group.[40] The variable findings suggest that the effects of exercise may differ with the amount of weight loss, the population and type of exercise. Overall, the positive results of aerobic exercise in two controlled trials[25,27] suggest that exercise will reduce bone loss at weight bearing sites during weight reduction.

## Calcium Level and Other Nutrients

During moderately low energy intake, women consume less calcium than recommended levels of intake.[29,51] This is important because the bone lost during energy restriction is, at least partially, due to inadequate calcium intake.[29,51] In postmenopausal women, our double-blind placebo controlled trial showed that 1 g of supplemental calcium/day prevented bone mobilization associated with caloric restriction.[51] In support of these findings, Jensen *et al.*[39] found that a 1 g calcium supplement/day in a group of pre- and post-menopausal women who lost 5.5% of their body weight prevented bone loss at the femoral neck. In contrast, others have found that postmenopausal women sustain significant bone loss from the spine (but not forearm or total body) despite supplementation with a nutrient formula and a total calcium intake of 1–1.2 g per day.[23] Consistent with this, our preliminary studies in overweight postmenopausal women[52] consuming a total of 1 g Ca per day shows an inadequate amount of calcium absorbed, and an increased activity of the calcium–PTH axis. It is possible that the current recommended intakes of calcium are not adequate during weight loss regimens. In addition, the level of vitamin D intake and its role with calcium during weight loss is an important consideration. Finally, other nutrients (such as protein, magnesium, vitamin K) and/or dietary induced changes in acid/base balance may be altered during energy restriction and could influence the response of bone to weight reduction.

## Measurement Errors of DXA and Body Weight

There are studies that question whether DXA is capable of accurately measuring bone in normal weight individuals[53,54] and that it may be less accurate in individuals at both extremes of body weight.[54,55] Weight reduction may alter the sensitivity of DXA due to changes in the soft tissue surrounding the bone.[32,38,56] For example, Svendsen[56] found that in obese women who lose weight (11 kg loss) there could be a theoretical false decrease in the anterior–posterior spine BMD of 1–2%. In a smaller clinical trial, it was found that a 5.5% weight loss resulted in total body BMC loss of 3%, but because bone area also decreased and could account for 89% of the BMC change, it was considered measurement error.[32] In some studies, lard has been placed on top of the individual to determine the degree of error in bone measurements by DXA.[38,57] Because large quantities of fat (lard) were used in these studies, it is likely that with moderate weight loss, the measurement error is much less or possibly insignificant.

In general, most investigators agree that the measurement error in the lumbar spine[54,56] due to excess abdominal fat in the obese makes this site less desirable than the femoral neck or other sites (*i.e.*, forearm, calcaneous). In addition, postmenopausal women are at greater risk of spinal osteoarthritis,[58] which can overestimate bone mass, and in obesity we found the prevalence to be 43%.[22,51] Osteoarthritis is another reason to avoid spinal bone measurements in older obese individuals. In our experience, the practicality of positioning an obese person within the scan area on the DXA bed can be challenging. In general, the accuracy of DXA measurements for the morbidly obese (> 40 BMI) is poor and the use of this method is not recommended for this population. Overall, there are intrinsic problems and/or limitations of DXA such as measuring bone mass for a projected bone area rather than measuring true volumetric BMD, and an inability to separate cortical and trabecular bone. Other instruments such as computerized tomography (CT) and magnetic resonance imaging (MRI) may reduce measurement error of bone by measuring true volumetric bone density, while also addressing bone quality, and should be considered for future trials, especially in those addressing body weight changes.

## Endocrine and Other Bone Regulators During Weight Reduction

It is likely that multiple endocrine changes during weight reduction will influence bone. Recent data have suggested several potential regulators, and oestrogen appears as one of the most relevant candidates. The influence of genetics on an individual's susceptibility should not be underestimated, and although this is outside the scope of the present review, this field currently needs further investigation to clarify the role of several candidate genes. The endocrine regulators of bone during weight reduction will be discussed individually; however, it must be kept in mind that there are multiple interactions (*i.e.*, between hormones and with the genetic background) that will modulate an individual's response.

# Parathyroid Hormone (PTH)

Serum PTH is typically within normal range in overweight and obese individuals, yet there is evidence that there is some secondary hyperparathyroidism in the morbidly obese.[59] During moderate energy restriction, our data suggest that serial measurements of serum PTH tend to increase compared to women maintaining their body weight.[22,51] Although we have found a small rise in serum PTH with values within the normal range, this could contribute to a rise in bone resorption. It is possible that the usually low dietary calcium intake in these women ($\sim 700$ mg day$^{-1}$) is exacerbated by the stress of energy restriction and causes a reduction in calcium absorption and a subsequent rise in serum PTH. Our preliminary studies examining calcium absorption during weight loss with 1 g calcium intake[52] suggest an elevation in the Ca–PTH axis when the rate of weight loss is approximately 0.7 kg loss per week.

# Oestrogen

With the onset of ovarian failure, serum oestrogen in women largely derives from the metabolism of circulating androstenedione by peripheral tissues, including fat.[60,61] Therefore, in obesity, the elevated serum oestrogen found in postmenopausal women is attributed to an increased conversion of androstenedione to estrone.[60,61] The direct relationship between fat mass and oestrogen levels may be a mechanism for the protective effect of obesity in reducing osteoporosis risk.[62] Low levels of oestrogen, present naturally in postmenopausal women, have a significant effect on BMD at several sites,[63,64] inhibit bone resorption and protect from fractures.[63–65] Therefore, within the low postmenopausal oestrogen levels, small variations in circulating oestrogen due to excess body fat or weight reduction are expected to be physiologically relevant to bone health.

Whether or not the decrease in body fat during moderate weight reduction affects serum oestrogens is important because it could provide a mechanism for weight reduction-associated bone loss. In postmenopausal women, serum oestrone decreases and sex hormone binding globulin increases with a loss in body fat during weight reduction, suggesting a decrease in bioavailable serum sex steroids.[22] In addition, weight loss in the rat results in a reduction in serum oestradiol levels.[44, 45,66] On the other hand, Hui *et al.*[67] found that weight gain can prevent the decrease in oestrogen levels in women between the ages of 31 and 50. Furthermore, we have shown that during energy restriction there is a greater oestrogen deficiency in lean compared with obese rats.[66] Local synthesis of oestrogen in the excess adipose tissue of the obese may preserve serum oestradiol levels and protect against the deleterious effects of weight reduction on bone.

In addition to the crucial and direct role of oestrogen in the maintenance of bone mass, evidence accumulated in the last few years has underscored its relevance in regulating intestinal Ca absorption.[68–70] Our data in the rat[66] strongly supports the hypothesis that the decrease in oestrogen due to energy restriction is an important mechanism in inducing a decrease in calcium absorption. Other physiological changes associated with decreased sex steroids may be affected during energy restriction, such as increased levels of cytokines and factors (*i.e.*, IL-1, IL-6, TNF-alpha) that promote osteoclastic activity. These events are observed with oestrogen deficiency due to ovariectomy, but have not been studied during weight loss. Together, these data suggest that a decrease in circulating oestrogens may regulate bone turnover by directly increasing osteoclast activity or indirectly by impairing intestinal calcium absorption.

# Leptin

Leptin, a soluble 16-kDa protein product of adipose tissue, has recently been shown also to be produced by osteoblasts.[71] Leptin has been shown to suppress body weight and appetite, whereas leptin-deficient (ob/ob) and leptin receptor deficient (db/db) mice are obese and hypogonadic, yet have a high bone mass phenotype. In humans, a genetically based leptin deficiency is associated with morbid obesity and hypogonadism, but does not occur with high bone mass.[72] Hence there is conflicting information about leptin and it is possible that the mouse differs from humans with respect to leptin deficiency.[73]

The effect of leptin on bone is conflicting because while its higher levels in the obese would be expected to mediate a greater bone mass, some investigators have suggested that it has a detrimental effect by inhibiting bone formation through a central mechanism.[74] Ducy *et al.*[75] treated ob/ob mice with leptin using an intracerebroventricular injection and found that leptin administration decreased bone mass. In contrast, Steppan *et al.*[76] showed that systemic leptin administration increases bone mass compared to vehicle treated ob/ob mice; whereas Burguera *et al.*[77] showed that ovariectomized rats respond to leptin by increasing OPG mRNA resulting in a reduced bone loss. It is possible that compared to a central hypothalamic control, the local anabolic effects of leptin predominates with systemic administration of the hormone. Leptin may directly effect osteoclast differentiation by stimulating OPG mRNA and inhibiting RANKL mRNA,[78] and redirect the differentiation of stem cells from adipocytes to osteoblasts.[79] Most studies show a positive effect of leptin on bone,[80–82] but not all.[83,84] The positive relationship between leptin and bone may only be present in non-obese individuals[81,85] who are not leptin resistant, and may be present in women but not men.[86] During weight reduction, it is possible that the predictable decrease in circulating leptin levels is responsible for bone loss, but this has not yet been shown.[22] Further studies are required to clarify whether leptin regulates bone during weight loss.

# Insulin-like Growth Factor-I (IGF-I)

The anabolic effect of IGF-I on bone is well-established,[87] and nutritional intake is considered a major regulator of circulating IGF-I. Both energy and protein deficiencies can independently reduce levels of IGF-I, as demonstrated in both human and animal studies.[88–91] During dieting, protein is usually adequate in most standard weight loss diets, and therefore, changes in IGF-I can be attributed to an energy deficit. In general, studies of fasting and severe energy restriction show a reduction in IGF-I,[90–94] whereas in studies of moderate energy restriction[51] and long-term energy restriction,[95] there is no change in IGF-I levels. Because there is a decrease in IGF-I with energy restriction that is associated with a decline in bone turnover markers,[92,94] administration of IGF-I has been used to increase bone mass in women with anorexia nervosa.[96] It is interesting that serum IGF-I is inversely related to abdominal fat before and during energy restriction,[97] suggesting that fat distribution may influence hormones that regulate bone density, but this area of research is still very limited.

# Glucocorticoids

It is well known that long-term exposure to glucocorticoids, used for the management of inflammatory disorders, is a risk factor for osteoporosis.[98] Acting *via* different mechanisms on bone cells, glucocorticoids promote bone loss by increasing bone resorption and inhibiting bone formation.[99] Excess endogenous cortisol can also lead to bone loss.[100] An example of such excess may be during periods of energy restriction. Most studies show that energy restriction is associated with hyperadrenocorticism.[66,101–103] In addition, McLean *et al.*[48] found that high dietary restraint in young women increases urinary cortisol levels. There is also evidence that acute fasting increases serum cortisol in healthy young women[104] and in women with anorexia nervosa,[105,106] but other studies of energy restriction show no change[107–109] or

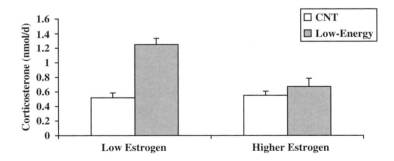

**Figure 27.3** *Corticosterone excretion increases due to energy restriction and low oestrogenic activity in mature rats (unpublished data). Differs from other groups, p < 0.002*

a decrease.[110] Differences in findings may be due to a single sample collection, rather than accounting for diurnal variation[111] of cortisol by measuring a 24 h urine sample or collecting multiple blood samples. In our laboratory, we found that the degree of hyperadrenocorticism is greater in a low compared to normal oestrogenic state (Figure 27.3). It is also suggested that the degree and type of energy restriction in different studies would result in a more or less pronounced cortisol response. Overall, it seems clear that most studies support a relationship between energy restriction and elevated cortisol levels. The detrimental effect of glucocorticoids on bone cells[99] and calcium absorption[111] may be one important mediator of the negative effects of weight reduction on calcium and bone metabolism.

# Conclusions

Overall, body weight is a significant predictor of bone density and fracture risk. The finding that bone is lost in many individuals during weight loss has major health implications. A large number of overweight and obese individuals are attempting to lose weight to reduce the co-morbid conditions associated with obesity. In young obese individuals, the risk of bone loss with moderate weight loss currently appears to be small, but the effect of more severe weight loss and of weight cycling is likely to be detrimental to bone health. For older individuals or women with oestrogen deficiency, involuntary, moderate or severe weight loss will increase bone turnover and loss and affect bone quality and fracture risk. The benefits of increased calcium, vitamin D, aerobic exercise and more moderate energy restriction during weight reduction could minimize this loss and should be encouraged for all individuals electing to lose weight. Further research is warranted on the safety and efficacy of oestrogen replacement therapy and optimal diets for the maintenance of bone mass in postmenopausal women who are attempting to lose weight. Newer methods to measure the response of BMD to weight loss will also increase the accuracy of assessing changes in body weight in this area.

# Acknowledgements

The authors appreciate the efforts of the staff and students in Dr. Shapses' laboratory who contributed to the work cited in this chapter, and to our illustrator, K. Mattejet.

# References

1. S.L. Edelstein and E. Barrett-Connor, Relation between body size and bone mineral density in elderly men and women, *Am. J. Epidemiol.*, 1993, **138**, 160–169.
2. D.T. Felson, Y. Zhang, M.T. Hannan and J.J. Anderson, Effects of weight and body mass index on bone mineral density in men and women: the Framingham study, *J. Bone Miner. Res.*, 1993, **8**, 567–573.

3. K.E. Ensrud, R.C. Lipschutz, J.A. Cauley, D. Seeley, M.C. Nevitt, J. Scott *et al.* Body size and hip fracture risk in older women: a prospective study. Study of Osteoporotic Fractures Research Group, *Am. J. Med.*, 1997, **103**, 274–280.

4. K.L. Margolis, K.E. Ensrud, P.J. Schreiner and H.K. Tabor, Body size and risk for clinical fractures in older women. Study of Osteoporotic Fractures Research Group, *Ann. Int. Med.*, 2000 **18**, 133, 123–127.

5. M.T. Hannan, D.T. Felson, B. Dawson-Hughes, K.L. Tucker, L.A. Cupples, P.W. Wilson and D.P. Kiel, Risk factors for longitudinal bone loss in elderly men and women: the Framingham Osteoporosis Study, *J. Bone Miner. Res.*, 2000, **15**, 710–720.

6. T.V. Nguyen, P.N. Sambrook and J.A. Eisman, Bone loss, physical activity, and weight change in elderly women: The Dubbo Osteoporosis Epidemiology Study, *J. Bone Miner. Res.*, 1998, **13**, 1458–1467.

7. K. Uusi-Rasi, H. Sievanen, M. Pasanen, P. Oja and I. Vuori, Maintenance of body weight, physical activity and calcium intake helps preserve bone mass in elderly women, *Osteoporosis Int.*, 2001, **12**, 373–379.

8. F. Wu, R. Ames, J. Clearwater, M.C. Evans, G. Gamble and I.R. Reid, Prospective 10-year study of the determinants of bone density and bone loss in normal postmenopausal women, including the effect of hormone replacement therapy, *Clin. Endocrinol. (Oxford)*, 2002, **56**, 703–711.

9. D.H. Taylor Jr. and T. Ostbye, The effect of middle- and old-age body mass index on short-term mortality in older people, *J. Am. Geriatr. Soc.*, 2001, **49**, 1319–1326.

10. NIH-NHLB obesity education initiative expert panel. Clinical guidelines on the identification, evaluation and treatment of overweight and obesity in adults – the evidence report. *Obes. Res.*, 1998, **6**, 51S–209S.

11. K.M. Flegal, M.D. Carroll, R.J. Kuczmarski and C.L. Johnson, Overweight and obesity in the United States: prevalence and trends, 1960–1994, *Int. J. Obes. Related Metab. Disord.*, 1998, **22**, 39–47.

12. S.A. Shapses, Body weight and the Skeleton, In *Nutritional Aspects of Osteoporosis*, P. Burckhardt, R. Heaney, and B. Dawson-Hughes (eds.), Academic Press, San Diego, 2001, pp. 341–354.

13. K.E. Ensrud, J. Cauley, R. Lipschutz and S.R. Cummings, Weight change and fractures in older women. Study of Osteoporotic Fractures Research Group, *Arch. Int. Med.*, 1997, **28**, 157 (See also pp. 857–863).

14. J.A. Langlois, M.E. Mussolino, M. Visser, A.C. Looker, T. Harris and J. Madans, Weight loss from maximum body weight among middle-aged and older white women and the risk of hip fracture: the NHANES I epidemiologic follow-up study, *Osteoporosis Int.*, 2001, **12**, 763–768.

15. M.E. Mussolino, A.C. Looker, J.H. Madans, J.A. Langlois and E.S. Orwoll, Risk factors for hip fracture in white men: the NHANES I Epidemiologic Follow-up Study, *J. Bone Miner. Res.*, 1998, **13**, 918–924.

16. G.R. Mundy, Metastasis to bone: causes, consequences and therapeutic opportunities, *Nat. Rev. Cancer*, 2002, **2**, 584–593.

17. T. Valdimarsson, O. Lofman, G. Toss and M. Strom, Reversal of osteopenia with diet in adult coeliac disease, *Gut*, 1996, **38**, 322–327.

18. D. Michelson, C. Stratakis, L. Hill, J. Reynolds, E. Galliven, G. Chrousos and P. Gold, Bone mineral density in women with depression, *N. Engl. J. Med.*, 1996, **17**, 335 (see also pp. 1176–1181).

19. J. Robbins, C. Hirsch, R. Whitmer, J. Cauley and T. Harris, The association of bone mineral density and depression in an older population, *J. Am. Geriatr. Soc.*, 2001, **49**, 732–736.

20. C.W. Bales and C.S. Ritchie, Sarcopenia, weight loss, and nutritional frailty in the elderly, *Annu. Rev. Nutr.*, 2002, **22**, 309–323.

21. A. Avenell, P.R. Richmond, M.E. Lean and D.M. Reid, Bone loss associated with a high fibre weight reduction diet in postmenopausal women, *Eur. J. Clin. Nutr.*, 1994, **48**, 561–566.

22. T.A. Ricci, S.B. Heymsfield, R.N. Pierson Jr., T. Stahl, H.A. Chowdhury and S.A. Shapses, Moderate energy restriction increases bone resorption in obese postmenopausal women, *Am. J. Clin. Nutr.*, 2001, **73**, 347–352.

23. O.L. Svendsen, C. Hassager and C. Christiansen, Effect of an energy-restrictive diet, with or without exercise, on lean tissue mass, resting metabolic rate, cardiovascular risk factors, and bone in overweight postmenopausal women, *Am. J. Med.*, 1993, **95**, 131–140.

24. D. Chao, M.A. Espeland, D. Farmer, T.C. Register, L. Lenchik, W.B. Applegate and W.H. Ettinger Jr., Effect of voluntary weight loss on bone mineral density in older overweight women, *J. Am. Geriatr. Soc.*, 2000, **48**, 753–759.

25. A.S. Ryan, B.J. Nicklas and K.E. Dennis, Aerobic exercise maintains regional bone mineral density during weight loss in postmenopausal women, *J. Appl. Physiol.*, 1998, **84**, 1305–1310.

26. S.A. Shapses, N.L. Von Thun, S.B. Heymsfield, T.A. Ricci, M. Ospina, R.N. Pierson Jr. and T. Stahl, Bone turnover and density in obese premenopausal women during moderate weight loss and calcium supplementation, *J. Bone Miner. Res.*, 2001, **16**, 1329–1336.

27. L.M. Salamone, J.A. Cauley, D.M. Black, L. Simkin-Silverman, W. Lang, E. Gregg, L. Palermo, R.S. Epstein, L.H. Kuller and R. Wing, Effect of a lifestyle intervention on bone mineral density in premenopausal women: a randomized trial, *Am. J. Clin. Nutr.*, 1999, **70**, 97–103.

28. J.C. Prior, S.L. Barr and Y.M. Vigna, The controversial endocrinology of the menopausal transition, *J. Clin. Endocrinol. Metab.*, 1996, **81**, 3127–3129.

29. S.J. Ramsdale and E.J. Bassey, Changes in bone mineral density associated with dietary-induced loss of body mass in young women, *Clin. Sci. (Colch)*, 1994, **87**, 343–348.

30. M.D. Van Loan, H.L. Johnson and T.F. Barbieri, Effect of weight loss on bone mineral content and bone mineral density in obese women, *Am. J. Clin. Nutr.*, 1998, **67**, 34–38.

31. G.M. Fogelholm, H.T. Sievanen, T.K. Kukkonen-Harjula and M.E. Pasanen, Bone mineral density during reduction, maintenance and regain of body weight in premenopausal, obese women, *Osteoporosis Int.*, 2001, **12**, 199–206.

32. P. Vestergaard, J. Borglum, L. Heickendorff, L. Mosekilde and B. Richelsen, Artifact in bone mineral measurements during a very low calorie diet: short-term effects of growth hormone, *J. Clin. Densitom.*, 2000, 3(1), 63–71.

33. R.E. Andersen, T.A. Wadden and R.J. Herzog, Changes in bone mineral content in obese dieting women, *Metabolism*, 1997, **46**, 857–861.

34. J.E. Compston, M.A. Laskey, P.I. Croucher, A. Coxon and S. Kreitzman, Effect of diet-induced weight loss on total body bone mass, *Clin. Sci.*, 1992, **82**, 429–432.

35. V.V. Gossain, D.S. Rao, M.J. Carella, G. Divine and D.R. Rovner, Bone mineral density (BMD) in obesity effect of weight loss, *J. Med.*, 1999, **30**, 367–376.

36. A. Gotfredsen, H. Westergren Hendel and T. Andersen, Influence of orlistat on bone turnover and body composition, _Int. J. Obes. Related Metab. Disord._, 2001, **25**, 1154–1160.

37. L. Hyldstrup, T. Andersen, P. McNair, L. Breum and I. Transbol, Bone metabolism in obesity: changes related to severe overweight and dietary weight reduction, _Acta Endocrinol. (Copenhagen)_, 1993, **129**, 393–398.

38. L.B. Jensen, F. Quaade and O.H. Sorensen, Bone loss accompanying voluntary weight loss in obese humans, _J. Bone Miner. Res._, 1994, **9**, 459–463.

39. L.B. Jensen, G. Kollerup, F. Quaade and O.H. Sorensen, Bone minerals changes in obese women during a moderate weight loss with and without calcium supplementation, _J. Bone Miner. Res._, 2001, **16**, 141–147.

40. J.E. Pritchard, C.A. Nowson and J.D. Wark, Bone loss accompanying diet-induced or exercise-induced weight loss: a randomised controlled study, _Int. J. Obes. Related Metab. Disord._, 1996, **20**, 513–520.

41. E.A. Jankowska, E. Rogucka and M. Medras, Are general obesity and visceral adiposity in men linked to reduced bone mineral content resulting from normal ageing? A population-based study, _Andrologia_, 2001, **33**, 384–389.

42. E. Kobyliansky, D. Karasik, V. Belkin and G. Livshits, Bone ageing: genetics versus environment, _Ann. Human Biol._, 2000, **27**, 433–451.

43. M. Fogelholm, H. Sievanen, A. Heinonen, M. Virtanen, K. Uusi-Rasi, M. Pasanen and I. Vuori, Association between weight cycling history and bone mineral density in premenopausal women, _Osteoporosis Int._, 1997, **7**, 354–358.

44. S.M. Talbott, M. Cifuentes, M.G. Dunn and S.A. Shapses, Energy restriction reduces bone density and biomechanical properties in aged female rats, _J. Nutr._, 2001, **131**, 2382–2387.

45. C. Wang, Y. Zhang, Y. Xiong and C.J. Lee, Bone composition and strength of female rats subjected to different rates of weight reduction, _Nutr. Res._, 2000, **20**, 1613–1622.

46. _NIH_ 2002 http://www.niddk.nih.gov/health/nutrit/pubs/wcycling.htm#further

47. H.E. Meyer, A. Tverdal and R. Selmer, Weight variability, weight change and the incidence of hip fracture: a prospective study of 39,000 middle-aged Norwegians, _Osteoporosis Int._, 1998, **8**, 373–378.

48. J.A. McLean, S.I. Barr and J.C. Prior, Cognitive dietary restraint is associated with higher urinary cortisol excretion in healthy premenopausal women, _Am. J. Clin. Nutr._, 2001, **73**, 7–12.

49. G.P. Bathalon, N.P. Hays, S.N. Meydani, B. Dawson-Hughes, E.J. Schaefer, R. Lipman _et al._, Metabolic, psychological, and health correlates of dietary restraint in healthy postmenopausal women, _J. Gerontol. Series A: Biol. Sci. Med. Sci._, 2001, **56**, M206–M211.

50. M.D. Van Loan and N.L. Keim, Influence of cognitive eating restraint on total-body measurements of bone mineral density and bone mineral content in premenopausal women aged 18–45 y: a cross-sectional study, _Am. J. Clin. Nutr._, 2000, **72**, 837–843.

51. T.A. Ricci, H.A. Chowdhury, S.B. Heymsfield, T. Stahl, R.N. Pierson Jr. and S.A. Shapses, Calcium supplementation suppresses bone turnover during weight reduction in postmenopausal women, _J. Bone Miner. Res._, 1998, **13**, 1045–1050.

52. S.A. Shapses, M. Cifuentes, R. Sherrell and C. Reidt, Rate of Weight Loss Influences Calcium Absorption, _J. Bone Miner. Res._, 2002, **17**, S471.

53. T.V. Nguyen, P.N. Sambrook and J.A. Eisman, Sources of variability in bone mineral density measurements: implications for study design and analysis of bone loss, _J. Bone Miner. Res._, 1997, **12**, 124–135.

54. H.H. Bolotin and H. Sievanen, Inaccuracies inherent in dual energy X-ray absorptiometry *in vivo* bone mineral density can seriously mislead diagnostic/prognostic interpretations of patient specific bone fragility, *J. Bone Miner. Res.*, 2001, **16**, 799–805.

55. T.N. Hangartner and C.C. Johnston, Influence of fat on bone measurements with dual-energy absorptiometry, *J. Bone Miner. Res.*, 1990, **9**, 71–81.

56. O.L. Svendsen, H.W. Hendel, A. Gotfredsen, B.H. Pedersen and T. Andersen, Are soft tissue composition of bone and non-bone pixels in spinal bone mineral measurements by DXA similar? Impact of weight loss, *Clin. Physiol. Funct. Imaging*, 2002, **22**, 72–77.

57. L.A. Milliken, S.B. Going and T.G. Lohman, Effects of variations in regional composition on soft tissue measurements by dual-energy X-ray absorptiometry, *Int. J. Obes. Related Metabol. Disord.*, 1996, **20**, 677–682.

58. G. Liu, M. Peacock, O. Eilam, G. Dorulla, E. Braunstein and C.C. Johnston, Effect of osteoarthritis in the lumbar spine and hip on bone mineral density and diagnosis of osteoporosis in elderly men and women, *Osteoporosis Int.*, 1997, **7**, 564–569.

59. T. Andersen, P. McNair, L. Hyldstrup, N. Fogh–Andersen, T.T. Nielsen, A. Astrup and I. Transbol, Secondary hyperparathyroidism of morbid obesity regresses during weight reduction, *Metabolism*, 1988, **37**, 425–428.

60. A.R. Glass, Endocrine aspects of obesity, In *Medical Clinics of North America*, G.A. Bray (ed.), W.B. Saunders Co., Philadelphia, PA, 1989, vol. 73, pp. 139–160.

61. R. Lindsay, Sex steroids in the pathogenesis and prevention of osteoporosis, In *Osteoporosis: Etiology, Diagnosis, and Management*, B.L. Riggs, L.J. Melton III, Raven Press, New York, 1988, pp. 333–358.

62. A.M. Frumar, D.R. Meldrum, F. Geola, I.M. Shamonki, I.V. Tataryn, L.J. Deftos and H.L. Judd, Relationship of fasting urinary calcium to circulating oestrogen and body weight in postmenopausal women, *J. Clin. Endocrinol. Metab.*, 1980, **50**, 70–75.

63. B. Ettinger, A. Pressman, P. Sklarin, D.C. Bauer, J.A. Cauley and S.R. Cummings, Associations between low levels of serum estradiol, bone density, and fractures among elderly women: the study of osteoporotic fractures, *J. Clin. Endocrinol. Metab.*, 1998, **83**, 2239–2243.

64. S. Khosla, L.J. Melton III, E.J. Atkinson, W.M. O'Fallon, G.G. Klee and B.L. Riggs, Relationship of serum sex steroid levels and bone turnover markers with bone mineral density in men and women: A key role for bioavailable oestrogen, *J. Clin. Endocrinol. Metab.*, 1998, **83**, 2266–2274.

65. H.M. Heshmati, S. Khosla, S.P. Robins, W.M. O'Fallon, L.J. Melton III and B.L. Riggs, Role of low levels of endogenous oestrogen in regulation of bone resorption in late postmenopausal women, *J. Bone Miner. Res.*, 2002, **17**, 172–178.

66. M. Cifuentes, A.B. Morano, H.A. Chowdhury and S.A. Shapses, Energy restriction reduces fractional calcium absorption in mature obese and lean rats, *J. Nutr.*, 2002, **132**, 2660–2666.

67. S.L. Hui, A.J. Perkins, L. Zhou, C. Longcope, M.J. Econs, M. Peacock, C. McClintock and C.C. Johnston Jr., Bone loss at the femoral neck in premenopausal white women: effects of weight change and sex-hormone levels, *J. Clin. Endocrinol. Metab.*, 2002, **87**, 1539–1543.

68. B.H. Arjmandi, B.W. Hollis and D.N. Kalu, *In vivo* effect of 17β-estradiol on intestinal calcium absorption in rats, *J. Bone Miner. Res.*, 1994, **26**, 181–189.

69. Y. Liel, S. Shany, P. Smirnoff and B. Schwartz, Oestrogen increases 1,25-dihydroxyvitamin D receptors expression and bioresponse in the rat duodenal mucosa, *Endocrinology*, 1999, **140**, 280–285.
70. P.D. O'Loughlin and H.A. Morris, Ooestrogen deficiency impairs intestinal calcium absorption in the rat, *J. Appl. Physiol.*, 1998, **511**, 313–322.
71. J.E. Reseland, U. Syversen, I. Bakke, G. Qvigstad, L.G. Eide, O. Hjertner *et al.*, Leptin is expressed in and secreted from primary cultures of human osteoblasts and promotes bone mineralization, *J. Bone Miner. Res.*, 2001, **16**, 1426–1433.
72. M. Ozata, I.C. Ozdemir and J. Licinio, Human leptin deficiency caused by a missense mutation: multiple endocrine defects, decreased sympathetic tone, and immune system dysfunction indicate new targets for leptin action, greater central than peripheral resistance to the effects of leptin, and spontaneous correction of leptin–mediated defects, *J. Clin. Endocrinol. Metab.*, 1999, **84**, 3686–3695.
73. M. Ozata, Different presentation of bone mass in mice and humans with congenital leptin deficiency, *J. Clin. Endocrinol. Metab.*, 2002, **87**, 951.
74. G. Karsenty, The central regulation of bone remodeling, *Trends Endocrinol. Metabol.*, 2000, **11**, 437–439.
75. P. Ducy, M. Amling, S. Takeda, M. Priemel, A.F. Schilling, F.T. Beil *et al.*, Leptin inhibits bone formation through a hypothalamic relay: a central control of bone mass, *Cell*, 2000, **100**, 197–207.
76. C.M. Steppan, D.T. Crawford, K.L. Chidsey-Frink, H. Ke and A.G. Swick, Leptin is a potent stimulator of bone growth in ob/ob mice, *Regul. Pept.*, 2000, **92**, 73–78.
77. B. Burguera, L.C. Hofbauer, T. Thomas, F. Gori, G.L. Evans, S. Khosla *et al.* Leptin reduces ovariectomy-induced bone loss in rats, *Endocrinology*, 2001, **142**, 3546–3553.
78. W.R. Holloway, F.M. Collier, C.J. Aitken, D.E. Myers, J.M. Hodge, M. Malakellis *et al.*, Leptin inhibits osteoclast generation, *J. Bone Miner. Res.*, 2002, **17**, 200–209.
79. T. Thomas, F. Gori, S. Khosla, M.D. Jensen, B. Burguera and B.L. Riggs, Leptin acts on human marrow stromal cells to enhance differentiation to osteoblasts and to inhibit differentiation to adipocytes, *Endocrinology*, 1999, **140**, 1630–1638.
80. H. Blain, A. Vuillemin, F. Guillemin, R. Durant, B. Hanesse, N. De Talance *et al.*, Serum leptin level is a predictor of bone mineral density in postmenopausal women, *J. Clin. Endocrinol. Metab.*, 2002, **87**, 1030–1035.
81. J.A. Pasco, M.J. Henry, M.A. Kotowicz, G.R. Collier, M.J. Ball, A.M. Ugoni and G.C. Nicholson, Serum leptin levels are associated with bone mass in nonobese women, *J. Clin. Endocrinol. Metab.*, 2001, **86**, 1884–1887.
82. M. Yamauchi, T. Sugimoto, T. Yamaguchi, D. Nakaoka, M. Kanzawa, S. Yano, *et al.*, Plasma leptin concentrations are associated with bone mineral density and the presence of vertebral fractures in postmenopausal women, *Clin. Endocrinol. (Oxford)*, 2001, **55**, 341–347.
83. G. Martini, R. Valenti, S. Giovani, B. Franci, S. Campagna and R. Nuti, Influence of insulin-like growth factor-1 and leptin on bone mass in healthy postmenopausal women, *Bone*, 2001, **28**, 113–117.
84. M. Sato, N. Takeda, H. Sarui, R. Takami, K. Takami, M. Hayashi *et al.*, Association between serum leptin concentrations and bone mineral density, and biochemical markers of bone turnover in adult men, *J. Clin. Endocrinol. Metab.*, 2001, **86**, 5273–5276.
85. M. Lee, J.M. Zmuda, S. Wisniewski, S. Krishnaswami, R.W. Evans and J.A. Cauley, Serum leptin concentrations and bone mass: differential association among obese and non-obese men, *J. Bone Miner. Res.*, 2002, **17**, S463.

86. T. Thomas, B. Burguera, L.J. Melton, 3rd, E.J. Atkinson, W.M. O'Fallon, B.L. Riggs and S. Khosla, Role of serum leptin, insulin, and oestrogen levels as potential mediators of the relationship between fat mass and bone mineral density in men versus women, *Bone*, 2001, **29**, 114–120.

87. C.J. Rosen, L.R. Donahue, S.J. Hunter, Insulin-like growth factors and bone: the osteoporosis connection, *Proc. Soc. Exp. Biol. Med.*, 1994, **206**, 83–102.

88. P. Ammann, S. Bourrin, J.P. Bonjour, J.M. Meyer and R. Rizzoli, Protein undernutrition-induced bone loss is associated with decreased IGF-I levels and oestrogen deficiency, *J. Bone Miner. Res.*, 2000, **15**, 683–690.

89. J.P. Bonjour, P. Ammann, T. Chevalley and R. Rizzoli, Protein intake and bone growth, *Can. J. Appl. Physiol.*, 2001, **26**(Suppl), S153–S166.

90. W.J. Smith, L.E. Underwood and D.R. Clemmons, Effects of caloric or protein restriction on insulin-like growth factor-I (IGF-I) and IGF-binding proteins in children and adults, *J. Clin. Endocrinol. Metab.*, 1995, **80**, 443–449.

91. J.P. Thissen, J.M. Ketelslegers and L.E. Underwood, Nutritional regulation of the insulin-like growth factors, *Endocrinol. Rev.*, 1994, **15**, 80–101.

92. S.K. Grinspoon, H.B. Baum, V. Kim, C. Coggins and A. Klibanski, Decreased bone formation and increased mineral dissolution during acute fasting in young women, *J. Clin. Endocrinol. Metab.*, 1995, **80**, 3628–3633.

93. A. Megia, L. Herranz, R. Luna, C. Gomez-Candela, F. Pallardo and P. Gonzalez-Gancedo, Protein intake during aggressive calorie restriction in obesity determines growth hormone response to growth hormone-releasing hormone after weight loss, *Clin. Endocrinol. (Oxford)*, 1993, **39**, 217–220.

94. C.L. Zanker and L.L. Swaine, Responses of bone turnover markers to repeated endurance running in humans under conditions of energy balance or energy restriction, *Eur. J. Appl. Physiol.*, 2000, **83**, 434–440.

95. N. Girard, G. Ferland, L. Boulanger and P. Gaudreau, Long-term calorie restriction protects rat pituitary growth hormone-releasing hormone binding sites from age-related alterations, *Neuroendocrinology*, 1998, **68**, 21–29.

96. S. Grinspoon, L. Thomas, K. Miller, D. Herzog and A. Klibanski, Effects of recombinant human IGF-I and oral contraceptive administration on bone density in anorexia nervosa, *J. Clin. Endocrinol. Metab.*, 2002, **87**, 2883–2891.

97. M.H. Rasmussen, A. Hvidberg, A. Juul, K.M. Main, A. Gotfredsen, N.E. Skakkebaek and J. Hilsted, Massive weight loss restores 24-hour growth hormone release profiles and serum insulin-like growth factor-I levels in obese subjects, *J. Clin. Endocrinol. Metab.*, 1995, **80**, 1407–1415.

98. J.A. Clowes, N. Peel and R. Eastell, Glucocorticoid-induced osteoporosis, *Curr. Opin. Rheumatol.*, 2001, **13**, 326–323.

99. E. Canalis and A.M. Delany, Mechanisms of glucocorticoid action in bone, *Ann. NY Acad. Sci.*, 2002, **966**, 73–81.

100. L. Tauchmanova, R. Rossi, V. Nuzzo, A. del Puente, A. Esposito-del Puente, C. Pizzi, F. Fonderico, G. Lupoli and G. Lombardi, Bone loss determined by quantitative ultrasonometry correlates inversely with disease activity in patients with endogenous glucocorticoid excess due to adrenal mass, *Eur. J. Endocrinol.*, 2001, **145**, 241–247.

101. E.S. Han, T.R. Evans, J.H. Shu, S. Lee and J.F. Nelson, Food restriction enhances endogenous and corticotropin-induced plasma elevations of free but not total corticosterone throughout life in rats, *J. Gerontol. Ser. A: Biol. Sci. Med. Sci.*, 2001, **56**, B391–B397.

102. A.S. Kitaysky, E.V. Kitaiskaia, J.C. Wingfield and J.F. Piatt, Dietary restriction causes chronic elevation of corticosterone and enhances stress response in red-legged kittiwake chicks, *J. Comp. Physiol [B]*, 2001. **171**, 701–709.
103. B.P. Yu and H.Y. Chung, Stress resistance by caloric restriction for longevity, *Ann. NY Acad. Sci.*, 2001, **928**, 39–47.
104. M. Bergendahl, A. Iranmanesh, T. Mulligan and J.D. Veldhuis, Impact of age on cortisol secretory dynamics basally and as driven by nutrient-withdrawal stress, *J. Clin. Endocrinol. Metab.*, 2000, **85**, 2203–2214.
105. B.M. Biller, V. Saxe, D.B. Herzog, D.I. Rosenthal, S. Holzman and A. Klibanski, Mechanisms of osteoporosis in adult and adolescent women with anorexia nervosa, *J. Clin. Endocrinol. Metab.*, 1989, **68**, 548–554.
106. J. Licinio, M.L. Wong, P.W. Gold, The hypothalamic-pituitary-adrenal axis in anorexia nervosa, *Psychiatr. Res.*, 1996, **62**, 75–83.
107. V. Hainer, M. Kunesova, A.J. Stunkard, J. Parizkova, B. Stich, R. Mikulova and L. Starka, The within-pair resemblance in serum levels of androgens, sex-hormonebinding globulin and cortisol in female obese identical twins-effect of negative energy balance induced by very low-calorie diet, *Hormone Metab. Res.*, 2001, **33**, 417–422.
108. E.F. Van Rossom, B.J. Nicklas, K.E. Dennis, D.M. Berman and A.P. Goldberg, Leptin responses to weight loss in postmenopausal women: relationship to sex-hormone binding globulin and visceral obesity, *Obes. Res.*, 2000, **8**, 29–35.
109. J.A. Yanovski, S.Z. Yanovski, P.W. Gold and G.P. Chrousos, Differences in corticotropin-releasing hormone-stimulated adrenocorticotropin and cortisol before and after weight loss, *J. Clin. Endocrinol. Metab.*, 1997, **82**, 1874–1878.
110. R. Buffenstein, A. Karklin and H.S. Driver, Beneficial physiological and performance responses to a month of restricted energy intake in healthy overweight women, *Physiol. Behav.*, 2000, **68**, 439–444.
111. S. Arnaud, M. Navidi, L. Deftos, M. Thierry-Palmer, R. Dotsenko, A. Bigbee and R.E. Grindeland, The calcium endocrine system of adolescent rhesus monkeys and controls before and after spaceflight, *Am. J. Physiol. Endocrinol. Metab.*, 2002, **282**, E514–E521.
112. M. Cifuentes, M.A. Johnson, R. Lewis, S.B. Heymsfield, H.A. Chowdhury, C. Modlesky and S.A. Shapses, Body weight reflects bone resorption in lean, but not overweight or obese postmenopausal women, *Osteoporosis Int.*, 2003, **14**, 116–122.

CHAPTER 28

# Clinical Eating Disorders and Subclinical Disordered Eating: Implications for Bone Health

CANDICE A. RIDEOUT, SUSAN I. BARR  and
JERILYNN C. PRIOR

The University of British Columbia, 2205 East Mall, Vancouver, British
Columbia, Canada, V6T 1Z4, Email: sibarr@interchange.ubc.ca

**Key Points**
- Eating disorders such as anorexia nervosa and bulimia nervosa are characterized by disturbed eating behaviours and distorted perceptions of body weight/size; evidence is accumulating to suggest that they, along with subclinical eating disturbances such as high levels of cognitive dietary restraint, have a negative impact on bone.
- Anorexia nervosa in particular is associated with reduced bone mass and increased risk for fracture, both at the time of active illness and many years later.
- Bone deficits associated with anorexia nervosa likely result from the combined effects of low body weight, poor nutritional status, and altered hormone levels (*e.g.*, reduced levels of insulin-like growth factor I and ovarian hormones, elevated cortisol).
- Normalisation of bone mass may occur following recovery from anorexia nervosa if the duration of the illness is relatively short and recovery occurs during adolescence when bone mineral density still has the potential to increase.
- Adequate intakes of calcium, vitamin D, and other key nutrients are important but not sufficient to maintain bone mass in the absence of weight gain and normal menstrual cycle activity.

- Although exogenous hormones are widely used in the treatment of anorexia nervosa, currently available data do not strongly support a bone-protective effect of oestrogen–progestin therapy for young women with anorexia nervosa.

## Clinical Scenario

A 34-year-old Caucasian woman was referred to the osteoporosis clinic in a tertiary care hospital after experiencing a low trauma fracture of the distal radius. Her intakes of calcium (approximately 500 mg day$^{-1}$ through diet and an additional 850 mg day$^{-1}$ through supplements) and vitamin D (400 IU day$^{-1}$ through supplements) were considered adequate. She exercised regularly, most frequently running (1 h, 6 days per week). Her weight was 48.3 kg and her height was 157 cm, resulting in a Body Mass Index (BMI; kg m$^{-2}$) of 19.6. She reported a first period at 12 years of age followed by irregular menstrual cycles throughout adulthood, including approximately five years during her late teens during which she was amenorrhoeic. She recalls losing a significant amount of weight around that time and reached a low weight of 37 kg, which she maintained for several months. However, her weight has been stable at its current level for approximately 6 years. Measurements of the lumbar spine and femoral neck were taken using dual energy X-ray absorptiometry (DXA) and revealed low bone mineral density (BMD) at both sites: $t$-scores (standard deviations (SD) below peak BMD) of $-2.1$ at the lumbar spine and $-1.6$ at the femoral neck were reported.

## Introduction

Several aspects of this young woman's history indicate that she may have had an eating disorder, specifically anorexia nervosa (AN), during her teenage years. Although AN is relatively uncommon in the general population, it is not the only pattern of disordered eating that could have adverse effects on bone health. In fact, many women may experience unwanted (and likely unexpected) consequences for bone as a result of their eating attitudes and behaviours. Like the woman whose situation was reported above, the eating attitudes and behaviours of many women fall along a continuum of disordered eating which, to varying degrees, could be detrimental to bone health.

Although the impact of AN on BMD has been recognized for many years,[1-8] questions remain regarding both the aetiology and the reversibility of the observed bone deficits. Moreover, the effects on bone density of other overt eating disorders (such as bulimia nervosa (BN)) and subclinical forms of disordered eating are unresolved and thus also topics of current research. Accordingly, this review will assess the impact of both clinical eating disorders and subclinical disordered eating (such as cognitive dietary restraint) on BMD and fractures in girls and young women. To do this, we first describe the continuum of disordered eating and briefly summarize the available data on BMD in women with eating disorders. Next, risk factors for reduced bone mass in this population are described, followed by an overview of the metabolic basis of bone loss. We subsequently review the extent to which low bone mass may be increased following recovery from AN and address the question of whether hormone replacement therapy (HRT) is an effective treatment. The chapter concludes with a discussion of the limitations of our current knowledge and possible directions for future research.

## Eating Disorders and the Continuum of Disordered Eating

Clinical eating disorders such as AN and BN are typically considered to be disease states affecting small numbers of individuals. However, rather than dichotomizing eating attitudes and behaviours as either 'healthy' or 'eating disordered,' it may be more appropriate to think of them as falling along a continuum, as depicted in Figure 28.1. As illustrated, an enjoyment of eating and self-acceptance of body weight and shape form one end of this continuum and the clinical eating disorders form the other end. Between these two extremes lie what has come to be known as disordered eating, or subclinical eating disturbances.

Although much less is known about the possible health consequences of these more subtle disturbances, they warrant attention because of their relatively high prevalence in the Western population of girls and women. Their frequency was illustrated by a cross-sectional retrospective study of a convenience sample of 239 young women in Norway (mean age = 22.7 years); not including those with a history of a clinical eating disorder, 14% ($n = 34$) of the women reported previous voluntary weight loss accompanied by missed menses.[9] Subclinical eating disturbances may exert profound effects on an individual's physiology, leading most subtly to disturbed ovulation and more obviously to skipped periods and amenorrhoea.[10] Thus, a discussion of the impact of disordered eating on bone health should not be limited to clinical eating disorders, but rather it should include other points along the continuum of disordered eating as well.

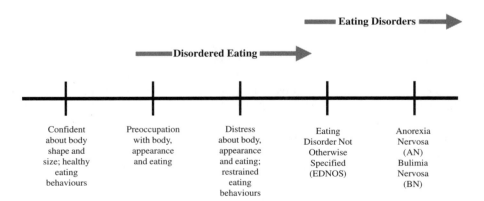

**Figure 28.1** *The continuum of disordered eating*

# Clinical Eating Disorders

Clinical eating disorders are characterized by marked disturbances in both eating behaviours and perceptions of body shape and weight.[11] They typically start during adolescence when body shape changes and both weight and proportion of body fat increase. Some reports indicate that the incidence of eating disorders appears to have increased in recent decades,[12,13] although others challenge this conclusion.[14] Such conflicting messages could be due to different trends observed in different age groups. In the past 40 years, the incidence of AN in teenagers has not increased significantly whereas the incidence in women in their 20 or 30s has increased roughly 3-fold.[15] Specific criteria for the diagnosis of the clinical eating disorders AN and BN have been established and they are listed in Table 28.1.

## *Anorexia Nervosa*

In AN, weight is not maintained at a minimally normal level (considered to be 85% of what is normal for age and height).[11] Although this low body weight typically results from weight loss, in children and young adolescents, it could also be due to a failure to gain weight despite an increase in height.[11]

More than 90% of the cases of AN occur in young women, for whom there is a lifetime prevalence of 0.5%.[11] It is not surprising that AN affects young women to a greater extent than young men given that they gain more fat weight during puberty and experience greater body dissatisfaction.[16] They are also more likely to perceive themselves to be overweight and they try to lose weight more often than boys and young men.[17,18] Teenage girls experience the highest incidence of AN. At 51 cases per 100 000 teenage girls per year, the incidence rate is roughly five times that of women over the age of 20 years.[15]

## *Bulimia Nervosa*

In contrast to those with AN, individuals with BN are typically within the normal range for weight.[11] BN is characterized by excessive, uncontrolled (binge) eating followed by feelings of guilt and inappropriate compensatory behaviours (such as purging through vomiting and/or excessive laxative use) intended to prevent weight gain.[11] In a clinical sample of young women with BN, the frequency of binge episodes prior to admission was between 2 and 23 per week.[19] Vomiting is the most commonly reported purging behaviour.[19,20] On average, women with BN recorded 1.25 purges for each of 3 days on which they maintained diet records.[20]

The lifetime prevalence of BN among women is roughly 1–3%.[11,21] However, a higher percentage of adolescents may demonstrate features of this disorder without meeting all of the criteria for the diagnosis of BN. For example, 7.5% of high school girls had vomited or used laxative pills in an attempt to lose weight in the month prior to completing a survey of various potentially risky behaviours.[17] Women with BN clearly demonstrate the dissatisfaction with their

**Table 28.1.** *Criteria for the diagnosis of eating disorders*

*Diagnostic Criteria for Anorexia Nervosa*

A. Refusal to maintain body weight at or above a minimally normal weight for age and height (*e.g.*, weight loss leading to maintenance of body weight less than 85% of that expected; or failure to make expected weight gain during period of growth, leading to body weight less than 85% of that expected).
B. Intense fear of gaining weight or becoming fat, even though underweight.
C. Disturbance in the way in which one's body weight or shape is experienced, undue influence of body weight or shape at self-evaluation, or denial of the seriousness of the current low body weight.
D. In postmenarcheal females, amenorrhea – *i.e.*, the absence of at least three consecutive menstrual cycles.

*Specify Type*

   *Restricting type:* during the current episode of Anorexia Nervosa, the person has not regularly engaged in binge-eating or purging behaviour (*i.e.*, self-induced vomiting or the misuse of laxatives, diuretics, or enemas).

   *Binge-Eating/Purging type:* during the current episode of Anorexia Nervosa, the person has regularly engaged in binge-eating or purging behaviour (*i.e.*, self-induced vomiting or the misuse of laxatives, diuretics, or enemas).

*Diagnostic Criteria for Bulimia Nervosa*

A. Recurrent episodes of binge eating. An episode of binge eating is characterized by both of the following: (1) eating, in a discrete period of time (*e.g.*, within any two-hour period), an amount of food that is definitely larger than most people would eat during a similar period of time and under similar circumstances, and (2) a sense of lack of control over eating during the episode (*e.g.*, a feeling that one cannot stop eating or control what or how much one is eating).
B. Recurrent inappropriate compensatory behaviour in order to prevent weight gain, such as self-induced vomiting; misuse of laxatives, diuretics, enemas, or other medications; fasting; or excessive exercise.
C. The binge eating and inappropriate behaviours both occur, on average, at least twice a week for three months.
D. Self-evaluation is unduly influenced by body shape and weight.
E. The disturbance does not occur exclusively during episodes of anorexia nervosa.

*Specify Type*

   *Purging type:* during the current episode of Bulimia Nervosa, the person has regularly engaged in self-induced vomiting or the misuse of laxatives, diuretics, or enemas.

   *Nonpurging type:* during the current episode of Bulimia Nervosa, the person has used other inappropriate compensatory behaviours, such as fasting or excessive exercise, but has not regularly engaged in self-induced vomiting or the misuse of laxatives, diuretics, or enemas.

Source: American Psychiatric Association. *Diagnostic and Statistical Manual of Mental Disorders*, 4th edn, Washington, DC, APA. 1994.[11]

weight and shape that is characteristic of eating disorders. When young women with BN were compared to non-eating disordered age-, height- and weight-matched controls, despite no difference in percent body fat, the women with BN perceived themselves to be more overweight.[20]

## Eating Disorder Not Otherwise Specified (EDNOS)

The fourth edition of the Diagnostic and Statistical Manual of Mental Disorders (DSM-IV) also includes a category entitled EDNOS.[11] This category is meant to include eating disorders that do not meet all of the criteria established for AN and BN. For example, if all criteria for AN were met in an adolescent girl except that she was >85% of her ideal body weight (IBW), the individual would be diagnosed with EDNOS. Binge-eating disorder (which is characterized by recurrent binges in the absence of the inappropriate compensatory behaviours that are symptomatic of BN) is also included in the EDNOS category.

## Subclinical Disordered Eating

Cultural promotion of an idealized and overly thin female shape is clearly demonstrated in the Western media.[22–24] and less obviously promoted in social interactions. Many individuals internalize the belief that a thin body is the ideal, and make efforts to change their body shape through strategies such as food restriction. For example, 37% of Canadian women classified as having a healthy body weight (BMI of 20.0–24.9) reported currently trying to lose weight.[25]

Dissatisfaction with one's body and attempts to lose weight occur even at young ages. Data from a recent USA multi-centre study of 2379 girls aged 9 and 10 years revealed that 40% were trying to lose weight, and 10% were classified as chronic dieters.[26] During adolescence, the preoccupation with weight continues and likely intensifies. In a representative sample of more than 16 000 high school students, roughly 60% were attempting to lose weight, the majority through restricting their food intake.[17] Thus, it appears that many girls and young women spend a large portion of their lives dissatisfied with their bodies and attempting to control their weight.

There are no diagnostic criteria for the subclinical eating disturbances. However, they have been measured and explored using self-report tests of eating attitudes and behaviours such as the Eating Attitudes Test,[27] the Children's Eating Attitudes Test,[28] the Restraint Scale,[29] the Three Factor Eating Questionnaire,[30] the Dutch Eating Behaviour Questionnaire,[31] and the Eating Disorders Inventory.[32]

## The Persistent Nature of Eating Disorders

Recovery from clinical eating disorders such as AN and BN is difficult to achieve. Recent studies report full recovery rates for AN ranging between 33[33] and 75%[34] after 7.5–15 years. In a review of the outcomes of AN from 119 studies in 5590 individuals, a recovery rate of approximately 47±19.7% was reported.[35] Of individuals with AN who did not recover, improvement in the symptoms of AN was reported in 33.5±17.8%, but in 21±12.8% of those with AN, the condition was chronic.[35] Although less is known about the long-term morbidity from BN, it is also a persistent disorder. A study of 221 patients with BN conducted 11.6±1.9 years after first presentation found that 30% still engaged in bingeing and purging; 11% still met all criteria for BN and 18.5% were classified with EDNOS.[36]

Even if recovery from an eating disorder occurs, it takes a long time. After two years of follow-up, none of 95 individuals with AN in one study demonstrated recovery, and after three years, the rate of recovery was only 1%.[34] The chronicity of eating disorders is further reflected by the fact that approximately 30% of patients relapse after discharge from a hospital treatment program[34] and up to 40% of patients with AN and 35% of those with BN relapse after a time of apparent full recovery.[33]

For a small but significant proportion of the eating disordered population, death results from complications of the disorder. AN is fatal in roughly 5% of cases.[35] The frequency of death as a consequence of severe AN is illustrated by the fact that longitudinal studies of bone health in adolescents with AN have lost participants because they died during the follow-up period.[37]

# The Impact of Eating Disorders and Disordered Eating on Bone

The consequences of eating disorders are numerous; both psychological and physical health can be severely affected and long-lasting adverse effects may occur. Compromised bone health is one possible result. Most notably, AN has been associated with reduced bone mass, osteopenia and osteoporosis, both at the time of active illness[3,38-45] and many years after recovery from the disorder.[46-48]

Eating disorders typically start during adolescence, which is a time of rapid growth and skeletal development. Between the ages of 9 and 15 years, the amount of bone mineral in the lumbar spine more than doubles in adolescent women[49,50] and women typically attain maximal adult bone density at the hip and spine by the age of 20 years.[49] Thus, if eating disorders during adolescence interfere with the ability to establish maximal peak bone density, risk for osteoporosis and fracture may be increased, especially later in life. Inadequate peak bone mass can cause the normal losses of bone occurring during perimenopause[51] and menopause[52] to produce risk for fragility fracture. Furthermore, as noted above, eating disorders typically follow a very protracted course and, as a consequence, their negative influence on bone could be exerted for quite some time.

## AN has Adverse Effects on Bone Health

A growing body of literature indicates that young people with AN have reduced BMD when compared to peers without the condition.[3,4,6-8,39,40,42-44,47,53-57] Because AN typically occurs when skeletal growth and bone mineral accrual are rapid, reduced bone mass may result from a combination of both the failure to attain what could have been the maximum bone density and the actual loss of bone tissue. Somewhat ironically, having been missed or hidden initially, a diagnosis of AN is sometimes made retrospectively during clinical investigation of osteoporosis and/or fracture.[47,58]

Lumbar and/or whole body BMD has been reported at $>2$ SD below normal in up to 50% of young women with AN.[3,5,39,59,60] Within the first year of diagnosis, those with AN have significantly reduced BMD, showing a 20–25% deficit in bone mass when compared to age- and sex-matched controls.[3] A recent study by Castro and colleagues[54] illustrated the adverse effects of AN on bone. Of 170 young patients with AN (mean age $= 15.2 \pm 1.4$ years), 44% ($n = 75$) were classified with osteopenia (defined as BMD $>1$ SD below the mean of healthy controls) at the lumbar spine and 25% ($n = 42$) with osteopenia at the femoral neck.[54] In another study of slightly older young women with AN (mean age $= 24.9 \pm 6.9$ years), approximately 85% ($n = 41$) had osteopenia.[39] In general, the incidence of osteopenia in adolescents and young women with AN seems to increase with increasing age (and thus, presumably, with the duration of illness).

## A History of AN is Associated with Increased Risk for Fracture

BMD measurements are useful because they can predict risk for fracture.[61] However, they cannot precisely identify who may experience a fracture, as other factors also contribute to fracture risk. Furthermore, BMD may not have the same predictive value at all ages.[61] Thus, it is informative to also examine rates of overt fracture among those with AN.

Fractures have been reported in association with AN in both relatively short-term prospective studies and longer retrospective investigations (some of which have included women who appear to have recovered from AN). In women with low bone mass associated with amenorrhoea and AN, fractures have occurred in the absence of significant trauma at young ages[7,39,47] and after many years of AN and/or apparent recovery.[40,58] Affected women may experience more than one fracture; for example, in four women with AN (aged 22–34 years), a total of 13 stress fractures were present (at the femoral neck, ribs, sacrum, and tibia).[62]

Recently, a population-based cohort study from Denmark estimated fracture risk for all Danes diagnosed with AN, BN or EDNOS over a 22-year period (from 1977 to 1998).[63] Each of 4385 individuals diagnosed with an eating disorder during that time was matched to three age- and gender-matched controls and incidence rate ratios (IRR) for fracture were calculated. An increased risk for fracture following a diagnosis of AN was reported (IRR $= 1.98$, 95% confidence interval $= 1.60$–$2.44$, $P < 0.01$).

In another retrospective cohort study, medical records were used to identify 208 cases of AN in the residents of Rochester, Minnesota between the years 1935 and 1989.[64] Incidence of fracture among those diagnosed with AN during this time was compared to the number of fractures that would have been expected based on age- and sex-specific rates of fracture derived from the larger local population. Forty years after a diagnosis with AN, the cumulative incidence of fracture at any site was 57%, compared to an expected rate of 42% ($P < 0.001$). In those previously diagnosed with AN, the standardized IRR for fracture was 3.9 for the distal forearm ($P < 0.001$), 4.4 for vertebrae ($P < 0.001$), and 3.0 for the proximal femur (not statistically significant).[64] This reflects an increased risk for fracture at sites most commonly affected by osteoporosis. Based on these data, the long-term risk for any fracture in women previously diagnosed with AN is 2.9 times greater than that for the general population (with slightly greater risks for fractures at sites typically associated with osteoporosis).[64]

Smaller prospective studies with shorter periods of follow-up have reported an even higher risk for fracture in women with AN. For example, a 7-fold increase in fracture risk was reported by Rigotti and colleagues.[8] However, other retrospective studies have concluded that while BMD may be reduced, fracture risk may not be elevated among those who have recovered from AN,[65] although sample sizes may not have been sufficient to detect an increase in fracture risk. Thus, the evidence from retrospective, population-based cohort studies is useful because it is possible to track a larger number of people for a longer time than would be possible in prospective studies.

## BN has Less Effect on Bone

Reduced BMD occurs less frequently with BN than with AN.[6] The BMD of women with BN is significantly higher than that of women with AN; in fact, it tends to be in the range for healthy women of comparable age.[20,66] However, while it appears that bone mass is typically conserved in those with BN, there are also reports of BMD reductions in connection with the disorder.

In those who experience reduced BMD in association with BN, it is likely due to factors such as amenorrhoea and low body weight which may be present in BN but are more commonly associated with AN.[45] For example, the two lowest values for BMD reported in a study by Howat and colleagues were in those with BN who also had a history of amenorrhoea.[20] However, somewhat surprisingly, normal BMD has also been reported in young women with BN who have amenorrhoea and/or a history of AN.[45,53] In the one patient with BN (age = 23 years, BMI = 17.8) included in a study by Mazess and colleagues, despite a 3-year history of amenorrhoea, total bone mineral content (BMC) and regional BMD measurements of the skeleton were normal (within 10% of the healthy age-matched controls).[40] Without an assessment of her BMD prior to the onset of BN, however, it is not possible to determine whether this patient had the genetic potential to have above average BMD, and lost bone into the normal range.

Unfortunately, most studies of eating disorders and bone health include few subjects with BN. Thus, while bone measurements for individuals with BN tend to fall within the normal range, the limited data set makes it difficult to draw firm conclusions.

## Subclinical Eating Disorders may have a Negative Impact on Bone

Although much less research has been done to assess whether subclinical disordered eating affects bone health, evidence is beginning to accumulate to suggest that it may. Much of the work in this area has focused on women with high levels of *cognitive dietary restraint*, which reflects the perception that one is constantly attempting to monitor and limit food intake in an effort to control body weight.[30] Researchers have consistently found high levels of cognitive dietary restraint in a substantial proportion of young women.[67,68] Given its prevalence, the possible health correlates of a restrained pattern of eating need to be determined. If data suggesting a negative impact of cognitive dietary restraint on bone health are confirmed, the implications could be profound, given that subtle disturbances of eating behaviour are so common that they are almost normative among Western girls and women.

There is evidence to suggest that physiological differences exist between women with high cognitive dietary restraint and those with low restraint. For example, normal weight young women with high scores for dietary restraint are more likely to experience irregular menstrual cycles.[69] In addition, subclinical disorders of ovulation within regular, apparently normal, menstrual cycles appear to be more common in women with high cognitive dietary restraint. These include anovulatory cycles or cycles with short luteal phases.[10,70–72] Such subclinical disorders of ovulation have been prospectively associated with loss of, and explain 20% of the variance in, spinal trabecular bone.[73] Also, women with high levels of cognitive dietary restraint have been reported to have elevated urinary cortisol excretion[68] and salivary cortisol levels.[74] This suggests that constantly monitoring and attempting to limit food intake may represent a subtle stressor, activating the hypothalamic–pituitary–adrenal (HPA) axis and leading to elevations in cortisol.[68] Strong prospective data from patients with Cushing's disease and high endogenous cortisol levels[75] as well as pharmacological dose glucocorticoid therapy[76] show that elevated cortisol is a risk factor for bone loss as well as for fracture. Moreover, even elevations of cortisol within the normal range have been associated with increased fracture risk in older adults.[77]

At this point, few studies have been conducted to directly assess whether subtle eating disturbances are negatively associated with bone health. In one report, however, high levels of cognitive dietary restraint were found to be an independent negative predictor of total body BMD and BMC in normal weight young women.[78] Although those with the highest levels of cognitive dietary restraint reported higher levels of physical activity, the bone benefit associated with exercise appeared to be eliminated by the high levels of restraint.[78] Also, in a 2-year prospective study of girls initially 10 years of age, scores on the oral control subscale of the children's Eating Attitudes Test negatively predicted baseline, 2-year, and 2-year change in total body and spinal BMC.[79]

More work is required in this area to elucidate relationships between these eating disturbances and bone health. Because subclinical eating disorders may share some characteristics with overt eating disorders such as AN (*e.g.*, menstrual disturbances, concern with weight and shape), it is reasonable to suspect that they may exert negative influences on bone through many of the same mechanisms, which are discussed below.

# The Multifactorial Aetiology of BMD Deficits Associated with AN: Risk Factors for Reduced BMD

The exact mechanisms through which AN and subclinical disturbances such as high cognitive dietary restraint may be associated with reduced BMD are uncertain. The negative impact of AN on bone is likely due to the combined effects of several factors, resulting in reductions in bone tissue through altered rates of bone accretion and resorption. The most likely contributors to the reduced bone mass observed in AN are listed in Table 28.2. Their association with, and potential contribution to, compromised bone health are summarized below.

## Factors Associated with Nutritional Status and Dietary Intake

### Low Body Weight

Reduced body mass, reflected by a low BMI or percentage ideal body weight (%IBW), has been associated with reduced BMD in many studies of individuals with AN.[3,6,7,37,42,45,47,48,53] Body weight is one of the most important contributors to peak bone mass and the duration of extreme thinness may contribute to the reduced BMD observed in those with AN.[80]

Because body weight is significantly reduced in AN, there is a concomitant reduction in the mechanical strain consistently applied to the skeleton. Mechanical strain, when applied to the skeleton, acts to sustain bone mass;[81] strain results from both the effects of one's body weight throughout normal activity as well as during physical exercise. It is reasonable to expect that this reduction in mechanical strain could contribute to reduced bone mass in AN.

**Table 28.2.** *Potential contributors to low bone density, bone loss and fracture in Anorexia Nervosa*

*Compromised Nutritional Status*
- Reduced body mass (low body weight, including low muscle mass and low fat mass)
- Malnutrition (including suboptimal intakes of calcium, vitamin D, protein)

*Alterations in Hormonal Status*
- Reduced insulin-like growth factor I (IGF-I)
- Menstrual disturbances such as long cycles, oligomenorrhea or amenorrhea (with low estradiol levels, often with concomicant low progesterone levels)
- Hypercortisolemia
- Low active thyroid hormone levels

In addition to reduced bone-loading, lower weight may also contribute to a slightly reduced oestrogen level (*via* lower levels of aromatisation in adipocytes and muscle cells) which would decrease antiresorptive activity.[82] As weight decreases in AN, body composition is also affected. Both fat mass and fat-free mass decrease, although adipose tissue is lost to a greater extent than lean tissues.[83] This has implications for bone health, given that a recent prospective twin study has shown that increases in muscle (lean body mass) are most important to bone change prior to menarche while afterwards, fat mass plays an important role in BMD.[84]

Weight may be a key determinant of BMD in women with AN, but it is clearly not the only factor contributing to BMD deficits. For example, when the regression coefficient of BMC on body weight was compared between patients with AN and healthy controls, it was found to be twice as high in the patients ($0.47 \pm 0.13$ *vs.* $0.23 \pm 0.04$ g kg$^{-1}$).[6] This suggests that other factors associated with low weight in those with AN also influence bone mass.

## *Malnutrition: Suboptimal Nutritional Status and Reduced Intake of Key Micronutrients*

Although it is reasonable to expect that reduced intakes of calcium and other key nutrients may play a role in the compromised bone mass associated with AN, few researchers have obtained estimates of calcium intake because they were considered to be too unreliable.[6] Those who have attempted to measure calcium intake through a 24-h recall,[38] 1-week food records[7] or food frequency questionnaires[45] have reported low calcium intakes among women with AN. For example, 8 of 13 women with AN reported intakes below 200 mg day$^{-1}$.[7] Secondary hyperparathyroidism can result from low calcium and vitamin D intakes, thus contributing to bone loss.[85]

Inadequate protein intakes may also have adverse effects on bone.[86] And, as discussed below, suboptimal nutritional status also affects bone metabolism indirectly through hormone-mediated effects.[87]

## Hormonal Factors

### *Reduced Insulin-like Growth Factor I (IGF-I)*

IGF-I is a bone trophic hormone which is associated with nutritional status. Several studies have demonstrated that serum IGF-I levels are lower in young women with AN than in healthy controls.[39,42,88,89] *In vitro*, IGF-I has been shown to stimulate osteoblastic proliferation and the synthesis of bone matrix proteins,[90] and levels of IGF-I have been significantly correlated with lumbar BMD in those with AN ($r = 0.5$, $P = 0.047$).[42] In bone, IGF-I is a modulator of hormonal input and local growth factor activity and influences the activity of osteoblasts and the rate of bone formation.[91] It is possible that malnutrition alters bone turnover in AN through the reduction of IGF-I stimulated bone formation.[42] Although it is not currently clear to what extent circulating levels of IGF-I reflect their level in bone,[87] it appears that IGF-I levels rise as weight is regained.[92] This could lead to an increase in bone mass as well.

## *Menstrual Cycle Disturbances (Combined Oestrogen and Progesterone Deficiencies)*

Amenorrhoea can be either primary (resulting in late menarche if AN developed prior to the onset of menstruation) or secondary (when menses cease after the post-menarcheal development of AN). It results in a deficiency of circulating oestrogen and progesterone levels through hypogonadotropic hypogonadism, which can exert a negative impact on bone. The duration of amenorrhoea has been identified as a significant predictor of bone mass in women with AN,[5,6,8,37,39,44,45,47,48,93,94] although this finding is not universally consistent.[3,89]

By definition, a young woman with AN must have experienced menstrual cycle disruption for a minimum of 3 months.[11] However, the duration of lack of flow associated with the condition is highly variable. Some women with chronic AN experience years, even decades, of amenorrhoea.[40] Although amenorrhoea is not one of the diagnostic criteria for BN, many young women with BN also experience abnormal menstrual function. For example, Pirke and colleagues[19] reported amenorrhoea in roughly half ($n = 7$) of their subjects with BN and Sundgot-Borgen and colleagues[56] found a history of menstrual dysfunction in 68% ($n = 30$). Other researchers have also reported amenorrhoea or menstrual cycle disturbances in a substantial number of their participants with BN.[20,53] And, as noted above, women with high scores for cognitive dietary restraint are also more likely to experience disruptions to normal menstrual cycle activity such as subclinical ovulatory disturbances[10,70–72] and irregular cycles.[69]

In women with AN, amenorrhoea is associated with lower serum oestrogen and progesterone levels than those measured in healthy controls. Lower oestrogen levels are due to both a reduced ovarian production and a decreased conversion of oestrogen in the periphery[87] and progesterone is essentially absent in amenorrhoea, since it is primarily produced by the corpus luteum following ovulation. Consistent with earlier reports,[7] Lennkh and colleagues[42] found levels of these reproductive hormones similar to what would be observed in the early follicular phase of normally menstruating women; in women with AN, mean serum oestrogen concentrations were $25.5 \pm 17.8$ U ml$^{-1}$ and progesterone levels were $0.95 \pm 3.7$ ng ml$^{-1}$.[42]

Reduced reproductive hormone levels associated with menstrual cycle disturbances undoubtedly contribute to BMD reductions in this population. This is supported by the fact that low BMD has been noted among women who do not have an eating disorder, but who do have low levels of reproductive hormones. For example, reduced bone mass has been observed in instances of premenopausal amenorrhoea such as hypothalamic amenorrhoea[95,96] and hyperprolactinemia,[97] although osteoporosis is associated with those conditions less frequently than it is with AN. In addition, healthy, exercising women between the ages of 20–40 years who were in the placebo group of a randomized controlled trial of cyclic progestin and calcium lost a mean of two percent of their spinal BMD during one year if they had subtle (anovulation or short luteal length cycles) or obvious (oligomenorrhoea or amenorrhoea) menstrual cycle disturbances.[98] This bone loss occurred despite normal BMI values, dietary calcium intakes that were adequate, and regular exercise.

Sex steroid hormones affect the production, longevity, and activity of osteoclasts and osteoblasts as well as other factors involved in bone remodelling.[99] Oestrogen has long been associated with bone health. It inhibits bone resorption through actions on Interleukin-1 and Interleukin-6.[99] It may possibly enhance bone formation through actions on transforming growth factor β and IGF-I,[99] although most evidence suggests that oestrogen suppresses bone formation through the coupling of resorption and formation.[100] Finally, oestrogen can also influence calcium balance by influencing parathyroid hormone secretion, calcitriol, calcitonin, intestinal absorption of calcium and calcium absorption/excretion in the kidney.[99]

Although the actions of progesterone and progestogens on bone are not as firmly established, there are several lines of evidence that support the importance of normal levels of progesterone in the bone health of premenopausal women. Using a nested case-control design, data from a population-based longitudinal study of bone health revealed that women with low BMD (< 10th percentile of the distribution) had significantly lower levels of sex steroids (both oestrogen and progesterone metabolites) in their urine when compared to women with normal BMD (50th to 75th percentile of the distribution).[101] In another study, luteal phase length (which reflects progesterone exposure) accounted for 20% of the variance in trabecular BMD change in a prospective study of ovulatory women.[73] Oestradiol levels, however, did not relate to bone change in this sample with normal levels and regular cycles.[73] In addition, spinal BMD was shown to increase in a 1-year randomized placebo-controlled trial of cyclic medroxyprogesterone (10 days per cycle) in otherwise healthy, normal weight women.[98]

It is not currently well understood exactly how gonadal hormones and body weight interact to influence bone health.[80] It is clear that as the duration of amenorrhoea increases, the risk of reduced bone mass also increases. Furthermore, because oestrogen and progesterone treatment of premenopausal menstrual cycle disturbances does not fully prevent the progression of bone loss in AN (as discussed below), we can conclude that while amenorrhoea and low circulating reproductive hormone levels contribute to a reduction in bone mass, there are additional factors involved in the development of osteopenia in AN.[44]

## Elevated Levels of Cortisol

AN, like any severe calorie restriction, is associated with hypercortisole-mia.[102] Increased cortisol is a fundamental physiological response to an energy intake that is inadequate to meet the metabolic requirement.[103] In patients with an eating disorder, the psychological stress associated with the condition also contributes to increased cortisol levels, because cortisol secretion is elevated in association with increased activity of the HPA axis. In AN, hypercortisolemia apparently results from both this increase in cortisol secretion per kg of body weight as well as a concomitant reduction in cortisol metabolism.[104]

Elevated cortisol is known to be associated with bone loss, both in those with AN[53,66] and in other populations characterized by elevated cortisol levels, such as those with Cushing's disease.[75] Negative effects on bone occur as a result of both increased bone resorption and decreased bone formation. Bone resorption is increased with elevated cortisol due to its interference with the action of vitamin D at the level of the intestine, and therefore through reduction in calcium absorption.[105,106] This, if severe, produces secondary hyperparathyroidism.[107] Bone formation appears to be affected by elevated cortisol levels through two processes: excess cortisol likely contributes to the central suppression of ovarian function (leading to decreased oestradiol and progesterone production)[108,109] and also directly interferes with osteoblastic bone formation.[110]

Elevations in cortisol among those with clinical or subclinical eating disorders may be subtle, resulting in values at the high end of what is considered normal. For example, in a study of college-aged women, participants with high levels of cognitive dietary restraint had higher levels of cortisol than those with low restraint, but cortisol levels remained within the normal range.[68] It may be for that reason that not all studies have found a difference in cortisol levels between young women with AN and healthy young women free of eating disorders.

## Reduced Thyroid Hormone Levels

Thyroid hormones are essential to maintaining balance in the process of bone remodelling,[99] and physiological levels of thyroid hormones may be important in bone growth and maintenance. With severe energy restriction such as that which occurs with starvation or AN, the body attempts to conserve energy by producing less active free triiodothyronine ($T_3$) and thyroxin ($T_4$) and more reverse $T_3$ which is metabolically inactive.[111] Both young women with AN and those with BN have $T_3$ levels that are significantly lower than those of healthy controls ($P<0.01$).[19] Furthermore, a significant correlation was reported between $T_3$ levels and BMD in young women with AN ($r=0.59$, $P<0.01$).[89]

## Summary of Factors Contributing to Reduced BMD Associated with AN

Clearly, the aetiology of reduced bone mass in AN is multifactorial. The main factors associated with reduced peak bone mass are all present in AN: alterations in reproductive and other hormone levels, changes in weight and bone-loading, and alterations in nutritional status.[82] It is possible that a synergistic effect of dietary and hormonal patterns occurs in AN which disrupts the balance of bone formation and resorption, leading to the development of osteopenia, secondary osteoporosis, and increased fracture risk.[43] There is evidence that the factors discussed above contribute in an independent and additive fashion to the reduction in BMD observed in association with AN. For example, studies demonstrating that the BMD of women with AN is even lower than would be predicted on the basis of weight loss alone[9] or amenorrhoea alone[112] illustrate the independent and additive effects of low weight and amenorrhoea on BMD in AN.

# Bone Metabolism in AN

Bone remodelling is normally balanced between processes of formation and resorption. Because these two aspects of bone turnover are linked, increased bone resorption is usually accompanied by increased bone formation.[113] However, markers of bone turnover in women with AN indicate that AN is characterized by elevated rates of bone resorption accompanied by normal rates of bone formation. This was supported by Lennkh and colleagues[42] who found significantly higher markers of bone resorption, specifically serum carboxyterminal crosslinked telopeptide and urinary deoxypyridinoline, in women with AN than in healthy controls.[42] C-terminal type 1 propeptide, a marker of bone formation, did not differ between groups.[42] These data suggest that osteopenia associated with AN is likely due to high rates of bone resorption in the absence of equally high rates of bone formation. This pattern of bone turnover appears to be rather unique to states of increased cortisol such as AN.

Alterations in bone metabolism associated with AN may vary, depending on the type of bone (trabecular *versus* cortical). There appear to be regional differences in the effect of AN on bone, with the axial skeleton (which is primarily composed of trabecular bone) affected to a greater extent than appendicular regions (which are largely compact cortical bone).[43] For example, it has been observed that the lumbar spine is more sensitive to short-term reduction in BMD in AN[54] and that lumbar spine BMD tends to be relatively lower than that of the femoral neck or total body BMD in patients with AN.[40,54] These results suggest a greater involvement or sensitivity of trabecular bone than cortical bone in the reduced bone mass associated with AN. However, due to the slower turnover of cortical bone, less recovery of BMD is observed in key potential fracture sites in the appendicular skeleton (such as the hip).[8] In other words, although deleterious effects for bone are typically first observed in areas of the skeleton primarily composed of trabecular bone (suggesting a more rapid influence on this type of bone), improvements in bone mass (*e.g.*, following weight recovery and return of menses, as discussed below) are also often first noted in these areas. In contrast, cortical bone does not appear to be negatively affected by AN as quickly as trabecular bone. However, if the condition progresses to a stage at which negative effects can be detected in cortical bone, subsequent improvements in these areas may take longer to observe.

# Can Bone Mass be Recovered Following Recovery from AN?

Whether or not it is possible to regain BMD that was lost (or not accrued) in patients with AN has not yet been clearly established. The answer to this question is of considerable clinical importance. If recovery from AN is *not* associated with a corresponding recovery of BMD to normal levels, a history of AN during adolescence could result in permanently reduced BMD and increased risk for fracture, especially after the accelerated bone loss that occurs during perimenopause and following menopause.

Definitions of recovery from AN vary (making comparisons between studies more difficult), but they tend to be based on the recovery of weight and menstrual status without consideration of disturbances of body image or other symptoms of AN.[45] Comparisons between studies are further hindered by the considerable variability among individuals with AN with respect to the extent of recovery achieved and the changes in BMD over time. Although these factors make it difficult to draw generalisations from the literature, despite some suggestions to the contrary, currently available data indicate that full recovery of bone mass does not necessarily accompany recovery from AN. For many women, bone deficits remain following recovery from AN despite normalisation of body weight and improved nutritional status.

## Cross-sectional Studies of Bone and Recovery from AN

Ward, Brown and Treasure[93] examined BMD in 18 women who had recovered from AN (recovery was defined as a BMI > 18.5 and resumption of menses for at least 6 months). Dual photon densitometry was used to assess BMD at the lumbar spine and the femoral neck. At the time of the study, subjects had met the criteria for recovery from AN for a mean of 6 years (range = 1–31 years) and had a mean BMI of 22.[93] Median *t*-scores for BMD at the hip and lumbar spine were −1.39 and −1.30, respectively. According to World Health Organisation guidelines,[114] 12 of the 18 women were classified as having osteopenia (*t*-score between −1 and −2.5) and two women were classified with osteoporosis (*t*-score < −2.5). Due to its cross-sectional nature, this study does not provide insight into how BMD may have changed since the time the women had AN. However, more than three quarters ($n = 14$) of those who had recovered from AN were found to have compromised BMD, suggesting that normalisation of BMD does not necessarily occur with weight restoration and recovery from AN.

Additional cross-sectional studies have demonstrated a persistence of BMD deficits following recovery from AN. For example, a case report of a young woman who recovered from severe AN revealed that after 6 years of recovery, despite a normal body weight and regular menses, BMD at the lumbar spine, femoral neck, and femoral trochanter was still > 2 SD below the mean for healthy women of the same age.[115] In addition, Rigotti and colleagues[8] found that cortical BMD was not recovered following weight restoration (defined in this study as at least 80% of IBW) in 18 recovered women with AN.[8]

The report of Hartman and colleagues[65] is particularly informative because the women who had recovered from AN had been weight-restored for a median of 21 years and they had very low scores on the Eating Attitudes Test, suggesting that their eating attitudes and behaviours had also normalized. The 19 women who had recovered from AN had comparable BMI to 13 age-matched controls (BMI = 21.1 and 22.4 for subjects and controls, respectively, $P = 0.065$). However, in the women who had recovered from AN, the median BMD at the proximal femur was significantly lower than that of controls (0.922 g cm$^{-2}$ for subjects *versus* 1.073 g cm$^{-2}$ for controls, $P = 0.004$).[65] Although the median lumbar spine BMD of subjects (1.245 g cm$^{-2}$) was also lower than that obtained in controls (1.303 g cm$^{-2}$), this finding was not statistically significant ($P = 0.374$). Data from this study suggest that BMD did not attain expected values in women who previously had AN, even after they had been recovered for many years.

There is some indication that while weight normalisation may not be sufficient for recovery of BMD at specific sites, total body BMD may improve as BMI increases. For example, when Siemers and colleagues[43] grouped their patients with AN by degree of weight recovery, there was no difference in lumbar spine BMD in severely malnourished (BMI < 17.2; $n = 6$), partially weight-restored (BMI = 17.2–19.9; $n = 8$), and weight-restored (BMI > 20.0; $n = 6$) individuals.[43] However, normalisation of body weight seemed to have a positive effect on whole body BMD. Total body BMD was significantly correlated with BMI ($r = 0.51$, $P < 0.001$) and it increased from 0.981 g cm$^{-2}$ in the severely malnourished patients to 0.999 g cm$^{-2}$ in the partially weight-restored to 1.046 g cm$^{-2}$ in the weight-restored.[43]

Another cross-sectional study suggested that bone mass could reach expected values following recovery from AN. Valla and colleagues found that in young women (aged 18–30 years) who had had AN but subsequently regained their weight (duration of recovery > 6 years), BMD values were what would be predicted on the basis of current anthropometric data.[9] Regression analyses indicated that time since AN accounted for 0.57 of the variation in BMD of the total femur, and weight gain accounted for 0.13 of the variance (model $r^2 = 0.70$).[9] However, the results of this study must be interpreted in light of the fact that only self-report data were used to ascertain the prior existence of AN.

The report of Treasure and colleagues[55] is frequently cited in support of the idea that it is possible to achieve expected values for bone mass once recovery from AN has occurred. Their study was designed to compare the BMD of three groups of women: young women with active AN, women who had recovered from AN, and age-matched healthy controls with no history of AN. They reported decreased BMD associated with active AN and a normalisation of BMD with recovery from AN.[55] However, this study has several methodological limitations. Notably, the precise definition of recovery used to classify women in this investigation was not specified. Also, due to the cross-sectional nature of this study, the women who had recovered from AN did not have a baseline BMD measurement. Thus, it is not possible to ascertain whether bone mass had, in fact, been recovered in those women. In general, although cross-sectional studies can provide valuable insights, in the absence of serial measurements of BMD, it is impossible to determine whether recovery of BMD is achievable.

## Longitudinal Studies of Bone and Recovery from AN

There is some indication from longitudinal studies that bone mass may increase in the short-term following recovery from AN, particularly if recovery occurs during adolescence. Castro and colleagues studied BMD changes in a relatively large sample of adolescent women with AN ($n = 108$) over the course of 6–30 months (mean = $15.4 \pm 6.1$ months).[116] In a subset of patients ($n = 64$) who were considered to have recovered from AN (*i.e.*, attained a BMI $> 19$ and experienced return of menses), BMD increased. In those who had normal lumbar spine BMD at the start of the study (during the time of active AN), BMD increased with recovery from AN at approximately the same rate that one would expect in healthy adolescents of the same age. Interestingly, those who had low lumbar spine BMD at the start of the study ($z$ score $< -1$), demonstrated an increase in BMD of 9.1% per year at the lumbar spine and 4.5% per year at the femoral neck following recovery from AN, approximately three times the gains normally seen in healthy adolescents at that age.[116]

In another longitudinal study, BMC was assessed by DXA and found to increase following short-term ($< 6$ months) weight gain in 26 patients with severe AN.[117] Before weight gain, mean total body BMC was $2.1 \pm 0.2$ kg compared to $2.2 \pm 0.2$ kg afterwards ($P < 0.01$). Also, Davies and colleagues reported results of serial measurements of BMD from a subset of their patients ($n = 17$) for a mean observation period of 1.4 years.[6] Although no statistically significant recovery of bone mass was demonstrated, the trend was favourable. However, extended periods of observation are required to establish conclusively whether bone mass is recoverable.

Another longitudinal study of patients with AN ($n = 24$) and BN ($n = 14$) was designed to measure BMD and indicators of bone turnover prospectively over a period of $3.6 \pm 1.5$ years.[118] At baseline, when compared to a group of age-matched healthy controls ($n = 42$), the lumbar spine BMD was reduced in those with AN (mean $t$-score = $-1.08 \pm 1.51$) but not in those with

BN (mean $t$-score $= 0.05 \pm 0.93$). During the course of the study, nine participants with AN and five with BN demonstrated recovery. Individuals with chronic AN had significantly lower lumbar spine BMD than those who had recovered at follow-up ($0.93 \pm 0.13$ g cm$^{-2}$ vs. $1.14 \pm 0.13$ g cm$^{-2}$, $P < 0.01$). Among the AN patients who did not recover, an annual loss of lumbar spine BMD of $-3.7 \pm 4.9\%$ was observed, whereas in the recovered AN subgroup a small but statistically insignificant annual increase of $0.7 \pm 1.7\%$ was reported.

## Normalisation of BMD Appears Possible if Recovery from AN Occurs during Adolescence

It is difficult to draw firm conclusions regarding the prognosis for BMD following recovery from AN. Most currently available data are ambiguous due to methodological limitations of the research. With few exceptions, reports provide data on small samples. In addition, there is usually insufficient information regarding the severity of the disease prior to recovery, and recovery itself is defined in a variety of ways. Furthermore, differences in bone size between those with AN, those who have recovered, and healthy controls may not be accounted for.[41]

The above points notwithstanding, current data suggest that early recovery from AN (*i.e.*, recovery during adolescence) is a key indicator of the subsequent likelihood of normalisation of bone mass. It is clear that even short periods of AN have negative effects on bone health. However, recovery during adolescence would likely be associated with a shorter duration of AN, a shorter duration of amenorrhoea, and a reduced length of exposure to the other possible contributors to reduced bone mass (*e.g.*, malnutrition, hypercortisolemia). Thus, if recovery from AN (including the normalisation of weight and return of menses) occurs during adolescence when BMD is increasing, the prognosis for bone health after recovery from AN would be positive.

# Promoting Recovery from AN: Therapeutic Goals and Treatment

Currently, the three main objectives of a therapeutic plan to promote recovery from AN (and thus stop the progression of osteopenia) are improved nutritional status, weight normalisation, and the resumption of menses. The primary focus should be on normalisation of body weight through a balanced, calcium-rich, vitamin D sufficient dietary intake.[60] While calcium and vitamin D intake should be augmented with supplementation if dietary intake is inadequate, such recommendations are clearly not sufficient in the absence of weight gain.[87]

The progressive deterioration of bone health in adolescents with AN and amenorrhoea has been likened to that observed in menopausal women, with the resultant clinical decision to administer HRT.[119] Typically, an oral contraceptive pill (OCP) is prescribed. Combinations of Premarin and Provera (oestrogen and progesterone) or Premarin alone are also used in treatment plans, albeit less frequently.[120] Evidence regarding the efficacy of HRT in the treatment of AN is discussed below.

# Is There a Role for HRT in the Treatment of AN-Induced Amenorrhoea and Bone Loss?

Exogenous oestrogens function as powerful antiresorptive agents in oestrogen-deficient conditions. Thus, it is logical to consider their use in the treatment of osteopenia associated with eating disorders. While recovery from AN and the natural resumption of menses are the ideal means for reducing risk for compromised bone health, they can be difficult to achieve. Normalisation of weight and menstruation occur in only $59.6 \pm 15.3\%$ and $57.0 \pm 17.2\%$ of patients with AN, respectively.[35] Thus, HRT is an intuitively tempting treatment option for the prevention of progressive osteopenia in adolescents with AN and amenorrhoea. However, while data support a bone-protective effect of HRT in menopausal women[121] and on fracture in the recent Women's Health Initiative randomized, double-blind placebo-controlled trial,[122] its efficacy in the treatment of bone loss associated with AN has not been established.

Although there is no doubt that reduced oestrogen levels and amenorrhoea play a role in BMD deficits associated with AN, the use of oestrogen and progesterone replacement in its treatment is controversial. Somewhat surprisingly, despite the current lack of clear evidence in support of its use, HRT is widely prescribed to treat amenorrhoea associated with AN in an attempt to alleviate negative consequences for bone. A recent report indicated that 78% of physicians treating adolescents with AN prescribe HRT;[120] most begin the administration of exogenous hormones after a patient has been amenorrhoeic for between 6 and 18 months.[119] However, few physicians reported testing BMD in their patients with AN, and BMD does not appear to strongly influence the decision to prescribe HRT.[120]

Both retrospective studies and prospective trials have been conducted in an attempt to determine the usefulness of HRT in alleviating possible adverse effects on bone in young women with AN. These studies are highlighted in Table 28.3 and discussed below.

## Observational Data Provide Weak Support for Benefits to Bone

Observational studies have provided some suggestion that HRT may alleviate the negative consequences of AN-induced amenorrhoea on bone at the lumbar spine, but not at the femoral neck. The femoral neck is primarily composed of cortical bone, which appears to be affected differently by oestrogen than the trabecular bone that is the main component of the spine.[123] For example, a case report of a woman (age = 26 years) with active AN (BMI = 17) and severe

**Table 28.3.** *Exogenous Hormone Treatment: Associations with Bone in Women with Anorexia Nervosa*

| Study | Subjects[a](age) | Study Design | Type HRT (number of subjects; duration) | Results |
|---|---|---|---|---|
| *Observational Studies* Seeman et al. (1992)[44] | 65 with AN 52 controls (27.6±1.9 years for those who had taken OCP) | Cross-sectional | Estrogen-progestin as OCP (n=16; 31.8±8.3 months) | Mean lumbar spine BMD and total-body BMC and BMD, but not femoral neck BMD, was greater for those with OCP use than those without. |
| Patel, (1999)[123] | 1 with AN (26 years) | Case report | Estradiol valerate + norethisterone (n=1; 1 year) | Lumbar spine BMD increased from 0.36 to 0.51 g cm$^{-2}$. Femoral neck BMD did not change. |
| Karlsson et al. (2000)[41] | 77 with untreated AN 58 with AN receiving ERT 26 recovered from AN 205 healthy controls (mean 27.12 years) | Cross-sectional | Estrogen as Premarin (n=58; mean 4.3 years, range 1–16 years) | ERT-treated women had greater vertebral body width, L3 BMC and L3 areal and volumetric BMD. Compared to untreated women. Femoral neck BMC and areal and volumetric BMD differences did not reach statistical significance (P=0.06). |
| Zipfel et al. (2001)[118] | 24 with AN 14 with BN 42 controls (mean 25.31 years for patients) | Prospective observational | Estrogen-progestin as OCP (n=12 of AN patients; not specified for full group) | HRT associated with higher BMI and lumbar spine BMD (but it was not an independent predictor of lumbar spine BMD). |
| *Intervention Studies* Klibanski et al. (1995)[39] | 48 with AN (24.9±6.9 years) | Randomized trial | Estrogen-progestin as Premarin and Provera | No difference between groups in absolute or net change lumbar spine BMD. |

| Study | Subjects | Study design | Intervention | Results |
|---|---|---|---|---|
| | | | or OCP ($n = 22$; $1.57 \pm 0.89$ years) | |
| Hergenroeder et al. (1997)[126] | 24 Women with menstrual dysfunction (14–28 years) six had current diagnosis of AN or BN | Randomized controlled trial | OCP ($n = 5$), medroxyprogesterone ($n = 10$), or placebo ($n = 9$) (1 year) | Among those <70% IBW at baseline, HRT resulted in small increase in mean lumbar spine BMD. Lumbar spine and total-body BMC/D in OCP group was greater at 12 months than the other two groups when controlled for baseline BMC/D, weight, age. No significant difference in femoral BMC/D. Placebo and medroxyprogesterone groups not statistically different. |
| Golden et al. (2002)[38] | 50 with AN ($16.8 \pm 2.3$ years) | Non-randomized prospective intervention | Estrogen-progestin as OCP ($n = 22$; variable) | After 1 year, no differences in absolute or net change BMD at lumbar spine or femoral neck. |
| Gordon et al. (2002)[125] | 61 with AN ($17.8 \pm 2.9$ years) | Randomized trial | Oral DHEA ($n = 31$) or estrogen-progestin as OCP ($n = 30$) (1 year) | After controlling for weight gain, no treatment effect was observed in either group in lumbar spine or total hip BMD. |
| Grinspoon et al. (2002)[124] | 60 with AN and osteopenia ($25.2 \pm 0.7$ years) | Randomized trial | Estrogen-progestin as OCP ($n = 15$), rhIGF-I ($n = 16$), both ($n = 14$) or neither ($n = 15$) (9 months) | OCP only did not increase lumbar spine BMD. rhIGF-I only increased lumbar spine BMD. BMD increased the most in combined (OCP + rhIGF-I) group. |
| Munoz et al. (2002)[89] | 38 with AN ($17.4 \pm 1.5$ years) | Non-randomized prospective intervention | 50 ug ethinyl estradiol per day and 0.5 mg norgestrel ($n = 22$; 1 year) | No difference in lumbar spine BMD after 1 year HRT, or 12 months later. |

AN, Anorexia Nervosa; BMC, Bone Mineral Content; BMD, Bone Mineral Density; BN, Bulimia Nervosa; DHEA, Dehydroepiandrosterone; ERT, Estrogen Replacement Therapy; HRT, Hormone Replacement Therapy; IBW, Ideal Body Weight; OCP, Oral Contraceptive Pill and rhIGF-I, recombinant human insulin-like growth factor I.
[a]number is given for the number of subjects originally enrolled; subjects were lost to follow-up.

osteoporosis (*t*-scores of −7.0 at the lumbar spine and −3.9 at the femoral neck) documented an increase in BMD at the lumbar spine from 0.36 to 0.51 g cm$^{-2}$ following 1 year of HRT, but there was no effect on BMD at the femoral neck.[123]

Data from larger observational studies have also provided some support for a beneficial effect of exogenous reproductive hormone therapy at the lumbar spine, but not the femoral neck.[41,44] Evidence that HRT was beneficial at the lumbar spine was also obtained in a prospective observational study; however, when duration of HRT was entered into a stepwise multivariate regression analysis to determine the contribution of various factors to lumbar spine BMD at follow-up, it was not an independent predictor of BMD.[118]

While such reports of individual cases or small groups of patients provide some insights, systematic evaluations of the effects of HRT on bone in patients with AN are clearly required for conclusions to be drawn. To determine the effects of HRT on bone in young women with AN, data from intervention studies must be reviewed (with an emphasis on data from randomized trials).

## Intervention Studies Suggest HRT does not Benefit Bone in AN

Experimental studies of the effect of exogenous hormone administration on the BMD of young women with AN have not provided a great deal of support for its use. Recently, a trial was conducted to examine the impact of combined oestrogen-progestin therapy on BMD in adolescent women with AN (aged 13–21 years).[38] The HRT group (*n* = 22) received an OCP with between 20 and 35 mcg of ethinyl oestradiol in addition to the standard treatment of medical, psychological, and nutritional counselling. The remaining participants (*n* = 28) received standard treatment alone (which included the administration of 400 IU of vitamin D and supplemental calcium to bring daily intake to 1200–1500 mg). The BMD at the lumbar spine and left femoral neck was measured using DXA at baseline and then annually for the mean follow-up time of 23.1 ± 11.4 months. There was no group difference in BMD at either site after one year of follow-up; in fact, in absolute terms, BMD was either unchanged or reduced.[38] Unfortunately, small numbers of subjects remained by the end of the second year, making group comparisons difficult at that point. Although this study is limited by its lack of random assignment to groups (patients selected their treatment condition with guidance from their physician and family, and the two groups differed in terms of age and duration of illness at baseline), the data do not support the use of OCP therapy to alleviate possible low BMD associated with AN-induced amenorrhoea.

Klibanski and colleagues[39] found that after a mean treatment duration of 1.6 ± 0.9 years, the change in BMD of women with AN who were treated with HRT was not significantly different from that of the control group. However, there was a 4% increase in mean BMD for those who weighed <70% of their IBW at baseline when HRT commenced.[39] They concluded that while a particular subset of patients may experience some improvement in BMD with exogenous oestrogen-progestin treatment, HRT does not uniformly prevent the progression of osteopenia observed in women with AN.[39] Interestingly, an improvement in spinal BMD was noted among patients who experienced natural return of menses, suggesting that normal menstrual cycle recovery may have a different effect on bone than exogenous oestrogen and progestin treatment.[39]

When the HRT intervention trials in Table 28.3 are examined collectively, it can be seen that in all but one study, BMD change in women treated with HRT did not differ from those treated with placebo[38,39,89,124] or oral dehydroepian-drosterone.[125] The only study reporting a beneficial effect of HRT was that of Hergenroeder and colleagues. They found that lumbar spine and total body BMD and BMC increased at 1 year in women treated with OCP, compared to those treated with placebo or medroxyprogesterone.[126] However, only six of their 24 subjects had a current diagnosis of AN or BN, so their results are not strictly comparable to those of the other studies. Finally, Grinspoon and colleagues, who observed no benefit of an OCP alone, found that women treated with both OCP and recombinant human IGF-I experienced a gain in BMD.[124]

## Interpreting Data from Studies of HRT and Bone in AN

There may be several reasons why, compared to results observed in menopausal women, exogenous oestrogen-progestin treatment in young women with AN does not appear to be as effective in preventing the progressive loss of bone tissue. First, in most cases, OCPs are not used throughout the duration of AN, so it is possible that bone loss occurring before or after their use is responsible for their apparent inability to completely prevent a reduction in bone mass.[44] Second, it has been suggested that adolescents with AN may require higher doses of these hormones due to the fact that they are in the process of reaching their peak bone mass[87,127] However, this explanation is weak because the oestradiol equivalent levels in OCPs may be up to 4–5 times higher than those found in the normal menstrual cycle. Third, a transdermal route of administration may be more effective than an oral one,[128] a possibility requiring further research. And fourth, the limited duration of follow-up in many studies precludes the possibility of observing improvement in bone mass over a long time period. In addition, OCP use by young women without AN in a prospective 5-year study was associated with no change in spinal BMD although the control group experienced a significant bone gain.[129] Overall, it is likely that HRT alone simply cannot overcome the many factors which likely contribute to the pathogenesis of reduced bone mass in AN.[89]

Controversy remains surrounding the use of exogenous ovarian hormones as a therapy for young women with AN. Opponents to its use argue that, in addition to the lack of evidence of a positive role in preventing or treating AN-induced osteopenia, potential hazards could accompany the use of HRT in a treatment plan for those with AN. For example, the resulting withdrawal bleed could provide false impressions of normal menstrual function, adequate weight gain, and protection against further loss of bone.[130] However, advocates of the use of HRT cite data in some studies[8,39,66] that, while not statistically significant due to small numbers of subjects or other methodological limitations, nevertheless suggest a protective effect of HRT.[127]

# Limitations of the Current Literature and Directions for Future Research

The evidence of negative associations of eating disorders with bone health continues to accumulate, but most reports are characterized by methodological limitations that preclude clear conclusions at this point. Most studies are of clinical samples of women with eating disorders; there is a paucity of population-based data which may be more representative of the larger group of young women affected by eating disorders and disordered eating patterns. Although the secretive nature of eating disorders can result in difficulty identifying cases in community settings and obtaining the participation of non-clinical samples,[12] investigators' focus on clinical populations leads to the omission of less severe cases. Thus, the impact of less serious but more prevalent eating disorders on bone health is still open to question. Also, the majority of studies report BMD data as group means, thus obscuring the large degree of individual variation in effects on bone. A group mean can be misleading if the group contains individuals who differ with respect to many key variables (*e.g.*, age at onset of disorder, duration of amenorrhoea, length of time since recovery, *etc.*). Furthermore, there are few prospective investigations of the effects of AN on bone; most reports are retrospective observational studies. This limitation is common to the literature on the outcomes of AN as a whole.[35]

Additional research is required to better understand the impact of eating disorders and disordered eating on bone. Cross-sectional studies have provided useful insights, but more prospective investigations with multiple measurements of bone are required. Although ethical considerations may pose challenges when conducting research with a severely ill population, randomized controlled trials with long follow-up periods are needed to determine the possible efficacy of treatments for reduced bone mass resulting from AN. For example, given the uncertainty regarding the effectiveness of treatment with reproductive hormones, additional research on this topic is warranted, perhaps conducted with transdermal oestrogen and cyclic bio-identical oral progesterone. To have adequate power, such studies may need to be conducted as multi-centre trials. Additional retrospective case-control studies of individuals who have recovered from AN may also provide further insights, particularly if recovery is carefully characterized and includes attaining normal eating attitudes and body size perceptions, in addition to restoration of weight restoration and normal menstrual function.

Given suggestive evidence that subclinical levels of disordered eating may adversely affect bone density, it is also imperative to clearly establish whether this does occur, and if so, to characterize the prevalence and impact of these subtle disturbances of eating attitudes and behaviour at the population level. Finally, formal evaluation of whether non-pharmaceutical measures such as mind-body therapy can reduce the stress associated with eating in these women, and thus normalize cortisol secretion and reproductive function, could lead to useful approaches to prevent adverse effects on bone.

# Conclusions

It is clear that AN has a negative impact on bone health among women treated clinically for the disorder. Its effects are compelling and long-lasting as evidenced by the large proportion of young women with AN who have sub-optimal bone mass at the time of active illness, and the fact that BMD deficits and increased risk for fracture remain many years after apparent recovery. Although questions regarding the exact mechanisms through which AN impairs BMD persist, the observed BMD deficits likely result from the combined effects of reduced body mass, malnutrition (including inadequate calcium, vitamin D and protein intakes), menstrual cycle disturbances (*i.e.*, combined oestrogen and progesterone deficiencies), and changes to other hormonal parameters (such as IGF-I and cortisol). Alterations in bone metabolism have been noted in association with AN; it appears to be associated with increased bone resorption without a concomitant increase in bone formation. And despite its frequent use, oestrogen-progestin therapy does not appear to alleviate negative effects on bone in AN. Further study is required to clarify these issues and to determine the possible effects of BN and subclinical disordered eating (such as cognitive dietary restraint) on bone.

# References

1. A.W. Brotman and T.A. Stern, Osteoporosis and pathologic fractures in anorexia nervosa, *Am. J. Psychiatry*, 1985, **142**, 495–496.
2. J. Ayers, G. Gidwani and I. Schmidt, Osteopenia in hypoestrogenic young women with anorexia nervosa, *Fertil. Steril.*, 1984, **41**, 224–228.
3. L.K. Bachrach, D. Guido, D. Katzman, I.F. Litt and R. Marcus, Decreased bone density in adolescent girls with anorexia nervosa, *Pediatrics*, 1990, **86**(3), 440–447.
4. L.K. Bachrach, D.K. Katzman, I.F. Litt, D. Guido and R. Marcus, Recovery from osteopenia in adolescent girls with anorexia nervosa, *J. Clin. Endocrinol. Metab.*, 1991, **72**, 602–606.
5. B.M.K. Biller, V. Saxe, D.B. Herzog, D.I. Rosenthal, S. Holzman and A. Klibanski, Mechanisms of osteoporosis in adult and adolescent women with anorexia nervosa, *J. Clin. Endocrinol. Metab.*, 1989, **68**, 548–554.
6. K.M. Davies, P.H. Pearson, C.A. Huseman, N.G. Greger, D.K. Kimmel and R.R. Recker, Reduced bone mineral in patients with eating disorders, *Bone*, 1990, **11**, 143–147.
7. N.A. Rigotti, S.R. Nussbaum, D.B. Herzog and R.M. Neer, Osteoporosis in women with anorexia nervosa, *N. Engl. J. Med.*, 1984, **311**, 1601–1606.
8. N.A. Rigotti, R.M. Neer, S.J. Skates, D.B. Herzog and S.R. Nussbaum, The clinical course of osteoporosis in anorexia nervosa: a longitudinal study of cortical bone mass, *J. Am. Med. Assoc.*, 1991, **265**(9), 1133–1138.

9. A. Valla, I.L. Groenning, U. Syversen and A. Hoeiseth, Anorexia nervosa: slow regain of bone mass, *Osteoporos. Int.*, 2000, **11**, 141–145.

10. S.I. Barr, J.C. Prior and Y.M. Vigna, Restrained eating and ovulatory disturbances: possible implications for bone health, *Am. J. Clin. Nutr.*, 1994, **59**, 92–97.

11. American Psychiatric Association, *Diagnostic and Statistical Manual of Mental Disorders*, American Psychiatric Association, Washington, DC, 2000.

12. A. Wakeling, Epidemiology of anorexia nervosa, *Psychiatry Res.*, 1996, **62**, 3–9.

13. S.J. Emans, Eating disorders in adolescent girls, *Pediatr. Int.*, 2000, **42**, 1–7.

14. E. Fombonne, Anorexia nervosa: no evidence of an increase, *Br. J. Psychiatry*, 1995, **166**(4), 462–471.

15. D.E. Pawluck and K.M. Gorey, Secular trends in the incidence of anorexia nervosa: integrative review of population-based studies, *Int. J. Eating Disord.*, 1998, **23**, 347-352.

16. G.B. Forbes, L.E. Adams-Curtis, B. Rade and P. Jaberg, Body dissatisfaction in women and men: the role of gender-typing and self-esteem, *Sex Roles*, 2001, **44**(7/8), 461–484.

17. L. Kann, S.A. Kinchen, B.I. Williams, J.G. Ross, R. Lowry, C.V. Hill, J.A. Grunbaum, P.S. Blumson, J. Collins and L.J. Koibe, Youth risk behavior surveillance – United States, 1997, *Morb. Mortal. Weekly Rep.*, 1998, **47**(SS-3), 1–89.

18. R. Lowry, D.A. Galuska, J.E. Fulton, H. Wechsler and L. Kann, Weight management goals and practices among U.S. high school students: associations with physical activity, diet, and smoking, *J. Adolesc. Health*, 2002, **31**(2), 133–144.

19. K.M. Pirke, J. Pahl, U. Schweiger and M. Warnhoff, Metabolic and endocrine indices of starvation in bulimia: a comparison with anorexia nervosa, *Psychiatry Res.*, 1985, **15**(1), 33–39.

20. P.M. Howat, L.M. Varner, M. Hegsted, M.M. Brewer and G.Q. Mills, The effect of bulimia upon diet, body fat, bone density, and blood components, *J. Am. Diet. Assoc.*, 1989, **89**, 929–934.

21. P.E. Garfinkel, E. Lin, P. Goering, C. Spegg, D.S. Goldbloom, S. Kennedy, A.S. Kapler and D.B. Woodside, Bulimia nervosa in a Canadian community sample: prevalence and comparison of subgroups, *Am. J. Psychiatry*, 1995, **152**(7), 1052–1058.

22. E.O. Guillen and S.I. Barr, Nutrition, dieting, and fitness messages in a magazine for adolescent women, 1970–1990, *J. Adolesc. Health*, 1994, **15**(6), 464–472.

23. L.M. Groesz, M.P. Levine and S.K. Murnen, The effect of experimental presentation of thin media images on body satisfaction: a meta-analytic review, *Int. J. Eating Disord.*, 2002, **31**, 1–16.

24. M.M. Morry and S.L. Staska, Magazine exposure: internalization, self-objectification, eating attitudes, and body satisfaction in male and female university students, *Can. J. Behav. Sci.*, 2001, **33**(4), 269–279.

25. Health and Welfare Canada, *Canada's Health Promotion Survey 1990: Technical Report*, Ottawa, Minister of Supply and Services Canada, 1993.

26. G.B. Schreiber, M. Robins, R. Striegel-Moore, E. Obarzanek, J.A. Morrison and D.J. Wright, Weight modification efforts reported by black and white preadolescent girls: National Heart, Lung, and Blood Institute Growth and Health Study, *Pediatrics*, 1996, **98**(1), 63–70.

27. D.M. Garner, M.P. Olmsted, Y. Bohr and P.E. Garfinkel, The eating attitudes test: psychometric features and clinical correlates, *Psychol. Med.*, 1982, **12**(4), 871–878.

28. M.J. Maloney, J. McGuire, S.R. Daniels and B. Specker, Dieting behavior and eating attitudes in children, *Pediatrics*, 1989, **84**(3), 482–489.

29. C.P. Herman and J. Polivy, Restrained eating, In *Obesity*, A.J. Stunkard (ed.), Saunders, Philadelphia, 1980, pp. 208–225.
30. A.J. Stunkard and S. Messick, The three factor eating questionnaire to measure dietary restraint, disinhibition, and hunger, *J. Psychosom. Res.*, 1985, **29**(1), 71–83.
31. T. van Strien, J.E.R. Frijters, G.P.A. Bergers and P.B. Defares, The Dutch eating behavior questionnaire (DEBQ) for assessment of restrained, emotional, and external eating behavior, *Int. J. Eating Disord.*, 1986, **5**, 295–315.
32. D.M. Garner, Development and validation of a multidimensional eating disorder inventory for anorexia nervosa and bulimia, *Int. J. Eating Disord.*, 1983, **2**, 15–34.
33. D.B. Herzog, D.J. Dorer, P.K. Keel, S.E. Selwyn, E.R. Ekeblad, A.T. Flores, D.N. Greenwood, B.A. Burwell and M.B. Keller, Recovery and relapse in anorexia and bulimia nervosa: a 7.5-year follow-up study, *J. Am. Acad. Child Adolesc. Psychiatry*, 1999, **38**(7), 829–837.
34. M. Strober, R. Freeman and W. Morrell, The long-term course of severe anorexia nervosa in adolescents: survival analysis of recovery, relapse and outcome predictors over 10–15 years in a prospective study, *Int. J. Eating Disord.*, 1997, **22**(4), 339–360.
35. H.-C. Steinhausen, The outcome of anorexia nervosa in the 20th century, *Am. J. Psychiatry*, 2002, **159**(8), 1284–1293.
36. P.K. Keel, J.E. Mitchell, K.B. Miller, T.L. Davis and S.J. Crow, Long-term outcome of bulimia nervosa, *Arch. Gen. Psychiatry*, 1999, **56**(1), 63–69.
37. D. Baker, R. Roberts and T. Towell, Factors predictive of bone mineral density in eating-disordered women: a longitudinal study, *Int. J. Eating Disord.*, 2000, **27**, 29–35.
38. N.H. Golden, L. Lanzkowsky, J. Schebendach, C.J. Palestro, M.S. Jacobson and I.R. Shenker, The effect of estrogen-progestin treatment on bone mineral density in anorexia nervosa, *J. Pediatr. Adolesc. Gynecol.*, 2002, **15**, 135–143.
39. A. Klibanski, B.M.K. Biller, D.A. Schoenfeld, D.B. Herzog and V.C. Saxe, The effects of estrogen administration on trabecular bone loss in young women with anorexia nervosa, *J. Clin. Endocrinol. Metab.*, 1995, **80**(3), 898–904.
40. R.B. Mazess, H.S. Barden and E.S. Ohlrich, Skeletal and body-composition effects of anorexia nervosa, *Am. J. Clin. Nutr.*, 1990, **52**, 438–441.
41. M.K. Karlsson, S.J. Weigall, Y. Duan and E. Seeman, Bone size and volumetric density in women with anorexia nervosa receiving estrogen replacement therapy and in women recovered from anorexia nervosa, *J. Clin. Endocrinol. Metab.*, 2000, **85**(9), 3177–3182.
42. C. Lennkh, M. de Zwaan, U. Bailer, A. Strnad, C. Nagy, N. El-Giamal, S. Wiesnagrotzki, E. Vytiska, J. Huber and S. Kasper, Osteopenia in anorexia nervosa: specific mechanisms of bone loss, *J. Psychiatr. Res.*, 1999, **33**, 349–356.
43. B. Siemers, Z. Chakmakjian and B. Gench, Bone density patterns in women with anorexia nervosa, *Int. J. Eating Disord.*, 1996, **19**(2), 179–186.
44. E. Seeman, G.I. Szmukler, C. Formica, C. Tsalamandris and R. Mestrovic, Osteoporosis in anorexia nervosa: the influence of peak bone density, bone loss, oral contraceptive use, and exercise, *J. Bone Miner. Res.*, 1992, **7**(12), 1467–1474.
45. P.J. Hay, J.W. Delahunt, A. Hall, A.W. Mitchell, G. Harper and C. Salmond, Predictors of osteopenia in premenopausal women with anorexia nervosa, *Calcif. Tissue Int.*, 1992, **50**, 498–501.
46. W. Herzog, H.-C. Deter, W. Fiehn and E. Petzold, Medical findings and predictors of long-term physical outcome in anorexia nervosa: a prospective, 12-year follow-up study, *Psychol. Med.*, 1997, **27**, 269–279.

47. Y.M. Maugars, J.-M.M. Berthelot, R. Forestier, N. Mammar, S. Lalande, J.-L. Venisse and A.M. Prost, Follow-up of bone mineral density in 27 cases of anorexia nervosa, *Eur. J. Endocrinol.*, 1996, **135**, 591–597.

48. E.R. Brooks, B.W. Ogden and D.S. Cavalier, Compromised bone density 11.4 years after diagnosis of anorexia nervosa, *J. Women's Health*, 1998, **7**(5), 567–574.

49. J.-P. Bonjour, G. Theintz, B. Buchs, D. Slosman and R. Rizzoli, Critical years and stages of puberty for spinal and femoral bone mass accumulation during adolescence, *J. Clin. Endocrinol. Metab.*, 1991, **73**, 555–563.

50. J.-P. Bonjour, G. Theintz, F. Law, D. Slosman and R. Rizzoli, Peak bone mass, *Osteoporos. Int.*, 1994, **4**(Suppl 1), S7–S13.

51. J.C. Prior, Perimenopause: The complex endocrinology of the menopausal transition, *Endocr. Rev.*, 1998, **19**, 397–428.

52. H. Okano, H. Mizunuma, M. Soda, I. Kagami, S. Miyamoto, M. Ohsawa, Y. Ibuki, M. Shiraki, M. Suzul and H. Shibata, The long-term effect of menopause on postmenopausal bone loss in Japanese women: results from a prospective study, *J. Bone Miner. Res.*, 1998, **13**, 303–309.

53. M.M. Newman and K.A. Halmi, Relationship of bone density to estradiol and cortisol in anorexia nervosa and bulimia, *Psychiatry Res.*, 1989, **29**(1), 105–112.

54. J. Castro, L. Lazaro, F. Pons, I. Halperin and J. Toro, Predictors of bone mineral density reduction in adolescents with anorexia nervosa, *J. Am. Acad. Child Adolesc. Psychiatry*, 2000, **39**(11), 1365–1370.

55. J.L. Treasure, G.F.M. Russell, I. Fogelman and B. Murby, Reversible bone loss in anorexia nervosa, *Br. Med. J.*, 1987, **295**, 474–475.

56. J. Sundgot-Borgen, R. Bahr, J.A. Falch and L. Sundgot Schneider, Normal bone mass in bulimic women, *J. Clin. Endocrinol. Metab.*, 1998, **83**, 3144–3149.

57. M.T. Munoz and J. Argente, Anorexia nervosa: hypogonadotrophic hypogonadism and bone mineral density, *Hormone Res.*, 2002, **57**(suppl 2), 57–62.

58. Y.M. Maugars, J.-M. Berthelot, S. Lalande, C. Charlier and A.M. Prost, Osteoporotic fractures revealing anorexia nervosa in five females, *Revue du Rhumatisme (English Edition)*, 1996, **63**(3), 201–206.

59. S. Grinspoon, E. Thomas, S. Pitts, E. Gross, D. Mickley, K. Miller, D. Herzog and A. Klibanski, Prevalence and predictive factors for regional osteopenia in women with anorexia nervosa, *Ann. Int. Med.*, 2000, **133**(10), 790–794.

60. S. Zipfel, W. Herzog, P.J. Beumont and J. Russell, Osteoporosis, *Eur. Eating Disord. Rev.*, 2000, **8**, 108–116.

61. D. Marshall, O. Johnell and H. Wedel, Meta-analysis of how well measures of bone mineral density predict occurrence of osteoporotic fractures, *Br. Med. J.*, 1996, **312**, 1254–1259.

62. M.M. LaBan, J.C. Wilkins, A.H. Sackeyfio and R.S. Taylor, Osteoporotic stress fractures in anorexia nervosa: etiology, diagnosis, and review of four cases, *Arch. Phys. Med. Rehabil.*, 1995, **76**, 884–887.

63. P. Vestergaard, C. Emborg, R.K. Stoving, C. Hagen, L. Mosekilde and K. Brixen, Fractures in patients with anorexia nervosa, bulimia nervosa, and other eating disorders – a nationwide register study, *Int. J. Eating Disord.*, 2002, **32**, 301–308.

64. A.R. Lucas, L.J. Melton, C.S. Crowson and W.M. O'Fallon, Long-term fracture risk among women with anorexia nervosa: a population-based cohort study, *Mayo Clin. Proc.*, 1999, **74**, 972–977.

65. D. Hartman, A. Crisp, B. Rooney, C. Rackow, R. Atkinson and S. Patel, Bone density of women who have recovered from anorexia nervosa, *Int. J. Eating Disord.*, 2000, **28**, 107–112.

66. K.A. Carmichael and D.H. Carmichael, Bone metabolism and osteopenia in eating disorders, *Medicine*, 1995, **74**(5), 254–267.
67. J.A. Alexander and B.J. Tepper, Use of reduced-calorie/reduced-fat foods by young adults: Influence of gender and restraint, *Appetite*, 1995, **25**, 217–230.
68. J.A. McLean, S.I. Barr and J.C. Prior, Cognitive dietary restraint is associated with higher urinary cortisol excretion in healthy premenopausal women, *Am. J. Clin. Nutr.*, 2001, **73**, 7–12.
69. J.A. McLean and S.I. Barr, Cognitive dietary restraint is associated with eating behaviors, lifestyle practices, personality characteristics, and menstrual irregularity in college women, *Appetite*, 2003, **40**, 185–192.
70. M. Lebenstedt, P. Platte and K.M. Pirke, Reduced resting metabolic rate in athletes with menstrual disorders, *Med. Sci. Sports Exerc.*, 1999, **31**, 1250–1256.
71. U. Schweiger, R.J. Tuschl, P. Platte, A. Broocks, R.G. Laessle and K.M. Pirke, Everyday eating behavior and menstrual function in young women, *Fertil. Steril.*, 1992, **57**(4), 771–775.
72. S.I. Barr, K.C. Janelle and J.C. Prior, Vegetarian *vs.* nonvegetarian diets, dietary restraint, and subclinical ovulatory disturbances: prospective 6-mo study, *Am. J. Clin. Nutr.*, 1994, **60**, 887–894.
73. J.C. Prior, Y.M. Vigna, M.T. Schechter and A.E. Burgess, Spinal bone loss and ovulatory disturbances, *N. Engl. J. Med.*, 1990, **323**, 1221–1227.
74. D.A. Anderson, J.R. Shapiro, J.D. Lundgren, L.E. Spataro and C.A. Frye, Self-reported dietary restraint is associated with elevated levels of salivary cortisol, *Appetite*, 2002, **38**, 13–17.
75. S. Hough, S.L. Teitelbaum, M.A. Bergfeld and L.V. Avioli, Isolated skeletal involvement in Cushing's syndrome: response to therapy, *J. Clin. Endocrinol. Metab.*, 1981, **52**, 1033–1038.
76. J.D. Adachi, W.G. Bensen, B.W. Brown, D.A. Hanley, A.B. Hodsman, R. Josse, D.L. Kendler, B.C. Lentle, W. Olszynski, L.G. Ste-Marie, A. Tenenhouse and A.A. Chines, Intermittent etidronate therapy to prevent corticosteroid-induced osteoporosis, *N. Engl. J. Med.*, 1997, **337**, 382–387.
77. G.A. Greendale, J.B. Unger, J.W. Rowe and T.E. Seeman, The relation between cortisol excretion and fractures in healthy older people: results from the MacArthur Studies, *J. Am. Geriatr. Soc.*, 1999, **47**, 799–803.
78. J.A. McLean, S.I. Barr and J.C. Prior, Dietary restraint, exercise, and bone density in young women: are they related? *Med. Sci. Sports Exerc.*, 2001, **33**(8), 1292–1296.
79. S.I. Barr, M.A. Petit, Y.M. Vigna and J.C. Prior, Eating attitudes and habitual calcium intake in prepubertal girls are associated with initial bone mineral content and its change over 2 years, *J. Bone Miner. Res.*, 2001, **16**(5), 940–947.
80. R.P. Heaney and V. Matkovic, Inadequate peak bone mass, In *Osteoporosis: Etiology, Diagnosis, and Management*, 2nd edn, B.L. Riggs and L.J. Melton (eds.), Lippincott-Raven Publishers, Philadelphia, 1995, pp. 115–131.
81. H.M. Frost, Bone "mass" and the "mechanostat": A proposal, *Anat Rec.*, 1987, **219**, 1–9.
82. M. Sowers, Lower peak bone mass and its decline, *Bailliere's Clin. Endocrinol. Metab.*, 2000, **14**(2), 317–329.
83. L. Scalfi, A. Polito, L. Bianchi, M. Marra, A. Caldara, E. Nicolai and F. Contaldo, Body composition changes in patients with anorexia nervosa after complete weight recovery, *Eur. J. Clin. Nutr.*, 2002, **56**, 15–20.
84. D. Young, J.L. Hopper, R.J. Macinnis, C.A. Nowson, N.H. Hoang and J.D. Wark, Changes in body composition as determinants of longitudinal changes in bone

mineral measures in 8 to 26-year-old female twins, *Osteoporosis. Int.*, 2001, **12**, 506–515.

85. Standing Committee on the Scientific Evaluation of Dietary Reference Intakes, Food and Nutrition Board, Institute of Medicine, *Dietary Reference Intakes for Calcium, Phosphorus, Magnesium, Vitamin D, and Fluoride*, National Academy Press, Washington, DC, 1997.

86. G.R. Cloutier and S.I. Barr, Protein and bone health: Literature review and counselling implications, *Can. J. Diet. Pract. Res.*, 2003, **64**, 5–11.

87. V. Bruni, M. Dei, I. Vicini, L. Beninato and L. Magnani, Estrogen replacement therapy in the management of osteopenia related to eating disorders, *Ann. NY Acad. Sci.*, 2000, **900**, 416–421.

88. S. Grinspoon, H. Baum, K. Lee, E. Anderson, D. Herzog and A. Klibanski, Effects of short-term recombinant human insulin-like growth factor I administration on bone turnover in osteopenic women with anorexia nervosa, *J. Clin. Endocrinol. Metab.*, 1996.

89. M.T. Munoz, G. Morande, J.A. Garcia-Centenera, F. Hervas, J. Pozo and J. Argente, The effects of estrogen administration on bone mineral density in adolescents with anorexia nervosa, *Eur. J. Endocrinol.*, 2002, **146**, 45–50.

90. C.A. Conover, In vitro studies of insulin-like growth factor I and bone, *Growth Horm. IGF Res.*, 2000, **10**(Suppl B), S107–S110.

91. B.Y. Reed, J.E. Zerwekh, K. Sakhaee, N.A. Breslau, F. Gottschalk and C.Y.C. Pak, Serum IGF 1 is low and correlated with osteoblastic surface in idiopathic osteoporosis, *J. Bone Miner. Res.*, 1995, **10**(8), 1218–1224.

92. D.R. Counts, H. Gwirtsman, L.M.S. Carlsson, M. Lesem and G.B. Cutler Jr., The effect of anorexia nervosa and refeeding on growth hormone-binding protein, the insulin-like growth factors (IGFs), and the IGF-binding proteins, *J. Clin. Endocrinol. Metab.*, 1992, **75**, 762–767.

93. A. Ward, N. Brown and J. Treasure, Persistent osteopenia after recovery from anorexia nervosa, *Int. J. Eating Disord.*, 1997, **22**, 71–75.

94. A.E. Andersen, P.J. Woodward and N. LaFrance, Bone mineral density of eating disorder subgroups, *Int. J. Eating Disord.*, 1995, **18**(4), 335–342.

95. B.M.K. Biller, J.F. Coughlin, V. Saxe, D. Schoenfeld, D.I. Spratt and A. Klibanski, Osteopenia in women with hypothalamic amenorrhea: a prospective study, *Obstet. Gynecol.*, 1991, **78**(6), 996–1001.

96. T. Csermely, L. Halvax, E. Schmidt, K. Zambo, G. Vadon, I. Szabo and A. Szilagyi, Occurrence of osteopenia among adolescent girls with oligo/amenorrhea, *Gynecol. Endocrinol.*, 2002, **16**, 99–105.

97. B.M.K. Biller, H.B. Baum, D.I. Rosenthal, V.C. Saxe, P.M. Charpie and A. Klibanski, Progressive trabecular osteopenia in women with hyperprolactinemic amenorrhea, *J. Clin. Endocrinol. Metab.*, 1992, **75**(3), 692–697.

98. J.C. Prior, Y.M. Vigna, S.I. Barr, C. Rexworthy and B.C. Lentle, Cyclic medroxyprogesterone treatment increases bone density: a controlled trial in active women with menstrual cycle disturbances, *Am. J. Med.*, 1994, **96**, 521–530.

99. A.M. Kenny and L.G. Raisz, Mechanics of bone remodeling: implications for clinical practice, *J. Reprod. Med.*, 2002, **47**(1), 63–70.

100. B.L. Riggs, J. Jowsey, P.J. Kelly, J.D. Jones and F.T. Maher, Effect of sex hormones on bone in primary osteoporosis, *J. Clin. Invest.*, 1969, **48**, 1065–1072.

101. M. Sowers, J.F. Randolph, M. Crutchfield, M.L. Jannausch, B. Shapiro, B. Zhang and M. La Pietra, Urinary ovarian and gonadotropin hormone levels in premenopausal women with low bone mass, *J. Bone Miner. Res.*, 1998, **13**, 1191–1202.

102. L. Douyon and D.E. Schteingart, Effect of obesity and starvation on thyroid hormone, growth hormone, and cortisol secretion, *Endocrinol. Metab. Clin.*, 2002, **31**(1), 173–189.
103. R. Alvero, L. Kimzey, N. Sebring, J. Reynolds, M. Loughran, L. Nieman and B.R. Olson, Effects of fasting on neuroendocrine function and follicle development in lean women, *J. Clin. Endocrinol. Metab.*, 1998, **83**, 76–80.
104. J. Licinio, M.-L. Wong and P.W. Gold, The hypothalamic-pituitary-adrenal axis in anorexia nervosa, *Psychiatry Res.*, 1996, **62**, 75–83.
105. E. Canalis, Mechanisms of glucocorticoid action in bone: implications to glucocorticoid-induced osteoporosis, *J. Clin. Endocrinol. Metab.*, 1996, **81**, 3441–3447.
106. I.R. Reid, Glucocorticoid osteoporosis – mechanisms and management, *Eur. J. Endocrinol.*, 1997, **137**, 209–217.
107. T.J. Hahn, L.R. Halstead, S.L. Teitelbaum and B.H. Hahn, Altered mineral metabolism in glucocorticoid induced osteopenia: effect of 25-hydroxyvitamin D administration, *J. Clin. Invest.*, 1979, **64**, 655–665.
108. S.L. Berga, T.L. Loucks-Daniels, L.J. Adler, G.P. Chrousos, J.L. Cameron, K.A. Matthews and M.D. Marcus, Cerebrospinal fluid levels of corticotropin-releasing hormone in women with functional hypothalamic amenorrhoea, *Am. J. Obstet. Gynecol.*, 2000, **182**(4), 776–784.
109. S.L. Berga, T.L. Daniels and D.E. Giles, Women with functional hypothalamic amenorrhea but not other forms of anovulation display amplified cortisol concentrations, *Fertil. Steril.*, 1997, **67**(6), 1024–1030.
110. T.L. Chen, L. Aronow and D.L. Feldman, Glucocorticoid receptors and inhibition of bone cell growth in primary culture, *Endocrinology*, 1977, **100**, 619–628.
111. E. Roti, R. Minelli and M. Salvi, Thyroid hormone metabolism in obesity, *Int. J. Obes. Related Metab. Disord.*, 2000, **24**(Suppl 2), S113–S115.
112. S. Grinspoon, K.K. Miller, C. Coyle, J. Krempin, C. Armstrong, S. Pitts, D. Herzog and A. Klibanski, Severity of osteopenia in estrogen-deficient women with anorexia nervosa and hypothalamic amenorrhea, *J. Clin. Endocrinol. Metab.*, 1999, **84**(6), 2049–2055.
113. H.K. Vaananen, Mechanism of bone turnover, *Ann. Med.*, 1993, **25**(4), 353–359.
114. World Health Organization, *Assessment of fracture risk and its application to screening for postmenopausal osteoporosis, WHO Technical Report Series 843*, Geneva, WHO, 1994.
115. L. Kotler, L. Katz, W. Anyan and F. Comite, Case study of the effects of prolonged and severe anorexia nervosa on bone mineral density, *Int. J. Eating Disord.*, 1994, **15**(4), 395–359.
116. J. Castro, L. Luisa, F. Pons, I. Halperin and J. Toro, Adolescent anorexia nervosa: the catch-up effect in bone mineral density after recovery, *J. Am. Acad. Child Adolesc. Psychiatry*, 2001, **40**(10), 1215–1221.
117. C.I. Orphanidou, L.J. McCargar, C.L. Birmingham and A.S. Belzberg, Changes in body composition and fat distribution after short-term weight gain in patients with anorexia nervosa, *Am. J. Clin. Nutr.*, 1997, **65**, 1034–1041.
118. S. Zipfel, M.J. Seibel, B. Lowe, P.J. Beumont, C. Kasperk and W. Herzog, Osteoporosis in eating disorders: a follow-up study of patients with anorexia and bulimia nervosa, *J. Clin. Endocrinol. Metab.*, 2001, **86**(11), 5227–5233.
119. T.J. Silber, Resumption of menses in anorexia nervosa: new research findings and their clinical implications, *Arch. Pediatr. Adolesc. Med.*, 1997, **151**(1), 14–15.

120. E. Robinson, L.K. Bachrach and D.K. Katzman, Use of hormone replacement therapy to reduce the risk of osteopenia in adolescent girls with anorexia nervosa, *J. Adolesc. Health*, 2000, **26**, 343–348.
121. The Writing Group for PEPI. Effects of hormone therapy on bone mineral density: results from the postmenopausal estrogen/progestin interventions (PEPI) trial, *J. Am. Med. Assoc.*, 1996, **276**(17), 1389–1396.
122. J.E. Rossouw, G.L. Anderson, R.L. Prentice, A.Z. LaCroix, C. Kooperberg, M.L. Stefanick, R.D. Jackson, S.A. Beresford, B.V. Howard, K.C. Johnson, J.M. Kotchen and J. Ockene, Risks and benefits of estrogen plus progestin in healthy postmenopausal women: principal results from the Women's Health Initiative randomized controlled trial, *J. Am. Med. Assoc.*, 2002, **288**(3), 321–333.
123. S. Patel, Effect of hormone replacement therapy on bone density in a patient with severe osteoporosis caused by anorexia nervosa, *Ann. Rheum. Dis.*, 1999, **58**(1), 66.
124. S. Grinspoon, L. Thomas, K.K. Miller, D. Herzog and A. Klibanski, Effects of recombinant human IGF-I and oral contraceptive administration on bone density in anorexia nervosa, *J. Clin. Endocrinol. Metab.*, 2002, **87**(6), 2883–2891.
125. C.M. Gordon, E. Goodman, S.J. Emans, E. Grace, K.A. Becker, C.J. Rosen, C.M. Gundberg and M.S. Le Boff, Physiologic regulators of bone turnover in young women with anorexia nervosa, *J. Pediatr.*, 2002, **141**(1), 64–70.
126. A.C. Hergenroeder, E. O'Brian Smith, R. Shypailo, L.A. Jones, W.J. Klish and K. Ellis, Bone mineral changes in young women with hypothalamic amenorrhea treated with oral contraceptives, medroxyprogesterone, or placebo over 12 months, *Am. J. Obstetr. Gynecol.*, 1997, **176**(5), 1017–1025.
127. B. Cromer, In support of HRT, *J. Pediatr. Adolesc. Gynecol.*, 2001, **14**(1), 42–45.
128. Z. Harel and S. Riggs, Transdermal *vs.* oral administration of estrogen in the management of lumbar spine osteopenia in an adolescent with anorexia nervosa, *J. Adolesc. Health*, 1997, **21**, 179–182.
129. F. Polatti, F. Perotti, N. Filippa, D. Gallina and R.E. Nappi, Bone mass and long-term monophasic oral contraceptive treatment in young women, *Contraception*, 1995, **51**, 221–224.
130. D.K. Katzman, In opposition of HRT, *J. Pediatr. Adolesc. Gynecol.*, 2001, **14**(1), 39–41.

# CHAPTER 29

# Nutrition and Bone Health—Effects of Pregnancy and Lactation

CAROLINE KARLSSON and MAGNUS K. KARLSSON

Department of Orthopedics, Malmo University Hospital, SE-205 02, Malmo, Sweden Email: magnus.karlsson@orto.mas.lu.se

**Key Points**
- Most studies report that pregnancy leads to a bone mineral density (BMD) loss of 2–5%, that lactation leads to a BMD loss of 2–5% and that bone loss following pregnancy and lactation is reversed after weaning.
- Women who have had multiple pregnancies and a long total duration of lactation have similar or higher BMD than nulliparous women and have no different or lower fracture risk than nulliparous women.

## Introduction

The skeleton is an organ with high metabolic activity, and 10–15% of the skeleton is remodelled in the course of 1 year.[1] The bone mass or the bone mineral density (BMD), *i.e.*, the total amount of bone in the skeleton, or in a specific skeletal region, for instance the hip, is one of the best predictors of bone strength. BMD is also known to correlate with the incidence of fractures, and the reduction in BMD by one standard deviation (SD) or about 10% is known to double the fracture risk.[2] A high BMD is generally regarded as a sign of good bone health, although diseases with extremely high BMD exist, for instance osteopetrosis and Morbus Paget. By contrast, low BMD is generally regarded as an impairment of bone health, and a 1.0–2.5 SD reduction in BMD compared to young individuals of the same gender is called osteopenia, while a more than 2.5 SD reduction in BMD is called osteoporosis.[3] However, this low BMD causes no actual problems until the clinically relevant end point for low BMD occurs, namely a fracture, and it is a fact that many individuals are unaware that their bone health is impaired until this event.

Reduced BMD may occur as a result of a inadequate accrual of BMD during growth, leading to a reduced peak bone mass (PBM), the highest BMD value during the lifecycle, which occurs at the completion of skeletal maturation, or an increased rate of age-dependent bone loss, which inevitably follows PBM.[4-6] As the accrual of BMD during growth is more than double in magnitude compared to the loss of BMD during the rest of the life-cycle,[4] PBM seems to be of major significance for bone health throughout life. This hypothesis is supported by reports proposing that as much as 60% of the variance in BMD at age 65 depends on the PBM attained.[7]

BMD is predominantly regulated by genetic factors and about 60–80% of the variance in BMD, both accrual of BMD during growth and bone loss later in life, is determined by genetic factors.[7,8] However, environmental factors may also contribute significantly to the variance in BMD. Pregnancy and lactation are examples of two episodes in many women's lives, often recurring, which are followed by changes in a variety of factors all of which may influence the BMD. Changes in hormones, such as oestrogen and prolactin, calcium metabolism, physical activity, weight, soft tissue composition, diet, smoking habits and alcohol consumption, are all factors that may influence the BMD.[9] These factors change during pregnancy and lactation, some will hypothetically increase BMD and some decrease BMD, making it almost impossible to predict the development of BMD during a pregnancy and a period of lactation.

There has never been, and will never be, a randomised, double-blind, placebo-controlled trial (RCT) demonstrating that pregnancy and/or lactation reduce bone health and increase fragility and/or osteoporosis-related fractures in old age. Blinded RCT, the highest level of 'evidence', will never be conducted because neither the investigator nor the participant can be blinded to pregnancy and lactation. It is only when we descend further, into observational, case-control studies, and prospective and retrospective cohort studies, that data become available supporting the inference that pregnancies and lactation lead to loss of BMD, *i.e.*, reduce bone health[10] and in the case of young mothers also reduce PBM.[11] Case reports exist of women who develop, not only osteoporosis but also the clinical end point, a fracture, during pregnancy.[12,13] However, these studies are never hypothesis-testing. They were all subjected to many systematic biases and should be interpreted with caution. The data are referred to as evidence but causality can never be proven by this approach, it is at best hypothesis-generating. This level of evidence may be seriously flawed and no matter how many studies demonstrate an association between a high BMD loss during pregnancy and lactation, the possibility remains that the high BMD loss shown in pregnant women compared to non-pregnant women in observational studies reflects a systemic bias in sampling. Meta-analyses should not even be performed in observational studies, much less interpreted, as meta-analyses cannot control for systemic bias. However, lack of evidence is not proof of lack of efficacy, and when making our inferences today we must search and interpret at the highest level of evidence available at the time when we make our inferences.

Furthermore, most published studies discuss reproductional history and changes in BMD, but BMD is only a surrogate end point for bone health and fractures. Other factors such as bone size, skeletal architecture and collagen orientation, also contribute to bone health and bone strength. It is only when data prove that women with a history of pregnancies and lactation have a higher incidence of fractures than women who have not given birth, we can finally state that pregnancy and lactation lead to negative clinical implications as regards bone health.

The purpose of this review is to present the changes in bone regulatory factors which occur during pregnancy and lactation, and speculate how changes in these factors may affect BMD. We also evaluate data in the literature, supporting or opposing the hypothesis that pregnancy and/or lactation lead to short- and long-term deficits in BMD with increased fracture incidence. When scrutinising the studies to be included in the review, we have found basically two types of study designs: (i) prospective studies, controlled or uncontrolled, following BMD from before or at the start of a pregnancy till after delivery and through the lactation period and (ii) retrospective and prospective epidemiological, observational and case-control studies addressing the relations between parity and low BMD, or at best the clinical end point of osteoporosis, *i.e.*, a fracture. The main clinical limitation of the first study design is that only short-term data are presented, leaving long-term changes in BMD without an answer, and in the second type of study, as in all studies with this type of design, the risk of selection bias. The studies may imply an association between pregnancy and low BMD, but the causal relationship could be between an undefined factor leading to a high probability to be pregnant and the same undefined factor leading to low BMD, not between pregnancy and low BMD. When inferences are made on the basis of observational and case-control studies, we must never forget that they are at best hypothesis-generating, not hypothesis-testing. Moreover, retrospective, epidemiological studies cannot distinguish between the effect of the pregnancy period and that of the lactation period. Even while studying mothers who have only bottle-fed their babies, a risk of selection bias exists, as some hormonal disturbances may be impairing the milk production and thus force the mothers to bottle feed, but also affect the BMD disadvantageously.

# Research Questions

- Do pregnancy and lactation lead to changes in environmental factors, possibly affecting BMD?
- Do pregnancy and lactation result in impaired bone health?
- Does an association exist between parity and low BMD?
- Could pregnancies and lactation periods be regarded as a risk factor for fractures?

# Methodological Problems when BMD is Evaluated in Pregnant and Lactating Women

A variety of factors, such as lean body mass, fat content and bone size influence the BMD value found by dual energy X-ray absorptiometry (DXA), the clinically most widely used method to evaluate BMD and bone health.[9,14–16] There are reports, which imply that weight reduction in pre-menopausal, overweight women is followed by a reduction in BMD[17] and that weight reduction in non-obese women is followed by even higher BMD loss.[18] However, it is unclear as to whether the discrepancy in BMD loss in the two cohorts is based on actual discrepancies in BMD loss, or whether a discrepancy in the changes in the soft tissue in obese and normal women leads to a different estimation of bone loss by the DXA technique.

Weight, lean body mass and fat content are all factors that change during pregnancy and lactation, thus being confounding factors when changes in BMD are evaluated by the DXA technique.[19,20] Any estimation of changes in BMD during pregnancy and lactation, evaluated by the DXA technique, may be the result of changes in soft tissue composition, with no actual or only minor changes in the BMD. Adjusting the BMD data for changes in soft tissue composition could solve the problem, but nevertheless, the data must be interpreted with caution, because there are controversies as to whether and how the data should be adjusted. No consensus exists on how the data should be presented, unadjusted for weight or soft tissue composition[21] adjusted for changes in weight[22] or adjusted for changes in fat and lean body mass separately.[9]

Another problem when BMD is to be interpreted during and after pregnancy and lactation by the DXA technique, is the fluid shift that occurs, most markedly just after delivery.[23] The increase in extra-cellular volume and glomerular filtration rate in pregnancy, together with the altered distribution of tissue volume, resulting from the development of the foetal, and changes in the placental and mammary compartments, makes measurements even more difficult to interpret.[24] All these changes could lead to a major difference in body water content during pregnancy, most obvious during the days preceding and following delivery, when a delivered mother often loses 5–10 kg in body weight. The changes in fluid content will inevitably influence the estimation of BMD, when using the DXA technique, as well as the evaluation of soft tissue composition and calcium metabolism.[9] The fluid shift often normalises within the first postpartum week, but if the end point measurements in a prospective study are conducted in close conjunction with the delivery, the fluid distribution will definitely influence our estimation of BMD.

A third problem to be mentioned is the ethical problem when BMD are evaluated by the DXA technique during pregnancy. The safety of the foetus when using ionising radiation must always be in focus, especially when measuring the axial skeleton of the mother. The radiation will then inevitably include the foetus, and such measurements should be avoided. The baseline measurements used in prospective studies are usually performed before conception, and in many studies, years before the pregnancy in question, since it is difficult to examine the mother just weeks before conception and inadvisable to perform the baseline measurement just after conception. For this reason, prospective studies are also usually presented with no measurements done during pregnancy, making the prospective data even more difficult to interpret.

# Changes in Determinants of BMD During a Pregnancy

## Nutrition

Nutrition is an important factor in the development and maintenance of BMD. Approximately 90% of BMD consists of calcium and phosphorus. Other dietary components such as protein, magnesium, zinc, copper, iron, fluoride, vitamins D, A, C and K are also included and required for normal bone metabolism, while some, as for example caffeine and alcohol, exert a negative influence on BMD.[25,26] Smoking and alcohol intake have been in epidemiological studies associated with low BMD[27] and during pregnancy, many women reduce both smoking frequency and alcohol consumption, changes that may affect BMD.

However, as the amount of calcium is so dominant in the skeleton and highly important for BMD, calcium is probably the most extensively studied nutritional element. Strong evidence exists that the amount of calcium in food is vital for skeletal development during growth[1] and two meta-analyses undertaken by Cummings *et al.*[2] and Welten *et al.*[28] conclude that calcium intake is positively associated with BMD in pre-menopausal women. Calcium seems to function as a threshold nutrient, *i.e.*, calcium intake is relevant up to a threshold intake only, and adding more calcium above this level will not improve BMD.[29] Adequate calcium intake has, in some studies, even been shown to not only prevent BMD loss, but also reduce the fracture risk in both pre- and post-menopausal women.[30]

During pregnancy, the mother provides the foetus with 25–30 g of calcium to support the development of the foetal skeleton.[31] The rate of calcium accretion in the foetus increases from approximately 50 mg per day at 20 weeks gestation to 330 mg per day at 35 weeks, *i.e.*, the calcium demand increases with the duration of the pregnancy, as most growth of the foetal skeleton takes place from mid-pregnancy and thereafter.[32] The increased calcium demand in women during pregnancy is met by major changes in calcium metabolism. The concentration of 1,25-dihydroxyvitamin (calictriol) increases, while the concentration of parathyroid hormone and calcitonin are almost unchanged during pregnancy.[24,32] As a result, calcium absorption is higher and urinary calcium excretion less during pregnancy than both before conception and after delivery. The higher absorption rate is more obvious in early-to-mid-pregnancy and precedes the period of highest calcium demand of the foetus.

Studies have been conducted on the influence of calcium intake on BMD in pregnancy. Most data infer that calcium supplements in pregnant women with normal or high calcium intake has no or little effect on BMD.[33] There is weak evidence that pregnant women with very low calcium intake may benefit from calcium supplementation.[34,35] As regards the foetus, most reports imply that physiological mechanisms ensure an adequate supply of calcium for the development of the foetal skeleton, even if the input from the maternal diet is low.[24,32]

In summary, the increased calcium demand of the foetus would hypothetically lead to lower maternal BMD during pregnancy. However, the maternal absorption of calcium adapts to the required level, and calcium supplementation seems to have a possible effect on BMD only in pregnant women with very low calcium intake. Reduced smoking and reduced alcohol consumption during pregnancy would, hypothetically, increase the BMD of the mother.

## Body Weight

During pregnancy there is an increase in maternal weight as well as an increase in fat content and lean body mass, partly due to changed nutritional habits.[9,36] As discussed in Introduction, changes in soft tissue composition during pregnancy also lead to methodological problems when the BMD is to be interpreted. The discrepancies in soft tissue composition between the two measurements can partly or totally explain discrepancies in the BMD values measured.[37] Moreover, increased weight, as well as increased fat content and lean body mass, are all factors known to increase BMD due to the increased mechanical load and increased production of hormones in the increased fat tissue.[38,39] High body weight is also associated with low fracture risk in post-menopausal women.[40,41] In summary, the weight and soft tissue changes during pregnancy would, hypothetically, lead to an increase in BMD.

## Changes in Hormones may Influence the BMD During Pregnancy

Oestrogen is probably the most important of the sex steroids, which regulate BMD in women. It not only plays a vital role in the regulation of the skeletal growth but also in the regulation of BMD in adult women.[42] As the bone remodelling cycle occurs on the surface of the skeleton, changes in bone turnover are usually more obvious in trabecular bone due to the higher surface/volume ratio in cancellous compared to cortical bone.[1] The oestrogen withdrawal at menopause is followed by an increased rate of bone remodelling. Bone formation increases, but as bone resorption increases even more, the result is a rapid loss of mainly trabecular BMD.[42,43] The mechanisms by which oestrogen exerts its effects are not fully understood, but the effects of oestrogen are mediated both through direct effects on oestrogen receptors[44] found on both osteoblasts[45] – the cells responsible for bone formation, and osteoclasts[46] – the cells responsible for bone resorption. Also indirect effects of oestrogen, such as an increased intestinal calcium absorption and increased renal calcium absorption are of importance. Oestrogen also stimulates the synthesis of 1,25 $(OH)_2$ D-vitamin in the kidney and reduces the PTH action on bone resorption, both conferring anabolic effects on the skeleton.[42] Furthermore, it seems that oestrogen interacts with other factors so that the bone preserving effect of oestrogen is enhanced by adequate nutrition and a high calcium intake[47] and high physical activity.[48]

The levels of oestrogen fluctuate during the normal menstrual cycle in women. It has been reported that the low oestradiol levels during the follicular phase in eumenorrheic women is associated with increased bone resorption, evaluated by bone metabolic markers.[49] Menstrual irregularities are also associated with reduced bone mass[50] and the most devastating condition, anorexia nervosa, is often followed by a major deficiency in BMD.[51] Thus, oestrogen is a central reproductive hormone associated with bone health, but progesterone and androgens may also affect BMD.[52] Among non-steroid hormones that may change during pregnancy, thyroid hormones,[53] growth hormone (GH), insulin-like growth factor (IGF-1)[54] and prolactin,[55] are all hormones that are assumed to influence the BMD.

Six weeks after conception, the placenta takes over the hormonal production from corpus luteum and progressively increases the production of oestrogens, progesterone, prolactin, placenta lactogen (hPL) and human choriongonadotropin (hCG). The most predominant oestrogen is oestriol, but oestradiol and oestron also increase during pregnancy. Pregnancy is thus characterised by higher levels of estrogens than in the inter-gravid state.[56] The levels of progesterone also increase at the beginning of a pregnancy, followed by a small decrease until the 11th week, after which a continuous increase occurs during the remainder of the pregnancy.

In summary, there are a variety of hormonal changes during pregnancy which may affect BMD, and it is all but possible to calculate the specific contribution of each hormone to the variance in BMD. As oestrogen is regarded as the most important regulatory sex hormone for the skeleton, changes during pregnancy would, hypothetically, increase BMD.

## Physical Activity

Physical activity is also an environmental factor known to influence BMD.[57] An increased level of physical activity increases BMD, predominantly during growth, but probably also to some extent during adult life.[57] Even if women with normal pregnancy are recommended to continue with physical activity, pregnancy often predisposes a woman to reduce her normal activity level, at least during the latter part of the pregnancy. In summary, the reduced physical activity seen in many pregnant women would, hypothetically, decrease BMD. Tissue changes during pregnancy would, hypothetically, lead to an increase in BMD.

# Pregnancy and Bone Health

## Bone Mineral Density

In all studies discussed in this review, relative changes are shown, with the significance level and confidence interval omitted for brevity, but all changes are significant unless otherwise stated. No RCTs exist to evaluate whether BMD is lost during pregnancy. Thus we have to rely on studies with a lower evidence level. Karlsson *et al.* reported in a cross-sectional, case-control study including 73 women, measured within a few days after delivery, a 7.6% lower lumbar spine BMD and a 3.9% lower total body BMD, after adjustment for differences in soft tissue composition, compared to 55 age- and gender-matched controls.[9] To our knowledge, the study presents the first BMD data adjusted separately for fat and lean soft tissue composition in an evaluation of the effect of pregnancy on BMD. The BMD in the delivered mothers was close to −1 SD, a reduction in BMD known to double the fracture incidence.

Several prospective publications have supported the contention that a reduction in BMD does occur during pregnancy, but at a biologically non-significant magnitude (Figure 29.1). Drinkwater *et al.* reported in a prospective, non-controlled case-control study including six pregnant women, a reduction in femoral neck BMD of 2.4% and in radial shaft BMD of 2.2% during a pregnancy.[58] However, as all women in this study were athletes who reduced their activity level during their pregnancy, these data are difficult to interpret. Black *et al.* reported that both spine BMD and total hip BMD decreased by 3.2% during pregnancy in 10 women.[59] More *et al.* reported that ultra-distal forearm BMD decreased by 4.9% during pregnancy in 38 women[60] and Holmberg-Mattila *et al.* reported that spine BMD decreased by 3% during pregnancy in five women.[61] Naylor *et al.* found that pelvis BMD decreased by 3.2% and spine BMD by 4.6% while arm BMD increased by 2.8% and leg BMD by 1.9% during the same period in 16 pregnant women, and explained the discrepancy by the fact that the BMD loss is found only in trabecular bone.[62] Ritchie *et al.* reported a BMD loss in 14 women far higher than in virtually all other published studies, with the finding that spine BMD decreased by 9% during pregnancy. However, this study used the computed tomography technique (CT), for the evaluation of the BMD loss, raising the question concerning methodology.[63]

In contrast, Sowers *et al.* in another, prospective, case-control study including 32 pregnant women, found no differences in BMD when the pre-pregnancy and postpartum values were compared, and no difference in BMD changes when pregnant and non-pregnant women during this period were compared.[21] However, the follow-up period was 475 days for the cases and 669 days for the controls, a fact that could influence the inferences.[21] Similar results were presented by Cross *et al.* whose results of a prospective, controlled study, including 10 pregnant women, showed that BMD changes did not differ between cases and controls during pregnancy.[33] However, the small sample size implies a high risk of making a type II error.

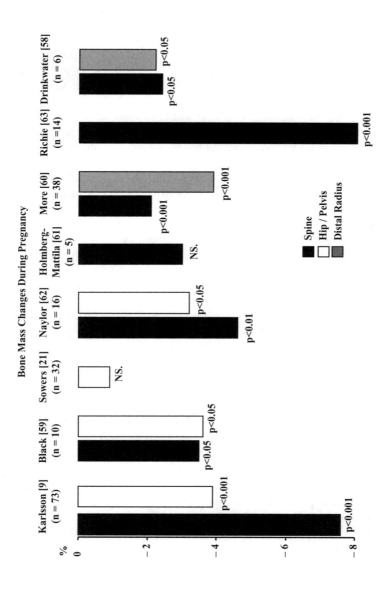

**Figure 29.1** *Relative changes in bone mass during a pregnancy evaluated in several original studies, presented with first author and reference number in this review, numbers of women included and level of significance when comparing the pre- and post-pregnancy values*

In the prospective studies cited above, the baseline measurements were performed up to 12 months before conception so that the baseline data at conception must be regarded with caution. The follow-up measurements were done 1–12 months after delivery, except the study by Karlsson *et al.* and Sowers *et al.* where the follow-up measurements were done within 3 and 15 days of parturition, respectively, but in all other studies it seems probable that the effect of lactation could confound the outcome. Furthermore, except for three of the cited studies, all others were uncontrolled, and only in the study by Karlsson *et al.* were the BMD values adjusted for changes in soft tissue composition during pregnancy in the comparison between the pre- and post-pregnancy values.

In summary, for practical reasons it is impossible to conduct a RCT, when the effects of a pregnancy on BMD are to be evaluated. Instead, we have to rely on results with a lower level of evidence within the evidence-based system. Most published studies are uncontrolled, include few cases, the baseline and the follow-up measurements are conducted months to years before or after pregnancy, and the measurements are not adjusted for changes in weight or soft tissue composition during pregnancy. However, when the latest published studies, together with a recent review are scrutinised, it seems feasible to conclude that during pregnancy there is a mobilisation of calcium from the maternal skeleton to the developing foetus so that a maternal loss of BMD occurs during the pregnancy.[10] The BMD loss is in most studies described to be 2–5%, a loss of minor biological significance (Figure 29.1). It also seems reasonable to conclude that the intervention of environmental factors, such as changes in nutritional habits, have little or no effect on the BMD loss during pregnancy, except in cases with a very low nutritional intake or in cases with other exceptional, abnormal, nutritional situations.

# Changes in Determinants of BMD During Lactation

## Nutrition

The metabolism of calcium is of utmost importance during lactation also. The homeostasis of calcium changes dramatically postpartum. Maternal absorption and excretion is also altered and the elevated calcium absorption and urinary calcium excretion found during pregnancy, shortly return to pre-pregnancy levels after.[22,33,63] An increased maternal demand for calcium during lactation is one factor that could also influence postpartum bone health. During lactation, and full breast-feeding, about 200 mg calcium are transferred in the mother's milk to the infant. The total calcium transfer via breast milk in one lactation period of 3–6 months, is greater than the calcium content transferred across the placenta during the whole of pregnancy.[32] The concentration of other minerals also changes during lactation, for example, lactating mothers have a reduced urinary phosphate excretion, which together with elevated serum phosphate concentrations implies a renal phosphorus conservation.[64] The importance of these changes in connection with bone health has, to date, not been well investigated.

Neither 1,25-dihydroxyvitamin nor parathyroid hormone (PTH) is elevated during lactation,[22,33,63] but at later stages during long periods of lactation and in the weaning period, the results of several studies imply that these hormones are elevated.[22,33,65] In contrast, calcitonin seems to remain at normal levels during the total lactating period as reported in most published studies,[24,32] albeit not in all.[63]

As during pregnancy, changes in nutritional habits during a period of lactation may affect the BMD. Smoking and alcohol are in many women postponed during lactation and may thus influence the BMD beneficially.[66] Also, changes in nutritional habits, associated with the change in family structure due to the arrival of a new baby, may influence the BMD.

A clinically relevant question then arises – is the calcium and the vitamin D intake of the mother of importance for BMD and could a supplement of extra calcium and vitamin D affect the BMD? The indications that calcium intake influences the biochemical response to lactation is weak.[67] Further, randomised, controlled intervention studies of lactating women with high and low calcium intake have shown no effects of an increased calcium supply on bone turnover markers.[22,24,32,33] There are studies on the influence of calcium intake on BMD during lactation. Polatti *et al.* inferred in a study, including 274 lactating mothers (of whom half were given calcium supplement and 153 acted as controls) that extra calcium possibly reduced the BMD loss during lactation by a few percent but that the effect was only transient, with no long-term benefits.[68] In contrast, Kalkwarf *et al.* reported on two prospective, controlled studies, no effect of calcium supplement on BMD loss during lactation, but possibly a small enhancement in the recovery of BMD during weaning[22,69] and Prentice *et al.* even strengthened the view by suggesting that women with low calcium intake do not benefit from increased calcium intake during lactation.[70] There is compelling evidence that the BMD changes that accompany lactation are independent of the concurrent calcium intake in food and independent of whether calcium is given as a supplement. The few studies which report a small and biologically non-significant effect of calcium postpartum have usually found the same beneficial effect in non-lactating mothers[33,70–74] *i.e.*, a calcium supplement does not seem to influence the BMD loss by lactation. Finally, there are no studies indicating that vitamin D requirements are greater in lactating compared to non-lactating women.[75]

In summary, the increased calcium loss from the mother to the child during lactation would, hypothetically, lead to lower maternal BMD during the pregnancy. However, the maternal absorption of calcium becomes adapted to the required level, and a calcium supplement appears to have no or only minor effects on BMD during lactation. Reduced smoking and alcohol consumption during lactation would, hypothetically, influence the BMD positively.

## Body Weight

Body weight is one of the more important factors that may determine the BMD level.[38,39] Reports exist which suggest that weight increase is of more importance for BMD than physical activity.[2] Also, the clinically relevant end point, namely a fracture, is lower in individuals with higher body weight.[40,41] A delivered mother usually rapidly loses weight, sometimes 10–20 kg during the first few weeks following delivery. The reduced weight is a factor known to decrease the BMD due to the reduced mechanical load on the skeleton. Also reduced fat content may occur, with a decreased production of peripheral hormones.[38,39] Thus, the weight and soft tissue changes taking place during lactation would, hypothetically, lead to a decreased BMD.

## Changes in Hormones may Influence the BMD During Lactation

During the first days postpartum, the levels of placental steroids rapidly decline, while the levels of follicle-stimulating hormone (FSH) and luteinising hormone (LH) remain at undetectable levels during the first week after delivery. Prolactin levels are high but decline to normal levels during the first month if lactation is not initiated. Furthermore, if lactation is not established, the synthesis of FSH and LH is resumed after approximately 10 days and the hypothalamus–pituitary axis is re-established after another 30 days. After approximately 2 months the menses resume and hypothetically the hormonal status is back to pre-pregnancy levels.[76]

By contrast, if lactation is initiated, the suppression of the hypothalamus–pituitary axis remains, together with a lactation-induced amenorrhea.[77] The re-establishment of a functional hypothalamus–pituitary axis at the end of lactation follows a well-defined pattern. Initially there is an almost complete inhibition of the pulsatile secretion; then, there is a return to erratic pulsatile secretion with some ovarian follicle development followed by a resumption of an apparently normal follicle growth, but most often an absence of ovulation and, finally, a return to normal ovulatory function.[78] However, a high prolactin level is also sustained and the stimulus responsible for this is the suckling reflex.[77–79] Prolactin concentrations remain elevated during the first 3–4 months of lactation, *i.e.*, until the introduction of solid food occurs, usually followed by reduced lactation frequency. This elevated prolactin level may also influence the BMD, as it has been suggested that prolactin itself interacts with oestrogen and suppresses the hypothalamus–pituitary axis and so plays an indirect role in calcium metabolism during lactation.[80]

Once more, oestrogen is probably the most important of the sex steroids, which regulate BMD in adult women.[42] Oestrogen levels are suppressed in lactating women and remain so for 3–6 months postpartum. The first menstruation is an indicator of recovery of ovulation.[77,81] The ovarian dysfunction has been proposed as one of the factors explaining reported bone loss during lactation.[82] Finding a dose–response relationship between oestrogen deprivation and BMD loss would indicate the importance of oestrogen. A short duration of lactational amenorrhea has been reported to be associated with a small bone loss during lactation[68,83] with a more pronounced increase in BMD after weaning.[72,83,84] Other authors have rejected that this relation exists.[9,68,85] However, the majority of studies propose that the bone loss during lactation and the recovery during weaning, are explained, at least partially, by variations in the oestrogen levels postpartum.[42] Positive associations between serum estradiol levels and BMD have also been described in postpartum women.[80,86]

In summary, there are a variety of hormonal changes during lactation that may affect BMD, but it is in practice not feasible to calculate the specific contribution of each hormone to the variance in BMD. As oestrogen is regarded as the most important regulatory sex hormone for the skeleton, the low oestrogen levels during lactation would, hypothetically, decrease the BMD.

## Physical Activity

Physical activity is also an environmental factor known to influence the skeleton.[9] The reduced physical activity, seen in many pregnant women, is often gradually increased with time elapsed since delivery, a circumstance which would be beneficial for bone health. By contrast, some women find the lack of spare time in their new family constellation an impediment to their taking part in extra physical activity. Those doing so are thus exceptions. Thus, it is difficult to predict activity level after delivery, and hypothetically the question of exercising may tend either way, active exercising resulting in increased BMD, or the opposite.

# Lactation and Bone Health

## Bone Mineral Density

Biochemical markers of bone resorption and formation, as markers for bone turnover, are elevated during the first months of lactation but decrease after 6–12 months, even in women who continue to breast-feed for 18 months or more.[24,32] The duration of lactation is an essential determinant of these markers; such women lactating for an extended period have a longer and more pronounced elevation, but some elevation is evident after delivery even in women who do not breast-feed.[24,32]

Comparable to the studies of the effect of pregnancy on BMD, the first studies of the effect of lactation on BMD were made using the single photon absorptiometry (SPA) and the dual photon absorptiometry (DPA) techniques[33,58,70,71,86–88] (Figure 29.2). A longitudinal study by Cross *et al.*[33] reported no BMD reduction with lactation, while another prospective study by Drinkwater *et al.*[58] reported a low postpartum BMD to be normalised in the lumbar spine, while a continued BMD loss of 6% was observed in the femoral neck after 6 months of lactation. Affinito *et al.* compared 36 mothers, for

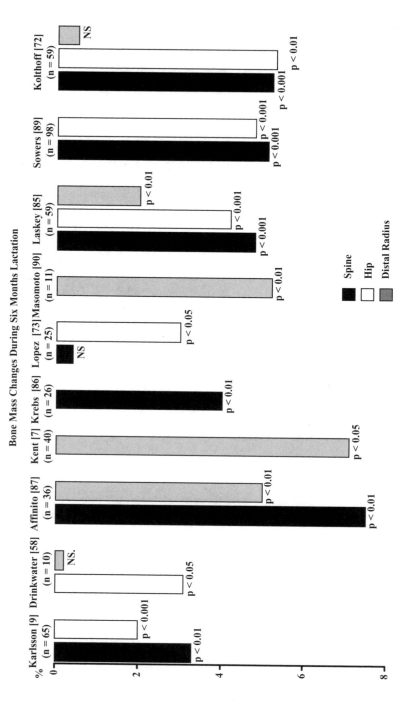

**Figure 29.2** *Relative changes in bone mass during 6 months of lactation evaluated in several original studies, presented with first author and reference number in this review, numbers of women included and level of significance when comparing the post-pregnancy and the 6 months post-pregnancy values*

whom lactation was inhibited pharmacologically, with 18 mothers lactating for 12 months after delivery. Lumbar spine BMD loss was in this cohort 7.5% higher and distal radius BMD loss 5% higher after 6 months of lactation compared to controls, with an incomplete recovery to be found 6 months after weaning.[87] In a study by Kent *et al.* including 40 mothers who breast-fed and 40 age-matched controls, a BMD loss of similar magnitude (7.1%) was found in the ultradistal radius after 6 months of lactation[71] and Krebs *et al.* including 26 lactating women and eight non-lactating controls, reported a 4% decline in spine BMD after 7 months of lactation.[86] However, BMD was almost restored post-weaning.

Several prospective, controlled studies using the DXA technique have demonstrated that lactation is followed by a significant reduction in maternal BMD[72,73,85,89,90] (Figure 29.2). The reductions are most pronounced in the axial skeleton and most published reports give a BMD loss of 3–6% after 3–6 months of lactation.[72,73,89,90] The clinically important question will then be whether BMD recovers after weaning (Figure 29.3). However, the time of the last measurement tends to vary greatly among studies, in some more than 12 months postpartum.[89,90] Matsumoto *et al.* reported a further decrease in BMD beyond the lactation period, namely, that the BMD of 22 women, lactating for 6–11 months, was still 8% lower 12 months postpartum compared with delivery, while for 11 non-lactating women the BMD was not reduced.[90] However, 24 months after delivery no difference remains in either of the groups compared to baseline. Sowers *et al.* reported BMD return to baseline already 12 months after delivery in 25 women who breast-fed for at least 6 months[89] while Kolthoff *et al.* reported a maximal BMD loss in the spine of 5.2% after 3 months in 59 lactating women, with an incomplete recovery after 12 months (still 3.3% lower). A complete recovery was observed 18 months postpartum.[72] A transient bone loss with lactation, at a magnitude similar to the deficits described above, but with recovery back to baseline levels after 6–18 months of weaning has been supported in a variety of other published works[73,85] (Figure 29.3). The studies cited defined and measured baseline postpartum, but three other studies also include pre-pregnancy BMD values.[60,61,63] However, the conclusions drawn above remain, when the latter authors confirmed that the lactation-induced BMD loss was found mainly in the lumbar spine, and that a dose–response relationship was seen with a longer period of lactation, leading to a larger BMD loss, a finding that strengthens the view that lactation does lead to BMD loss.[60,63,85]

One confounding factor in the measurement of women postpartum is the changes in weight, soft tissue composition and the fluid flux that occur during the first days to weeks after delivery, changes that imply methodological problems when using the DXA technique, as previously described. However, in an investigation, including 40 lactating women and 36 non-lactating women in a 24-month prospective study, Hopkins *et al.* supported the established view by reporting that, after 3–6 months of breast-feeding, BMD was reduced, in particular in the lumbar spine but that the BMD loss was recovered after the onset of menses. All BMD data in this study were adjusted for changes in bone

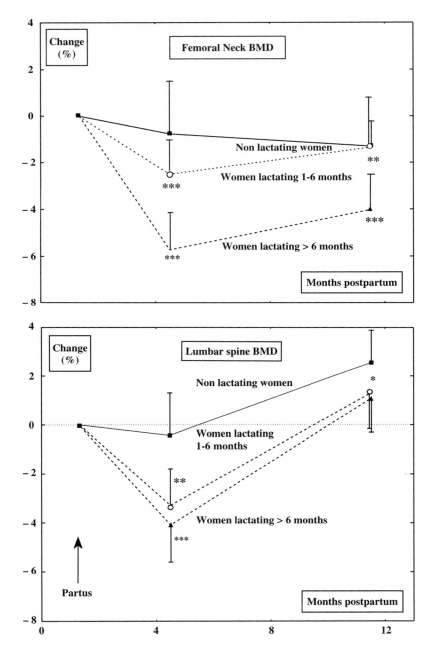

**Figure 29.3**   *Relative changes in bone mass during the first year following delivery in women with no breastfeeding and in women with breastfeeding below or above 6 months duration*
Adapted from Karlsson *et al.*[9]

size, weight and height.[84] This notion was supported by Karlsson *et al.* in a report showing that the BMD of the lumbar spine of 65 breast-feeding mothers was reduced by 4.1% and that of the femoral neck by 2.0% after 5 months of lactation following delivery, compared with 55 non-lactating mothers. Lactation beyond this period, however, did not decrease the BMD further.[9] Moreover, BMD of the spine 12 months after delivery was completely recovered, while the BMD in the femoral neck was only partly recovered (Figure 29.3). The strength of this study was that all BMD data were adjusted for changes in fat and lean mass during the study period.[9] Finally, an extensive review by Ensom *et al.* supports the suggestions described, finding that most studies imply that a reduction in BMD occurs with lactation, but that this reduction is reversible after weaning.[10] There is, however, a controversy concerning whether the recovery is related to the cessation of breast-feeding, the length of lactation or the resumption of menses.[79] As these three variables are associated, it is extremely difficult to draw any conclusions as regards the causality.

It has also been proposed that closely spaced pregnancies with a new pregnancy during or soon after a lactation period would be a risk factor for osteoporosis later in life, due to additive periods with BMD loss in quick succession.[87] There are studies which investigate mothers with narrow childbirth and lactation periods longitudinally, with BMD measurements performed during lactation, post-lactation and after delivery of the subsequent infant.[91,92] However, virtually all studies imply that women with the dual calcium demands of extended lactation and a subsequent pregnancy do not risk failure of bone recovery to pre-lactation levels.

In summary, there are no RCT which evaluate the effect of lactation on BMD. Instead, we have to rely on published results with a lower level of evidence. Several controlled or uncontrolled, prospective studies infer that lactation after childbirth produces a BMD loss of 2–5% (Figure 29.2), and most studies also conclude that recovery follows weaning (Figure 29.3). It also seems reasonable to conclude that intervention with environmental factors, such as changes in nutritional habits, have little or no effect on the BMD loss during lactation, except maybe in cases with a very low nutritional intake.

# Pregnancy, Lactation and Long-term Effects on BMD

Maybe the clinically most significant question as regards reproductive history and its association with BMD, is whether any reduction in BMD during pregnancy and lactation is irreversible. If all mothers achieve a slightly reduced BMD after every period of pregnancy and lactation, and if the women do not recover postpartum, this will leave the individuals with a residually low BMD compared to their female counterparts who never experienced a pregnancy. If this is so, it would produce a cohort of women with reduced bone health and an increased risk to sustain fractures later in life, and pregnancy and lactation could then be regarded as a risk factor for future fractures.

Karlsson *et al.* have presented a cross-sectional study, including 39 pre-menopausal women with a minimum of four pregnancies and with a long total duration of lactation, in which the BMD was found to be no lower for these women than for 58 age-matched controls with a maximum of two pregnancies and a short duration of lactation[9] (Figure 29.4). Furthermore, the total duration of lactation was not correlated with the current BMD. All BMD values in this study were adjusted for differences in soft tissue composition, and the authors concluded that neither an extended lactation period nor multiple pregnancies could be used as a risk factor for the prediction of women at risk for future osteoporosis. The same conclusion was drawn by Kojima *et al.* in a cross-sectional study including 465 pre- and 713 post-menopausal Japanese women, when they stated that a history of lactation and parity was not a major determinant of BMD later in life.[93] The authors reported a small but significant inverse association with total lactation period and BMD ($r = -0.29$) in pre-menopausal women aged 40–44 years, but not in post-menopausal women aged 60–64 years, the years when fragility fractures begin to be a problem of magnitude. Another Swedish study, including 70 year-old women, once more stated that the BMD of women with a history of pregnancies was not different from that of nulliparous women[94] and a recently published review, including 23 citations, supports the notion, as summarised by the authors, that none of the studies could verify an association linking a greater number of pregnancies to a greater decrease in BMD.[10]

In contrast, studies also exist which propose an association between pregnancy and high BMD (Figure 29.4). Cure-Cure *et al.* studied 1855 post-menopausal women and reported that total body BMD, hip BMD and leg BMD were higher in women with at least one delivery in comparison with nulliparous women.[95] Women with two deliveries or more had 3% higher total body BMD, 8% higher femoral neck BMD and 4% higher leg BMD in comparison to women with no children. The same association was reported by Forsmo *et al.* in a cross-sectional study including 1652 peri- and post-menopausal Norwegian women,[96] in one by Grainge *et al.* including 580 English women aged 45–61 years,[97] in one by Tuppuvainen *et al.* including 3126 Finnish women aged 47–56 years,[98] in one by Murphy *et al.* including 825 English women aged 41–76 years,[99] in one by Sowers *et al.* including 217 white American women aged 22–54 years,[66] and in one study by Mariconda *et al.* including Italian women.[100] All of these studies concluded that women with a history of one or several children had in general 2–5% higher BMD than nulliparous women. One of the most cited population-based, prospectively followed cohorts, the Dubbo Osteoporotic Study in Australia, including 1091 women with a mean age of 70 years, showed that women with a history of one or more pregnancies had 5–6% higher femoral neck BMD than their peers with no children.[101] Reports also exist, which propose that a dose relationship does occur between number of pregnancies and BMD, a fact that strengthens the view that pregnancies are associated with high BMD. A large British study by Murphy *et al.* reported the lower BMD of the spine and femoral neck in

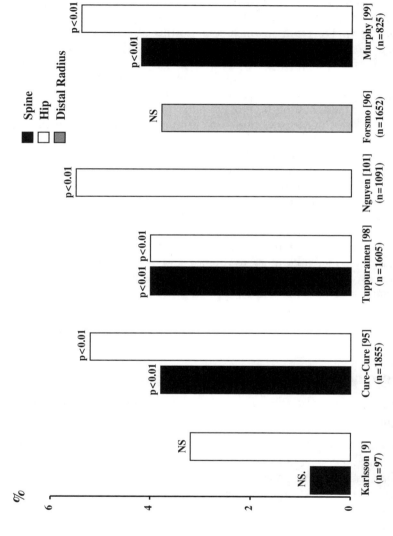

**Figure 29.4** *Relative difference in bone mass in parous women compared with nulliparous women evaluated in several original studies, presented with first author and reference number in this review, numbers of women included and level of significance when comparing the parous and the nulliparous women*

nulliparous women, intermediate values in primiparas and the highest value in women with two or more children.[99]

In summary, it seems probable that both pregnancy and lactation result in a reduced BMD. Hypothetically this could, after several pregnancies with a long total period of lactation, lead to a more serious residual deficit in BMD with increased risk of fractures. However, virtually all existing data in the literature oppose this hypothesis. The BMD of women with a history of pregnancies and lactation is the same or higher than that of their peers who have not given birth (Figure 29.4). The reports, which come to this conclusion, are based on study designs with a lower level of evidence within the evidence-based system. Observational studies and cross-sectional case-control studies, as discussed above, must always be treated with caution, as selection bias may be producing the inferences. The study design can only be regarded as hypothesis-generating, not as hypothesis-testing, as the causality can be interpreted to mean that a disease or sub-clinical condition is preventing a pregnancy, and that the same disease or condition is leading to low BMD among the nulliparous controls, not that pregnancies lead to high BMD. However, the data raise interesting hypotheses. A woman with several children is often forced to change her lifestyle. The housework increases with many family members, probably forcing the woman to perform more work in the home environment. More weight loaded work leads to higher BMD, more work leads to higher muscle mass and a higher body weight are factors that may also increase the BMD. Also, with many children, the nutrition habits of the family may change. Milk consumption increases in many homes and the eating habits become more focused on a healthy diet. Thus, even if pregnancy and lactation reduce BMD, the changes in lifestyle associated with being a parent might outweigh these negative BMD changes, and, in fact, leave mothers with a higher BMD than their peers with no children.

# Pregnancy, Lactation and Fractures

## Pregnancy-related Osteoporosis or Transient Osteoporosis of the Hip with or Without Fractures

Osteoporosis associated with pregnancy is considered a rare complication of pregnancy and lactation, but the actual incidence is unknown.[102] Another diagnosis, namely pregnancy-related, transient osteoporosis of the hip, probably reflects the same condition, albeit localised to only one specific skeletal region.[103–106] Our knowledge regarding the condition is based mainly on case reports. It affects predominantly slightly built, primigravid, lactating women and usually does not recur during subsequent pregnancies. Little is known about its pathogenesis or the dynamics of BMD decrement when the disease develops and the BMD increment during the recovery phase.[12,102,107–110] The disease becomes obvious most often at the beginning of the third trimester,[12,109,110] but the condition is relatively benign and patients are expected to return to normal BMD 6–12 months after weaning, without treatment. The condition may also lead to vertebral fractures, but also sacrum fractures and hip fractures have been described, and in these cases long-standing clinical problems could develop due to the fracture.[107–109,111]

The two published studies with the largest sample size included 24 patients in a retrospective, observational study and 35 patients and 35 matched controls in a case-control study.[12,110] Both studies supported the descriptions above as being a disease that recovered without treatment, and Smith *et al.* further concluded that most of the patients were without any long-term complaints 24 years after their pregnancy.[12] Apart from these studies our knowledge concerning the disease comes mainly from case reports.

In summary, pregnancy related osteoporosis or pregnancy-related transient osteoporosis of the hip is a rare condition that may occur predominantly during the first pregnancy, and is a disease that does not recur with subsequent pregnancies. The disease is self-limiting so that most affected women are usually without a BMD deficit 6 months after weaning. The condition is sometimes complicated with a fracture, but if no fracture with a sequel develops during the period with osteoporosis, the disease does not lead to long-term complications.

## Pregnancy, Lactation and the Long-term Fracture Risk

Most studies evaluate the effect of parity and lactation on BMD as a surrogate end point for fractures. It is only when studies can prove that an association exists between the number of pregnancies and fractures and/or duration of lactation and fractures, that we can draw the conclusion that these conditions confer clinically relevant, impaired bone health. However, few studies use fractures as the end point (Figure 29.5) The incidence of hip and forearm fractures was not higher in women who had given birth to four or more children than in women who had not given birth, in a cross-sectional case-control study including 355 post-menopausal women with a fracture and 562 matched women with no history of fracture.[112] Additionally, women who had breast-fed for more than 2 years did not have a higher fracture risk compared to women who had never breast-fed.[112] A Swedish study, including 70-year old women supported the conclusion that there was no difference in fracture incidence in a comparison between women with a history of pregnancies and women with no pregnancies.[94] Another case-control study from Australia supports this view. The authors reported no association between hip fracture risk and parous women in an evaluation of 174 cases and 137 controls aged 65 years and over, but a protective effect of breast-feeding, which reduced the incidence of hip fractures by 53%.[2] No association was found between pregnancy and hip fracture risk in the Study of Osteoporotic Fractures (SOF), a longitudinal study of 9704 women over 65 years of age, followed for ~8 years,[2] in the Mediterranean Osteoporosis study (MEDOS), a cross-sectional study including 2086 women from 14 centres in six countries in Southern

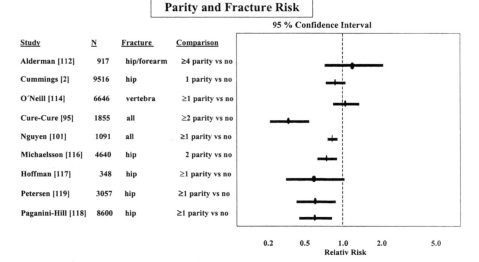

**Figure 29.5**   *Relative risk of sustaining a fracture in parous compared with nulliparous women, presented with first author and reference number in this review, numbers (N), type of fracture evaluated and numbers of pregnancies with 95% confidence interval for the risk ratio presented*

Europe with a hip fracture and 3532 non-fractured controls[113] or in the European Vertebral Osteoporosis Study (EVOS), a longitudinal study including 6646 women aged 50–79 years, of whom 884 had a vertebral deformity.[114]

In contrast, Cure-Cure *et al.* reported in two cross-sectional studies including 1855 post-menopausal women that the relative risk of fracture was 59% lower in women with two or more deliveries in comparison with nulliparous women[95,115] (Figure 29.5). The prospective Dubbo Epidemiological Osteoporosis Study, previously cited, also reported that there was a 6% lower incidence of fractures in parous than in nulliparous women.[101] There are also case-control studies which support the notion that pregnancies are associated with reduced fracture risk. A case-control study including 1328 post-menopausal Swedish women aged 50–81 years with a hip fracture and 3312 matched controls reported in an age-adjusted analysis that the risk of hip fractures among all women was reduced by 10% per child.[116] After adjustment for body mass index the risk reduction was 5% per child, but no association was found between the duration of lactation and fracture risk after parity in this study. An even higher protective effect against hip fractures was suggested in another case-control study, including 174 cases aged 54 years and over and 74 matched controls, where the authors implied that a live birth was associated with a 35% reduction in the risk of hip fractures in comparison with having no births.[117]

A reduction in hip fracture incidence in relation to high parity has also been reported in at least two large, prospective, observational studies, the Leisure

World Study, a 7-year prospective study of 8600 post-menopausal women[118] and in a Danish twin study including 3057 twins aged 66–99 years and followed for a total of 29 112 person-years. The results of the studies implied that having one or more child was associated with a lower hip fracture risk than having no children.[119]

In summary, data in the literature are conflicting when the association between parity, lactation and fracture is scrutinised. About half of the studies suggest that there is no association between parity and fractures, while the other half imply that a history of one or several pregnancies is associated with a reduced fracture risk (Figure 29.5). Most studies also conclude that there is no association between lactation and fractures. Once more, as data inferring these conclusions are derived from observational studies and cross-sectional case-control studies, the risk of selection bias producing the outcome must be included as a possible explanation. However, even if studies within the lower level of the evidence-based hierarchy are evaluated, virtually no studies exist which imply that pregnancies and lactation are associated with an increased fracture risk. This is strong, indirect evidence, that the previously described BMD reduction following pregnancy and lactation periods, do not lead to negative, clinically relevant, long-term effects, with an increased fracture incidence. Contrary, it seems as having many children leads to a situation that, if anything, leads to high BMD and reduced fracture risk.

## Conclusions

A variety of factors usually change during a pregnancy and lactation as smoking habits, alcohol consumption, level of physical activity, body weight, soft tissue composition, hormones, some factors increase BMD, some decrease BMD, making it nearly impossible to predict the changes in BMD with a pregnancy and lactation. Both a pregnancy and a period of lactation is usually described to be followed with a BMD loss of 2–5%, but most studies suggest that recovery follows weaning. This is further supported when data infer that BMD of women with a history of multiple pregnancies and lactation is the same or higher than that of their peers who have not given birth, and epidemiological data indicate no association or that pregnancies is associated with a reduced fracture risk. However, as no published studies are blinded randomised controlled trails, but instead observational and cross-sectional case-control studies, no studies can be regarded as true hypothesis proven, and the data must be regarded with scepticism.

## References

1. A.M. Parfitt, A.R. Villanueva, J. Foldes and D.S. Rao, Relations between histologic indices of bone formation: implications for the pathogenesis of spinal osteoporosis, *J. Bone Miner. Res.*, 1995, **10**(3), 466–473.

2. S.R. Cummings, M.C. Nevitt, W.S. Browner, K. Stone, K.M. Fox, K.E. Ensrud, J. Cauley, D. Black and T.M. Vogt, Risk factors for hip fracture in white women. Study of Osteoporotic Fractures Research Group, *N. Engl. J. Med.*, 1995, **332**(12), 767–773.

3. WHO. Assessment of fracture risk and its application to sceening for postmeno-pausal osteoporosis. *WHO Technical Report Series 843*, Geneva. 1994.

4. S. Bass, G. Pearce, M. Bradney, E. Hendrich, P.D. Delmas, A. Harding and E. Seeman, Exercise before puberty may confer residual benefits in bone density in adulthood: studies in active prepubertal and retired female gymnasts, *J. Bone Miner. Res.*, 1998, **13**(3), 500–507.

5. J.P. Bonjour, G. Theintz, F. Law, D. Slosman and R. Rizzoli, Peak bone mass, *Osteoporosis Int.*, 1994, **4**(Suppl. 1), 7–13.

6. C.C. Johnston Jr. and C.W. Slemenda, Peak bone mass, bone loss and risk of fracture, *Osteoporosis Int.*, 1994, **4**(Suppl. 1), 43–45.

7. P.J. Kelly, N.A. Morrison, P.N. Sambrook, T.V. Nguyen and J.A. Eisman, Genetic influences on bone turnover, bone density and fracture, *Eur. J. Endocrinol.*, 1995, **133**(3), 265–271.

8. S.L. Hui, C.W. Slemenda and C.C. Johnston Jr., The contribution of bone loss to postmenopausal osteoporosis, *Osteoporosis Int.*, 1990, **1**(1), 30–34.

9. C. Karlsson, K.J. Obrant and M. Karlsson, Pregnancy and lactation confer reversible bone loss in humans, *Osteoporosis Int.*, 2001, **12**(10), 828–834.

10. M.H. Ensom, P.Y. Liu and M.D. Stephenson, Effect of pregnancy on bone mineral density in healthy women, *Obstet. Gynecol. Surv.*, 2002, **57**(2), 99–111.

11. T. Lloyd, H.M. Lin, D.F. Eggli, W.C. Dodson, L.M. Demers and R.S. Legro, Adolescent Caucasian mothers have reduced adult hip bone density, *Fertil. Steril.*, 2002, **77**(1), 136–140.

12. R. Smith, N.A. Athanasou, S.J. Ostlere and S.E. Vipond, Pregnancy-associated osteoporosis, *QJM*, 1995, **88**(12), 865–878.

13. A.J. Phillips, S.J. Ostlere and R. Smith, Pregnancy-associated osteoporosis: does the skeleton recover? *Osteoporosis Int.*, 2000, **11**(5), 449–454.

14. H. Sievanen, A physical model for dual-energy X-ray absorptiometry–derived bone mineral density, *Investig. Radiol.*, 2000, **35**(5), 325–330.

15. M.K. Karlsson, P. Gardsell, O. Johnell, B.E. Nilsson, K. Akesson and K.J. Obrant, Bone mineral normative data in Malmo, Sweden. Comparison with reference data and hip fracture incidence in other ethnic groups, *Acta Orthop. Scand.*, 1993, **64**(2), 168–172.

16. E. Seeman, M.K. Karlsson and Y. Duan, On exposure to anorexia nervosa, the temporal variation in axial and appendicular skeletal development predisposes to site-specific deficits in bone size and density: a cross-sectional study, *J. Bone Miner. Res.*, 2000, **15**(11), 2259–2265.

17. G.M. Fogelholm, H.T. Sievanen, T.K. Kukkonen-Harjula and M.E. Pasanen, Bone mineral density during reduction, maintenance and regain of body weight in premenopausal, obese women, *Osteoporosis Int.*, 2001, **12**(3), 199–206.

18. L.M. Salamone, J.A. Cauley, D.M. Black, L. Simkin-Silverman, W. Lang, E. Gregg, L. Palermo, R.S. Epstein, L.H. Kuller and R. Wing, Effect of a lifestyle intervention on bone mineral density in premenopausal women: a randomized trial, *Am. J. Clin. Nutr.*, 1999, **70**(1), 97–103.

19. A. Martin, M.A. Brown and A.J. O'Sullivan, Body composition and energy metabolism in pregnancy, *Aust. NZ J. Obstet. Gynaecol.*, 2001, **41**(2), 217–223.

20. N.G. Pipe, T. Smith, D. Halliday, C.J. Edmonds, C. Williams and T.M. Coltart, Changes in fat, fat-free mass and body water in human normal pregnancy, *Br. J. Obstet. Gynaecol.*, 1979, **86**(12), 929–940.
21. M. Sowers, M. Crutchfield, M. Jannausch, S. Updike and G. Corton, A prospective evaluation of bone mineral change in pregnancy, *Obstet. Gynecol*, 1991, **77**(6), 841–845.
22. H.J. Kalkwarf and B.L. Specker, Bone mineral changes during pregnancy and lactation, *Endocrine.*, 2002, **17**(1), 49–53.
23. D.A. Southgate, Body content and distribution of water in healthy individuals, *Bibl. Nutr. Dieta.*, 1987(40), 108–116.
24. A. Prentice, Maternal calcium metabolism and bone mineral status, *Am. J. Clin. Nutr.*, 2000, **71**(5 Suppl), 1312S–1316S.
25. J.J. Anderson, The role of nutrition in the functioning of skeletal tissue, *Nutr. Rev.*, 1992, **50**(12), 388–394.
26. J.Z. Ilich and J.E. Kerstetter, Nutrition in bone health revisited: a story beyond calcium, *J. Am. Coll. Nutr.*, 2000, **19**(6), 715–737.
27. B. Jonsson, *Life style and fracture risk*. Thesis, University of Lund. 1993
28. D.C. Welten, H.C. Kemper, G.B. Post and W.A. van Staveren, A meta-analysis of the effect of calcium intake on bone mass in young and middle aged females and males, *J. Nutr.*, 1995, **125**(11), 2802–2813.
29. V. Matkovic and R.P. Heaney, Calcium balance during human growth: evidence for threshold behavior, *Am. J. Clin. Nutr.*, 1992, **55**(5), 992–996.
30. M.B. Andon, T. Lloyd and V. Matkovic, Supplementation trials with calcium citrate malate: evidence in favor of increasing the calcium RDA during childhood and adolescence, *J. Nutr.*, 1994, **124**(8 Suppl), 1412S–1417S.
31. S.A. Abrams, Bone turnover during lactation–can calcium supplementation make a difference? *J. Clin. Endocrinol. Metab.*, 1998, **83**(4), 1056–1058.
32. A. Prentice, Calcium in pregnancy and lactation, *Annu. Rev. Nutr.*, 2000, **20**, 249–272.
33. N.A. Cross, L.S. Hillman, S.H. Allen, G.F. Krause and N.E. Vieira, Calcium homeostasis and bone metabolism during pregnancy, lactation, and postweaning: a longitudinal study, *Am. J. Clin. Nutr.*, 1995, **61**(3), 514–523.
34. A. Prentice, T.J. Parsons and T.J. Cole, Uncritical use of bone mineral density in absorptiometry may lead to size-related artifacts in the identification of bone mineral determinants, *Am. J. Clin. Nutr.*, 1994, **60**(6), 837–842.
35. L. Raman, K. Rajalakshmi, K.A. Krishnamachari and J.G. Sastry, Effect of calcium supplementation to undernourished mothers during pregnancy on the bone density of the bone density of the neonates, *Am. J. Clin. Nutr.*, 1978, **31**(3), 466–469.
36. S.V. Jaque-Fortunato, N. Khodiguian, R. Artal and R.A. Wiswell, Body composition in pregnancy, *Semin. Perinatol.*, 1996, **20**(4), 340–342.
37. G.M. Blake, R.J. Herd, R. Patel and I. Fogelman, The effect of weight change on total body dual-energy X-ray absorptiometry: results from a clinical trial, *Osteoporosis Int.*, 2000, **11**(10), 832–839.
38. M. Sowers, A. Kshirsagar, M. Crutchfield and S. Updike, Body composition, age and femoral bone mass of young adult women, *Ann. Epidemiol.*, 1991, **1**(3), 245–254.
39. R. Lindsay, F. Cosman, B.S. Herrington and S. Himmelstein, Bone mass and body composition in normal women, *J. Bone Miner. Res.*, 1992, **7**(1), 55–63.
40. G.M. Wardlaw, Putting body weight and osteoporosis into perspective, *Am. J. Clin. Nutr.*, 1996, **63**(3 Suppl), 433S–436S.
41. K.E. Ensrud, R.C. Lipschutz, J.A. Cauley, D. Seeley, M.C. Nevitt, J. Scott, E.S. Orwoll, H.K. Genant and S.R. Cummings, Body size and hip fracture risk in

older women: a prospective study. Study of Osteoporotic Fractures Research Group, *Am. J. Med.*, 1997, **103**(4), 274–280.

42. R.T. Turner, B.L. Riggs and T.C. Spelsberg, Skeletal effects of estrogen, *Endocrine Rev.*, 1994, **15**(3), 275–300.

43. B.L. Riggs, S. Khosla and L.J. Melton III, A unitary model for involutional osteoporosis: estrogen deficiency causes both type I and type II osteoporosis in postmenopausal women and contributes to bone loss in aging men, *J. Bone Miner. Res.*, 1998, **13**(5), 763–773.

44. J.M. Zmuda, J.A. Cauley and R.E. Ferrell, Recent progress in understanding the genetic susceptibility to osteoporosis, *Genet. Epidemiol.*, 1999, **16**(4), 356–367.

45. E.F. Eriksen, D.S. Colvard, N.J. Berg, M.L. Graham, K.G. Mann, T.C. Spelsberg and B.L. Rigs, Evidence of estrogen receptors in normal human osteoblast-like cells, *Science.*, 1988, **241**(4861), 84–86.

46. M.J. Oursler, P. Osdoby, J. Pyfferoen, B.L. Riggs and T.C. Spelsberg, Avian osteoclasts as estrogen target cells, *Proc. Natl Acad. Sci. USA*, 1991, **88**(15), 6613–6617.

47. J.W. Nieves, L. Komar, F. Cosman and R. Lindsay, Calcium potentiates the effect of estrogen and calcitonin on bone mass: review and analysis, *Am. J. Clin. Nutr.*, 1998, **67**(1), 18–24.

48. W.M. Kohrt, D.B. Snead, E. Slatopolsky and S.J. Birge Jr., Additive effects of weight-bearing exercise and estrogen on bone mineral density in older women, *J. Bone Miner. Res.*, 1995, **10**(9), 1303–1311.

49. K.M. Chiu, J. Ju, D. Mayes, P. Bacchetti, S. Weitz and C.D. Arnaud, Changes in bone resorption during the menstrual cycle, *J. Bone Miner. Res.*, 1999, **14**(4), 609–615.

50. S.E. Tomten, J.A. Falch, K.I. Birkeland, P. Hemmersbach and A.T. Hostmark, Bone mineral density and menstrual irregularities. A comparative study on cortical and trabecular bone structures in runners with alleged normal eating behavior, *Int. J. Sports Med.*, 1998, **19**(2), 92–97.

51. M.K. Karlsson, S.J. Weigall, Y. Duan and E. Seeman, Bone size and volumetric density in women with anorexia nervosa receiving estrogen replacement therapy and in women recovered from anorexia nervosa, *J. Clin. Endocrinol. Metab.*, 2000, **85**(9), 3177–3182.

52. G. Saggese, S. Bertelloni and G.I. Baroncelli, Sex steroids and the acquisition of bone mass, *Hormone Res.*, 1997, **48**(Suppl. 5), 65–71.

53. S. Mora, G. Weber, K. Marenzi, E. Signorini, R. Rovelli, M.C. Proverbio and G. Chiumello, Longitudinal changes of bone density and bone resorption in hyperthyroid girls during treatment, *J. Bone Miner. Res.*, 1999, **14**(11), 1971–1977.

54. S.E. Inzucchi and R.J. Robbins, Clinical review 61: Effects of growth hormone on human bone biology, *J. Clin. Endocrinol. Metab.*, 1994, **79**(3), 691–694.

55. A. Colao, C. Di Somma, S. Loche, A. Di Sarno, M. Klain, R. Pivonello, M. Pietrosante, M. Salvatore and G. Lombardi, Prolactinomas in adolescents: persistent bone loss after 2 years of prolactin normalization, *Clin. Endocrinol. (Oxford)*, 2000, **52**(3), 319–327.

56. K.J. Catt IV. Reproductive endocrinology, *Lancet*, 1970, **1**(7656), 1097–1104.

57. M. Karlsson, S. Bass and E. Seeman, The evidence that exercise during growth or adulthood reduces the risk of fragility fractures is weak, *Best Pract. Res. Clin. Rheumatol.*, 2001, **15**(3), 429–450.

58. B.L. Drinkwater and C.H. Chesnut III, Bone density changes during pregnancy and lactation in active women: a longitudinal study, *Bone Miner.*, 1991, **14**(2), 153–160.
59. A.J. Black, J. Topping, B. Durham, R.G. Farquharson and W.D. Fraser, A detailed assessment of alterations in bone turnover, calcium homeostasis, and bone density in normal pregnancy, *J. Bone Miner. Res.*, 2000, **15**(3), 557–563.
60. C. More, P. Bettembuk, H.P. Bhattoa and A. Balogh, The effects of pregnancy and lactation on bone mineral density, *Osteoporosis Int.*, 2001, **12**(9), 732–727.
61. D. Holmberg-Marttila, H. Sievanen and R. Tuimala, Changes in bone mineral density during pregnancy and postpartum: prospective data on five women, *Osteoporosis Int.*, 1999, **10**(1), 41–46.
62. K.E. Naylor, P. Iqbal, C. Fledelius, R.B. Fraser and R. Eastell, The effect of pregnancy on bone density and bone turnover, *J. Bone Miner. Res.*, 2000, **15**(1), 129–137.
63. L.D. Ritchie, E.B. Fung, B.P. Halloran, J.R. Turnlund, M.D. Van Loan, C.E. Cann and J.C. King, A longitudinal study of calcium homeostasis during human pregnancy and lactation and after resumption of menses, *Am. J. Clin. Nutr.*, 1998, **67**(4), 693–701.
64. C.S. Kovacs and H.M. Kronenberg, Maternal-fetal calcium and bone metabolism during pregnancy, puerperium, and lactation, *Endocr. Rev.*, 1997, **18**(6), 832–872.
65. M. Sowers, D. Zhang, B.W. Hollis, B. Shapiro, C.A. Janney, M. Crutchfield, M.A. Schork, F. Stanczyk and J. Randolph, Role of calciotrophic hormones in calcium mobilization of lactation, *Am. J. Clin. Nutr.*, 1998, **67**(2), 284–291.
66. M.R. Sowers, M.K. Clark, B. Hollis, R.B. Wallace and M. Jannausch, Radial bone mineral density in pre- and perimenopausal women: a prospective study of rates and risk factors for loss, *J. Bone Miner. Res.*, 1992, **7**(6), 647–657.
67. A. Prentice, B. Dibba, L.M. Jarjou, M.A. Laskey and A.A. Paul, In breast milk calcium concentration influenced by calcium intake during pregnancy, *Lancet*, 1994, **344**(8919), 411–412.
68. F. Polatti, E. Capuzzo, F. Viazzo, R. Colleoni and C. Klersy, Bone mineral changes during and after lactation, *Obstet. Gynecol.*, 1999, **94**(1), 52–56.
69. H.J. Kalkwarf, B.L. Specker, D.C. Bianchi, J. Ranz and M. Ho, The effect of calcium supplementation on bone density during lactation and after weaning, *N. Engl. J. Med.*, 1997, **337**(8), 523–528.
70. A. Prentice, L.M. Jarjou, T.J. Cole, D.M. Stirling, B. Dibba and S. Fairweather-Tait, Calcium requirements of lactating Gambian mothers: effects of a calcium supplement on breast-milk calcium concentration, maternal bone mineral content, and urinary calcium excretion, *Am. J. Clin. Nutr.*, 1995, **62**(1), 58–67.
71. G.N. Kent, R.I. Price, D.H. Gutteridge, M. Smith, J.R. Allen, C.I. Bhagat, M.P. Barnes, C.J. Hickling, R.W. Retallack and S.G. Wilson, Human lactation: forearm trabecular bone loss, increased bone turnover, and renal conservation of calcium and inorganic phosphate with recovery of bone mass following weaning, *J. Bone Miner. Res.*, 1990, **5**(4), 361–369.
72. N. Kolthoff, P. Eiken, B. Kristensen and S.P. Nielsen, Bone mineral changes during pregnancy and lactation: a longitudinal cohort study, *Clin. Sci. (London)*, 1998, **94**(4), 405–412.
73. J.M. Lopez, G. Gonzalez, V. Reyes, C. Campino and S. Diaz, Bone turnover and density in healthy women during breastfeeding and after weaning, *Osteoporosis Int.*, 1996, **6**(2), 153–159.

74. M.A. Laskey, A. Prentice, L.A. Hanratty, L.M. Jarjou, B. Dibba, S.R. Beavan and T.J. Cole, Bone changes after 3 mo of lactation: influence of calcium intake, breast-milk output, and vitamin D-receptor genotype, *Am. J. Clin. Nutr.*, 1998, **67**(4), 685–692.

75. B.L. Specker, Do North American women need supplemental vitamin D during pregnancy or lactation? *Am. J. Clin. Nutr.*, 1994, **59**(Suppl. 2), 484S–490S; discussion 490S–491S.

76. A. Zarate and E.S. Canales, Endocrine aspects of lactation and postpartum infertility, *J. Steroid Biochem.*, 1987, **27**(4–6), 1023–1028.

77. A.S. McNeilly, C.C. Tay and A. Glasier, Physiological mechanisms underlying lactational amenorrhea, *Ann. NY Acad. Sci.*, 1994, **709**, 145–155.

78. A.S. McNeilly, Lactational control of reproduction, *Reprod. Fertil. Dev.*, 2001, **13**(7–8), 583–590.

79. A.S. McNeilly, Neuroendocrine changes and fertility in breast-feeding women, *Prog. Brain Res.*, 2001, **133**, 207–214.

80. M.F. Sowers, B.W. Hollis, B. Shapiro, J. Randolph, C.A. Janney, D. Zhang, A. Schork, M. Crutchfield, F. Stanczyk and M. Russell-Aulet, Elevated parathyroid hormone-related peptide associated with lactation and bone density loss, *JAMA.*, 1996, **276**(7), 549–554.

81. A.S. McNeilly, Lactational amenorrhea, *Endocrinol. Metab. Clin. N. Am.*, 1993, **22**(1), 59–73.

82. A. Honda, T. Kurabayashi, T. Yahata, M. Tomita, K. Takakuwa and K. Tanaka, Lumbar bone mineral density changes during pregnancy and lactation, *Int. J. Gynaecol. Obstet.*, 1998, **63**(3), 253–258.

83. H.J. Kalkwarf and B.L. Specker, Bone mineral loss during lactation and recovery after weaning, *Obstet. Gynecol.*, 1995, **86**(1), 26–32.

84. J.M. Hopkinson, N.F. Butte, K. Ellis and E.O. Smith, Lactation delays postpartum bone mineral accretion and temporarily alters its regional distribution in women, *J. Nutr.*, 2000, **130**(4), 777–783.

85. M.A. Laskey and A. Prentice, Bone mineral changes during and after lactation, *Obstet. Gynecol.*, 1999, **94**(4), 608–615.

86. N.F. Krebs, C.J. Reidinger, A.D. Robertson and M. Brenner, Bone mineral density changes during lactation: maternal, dietary, and biochemical correlates, *Am. J. Clin. Nutr.*, 1997, **65**(6), 1738–1746.

87. P. Affinito, G.A. Tommaselli, C. di Carlo, F. Guida and C. Nappi, Changes in bone mineral density and calcium metabolism in breastfeeding women: a one year follow-up study, *J. Clin. Endocrinol. Metab.*, 1996, **81**(6), 2314–2318.

88. G.N. Kent, R.L. Price, D.H. Gutteridge, J.R. Allen, K.J. Rosman, M. Smith, C.I. Bhagat, S.G. Wilson and R.W. Retallack, Effect of pregnancy and lactation on maternal bone mass and calcium metabolism, *Osteoporosis Int.*, 1993, **3**(Suppl. 1), 44–47.

89. M. Sowers, G. Corton, B. Shapiro, M.L. Jannausch, M. Crutchfield, M.L. Smith, J.F. Randolph and B. Hollis, Changes in bone density with lactation, *JAMA*, 1993, **269**(24), 3130–3135.

90. I. Matsumoto, S. Kosha, S. Noguchi, N. Kojima, T. Oki, T. Douchi and Y. Nagata, Changes of bone mineral density in pregnant and postpartum women, *J. Obstet. Gynaecol.*, 1995, **21**(5), 419–425.

91. M. Sowers, J. Randolph, B. Shapiro and M. Jannausch, A prospective study of bone density and pregnancy after an extended period of lactation with bone loss, *Obstet. Gynecol.*, 1995, **85**(2), 285–289.

92. M.A. Laskey and A. Prentice, Effect of pregnancy on recovery of lactational bone loss, *Lancet*, 1997, **349**(9064), 1518–1519.

93. N. Kojima, T. Douchi, S. Kosha and Y. Nagata, Cross-sectional study of the effects of parturition and lactation on bone mineral density later in life, *Maturitas*, 2002, **41**(3), 203–209.

94. C. Johansson, D. Mellstrom and L. Milsom, Reproductive factors as predictors of bone density and fractures in women at the age of 70, *Maturitas*, 1993, **17**(1), 39–50.

95. C. Cure-Cure, P. Cure-Ramirez, E. Teran and P. Lopez-Jaramillo, Bone-mass peak in multiparity and reduced risk of bone-fractures in menopause, *Int. J. Gynaecol. Obstet.*, 2002, **76**(3), 285–291.

96. S. Forsmo, B. Schei, A. Langhammer and L. Forsen, How do reproductive and lifestyle factors influence bone density in distal and ultradistal radius of early postmenopausal women? The Nord-Trondelag Health Survey, Norway, *Osteoporosis Int.*, 2001, **12**(3), 222–229.

97. M.J. Grainge, C.A. Coupland, S.J. Cliffe, C.E. Chilvers and D.J. Hosking, Reproductive, menstrual and menopausal factors: which are associated with bone mineral density in early postmenopausal women? *Osteoporosis Int.*, 2001, **12**(9), 777–787.

98. M. Tuppurainen, H. Kroger, S. Saarikoski, R. Honkanen and E. Alhava, The effect of gynecological risk factors on lumbar and femoral bone mineral density in peri- and postmenopausal women, *Maturitas*, 1995, **21**(2), 137–145.

99. S. Murphy, K.T. Khaw, H. May and J.E. Compston, Parity and bone mineral density in middle-aged women, *Osteoporosis Int.*, 1994, **4**(3), 162–166.

100. M. Mariconda, M. Pavia, A. Colonna, I.F. Angelillo, O. Marsico, F. Sanzo, C. Mancuso and C. Milano, Appendicular bone density, biochemical markers of bone turnover and lifestyle factors in female teachers of Southern Italy, *Eur. J. Epidemiol.*, 1997, **13**(8), 909–917.

101. T.V. Nguyen, G. Jones, P.N. Sambrook, C.P. White, P.J. Kelly and J.A. Eisman, Effects of estrogen exposure and reproductive factors on bone mineral density and osteoporotic fractures, *J. Clin. Endocrinol. Metab.*, 1995, **80**(9), 2709–2714.

102. Y. Liel, D. Atar and N. Ohana, Pregnancy-associated osteoporosis: preliminary densitometric evidence of extremely rapid recovery of bone mineral density, *South Med. J.*, 1998, **91**(1), 33–35.

103. W.G. Boissonnault and J.S. Boissonnault, Transient osteoporosis of the hip associated with pregnancy, *J. Orthop. Sports Phys. Ther.*, 2001, **31**(7), 359–365; discussion 366–367.

104. R. Axt-Fliedner, G. Schneider, R. Seil, M. Friedrich, D. Mink and W. Schmidt, Transient bilateral osteoporosis of the hip in pregnancy, A case report and review of the literature, *Gynecol. Obstet. Investig.*, 2001, **51**(2), 138–140.

105. A. Samdani, E. Lachmann and W. Nagler, Transient osteoporosis of the hip during pregnancy: a case report, *Am. J. Phys. Med. Rehab.*, 1998, **77**(2), 153–156.

106. N. Uematsu, Y. Nakayama, Y. Shirai, K. Tamai, H. Hashiguchi and Y. Banzai, Transient osteoporosis of the hip during pregnancy, *J. Nippon Med. Sch.*, 2000, **67**(6), 459–463.

107. H.A. Tran and N. Petrovsky, Pregnancy-associated osteoporosis with hypercalcaemia, *Intern. Med. J.*, 2002, **32**(9–10), 481–485.

108. T. Anai, T. Tomiyasu, K. Arima and L. Miyakawa, Pregnancy-associated osteoporosis with elevated levels of circulating parathyroid hormone-related protein: a report of two cases, *J. Obstet. Gynaecol. Res.*, 1999, **25**(1), 63–67.

109. R. Smith and A.J. Phillips, Osteoporosis during pregnancy and its management, *Scand. J. Rheumatol. Suppl.*, 1998, **107**, 66–67.
110. F. Dunne, B. Walters, T. Marshall and D.A. Heath, Pregnancy associated osteoporosis, *Clin. Endocrinol. (Oxford)*, 1993, **39**(4), 487–490.
111. V. Breuil, O. Brocq, L. Euller-Ziegler and A. Grimaud, Insufficiency fracture of the sacrum revealing a pregnancy associated osteoporosis. First case report, *Ann. Rheum. Dis.*, 1997, **56**(4), 278–279.
112. B.W. Alderman, N.S. Weiss, J.R. Daling, C.L. Ure and J.H. Ballard, Reproductive history and postmenopausal risk of hip and forearm fracture, *Am. J. Epidemiol.*, 1986, **124**(2), 262–267.
113. O. Johnell, B. Gullberg, J.A. Kanis, E. Allander, L. Elffors, J. Dequeker, G. Dilsen, C. Gennari, A. Lopes Vaz, G. Lyritis *et al.*, Risk factors for hip fracture in European women: the MEDOS Study. Mediterranean Osteoporosis Study, *J. Bone Miner. Res.*, 1995, **10**(11), 1802–1815.
114. T.W. O'Neill, A.J. Silman, M. Naves Diaz, C. Cooper, J. Kanis and D. Felsenberg, Influence of hormonal and reproductive factors on the risk of vertebral deformity in European women. European Vertebral Osteoporosis Study Group, *Osteoporosis Int.*, 1997, **7**(1), 72–78.
115. C. Cure-Cure and P. Cure-Ramirez, Hormone replacement therapy for bone protection in multiparous women: when to initiate it, *Am. J. Obstet. Gynecol.*, 2001, **184**(4), 580–583.
116. K. Michaelsson, J.A. Baron, B.Y. Farahmand and S. Ljunghall, Influence of parity and lactation on hip fracture risk, *Am. J. Epidemiol.*, 2001, **153**(12), 1166–1172.
117. S. Hoffman, J.A. Grisso, J.L. Kelsey, M.D. Gammon and L.A. O'Brien, Parity, lactation and hip fracture, *Osteoporosis Int.*, 1993, **3**(4), 171–176.
118. A. Paganini-Hill, A. Chao, R.K. Ross and B.E. Henderson, Exercise and other factors in the prevention of hip fracture: the Leisure World study, *Epidemiology*, 1991, **2**(1), 16–25.
119. H.C. Petersen, B. Jeune, J.W. Vaupel and K. Christensen, Reproduction life history and hip fractures, *Ann. Epidemiol.*, 2002, **12**(4), 257–263.

# CHAPTER 30

# Can Nutrition Alter the Population Burden of Fractures?

SANDRA IULIANO-BURNS[1] and EGO SEEMAN

[1]Department of Endocrinology, Austin Health, Studley Rd, Heidelberg, Victoria, Australia, 3084
Email: sandraib@unimelb.edu.au

## Key Points

- The greatest burden of fragility fractures comes from the large low-risk segment of the population with bone mineral density (BMD) in the normal and osteopenic range. Drug therapy is not an option for treating this population.
- Non-drug related approaches are needed to reduce this burden of fractures. Improving nutritional intake may be a safe, inexpensive and effective way to reduce the population burden of fractures. Whether this can be achieved remains to be determined.
- The most rigorous means of determining the fracture efficacy of nutritional interventions is using randomized, placebo-controlled trials. However, this approach is rarely undertaken for macronutrients, but has been studied for nutrients such as calcium.
- Evidence supporting sustained benefits of calcium supplementation on bone size and bone mass in growing children does exist. Calcium and vitamin D supplementation reduce the risk of fractures in institutionalized elderly, but not in ambulant community dwellers.

## Introduction

Fragility fractures may effect up to 50% of women and 20% of men.[1] The burden of fractures in absolute terms will increase because the population is ageing. Drug therapy halves the risk of fracture in high-risk individuals with osteoporosis (BMD 2.5 or more standard deviations below the young normal mean). However, the *population* burden of fractures does not come from this tail of the population distribution for BMD. The majority of fractures come from the larger segment of the population at low or moderate risk for fracture with BMD in the lower part of the normal distribution.[2–4]

Drug therapy is not an option in this low-risk population because drugs such as Alendronate, Risedronate and Raloxifene have not been proven to be efficacious in this group. Even if treatment is efficacious in this low-risk population, the cost of treatment is higher than the cost of the fractures as a huge segment of the community will need to be exposed to the cost, inconvenience and side-effects of drug therapy with little or no benefit, to capture the cases coming to fracture.[5]

Thus, the challenge is to devise means of reducing the population burden of fractures incurred from this larger segment of the population. One theoretical option is to shift the whole of the population distribution for BMD to a higher level.[5] For this approach to be successful, the modification must be efficacious, accessible to all persons, safe and easy to administer. Compliance must be high, and safety is critical, as many individuals must be treated to avert fractures in a few.[5]

There is limited evidence that this approach has met with some success in the cardiovascular field. For example, community-based dietary interventions reduced blood pressure (BP) in mildly hypertensive adults and improved lipid profiles.[6–8] Favourable health outcomes translate into reduced cardiovascular disease (CVD) mortality rates with an estimated 19% reduction in the probability of CVD through community-based programs.[9,10] Appel *et al.*[6] reported a 2–6% reduction in systolic and diastolic BP in mildly hypertensive adults consuming a low-fat, high-fruit and vegetable diet or high-fruit and vegetable diet compared to a typically 'Western' control diet. The diet was also effective in reducing total, low-density and high-density cholesterol.[7] Systolic and diastolic BP decreased with reduced sodium intake.[8] Nutritional changes may fulfil many of the above criteria in the field of osteoporosis.

In this chapter, we review nutritional modifications that may influence the pathogenetic mechanisms operative during skeletal growth and ageing and that influence bone fragility in old age. Where possible we will provide one or two examples from the literature. A more exhaustive review of nutritional factors is provided in other chapters of this book.

# Constraints Imposed by Limitations in Study Design, Execution and Measurement Methods

Most studies are observational prospective or retrospective cohort studies, or case control studies, which are at best hypothesis-generating rather than hypothesis-testing. This is a serious limitation that cannot be overstated as sampling bias, socio-economic factors, confounding factors such as tobacco or alcohol use, that influence the outcome variable cannot be easily measured and so cannot be convincingly taken into account, even using statistical methods of 'adjustment'. In addition, most of these studies do not have anti-fracture efficacy as an endpoint, but rather use surrogates, such as group differences in BMD, changes in BMD or biochemical measurements of bone remodelling in response to deficiency or repletion, as the phenotypic endpoint – surrogates that either do not predict fracture incidence or do so poorly.

Moreover, if up to 80% of the variance in bone traits is genetically determined[11] *i.e.*, if the majority of differences in bone size, mass or architecture are attributable to differences in inherited genetic factors, then the remaining 20% of the variance would be due to individual-specific differences in lifestyle factors such as physical activity, occupation, sunlight exposure, smoking and alcohol consumption. To divide this 'environmental' component of the variance into influential lifestyle factors including nutrition, then to further divide this component into bone specific nutrients, of which many demonstrate co-linearity, makes the task of identifying the contribution of one nutrient formidable. For example, Lunt *et al.*[12] reported that between 0.6 and 1.1% of the variance in BMD at various sites could be accounted for by differences in dairy calcium intake. The power to detect small effects like this is low unless studies have very large sample sizes, well in excess of the usual study size in which BMD and dietary intake have been measured.

# Nutritional Factors and Skeletal Growth

BMD in old age is a function of net gain in BMD during growth and the net loss of BMD during ageing. The net gain in BMD during growth is the result of the increase in the external dimensions of bone, the accrual of cortical and trabecular bone tissue within the growing bone, and the completion of primary and secondary mineralization of the newly deposited bone tissue. The net loss of bone with its mineral content during ageing is a function of the slower continued periosteal apposition of bone and the net loss of bone and bone mineral from the endocortical (inner) surface, the latter a function of the rate of bone remodelling and the size of the negative bone balance in each of the remodelling foci on the endocortical surfaces of bone.

## Appendicular Skeleton

Peak young adult appendicular bone size is determined by the extent of longitudinal growth and periosteal apposition, which together form the external dimensions of the bone, while cortical thickness is determined by the net movement of the endocortical surface produced by endocortical remodelling (formation and resorption) relative to the periosteal surface. Together, the surface extent of periosteal modelling and endocortical remodelling determine cortical thickness and the distance this cortical mass is placed from the neutral or long axis of the bone.

Nutritional factors that (directly or indirectly) modify periosteal bone formation and endocortical bone remodelling may then produce a long bone that differs in its diameter and cortical thickness, while nutritional factors effecting the cellular proliferation or life span of the cells forming the trabecular bone at the metaphysis modify bone length. In boys, 90% of the cortical thickness achieved at peak young adulthood is the result of periosteal apposition, while in girls about 75–80% is the result of periosteal apposition with the remainder the result of endocortical bone formation (or reduced endocortical bone resorption).[13–16]

The effect of nutritional deficiency or repletion states are likely to be sex- race- site- and surface specific depending on the systemic and local factors (growth hormone, insulin-like growth factor 1, sex steroids) responsible for mediating their effects. For example, nutritional deficiencies that effect oestrogen production in women may result in a thinner but wider cortex as the inhibitory effect of oestrogen on periosteal apposition is removed, but concomitantly endocortical apposition fails to occur. The same nutritional deficiency in men may result in a smaller bone with a thinner cortex due to loss of periosteal apposition.

## Axial Skeleton

Trabecular numbers are determined largely at the growth plate although trabecularization of the expanding cortex also contributes.[17] Trabecular thickening, during the pre- and peri-pubertal years by net bone formation to reach a peak in young adulthood, is responsible for the increase in trabecular BMD in the peri-pubertal years. Peak trabecular BMD is similar in boys and girls of the same race, but differs by race.[17–20] So, nutritional factors *in utero* and in the early postnatal period may influence trabecular numbers while nutritional factors during infancy, childhood and puberty may influence trabecular thickness. Whether the higher peak trabecular BMD in African Americans (AA) and Chinese than Caucasians is the result of differences in nutritional factors is unknown.

## Sex and Racial Differences in Upper and Lower Body Size

Growth in body length is region specific. Growth of the appendicular, not axial, skeleton accelerates at 12 months of age after both reach a nadir after birth and proceeds more rapidly than the growth of the axial skeleton before puberty. Sex- and racial-differences in growth of the legs account for more of the sex- and race-specific differences in final height than does sex- and race-specific differences in axial growth. The sex specific difference is the result of the later onset of puberty in boys than girls so that there is more time available for longitudinal growth of the legs in boys than girls. Nutritional deficiencies that effect oestrogen production or action during growth enable epiphyseal growth to continue producing longer leg length; however, axial growth is inhibited resulting in a shorter trunk.[21]

The mechanisms responsible for racial differences in upper and lower body lengths include shorter leg length in Asians than Caucasians, greater leg length in AA than Caucasians, similar axial length in Asians and Caucasians, and shorter axial length in AA than the other groups. These differences in upper and lower body length may partly account for sex- and race-specific differences in fracture rates as the load imposed on bone during a fall or during bending. The potential energy exerted on a bone during a fall may vary according to the proportion of upper body mass, and the distance from ground level. Moreover, these differences in vertebral height and long bone length are accompanied by differences in the cross-sectional area of the whole bone; the cross-sectional area of the bone tissue, and the distance the cortical shell is placed from the long or neutral axis of the bone. These structural and geometric features are sex- and race-specific, and may be determined by nutritional factors, which therefore produce sex- and race-specific differences in the relationship between the imposed load on the bone and its strength.[18,22,23] AA have a shorter but wider vertebral body with a higher volumetric BMD than Caucasians while the long bones are more narrow, but have a relatively thicker cortex.[20] Chinese have a smaller vertebral body with a higher volumetric BMD than Caucasians while the femoral neck (FN) is more narrow in absolute terms, but has a relatively thicker cortex.[23]

The relevance of the above to nutrition may seem remote, but secular trends in upper and lower body lengths are well documented and are likely to be accompanied by secular trends in accompanying skeletal proportions. The most likely factors producing these secular trends are changes in nutrition and physical activity. The heterogeneity in secular trends is fascinating and virtually unstudied in the bone field. Secular increases occur in one or both of leg and trunk length in some races, but not others, and in one or both sexes. These are all documented in studies in children around puberty. Nutritional changes during ageing cannot account for the differences, as height changes are not feasible after epiphyseal closure.[24–26]

Unlike the situation during ageing in which the effect of nutrition or a risk factor depends on the 'dose' and duration of the exposure, during growth, the effect of nutrition or risk factors depend on the maturational stage at the time of exposure. The rate of growth of the appendicular skeleton is greater than the rate of growth of the axial skeleton before puberty so that nutritional factors operating at this time may have greater effects on the appendicular skeleton. Women with hip fractures have a wider FN, a structural difference that is present in their daughters and so is likely to have its origin during growth.[27]

During puberty the rate of growth of the appendicular skeleton slows down as the rate of growth of the axial skeleton accelerates. At this time, nutritional factors may have effects on the axial skeleton, reducing its growth rate, or may influence (*e.g.*, slow down) the rate of *deceleration* of growth of the appendicular skeleton (especially if the nutritional factor influences sex steroids.[14] Anorexia nervosa of early onset produces deficits in bone mass and size of the appendicular skeleton while later onset of this disease produces changes confined to the axial skeleton.[28] Overall, nutritional deficiencies before and during puberty will result in *greater* deficits at skeletal sites than deprivation after puberty when sites are closer to their mature size and mass.[14]

# Nutritional Factors and the Ageing Skeleton

At the completion of linear growth, signalled by the closure of the epiphyses, periosteal apposition continues in both sexes, but much more slowly than during growth. Periosteal apposition appears to be greater in males than females and varies from site to site.[29,30] The importance of periosteal apposition lies in the unique biomechanical advantage conferred by bone formation distant from the neutral or long axis of a bone.[29] The regulators of periosteal apposition are not well defined in either sex nor from race to race. Whether periosteal apposition is largely the result of mechanical loads acting at this surface, and whether nutritional factors operate here is not known. There is evidence to suggest that exercise and nutrition may interact to confer a larger and potentially stronger bone[31–33]. Secondary hyperparathyroidism is accompanied by periosteal remodelling and this process may be modified by calcium supplementation.[34,35] Periosteal apposition may also contribute to the final volumetric BMD of the whole bone in old age as this process is likely to partly offset the loss of bone occurring on the endocortical surfaces of the skeleton.[22,30]

During growth, remodelling on the endocortical surface is associated with a positive bone balance so that each remodelling event deposits a small but net positive amount of bone. The reason remodelling rate is very high in the first year of growth is to facilitate the deposition and growth of bone mass. The rate of remodelling declines and it is likely, but unproven, that the net balance between the volumes of bone resorbed and formed in each remodelling site or basic multicellular unit (BMU) becomes less positive and then zero as bone formation in each BMU declines, and eventually becomes negative. The negative bone balance is probably present well before the menopause in women and well before midlife in men and is the result of a fall in bone formation in the BMU, not an increase in bone resorption. [17,36] The anti-resorptive effect of nutritional interventions such as calcium supplementation may have minimal effect during this phase in skeletal development, which is evident by the small effect or lack of effect of calcium supplementation on BMD in pre-menopausal women.[37]

By the time of menopause, BMU balance is negative, but the loss of bone and the accompanying structural damage is slight because the rate of bone remodelling is slow. It is the rate of remodelling rather than the size of the negative BMU balance that drives the bone loss. At menopause the rate of remodelling accelerates and BMU balance becomes more negative as oestrogen deficiency increases the life span of the osteoclast and reduces the life span of the osteoblast.[38]

If nutritional factors are important in modifying the rate of bone loss, then the most effective means to achieve this is by reducing the rate of remodelling. Whether dietary factors such as calcium supplementation, phyto-oestrogens or protein intake can achieve this is uncertain, but possible and more plausible than an effect on the amount of bone formed or resorbed in each BMU. Nutritional factors could influence bone modelling outside the remodelling unit by increasing bone formation through an anabolic effect directly on the bone lining cells like anabolic agents. There is some data examining the effects of nutritional factors on one or more of these components responsible for bone loss (remodelling rate, BMU based bone formation, resorption or balance). For example, calcium supplements are likely to reduce the rate of bone remodelling as measured by circulating biochemical markers of remodelling, however, variable changes in remodelling are reported using vitamin D supplements.[34,39–48]

## Nutrition during Growth and the Incidence of Fractures in Old Age

Studies of the effects of nutrient intake during growth and fracture rates in old age cannot be readily undertaken given the many years between the supplement during growth and fracture incidence some 50 years later. Indirect inferences based on surrogates of anti-fracture efficacy are feasible, but lack credibility. Data from the EVOS study suggest that milk consumption during growth and adulthood is associated with a reduced risk of vertebral fractures in women.[12] The study by Matkovic *et al.*[49] purporting a lower incidence of fractures in a high calcium consuming community compared to a low calcium community may not reflect the effects of calcium nutrition as the high calcium consumers also had greater energy, protein and fat intakes and were also more active given there were no differences in body weight despite the higher calorie intake. The difference in bone mass measured at the metacarpals was found at peak rather than during ageing where the diminution in bone mass was no less than in the low calcium intake group. These observations suggest that if calcium was responsible for the higher peak bone mass and lower fracture rates, the effect was growth-related rather than age-related.

## Nutrition during Ageing and the Incidence of Fractures in Old Age

Very few long-term randomized trials have been done examining the effects of changes in nutrients such as protein or calcium intake in the community. Although the effects of calcium intake with vitamin D supplementation in elderly institutionalized women provide replicable and consistent evidence for a reduction in non-vertebral fracture risk, this has not been demonstrated in ambulant community dwellers.[47,50,51] Studies of the effects of calcium intake, milk intake and forearm or hip fractures are variably negative or positive and no conclusive view can be expressed from these studies.[52–71]

In studies of patients with osteoporosis, a reduction in vertebral fractures with calcium supplementation has been reported in some, but not in all, studies.[34,72–74] Chevally *et al.*[72] compared the effectiveness of 18 months of supplementation with 800 mg of calcium to a placebo on vertebral fracture rates in 93 elderly healthy subjects. No difference was observed in the rate of new vertebral fractures (defined as a >20% reduction in vertebral height). However, if new fractures were defined as a >20% decrease in the ratio of anterior:posterior height or mid:posterior height, fewer fractures occurred in the calcium treated groups (11 fractures in 54 people) compared to the placebo group (11 fractures in 25 people).

Recker et al.[73] compared the effects of a supplement of 1.2 g of calcium on fracture rates in elderly women with and without vertebral fractures. Fewer women with existing fractures supplemented with calcium incurred vertebral fractures (15 of 53) compared to non-supplemented women with prior fractures (21 of 41). No differences in fracture rates were reported between calcium treated and control women without prior fractures. This observation was almost certainly discovered during a post hoc analysis. This is apparent by examination of the uneven sample size in each of the four groups. Analysis of the whole group did not show a significant effect of the supplement. Stratification into those with and without a baseline fracture was not done prior to randomization to placebo or calcium.

Several other studies have been conducted in women with osteoporosis, but problems in design with small sample sizes and brief follow-up fail to convincingly demonstrate whether calcium supplementation has, or has no beneficial effect on fracture rates.[34,46,75] No association was reported between milk consumption during any phase of life and vertebral deformities in men.[12] Chan et al.[76] reported a mean lower calcium intake in elderly Chinese women with vertebral fractures compared to women without. The relative risk for fractures in the lowest quartile for calcium was 2.1. Cumming et al.[57] reported no association between dietary calcium or milk intake and vertebral fracture risk in women, but did report a greater risk of vertebral fractures with increasing total calcium intake (including that from supplementation; $RR = 2.5$ for top vs bottom quintile for calcium intake). Women with a family history of osteoporosis and a higher risk of fractures were more likely to take calcium supplements.

Decreased risk of hip fracture with increased protein intakes has been reported in women, while a non-significant trend for a decrease in fracture risk with increasing protein intake was observed in men[77-80]. Abelow et al.[81] and Sellmeyer et al.[82] reported positive associations between animal protein and animal:vegetable protein ratio and hip fracture risk. Meyer et al.[59] and Michaelsson et al.[60] both reported no association between protein or non-dairy animal protein intakes and hip fracture risk. High intakes of total and animal protein, but not vegetable protein were reported to be a risk factor for forearm fractures with $RR$s of 1.18 and 1.21 for the highest quintiles, respectively.[79]

The effect of protein intake on bone loss is conflicting. While some authors have reported lower BMD and greater bone loss in low protein consumers, others have reported greater bone loss in postmenopausal women consuming a high animal:vegetable ratio of protein intake.[82-84] A high animal protein intake was proposed to contribute to bone loss because of the acid load brought upon by the protein.[85] Calcium excretion was positively associated with animal but not vegetable protein intake in adults.[86] Short-term manipulation of the acid content of the diet results in greater rates of urinary calcium and *C*-telopeptide excretion.[87] Dietary manipulation to induce a high-compared to low-protein diet augmented urinary calcium and *N*-telopeptide excretion rates.[88] Whether the observed changes are of detriment to bone mass and fracture risk is yet to be determined.

One difficulty in establishing a relationship between dietary vitamin D intake and fracture risk is that the major source of vitamin D in healthy people is cutaneous production from sunlight exposure. Provided there is adequate sunlight exposure (*e.g.*, ability to move outside) and ultraviolet radiation (*e.g.*, latitude of inhabitants) nutrition is likely to be a minor contributor to vitamin D levels. Therefore, lack of sunlight exposure may be a more important contributor to vitamin D deficiency with lack of sunlight exposure reported to be a risk factor for hip fractures in women and men.[67,71,89]

The principal dietary sources of vitamin D are cod liver oil and fatty fish such as salmon, sardines, tuna, herring and mackerel. Small amounts are found in diary products and eggs and, depending on the country, foods such as margarine and milk are fortified with vitamin D. The elderly, who are at the highest risk of fracture, are also more prone to vitamin D deficiency due to reduced capacity for the skin to synthesize vitamin D.[90,91] This state may be exacerbated by restricted mobility, resulting in lack of sunlight exposure. Munger *et al.*[77] reported no association between vitamin D intake and fracture risk, while Ranstam *et al.*[92] reported a decreased risk of fracture with increasing use of vitamin D supplements.

Two large randomized trials involving elderly men and women supplemented from 2 to 3.5 years with a daily dose of approximately 400 IU per day of vitamin D or placebo did not report a reduction in hip or peripheral fracture rates with vitamin D supplementation using either active treatment or intention to treat analysis.[43,50] Five years of treatment using a daily dose of 300 IU of vitamin D, with or without HRT did not improve non-vertebral fracture rates in postmenopausal women.[93] A daily dose of 800 IU of Vitamin D combined with 1.2 g of calcium taken for 18 months was effective in reducing hip fracture rates in elderly women.[47] Improved total and upper body but not lower body fracture rates were reported in elderly men and women randomly assigned to receive a yearly injection of 150 000–300 000 IU of vitamin D for a mean of approximately 3.5 years compared to controls.[94]

# The Effect of Nutrition on BMD, Bone Size and Structure during Growth

Given the complexities of study design needed to demonstrate the effect of a nutrient given during adulthood and anti-fracture efficacy in old age, an alternative approach is to examine the effect of a nutrient on a surrogate of anti-fracture efficacy such as BMD, a bone remodelling marker or a structural change. BMD is most often used as a surrogate for bone strength and fracture risk.

Earlier studies reported greater gains in height in children supplemented with milk, effects likely be the result of the additional energy or protein rather than the calcium content of the milk.[95] Calcium supplementation does not appear to enhance height gains in Asian children or Gambian children accustomed to low calcium intakes.[96,97] Gilsanz et al.[98] and Lento-Axtelius et al.[99] reported no reductions in bone length with calcium depletion although reductions in BMD and bone area were observed. No effect on trabecular number, thickness or spacing was reported with moderate deficits in calcium although others have reported reductions in trabecular numbers and thickness with calcium deficiency.[98–101] Reduced cortical thickness and elevated rates of bone resorption were observed with calcium deprivation.[101,102]

Calcium supplementation has been reported to enhance bone mass accrual by up to 5% in children.[96,97,103–105] However, residual benefits following the cessation of supplementation are usually not found.[106–108] The most compelling evidence to support the notion that calcium supplementation increases peak bone size and bone mass is based on a study in pre-pubertal girls. Bonjour et al.[104] reported greater gains in BMD with 850 mg per day of calcium supplementation at the radius and femoral diaphyses using both the intention to treat and active treatment analysis, and at the radius metaphysis and trochanter using active treatment analysis only. No effect of supplementation was observed for lumbar spine (LS) or FN BMD. The mean gain from all sites was only significant for the participants with calcium intakes below the median. In the girls with low calcium intakes the greater gains observed for LS height and femoral shaft bone area, width, BMC and BMD were evident one year after the cessation of supplementation

Even if a supplement does produce a higher peak bone size or bone mass, it is essential that evidence is provided for the maintenance of this effect into adulthood. This has never been demonstrated. In a recent follow-up study, Bonjour et al.[109] reported that the differences in bone mass and bone size documented at the completion of the one year study persisted and indeed were greater 3.5 years after completion of the study than at the end of the supplementation period.[109,110] In this follow up, 116 of the original 144 girls were available for analysis. In the girls who had previously received calcium supplementation, greater gains in LS, radial and femoral diaphyses, FN and trochanteric BMD were observed compared to controls. The mean gains in BMD, BMC and bone area for all sites were greater in this supplemented group. Furthermore, despite no differences in standing and LS heights and LS BMC being observed between the two groups at the end of the supplementation period, or a year after supplementation, a difference was reported at the end of the 3.5-year follow-up period.

The interpretation of this study is difficult. Firstly, in the initial study the benefits of supplementation were only observed in those girls with below median calcium intakes. The supplementation may have accelerated the rate of growth along the same trajectory (percentile) and so when the study stopped the growth in the controls may 'catch up' leaving no difference in bone size and bone mass. This interpretation would fit the follow-up results of the study in twins reported by Slemenda *et al.*[106] Alternatively, and as proposed by Bonjour *et al.*[109] the intervention may have altered the trajectory of growth producing the observed persistent benefit. Another interpretation discussed and examined by the authors was that the dropouts produced an imbalance in the level of maturity in the two groups so that, by chance alone, those receiving the supplement were more maturationally advanced.[110] In a sub-analysis matching by maturational status the data did indicate a persisting effect. Of the girls who remained in Tanner stages 1–3 for the follow up period the girls in the calcium group gained more height, and BMD at the FN and femoral diaphysis compared to the placebo group. However, again it is not clear what proportion of these less mature girls were in the initial 'below median calcium intake' group where the reported benefits were observed.

The observations by Bonjour *et al.*[109] are potentially very important and if the result is confirmed then the question arises as to the mechanisms responsible for the greater bone size and bone mass produced by calcium supplementation in the pre-pubertal years. Does the calcium supplement increase the rate of longitudinal and radial growth of bone by influencing the GH/IGF-1 axis? If cortical thickness and increased trabecular thickness is the structural change responsible for the higher bone mass in the bigger bone, is this mediated by sex steroids? Are the effects mediated by increased bone formation and reduced bone resorption on the endosteal and trabecular envelopes?

Protein deficiency is associated with reduced linear growth, perhaps due to its effect on the growth hormone-IGF-1 system.[111] IGF-1 is required for the stimulation and proliferation of chondrocytes in the epiphyseal plate.[112] Reductions in cortical and trabecular bone, but not bone width, have been reported in protein deficient children.[111] Reductions in cortical area but not total area, and in trabecular number and thickness have been reported in protein depleted mature rats.[113–115]

# The Effect of Nutrition on BMD, Bone Size and Structure during Adulthood

Significantly, lower spine BMD is reported in women with a low intake of milk during early adulthood. Current calcium intake did not correlate with spine BMD after adjusting for age, anthropometry and other lifestyle variables.[116] Michaelsson *et al.*[117] compared BMD in women with high ( > 1400 mg per day) and low (400–550 mg per day) calcium intakes. Unadjusted BMD did not differ between the two groups. Following adjustment, total body and FN BMD, but not spine BMD was greater in the high- compared to low-calcium consumers. Others have observed no association between calcium intake and BMD or bone loss [118–120].

Many authors reported that the benefits of supplementation were evident in the first, but not during subsequent years.[40,45,46,75,121,122] This observation is in keeping with the notion that dietary interventions such as calcium supplementation reduce the rate of bone remodelling and then cease to be detectable after steady-state is restored.[123]

The increase in BMD reported using calcium supplements is the result of a reduction in the remodelling space. This reduction in the remodelling space consists of: bone formation in the many more resorption cavities formed during the remodelling cycle *before* treatment was begun, and a continued increase in secondary mineralization of the newly laid down bone and the older bone. The more complete secondary mineralization occurs because the bone is being replaced (turned over) more slowly as a result of the slower remodelling rate during therapy.

The existing bone tissue mass becomes more fully mineralized, an effect that may restore the strength of bone and may therefore reduce the occurrence of further fractures, but there is only a weak relationship between a change in BMD and a reduction in fracture rate. Indeed, in the studies of Raloxifene treatment, fracture rates decline even in subjects losing bone during therapy.[124,125] Therefore, the increase in BMD observed is not a valid surrogate of anti-fracture efficacy. Although some studies that reported positive changes to BMD with vitamin D supplementation also reported reductions in fracture rates.[47] Studies that reported no difference in BMD at fracture sites also reported non-significant changes to fracture rates at those sites.[34,44,72]

The limited data available of the structural changes to bone resulting from dietary interventions indicate that calcium slows the reduction in cortical thickness in peri- and post-menopausal women, and men.[39,40,126,127] This resulted from a slowing of the expansion of the medullary cavity, with no effect on periosteal width.[39] The bending strength of bone may not be altered as a unit of bone preserved at the endocortical surface offers less biomechanical advantage than the same (or less) bone deposited distant from the neutral axis of the bone on the periosteal surface. There is no evidence that calcium supplements increase periosteal apposition. Nevertheless, if calcium supplements reduce the rate of remodelling and preserve endocortical bone and slows down the appearance and size of intracortical porosity, the compressive strength of bone may be better preserved as loads on bone causing compressive stress may be reduced as the load will be imposed on a larger cross-sectional area than would occur had no supplement been given. The effect of calcium supplements on intracortical porosity or the maintenance of trabecular connectivity has not been studied.

# Conclusion

There are many areas of ignorance regarding the effects of nutrients on skeletal growth and ageing. In part, this is because nutrients are difficult to study in isolation, effects are likely to be small and so, difficult to detect. In addition, effects of nutrients are likely to be indirect and expressed as changes on the periosteal and endosteal surfaces. These surface specific changes are the result of modelling and remodelling and are not easily studied. Of all periods of life, the most opportune time for intervention appears to be during growth. Calcium supplementation appears effective during growth; however, the optimal time period to intervene is not clearly established. The anti-fracture efficacy of calcium and vitamin D interventions has been demonstrated in high-risk populations, but remains to be confirmed in low-risk groups.

# References

1. J.A. Kanis, O. Johnell, A. Oden *et al.*, Long-term risk of osteoporotic fracture in Malm, *Osteoporosis Int.*, 2000, **11**, 669–674.
2. K.M. Sanders, G.C. Nicholson, M.A. Kotowicz *et al.*, The inadequacy of the T = −2.5 SD as a threshold for reducing the population burden of all fractures: Geelong Osteoporosis Study (GOS), In *Third International Symposium on Clinical and Economic Aspects of Osteoporosis and Osteoarthritis*, 2002, Barcelona, Spain.
3. M. van der Klift, Osteoporosis, more than fractures alone. An epidemiological approach, In *Institute of Medical Technology Assessment, Department of Internal Medicine and Department of Epidemiology and Biostatistics*, 2002, Erasmus University: Rotterdam.
4. S.A. Wainwright, K.R. Phipps, J.V. Stone *et al.*, A large proportion of fractures in postmenopausal women occur with baseline bone mineral density *T*-score > −2.5, In *Twenty-third annual meeting of the American Society for Bone and Mineral Research*, 2001, Phoenix, Arizona, USA.
5. G. Rose, Sick individuals and sick populations, 1985, [see comments.], *Bull. WHO*, 2001, **79**, 990–996.
6. L.J. Appel, T.J. Moore, E. Obarzanek *et al.*, A clinical trial of the effects of dietary patterns on blood pressure. DASH collaborative research group, [see comments.], *N. Engl. J. Med.*, 1997, **336**, 1117–1124.
7. E. Obarzanek, F.M. Sacks, W.M. Vollmer *et al.*, Effects on blood lipids of a blood pressure-lowering diet: the dietary approaches to stop hypertension (DASH) trial, [see comments.], *Am. J. Clin. Nutr.*, 2001, **74**, 80–89.
8. F.M. Sacks, L.P. Svetkey, W.M. Vollmer *et al.*, Effects on blood pressure of reduced dietary sodium and the dietary approaches to stop hypertension (DASH) diet. DASH-sodium collaborative research group, [see comments.], *N. Engl. J. Med.*, 2001, **344**, 3–10.
9. P. Puska, E. Vartiainen, J. Tuomilehto *et al.*, Changes in premature deaths in Finland: successful long-term prevention of cardiovascular diseases, *Bull. World Health Org.*, 1998, **76**, 419–425.
10. L. Weinehall, G. Westman, G. Hellsten *et al.*, Shifting the distribution of risk: results of a community intervention in a Swedish programme for the prevention of cardiovascular disease, *J. Epidemiol. Community Health*, 1999, **53**, 243–250.
11. Seeman, E., J.L. Hopper, N.R. Young *et al.*, Do genetic factors explain associations between muscle strength, lean mass, and bone density? A twin study, *Am. J. Physiol.*, 1996, **270**, E320–E327.
12. M. Lunt, P. Masaryk, C. Scheidt-Nave *et al.*, The effects of lifestyle, dietary dairy intake and diabetes on bone density and vertebral deformity prevalence: the EVOS study, *Osteoporosis Int.*, 2001. **12**, 688–698.
13. S. Garn, In *The earlier gain and later loss of critical bone: nutritional perspectives*, Charles C. Thomas, Springfield, IL, 1970, pp. 3–120.

14. S. Bass, P.D. Delmas, G. Pearce *et al.*, The differing tempo of growth in bone size, mass, and density in girls is region-specific, [see comments.], *J. Clin. Invest.*, 1999, **104**, 795–804.

15. M. Bradney, M.K. Karlsson, Y. Duan *et al.*, Heterogeneity in the growth of the axial and appendicular skeleton in boys: implications for the pathogenesis of bone fragility in men, *J. Bone Miner. Res.*, 2000, **15**, 1871–1878.

16. E. Schoenau, The development of the skeletal system in children and the influence of muscle strength, *Hormonal Res.*, 1989, **49**, 27–31.

17. A.M. Parfitt, R. Travers, F. Rauch *et al.*, Structural and cellular changes during bone growth in healthy children, *Bone*, 2000, **27**, 487–494.

18. Y. Duan, X.-F. Wang, C.H. Turner *et al.*, Racial but not sex differences in vertebral body volumetric bone mineral density in Asian and Caucasian women and men, *J. Bone Miner. Res.*, 2002, **17**, SA315 (Abstract).

19. E. Seeman, Growth in bone mass and size–are racial and gender differences in bone mineral density more apparent than real? [letter; comment], *J. Clin. Endocrinol. Metabol.*, 1998, **83**, 1414–1419.

20. V. Gilsanz, T.F. Roe, S. Mora *et al.*, Changes in vertebral bone density in black girls and white girls during childhood and puberty, *N. Engl. J. Med.*, 1991, **325**, 1597–1600.

21. E. Seeman, Pathogenesis of bone fragility in women and men, *Lancet*, 2002, **359**, 1841–1850.

22. Y. Duan, E. Seeman and C.H. Turner, The biomechanical basis of vertebral body fragility in men and women, *J. Bone Miner. Res.*, 2001, **16**, 2276–2283.

23. X.-F. Wang, T.J. Beck, Y. Duan *et al.*, Racial differences in hip fracture risk are established during growth: structural and biomechanical determinants of strength at the proximal femur in Asians and Caucasians, *J. Bone Miner. Res.*, 2002, **17**, 1167 (Abstract).

24. N. Cameron, J.M. Tanner and R.H. Whitehouse, A longitudinal analysis of the growth of limb segments in adolescence, *Ann. Human Biol.*, 1982, **9**, 211–220.

25. A.M. Fredriks, S. van Buuren, R.J. Burgmeijer *et al.*, Continuing positive secular growth change in The Netherlands 1955–1997, *Ped. Res.*, 2000, **47**, 316–323.

26. D.S. Freedman, L.K. Khan, M.K. Serdula *et al.*, Secular trends in height among children during 2 decades: The Bogalusa Heart Study, *Arch. Ped. Adol. Med.*, 2000, **154**, 155–161.

27. A. Tabensky, Y. Duan, J. Edmonds *et al.*, The contribution of reduced peak accrual of bone and age-related bone loss to osteoporosis at the spine and hip: insights from the daughters of women with vertebral or hip fractures, *J. Bone Miner. Res.*, 2001, **16**, 1101–1107.

28. E. Seeman, M.K. Karlsson and Y. Duan, On exposure to anorexia nervosa, the temporal variation in axial and appendicular skeletal development predisposes to site-specific deficits in bone size and density: a cross-sectional study, *J. Bone Miner. Res.*, 2000, **15**, 2259–2265.

29. C.B. Ruff and W.C. Hayes, Sex differences in age-related remodeling of the femur and tibia, *J. Orthopaed. Res.*, 1988, **6**, 886–896.

30. Y. Duan, C.H. Turner, B.T. Kim *et al.*, Sexual dimorphism in vertebral fragility is more the result of gender differences in age-related bone gain than bone loss, *J. Bone Miner. Res.*, 2001, **16**, 2267–2275.

31. S. Iuliano-Burns, L. Saxon, G. Naughton *et al.*, Regional specificity of exercise and calcium during skeletal growth in girls: A randomised controlled trial, *J. Bone. Miner. Res.*, 2003, **18**, 156–162.

32. E.M. Lau, J. Woo, P.C. Leung *et al.*, The effects of calcium supplementation and exercise on bone density in elderly Chinese women, *Osteoporosis Int.*, 1992, **2**, 168–173.

33. R. Prince, A. Devine, I. Dick *et al.*, The effects of calcium supplementation (milk powder or tablets) and exercise on bone density in postmenopausal women, *J. Bone Miner. Res.*, 1995, **10**, 1068–1075.

34. B.L. Riggs, W.M. O'Fallon, J. Muhs *et al.*, Long-term effects of calcium supplementation on serum parathyroid hormone level, bone turnover, and bone loss in elderly women, *J. Bone Miner. Res.*, 1998, **13**, 168–174.

35. B. Dawson-Hughes, G.E. Dallal, E.A. Krall *et al.*, Effect of vitamin D supplementation on wintertime and overall bone loss in healthy postmenopausal women, *Ann. Intern. Med.*, 1991, **115**, 505–512.

36. P. Lips, P. Courpron and P.J. Meunier, Mean wall thickness of trabecular bone packets in the human iliac crest: changes with age, *Calcif. Tissue Res.*, 1978, **26**, 13–17.

37. E.L. Smith, C. Gilligan, P.E. Smith *et al.*, Calcium supplementation and bone loss in middle-aged women, *Am. J. Clin. Nutr.*, 1989, **50**, 833–842.

38. S.C. Manolagas, Birth and death of bone cells: basic regulatory mechanisms and implications for the pathogenesis and treatment of osteoporosis, *Endocr. Rev.*, 2000, **21**, 115–137.

39. M. Peacock, G. Liu, M. Carey *et al.*, Effect of calcium or 25OH vitamin D3 dietary supplementation on bone loss at the hip in men and women over the age of 60. [see comments.], *J. Clin. Endocrinol. Metabol.*, 2000, **85**, 3011–3019.

40. P.J. Elders, P. Lips, J.C. Netelenbos *et al.*, Long-term effect of calcium supplementation on bone loss in perimenopausal women, *J. Bone Miner. Res.*, 1994, **9**, 963–970.

41. D. Storm, R. Eslin, E.S. Porter *et al.*, Calcium supplementation prevents seasonal bone loss and changes in biochemical markers of bone turnover in elderly New England women: a randomized placebo-controlled trial, *J. Clin. Endocrinol. Metabol.*, 1998, **83**, 3817–3825.

42. B. Riis, K. Thomsen and C. Christiansen, Does calcium supplementation prevent postmenopausal bone loss? A double-blind, controlled clinical study, *N. Engl. J. Med.*, 1987, **316**, 173–177.

43. H.E. Meyer, G.B. Smedshaug, E. Kvaavik *et al.*, Can vitamin D supplementation reduce the risk of fracture in the elderly? A randomized controlled trial, *J. Bone Miner. Res.*, 2002, **17**, 709–715.

44. P.R. Ebeling, J.D. Wark, S. Yeung *et al.*, Effects of calcitriol or calcium on bone mineral density, bone turnover, and fractures in men with primary osteoporosis: a two-year randomized, double blind, double placebo study, *J. Clin. Endocrinol. Metabol.*, 2001, **86**, 4098–4103.

45. M.E. Ooms, J.C. Roos, P.D. Bezemer *et al.*, Prevention of bone loss by vitamin D supplementation in elderly women: a randomized double-blind trial, *J. Clin. Endocrinol. Metabol.*, 1995, **80**, 1052–1058.

46. B. Dawson-Hughes, S.S. Harris, E.A. Krall *et al.*, Effect of calcium and vitamin D supplementation on bone density in men and women 65 years of age or older, *N. Engl. J. Med.*, 1997, **337**, 670–676.

47. M.C. Chapuy, M.E. Arlot, F. Duboeuf *et al.*, Vitamin D3 and calcium to prevent hip fractures in the elderly women, *N. Engl. J. Med.*, 1992, **327**, 1637–1642.

48. E.S. Orwoll, S.K. Oviatt, M.R. McClung *et al.*, The rate of bone mineral loss in normal men and the effects of calcium and cholecalciferol supplementation, *Ann. Intern. Med.*, 1990, **112**, 29–34.

49. V. Matkovic, K. Kostial, I. Simonovic *et al.*, Bone status and fracture rates in two regions of Yugoslavia, *Am. J. Clin. Nutr.*, 1979, **32**, 540–549.
50. P. Lips, W.C. Graafmans, M.E. Ooms *et al.*, Vitamin D supplementation and fracture incidence in elderly persons. A randomized, placebo-controlled clinical trial, *Ann. Internal Med.*, 1996, **124**, 400–406.
51. M.C. Chapuy, R. Pamphile, E. Paris *et al.*, Combined calcium and vitamin D3 supplementation in elderly women: confirmation of reversal of secondary hyperparathyroidism and hip fracture risk: the Decalyos II study, *Osteoporosis Int.*, 2002, **13**, 257–264.
52. R.J. Honkanen, K. Honkanen, H. Kroger *et al.*, Risk factors for perimenopausal distal forearm fracture, *Osteoporosis Int.*, 2000, **11**, 265–270.
53. N. Kreiger, A. Gross and G. Hunter, Dietary factors and fracture in postmenopausal women: a case-control study, *Int. J. Epidemiol.*, 1992, **21**, 953–958.
54. T.V. Nguyen, J.R. Center, P.N. Sambrook *et al.*, Risk factors for proximal humerus, forearm, and wrist fractures in elderly men and women: the Dubbo Osteoporosis Epidemiology Study, *Am. J. Epidemiol.*, 2001, **153**, 587–595.
55. D. Feskanich, W.C. Willett, M.J. Stampfer *et al.*, Milk, dietary calcium, and bone fractures in women: a 12-year prospective study, *Am. J. Pub. Health*, 1997, **87**, 992–997.
56. J. Huopio, H. Kroger, R. Honkanen *et al.*, Risk factors for perimenopausal fractures: a prospective study, *Osteoporosis Int.*, 2000, **11**, 219–227.
57. R.G. Cumming, S.R. Cummings, M.C. Nevitt *et al.*, Calcium intake and fracture risk: results from the study of osteoporotic fractures, *Am. J. Epidemiol.*, 1997, **145**, 926–934.
58. W. Owusu, W.C. Willett, D. Feskanich *et al.*, Calcium intake and the incidence of forearm and hip fractures among men, *J. Nutr.*, 1997, **127**, 1782–1787.
59. H.E. Meyer, J.I. Pedersen, E.B. Loken *et al.*, Dietary factors and the incidence of hip fracture in middle-aged Norwegians. A prospective study, *Am. J. Epidemiol.*, 1997, **145**, 117–123.
60. K. Michaelsson, L. Holmberg, H. Mallmin *et al.*, Diet and hip fracture risk: a case-control study. Study Group of the Multiple Risk Survey on Swedish Women for eating assessment, *Int. J. Epidemiol.*, 1995, **24**, 771–782.
61. A.C. Looker, T.B. Harris, J.H. Madans *et al.*, Dietary calcium and hip fracture risk: the NHANES I Epidemiologic Follow-Up Study, *Osteoporosis Int.*, 1993, **3**, 177–184.
62. A. Tavani, E. Negri and C. La Vecchia, Calcium, dairy products, and the risk of hip fracture in women in northern Italy, *Epidemiology*, 1995, **6**, 554–557.
63. C.A. Wickham, K. Walsh, C. Cooper *et al.*, Dietary calcium, physical activity, and risk of hip fracture: a prospective study, *Br. Medical J.*, 1989, **299**, 889–892.
64. C. Cooper, D.J. Barker and C. Wickham, Physical activity, muscle strength, and calcium intake in fracture of the proximal femur in Britain, *Br. Medical J.*, 1988, **297**, 1443–1446.
65. E. Lau, S. Donnan, D.J. Barker *et al.*, Physical activity and calcium intake in fracture of the proximal femur in Hong Kong, *Br. Medical J.*, 1988, **297**, 1441–1443.
66. E.M. Lau, P. Suriwongpaisal, J.K. Lee *et al.*, Risk factors for hip fracture in Asian men and women: the Asian osteoporosis study, *J. Bone Miner. Res.*, 2001, **16**, 572–580.
67. O. Johnell, B. Gullberg, J.A. Kanis *et al.*, Risk factors for hip fracture in European women: the MEDOS Study. Mediterranean Osteoporosis Study, *J. Bone Miner. Res.*, 1995, **10**, 1802–1815.

68. R. Perez Cano, F. Galan Galan and G. Dilsen, Risk factors for hip fracture in Spanish and Turkish women, *Bone*, 1993, **14** (Suppl. 1), S69–S72.

69. C. Ribot, F. Tremollieres, J.M. Pouilles *et al.*, Risk factors for hip fracture. MEDOS study: results of the Toulouse Centre, *Bone*, 1993, **14** (Suppl. 1), S77–S80.

70. T.L. Holbrook, E. Barrett-Connor and D.L. Wingard, Dietary calcium and risk of hip fracture: 14-year prospective population study, *Lancet*, 1988, **2**, 1046–1049.

71. J. Kanis, O. Johnell, B. Gullberg *et al.*, Risk factors for hip fracture in men from southern Europe: the MEDOS study. Mediterranean Osteoporosis Study, *Osteoporosis Int.*, 1999, **9**, 45–54.

72. T. Chevalley, R. Rizzoli, V. Nydegger *et al.*, Effects of calcium supplements on femoral bone mineral density and vertebral fracture rate in vitamin-D-replete elderly patients, *Osteoporosis Int.*, 1994, **4**, 245–252.

73. R.R. Recker, S. Hinders, K.M. Davies *et al.*, Correcting calcium nutritional deficiency prevents spine fractures in elderly women, *J. Bone Mineral Res.*, 1996, **11**, 1961–1966.

74. B.L. Riggs, E. Seeman, S.F. Hodgson *et al.*, Effect of the fluoride/calcium regimen on vertebral fracture occurrence in postmenopausal osteoporosis. Comparison with conventional therapy, *N. Engl. J. Med.*, 1982, **306**, 446–450.

75. I.R. Reid, R.W. Ames, M.C. Evans *et al.*, Long-term effects of calcium supplementation on bone loss and fractures in postmenopausal women: a randomized controlled trial, *Am. J. Med.*, 1995, **98**, 331–335.

76. H.H. Chan, E.M. Lau, J. Woo *et al.*, Dietary calcium intake, physical activity and the risk of vertebral fracture in Chinese, *Osteoporosis Int.*, 1996, **6**, 228–232.

77. R.G. Munger, J.R. Cerhan and B.C. Chiu, Prospective study of dietary protein intake and risk of hip fracture in postmenopausal women, *Am. J. Clin. Nutr.*, 1999, **69**, 147–152.

78. Z. Huang, J.H. Himes and P.G. McGovern, Nutrition and subsequent hip fracture risk among a national cohort of white women, *Am. J. Epidemiol.*, 1996, **144**, 124–134.

79. D. Feskanich, W.C. Willett, M.J. Stampfer *et al.*, Protein consumption and bone fractures in women, *Am. J. Epidemiol.*, 1996, **143**, 472–479.

80. M.E. Mussolino, A.C. Looker, J.H. Madans *et al.*, Risk factors for hip fracture in white men: the NHANES I epidemiologic follow-up study, *J. Bone Miner. Res.*, 1998, **13**, 918–924.

81. B.J. Abelow, T.R. Holford and K.L. Insogna, Cross-cultural association between dietary animal protein and hip fracture: a hypothesis, *Calcif. Tissue Int.*, 1992, **50**, 14–18.

82. D.E. Sellmeyer, K.L. Stone, A. Sebastian *et al.*, A high ratio of dietary animal to vegetable protein increases the rate of bone loss and the risk of fracture in postmenopausal women. Study of Osteoporotic Fractures Research Group, *Am. J. Clin. Nutr.*, 2001, **73**, 118–122.

83. J.H. Promislow, D. Goodman-Gruen, D.J. Slymen *et al.*, Protein consumption and bone mineral density in the elderly: the Rancho Bernardo Study, *Am. J. Epidemiol.*, 2002, **155**, 636–644.

84. M.T. Hannan, K.L. Tucker, B. Dawson-Hughes *et al.*, Effect of dietary protein on bone loss in elderly men and women: the Framingham Osteoporosis Study, *J. Bone Miner. Res.*, 2000, **15**, 2504–2512.

85. A. Wachman and D.S. Bernstein, Hypothesis: Diet and osteoporosis, *Lancet*, 1968, 958–959.

86. R. Itoh, N. Nishiyama and Y. Suyama, Dietary protein intake and urinary excretion of calcium: a cross-sectional study in a healthy Japanese population, [see comments.], *Am. J. Clin. Nutr.*, 1998, **67**, 438–444.

87. T. Buclin, M. Cosma, M. Appenzeller *et al.*, Diet acids and alkalis influence calcium retention in bone, *Osteoporosis Int.*, 2001, **12**, 493–499.

88. J.E. Kerstetter, M.E. Mitnick, C.M. Gundberg *et al.*, Changes in bone turnover in young women consuming different levels of dietary protein, *J. Clin. Endocrinol. Metabol.*, 1999, **84**, 1052–1055.

89. J.E. Compston, The role of vitamin D and calcium supplementation in the prevention of osteoporotic fractures in the elderly, *Clin. Endocrinol.*, 1995, **43**, 393–405.

90. M.F. Holick, L.Y. Matsuoka and J. Wortsman, Age, vitamin D, and solar ultraviolet, *Lancet*, 1989, **2**, 1104–1105.

91. J. MacLaughlin and M.F. Holick, Aging decreases the capacity of human skin to produce vitamin D3, *J. Clin. Invest.*, 1985, **76**, 1536–1538.

92. J. Ranstam and J.A. Kanis, Influence of age and body mass on the effects of vitamin D on hip fracture risk, *Osteoporosis Int.*, 1995, **5**, 450–454.

93. M.H. Komulainen, H. Kroger, M.T. Tuppurainen *et al.*, HRT and Vit D in prevention of non-vertebral fractures in postmenopausal women; a 5 year randomized trial, *Maturitas*, 1998, **31**, 45–54.

94. R.J. Heikinheimo, J.A. Inkovaara, E.J. Harju *et al.*, Annual injection of vitamin D and fractures of aged bones, *Calcif. Tissue Int.*, 1992, **51**, 105–110.

95. J.B. Orr, Milk consumption and the growth of school children, *Lancet*, 1928, **28**, 202–203.

96. B. Dibba, A. Prentice, M. Ceesay *et al.*, Effect of calcium supplementation on bone mineral accretion in gambian children accustomed to a low-calcium diet, *Am. J. Clin. Nutr.*, 2000, **71**, 544–549.

97. W.T. Lee, S.S. Leung, D.M. Leung *et al.*, A randomized double-blind controlled calcium supplementation trial, and bone and height acquisition in children, *Br. J. Nutr.*, 1995, **74**, 125–139.

98. V. Gilsanz, T.F. Roe, J. Antunes *et al.*, Effect of dietary calcium on bone density in growing rabbits, *Am. J. Physiol.*, 1991, **260**, E471–E476.

99. D. Lehto-Axtelius, V.V. Surve, O. Johnell *et al.*, Effects of calcium deficiency and calcium supplementation on gastrectomy-induced osteopenia in the young male rat, *Scand. J. Gastroenterol.*, 2002, **37**, 299–306.

100. C.H. Turner, W.R. Hinckley, M.E. Wilson *et al.*, Combined effects of diets with reduced calcium and phosphate and increased fluoride intake on vertebral bone strength and histology in rats, *Calcif. Tissue Int.*, 2001, **69**, 51–57.

101. P. Mocetti, P. Ballanti, S. Zalzal *et al.*, A histomorphometric, structural, and immunocytochemical study of the effects of diet-induced hypocalcemia on bone in growing rats, *J. Histochem. Cytochem.*, 2000, **48**, 1059–1078.

102. W. Geng, D.L. DeMoss and G.L. Wright, Effect of calcium stress on the skeleton mass of intact and ovariectomized rats, *Life Sciences*, 2000, **66**, 2309–2321.

103. C.C. Johnston Jr., J.Z. Miller, C.W. Slemenda *et al.*, Calcium supplementation and increases in bone mineral density in children, [see comments.], *N. Engl. J. Med.*, 1992, **327**, 82–87.

104. J.P. Bonjour, A.L. Carrie, S. Ferrari *et al.*, Calcium-enriched foods and bone mass growth in prepubertal girls: a randomized, double-blind, placebo-controlled trial, *J. Clin. Invest.*, 1997, **99**, 1287–1294.

105. W.T. Lee, S.S. Leung, S.H. Wang *et al.*, Double-blind, controlled calcium supplementation and bone mineral accretion in children accustomed to a low-calcium diet, [see comments.], *Am. J. Clin. Nutr.*, 1994, **60**, 744–750.

106. C.W. Slemenda, T.K. Reister, M. Peacock *et al.*, Bone growth in children following the cessation of calcium supplementation, *J. Bone Miner. Res.*, 1993, **8**, S154.

107. W.T. Lee, S.S. Leung, D.M. Leung *et al.*, A follow-up study on the effects of calcium-supplement withdrawal and puberty on bone acquisition of children, *Am. J. Clin. Nutr.*, 1996, **64**, 71–77.

108. W.T. Lee, S.S. Leung, D.M. Leung *et al.*, Bone mineral acquisition in low calcium intake children following the withdrawal of calcium supplement, *Acta Paediatr.*, 1997, **86**, 570–576.

109. J.P. Bonjour, T. Chevalley, P. Ammann *et al.*, Gain in bone mineral mass in prepubertal girls 3.5 years after discontinuation of calcium supplementation: a follow-up study, [see comments.], *Lancet*, 2001, **358**, 1208–1212.

110. T. Remer, K.R. Boye and F. Manz, Long-term increase in bone mass through high calcium intake before puberty, *Lancet*, 2002, **359**, 2037–2038; discussion 2038.

111. Adams, P. and F.R. Berridge, Effects of kwashiorkor on cortical and trabecular bone, *Arch. Dis. Childhood*, 1969, **44**, 705–709.

112. E. Canalis, T.L. McCarthy and M. Centrella, Growth factors and cytokines in bone cell metabolism, *Annu. Rev. Med.*, 1991, **42**, 17–24.

113. S. Bourrin, A. Toromanoff, P. Ammann *et al.*, Dietary protein deficiency induces osteoporosis in aged male rats, *J. Bone Miner. Res.*, 2000, **15**, 1555–1563.

114. P. Ammann, S. Bourrin, J.P. Bonjour *et al.*, Protein undernutrition-induced bone loss is associated with decreased IGF-I levels and estrogen deficiency, *J. Bone Miner. Res.*, 2000, **15**, 683–690.

115. P. Ammann, A. Laib, J.P. Bonjour *et al.*, Dietary essential amino acid supplements increase bone strength by influencing bone mass and bone microarchitecture in ovariectomized adult rats fed an isocaloric low-protein diet, *J. Bone Miner. Res.*, 2002, **17**, 1264–1272.

116. S.A. New, C. Bolton-Smith, D.A. Grubb *et al.*, Nutritional influences on bone mineral density: a cross-sectional study in premenopausal women, *Am. J. Clin. Nutr.*, 1997, **65**, 1831–1839.

117. K. Michaelsson, R. Bergstrom, L. Holmberg *et al.*, A high dietary calcium intake is needed for a positive effect on bone density in Swedish postmenopausal women, *Osteoporosis Int.*, 1997, **7**, 155–161.

118. S.A. Earnshaw, A. Worley and D.J. Hosking, Current diet does not relate to bone mineral density after the menopause. The Nottingham Early Postmenopausal Intervention Cohort (EPIC) Study Group, *Br. J. Nutr.*, 1997, **78**, 65–72.

119. C. Cooper, E.J. Atkinson, D.D. Hensrud *et al.*, Dietary protein intake and bone mass in women, *Calcif. Tissue Int.*, 1996, **58**, 320–325.

120. E.C. van Beresteijn, M.A. van't Hof, G. Schaafsma *et al.*, Habitual dietary calcium intake and cortical bone loss in perimenopausal women: a longitudinal study, *Calcif. Tissue Int.*, 1990, **47**, 338–344.

121. B. Dawson-Hughes, G.E. Dallal, E.A. Krall *et al.*, A controlled trial of the effect of calcium supplementation on bone density in postmenopausal women, *N. Engl. J. Med.*, 1990, **323**, 878–883.

122. D. Mackerras and T. Lumley, First- and second-year effects in trials of calcium supplementation on the loss of bone density in postmenopausal women, *Bone*, 1997, **21**, 527–533.

123. J.A. Kanis, The use of calcium in the management of osteoporosis, *Bone*, 1999, **24**, 279–290.
124. S.R. Cummings, D.B. Karpf, F. Harris *et al.*, Improvement in spine bone density and reduction in risk of vertebral fractures during treatment with antiresorptive drugs, *Am. J. Med.*, 2002, **112**, 281–289.
125. S. Sarkar, B.H. Mitlak, M. Wong *et al.*, Relationships between bone mineral density and incident vertebral fracture risk with raloxifene therapy, [see comments.], *J. Bone Miner. Res.*, 2002, **17**, 1–10.
126. B. Ettinger, H.K. Genant and C.E. Cann, Postmenopausal bone loss is prevented by treatment with low-dosage estrogen with calcium, *Ann. Intern. Med.*, 1987, **106**, 40–45.
127. R.R. Recker, P.D. Saville and R.P. Heaney, Effect of estrogens and calcium carbonate on bone loss in postmenopausal women, *Ann. Intern. Med.*, 1977, **87**, 649–655.

# CHAPTER 31

# Cost-effectiveness of Nutritional Supplements for Osteoporosis Prevention

R.L. FLEURENCE, C.P. IGLESIAS and D.J. TORGERSON

York Trials Unit, Department of Health Sciences, University of York,
Email: djt6@york.ac.uk

**Key Points**

- Nutritional supplements constitute promising options for the treatment of osteoporosis, which is an important source of morbidity and cost to society.
- Evaluating the cost-effectiveness of these treatments is important when allocating scarce health resources for the prevention of fractures.
- A systematic review of the literature identified a number of economic evaluations of vitamin D and calcium with or without calcium supplementation.
- The evidence suggests that vitamin D with or without calcium supplementation is probably cost-effective in the UK, in particular in high-risk groups, such as older women and women with previous fractures.
- Further empirical evidence is needed on the effectiveness of the interventions and their applicability to populations at different risks of fractures.
- Three ongoing clinical trials will be reporting in 2004 and will address these important questions.

## Background

Osteoporotic fractures are an important source of morbidity and cost to society. The burden of fractures will increase due to an increasingly elderly population. Improving the nutrition of the older population either through diet or by the use of supplements may be a useful approach in helping to reduce this fracture burden. Vitamins and minerals play an integral role in bone development. These compounds may contribute to the prevention of osteoporosis and of osteoporotic fractures.[1] Consequently, dietary and supplemental intake, particularly in populations at risk of osteoporosis, is receiving increasing attention. Some clinical studies indicate that vitamin K may decrease fracture rates, while on the contrary, high intake of vitamin A may cause fractures.[2,3] More recently a large trial of anti-oxidant vitamins (*e.g.*, Vitamin C) showed no beneficial effect on fracture rates.[4] In contrast, however, there is some evidence that vitamin D with or without calcium supplementation appears to have a positive effect in decreasing fracture rates.[5–8] Indeed, a recent trial of supplementing people with 100 000 IU of vitamin D every 4 months showed a statistically significant reduction in fractures.[9]

Because of the relative low cost of nutritional supplements, compared to other available treatments for osteoporosis such as hormone replacement therapy (HRT) and bisphosphonates, and the relatively high compliance with such treatments, analyzing the cost-effectiveness of such strategies is likely to play an important role in the allocation of scarce health resources for the prevention of fractures.[10] Nevertheless, even if nutritional supplements are inexpensive, if they are ineffective, this is a waste of scarce resources. Therefore, it is important to consider the costs and benefits of any health care intervention.

# Economic Evaluation

In an economic evaluation we wish to ascertain whether a change in a health care policy is worthwhile. A full economic evaluation addresses the comparison of significant expected benefits and costs associated to them. Depending on the way the expected benefits associated with each intervention are measured the analysis can be classified as:

- Cost analysis
- Cost-effectiveness analysis
- Cost-utility analysis
- Marginal cost analysis
- Cost-benefit analysis

In economic evaluation, the definition of costs is much broader than the financial one; it does not only comprise the actual cost related to the retail prices of drugs, equipment, wages, and any other required resources. But most relevantly, an economic evaluation should include an economically consistent estimation of costs, thus measure the benefit forgone by having used the resources in one particular way and not in another. Opportunity cost evaluation requires comprehensive, disaggregated data at the individual patient level. Complications in obtaining such information compel the use of accounting cost data, retail prices, *etc.*, which in general are poor estimators of the true opportunity cost.

The point of view of the analysis is a key element in the identification of all relevant costs related to the implementation of any health care programme. Suppose calcium supplementation reduces admissions to hospital wards. In an analysis with a health care provider perspective, direct costs;[*] such as the new drug's retail price, and the 'hotel costs' prevented from having delayed the admission of the patient to a specialised institution, would constitute main components of the cost analysis. Whereas from the societal point of view the estimation of costs will not only include direct costs associated with the interventions but also an estimation of the associated indirect cost,[†] such as all the costs borne by the patient's relatives, would also need to be included. For example, patients who have sustained a hip fracture often require extra help around the house. Typically this is often provided 'free' by the patient's relatives. The extra time spent by sons or daughters assisting their parent is not without its cost. Ideally, such time should be valued and counted as part of the cost of caring for a person with a hip fracture.

Having identified and estimated all relevant costs and benefits of the alternative programmes/services of interest, the following step is to perform a comparison of the additional costs and benefits that a programme/service imposes over another. The most commonly used incremental approach is that based on incremental cost-effectiveness ratios (ICER).

The decision rule is straightforward. If intervention A is more costly and less effective than B, its implementation should not be recommended. If intervention A is more effective and less costly than intervention B, in economic evaluation jargon this is described as a case of dominance, where A dominates B. The decision rule is more contentious if intervention A is both more costly and more effective than the alternative B, or intervention A is less costly but also less effective than B. In these cases the final decision regarding the implementation of the new intervention, in our example intervention A, will in both cases strongly depend on the maximum cost-effectiveness ratio that the decision makers are willing to accept; in other words, the implicit value the decision maker assigns to one additional unit of health outcome.

## Cost Analysis

Cost analyses can be useful in that they can be used to estimate the cost savings of any potential intervention. The cost components of a fracture are complex. Clearly there is the cost of surgical treatment of fracture, if this is necessary, with the commensurate hospital stay. In addition to hospital costs there are other costs that fall on primary care, the patient and their family and social services costs.

---

[*]Direct cost: those cost borne by the health care provider.
[†]In direct costs: are defined as those costs borne by the society.

## Costs of Hip Fracture

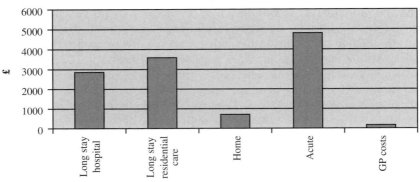

**Figure 31.1**   *Breakdown of hip fracture costs by location*
(Source, Dolan & Torgerson, 1998)

The cost breakdown for a typical hip fracture patient for the first 12 months after fracture is shown in Figure 31.1.[11] Whilst Figure 31.1 shows that the single largest cost is the acute hospital costs due to the surgical treatment of hip fracture and the associated bed day costs, the other post fracture costs are actually greater than these. A significant proportion of hip fracture patients cannot be discharged back into the community. For these patients the costs of residential or long stay hospital care is substantial. Even for those patients who can return to home, many, if not the majority, require social service support to enable them to continue living in the community.

The acute hospital costs, therefore, represent only about 40% of the total costs of hip fracture in this UK study.

Whilst the costs outlined in Figure 31.1 seem substantial, they do not include other relevant costs of care. For example, many patients may have to go and live with a relative (*e.g.*, son or daughter) who will have to give up time to care for the person. Whilst these time costs are difficult to measure financially they do represent a real opportunity cost of a hip fracture.

Other studies have shown that hip fractures incur substantial health care costs. Gabriel and colleagues in a study in the USA found that hip fracture costs were about $11 000 in the year after injury.[12] These costs were mainly health care costs and the authors did not include additional non health care support costs, which as noted above form a substantial part of a hip fracture cost profile. On the other hand, that study did note significantly higher costs attributable to vertebral fracture than did the Dolan and Torgerson, UK study. In the UK study vertebral fracture costs was only about 4% of a hip fracture cost. In contrast, Gabriel *et al.* noted that vertebral fractures in the USA cost approximately 17% of a hip fracture. The differences may be due to the different treatment strategies that occur in the different countries. Or it may be that, whilst the UK study, had access to a fairly comprehensive data set relating to hip fractures the data were less robust to the cost consequences of vertebral fractures and therefore may have significantly underestimated vertebral fracture costs.

In contrast, other fractures, such as wrist fracture, tend to be less disabling and therefore have fewer resource implications both within the health service sector and outside.

Cost analyses to put a financial estimate of a disease burden are quite commonly undertaken. Unfortunately, whilst they can give an idea of the scale of the problem they tell us nothing about the most effective method of reducing fractures. For example, Dolan and Torgerson estimated the cost of fractures within a UK setting of about £942 million.[11] Whilst this sum is considerable it does not give us any indication as to whether intervening in the problem is cost-effective or not. To assess whether any intervention is worthwhile, more sophisticated economic methods are required.

## Cost-Effectiveness Analysis (CEA)

One way of evaluating two interventions is to ascertain the costs and compare their benefits in terms of a clinical unit of outcome. For example, one might compare the cost-effectiveness of HRT versus vitamin D and calcium in terms of a cost per fracture averted. Or, alternatively, one might compare two programmes in cost per life year gained. Unfortunately, this economic method tends to be rather limited in that the unit of outcome may not capture all the clinical benefits and side-effects of treatment. For example, a cost per hip fracture averted will not include the possible disbenefit of developing breast cancer for the HRT user. One method of addressing these shortcomings is to use the Cost Utility method.

## Cost Utility Analysis (CUA)

A cost utility analysis (CUA) aims to capture all the health benefits of treatment within a single index value. Thus, in a 0–1 scale, 0 represents death and 1 full health. CUA therefore will incorporate all the health benefits and side effects of treatment. Importantly, CUA facilitates comparisons between different interventions, such as an osteoporosis prevention programme or a breast cancer intervention.

## Cost Benefit Analysis

Theoretically this is the most powerful economic technique. Both benefits and costs are expressed in monetary terms and if benefits are greater than costs the programme should be adopted. However, the difficulty of valuing health benefits in monetary terms has, to date, not been satisfactorily overcome; therefore, there are few cost benefit analyses in health care.

Whilst there is a large focus on pharmaceutical treatments for osteoporosis, nutritional methods are likely to be relatively more cost-effective due to their low cost. To assess the state of the economic literature in this field we have undertaken a critical review of published economic evaluations of nutritional supplements for the prevention of osteoporotic fractures.

## Search Strategy

A systematic review of published economic evaluations of interventions for osteoporosis was carried out. Two reviewers (RF and CI) independently conducted a search on Medline for papers published up to May 2002, according to predefined inclusion and exclusion criteria.[‡] Bibliographies of the retrieved evaluations as well as those of review articles were also searched. Thirty economic evaluations of interventions relating to osteoporosis were identified, six of which assessed the cost-effectiveness of either vitamin D, vitamin D with calcium or calcium alone, compared to a range of treatments available for osteoporosis or to no intervention.[13–18] There were no economic evaluations analyzing the cost-effectiveness of vitamins A, C or K. Given the evidence for no effect of vitamins A or C this may be appropriate. For simplicity, nutritional supplements, in this chapter, will refer to vitamin D with or without calcium supplementation.

## Economic Evaluations

In all six of the studies, presented in Table 31.1, the cost-effectiveness of the nutritional supplements of calcium or vitamin D was modelled using clinical evidence available from the literature. We did not identify economic evaluations of any other nutritional supplement.

To date, no economic evaluation of Vitamin D and calcium supplementation has been carried out alongside a clinical trial. In the reviewed studies, nutritional supplements were compared to a range of treatments available for osteoporosis, such as HRT, bisphosphonates or to no intervention. The populations studied were diverse, ranging from young postmenopausal women to elderly male and female patients. Some studies looked at high-risk populations, such as elderly women with low bone density measurement (BMD) or low body mass index (BMI). Community as well as institutional settings was considered. All studies considered the prevention of hip fractures as a primary outcome, since these fractures are the most costly to the health service and are associated with significant mortality and loss of quality of life to the patient. Some studies also included other prevalent fractures in the analysis, such as forearm and vertebral fractures. The time horizons varied from three

---

[‡]To be included a study had to be an economic evaluation of interventions to prevent fractures in populations at risk of osteoporosis. Interventions included ERT, HRT, calcium, vitamin D, bisphosphonates, alendronate, etidronate, nasal calcitonin and raloxifene. Exclusion criteria for the studies were cost of illness studies, burden of disease studies, editorials and reviews.

**Table 31.1.** *Main characteristics of economic studies of supplementation treatments*

| Study | Intervention and comparator | Study population and setting | Type of fracture prevented | Time horizon of the study | Perspective of the study | Main source of clinical data | Main source of cost data |
|---|---|---|---|---|---|---|---|
| Geelhoed (1994) | Different strategies with ORT, calcium and exercise, compared to no intervention | Healthy Caucasian women from age 50 onwards | Hip fractures | 50 years | Australian health service | Calcium was assumed to halve the rate of bone loss without treatment | Royal Perth Hospital and Australian published drug prices |
| Torgerson (1995) | D by injection and VDC compared to no prevention strategy | Women with and without low BMI in community and institutional settings | All fractures and hip fractures | 3 or 4 years | National Health Service, UK | 1 RCT, 2 observational studies | Published drug costs in the UK and NHS fees |
| Torgerson (1996) | D by injection, thiazide, HRT, VDC, calcium and calcitonin | Hypothetical cohort of women aged 80 | Hip fractures | 5 years | National Health Service, UK | 2 RCTs, 2 case-control studies, 1 meta-analysis of observational studies | Published drug costs in the UK |
| Ankjaer-Jensen (1996) | Calcium, etidronate and calcitonin for 5 years, HRT for 10 years | Women aged 70 years at onset of treatment and women screened for low BMD | Hip, forearm and vertebral fractures | 5 years for calcium, 10 years for HRT | Danish health service. | 3 published studies were used. Study designs were not reported | Not reported in detail |
| Rodriguez (1999) | Alendronate compared to VDC | Women with established osteoporosis and a previous vertebral fracture | Hip fractures | 3 years | Spanish health service. | Previously published study | Spanish hospital and drug prices |
| Bendich (1999) | Daily calcium supplementation | Individuals aged 50+ in the United States | Hip fractures | 34 months | Societal, United States | 3 RCTs | US Congress Office of Technology Assessment |

D = vitamin D, VDC = oral vitamin D and calcium, BMI = body mass index, BMD = bone mineral density, RCT = randomized controlled trial, ORT = Oestrogen Replacement Therapy.

years to the patients' lifetime. Two studies were done in the UK, one in Denmark, one in Australia, one in Spain and one in the US. Five studies chose the perspective of the national health service in the country to which the data referred, whilst one looked at the costs from a societal perspective (the US study). Most evaluations used published clinical studies as the source of the effectiveness data (one study used the authors' assumptions), although the quality of reporting varied from study to study. The cost data was generally obtained from a local hospital accounting system and from official drug price lists. Again the quality in reporting the source of the costing data was uneven. The main benefits, costs and cost-effectiveness results are summarized in Table 31.2, along with the authors' conclusions with respect to the cost-effectiveness of the interventions.

# Discussion

Osteoporosis represents a significant cost and clinical burden of disease. The ideal method of prevention would be relatively inexpensive, have few, or no, side effects and be acceptable to a wider population. Nutritional supplementation fulfils these criteria. The main issue with promoting their wider use is poor evidence of effectiveness. Calcium with vitamin D supplementation has been shown in trials, mainly among women in residential care, to reduce fractures significantly. Until very recently there was no robust evidence for the use of vitamin D alone. The evidence for vitamin D is contradictory with one large trial showing no effect[6] and another showing a positive benefit.[9] These contradictory data may be due to the population types recruited to the trials or to the different dosing schedules used. Trials due to be completed in 2004 should answer the question of whether the use of vitamin D is an effective, and cost-effective, method of fracture prevention on a population basis.

A similar situation applies to the use of calcium supplements. These are widely used on the basis of very weak evidence. Their effectiveness, on fracture rates, is mainly confined to a high risk population that may have been malnourished, although a small trial has shown a benefit on a healthier community.[8] As with vitamin D, large trials currently underway in the UK will report in 2004 and should answer the effectiveness and cost-effectiveness questions of calcium supplementation in community dwelling older people.

Existing economic evaluations are hampered by the lack of good effectiveness evidence. Furthermore, whilst economic evaluations do constitute essential tools for decision-makers when allocating limited resources for health care their quality, and the validity of their conclusions should be assessed. Several guidelines have been developed and are available for that purpose.[2] The quality of the studies reviewed in this article, is discussed in the light of a selected number of these criteria. A summary is presented in Table 31.3.

**Table 31.2.** *Economic evaluation results*

| Study | Benefit results | Cost results | Cost-effectiveness results* | Cost-effectiveness conclusions |
|---|---|---|---|---|
| Geelhoed (1994) | The number of QALYs per woman was 32 for no intervention, between 32 and 33 for ERT regimens and 32 for calcium and exercise. | The total cost (in Aus $ 000) for no intervention was 74, between 77 and 86 for ERT regimens and 81 for calcium and exercise. | The cost per QALY gained was between $8,500 and $16,500 for ERT regimens and $28,500 for calcium and exercise. | Lifetime estrogen therapy from age 50 years was the most cost-effective treatment. |
| Torgerson (1995) | Vit D injection: all fractures □ 25%, hip fractures □ 22% Oral Vit D+C all fractures □ 21%, hip fractures □ 28%. | Total costs were not reported. | Net cost per averted hip fracture ranged from £2,221 to £17,379 for VD injection and from £4,781 to £2,573 for VDC. | VD injection is potentially cost-effective in the elderly population. VDC may be cost-effective in targeted high-risk populations. |
| Torgerson (1996) | The number of hip fractures avoided for VD injection, thiazide, HRT, VDC, calcium and calcitonin were 1867, 1527, 2546, 2291, 1527 and 3140, respectively. | The total discounted costs (5 years for 100,000 women) for VD injection, thiazide, HRT, VDC, calcium and calcitonin were in millions £1.1, £3.6, £33, £38, £30 and £56, respectively. | ICERs (including averted costs) were a cost-saving of £9,176,496 for VD injection, a cost-saving of £4,775,704 for thiazide (compared to do-nothing), £41,493 for HRT, £81,547 for VDC and £433,548 for calcitonin. Calcium was dominated. | The study concluded that vitamin D was potentially the most cost-effective treatment, given the assumptions of the model. Further research into the effectiveness of VD and VDC should be conducted. |
| Ankjaer-Jensen (1996) | For all women over 70, under optimistic assumptions, the number of hip fractures avoided was 0.17 with calcitonin, 0.11 with etidronate, 0.11 with calcium and 0.11 with HRT. Under pessimistic assumptions, the number of hip fractures avoided was 0.04 with calcitonin, 0.07 with etidronate, 0.03 with calcium and 0.03 with HRT. | For all women over 70, under optimistic assumptions, the net costs (in Danish Kroner) were 24,191 for calcitonin, 1,195 for etidronate, 837 with calcium and -4,280 with HRT. Under pessimistic assumptions, the net costs were 34,678 with calcitonin, 1,709 with etidronate, 6,849 with calcium and 5,343 with HRT. | The cost-effectiveness ratio for calcitonin ranged from 140,000 to 860,000 DKK, from 10,864 to 24,400 for tidronate, from 7,600 to 220,000 for calcium and from cost-saving to 178,000 for HRT, depending on optimistic and pessimistic assumptions. | Methodological problems in this study do not permit valid conclusions to be drawn regarding the cost-effectiveness of calcium compared to the other interventions. |

cont.

**Table 31.2.** cont.

| Study | Benefit results | Cost results | Cost-effectiveness results* | Cost-effectiveness conclusions |
|---|---|---|---|---|
| Rodriguez (1999) | 1.1% of women taking alendronate had a hip fracture, compared to 2.2 % taking VDC. | The total cost of hip fractures (pesetas) was 668,616, 3-year costs of alendronate were 279,121, 3-year costs of calcium and vitamin D were 8,079. | Incremental cost per fracture averted was 25.6 million pesetas. | Alendronate is not a cost-effective intervention. |
| Bendich (1999) | Daily intake of calcium would prevent 134 764 hip fractures in individuals aged 50+ in the US. | $2.6 billion in direct medical costs would be averted with daily intake of calcium. | Not calculated. | Calcium is cost effective for women aged 75+ and possibly for all individuals aged 65+. |

ICER = incremental cost-effectiveness ratio, ERT = estrogen replacement therapy.
* negative cost-effectiveness ratios indicate cost-savings according to the authors, however, there are methodological problems in interpreting negative cost-effectiveness ratios and they should be used with caution.[15]

**Table 31.3.** *Summary of quality criteria applied to the reviewed studies*

| Study | Quality of clinical data used | Appropriate use of cost-effectiveness decision rules | Analysis of uncertainty | |
|---|---|---|---|---|
| | | | Deterministic | Stochastic |
| Geelhoed (1994) | ✗ | ✗ | ✓ | ✗ |
| Torgerson (1995) | ✓ | ✓ | ✓ | ✗ |
| Torgerson (1996) | | | | ✗ |
| Ankjaer-Jensen (1996) | ✗ | ✗ | ✓ | ✗ |
| Rodriguez (1999) | ✗ | ✓ | ✓ | ✗ |
| Bendich (1999) | ✗ | ✗ | ✓ | ✗ |

## Quality of Clinical Data

One important consideration is the quality and validity of the clinical data used to populate the cost-effectiveness model. The scarcity of clinical evidence available on the effectiveness of Vitamin D and/or calcium should not in itself be construed as an indication of poor quality of the studies, or necessarily reduce the validity of their results. Indeed, economic models unavoidably rely on assumptions to assess the cost-effectiveness of interventions. Moreover, it is in cases where the uncertainty surrounding many parameters is high that such studies constitute powerful tools for decision-making.[21] However, it is important that the economic evaluation reports in detail the source of the data used in the model, and gives some indication of the quality of the studies from which the data were obtained.

In the study by Geelhoed (1994), the assumptions concerning the cardio-protective effect of HRT are now out of date, so the conclusions drawn in this study should be viewed with the utmost caution.[22] Both studies by Torgerson (1995, 1996) contain the most recent evidence and describe in detail the studies used to obtain estimates of effectiveness. In the study by Ankjaer-Jensen (1996) and the one by Rodriguez (1999), the source of the clinical data is reported, although little detail is provided on the rationale for selecting particular studies. Finally, in the study by Bendich (1999), the source of the estimates of effectiveness is reported in great detail, as a meta-analysis was done to obtain the final estimate. However, although the intervention studied was calcium supplementation alone, the authors actually use some studies with vitamin D as well as calcium without further investigation, casting some doubt on the results of the study.

## Appropriate Use of Cost-effectiveness Rules

A second consideration when assessing the quality of an economic evaluation is the appropriate use of cost-effectiveness methodology and decision rules. There are now a number of textbooks and published articles that set out their use.[23,24]

However, it should be noted that there were methodological problems in the use of cost-effectiveness ratios in the studies by Geelhoed (1994)[18] and Ankjaer-Jensen (1996).[25] The study by Bendich (1999)[17] is not a cost-effectiveness study from a methodological point of view, but is rather a calculation of the total cost savings and total hip fractures avoided. Both studies by Torgerson[13,14] and the study by Rodriguez (1999)[15] appropriately apply the rules of cost-effectiveness analysis.

## Analysis of Uncertainty

Because of the range of uncertainties within the effectiveness and costs used in the economic models it is important to account for such uncertainty. Accounting for the uncertainty that necessarily surrounds the parameters used in a model, provides information on which to assess the sensitivity of the results to a number of assumptions. It also provides important information on the generalisability of the model to other settings and other populations. While all studies conducted deterministic one-way sensitivity analyses and scenario analyses (such as best case worst case scenarios), none of the studies conducted a stochastic analysis, which allows all parameters to vary jointly.

Caution should be exercised before drawing conclusions on the cost-effectiveness of nutritional supplements, because of the uneven quality of economic evaluations. This critical review of the evidence would suggest, based on the studies by Torgerson (1995, 1996),[13,14] that vitamin D with or without calcium supplementation is probably cost-effective in the UK, in particular in high-risk groups, such as older women and women with previous fractures. The results of the studies by Geelhoed,[18] Ankjaer-Jensen,[16] Rodriguez[15] and Bendich[17] should be viewed with some caution.

However, stronger evidence on the effectiveness of both treatments is needed. In particular it is necessary to know whether vitamin D can act alone without calcium supplementation, since the cost of vitamin D injection would be significantly lower than a daily vitamin supplement with added calcium. Or alternatively, as has been described recently, intermittent large doses of oral vitamin D would appear to be a cost-effective choice.

## The Future?

There are currently three ongoing clinical trials of vitamin D with and without calcium taking place in the UK, that are also collecting economic as well as clinical data. The three, all of which are due to report in 2004, will make a substantial contribution to the health economics of this area. The studies are quite different: one study examines the effectiveness of an annual injection of vitamin D; another looks at vitamin D alone or with calcium in the field of secondary prevention; whilst the third study examines the use of calcium with vitamin D supplements for the prevention of fractures in primary care.

The study of an annual vitamin D injection, if shown to be effective, will demonstrate a potentially very cost-effective treatment. The annual injection could be given alongside influenza vaccinations and therefore the marginal or incremental cost of delivering the therapy would virtually be only the cost of the injection itself as the cost of the practice nurse time, and the patients travel cost and time, has already been incurred for the influenza vaccination.

The evaluation of calcium with or without vitamin D for secondary prevention is, again, evaluating a potentially very cost-effective treatment. Routinely prescribing supplements to older men and women who have sustained a recent fragility fracture will automatically target a relatively high risk population as any fracture among people over the age of 70 years is usually a marker for increased risk of further fractures. If vitamin D supplementation alone is ineffective then the additional cost of using a calcium supplement (if the combination were shown to be effective) will be more than balanced out by the relatively high absolute risk of the target population.

Finally, there is the trial of calcium with vitamin D supplements in primary care. The target population in this study are women, aged over 70 years, with one or more risk factors for hip fracture, which actually accounts for about 60% of the women over 70 years of age.[26] These women are at an elevated risk and therefore as long as supplementation is demonstrated to be effective it is highly likely that it will prove to be a cost-effective method of prevention.

At present modelling studies of vitamin D show that though it is a very cost-effective intervention there is still a lack of empirical data. Similarly, clinical data supporting the widespread use of calcium supplementation is mainly confined to high-risk older people in residential care. What is required is data from large, well designed, clinical trials with an economic evaluation built into the design.

## Conclusions

The cost-effectiveness of nutritional supplements for the treatment of osteoporosis has been confined to studies of vitamin D with or without calcium supplementation.

There have been no studies involving other nutritional supplements.

All published economic evaluations have been based on modelling techniques, which require a number of assumptions to be made concerning the effectiveness and costs of the treatments under consideration. To date there has been no economic evaluation carried out alongside a clinical trial.

The quality of the published economic evaluations is uneven and caution should be exercised in drawing conclusions from them.

A critical review of the current economic evidence would suggest that vitamin D with or without calcium supplementation is probably cost-effective in the UK, in particular in high-risk groups, such as older women and women with previous fractures.

However, further empirical evidence on the effectiveness of vitamin D with and without calcium is needed. Three ongoing clinical trials that will report in 2004 will provide important answers to these questions.

# References

1. K.M. Fairfield and R.H. Fletcher, Vitamins for chronic disease prevention in adults: scientific review, *JAMA*, 2002, **287**, 3116–3126.
2. D. Feskanich, V. Singh, W.C. Willett and G.A. Colditz, Vitamin A intake and hip fractures among postmenopausal women, *JAMA*, 2002, **287**, 47–54.
3. D. Feskanich, P. Weber, W.C. Willett, H. Rockett, S.L. Booth and G.A. Colditz, Vitamin K intake and hip fractures in women: a prospective study, *Am. J. Clin. Nutr.*, 1999, **69**, 74–79.
4. MRC/BHF Heart Protection Study of antioxidant vitamin supplementation in 20 536 high-risk individuals: a randomised placebo-controlled trial, *Lancet*, 2002, **360**, 23–33.
5. M.C. Chapuy, M.E. Arlot, F. Duboeuf, J. Brun, B. Crouzet, Arnaud S, P.D. Delmas and P.J. Meunier, Vitamin D3 and calcium to prevent hip fractures in elderly women, *N. Engl. J. Med.*, 1992, **327**, 1637–1642.
6. P. Lips, W.C. Graafmans, M.E. Ooms, P.D. Bezemer and L.M. Bouter, Vitamin D supplementation and fracture incidence in elderly persons. A randomized, placebo-controlled clinical trial, *Ann. Intern. Med.*, 1996, **124**, 400–406.
7. R.J. Heikinheimo, J.A. Inkovaara, E.J. Harju, M.V. Haavisto, R.H. Kaarela, J.M. Kataja, *et al.*, Annual injection of vitamin D and fractures of aged bones, *Calcif. Tissue Int.*, 1992, **51**, 105–110.
8. B. Dawson-Hughes, S.S. Harris, E.A. Krall and G.E. Dallal, Effect of calcium and vitamin D supplementation on bone density in men and women 65 years of age or older, *N. Engl. J. Med.*, 1997, **337**, 670–676.
9. D.P. Trivedi, R. Doll and K.T. Khaw, Effect of four monthly oral vitamin D3 (cholecalciferol) supplementation on fractures and mortality in men and women living in the community: randomised double blind controlled trial, *BMJ*, 2003, **326**, 469.
10. D.J. Torgerson and D.M. Reid, The economics of osteoporosis and its prevention. A review, *Pharmacoeconomics*, 1997, **11**, 126–138.
11. P. Dolan and D.J. Torgerson, The cost of treating osteoporotic fractures in the United-Kingdom female population, *Osteoporosis Int.*, 1998, **8**, 611–617.
12. S.E. Gabriel, A.NA. Tosteson, C.L. Leibson, C.S. Crowson, G.R. Pond, C.S. Hammand and L.J. Melton, Direct medical costs attributable to osteoporotic fractures, *Osteoporosis Int.*, 2002, **13**, 323–330.
13. D.J. Torgerson and J.A. Kanis, Cost-effectiveness of preventing hip fractures in the elderly population using vitamin D and calcium, *QJM*, 1995, **88**, 135–139.
14. D.J. Torgerson, C. Donaldson and D.M. Reid, Using economics to prioritize research: a case study of randomized trials for the prevention of hip fractures due to osteoporosis, *J. Health Services Res.*, 1996, **1**, 141–146.
15. E.C. Rodriguez, M.L. Fidalgo Garcia and C.S. Rubio, A cost-effectiveness analysis of alendronate compared to placebo in the prevention of hip fracture, *Aten Primaria*, 1999, **24**, 390–396.
16. A. Ankjaer-Jensen and O. Johnell, Prevention of osteoporosis: cost-effectiveness of different pharmaceutical treatments, *Osteoporosis Int.*, 1996, **6**, 265–275.

17. A. Bendich, S. Leader and P. Muhuri, Supplemental calcium for the prevention of hip fracture: potential health economic benefits, *Clin. Therapeut.*, 1999, **21**, 1058–1072.

18. E. Geelhoed, A. Harris and R. Prince, Cost-effectiveness analysis of hormone replacement therapy and lifestyle intervention for hip fracture, *Aust. J. Public Health*, 1994, **18**, 153–160.

19. A. Stinnett and A.D. Paltiel, Estimating CE ratios under second order uncertainty, *Med. Decision Making*, 1997, **7**, 483–489.

20. M. Sculpher, E. Fenwick and K. Claxton, Assessing quality in decision analytic cost-effectiveness models. A suggested framework and example of application, *Pharmacoeconomics*, 2000, **17**, 461–477.

21. M.J. Buxton, M.F. Drummond, Van B.A. Hout, R.L. Prince, T.A. Sheldon, T. Szucs *et al.*, Modelling in economic evaluation: an unavoidable fact of life, *J. Health Econ.*, 1997, **6**, 217–227.

22. Risks and benefits of estrogen plus progestin in healthy postmenopausal women: principal results from the Women's Health Initiative randomized controlled trial, *JAMA*, 2002, **288**, 321–333.

23. M.F. Drummond, G.W. Torrance and G.L. Stoddart, *Methods for the Economic Evaluation of Health Care Programmes*, Oxford University Press, Oxford, 1995.

24. M. Johannesson and M.C. Weinstein, On the decision rules of cost-effectiveness analysis, *J. Health Econ.*, 1993, **12**, 459–467.

25. M. Sculpher, D. Torgerson, R. Goeree and B.J. O'Brien, *A Critical Structured Review of Economic Evaluations of Interventions for the Prevention and Treatment of Osteoporosis*. Discussion Paper (169), 1999, York, University of York. CHE discussion paper. Centre for Health Economics.

26. A. Stewart, L.D. Calder, D.J. Torgerson, D.G. Seymour, L.D. Ritchie, C.P. Iglesias and D.M. Reid, Prevalence of hip fracture risk factors in women aged 70 years and over, *QJM*, 2000, **93**, 677–680.

CHAPTER 32

# Nutritional Strategies for Prevention and Treatment of Osteoporosis in Populations and Individuals

AILSA GOULDING

Department of Medical and Surgical Sciences, University of Otago,
PO Box 913, Dunedin, New Zealand,
Email: ailsagoulding@stonebow.otago.ac.nz

> *Bone formation equals bone resorption,*
> *result happiness.*
> *Bone resorption exceeds bone formation,*
> *result osteoporosis, fractures and misery.*
>                 With apologies to Charles Dickens and Mr Micawber.

**Key Points**

- It is important that we grow a good skeleton, achieve our optimal genetic peak bone mass and hang on to this as long as possible later in life.
- The value of sensible nutrition, regular physical activity and a healthy lifestyle to maintenance of good bone health needs a better press.
- Simple dietary guidelines for optimal bone health will be most effective.
- More emphasis/funding should be given to nutritional education – both for health professionals and the general public.
- Earlier identification of groups at high risk of osteoporosis should be a priority.
- Individuals in high risk groups should have bone density measurements.
- Interventions to counter nutritional inadequacies should be used more widely (evidence supports the value of vitamin D, protein, and calcium interventions).

It is a truth universally acknowledged that we are what we eat. Our diets supply the raw materials needed to build and keep strong bones. Accordingly, nutrition plays an important role in enabling us to achieve our optimal genetic peak bone mass during growth[1] and in curbing unnecessary losses of bone thereafter. Unfortunately, bad dietary habits, hormonal imbalances, physical inactivity and smoking can each accelerate the thinning of bone. Yet poor nutritional habits can be modified. We know that thin bones with low mineral density are weak and that they break much more easily than strong ones at every age.[2-4] Given the high rates of bone fracture currently occurring worldwide, both in youth and old age, it is imperative that sound, sensible and simple nutritional advice for optimising bone health should be available *throughout the life course* (Table 32.l). After all, elevating bone mineral density by one standard deviation unit strengthens the skeleton sufficiently to reduce fracture rates by 50%.[5] An effect of this magnitude would have an important effect on fracture rates.

Earlier chapters in this book have set out and discussed the latest evidence regarding the effects of different individual nutrients, behaviour and conditions upon the skeleton. The purpose of the present chapter is to consider ways to put into practice in the public health arena up-to-date nutritional advice. Three broad strategy levels of different intensity are recommended to facilitate this (Figure 32.1, Table 32.2). Firstly, *everyone in the community* should be encouraged to eat a healthy diet lifelong, regardless of their particular risks of osteoporosis (this is universal primary prevention). The success of this strategy requires provision of, and access to, an adequate food supply, good nutrition policy and education, an informed workforce of health professionals and good communication to inform all age sectors of the population.

**Table 32.l.** *Universal needs through the life course*

| Stage of life | Nutritional and lifestyle recommendations |
| --- | --- |
| Pregnancy and lactation | Consider needs of both mother and child, advocate breastfeeding |
| Infancy and childhood | Satisfy needs for linear growth, instil good habits of nutrition and physical activity from a young age |
| Pubertal growth and adolescence | Provide the higher intakes of minerals and nutrients needed now, encourage outdoor sports and weight-bearing physical activity to optimise strong skeletal structure and minimize excess adiposity. |
| Young adulthood | Avoid bone-wasting behaviours (smoking, amenorrhoea, excessive consumption of alcohol, sodium, caffeine) |
| Older adult life | Requirements for vitamin D are higher, muscle mass is dwindling and greater attention is needed to maintain 25(OH)D stores, mobility and balance and to avoid falls |
| All stages of life | Ensure diet meets nutritional requirements for age and gender, maintain healthy body weight, undertake daily weight-bearing physical activity, and protect vitamin D adequacy |

**Figure 32.1**    *The three broad strategies of prevention advocated to optimise bone health via nutrition*

**Table 32.2.** *Nutritional strategies for populations and individuals*

| Strategic level of prevention | Nutritional recommendations |
|---|---|
| **Universal primary prevention**<br>Males and females of all ages | A balanced diet satisfying all nutritional needs for the major food groups, minerals and micronutrients should be provided throughout the life course |
| **High-risk groups**<br>Examples: those with a family history of osteoporosis, some ethnic populations, the oestrogen-deficient, the physically inactive, the housebound, the lactose-intolerant, those with low energy intakes or eating disorders, alimentary malabsorption problems, renal failure, and patients requiring corticosteroids or anticonvulsants | People identified as having above-average risks of developing thin bones for reasons of genetic inheritance, ethnicity, disease or behaviour, should be encouraged to pay special attention to nutrition; they may require specific supplements or foods fortified with particular nutrients |
| **Individuals with osteoporosis**<br>Particularly those with any history of minimal trauma fractures, poor balance, musculoskeletal problems or a history of frequent falling | Dietary interventions should be instituted to correct undernutrition, particular deficiencies and to counter excessive bone loss. Over consumption of urinary calcium-wasting agents (e.g. sodium chloride, caffeine) should be avoided. For those on bone-sparing anabolic therapies extra nutrients may be required to meet the demands of new bone (calcium, vitamin D, protein) |

Secondly, *subgroups of the population possessing above average risks of developing osteoporosis* for economic, genetic, disease or behavioural reasons, require more intensive nutritional advice and action, to cater for their special needs. People in these higher risk groups may need to make particular food and lifestyle choices to counter their inherent risks of osteopenia and some may require specific nutritional supplements. Early identification of people belonging to these high risk groups and efforts to educate them about ways to improve their diets should therefore be a priority. Individuals belonging to these high risk groups should also consider having their bone density measured by dual energy X-ray absorptiometry in order to establish whether that they do exhibit osteopenia or osteoporosis.[5] Thirdly, *people with fragility fractures or documented osteoporosis* should be targeted for mandatory rigorous dietary evaluation and dietary interventions, since these individuals have the highest risk of suffering new fractures within the short term. Dietary improvements are especially important for people with recent fractures, as optimising nutrition can act to speed recuperation and slow bone loss[6–8] and it is known that people admitted to hospital with fractures are at a high risk of sustaining more fractures soon after discharge.[9] Considering nutritional needs is also crucial for osteoporotic patients receiving prophylactic therapeutic drugs which promote the rapid rebuilding of lost bone[10] because their nutritional requirements for the raw materials of new bone can rise sharply. Indeed, anabolic agents can stimulate gains of bone mass comparable in magnitude to those occurring at the growth spurt.[11]

In an ideal world every individual would enjoy a healthy balanced diet, engage in regular weight-bearing physical activity, maintain a healthy body weight and avoid unnecessary bone-thinning lifestyle options, so that they would achieve their optimal genetic peak bone mass during youth, and subsequently conserve bone mass above the fracture threshold lifelong. Unfortunately we do not live in an ideal world. Many people, both young and old, develop weak skeletons and are in consequence fracture-prone. Although some of this skeletal weakness is undoubtedly due to genetic influences,[12,13] and some fractures are the result of high trauma, or the use of medications which rarefy bone,[14,15] poor nutritional habits and excessive calorie restriction can also make a major contribution to the pathogenesis of low bone density even in young people (coeliac disease, milk avoidance, lack of vitamin D, anorexia nervosa, athletic amenorrhoea). Inactivity and malnutrition undoubtedly contribute to the genesis of musculoskeletal weakness in the elderly.[16,17] In consequence inadequate or inappropriate nutrition contribute significantly to the high fracture burden currently faced by societies throughout the world.[18] Biomarkers of bone resorption and formation are already being used to investigate short-term skeletal responses to nutritional interventions.[19] Another future research area that deserves urgent attention is study of the direct influences of specific nutrients upon bone cells *via* the RANK ligand/RANK/osteoprotegerin cascade, which plays such a critical role in building and maintaining skeletal mass.[20]

# The Importance of Safeguarding Good Nutrition in Early Life

Preventing early malnutrition *via* the provision of a sound and balanced diet is particularly important because the adverse effects of inadequate nutrition during early growth can be long-lasting. Infants born preterm have small skeletons[21] and research shows that their low bone mass persists at least up to the age of puberty.[22,23] Indeed, some epidemiological evidence indicates that low birth weight is associated with reduced bone mineral content[24] and an increased incidence of hip fracture among elderly people in their seventies.[25]

In some countries malnutrition is rife, energy intakes are limited and habitually low intakes of protein, minerals and vitamins are responsible for impaired mineralization in infancy and childhood[26]: in these circumstances supplements of single minerals and vitamins can increase bone density usefully.[27–29] On the other hand, even when ample nutrients are widely available, inadequate nutrition can occur as a consequence of poor food choices. Substitution of nutrient-poor energy-dense soft drinks for fluid milk by children and adolescents is a salient example of this.[30] There are concerns that the current trend towards rising consumption of soda drinks and reduced intakes of milk is likely to compromise dietary calcium intakes and be detrimental to bone health.[31,32] Excessive consumption of energy-dense soft drinks can also promote obesity,[33] which in turn can raise the susceptibility of children and adolescents to bone fractures and other orthopaedic problems during growth because bone growth lags behind increases of body weight,[34,35] as demonstrated by the case illustrated in Figure 32.2.

Wise beverage choices can have a considerable influence upon bone health (Table 32.3). Evidence supporting the view that regular consumption of milk during childhood and adolescence augments adult bone mass appears to be growing.[36] Consumption of dairy foods should therefore be encouraged. Stepping up milk consumption is also acceptable for older subjects.[37] Trials of milk/dairy supplementation in children and adolescents also demonstrate enhanced bone gain,[38] at least in the short term, though some of this benefit does appear to be transient.[39,40] By contrast, research suggests that children who avoid milk have low calcium intakes, small stature, poor skeletons and are fracture-prone at a young age.[41,42] Interestingly, information collected from 3251 women participating in the NHANES III study indicated that women over 50 who had low intakes of milk in childhood (less than a glass per week), had not only less bone as adults, but also significantly more osteoporotic fractures, than women who consumed a glass of milk or more daily in childhood.[43] The effect was substantial, with low milk consumption accounting for 11% of the osteoporotic fractures reported by their subjects.[43]

Diets which have a low calcium content are often nutritionally poor.[44] By contrast, dairy foods, which supply a high proportion of the total calcium intake in western countries, are naturally rich in many different nutrients and minerals and also high in protein.[37] The beneficial effects of dairy foods for

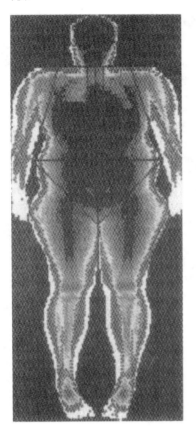

**A Case History. Distal forearm fracture in an overweight boy with habitual low physical activity levels and low dietary calcium intake.**

This 13-year-old fell while standing still on his roller blades and fractured his right radius. His current calcium intake was poor (392 mg day$^{-1}$) and he rated his physical activity relative to other boys of his own age as low (1 on a scale of 1–5). He had avoided participating in sports for 4 years. His pubertal development was Tanner Stage 2. Dual energy X-ray absorptiometry (DXA) scans were performed.

For his age, height was normal 155 cm, 25th centile), but both weight (74 kg) and body mass index (30.8 kg m$^{-2}$) values were above the 95th centile. He had a high body fat mass (32.7 kg) and a high body fat percentage (53.1%) and although lean tissue mass (39.3 kg) and bone mineral content (1.7 kg) were appropriate for his age, his total bone mineral was extremely low for his body weight (see graph below). His bone mineral density was very low at several sites, being only 79.4% of New Zealand norms for age at the femoral neck and 83.7% at the ultradistal radius.

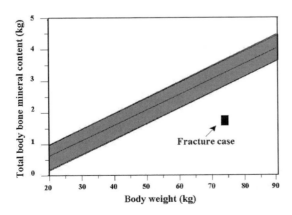

**Low total skeletal bone mass in relation to total body weight.**

Shaded diagonal area shows the normal range for fracture-free New Zealand boys aged 3–19 years. Our case has appropriate total body bone mineral content (1.7 kg) and bone area (1.783 cm$^2$) for his age (13 years) but very low bone mineral content for his high body weight (74 kg). Normal body weight (50th centile) for a 13-year-old boy is 48 kg. Older boys weighing 74 kg have about 3.4 kg of bone mineral in the total skeleton.

**Figure 32.2**   *Over weight children and adolescents are prone to fracture.*

**Table 32.3.** *Beverage choices can influence bone health*

Drinks
- Remember that milk (plain or flavoured) is nutrient-rich: an important source of protein, vitamins and minerals – a meal in itself
- Mineral-rich drinking waters can provide an alternative to dairy and boost intakes of calcium and magnesium without raising calorie intakes
- Soymilks also provide useful calcium and may have weak oestrogenic actions
- Although coffee increases urinary calcium, adding milk counters the calciuria
- Tea-drinkers may have higher bone density than those who drink no tea
- Mineral-fortified fruit juices may benefit bone but are high in calories and generally nutrient poor
- Excessive consumption of high-calorie, mineral-poor carbonated or soft drinks should not be encouraged
- Consumption of alcohol should be moderate, not excessive

bone health may relate to these properties, not just to their high calcium content, since increases in bone associated with dairy food supplementation appear to increase bone mass during growth more effectively than mineral supplements of calcium salts. Adequate dietary protein and phosphorus are undoubtedly important to skeletal health and adults with low protein intakes exhibit rapid bone loss.[45] The greater anabolic effects of dairy products upon bone may also be due to the influence of certain bone-active proteins and cytokines contained in milk[38] such as osteoprotegerin,[46] milk basic protein[47] and lactoferrin,[48] as well as to the important rises in the blood levels of insulin-like growth factor 1 (IGF-1) which follow milk consumption.[49,50] Growth hormone influences bone size and there is strong evidence to show that IGF-I expands the outer dimensions of bone.[51,52] Any nutritional factors that result in the enlargement of individual bones will increase their strength for biomechanical reasons.

In developed countries, low socio-economic status is often associated with consumption of poor diets. Regrettably, socio-economic disadvantages of childhood appear to exert long-lasting ill-effects on dental and general health in adult life.[53] Inadequate nutrition in childhood can impair linear growth,[41,54] leading to shorter leg length[55] and reduced bone mass.[54] Moreover, some authorities consider that brief periods of hormonal disturbance, inactivity, disease or inadequate nutrition during growth, may influence skeletal development adversely and explain the presence of *regional* osteoporosis in later life.[56]

Regular weight-bearing exercise is necessary for good bone health in people of all ages since mechanical loading is of critical importance to osteogenesis.[57] Intermittent loading has a strong anabolic effect on the bone cells, whereas lack of weight-bearing initiates rapid bone loss. Levels of physical activity in youth influence both the magnitude of peak bone mass attained and the rapidity of bone gain during growth. Greater participation in weight-bearing exercise may also explain the higher bone density values and lower fracture rates seen in people from rural communities *vs.* those from cities.[58] Unfortunately, lifestyles in the 21st century are becoming increasingly sedentary and many people fear that inactive children may fail to achieve their genetic peak bone mass. Well-controlled trials certainly demonstrate that rapid gains in bone mass can occur within one school year in groups of prepubertal children given weight-bearing exercise programmes.[59,60] Moreover, the effects of inadequate mineral intakes and low exercise can interact.[61,62] In addition, recent research suggests that early dietary intakes and weight-bearing physical activity influence bone shape and strength as well as bone mass.[63–65]

When dietary energy intakes are low it is hard to consume enough calcium from food. Thus dietary energy deprivation *per se* can induce malnutrition and so contribute to increased fracture risk. This scenario is seen in younger people with eating disorders who slim to excess[66] and in some athletes.[67] It is also evident in many frail and hospitalised elderly patients.[68] The institutionalised elderly are a particularly vulnerable group for low energy consumption and greater vigilance is needed to ensure adequate nutritional support is provided for them. Health authorities could improve care in this respect by requiring regular documentation of simple anthropometry (weights, heights, body mass index) and record taking of the use of nutritional supplementation by patients in their care. Regular review of the dietary intakes of calories, protein, minerals and vitamins should be mandatory for all institutionalised patients. Attention should also be paid to their oral health[69] since edentulous subjects and those with ill-fitting dentures may lose appetite and have difficulty consuming certain foods. Doctors should consider the possible adverse effects of vitamin D deficiency and malnutrition upon bone health in *every* frail elderly patient in their care and take action to remedy deficiencies when these are detected. In addition well-monitored home-based nutrition intervention schemes should play a useful role in improving the dietary intakes and general health of functionally impaired, community-dwelling older people. Trials have shown this approach is effective in increasing intakes of both fruit and vegetables and calcium-rich foods.[70]

## Childhood Fracture Prevention

Although many studies have been undertaken in children to examine the relationships of nutrition to bone growth, there is to date a paucity of research regarding the relationships of poor nutrition to bone *fracture* in childhood. This is somewhat surprising since fractures are extremely common at this time of life, with 40% of girls and 51% of boys reporting at least one fracture between birth and age 18 years.[71] The highest rates of fracture occur around the growth spurt[72] (9–13 years in girls and 11–14 years in boys) when demands for the building components of bone are high and rapid skeletal remodelling increases bone fragility.[73] Thus this is a time when it is particularly important for every parent to ensure their children eat healthily and take plenty of exercise.[74] Our own research has established that girls and boys who fracture their forearms have lower bone density than those who remain fracture-free.[75,76] Because overweight adults have stronger bones and higher bone mineral density than lighter people[77] we were surprised to note that obese and overweight children seem to be especially fracture prone: we postulate that this effect occurs because fat mass is gained faster than bone mass during growth. Thus a mismatch of bone mass to body mass develops in obese children and adolescents which increases their susceptibility to fracture[34,78] (see case report in Figure 32.2). This is a worrying observation since obesity in childhood is escalating sharply as a result of overeating and under-exercising.[35] Overweight children also fall more heavily than children of healthy weight, and they may have more difficulty balancing during everyday activities of sport and play,[79] possibly because their musculature relative to weight is lower than in children of healthy weight. In girls, we have shown that low bone mineral density, high body weight and a previous history of fracture are independent and important predictors of new fractures during growth.[4] Many children and adolescents break bones repeatedly and we believe that public health programmes designed to improve childhood bone mass, maintain healthy body weight and prevent obesity have considerable potential to lower rates of fracture during childhood and adolescence. Randomised control trials have already demonstrated that the bones of children can be strengthened and bone density augmented, at least in the short term, by dietary supplements of milk and/or calcium, and by increased participation in weight-bearing physical activity,[80] especially when physical activity is increased before puberty.[64,81–83] Future trials designed to increase the bone mass and promote healthy body weight in children presenting with a first fracture could be rewarding in lowering paediatric fracture rates. These interventions should also improve peak bone mass and have a favourable influence upon osteoporotic fracture rates later in life. As our understanding of gene/nutrient interactions improves, it should be possible to implement strategies to build and maintain good peak bone mass in high risk individuals earlier, and more effectively.

# Fracture Prevention in Adults

There is now strong evidence from randomised controlled trials to support the view that bone-sparing medications[18] and use of hip-protectors[84] can reduce fractures in adults. Rather fewer well-designed trials with sufficient statistical power to evaluate the effects of *nutritional change* on bone fractures have been undertaken. However, a major randomised clinical trial demonstrated convincing benefits of calcium plus vitamin D in reducing hip fractures among institutionalised very elderly women.[85,86] Younger men and women given calcium/vitamin D therapy also suffered fewer fractures in another study.[87] Defining the optimal dietary intakes of calcium is still a subject of hot debate and evidence for a positive effect of calcium supplementation alone in fracture prevention is relatively weak. Similarly, the efficacy of vitamin D alone in this regard[88] continues to be controversial.[89] Nevertheless, a recent large British trial[90] of 2037 men and 649 women aged 65–85 years living in the community, demonstrated that 4-monthly supplementation with 100,000 IU of oral cholecalciferol for 5 years without giving supplementary calcium, lowered the occurrence of major new osteoporotic fractures (forearm, vertebrae and hip) by 33%: this highly favourable outcome was achieved very economically, with the medication costing less than one British pound per patient per year. Cheap indeed! The importance of adequate nutrition in speeding post-operative recovery and slowing bone loss immediately *after fracture* also deserves greater emphasis and should be practised more widely. Trials have established that improving the general nutrition of patients treated for recent hip fractures *via* dietary protein and multinutrient supplementation can shorten the duration of inpatient hospital care and slow subsequent bone loss.[7,8] Finally it should be noted that although the evidence that increased physical activity lowers actual fracture risk in adult life is weak,[91] exercise programmes can reduce falls.[92] In addition bone loss will be slower when older patients remain active than when they become less mobile. Thus regular daily weight-bearing activity (>60 min per day where possible) should be encouraged for all older people.

# Food Fortification/Use of Supplements

People who avoid certain foods that supply necessary nutrients need to make appropriate alternative food choices,[93] or use fortified foods/dietary supplements, in order to achieve a nutritionally balanced diet. For some consumers the monetary costs of mineral supplements *vs.* food sources of minerals may be a consideration in this choice.[94]

It is feasible to achieve adequate dietary intakes of calcium without consuming dairy products, but this option is considerably more difficult because greater quantities of other foods are required to ensure the recommended daily intake of calcium is met. Bulkier serving sizes must be consumed.[30] However, today many processed foods are fortified with calcium salts, and these products can provide useful sources of calcium alternative to dairy products or soy milks: they have a place for individuals experiencing difficulty in consuming sufficient minerals, proteins or vitamins from their usual diets. Vegetarians, people avoiding dairy foods, having lactose malabsorption,[95,96] allergies to particular foods such as gluten, those with renal failure, malabsorptive syndromes and people with habitually low caloric intakes may use fortified foods to help them to meet their dietary needs adequately. These people find fortified foods are more acceptable than consuming pills regularly. Others, such as those with low dietary energy intakes, have difficulty in consuming a nutritionally adequate diet from the foods they eat, and prefer to take specific dietary supplements.

# Role of Education/Advertising/Industry

## Behaviour

It is essential that healthy foods meeting the needs of skeletal development should be provided for youngsters throughout the growth period. Good early dietary habits should be encouraged by parental example.[97,98] This early education by example is critically important because bad eating habits of early onset may persist into adult life. For example although many children with milk allergies grow out of their adverse reactions to milk, early avoidance of dairy products can lead to lifelong residual food aversions which can severely limit the choices of calcium-rich foods later in life.[99]

Desirable dietary habits undoubtedly have the potential both to improve bone growth in youth and to improve general health and reduce osteoporotic bone loss in older people.[100] However, what we eat is strongly influenced by economic circumstances, beliefs and culture. Many people with poor nutritional habits cannot afford to consume adequate diets. Others have no intention of changing their eating patterns. People will only follow advice for dietary change when they are convinced this will help their bone health. Thus stimulating motivation for change in dietary habits may be a critically important step towards the success of both educational and direct dietary interventions.[101] Collaborative efforts involving health professionals and the mass media can play a useful role here. Accenting the value of a slogan such as "3 daily servings of calcium-rich foods for optimal bone health" might improve calcium intakes in much the same way as the strategy of "five servings of fresh fruit and vegetables for cancer prevention" has proved to be acceptable in the community. The challenge for nutritional osteoporosis prevention is to provide health messages, which will promote positive changes in dietary behaviour in a cost-effective manner.

There is a widespread public perception that dairy foods are fattening. The converse may be the reality[102] as some postulate that high calcium intakes simultaneously suppress adipogenesis and inhibit lipolysis.[103] More work is needed to validate this hypothesis. However, we should not forget that low-fat dairy foods and alternative sources of essential minerals to dairy products are available (Table 32.3). Soy products, bony fish and nuts are rich in calcium.[93] Mineral-rich drinking waters can be an especially useful source of calcium for the calorie-conscious, those intolerant to dairy products, and people obtaining calcium from solely vegetarian diets. Calcium provided in mineral-rich drinking water is absorbed well and mineral-rich drinking waters can make a significant and substantial contribution to total dietary intakes of calcium and magnesium without interfering with the consumption of food sources of calcium. A French study of 624 subjects showed that drinking water rich in these minerals (supplying 486 mg of calcium per litre and 84 mg of magnesium per litre) contributed a quarter of the total daily calcium and 17% of the magnesium intake.[104]

Fortunately, most people with mild malabsorption of lactose can tolerate moderate amounts of dairy foods.[105] Hard cheese and yoghurt consumption should be encouraged too, since these are better tolerated than fluid milk by malabsorbers. In older women yoghurt improves diet quality and lowers biochemical markers of bone resorption.[106] In addition many people find yoghurt becomes preferable to milk itself as they get older, perhaps because lactose malabsorption increases sharply in the seventh decade of life, causing mild GI discomfort to follow milk consumption.[107] Importantly, the evidence suggests that reductions in peak bone mass[96] and accelerated rates of bone loss[95] are associated with reductions in total dietary calcium intake, not with lactose malabsorption *per se*.

Excessive consumption of caffeine and high intakes of alcohol can be detrimental to bone. The calciuric effects of caffeine are modest and are readily offset however, by taking milk in coffee.[108] The importance of increases in urinary calcium caused by high protein diets to overall calcium balance and to bone health is still debated. Moderate to high protein intakes seem best, since low intakes (below 50 g per day) are associated with greater bone loss in the spine and hip and higher protein intakes to smaller reductions in bone density.[45] Cigarette smoking should be strongly discouraged as this has detrimental effects on bone mass. Smokers undergo menopause earlier than non-smokers[109] and smoking is linked with increased fracture risk in older men and women.[110] The public message that smokers have less bone than non-smokers should be promoted more widely. Finally, we are all aware that people in many countries are today becoming unduly sedentary. Because weight-bearing activity promotes osteogenesis, whereas inactivity causes rapid bone loss, people of all ages should be encouraged to undertake weight-bearing activity every day to help build and keep their bones strong and their muscles co-ordinated.

## Vitamin Supplementation

Most vitamin D is derived from cutaneous synthesis rather than from dietary sources[111] so regular exposure to adequate ultraviolet light is necessary to ensure adequate vitamin D status. When this is not achieved, supplementation should be initiated in order to enhance alimentary calcium absorption, suppress high secretion of parathyroid hormone and maintain good muscle function. We are starting to recognize that our usual vitamin D requirements are considerably higher than was hitherto thought.[112] In developed countries rickets is still seen in dark-skinned children[113] while seasonal vitamin D insufficiency is also common in childhood even in sunny countries,[114,115] as well as being virtually the norm among the elderly in winter, particularly at high latitudes. Wider use of vitamin D supplementation appears justified by present evidence.[116] It is safe and cheap and should be mandatory for the housebound, and for those who generally cover their skin when outdoors for cultural reasons. Vitamin D supplementation should be considered in pregnancy and for infants, a practice already followed in some countries.[117] Nor should we forget that low dietary calcium intakes can exacerbate vitamin D deficiency.[27] Supplementation can be provided either orally or by injection. Although many patients like to take a daily multivitamin supplement to get their additional vitamin D, intermittent use of higher oral doses of vitamin D 4-monthly under medical supervision[90] might make it easier to treat vitamin D insufficiency in the frail elderly.

While daily use of small amounts of multivitamins is quite safe it should be borne in mind that megadoses of vitamins A and D can be harmful. Thus caution should be used when fortifying *foods* with specific vitamins. When milk is fortified with vitamin D this process needs to be carefully monitored.[118] Several studies suggest that excessive dietary vitamin A is associated with lower bone density and increased risk of hip fracture.[119,120] High levels of retinol in the serum were associated with higher fracture rates in a recent population-based study of 2322 Swedish males aged 49–51 years living in Uppsala who were followed for up to 30 years.[121] Fracture risk rose sharply in men with serum retinol levels within the highest quintile($>2.64$ µmol $L^{-1}$). Vitamin A in excess has long been known when administered to animals to induce high rates of bone resorption, skeletal abnormalities and fractures. This is seen for example in cats and dogs fed solely on raw liver.

## The Importance of Safeguarding Oestrogen Status

Dietary energy intakes need to be adequate for energy expenditure. Unfortunately, many people slim and exercise to excess. Their body fat stores fall, reducing circulating levels of leptin, a hormone which affects both bone formation and pubertal maturation.[122] Levels of cortisol may also rise, favouring breakdown of bone. But most importantly, consumption of inadequate calories for energy expenditure leads to oestrogen deficiency (anorexia, athletic amenorrhoea),[123] which causes accelerated bone turnover, rapid loss of bone mass and fractures, even in men.[124] Thin elderly people, with low aromatisation of androgen to oestrogen in adipose tissue, often have lower endogenous oestrogen levels and higher bone loss than people with higher stores of body fat.

## Stepping up Consumption of Fruit and Vegetables

Most nutritionists agree that a varied diet is desirable for everyone, because this will help to maintain intakes of essential vitamins, minerals, trace elements and micronutrients. This advice applies to bone health. The possible adverse influences of vitamin K deficiency upon bone metabolism and mass have been studied in the last few years because we know that vitamin K is an essential co-factor for the synthesis of osteocalcin, a protein required for osteoblastic activity and bone matrix formation. Lower bone density and higher hip fracture rates are seen in patients with low vitamin K levels, possibly because of reduced γ-carboxylation of osteocalcin.[125] By contrast, the higher vitamin K levels of Asian populations who consume large amounts of fruit and vegetables may help to explain their lower fracture rates in comparison with Caucasians.

There has also been a considerable resurgence of interest in the effects of acid–base status to the pathogenesis of osteoporosis. Bone certainly plays an important role in buffering endogenous acid, as can be seen by the effects of blocking bone resorption with bisphosphonate drugs.[126] High acid loads increase the need for skeletal buffering, stimulate osteoclastic bone breakdown and augment urinary calcium excretion. Because vegetable diets contribute excess base, the endogenous load of acid which must be buffered by the skeleton is lower on diets supplying plenty of fruit and vegetables than on diets high in meat, which have a high *S*-amino acid-content and confer a greater load of endogenous acid to be buffered by the skeleton.[127–129] The higher potassium intakes and the production of a more alkaline urine produced by vegetarian diets *vs.* meat diets, will also help to lower obligatory urinary calcium losses.[130] The hypercalciuric response to acidifying agents and diets high in *S*-amino acids may reflect actions of *S*-amino acids on calcium-ion sensing receptors in the kidney.[131]The extreme hypercalciuric effects produced by *S*-amino acid supplementation studies *vs.* diets supplemented with whole protein,[132] may have a similar explanation.

The presence of natural antioxidants and phytoestrogen components in some plant foods (such as soy products) may also act to slow postmenopausal bone loss, though the evidence for this is not compelling.[133] In addition some plant foods such as onions, appear to contain bone-active compounds with the ability to suppress bone resorption.[134] Another interesting epidemiological study demonstrated that tea-drinkers had considerably higher bone density than those avoiding tea, an effect which could have been related to the fluoride and flavonoid content of tea.[135] However, in my view evidence to date of bone benefits from stepping up fruit and vegetables, antioxidants and flavonoids may not yet be sufficiently strong to justify these dietary changes on the grounds of osteoporosis prevention.

On the other hand reducing the amounts of sodium chloride (NaCl) ingested in processed foods and avoiding liberal use of salt as a condiment, will lower obligatory losses of calcium in the urine usefully, since urinary calcium increases by 1 mmol for every 100 mmol NaCl (approximately one teaspoonful of salt) consumed.[136] The strategy of avoiding too many salty foods will enable calcium requirements to be met more economically than is the case when discretionary salt intake is high and obligatory salt-mediated urinary calcium losses are raised. Advice to limit dietary sodium chloride intakes is probably especially relevant to people with kidney stones and to salt-sensitive hypertensive patients, possibly because many of these individuals display exaggerated hypercalciuric responses to sodium.

## Providing Nutritionally Balanced Meals may Improve Bone Health More Effectively than Supplementation to Boost Single Nutrients

We cannot consciously control the absorption or excretion of nutrients we ingest and must rely upon the efficiency of our metabolism to absorb and retain the appropriate quantities of ingredients our skeletons require daily from the foods we consume. This is probably just as well, because achieving a balanced nutritional diet throughout life by choosing a suitable menu from a complex assortment of single nutrients would be an impossibly difficult task. People choose to eat foods and drinks they like: they are much less likely to self-select a correctly balanced healthy mixture of calories, minerals and vitamins from the large variety of food items on display. Regrettably, the majority of human studies investigating the effects of different nutrients on bone metabolism and skeletal mass have concentrated to date upon the effects of manipulating consumption of single nutrients and examining the bioavailability of these, rather than upon the effects of more general food patterns and/or habitual use

**Table 32.4.** *Innovative ways to increase consumption of foods augmenting bone health*

*Meals*
- Select good role-models to promote bone-healthy eating habits in advertising and the media
- Use TV cooking programmes to prepare bone-healthy nutritionally balanced whole meals (encourage children and adults to participate)
- Industry to produce attractive designer meals for good bone health suitable for different age groups
- Market these nutritionally balanced meals as bone-healthy
- Offer price-reductions of bone-healthy meals for low socio-economic status groups
- Design bone-healthy menus for groups with specific health needs (milk-allergy, diabetics, coeliac disease, renal failure, the calorie conscious)

of nutritionally balanced menus. By contrast, in the fields of hypertension and cardiovascular disease, clinical trials show that providing high risk patients with tasty *nutritionally balanced meals* meeting the full recommendations of national health organizations, ameliorates disease risk factors far more efficaciously, than attempts to modify intakes of single nutrients (such as sodium) or food groups (such as fats).[137,138] It appears likely that a similar approach could also prove beneficial for osteoporosis prevention (Table 32.4). The 'whole meal' strategy could be employed to promote good bone health and to prevent fractures in both youngsters and in older people. Balanced 'bone-healthy' meals could be purchased by families, schools, hospitals and institutions caring for the elderly. Such meals could also be designed specifically for high risk groups, such

**Table 32.5.** *Nutritional/lifestyle strategies to prevent osteoporosis throughout the life course*

*For individuals*
- Maintain a healthy body weight and match energy intake to energy expenditure
- Consume a varied balanced diet providing calcium-rich foods daily. If avoiding dairy foods, make strenuous efforts to keep up calcium intakes from other dietary sources (including fortified foods and mineral-rich water), or use mineral supplements
- Spend 30 min outside daily to safeguard vitamin D status *via* cutaneous synthesis. If housebound take vitamin D supplements under the advice of a health professional
- Undertake appropriate weight-bearing physical activity every day to stimulate osteogenesis and maintain good muscular co-ordination. Encourage enjoyment in walking and sport
- Avoid jeopardizing oestrogen-status through undernutrition and over-exercise (Do not diet or exercise to excess)
- Improve the calcium economy of the body by avoiding excessive urinary calcium wasting. (Limit discretionary intakes of sodium chloride, consume moderate protein and offset the calciuric effects of caffeine, alcohol and high *S*-amino acid intakes)
- Do not smoke cigarettes
- Do not consume excessive vitamin A
- Consume fruit and vegetables daily to boost intakes of potassium, magnesium, phosphorus, vitamin K, micronutrients, antioxidants and flavonoids
- Keep the environment safe to avoid falls

**Table 32.6.** *Strategies for professional groups to prevent osteoporosis and improve its management*

*For health professionals*
1. Know the nutrients important to bone health for every life stage
2. Be aware of groups with an increased risk of osteoporosis and offer education, advice and nutritional intervention to these more widely
3. Document fracture histories of all individuals (children and adults)
4. Promote the value of bone densitometry to detect osteoporosis
5. Prescribe supplementary vitamin D, calcium and protein when this is needed

*For industry and agriculture*
1. Provide and promote bone-healthy foods
2. Develop a wider range of mineral-enriched waters for daily use
3. Market appropriate doses of oral vitamin D supplements for 4-monthly use
4. Ensure advertising meets current nutritional and ethical guidelines

*For policy makers*
1. Ensure a safe and affordable food supply for all people
2. Teach the value of maintaining good bone health in the school curriculum
3. Promote nutrition education for health professionals and the general public
4. Ensure media messages are simple, sound and comprehensible
5. Discourage cigarette smoking and physical inactivity
6. Promote the value of bone densitometry to document bone status in those vulnerable to osteoporosis
7. Require yearly statistics on fracture rates and the use of bone-sparing medications to be reported
8. Require hospitals, nursing homes to document use of supplementary nutritional support (vitamin D, protein and calcium supplements)
9. Fund research (basic, molecular and clinical) to improve understanding of the role of nutrition in prevention and management of osteoporosis
10. Maintain a trained workforce of health professionals with expertise in bone health

as those with allergies, as well as for specific age groupings. Use of these products could be facilitated by government agencies and by industry. A catchy logo for approved bone-healthy meals and foods could help marketing. Judicious advertising could offer price concessions for bone-healthy meals. This procedure could lower public health costs substantially by improving the bone health of the population and helping to reduce fractures.

# Conclusions

To conclude, ways advocated in this chapter for individuals to optimise their bone gain in youth and to slow bone loss later in life, by consuming a balanced healthy diet and following a healthy lifestyle throughout the life course are summarized in Table 32.5. Table 32.6. list some key advice for health professionals, for industry, for agriculture and for policy makers that is directed towards preventing osteoporotic fractures *via* improved nutrition education and by making healthy food more widely available. Finally, when asked the question "who needs nutritional osteoporosis prevention?" we should remember that "we all do, throughout life" because strong bones are harder to break than weak ones.

# References

1. R.P. Heaney, S. Abrams, B. Dawson-Hughes, A. Looker, R. Marcus, V. Matkovic and C. Weaver, Peak bone mass, *Osteoporosis Int.*, 2000, **11**, 985–1009.
2. S.L. Hui, C.W. Slemenda and C.C. Johnston, Baseline measurement of bone mass predicts fracture in white women, *Ann. Intern. Med.*, 1989, **111**, 355–361.
3. D. Marshall, O. Johnell and H. Wedel, Meta-analysis of how well measures of bone mineral density predict occurrence of osteoporotic fractures, *Br. Med. J.*, 1996, **312**, 1254–1259.
4. A. Goulding, I.E. Jones, R.W. Taylor, P.J. Manning and S.M. Williams, More broken bones: a 4-year double cohort study of young girls with and without distal forearm fractures, *J. Bone Miner. Res.*, 2000; **15**(10): 2011–2018.
5. J.A. Kanis, L.J. Melton, C. Christiansen, C.C. Johnston and N. Khaltaev, The diagnosis of osteoporosis, *J. Bone Miner. Res.*, 1994, **9**, 1137–1141.
6. M. Delmi, C-H. Rapin, J.-M. Bengoa, P.D. Delmas, H. Vasey and J-P. Bonjour, Dietary supplementation in elderly patients with fractured neck of femur, *Lancet*, 1990, **335**, 1013–1016.
7. J.-P. Bonjour, M.A. Schurch, and R. Rizzoli, Nutritional aspects of hip fracture, *Bone*, 1996, **18**, S139-S144.
8. M.A. Schurch, R. Rizzoli, D. Slosman, L. Vadas, P. Vergnaud and J.P. Bonjour, Protein supplements increase serum insulin-like growth factor-I levels and attenuate proximal femur bone loss in patients with recent hip fracture – a randomized, double-blind, placebo-controlled trial, *Ann. Intern. Med.*, 1998, **128**, 801–809.
9. O. Johnell, S. Oden, F. Caulin, J. Kanis, Acute and long-term increase in fracture risk after hospitalization for vertebral fracture, *Osteoporosis Int.*, 2001, **12**, 207–214.
10. R.P. Heaney, Constructive interactions among nutrients and bone-active pharmacologic agents with principal emphasis on calcium, phosphorus, vitamin D and protein, *J. Am. Coll. Nutr.*, 2001, **20**, 403S-405S.
11. R.M. Neer, C.D. Arnaud, J. Zanchetta, R. Prince, G.A. Gaich, J. Reginster, A.B. Hodsman, E.F. Erikson, S. Ish-Shalom, H.K. Genan, O.H. Wang and B.H. Mitlak, Effect of parathyroid hormone (1–34) on fractures and bone mineral density in postmenopausal women with osteoporosis, *N. Engl. J. Med.*, 2001, **344**, 1434–1441.
12. F.H. Glorieux, N.J. Bishop, H. Plotkin, G. Chabot, G. Lanoue and R. Travers, Cyclic administration of pamidronate in children with severe osteogenesis imperfecta, *N. Engl. J. Med.*, 1998, **339**, 947–952.
13. E.E. Hobson and S.H. Ralston. Role of genetic factors in the pathophysiology and management of osteoporosis, *Clin. Endocrinol.*, 2001, **54**, 1–9.
14. S. Epstein, Post-transplantation bone disease: the role of immunosuppressive agents and the skeleton, *J. Bone Miner. Res.*, 1996, **11**, 1–7.
15. I.M. van der Sluis, M.M. van den Heuvel-Eibrink, K. Hablen, E.P. Krenning and S.M.P.F.D. Kesizer-Schrama, Altered bone mineral density and body composition, and increased fracture risk in childhood acute lymphoblastic leukemia, *J. Pediatr.*, 2002, **141**, 204–210.
16. M. Pfeifer, B. Begerow, H.W. Minne, T. Schlotthauer, M. Pospeschill, M. Scholz, A.D. Lazarescu and W. Pollahne, Vitamin D status, trunk muscle strength, body sway, falls and fractures among 237 postmenopausal women with osteoporosis *Exp. Clin. Endocrinol. Diab.*, 2001, **109**, 87–92.
17. H.A. Bishcoff, H.B. Stahelin, W. Dick, R. Akos, M. Knecht, C. Salis, S.M. Cadarette, M. Nebike, R. Theiler, M. Pfeifer, B. Begerow, R.A. Lew and

M. Conzelmann, Effects of vitamin D and calcium supplementation on falls: a randomized controlled trial, *J. Bone Miner. Res.*, 2003, **18**, 343–351.

18. P.D. Delmas, Treatment of postmenopausal osteoporosis, *Lancet*, 2002, **359**, 2018–2026.

19. C.M. Weaver and M. Liebman, Biomarkers of bone health appropriate for evaluating functional foods designed to reduce risk of osteoporosis, *Br. J. Nutr.*, 2002, **88**(Suppl. 2):S225-S232.

20. L.C. Hofbauer and M. Schoppet, Editorial: Osteoprotegerin gene polymorphism and the risk of osteoporosis and vascular disease, *J. Clin. Endocrinol. Metab.*, 2002, **87**, 4078–4079.

21. W.W.K. Koo, R. Sherman, P. Succop, A.E. Oestreich, R.C. Tsang, S.K. Krug-Wispe and J.J. Steichen, Sequential bone mineral content in small preterm infants with and without fractures and rickets, *J. Bone Miner. Res.*, 1988, **3**, 193–197.

22. M.S. Fewtrell, T.J. Cole, N.J. Bishop and A. Lucas, Neonatal factors predicting childhood height in preterm infants: evidence for a persisting effect of early metabolic bone disease? *J. Pediatr.*, 2000, **137**, 668–673.

23. S.A. Zamora, D.C. Belli, R. Rizzoli, D.O. Slosman and J.-P. Bonjour, Lower femoral neck bone mineral density in prepubertal former preterm girls, *Bone*, 2001, **29**, 424–427.

24. D.E. Yarbrough, E. Barrett-Connor and D.J. Morton, Birth weight as a predictor of adult bone mass in postmenopausal women: the Rancho Bernardo Study, *Osteoporosis Int.*, 2000, **11**, 626–630.

25. C. Cooper, M.K. Javaid, P. Taylor, K. Walker-Bone, E. Dennison and N. Arden, The fetal origins of osteoporotic fracture, *Calcif. Tissue Int.*, 2002, **70**, 391–394.

26. J.M. Pettifor and G.P. Moodley, Appendicular bone mass in children with a high prevalence of low dietary calcium intakes, *J. Bone Miner. Res.*, 1997, **12**, 1824–1832.

27. T.D. Thacher, P.R. Fischer, J.M. Pettifor, J.O. Lawson, C.O. Isichei and G.M. Chan, Case-control study of factors associated with nutritional rickets in Nigerian children, *J. Pediatr.*, 2000, **137**, 367–373.

28. B. Dibba, A. Prentice, M. Ceesay, D.M. Stirling, T.J. Cole and E.M.E. Poskitt, Effect of calcium supplementation on bone mineral accretion in Gambian children accustomed to a low-calcium diet, *Am. J. Clin. Nutr.*, 2000, **71**, 544–549.

29. X.Q. Du, H. Greenfield, D.R. Fraser, K.Y. Ge, Z.H. Liu and W. He, Milk consumption and bone mineral content in Chinese adolescent girls, *Bone*, 2002, **30**, 521–528.

30. J.P. Goldberg, S.C. Folta and A. Must, Milk: can a "good" food be so bad? *Pediatrics*, 2002, **110**, 826–831.

31. S.J. Whiting, A. Healey, S. Psiuk, R.L. Mirwald, K. Kowalski and D.A. Bailey, Relationship between carbonated and other low nutrient dense beverages and bone mineral content of adolescents, *Nutr. Res.*, 2001, **21**, 1107–1115.

32. S.A. Bowman, Beverage choices of young females: changes and impact on nutrient intakes, *J. Am. Diet. Assoc.*, 2002, **102**, 1234–1239.

33. D.S. Ludwig, K.E. Peterson and S.L. Gortmaker, Relation between consumption of sugar-sweetened drinks and childhood obesity: a prospective, observational study, *Lancet*, 2001, **357**, 505–508.

34. A. Goulding, R.W. Taylor, I.E. Jones, K.A. McAuley, P.J. Manning and S.M. Williams, Overweight and obese children have low bone mass and area for their weight, *Int. J. Obesity*, 2000, **24**, 627–632.

35. C.B. Ebbeling, D.B. Pawlak and D.S. Ludwig, Childhood obesity: public-health crisis, common sense cure, *Lancet*, 2002, **360**, 473–482.

36. K.L. Tucker, Does milk intake in childhood protect against later osteoporosis? *Am. J. Clin. Nutr.*, 2003, **77**, 10–11.
37. S.I. Barr, D.A. McCarron, R.P. Heaney, B. Dawson-Hughes, S.L. Berga, J.S. Stern and S. Oparil, Effects of increased consumption of fluid milk on energy and nutrient intake, body weight, and cardiovascular risk factors in healthy older adults, *J. Am. Diet. Assoc.*, 2000, **100**, 810–817.
38. R. Eastell and H. Lambert, Diet and healthy bones, *Calcif. Tissue Int.*, 2002, **70**, 400–404.
39. J.-P. Bonjour, T. Chevalley, P. Ammann, D. Slosman and R. Rizzoli, Gain in bone mineral mass in prepubertal girls 3–5 years after discontinuation of calcium supplementation: a follow-up study, *Lancet*, 2001, **358**, 1208–1212.
40. B. Dibba, A. Prentice, M. Ceesay, M. Mendy, S. Darboe, D.M. Stirling, T.J. Cole and E.M.E. Poskitt, Bone mineral contents and plasma osteocalcin concentrations of Gambian children 12 and 24 mo after the withdrawal of a calcium supplement, *Am. J. Clin. Nutr.*, 2002, **76**, 681–686.
41. R.E. Black, S.M. Williams, I.E. Jones and A. Goulding, Children who avoid drinking cow milk have low dietary calcium intakes and poor bone health, *Am. J. Clin. Nutr.*, 2002, **76**, 675–680.
42. A. Goulding, J.E.P. Rockell, R.E. Black, A.M. Grant, I.E. Jones and S.M. Willams, Children who avoid drinking cow milk are at increased risk for prepubertal bone fractures, *J. Am. Diet. Assoc.*, 2003, in press.
43. H.J. Kalkwarf, J.C. Khoury and B.P. Lanphear, Milk intake during childhood and adolescence, adult bone density, and osteoporotic fractures in US women, *Am. J. Clin. Nutr.*, 2003, **77**, 257–265.
44. M.J. Barger-Lux, R.P. Heaney, P.T. Packard, J.M. Lappe and R.R. Recker, Nutritional correlates of low calcium intake, *Clin. Appl. Nutr.*, 1992, **2**, 39.
45. M.T. Hannan, K.L. Tucker, B. Dawson-Hughes, L.A. Cupples, D.T. Felson and D.P. Keil, Effect of dietary protein on bone loss in elderly men and women: the Framingham Osteoporosis Study, *J. Bone Miner. Res.*, 2000, **15**, 2504–2512.
46. J.M. Kanczler, T. Bodamyali, T.M. Millar, J.G. Clinch, C.R. Stevens and D.R. Blake, Human and bovine milk contains the osteoclasto-genesis inhibitory factor, osteoprotegerin, *J. Bone Miner. Res.*, 2001, **16**(6), 1176.
47. J. Yamamura, S. Aoe, Y. Toba, M. Motouri, H. Kawakami, M. Kumegawa, A. Itabashi and Y. Takada, Milk basic protein (MBP) increases radial bone mineral density in healthy adult women, *Biosci. Biotechnol. Biochem.*, 2002, **66**, 702–704.
48. J. Cornish, K. Callon, T. Banovic, U. Bava, M. Watson, Q. Chen, K. Palmano, W. Haggarty, A.B. Grey and I.R. Reid, Lactoferrin is a potent osteoblast/chondrocyte growth factor, inhibits osteoclastogenesis and is anabolic to bone *in vivo*, *J. Bone Miner. Res.*, 2002, **17**(Suppl. l), SU151.
49. J. Cadogan, R. Eastell, N. Jones and M.E. Barker, Milk intake and bone mineral acquisition in adolescent girls: randomised, controlled intervention trial, *BMJ*, 1997, **315**, 1255–1260.
50. R.P. Heaney, D.A. McCarron, B. Dawson-Hughes, S. Oparil, S.L. Berga, J.S. Stern, S.I. Barr and C.J. Rosen, Dietary changes favorably affect bone remodeling in older adults, *J. Am. Diet. Assoc.*, 1999, **99**(10), 1228–1233.
51. C. Libanati, D.J. Baylink, E. Lois-Wenzel, N. Srinivasan and S. Mohan, Studies on the potential mediators of skeletal changes occurring during puberty in girls, *J. Clin. Endocrinol. Metab.*, 1999, **84**, 2807–2814.
52. S. Yakar, C.J. Rosen, W.G. Beamer, C.L. Ackert-Bicknell, Y. Wu, J-L. Liu, G.T. Ooi, J.F. Setser, J. Frystyk, Y.R. Boislair and D. LeRoith, Circulating levels

of IGF-1 directly regulate bone growth and density, *J. Clin. Invest.*, 2002, **110**, 771–781.

53. R. Poulton, A. Caspi, B.J. Milne, W.M. Thomson, A. Taylor, M.R. Sears and T.E. Moffitt, Association between children's experience of socioeconomic disadvantage and adult health: a life-course study, *Lancet*, 2002, **360**, 1640–1645.

54. T. Parsons, M. Van Dusseldorp, M. Van der Vliet, K. Van de Werken, G. Schaafsma and W. Van Staveren, Reduced bone mass in Dutch adolescents fed a macrobiotic diet in early life, *J. Bone Miner. Res.*, 1997, **12**, 1486–1494.

55. C. Power, Childhood adversity still matters for adult health outcomes, *Lancet*, 2002, **360**, 1619–1620.

56. E. Seeman, Pathogenesis of bone fragility in women and men, *Lancet*, 2002, **359**, 1841–1850.

57. H.M. Frost, Muscle, bone and the Utah paradigm a 1999 overview, *Med. Sci. Sports Exerc.*, 2000, **32**, 911–917.

58. K.M. Sanders, G.D. Nicholson, A.M. Ugoni, E. Seeman, J.A. Pasco and M.A. Kotowicz, Fracture rates lower in rural than urban communities: the Geelong Osteoporosis Study, *J. Epidemiol. Commun. Health*, 2002, **56**, 466–470.

59. F.L. Morris, G.A. Naughton, J.L. Gibbs, J.S. Carlson and J.D. Wark, Prospective ten-month exercise intervention in premenarcheal girls: positive effect on bone and lean mass, *J. Bone Miner. Res.*, 1997, **12**, 1453–1462.

60. R.K. Fuchs, J.J. Bauer and C.M. Snow, Jumping improves hip and lumbar spine bone mass in prepubescent children: a randomized controlled trial, *J. Bone Miner. Res.*, 2001, **16**, 148–156.

61. B. Specker, Evidence for an interaction between calcium intake and physical activity on changes in bone mineral density, *J. Bone Miner. Res.*, 1996, **11**, 1539–1544.

62. S. Iuliano-Burns, L. Saxon, G. Naughton, K. Gibbons and S.L. Bass, Regional specificity of exercise and calcium during skeletal growth in girls: a randomised controlled trial, *J. Bone Miner. Res.*, 2003, **18**, 156–162.

63. M.A. Petit, H.A. McKay, K.J. MacKelvie, A. Heinonen, K.M. Khan and T.J. Beck, A randomized school-based jumping intervention confers site and maturity-specific benefits on bone structural properties in girls: a hip structural analysis study, *J. Bone Miner. Res.*, 2002, **17**, 363–372.

64. S.L. Bass, L. Saxon, R.M. Daly, C.H. Turner, A.G. Robling, E. Seeman and S. Stuckey, The effect of mechanical loading on the size and shape of bone in pre-, peri- and postpubertal girls: a study in tennis players, *J. Bone Miner. Res.*, 2002, **17**, 2274–2280.

65. S. Kontulainen, H. Sievanen, P. Kannus, M. Pasanen and I. Vuori, Effect of long-term impact-loading on mass, size and estimated strength of humerus and radius of female racquet-sports players: a peripheral quantitative computed tomography study between young and old starters and controls, *J. Bone Miner. Res.*, 2002, **17**(12), 2281–2289.

66. P. Vestergaard, C. Emborg, R.K. Stoving, C. Hagen, L. Mosekilde and K. Brixen, Fractures in patients with anorexia nervosa, bulimia nervosa, and other eating disorders – a nationwide register study, *Int. J. Eat. Disord.*, 2002, **32**, 301–308.

67. M.M. Manore, Dietary recommendations and athletic menstrual dysfunction, *Sports Med.*, 2002, **32**, 887–901.

68. D.H. Sullivan, S. Sun and R.C. Walls, Protein-energy undernutrition among elderly hospitalized patients, *JAMA*, 2002, **281**, 2013–2019.

69. M. Lamy, P. Mojon, G. Kalykakis, R. Legrand and E. Butz-Jorgensen, Oral status and nutrition in the institutionalized elderly, *J. Dentistry*, 1999, 443–448.

70. M.A. Bernstein, M.E. Nelson, K.L. Tucker, J. Layne, E. Johnson, A. Nuernberger, C. Castaneda, J.O. Judge, D. Buchner and M.F. Singh, A home-based nutrition intervention to increase consumption of fruits, vegetables, and calcium-rich foods in community dwelling elders, *J. Am. Diet. Assoc.*, 2002, **102**, 1421–1427.

71. I.E. Jones, S. Williams, N. Dow and A. Goulding, How many children remain fracture-free during growth? A longitudinal study of children and adolescents participating in the Dunedin Multidisciplinary Health and Development Study, *Osteoporosis Int.*, 2002, **13**, 990–995.

72. G. Jones and H.M. Cooley, Symptomatic fracture incidence in those under 50 years of age in southern Tasmania, *J. Paediatr. Child Health*, 2002, **38**, 278–283.

73. V. Matkovic, Nutrition, genetics and skeletal development, *J. Am. Coll. Nutr.*, 1996, **15**, 556–569.

74. Z. Kun, H. Greenfield, X. Du and D.R. Fraser, Improvement of bone health in childhood and adolescence, *Nutr. Res. Rev.*, 2001, **14**, 119–151.

75. A. Goulding, R. Cannan, S.M. Williams, E.J. Gold, R.W Taylor and N.J. Lewis-Barned, Bone mineral density in girls with forearm fractures, *J. Bone Miner. Res.*, 1998, **13**, 143–148.

76. A. Goulding, I.E. Jones, R.W. Taylor, P.J. Manning and S.M. Williams, Bone mineral density and body composition in boys with distal forearm fractures: a dual energy x-ray absorptiometry study, *J. Pediatr.*, 2001, **139**, 509–515.

77. I.R. Reid, Relationships among body mass, its components, and bone, *Bone*, 2002, **31**, 547–555.

78. A. Goulding, R.W. Taylor, I.E. Jones, P.J. Manning and S.M. Williams, Spinal overload – a concern for obese children and adolescents? *Osteoporosis Int.*, 2002, **13**, 835–840.

79. A. Goulding, I.E. Jones, R.W. Taylor, J.M. Piggot and D. Taylor, Dynamic and static tests of balance and postural sway in boys: effects of previous wrist bone fractures and high adiposity, *Gait Posture*, 2003, **17**, 136–141.

80. S.A. French, J.A. Fulkerson and M. Story, Increasing weight-bearing physical activity and calcium intake for bone mass growth in children and adolescents: a review of intervention trials, *Prev. Med.*, 2000, **31**, 722–731.

81. P. Kannus, H. Haapasalo, M. Sankelo, H. Sievanen, M. Pasanen, A. Heinonen, P. Oja and I. Vuori, Effect of starting age of physical activity on bone mass in the dominant arm of tennis and squash players, *Ann. Intern. Med.*, 1995, **123**, 27–31.

82. S. Bass, G. Pearce, M. Bradney, E. Hendrich, P.D. Delmas, A. Harding and E. Seeman, Exercise before puberty may confer residual benefits in bone density in adulthood: studies in active prepubertal and retired female gymnasts, *J. Bone Miner. Res.*, 1998, **13**, 500–507.

83. M. Sundberg, P. Gardsell, O. Johnell, M.K. Karlsson, E. Ornstein, B. Sandstedt and I. Sernbo, Physical activity increases bone size in prepubertal boys and bone mass in prepubertal girls: a combined cross-sectional and 3-year longitudinal study, *Calcif. Tissue Int.*, 2002, **71**, 406–415.

84. P. Kannus, J. Parkkari, S. Niemi, M. Pasanen, M. Palvanen, M. Jarvinen and I. Vuori, Prevention of hip fracture in elderly people with use of a hip protector, *N. Engl. J. Med.*, 2000, **343**, 1506–1513.

85. M. Chapuy, M. Arlot, F. Duboeuf, J. Brun, B. Crouzet, S. Arnaud, P.D. Delmas and P.J. Meunier, Vitamin D3 and calcium to prevent hip fractures in elderly women, *N. Engl. J. Med.*, 1992, **327**, 1637–1642.

86. M. Chapuy, M. Arlot, P. Delmas and P. Meunier, Effect of calcium and cholecalciferol treatment for three years on hip fractures in elderly women, *Br. Med. J.*, 1994, **308**, 1081–1082.

87. B. Dawson-Hughes, S. Harris, E. Krall and G. Dallal, Effect of calcium and vitamin D supplementation on bone density in men and women 65 years of age or older, *N. Engl. J. Med.*, 1997, **337**, 670–676.

88. R.J. Heikinheimo, J.A. Inkovaara, E.J. Harju, M.V. Haavisto, R.H. Kaarela, J.M. Kataja, A.M.L. Kokko, L.A. Kohlo and S.A. Rajala, Annual injection of vitamin D and fractures of aged bones, *Calcif. Tissue Int.*, 1992, **51**, 105–110.

89. P. Lips, W. Graafmans, M. Ooms, P. Bezemer and L. Bouter, Vitamin D supplementation and fracture incidence in elderly persons: a randomized, placebo-controlled clinical trial, *Ann. Intern. Med.*, 1996, **124**, 400–406.

90. D.P. Trivedi, R. Doll and K.T. Khaw, Effect of four monthly oral vitamin D3 (cholecalciferol) supplementation on fractures and mortality in men and women living in the community: randomised double blind controlled trial, *BMJ*, 2003, **326**, 469–472.

91. M. Karlsson, S. Bass and E. Seeman, The evidence that exercise during growth or adulthood reduces the risk of fragility fractures is weak, *Best Pract. Res. Clin. Rheumatol.*, 2001, **15**, 429–450.

92. A.J. Campbell, M.C. Robertson, M.M. Gardner, R.N. Norton, M.W. Tilyard and D.M. Buchner, Randomised controlled trial of a general practice programme of home based exercise to prevent falls in elderly women, *Br. Med. J.*, 1997, **315**(7115), 1065–1069.

93. C. Weaver, W. Proulx and R. Heaney, Choices for achieving adequate dietary calcium with a vegetarian diet, *Am. J. Clin. Nutr.*, 1999, **70**, 543S–548S.

94. J.L. Keller, A.J. Lanou and N.D. Barnard, The consumer cost of calcium from food and supplements, *J. Am. Diet. Assoc.*, 2002, **102**, 1669–1671.

95. A. Goulding, R.W. Taylor, D. Keil, E. Gold, N.J. Lewis-Barned and S.M. Williams, Lactose malabsorption and rate of bone loss in older women. *Age Ageing*, 1999, **28**, 175–180.

96. M. Di Stefano, G. Veneto, S. Malservisi, L. Cecchetti, L. Minguzzi, A. Strocchi and G.R. Corazza, Lactose malabsorption and intolerance and peak bone mass, *Gastroenterology*, 2002, **122**, 1793–1799.

97. L.L. Birch and J.O. Fisher, Development of eating behaviors among children and adolescents, *Pediatrics*, 1998, **101**(3), 539–549.

98. J.O. Fisher, D.C. Mitchell, H. Smiciklas-Wright and L.L. Birch, Maternal milk consumption predicts the tradeoff between milk and soft drinks in young girls' diets, *J. Nutr.*, 2001, **131**, 246–250.

99. L. Christie, R.J. Hine, J.G. Parker and W. Burks, Food allergies in children affect nutrient intake and growth, *J. Am. Diet. Assoc.*, 2002, **102**, 1648–1651.

100. M.L. Wahlqvist and G.S. Savige, Interventions aimed at dietary and liefestyle changes to promote healthy aging, *Eur. J. Clin. Nutr.*, 2000, **54**(Suppl. 3), S148–S156.

101. P. Gulliver and C.C. Horwath, Assessing women's perceived benefits, barriers, and stage of change for meeting milk product consumption recommendations, *J. Am. Diet. Assoc.*, 2001, **101**, 1354–1357.

102. S.J. Parikh and J.A. Yanovski, Calcium intake and adiposity, *Am. J. Clin. Nutr.*, 2003, **77**, 281–287.

103. M.B. Zemel, H. Shi, B. Greor, D. Dirienzo and P.C. Zemel, Regulation of adiposity by dietary calcium, *FASEB J.*, 2000, **14**, 1132–1138.

104. P. Galan, M.J. Arnaud, S. Czernichow, A-M. Delabroise, P. Preziosi, S. Bertrais, C. Franchisseur, M. Maurel, A. Farier and S. Hercberg, Contribution of mineral waters to dietary calcium and magnesium intake in a French adult population, *J. Am. Diet. Assoc.*, 2002, **102**, 1658–1662.

105. F. Suarez, J. Adshead, J. Furne and M. Levitt, Lactose maldigestion is not an impediment to the intake of 1500 mg calcium daily as dairy products, *Am. J. Clin. Nutr.*, 1998, **68**, 1118–1122.

106. R.P. Heaney, K. Rafferty and M.S. Dowell, Effect of yoghurt on a urinary marker of bone resorption in postmenopausal women, *J. Am. Diet. Assoc.*, 2002, **102**, 1672–1674.

107. M. Wheadon, A. Goulding, G.O. Barbezat and A.J. Campbell, Lactose malabsorption and calcium intake as risk factors for osteoporosis in elderly New Zealand women, *NZ Med. J.*, 1991, **104**, 417–419.

108. E. Barrett-Connor, J.C. Chang and S.L. Edelstein, Coffee-associated osteoporosis offset by daily milk consumption – the Rancho-Bernado study, *J. Am. Med. Assoc.*, 1994, **271**, 280–283.

109. A. Goulding, Smoking, the menopause and biochemical parameters of bone loss and bone turnover, *NZ Med. J.*, 1982, **95**, 218–220.

110. D. Jesudason and A.G. Need, Effects of smoking on bone and mineral metabolism, *The Endocrinol.*, 2002, **12**,199–209.

111. R.P. Heaney, K.M. Davies, T.C. Chen, M.F. Holick and M.J. Barger-Lux, Human serum 25-hydroxycholecalciferol response to extended oral dosing with cholecalciferol, *Am. J. Clin. Nutr.*, 2003, **77**, 204–210.

112. M. Holick, Vitamin D requirements for humans of all ages: new increased requirements for women and men 50 years and older, *Osteoporosis Int.*, 1998, (Suppl. 8), S24–S29.

113. B.H. Blok, C.C. Grant and I.R. Reid, Characteristics of children with florid vitamin D deficient rickets in the Auckland region in 1998, *NZ Med. J.*, 2000, **113**, 374–376.

114. S. Docio, J. Riancho, A. Perez, J. Olmos, J. Amado and J. Gonzalez-Macias, Seasonal deficiency of vitamin D in children: a potential target for osteoporosis-preventing strategies? *J. Bone Miner. Res.*, 1998, **13**, 544–548.

115. G. Jones, C. Blizzard, M.D. Riley, V. Parameswaran, T.M. Greenaway and T. Dwyer, Vitamin D levels in prepubertal chldren in Southern Tasmania: prevalence and determinants, *Eur. J. Clin. Nutr.*, 1999, **53**, 824–829.

116. A. Goulding, Lightening the fracture load: growing evidence suggests many older New Zealanders would benefit from more vitamin D, *NZ Med. J.*, 1999, **112**, 329–330.

117. N.J. Shaw and B.R. Pal, Vitamin D deficiency in UK Asian families: activating a new concern, *Arch. Dis. Childhood*, 2001, **86**, 147–148.

118. C.H. Jacobus, M.F. Holick, Q. Shao, T.C. Chen, I.A. Holm, J.M. Kolodny, G.E. Fuleihan and E.N. Seely, Hypervitaminosis-D associated with drinking milk, *N. Engl. J. Med.*, 1992, **326**, 1173–1177.

119. D. Feskanich, V. Singh, W.C. Willett and G.A. Colditz, Vitamin A intake and hip fractures among postmenopausal women, *J. Am. Med. Assoc.*, 2002, **287**, 47–54.

120. D. Feskanich, W.C. Willett and G.A. Colditz, Calcium, vitamin D, milk consumption, and hip fractures: a prospective study among postmenopausal women, *Am. J. Clin. Nutr.*, 2003, **77**, 504–511.

121. K. Michaelsson, H. Lithell, B. Vessby and H. Melhus, Serum retinol levels and the risk of fracture, *N. Engl. J. Med.*, 2003, **348**, 287–294.

122. S. Khosla, Leptin – Central or peripheral to the regulation of bone metabolism? *Endocrinology*, 2002, **143**, 4161–4164.

123. M.P. Warren and S. Shantha, The female athlete. *Best Pract. Res. Clin. Endocrinol. Metab.*, 2000, **14**, 37–53.

124. B.L. Riggs, S. Khosla and L.J. Melton, Sex steroids and the construction and conservation of the adult skeleton, *Endocrine Rev.*, 2002, **23**, 279–302.

125. D. Feskanich, P. Weber, W.C. Willett, H. Rockett, S.L. Booth and G.A. Colditz, Vitamin K intake and hip fractures in women: a prospective study, *Am. J. Clin. Nutr.*, 1999, **77**, 74–79.

126. A. Goulding and M.F. Broom, Effects of diphosphonate and colchicine administration upon acid–base changes induced by bilateral nephrectomy in rats, *Clin. Sci.*, 1979, **57**, 19–23.

127. A. Sebastian, D.E. Sellmeyer, K.L. Stone and S.R. Cummings, Dietary ratio of animal to vegetable protein and rate of bone loss and risk of fracture in postmenopausal women, *Am. J. Clin. Nutr.*, 2001, **74**, 411–412.

128. D.E. Sellmeyer, K.L. Stone, A. Sebastian and S.R. Cummings, Dietary ratio of animal to vegetable protein and rate of bone loss and risk of fracture in postmenopausal women, *Am. J. Clin. Nutr.*, 2001(73), 118–122.

129. S.A. New, The role of the skeleton in acid–base homeostasis, *Proc. Nutr. Soc.*, 2002, **61**, 151–164.

130. A. Goulding and D.R. Campbell, Hypocalciuric effects of hydrochlorothiazide in the rat during $NaHCO_3$, NaCl, and $NH_4Cl$ loading. *Renal Physiol.*, 1984, **7**, 185–191.

131. A.D. Conigrave, A.H. Franks, E.M. Brown and S.J. Quinn, L-amino acid sensing by the calcium-sensing receptor: a general mechanism for coupling protein and calcium metabolism? *Eur. J. Clin. Nutr.*, 2002, **56**, 1072–1080.

132. H. Spencer, L. Kramer, D. Osis and C. Norris, Effect of a high protein (meat) intake on calcium metabolism in man, *Am. J. Clin. Nutr.*, 1978, **3**, 2167–2180.

133. K.E. Wangen, A.M. Duncan, B.E. Merz-Demlow, X. Xu, R. Marcus, W.R. Phipps and M.S. Kurzer, Effects of soy isoflavones on markers of bone turnover in premenopausal and postmenopausal women, *J. Clin. Endocrinol. Metab.*, 2000, **85**, 3043–3048.

134. R.C. Muhlbauer, A. Losano and A. Reinli, Onion and a mixture of vegetables, salads and herbs affect bone resorption in the rat by a mechanism independent of their base excess, *J. Bone Miner. Res.*, 2002, **17**, 1230–1236.

135. V.M. Hegarty, H.M. May and K.-T. Khaw, Tea drinking and bone mineral density in older women, *Am. J. Clin. Nutr.*, 2000, **71**, 1003–1007.

136. A. Goulding, Osteoporosis: why consuming less sodium chloride helps to conserve bone, *NZ Med. J.*, 1990, **103**, 120–122.

137. W.M. Vollmer, F.M. Sacks, J. Ard, L.J. Appel, G.A. Bray, D.G. Simons-Morton, P.R. Conlin, L.P. Svetkey, T.P. Erlinger, T.J. Moor and N. Karanja for the Dash–Sodium trial collaborative research group, Effects of diet and sodium intake on blood pressure: subgroup analysis of the DASH-sodium trial, *Ann. Intern. Med.*, 2001, **135**, 1019–1028.

138. D.A. McCarron, S. Oparil, A. Chait, R.B. Haynes, P. KrisEtherton, J.S. Stern, L.M. Resnick, S. Clark, C.D. Morris, D.C. Hatton, J.A. Metz, M. McMahon, S. Holcomb, G.W. Snyder and X. Pisunyer, Nutritional management of cardiovascular risk factors – a randomized clinical trial, *Arch. Intern. Med.*, 1997, **157**, 169–177.

# Subject Index